Neurobehavioral of Childhood

An Evolutionary Perspective

Robert Melillo • Gerry Leisman

Neurobehavioral Disorders of Childhood

An Evolutionary Perspective

Robert Melillo
Department of Clinical Neurology
Carrick Institute for Graduate Studies
203-8941 Lake Drive
Cape Canaveral, FL 32920
USA

Gerry Leisman
Department of Psychology
College of Staten Island
City University of New York
2800 Victory Blvd.
Staten Island, NY 10314
USA
drgersh@yahoo.com

ISBN 978-1-4419-1232-9   e-ISBN 978-1-4419-1231-2
DOI 10.1007/978-1-4419-1231-2
Springer Dordrecht Heidelberg London New York

Library of Congress Control Number: 2009935459

© Springer Science+Business Media, LLC 2009
All rights reserved. This work may not be translated or copied in whole or in part without the written permission of the publisher (Springer Science+Business Media, LLC, 233 Spring Street, New York, NY 10013, USA), except for brief excerpts in connection with reviews or scholarly analysis. Use in connection with any form of information storage and retrieval, electronic adaptation, computer software, or by similar or dissimilar methodology now known or hereafter developed is forbidden.
The use in this publication of trade names, trademarks, service marks, and similar terms, even if they are not identified as such, is not to be taken as an expression of opinion as to whether or not they are subject to proprietary rights.

Printed on acid-free paper

Springer is part of Springer Science+Business Media (www.springer.com)

To
My wife, Carolyn, and my children, Robby, Ellie, and Ty.
I love all of you. You are my life's inspiration.
To my parents, Joseph and Catherine, my brother, Domenic,
my sister, Susan, and their children for all of your love and support.
To Ted Carrick for your friendship and professional inspiration.
To Janet Groschel for your friendship and support; this book
would not have been possible without all of your help.
Thank you.

To my wife and best friend
Yael
And all of our children
Yael, Amit, Akiba, Michal, and Daniel

Foreword

Attention deficit/hyperactivity disorder (ADHD) affects 3–8 percent of children and is associated with social, cognitive, and academic impairments. The DSM-IV estimates the prevalence in children closer to 10 percent with a predominantly hyperactive impulsive type or a predominantly inattentive type. With prevalence estimates of ADHD ranging from 1.9 percent to 17.8 percent and the persistence of the disorder into adulthood in 10–60 percent of the child onset cases, the consequences of this disorder are far reaching.

Over the last decade, there has been a dramatic increase in the number of ADHD-related visits to health care providers with an associated increase in pharmaceutical intervention in children. Conservative estimates of greater than 10 million prescriptions for methylphenidate in the United States alone support the increased incidence or recognition of a serious developmental problem. The personal, familial, and societal consequences associated with inappropriate over activity, distractibility, inattention, and impulsive behavior in children and adults are alarming.

Educators have identified problems in test-taking and difficulty completing homework that reflect the poor organizational skills which carries over into adulthood. Delinquent and antisocial behaviors are commonly associated in the male ADHD sufferer. Females make up approximately 10–25 percent of cases with a male to female ratio ranging from 4 : 1 for the predominantly hyperactive impulsive type to 2 : 1 for the predominantly inattentive type.

The DSM-IV criteria for ADHD require that a minimum of six symptoms of inattention or a minimum of six symptoms of hyperactivity–impulsivity occurances are noted. The symptoms must be present before the age of 7 years and must have persisted for a minimum of 6 months before a diagnosis might be made. The diagnosis of ADHD is largely subject to an individual fitting DSM-IV criteria leaving the clinical examination to act as a mechanism that might identify other medical or neurological disorders.

Physicians, educators, therapists, parents, and sufferers are searching for answers to their many questions regarding ADHD. In their textbook *Neurobehavioral Disorders of Childhood: An Evolutionary Perspective*, Drs Melillo and Leisman present such answers. The book is uniquely different as it is written by a clinician and a neuroscientist who have embraced a concern for society with an expression of humanism.

Rob Melillo and Gerry Leisman have specialized in the diagnosis, treatment, and science of neurological disorders for over 30 years. Their extensive clinical knowledge of neurology is evident and facilitates an understanding of the disorder of autistic spectrum disorders. Refreshingly, their work is not limited to the neurological but encompasses a unique historical or developmental approach to the subject that promotes an awareness of autistic spectrum disorders in a manner not before presented.

This text will promote change and better understanding of a complex dilemma. It will be well used by health care providers, educators, parents, and all of us concerned with the optimally developing child. I was honored to have been asked to write a foreword to this work, but blessed to have been fortunate enough to read it.

Frederick Carrick
Professor Emeritus of Neurology, Parker College
Distinguished Post Graduate Professor of Clinical Neurology, Logan College

Contents

Chapter 1	**Introduction**	1
	Comorbidities of Neurobehavioral Disorders of Childhood	9
	ADD/ADHD	9
	Learning Disabilities	10
	Obsessive–Compulsive Disorder (OCD)	10
	Tourette's Syndrome	10
	Pervasive Developmental Disorder (PDD)	10
	Autism/Asperger's Syndrome	11
	The Dopamine Connection	13
Chapter 2	**Evolution of the Human Brain**	15
	Evolutionary Perspectives	15
	The Mechanics of Walking	18
	Changes to the Vestibular System in Upright Posture	20
	Bipedalism and the Growth of the Human Brain	21
	Basic Structure of the Brain	21
	Neuron Theory	23
	Plasticity	25
	Evolutionary Growth of the Human Brain	28
Chapter 3	**Why the Brain Works the Way it Does: Evolution and Cognition from Movement**	33
	The Evolution of Cognition from Movement	39
Chapter 4	**The *Cerebellum* and *Basal Ganglia***	47
	Anatomy and Function of the Cerebellum in the Context of Neurobehavioral Disorders of Childhood	47
	Structure of the Cerebellum	47
	Functions of the Cerebellum	50
	The Cerebellum and Motor Learning	56
	Non-Motor Functions of the Cerebellum: Evolutionary Implications for Cognitive Function	58
	The Cerebellum's Role in Information Processing	59

Chapter 5	**The Basal Ganglia**	**69**
	Direct and Indirect Pathways	72
	The Dopamine System	75
Chapter 6	**The Thalamus**	**81**
	Anatomy and Function	81
	Closed vs. Open Systems and the Thalamus: The Binding Problem	84
	Subcortical Neurotransmitter Systems	
	Facilitating Interregional Communication	88
Chapter 7	**The *Limbic System* and the Evolution of the Cerebral Cortex**	**93**
	Anatomy and Function of the Limbic System	93
	The Anatomy of the Amygdaloid Nucleus (Amygdala)	95
	Functions of the Amygdala	97
	Anatomy and Function of the Septal Nuclei	98
	Anatomy and Function of the Mammillary Bodies	99
	Anatomy of the Hippocampus	100
	Functions of the Hippocampus	101
	The Hypothalamus	103
Chapter 8	**The Cerebral Cortex**	**105**
	Cerebral Cortex: Function and Development	105
	Frontal Lobes	108
	The Parietal Lobes	109
	The Occipital Lobes	109
	The Temporal Lobes	110
	The Prefrontal Cortex	110
	Brain Asymmetry	114
	Evolution and Development of Cerebral Lateralization:	
	A Focus on the Adaptation for Movement	116
	Behavior	119
	Non-Human Primates	120
	Ontogeny of Human Lateralization	121
	Motor Asymmetries	127
	Cerebellar Components of Speech and Language Asymmetries	132
	Handedness and Abilities	132
	Visual Processing	133
	Auditory Processing	137
	Face Processing	139
	Autonomic Asymmetry	142
	Neurotransmitters and Neuroendocrine Function	147
	Serotonin	148
	Dopamine	149
	Neuroendocrine System	149
	Asymmetry and Emotions	152
	Attentional Asymmetries	164
	Tactile Inattention	165
	Arousal	166

	Learning	167
	The Functions of the Cortico-Cortico Fibers	169
	Direct Transfer	174
	Third Party Convergence	174
	Nonconvergent Temporal Integration	174
Chapter 9	**Signs and Symptoms of Neurobehavioral Disorders of Childhood**	**177**
	Balance is Essential	183
	Tactile, Visual, and Auditory Symptoms	190
	Tactile	190
	Emotionally Significant	193
	Vision and Hearing	195
	Motor Symptoms	204
	Cognitive Disabilities	215
	Language	220
	Memory	224
	Emotional and Affective Symptoms	225
	Attention Deficit Hyperactivity Disorder	231
	Cortical–Subcortical Circuits and Affective Disorders	233
	Fear, Anger, and Violent Behavior	234
	Overlap of Symptoms	237
Chapter 10	**Causation from an Evolutionary Perspective**	**243**
	Voluntary Reduction in Physical Activity	245
	Neuromuscular Adaptations to Inactivity	249
	Neurological Adaptations	251
	Sedentary Activity and the Stress Response	252
	Gravity and it Effects on the Brain and Performance	253
	Gravity, Posture, Oxygen, and Performance	254
	Plasticity and Activity Dependence	257
	Involuntary Lack of Physical Activity	261
	Injury and Illness	264
	Prenatal Injury or Illness	264
	Prenatal Stress	267
	The Birthing Process	268
	The Neck and Movement in Relation to Brain Development	270
	Infection and Immunity	273
	Parental Physical and Environmental Influences	275
	Genetics and Neurotransmitters	279
	Catecholamine Theory	283
	Autonomic, Hormonal, and Immune System Involvement	285
	Nutrition and Autistic Spectrum Disorders in an Evolutionary Context	291
Chapter 11	**Therapeutic Theory and Strategy**	**305**
	Treatment Rationale	305
	Behavioral Intervention Strategies	305
	Sensory-Motor Intervention Strategies	318

Integrated Sensory-Motor Intervention Strategies 329
Theories of Physical-Mechanical Interventions 332
Metabolic-Physiological Intervention Strategies 339
Psychopharmacology 348

References 367

Index 441

1

Introduction

Childhood neurobehavioral disorders share many features in common. While often referred to as *Learning Disabilities*, the implication of such a term would indicate that the primary manifestation of these childrens' condition affects classroom behavior exclusively and that the rest of their development proceeds smoothly and without incident.

Attention deficit disorder (ADD), *attention deficit hyperactive disorder* (ADHD), *pervasive developmental disorder* (PDD), *obsessive–compulsive disorder* (OCD), *Asperger's* syndrome, and *Autism* to name but a few, may be viewed as points on a spectrum of developmental disabilities in which those points share features in common and possibly etiology as well, varying only in severity and in the primary anatomical region of dysfunctional activity.

As this text focuses on alterations in the normal development of the child, it would be natural and useful to develop a working theory based on what we know of the development and evolution of the human species and its brain.

In outlining our theory of developmental disabilities in evolutionary terms, we will offer evidence to support the following notions:

1. *Bipedalism was the major reason for human neocortical evolution.* We propose that bipedal locomotion is phylogenetically the most sophisticated and complex movement. It is unique to humans and was responsible for the development of the large human brain. This occurred, we postulate, because of the bipedal posture's unique ability to harness gravitational forces through the now upright postural motor system. Bipedalism has been utilized as a power source to maintain a genetic mutation, having created larger pools of neurons. The same evolutionary process has allowed us to develop the binding of the motor system into a synchronous, rhythmic, purposeful movement, which expanded to eventually allow for cognitive binding or consciousness.

Postural muscles, we claim, were the main conduit for this motor and cognitive binding to evolve and continue to exist. Deviations from normal postural development or from normal levels of postural activity can disrupt or delay *cerebellar* and cortical maturation and may disrupt the underlying oscillatory timing mechanisms on which both motor and cognitive binding is based.

2. *Cognition evolved secondary and parallell to evolution of motricity.* We will explore the relationship between cognitive and motor functions from an evolutionary perspective.

We will demonstrate that cognitive and motor functions are actually part of the same function, even though they have been historically viewed as separate. They both evolved in parallel as a product of the evolution of sophisticated complex movement. The same underlying mechanisms that evolved to enable more complex coordinated movements were adapted and utilized to effect more sophisticated cognitive processes.

3. *There exists an overlap of cognitive and motor symptoms.* Another major theme to be explored includes the articulation of *cerebellum*, *basal ganglia*, and *frontal lobes*, which are areas of the brain recognized to control motor and nonmotor intentional and executive function. Most developmental disabilities have as their most common symptom, motor incoordination or clumsiness, especially of posture and gait. Impulse control, either inhibited or facilitated, and judgment disorders can all be attributed to dysfunction of this network and its control of motor and nonmotor behavior. This spectrum of disorders all involve disruption primarily of what is known as executive functions which are functions attributed to the *frontal lobe*.

4. *Lack of thalamo—cortical stimulation not overstimulation is a fundamental problem of developmental disabilities.* We postulate that under stimulation of the *cerebellum* and *thalamus* produces a disruption of normal thalamocortical oscillations disrupting the ability to bind motor and cognitive functions in space and time. This temporal incoherence particularly affects the higher *frontal lobe* functions or executive functions and its reciprocal connections to the *basal ganglia*. This results in disruption of both executive and motor functions since these areas (the *frontal lobe, basal ganglia,* and *cerebellum*) are primarily responsible for both. Various studies show decreased activity and size of the *cerebellum, basal ganglia,* and *frontal lobe*, which is consistent with functional losses seen. Stimulant medication especially medications that increase *dopamine*, in children with developmental disabilities, have been shown to improve symptoms by increasing stimulation to those areas. *Dopamine* is the most common neurotransmitter in this motor system network and its dysfunction has been implicated in many neurobehavioral disorders in adults.

5. *A primary problem is dysfunction of hemisphericity.* We will explain why despite these conditions having a possible common neurological basis, children still present different symptoms. The explanation is that a primary problem relates to the failure to achieve equal activity and resonance in both right and left cerebral hemispheres. There exists an asymmetric distribution of almost all human functions within the cerebral cortex including cognitive, motor, sensory, neurohormonal, immune, autonomic, and endocrine. Failure to develop and achieve temporal coherence between the two hemispheres results in a form of functional independence of each hemisphere, resulting in adaptive functioning of the brain and its control properties. This may sometimes result in extremely high degrees of functioning in specific tasks. An example is the *savant* syndrome, but the same process is seen throughout all disorders in the spectrum and may even be a basis for individual differences in cognitive style. We will discuss the ramifications of viewing many of the cognitive and motor effects of developmental disabilities as based on a functional disconnection syndrome. Additionally, it will be noted that in these children, global functions are not affected, but many specific functions are decreased while others are enhanced, demonstrating region specific brain effects.

6. *Most conditions in this spectrum of disorders are the result of a right hemisphericity.* Most developmental disability syndromes can be clearly related to dysfunction or delay in development of the right hemisphere. The right hemisphere is understimulated resulting in slower temporal processing within that hemisphere, especially in the *frontal lobe*. This slower temporal processing results, in turn, in decreased effectiveness of the right hemisphere's normal executive functions. This decrease in activity has been shown with modern functional imaging of the brain, which has noted a decreased activity in the

right frontal cortex with an asymmetric distribution of activity in the *basal ganglia* and *cerebellum*. This right hemisphericity may also explain why males are affected more than females. Almost all of the specific disorders described earlier are found with significantly greater frequency in males. The frequency ranges from approximately 6 to 1 in *ADD* to 50 to 1 in high functioning *autistic* individuals. Male brains are more asymmetrical than female brains. Male brains are more susceptible to prenatal and postnatal influences; these influences, which primarily consist of maternal prenatal levels of estrogen, create this greater right cortical development than left characteristic of male brains. It has been further noted that *dopamine* decreases have a greater negative effect on right frontal cortex function than left due to the asymmetrical distribution of *dopamine* receptors in the brain.

7. *Environment is a fundamental problem.* Although genetic predispositions are likely to be present, the main factors in causation of developmental disabilities are hypothesized to be environmental, especially in the more severely afflicted. The dramatic rise in the diagnosis of these problems is not consistent with a purely genetic cause. We intend to demonstrate that current socially acceptable childhood behaviors, primarily those which are sedentary, such as a high proportion of time spent by the child watching television or playing video games, is a primary factor for the dramatic increase in neurobehavioral problems of childhood. The human brain is extremely plastic allowing us to adapt to the environment in which we develop. The long postnatal development in humans is primarily responsible for this flexibility. Although this has some obvious advantages from a evolutionary perspective it also makes the human brain more susceptible to negative environmental influences or lack of appropriate environmental stimuli. The window of time for the greatest development is between conception and the age of 6. Motor activities facilitate this brain development, particularly in males. A dramatic decrease in early motor activity will affect development of gross motor behavior, which is more specific to right hemisphere development. Sedentary behavior is pervasive amongst children and has been witnessed by dramatic increases in their obesity rates. These increased rates are comorbid with the rates and time frame of the increase in learning disabilities and neurobehavioral disorders. The primary social influences negatively affecting cognitive and motor functions in childhood consist of the increased use of television, VCR, computers, working parents, and parental fears for their children. Other environmental factors such as poor nutrition, increased caloric intake, environmental toxins, and early sensory deprivation are other important factors but are not as significant as sedentary behavior.

8. *All of these conditions are variations of the same problem.* We conclude that most developmental disabilities are of similar etiology and are variations of the same underlying problem. A high rate of comorbidity exists for all of these conditions. The *frontal lobe*s, *cerebellum*, *basal ganglia*, and *thalamus* have been implicated in all of these conditions. This has been documented on static imaging such as CT scans and MRI, as well as functional imaging such as PET scans and fMRI. In addition, *dopamine*rgic system involvement is highly related to each of these conditions in a way similar to that found in adult neurobehavioral disorders such as schizophrenia.

9. *These problems are correctable.* An additional theme propounded in the text concerns intervention strategies to remediate these conditions. As brain organization is plastic, many aspects of neurobehavioral disorders do not have to result in permanent impairment. Appropriate forms of environmental stimulation and behavioral modifications can significantly improve or completely correct the underlying problem. Since motor and cognitive dysfunction often coexist, improving the function of one effects changes in the other. In children, we will demonstrate that the main focus of treatment should be on improving motor performance combined with some cognitive training and behavior modification as necessary. We will demonstrate that motor and cognitive early intervention strategies will effect objective change in an electrical

asynchrony of the two cerebral hemispheres that is associated with positive change in both the cognitive and motor symptoms.

10. *Hemisphere specific treatment is the key to success.* Besides increasing motor performance, timing, endurance, and posture, we will finally address the need for hemisphere specific treatment modalities. Motor activity, sensory stimulation, and cognitive functions directed toward the under-functioning hemisphere is the most important consideration in treatment. Achieving temporal coherence or a balance of activity between the two hemispheres is critical for allowing cognitive and bilateral motor binding to occur, which would reduce hemispheric neglect. As the hemispheres achieve a normal coherence and synchronization, motor and cognitive performance will improve.

We consider this book the beginning and not the end of our work. We present this information so that others may reproduce our results and hopefully improve on them over time. We have used widely-accepted assessment tools whenever possible to help confirm and perfect our diagnosis and to help document the results. Hopefully, we may combine our results with others who will use our program, and we will be able to publish a vast body of information in the coming years. Our children and our society's future may depend on it.

Numerous investigators have reported that the incidence of *autism* spectrum disorders and neurobehavioral developmental disabilities are on the rise (Case-Smith and Miller, 1999; Fombonne et al., 1999; Chakrabarti and Fombonne, 2001). It also has been reported that today's children, in general, seem to have shorter attention span (Chakrabarti and Fombonne, 2001; Keren et al., 2001), more impulsive behavior (Chakrabarti and Fombonne, 2001), and decreasing scores in reading and language skills (Chakrabarti and Fombonne, 2001) than they had as recently as 10 years ago. It appears that these problems are becoming epidemic.

We have also been noting, both clinically and as reported in the literature, an increasing number of adult patients being diagnosed with *ADD/ADHD*, often following physical trauma, such as an automobile accident (Arcia and Gualtieri, 1994). There exists a similarity in the neurological symptoms between adult patients and children diagnosed as *ADD/ADHD*.

Autistic spectrum disorders including *ADD*, *ADHD*, and *autism* itself as well as other learning disabilities are growing at a staggering rate in the United States. In January, 2001, it was reported (Newsday, 2001) that 1 in 10 U.S. children have some sort of mental health problem, but fewer than 1 in 5 of them are being treated. This claim was made by the then Surgeon General of the United States, David Satcher, and reported in the same article. Among the conditions that are reportedly undertreated, include major depression, considered one of the most common, *ADHD*, and *OCD*. Satcher stated, "Short of those diagnosable problems, are problems that children have in their development and functioning very early." Satcher proposed a complete overhaul of how mental health in children is handled from training teachers and doctors, to better recognizing and understanding these disorders, to doing more research and translating that research into effective treatment programs. Satcher is also quoted as noting, "In any given year it is estimated that fewer than one in five of the children suffering from mental illness receive needed treatment."

In an article in *U.S. News and World Report* (2000), it was stated that 1 in every 6 children in America suffer from problems such as *autism*, aggression, dyslexia, and ADHD. In California, reported cases of *autism* rose 210 percent from 3,864 to 100,995 between 1987 and 1998. In New York, the number of children purportedly with learning disabilities jumped 55 percent from 132,000 to 204,000 between 1983 and 1996. "In the past decade, there has been a significant surge in the number of children diagnosed with *autism* throughout California," states the article. In August 1993 the article continues,

> there were 4,911 cases of so-called level-one *autism* logged in the state's Department of Developmental Services client-management system. This figure does not include children

with *Asperger's* syndrome, but only those who have received a diagnosis of classic *autism*. In the mid-'90s, the caseload started spiraling up. In 1999, the number of clients was more than double what it had been six years earlier. Then the curve started spiking. July 2001, there were 15,441 clients in the DDS database. Now there are more than seven new cases of level-one *autism*—85 percent of them children—entering the system every day.

California is not alone. Rates of both classic *autism* and *Asperger's* syndrome are reportedly rising all over the world, which is certainly a cause for alarm and for the urgent mobilization of research. *Autism* was once considered a rare disorder, occurring in 1 out of every 10,000 births (Fombonne, 2003). Now it is reportedly more common—perhaps 20 times more (Chakrabarti and Fombonne, 2001). However, according to local authorities, the picture in California is particularly bleak in Santa Clara County. Here in Silicon Valley, family support services provided by the DDS are brokered by the San Andreas Regional Center, one of 21 such centers in the state.

The total number of children affected at this point, we estimate to be approximately 15 million in the United States alone. No comprehensive treatment methodology exists at present that can effectively ameliorate the broad range of symptoms characteristic of children of *autistic* and neurobehavioral disorders. Drug treatment is not a panacea neither is educational or psychological remediation approaches. Some parents do not want to use medication because they are rightly afraid of possible long-term side effects. For others; the medication does not work very well or at all. While the vast majority of parents feel that psychological therapy is useful, many report that behavioral interventions did not help their children as much as they had hoped. They all feel there should be an alternative.

Most parents of children with developmental neurobehavioral disorders do not have a good understanding of the nature of the etiology of their children's dysfunction. They do not understand what is "wrong" with their children and often blame themselves, feel inadequate, and think of themselves as "bad" parents.

Subjective reports from experienced teachers have consistently yielded opinions that while the *ADD/ADHD* problem might be over diagnosed, there exists a growing problem among significant numbers of children in their classrooms. They have indicated that today's children are different now than children in their classes 20 or 30 years ago. The teachers claim to see a steady decline in the quality of the children's work over the last three decades and this trend seems to have been accelerating within the past 10 years. They find it much more difficult to teach because in today's classroom the children have a harder time paying attention or following instructions. The percentage of children with severe behavioral and language problems is reportedly increasing (U.S. Department of Health and Human Services [USDHHS], 1997). Teachers complain that children do not read as much as earlier generations and their comprehension, when they did read, had decreased significantly. The most frequent and consistent report of the many teachers that have been interviewed is that it is harder to get children today to sit still.

The teachers report that at lunchtime they see children line up in the hallway outside the nurse's office for their medication. "Years ago," one of the veteran teachers said, "… maybe you had one or two children who were hyperactive or with attention problems per class. Now in many instances, it is almost half the children who have problems." They all noted that many of their colleagues were opting for early retirement because they could not take it any longer and felt ineffectual in their careers. They also believe that the problem is getting worse and they are powerless to change it.

When asked if they knew the basis of these subjectively noted changes in school behavior the teachers, like many parents report that they do not really understand the nature of the physiological problems. Teachers appear to be consistently in agreement with each other. One teacher's response characteristic of so many others is,

> I don't know how it works, but I feel children don't use their muscles as much as they used to. At lunchtime, children do not play on the playground as they used to. They do not run around, they just sit or stand there. They seem less physically active. If you drive around any neighborhood, you do not see children outside. You used to see children out on their bikes or roller skates, climbing trees or playing ball. You don't see that any more. It's not that there are fewer children, in many cases there are more. They are inside watching TV, playing Nintendo or computer games.

She continued, "I don't know exactly what the relationship is or how it works, but I believe that there is definitely some connection."

Another common response from teachers questioned was that teachers reported a conjectural "cause and effect" relationship between increased developmental problems and an increased number of children coming from families where both parents work. They felt that increasing numbers of children from these families have less parental stimulation and fewer meaningful interpersonal relationships. Due to time constraints, parents do not talk to their children as much as they did 10 years ago, or read and spend time just being together. Because of this, the baby-sitter is often the television, video games or the computer, intrinsically solitary activities, so even talking with other people is greatly decreased. Parents attempt to compensate for less time with their children by providing more stimulating activities, or "quality" time between parent and child—just being together and talking and relating with one another.

Healey (1990) in her book *Endangered Minds*," quotes many interviews with teachers and some interesting studies with startling statistics. As an educator and administrator with 30 years experience, she comments that modern children seem to have "changing brains." Healey notes that youngsters today seem different than those she used to teach, even though the average IQ score has remained fairly stable. She became convinced that the changes are due to the way children are now absorbing and processing information.

Healey writes, "children were likable, fun to be with, intuitive, and often amazingly self aware." Today, in her estimation, they seem harder to teach, less attuned to verbal material both spoken and written. She discovered, "many admitted they didn't read very much, sometimes even the required homework." They struggle with or avoid writing assignments while teachers anguish over the results. One of the teachers she interviewed stated, "I feel like children have one foot out the door with whatever they are doing, they are incredibly easily distracted. I think there may have been a shift in the last five years." Healey (1990) developed a questionnaire requesting anecdotal information of cognitive changes observed in students. She handed it out at national meetings and conferences to experienced teachers working in schools where demographics had remained relatively stable. Approximately 300 teachers responded and there was unanimity in their response. The consensus was that attention spans had become perceptibly shorter and reading, writing, and oral language skills seemed to be declining, even in neighborhoods that were more affluent. Additionally, Healy indicated that no matter how "bright," students are less able to "bend their minds" around difficult problems in math, science, and other subjects as they had been able to previously.

Statistics comparing math and science performance of students from Asia, Europe, and the United States, show that students in the United States are at the "rock bottom," particularly in understanding complex interpretations of data (Healy, 1990). Healey goes on to quote a cover story in Fortune magazine indicating, "... in a high tech age where nation's increasingly compete on brain power, American schools are producing an army of illiterates."

The scholastic aptitude test (SAT) taken by students who intend to apply for college in the United States has shown drastically declining scores (Choy, 2002), particularly in the areas of higher level verbal and reasoning skills. Starting in 1965, the average SAT verbal and math scores have declined steadily until the mid-1980s, where they leveled off and then experienced a slight rise. Subsequent math scores have remained stable, but verbal scores have begun another gradual decline. Overall, verbal declines have been considerably greater, 47 points by 1988 as opposed to 22 for math (Choy, 2002). In the past, it was shown that children from less priviledged educational backgrounds had poor scores on standardized tests. Recently, scores of minority populations are the only ones showing consistent improvement, with African American students in particular making the most impressive gains (Choy, 2002). This steady decline in standardized testing scores has occurred in spite of the proliferation of commercial courses that claim to successfully prepare students for standardized tests like the SATs, and it happened especially in the privileged groups, whose double income parents can afford the additional expense. Educators agree that for all students, increased television watching and decreased reading negatively influence verbal and attentional performance of children (Kagan, 1971; White, 1975; Woody-Ramsey and Miller, 1988; Ruff and Lawson, 1990). The National Assessment of Education Progress (NCES, 2001) has reported deficiencies in higher order reasoning skills, including those necessary for advanced reading, comprehension, math, and science. The National Assessment of Educational Progress's (NCES, 2001) most recent report found that only 5 percent of high school graduates could satisfactorily master material traditionally used at college level. Perhaps, it is no surprise that 80 percent of the books in the United States are read by about 10 percent of the people. Cullinen (1997) asked a group of typical fifth graders what percentage of time they spent outside of school reading. Fifty percent read 4 minutes a day or less; 30 percent read two minutes a day or less, and 10 percent read nothing. Cullinen remarked that our society is becoming increasingly "alliterate," meaning that people know how to read, but choose not to. Most alliterates watch television for their news and achieve only a superficial level of understanding. Healey (1990) cites a report of a survey of 443 students entering community college revealed that 50 percent were reading below ninth grade level. The *New York Times* reported in March, 1988 (*New York Times*, 1988), that youngsters may "sound out" words better than they used to, but are actually understanding less (cf. Pearson, 1986; Pappas and Brown, 1987; Kamil et al., 2000). Several of the teachers interviewed by Healy felt that children from every neighborhood come to class with fewer social skills, less language ability, less ability to listen, and less motor ability and that in general, a frightening majority of children's attention spans have degenerated over recent years. Objective support for these reported conclusions is supported in the literature by numerous studies (Woody-Ramsey and Miller, 1988; Ruff and Lawson, 1990). We conclude that sedentary lifestyles, increased television viewing, and busy parents are major causes of these dramatic statistical changes in cognitive, motor, and academic performance of present day western school aged children.

Recent estimates (Nelson, 2002) state that three quarters of all school age children and two thirds of pre-school children have mothers in the labor force. A large percentage of those children come home to a house without a parent or other adult. Fifty percent of those parents do not have adequate day care available to them. More than half of American 1 year olds are spending their days with someone other than their mothers. Additionally, English is frequently a second language for many daycare workers. Therefore, the children in their care are initially learning to speak English from non-English speaking adults. When parents are not around and the availability and quality of daycare are not standardized, the television, VCR, and video-computer games become the surrogate babysitter. These are sedentary activities, which tend to breed increasingly sedentary behavior. We will later outline why we think the most important factor in children's "changing brains" is lack of physical activity.

There has been a significant lifestyle change in the last two decades where rates of obesity among children and adolescents have jumped 45 percent between 1960 and the 1980s

(Campbell et al., 2001; Steinbeck, 2001; Child Health Alert, 2002). A number of studies (Barlow et al., 2000; Kiess et al., 2001; Morgan et al., 2002) have shown that a significant number of American children are overweight and cannot pass basic physical tests of strength, endurance, and agility. In 1984, only 2 percent of 18 million who took the presidential physical fitness test received an award. The American Academy of Pediatrics recently issued a report declaring that up to 50 percent of the nation's school children are not getting enough exercise to develop healthy hearts and lungs and that 40 percent of youngsters between ages 5 and 8 exhibit at least one risk factor for heart disease (Story et al., 2002). In 1989, the United States Army was forced to modify the physical requirements in basic training. Lt. Col. John Anderson was quoted as stating, "it's our opinion that the young people coming into the military now have spent more time in front of the television than on the tennis court or softball field." (*Associated Press*, April 18, 1989). Newsday (November 17, 1997) reported that Dr Pat Vehrs, an adolescent and health expert at Baylor College of Medicine and Texas Children's Hospital in Houston had said that children are increasing their body fat and decreasing their exercise time. "Children and teenagers are increasingly obese and are not as physically active as their counterparts in previous decades. National surveys have shown that since the 1960s, *but especially in the last decade* (italics ours), there has been an increase in the percentage of body fat and a decrease in physical activity among youth." Not coincidentally, it has been during the past decade that we have also seen a sharp increase in the percentage of children with learning disabilities, attention problems, and subsequently increased use of Ritalin. Vehrs claims that 21 percent of children aged 6–17 may be classified as obese. He believes that there are many reasons for this trend, among which are the following:

1. Parents who work may ask their children to stay at home alone until their parents return. This encourages inactivity by limiting children to indoor pursuits.
2. Family members may participate in few physical activities together. Often parents are too tired when they arrive home to walk, shoot baskets, or ride bicycles with their children.
3. Fear of neighborhood crime.
4. Lack of sidewalks, well-lit streets, access to parks, gyms, or pools in the suburbs.

Vehrs also commented, "Parents shouldn't be complacent because their children are taking physical education classes at school." Physical education classes usually last less than an hour a day. "By the time (the children have) changed clothes, gotten into the gym, and taken roll, there is very little actual time left." The same is true for children's weekend sports like soccer where the child spends more time actually standing around than doing any real continuous physical activity.

The reasons that Vehrs provides for increasing obesity in childhood are the same reasons teachers give for decreased verbal and reading scores, and increases in sustaining attention. There seems to be a connection between the two, that decreased physical activity is related in some way to decreased efficiency of brain function and decreased scholastic and behavioral performance. Healey (1990) poses the question, "If young bodies are in bad shape, what about the brains attached to them."

Surprisingly, we know little about the relationship between motor performance and physical exercise on the one hand, and school success motivation, and subsequent abilities to concentrate on the other. Surprisingly, there are few studies in the literature that have explored the effects of a sedentary behavior on the child's capacity to learn.

We intend to thoroughly explore this relationship between motor activities and learning. We will examine the relationship between less physically active children and increased brain dysfunction. Studies examining alterations of neurochemistry, and clinical outcome studies are not sufficient to explain the measured and reported changes in child classroom behavior. While a significant portion of the explained variance of neurobehavioral disorders of childhood that affect learning ability has a genetic component affecting the child's neurochemistry, we posit that environmental factors are as significant in explaining and in possibly remediating this class of disorders. While it may be accurate that neurotransmitter function may be altered in some way in neurobehaviorally involved children but not necessarily because these chemicals cannot be adequately produced or because they are unbalanced. Neurotransmitters are simply messengers; the problem is more likely based with the message rather than with the messenger.

Current theory and approaches to treatment of neurobehavioral disorders in childhood are failing. We do require some comprehensive theory to account for these increasingly frequent disorders of childhood that can be translated into meaningful strategies for better addressing management at home and in school.

We first started to conceive of the text as an overview of attention deficit hyperactivity disorder. In the literature, and in the practices of numerous Education and Behavioral specialists we have been noting a more rapid increase of *autistic spectrum disorders*.

Before even beginning to speculate about commonalities in various neurobehavioral disorders, we have noticed both in the literature and in clinical practices that there seems to be a high rate of comorbidity between *ADHD* and *autism* (Bonde, 2000; Luteijn et al., 2000; Richardson and Ross, 2000; Noterdaeme et al., 2001) and other disorders such as *obsessive–compulsive disorder*, *Tourette's* syndrome, learning disability, dyslexia, *pervasive developmental disorder*, and *Asperger's* syndrome (Adesman, 1996; Gartner et al., 1997; Ghaziuddin et al., 1998; Bonde, 2000; Kadesjo and Gillberg, 2000). Many studies have found that over 50 percent of children diagnosed with an ADD, also meet the diagnostic criteria for one or more additional psychiatric disorders such as mood, anxiety, substance abuse, learning, or behavior disorders (Gartner et al., 1997; Ghaziuddin et al., 1998; Gralton et al., 1998; Bonde, 2000; Dykens, 2000; Kadesjo and Gillberg, 2000). It seems unlikely that a child would have two or three different psychiatric conditions simultaneously. Many experts theorize that the comorbidity of these conditions exist because these disorders have similar underlying genotypes, although there is no evidence to support such a contention at present. In *Tourette's* syndrome, the early developmental manifestations include not the characteristic vocal tics but rather the diagnosis of *ADHD* (Thomsen, 2000; Cohen, 2001; State et al., 2001; Zappella, 2002). Well then why is it not diagnosed as *ADHD* with a vocal tic rather than a whole separate diagnosis? If all of these conditions are usually present together in one child perhaps, they are really all in the same condition. We believe that this in fact is the case. Many researchers no longer look at these disorders as discrete separate conditions but rather *comorbidities*.

COMORBIDITIES OF NEUROBEHAVIORAL DISORDERS OF CHILDHOOD

Many researchers no longer look at these disorders as a discrete separate condition, but rather as a spectrum of disorders. They are more frequently being viewed as related clusters, spectrum, or dimensional groupings of slightly varying dysfunctions of a related functional system. Examples include Schizophrenic Spectrum Disorder (Bellak, 1994), Compulsive–Impulsive Spectrum Disorders (Oldham et al., 1996), *autistic spectrum disorder* (Towbin, 1994), and *depressive spectrum disorders* (Angst and Merikangas, 1997).

ADD/ADHD

Tannock and Schachar (1996) note, "that there is a growing consensus that the fundamental problems in (*ADHD*) are in self-regulation and that *ADHD* is better conceptualized as an impairment of higher-order cognitive processing known as (executive function)."

Castellnnos (1999) also noted that "unifying abstraction that currently best encompasses the faculties principally affected in *ADHD* has been termed executive function (EF) which is an evolving concept. ... There is no impressive empirical support for its importance in *ADHD*." What is clear in the literature is that the main functions that are affected have been termed executive functions and it is known that executive functions seem to primarily reside in the *frontal lobes*. In fact, *ADD* is considered a name for a spectrum of deficits of cognitive executive functions that may respond to similar treatments and are often comorbid with a wide variety of psychiatric disorders, many of which may also be spectrum disorders.

According to Brown (1991), this view of *ADD* as a cluster of attentional/executive impairments that appear and may persist with or without psychiatric comorbidity is consistent with Seidmans' findings from neuropsychological testing of children and adults with *ADD* (Seidman et al., 1995a, 1995b, 1997a, 1997b, 1998).

Hudziak and Todd (1993) also noted that the rates of comorbidity in children for *ADHD* and ODD was 35 percent, CD was 50 percent, mood disorders 15–75 percent, anxiety disorder 25 percent, and learning disabilities 10–92 percent. It has also been noted that individuals with ADD have a significantly increased probability of having increasingly additional psychiatric disorders (Biederman, J. et al., 1991; Jensen et al., 1997).

Learning Disabilities

Although the relationship between learning disabilities and *ADD* is not well understood, there is nonetheless significant resource that shows significant elevation of specific learning disorders such as reading disorder, math disorder, and disorders of written expression in individuals who are diagnosed with *ADD* (Cantwell and Baker, 1991).

Obsessive–Compulsive Disorder (OCD)

Numerous authors have noted varying degrees of overlap between *OCD* and *ADHD*. Percentage overlaps range from 6 percent (Toro et al., 1992) up to 32 percent and 33 percent (Geller et al., 1995, 1996) respectively.

Tourette's Syndrome

Most studies developmentally examining *Tourette's* syndrome and its comorbidity with *ADHD* demonstrate that between 25 and 85 percent of *Tourette's* syndrome probands have comorbid *ADHD* or *ADD* (Comings and Comings, 1984, 1985, 1986, 1987, 1988, 1998; Shapiro et al., 1988). Another interesting finding is that in *Tourette's* syndrome, as the severity of symptoms increases, the frequency of comorbid *ADD* also increases. It has also been noted that the combined prevalence of *Tourette's* syndrome in males was 1 in 1,400 and that males with *Tourette's* syndrome 27 percent had *ADHD*, 27 percent had sleep disorders, 17 percent had conduct disorders, 7 percent had *obsessive-compulsive disorder*, 27 percent had repeated a grade, and 24 percent had learning disorder (Caine et al., 1998). As we had noted earlier, the first signs of *Tourette's* syndrome are not necessarily vocal tics, but rather the diagnosis of *ADHD*.

Pervasive Developmental Disorder (PDD)

It has been noted that there also is a relationship between *ADD* and severe *autistic* and/or schizophrenic spectrum disorders (Luteijn et al., 2000). Luteijn and colleagues examined differences and similarities between social behavior problems in children with problems classified as *pervasive developmental disorder* not otherwise specified (*PDD-NOS*) and a group of children with problems classified as *ADHD*, as measured by parent questionnaires. In comparing the *PDD-NOS* group and the *ADHD* group, the results demonstrated that both groups have severe problems in executing appropriate social behavior. The two groups could be distinguished only by the nature and the extent of these problems. Roeyers and colleagues (1998) also investigated early clinical differences between children with a diagnosis of *PDD-NOS* and children with *ADHD*. A differential diagnosis between the two disorders is often difficult in infancy or early childhood. Twenty-seven children with *PDD-NOS* were matched with 27 children with *ADHD* as to IQ and chronological age. Their parents were retrospectively questioned on pre-, peri-, and postnatal complications and on atypical or delayed development of the children between 0 and 4 years of age. This exploratory study revealed almost no differences between both groups with respect to pregnancy or birth complications.

Autism/Asperger's Syndrome

The similarities between the symptoms and *autistic spectrum disorders* are actually significant when one looks at the symptoms associated with *ADHD*. In fact, when we examine them, they seem almost identical. It has been noted that *autistic* individuals maybe hyperactive, but that they also present with executive dysfunction in attention, impulsivity, and distractibility. It has also been noted that there is a similarity between *autistic* disorder and *Asperger's* syndrome and that *Asperger's* syndrome goes under many different types of names, some of the names are semantic-pragmatic disorder, right hemisphere learning disability, nonverbal learning disability, and schizoid disorder.

Much of this confusion has come about by the way we diagnose these problems. We would like to believe that there is a lab test or an objective test somewhere that confirms the diagnosis of *ADHD*, *OCD*, or *Tourette's*; but in fact, the diagnosis is purely subjective. There are no consistent anatomic or physical markers for these conditions. Most often, these disorders are diagnosed by a professional sitting down with a parent or teacher and reading to them a list of symptoms and checking off if the parent or teacher believes that the child manifests the relevant symptoms. However, even this process is not as clear-cut as it sounds. The list of symptoms is extremely vague and many of these conditions are hard if not impossible to distinguish.

One problem, according to Linda Lotspeich (Lotspeich and Ciaranello, 1993; personal communication, 2001), Director of the Stanford *Pervasive Developmental Disorders* Clinic, is that the rules in the *DSM-IV* do not work. "The diagnostic criteria are subjective, like marked impairment in the use of nonverbal behaviors such as eye-to-eye gaze, facial expression, body posture, and gestures to regulate social interaction." "How much 'eye-to-eye gaze' do you have to have to be normal?" asks Lotspeich. "How do you define what 'marked' is? in shades of gray, when does black become white? What is happening is that a group of symptoms is being called a disorder and if we add or subtract a few symptoms or make a few more severe, then it is called a different condition or syndrome. However, when we look at the areas of the brain involved in all of these conditions, and the neurotransmitter systems involved, they are all basically the same. Therefore, in reality, these are all possibly the same problem along a spectrum of severity. The most common of all comorbidities is *OCD*, developmental coordination disorder or more simply put "clumsiness" or motor incoordination. In fact, practically all children in this spectrum have some degree of motor incoordination. The type of incoordination is also usually the same. It involves primarily the muscles that control gait and posture or gross motor activity. Sometimes to a lesser degree, we find fine motor coordination also affected. Although it has been fairly well known that attention deficit disorders are comorbid with psychiatric disorders such as the ones described above, what is less known and what is more significant is the association between *ADD* and motor controlled dysfunction (clumsiness) or what has been termed as developmental coordination disorder (*American Psychiatric Association*, 1994). In the past, motor clumsiness or *OCD* have not been looked at as being psychiatric in nature, but rather being neurological and falling more under the realm of the pediatric neurologist. Motor control problems were first noted in what was then called the minimal brain dysfunction syndromes or MBD. Minimal Brain dysfunction was the term denoting children who had normal intelligence, but who had comorbidity of attention deficit and motor dysfunction or "soft" neurological signs.

Several studies by Denckla and others (Denckla and Rudel, 1978; Gillberg et al., 1982, 1993; Denckla et al., 1985; Wolffet et al., 1990; Landgren et al., 1996; Kadesjo and Gillberg, 1998) have shown that comorbidity exists between *ADHD* and *OCD*, dyscoordination or motor perceptual dysfunction. Several Swedish studies have shown that 50 percent of children with *ADHD* also had *OCD* (Brown, 2001).

In a Dutch study (Hadders-Algra and Towen, 1992), 15 percent of school age children were judged to have mild minor neural developmental deviations and another 6 percent demonstrated severe neural developmental deviations (occurring in boys twice as often as in girls). Minor developmental deviations were noted to consist of dyscoordination, fine motor deviations, choreiform movements, and abnormalities of muscle tone. Researches that have dealt with these minor neural developmental deviations tend to look at motor dysfunction as a sign of neurological disorder that may be associated with other problems such as language and perception dysfunction. Motor dyscoordination has also been noted as a significant sign in *autistic spectrum disorders* and in *Asperger's* syndrome. In fact, it has been speculated that the type of motor incoordination might be able to differentiate high functioning *autistic* HSA individuals from *Asperger's* syndrome individuals (Gepner and Mestre, 2002; Green et al., 2002; Rutherford et al., 2002). In *Asperger's* syndrome, it has been noted that individual's have significant degrees of motor incoordination. In fact, in Wing's original paper, she noted that the 34 cases that she had diagnosed based on *Asperger's* description, "90 percent were poor at games involving motor skill, and sometimes the executive problems affect their ability to write or draw." Although, gross motor skills are most frequently affected, fine motor and specifically graphomotor skills were sometimes considered significant in *Asperger's* syndrome" (Wing and Attwood, 1987; Wing, 1988).

Wing (1981) noted that posture, gait, and gesture incoordination was most often seen in *Asperger's* syndrome and that children with classic *autism* seem not to have the same degree of balance and gross motor skill deficits. However, it was also noted that the agility and gross motor skills in children with *autism* seem to decrease as they get older and may eventually present in similar or at the same level as *Asperger's* syndrome.

Gillberg (1989) reported clumsiness to be almost universal among children that she had examined for *Asperger's* syndrome. The other symptoms she noted that were associated with *Asperger's* syndrome consisted of severe impairment and social interaction difficulties, preoccupation with a topic, reliance on routines, pedantic language, comprehension, and dysfunction of nonverbal communication. In subsequent work, Gillberg included clumsiness as an essential diagnostic feature of *Asperger's* syndrome (Gillberg and Gillberg, 1989; Ehlers and Gillberg, 1993).

Tantam (1991) noted that 91 percent of the *Asperger's* individuals in his study were deemed clumsy and he reported that the most significant difference between *Asperger's* and non-*Asperger's* individuals was that ball catching was significantly poor in *Asperger's* individuals. Kline and colleagues (1995) noted that a significantly higher percentage of *Asperger's* rather than non-*Asperger's* autistic individuals showed deficits in both fine and gross motor skills either relative to norms or by clinical judgment. They further noted that all 21 *Asperger's* cases showed gross motor skill deficits, but 19 of these also had impairment in manual dexterity which seem to suggest that poor coordination was a general characteristic of *Asperger's*. With studies like this, many researchers have looked at fine motor coordinative skills as being disrupted as a general feature of *autistic spectrum disorders*. However, when we examine the condition from a hemispheric perspective, as we will note later, gross motor skill dysfunctions are more typical of right hemisphere involvement whereas fine motor skill dysfunctions are more typical of left hemisphere involvement. We will demonstrate later that both classic *autism* and *Asperger's syndrome* are associated with right hemisphere deficits, and thereby, would be expected to show a greater involvement of gross motor skill deficits. It might seem somewhat confusing initially when fine motor skills seem to be disrupted at almost equal levels. According to a neuropsychological model, this type of weakness would be more indicative of a left hemisphere deficit. However, when examining the literature closely, it has been noted (Szatmari et al., 1990) that manual dexterity is less effective for high functioning *autistic*s than for *Asperger's*, but only for the nondominant hand. This suggests a lateralized difference. This would show that although fine motor coordinative skill is decreased, it is decreased in the left hand more specifically, which is associated with right hemisphere function. This is consistent with a hemispheric imbalance model and specifically a right hemisphericity.

Manjiviona and Prior (1995) noted that 50 percent of *autistics* and 67 percent of their *Asperger's* group presented with significant motor impairment as defined by norms on a test of motor impairment. However, the two *autistic* subgroups did not differ significantly.

Szatmari and colleagues (1990) also noted that *autistic* groups did not differ from *Asperger's* groups with respect to dominant hand speeds on type boards although both were slower than psychiatric controls. Vilensky and associates (1981) analyzed the gait pattern of a group of children with *autism*. They used film records and identified gait abnormalities in these children that were not observed in a controlled group of normally developing children or in small groups of "hyperactive/aggressive children." Reported abnormalities were noted to be similar to those associated with *Parkinson's*. Hallet and colleagues (1993) assessed the gait of five high functioning adults with *autism* compared with age matched normal controls. Using a computer assisted video kinematic technique, they found that gait was atypical in these individuals. The authors noted that the overall clinical findings were consistent with a *cerebellar* rather than a basal ganglionic dysfunction.

Kohen-Raz and colleagues (1992) noted that postural control of children with *autism* differs from that of matched mentally handicapped and normally developing children and from adults with vestibular pathology. These objective measures were obtained using a computerized posturographic technique. It has been also noted that the pattern of atypical postures in children with *autism* is more consistent with a mesocortical or *cerebellar* rather than vestibular pathology. Numerous investigators (Howard et al., 2000) have shown independently empirical evidence that basic disturbances of the motor systems of individuals with *autism* are especially involved in postural and lower limb motor control.

THE *DOPAMINE* CONNECTION

Neural substrates, which may be especially important in executive function, working memory, and *ADD*, are those of the nigro*striatal* structures. Crinella and associates (1997) reported findings from organism studies suggesting that nigro*striatal* structures contribute essential, superordinate control of functions such as shifting mental set, planning action, and sequencing (i.e., executive functions). As Pennington and colleagues (1996) pointed out, many developmental disorders may result from a general change in some aspect of brain development such as neuronal number, structure, connectivity, neurochemistry, or metabolism. Such a general change could have a differential impact across different domains of cognition, with more complex aspects of cognition, such as executive functions, being most vulnerable and other aspects being less vulnerable. In this same context, Pennington and colleagues noted that the executive function impairments associated with *ADHD* and some other developmental disorders may all involve varying degrees of *dopamine* depletion in the *prefrontal* cortex and in related areas (p. 330).

In a review of findings from neuroimaging studies of the human brain, Posner and Raichle (1994, pp. 154–179) showed evidence of at least three anatomic networks that function separately and together to support various aspects of attention. The possibility that attention impairments resulting from *ADD* may be closely related to *dopamine* decreases in certain areas of the brain finds support in the numerous studies that have demonstrated *dopamine*rgic medications (e.g., *methylphenidate, dextroamphetamine*) to be effective in alleviating a wide variety of inattention symptoms (see Levy, 1991). Although *noradrenergic* medications (e.g., *desipramine, nortriptyline*) and alpha2-agonist medications (e.g., *clonidine, guanfadine*) have been demonstrated to be effective in alleviating hyperactivity–impulsivity symptoms of *ADHD*, there is some evidence that these nonstimulant medications are less effective in alleviating inattention symptoms (Spencer et al., 1996; Levy and Hobbes, 1988; *American Academy of Child and Adolescent Psychiatry*, 1997). These findings suggest that a specific neurotransmitter system, the *dopamine*rgic system, may play a particularly important role in inattention symptoms of *ADD*. Servan-Schreiber and associates (1998) summarized the research literature on the impact of *dopamine* on specific neural networks in human information processing. They developed and tested a model demonstrating that *dopamine* has a direct positive effect on the gain in the activation function of the neural networks underlying attentional processing. Additional evidence for the critical role of *dopamine* in management of cognition comes from recent laboratory studies summarized by Wickelgren (1997), which indicate that in many species *dopamine* plays a critical role in mobilizing attention, facilitating learning, and motivating behavior that is critical for adaptation. The role of *dopamine* in facilitating these functions may be far broader, subtle, and complex than had previously been thought. Inattention symptoms of *ADD* may be reflecting impairments resulting primarily from insufficient functioning of aspects of *dopamine*rgic transmission in the human brain.

What is the connection between the motor and the cognitive/emotional systems? In the past, motor areas of the brain were thought to be distinct from areas that control cognitive functions. However, over the last few years, those lines have blurred significantly and it is now recognized that areas like the *cerebellum* and the *basal ganglia* influence both motor function and nonmotor function as well. Motor and cognitive functions are closely related. In fact, it is thought that cognitive function, or what we call thinking, is the internalization of movement and that cognition and movement are really the same. We will attempt to better understand the connection between motor control, cognition, and posture and how these connectivities may be involved in learning and its dysfunction as well as in neurobehavioral disorders of childhood.

To fully understand the connection between motor and cognitive function and how they are connected in dysfunctioning systems we will examine these processes in evolutionary terms. We will explore the evolution of movement and how it relates to the evolution of nervous systems and ultimately brains and in particular the human brain. There are three elements that are important in facilitating an understanding of the growth of the human brain: (1) environmental stimulus and its effects on the brain, (2) plasticity, and (3) Darwin's theory of natural selection. With these three elements better understood, we can better understand why and how the human brain developed as it did.

2

Evolution of the Human Brain

EVOLUTIONARY PERSPECTIVES

"Nothing in biology makes sense except in the light evolution..." says Theodosius Dobzhansky, distinguished geneticist in Scientific American (Ewald, 1993). "Evolutionary biology is, of course, the scientific foundation for all biology, and biology is the foundation for all medicine. To a surprising degree, however, evolutionary biology is just now being recognized as a basic medical science.... The enterprise of studying medical problems in an evolutionary context has been termed *Darwinian medicine*.... Darwinian medicine asks why the body is designed in a way that makes us all vulnerable to problems like cancer, atherosclerosis, depression, and choking, thus offering a broader context in which to do research."

Modern science tells us that the earth was formed approximately 4.5 billion years ago. It is also thought that life started on earth about 3.85 billion years ago. How it started is still a mystery but it is thought that a microbe either arose spontaneously on earth, or was transported from space. However, either way life started with simple single cell creatures or *prokaryotes* and eventually *eukaryotes*. From there, life developed in the salty oceans and the first multicellular creatures were thought to be sessile and more plant-like. They were implanted on the floor of the ocean and would filter-feed, they did not move. However, it is thought that through a process known as *paedogenesis*, the foundation of vertebrates was formed. *Paedogenesis* is the process whereby the larval form of a creature becomes sexually active and thereby reproduces at that stage of development. This is what is thought to have happened several billion years ago. We can use a sessile creature such as a sea squirt as an example of how this may have occurred. Sessile creatures like the sea squirt were probably some of the earliest forms of life. These organisms do not actually have a nervous system so to speak; they have a primitive nerve net. However, the larval form of the sea squirt has a very different appearance. It resembles a tadpole. It has a muscular tail, a primitive *notochord*, and a form of nervous system, and brain. What happens in the sea squirt is that once the larva is formed it swims around and finds a hospitable place to feed and implant itself into the ground, the tail and nervous system dissolve and the creature becomes sessile once again without a nervous system. The lesson we learn from this is that a brain and nervous system is only necessary in moving creatures.

What is thought to have happened is that at some point, this larval form became sexually active and reproduced at that stage and life on earth was never the same. This scenario is very feasible. In fact, 70 percent of evolution is thought to have happened by the addition of a step in the beginning or subtraction of a step from the end of the development of the organism thereby creating a different form. The late Steven J. Gould thought that this is because there are different timing mechanisms for development in organisms. He thought that there are separate timing mechanisms for sexual and physical development and one may slow down or speed up while the other remains the same. By the time an organism reaches sexual maturity and stops developing, it may have achieved only the body plan of a juvenile ancestor. Alternatively, an organism may go through the entire development of its forerunners while it is still young and then simply continue the program, growing bigger horns, or more shell coils. Researchers now think that there are many different timers in a single organism, each controlling the growth of a step (Soll, 1983; Lloyd and Edwards, 1987). It has been speculated that the retardation of the development of the skull may have been connected with an expanded period of neuroblast proliferation in the cerebral cortex thereby increasing cell number (Noden, 1992). In this larva form, the sexually matured mechanism sped up allowing it to reproduce while the physical mechanism remained unchanged. Most importantly, we see that the brain was born out of movement. Even though these creatures could now move and swim all over, they still were filter-feeders.

The next major step in evolution occurred with the development of biting jaws, arising from modified gills and denticles. Armed with jaws, organisms were now free to roam anywhere in search of food and were no longer lowly bottom feeders. This further increased mobility and allowed creatures to inhabit, not only seas and oceans but also fresh water streams and rivers. As a result of frequent periods of drought throughout the world, fish inhabiting these streams and rivers would see these bodies of water significantly reduced in size and the oxygen content become dangerously low. Organisms that could develop an alternate source of respiration would be better fitted for survival and we know that some fish developed primitive lungs and were able to breathe air. A further stage in evolution was the development of the typical vertebrate shape, the development of pectoral and pelvic fins, and the lateral swim line. With more sophisticated movement, we see the development of a brain to control that movement and the first appearance of the *cerebellum*. Some organisms developed the pectoral and pelvic fins further to move around in the bottom of the water by means of fleshy lobed fins. These also provided an advantage during times of drought. When streams and ponds dried up and the oxygen content was low, fish with fleshy lobed fins and lungs could skirt along the land to another pond or stream and thereby enter that stream and survive. Although we start to see the ability to move on land, it was not until sometime later after insects proliferated and there was a food source on land that we see a development of amphibians.

This first part-time land dwellers had to be able to move equally as well in land and water so they maintained the basic shape of fish for movement. The midline trunk muscles provided the main thrust with oscillations back and forth. The limbs were used as mere anchors to pull forward. Eventually the limbs were adducted and were placed under the organism more often. This allowed the organism to easily lift itself off the ground reducing friction and allowing the organism to move quickly improving its ability to hunt, prey, or retreat from predators. It also allowed better respiratory function allowing the ribs to expand more efficiently and provided the organism with more oxygen to improve endurance. This also increased the available oxygen for the organism's brain, which is critical for brain growth especially in the areas of the *cerebellum* and cortex, which use the most oxygen. With the *thecodonts*, there developed true locomotion with a trend toward bipedalism. From the *thecodonts* developed the bipedal dinosaur and pterosaurs, the bird ancestors. It is thought that these organisms first became arboreal before they developed flight with a probable intermediate stage of gliding.

Although reptiles were the ancestors of mammals there was an early split close in time to the development of the class. As early as the *Pennsylvanian era*, there is evidence of the existence of a distinct line leading to the development of mammals (MacLean, 1985). This early stock consisted of the *synapsids*, which led to the more advanced *therapsids*. These were mammal-like reptiles with a marked change in locomotion. Their elbows were turned back and the knees turned forward. The arms were placed almost directly under the body where their weight was supported by the bones rather than by continuous muscular effort. This conserved energy and allowed for longer strides. With this limb structure, *therapsids* resembled the limb structure of mammals. It is thought that temperature regulation was more advanced in these creatures, and scales were replaced by hair. However, there does not appear to be an increase in brain size characteristic of mammals. *Therapsids* were dominant in later *Permian* and early *Triassic* times. In the later *Triassic era*, however, we see a sharp reduction in *therapsids* with only a scarce few remaining. Apparently, the ruling reptiles, possibly due to the development of bipedalism as a more efficient form of locomotion, temporarily overtook *therapsids*.

The *Jurassic* and *Cretaceous* periods belong to the dinosaurs while *therapsids* disappeared; however, they left behind descendants, which would once again eventually dominate as mammals. It is thought that the main characteristic that was unique to mammals was intelligence. In mammals, we also see the retention of the advanced locomotor activity of the *therapsids* with further development of a more efficient temperature regulation mechanism. Advanced mammals began domination again after the late cretaceous extinction of the dinosaurs and are most notable for the size and complexity of their brain. It is also thought that improvements in the reproduction of mammals are related to allowing the brain as long a time as possible to reach maturity before being used most efficiently. The retention of the embryo and the fetus in the mother for a longer period was a major step toward this end. Also, development of a nursing habit is thought to not only postpone the time when the youngster must go it alone, but also allows for a period of training to take place. Eventually, we see the development of small primates, similar to lemurs, which were mostly arboreal at the time. The development of *brachiation* also was an advanced form of locomotion, which would also require accompanying cognitive advances in hand–eye coordination and in predictive ability. We then see the development of advanced primates, which appear to have had an arboreal and partly bipedal existence, which has been postulated to exist in *Australopithecus* and *Paranthropus* species, the species that predate hominids.

Eventually, this development progressed further and produced various species of primates that were primarily tree-dwelling creatures. It was previously thought that approximately five million years ago with the beginning of the ice age, certain catastrophic environmental changes occurred in Southeast Africa, which destroyed most of the forests and left the *Australopithecus* in its wake. The story goes that some inhabitants stayed in trees and developed into chimpanzees and some were forced to remain on the ground. It was also previously thought that the first being known to walk upright was *Australopithecus afarensis* or Lucy, who was thought to have appeared about three or four million years ago. She was thought to have been the first bipedal primate. However, in 2001, it was reported (Haile-Selassie, 2001) that a graduate student named Yohannes Haile-Selassie had found what appeared to be the most ancient human ancestor ever discovered. It was a chimp-size creature that lived in the Ethiopian forest between 5.8 and 5.2 million years ago, which is nearly $1-\frac{1}{2}$ million years earlier than the previous Lucy. This new human ancestor had been listed as a subspecies variant of *Ramidus*, and has been given the name of *Ardipithecus ramidus kadabba*. One of the unique features of *Kadabba* is that it is thought to have walked upright much of the time. Although it is also thought that it probably spent some time in trees. It is thought that this creature lived in large social groups that would include both sexes.

Paleontologists have suspected for nearly 200 years that bipedalism was probably the key to the evolutionary transition that split the human line off from the apes. Fossil discoveries as far back as *Java Man* in the 1890s supported this belief. However, as is typical in this rapidly changing evolutionary environment, in July, 2002 it was reported that an even older hominid species was found. This new species named *Sahelanthropus tchadensis* was found by French paleontologist Michael Brunet and his team in the central African region of Chad (Brunet et al., 2002). This species is thought to be between 6 and 7 million years old and is thought to be the oldest known bipedal hominid species. This is also thought to be close to the time when humans and chimps first separated in evolution, a finding making things somewhat more confusing as the circumstances under which bipedalism was thought to have arisen has changed from the previously held view.

It had been thought that as Africa became significantly drier, the grasslands favored bipedalism so that early hominids could see over tall grasses to spot potential predators. Bipedalism, it is thought, also would have allowed the dissipation of heat from the grassland sun. However, this recent find changes these theories because, as it turns out, the earliest humans may have not developed in grasslands at all. In a companion paper to the one published in July, 2001 in *Nature* (Vignaud et al., 2002), it is reported that the most recent *Ardipithecus ramidus kadabba* and *Sahelanthropus tchadensis*, as did other then contemporary ancient hominids, all lived in a well-forested environment. At this time, scientists do not understand what the main advantage to walking upright was. According to evolution, we now have a picture of how human development may have progressed. About 6–7 million years ago, we have the newest member of the hominid species *Shadantropus tchadereniss* then approximately 5.8 million years ago, the *Ardipithecus ramidus kadabba*. Then, a million years later, its descendent, the renamed *Ardipithecus ramidus* appears. After that arises a new genus, *Australopithecus*, where Lucy belongs and it is only about two million years ago that the first member of the human genus *homo* arises. For most of the past six million years, multiple hominid species roamed the earth until approximately 30,000 years ago when modern humans and neantherdals coexisted.

Overall, these new findings raise the question as to the main advantage of bipedalism. We believe bipedalism to be associated not with the biomechanical or social development of *Homo sapiens*, but rather with changes in the development of man's nervous system that occurred because of walking upright. Nevertheless, before we explore this further let us understand the mechanics of walking.

The Mechanics of Walking

From a biomechanical perspective, Lovejoy and colleagues (1982), contend that the advantages of walking upright were somehow so great that the behavior endured throughout thousands of generations. It is well known that the anatomy of our ancestors underwent all sorts of basic changes to accommodate bipedalism. Some of these changes helped the body stay balanced by stabilizing the weight-bearing leg and in keeping the upper torso centered over the feet. Much of the changes associated with bipedalism may have improved coordination as well. They state "to walk upright in an habitual way you have to do so in synchrony." They state further, "if the ligaments and muscles are out of sync, that leads to injuries and then you would be chetah-meat." In addition, according to Lovejoy, by far the most crucial changes were those in the spine. The distance between the chest and pelvis is longer in humans than apes allowing the lower spine to curve, which locates the upper body over the pelvis for balance. This is an interesting statement when we understand that the main basis of consciousness itself, which is thought to be developed uniquely in the human brain and possibly as a byproduct of bipedalism, are the oscillations that occur in the brain and the ability to keep the brain in synchrony due to these oscillations (cf. see Leisman, 1976a; Koch and Leisman, 1990). It is also important that one half of our brain be in synchrony or coherent with the other and that possibly bipedalism may have helped to create such an environment in the brain.

In evolution, the laws of biomechanics are simple. Two factors determine adaptation. One is safety and the second is the conservation of energy. This means that biomechanically, if an adaptation is safe and more energy efficient, it will most likely be selected. Walking and standing upright, both fit these criteria. Bipedal walking is efficient; it uses two limbs instead of four, and in doing so conserves energy. We can still run over a short distance as fast as some organisms with four legs. There are stories of American Indians, for example, being able to chase and catch horses on foot. Bipedalism did not compromise our speed or agility to hunt. We can swim and climb trees. Our hands are freed to make tools and allow greater mobility to search for food in distant locations. Humans have been hunter–gatherers for 99 percent of their history, so the physical adaptation to hunt and gather mechanically makes the most sense. From a biomechanical standpoint, the key to standing and walking upright involves the architecture of the spine.

During the evolution of erect posture, the lumbar or lower back joints, and the lumbosacral joints developed the ability to achieve a position of pronounced extension. This allows for a marked lumbar curvature or *lordosis* of the spinal column. This curve is present and developed not only in the lower back, but also in the neck. This adaptation puts the spine and head in an upright position. Except for the tailbone or sacrum, the spinal column has no curves at birth. The thoracic or mid-back part of the spine gradually develops a relatively fixed curve in the young child. A flexible cervical or neck curve appears when the infant is able to raise its head.

The flexible lumbar curve or lower back curve appears at the end of the first year when the child starts to walk. The lower back curve is necessary to obtain the erect posture because the pelvis remains essentially in the same position as that in standing all fours in the quadruped position. The fact that the pelvis did not shift from its quadruped position during evolution of the erect posture also necessitated placing the hip and knee joints into full extension.

Additionally, the arch of the foot developed so that the bones are structurally arranged to support the body weight with reduced muscular activity. In humans, where the ligaments are passive and do not use energy, the bones and fully extended hip and knee joints bear the brunt of the forces involved in standing erect. Only humans stand perfectly erect. Quadrupeds, including knuckle-walking apes can mimic the human erect posture. However, they do this with a great expenditure of energy because their hips and knee joints cannot be fully extended; extended so that the passive ligaments and joints can withstand the forces involved in standing erect.

Evolution must be energy efficient or it will not proceed optimally. We see this same expenditure of energy when a child first starts to stand with partially flexed hips and knee joints. An erect posture appears to be most awkward when compared to a quadrupedal posture; however, it is the most efficient and economic posture to have evolved. Once man rose by muscular activity to attain the fully erect position, the contraction of small "postural" muscles in the spine, hips, and legs are required to keep the head, trunk, and limbs aligned vertically in line with the center of gravity. Evolving to this position took some effort.

The driving force behind the spinal curve and the changes in the hip and knee joints are muscular activity. The physiological principle known as *Wolfe's Law* states that bone or connective tissue will adapt based on the forces that are applied to it. Those forces are generally muscles and gravity. The contraction or activation of muscle would have had to promote an upright posture, and relaxation or inhibition of muscle would have been required to promote a quadrupedal position. We think that these principles are associated with the uniqueness of our spine, pelvis, knee, and foot. Although the final achievement of standing is efficient, there must have been a greater incentive than purely biomechanics.

What controls the muscles that coordinate muscular contraction that supports locomotion? These neuro-anatomical changes developed rapidly in the evolution of man, with the flexible muscles and ligaments adapting rather than the bones and joints which would have required a longer time to adapt. Efficient and complex movement is best accomplished by efficient brain stimulation. Biomechanically and from the standpoint of survival of the human species, bipedalism makes sense. Ultimately, the brain's adaptation allowed for greater motor control and upright locomotion.

Changes to the Vestibular System in Upright Posture

The unique posture and obligatory bipedalism of modern humans are unique among living primates. According to Gracovetsky and Farfan (1986), bipedalism makes sense from a mechanical perspective because it is energy efficient without losing much in potential for speed. However, there must also have been advancements in the mechanisms to provide for the unconscious perception of movement, especially the vestibular apparatus and the *cerebellum*. Modern human locomotor behavior makes specific demands on the vestibular apparatus due to the fact, that it requires an upright body posture balancing on a very small area of support. In humans, the morphology of the vestibular apparatus comprises three bony semicircular canals and their dimensions are thought to reflect the arc size of the enclosed membranous semicircular ducts.

It is thought that there is a direct relationship between duct size and the sensitivity and time constants of the semicircular canals and monitoring systems and that canal size and locomotor behavior are interrelated (Spoor and Zonneveld, 1998; Spoor et al., 1999). It has been shown that modern humans have large anterior and posterior canals and a smaller lateral canal than the great apes. In contrast, fossil hominids such as *Australopithecine* show great ape-like proportions and *Homo erectus* shows modern human-like proportions. Increased sensitivity of the vertically oriented anterior and posterior canals in humans would make sense due to the role of the vestibular system in coordinating upright bipedal posture through the vestibular reflexes which monitor body movements in the vertical plane. This would be especially true in more complex movements such as running and jumping which is thought to be absent in less advanced species such as *Australopithecine paranthropus*, and earlier hominids. What we see here is a clear example of how biomechanical changes in movement and posture are accompanied by structural and functional changes in the nervous system areas that control and coordinate that movement and this is the first example in humans that has been noted.

The cortex, especially the neo-cortex and neo-*cerebellum*, is the most sophisticated nervous system. Our cortices support and control the ability of the human to stand and walk upright. How did standing upright and walking affect the unique development of our brain and what is the basis for assuming a relationship?

Fossil records indicate that *Homo rudolforensis* may have descended from *A. faraensis* between 2.3 and 1.3 million years ago. These beings had a progressive upright stance, increased use of tools, greater growth of the brain, and more socialization than contemporaneous hominids. *Homo rudolferensis* is also known to have hunted in groups. This required more interaction and better communication skills in order to make plans and cooperate with one another to achieve a common goal. This also probably forced or framed some of the groundwork to develop sophisticated language skills and the ability to think and reason—characteristics of an intelligent being. The progression from *Australopithecus* to *Homo rudolforensis* to *Homo sapiens* probably happened over a period of 1 to 1.25 million years.

How this last step of evolution progressed is a mystery of surpassing consequence. One of the most baffling questions that has confounded scientists and philosophers is, "Why did the human brain grow much larger than its nearest relatives' brains and why did it happen in such a short period of time, by evolutionary standards?" The 400 cm³ brain of *australophitecines* was the product of hundreds of millions of years of evolution. However, it took only several million years, literally the "blink of an eye" in evolutionary terms, for the brain to triple its size and become capable of complex thought which led to further physical adaptations.

Early humans did not need great intellect to deal with the hunter–gatherer life style. Intelligence was obviously a later development. *Homo erectus*' brain was roughly the same size as ours. *Homo erectus* with the same size brain as ours did not speak or write. We today, on the other hand, still have not even come close to our intellectual potential. Why did we "build" a bigger brain than we needed to survive? What kind of adaptation might have increased the brain size and more importantly, what does this tell us about our brain now?

Bipedalism and the Growth of the Human Brain

There has been much speculation over what may have been the driving force behind the physical growth of the brain. Many theories, including tool use, throwing, language, and culture have been given as possibilities. It is clear that all these circumstances developed after the massive growth of the brain was achieved and all of the speculation leads to what is known among biologists as *exaptation*. That is, an evolutionary change that adapts organisms to one set of environmental conditions, but additionally positions them for a new surge in adaptive evolution. One example of exaptation becoming adaptation are birds' feathers. These structures are essential for flight, before flight they were used as insulators. New structures do not necessarily arise for a particular reason. They simply develop spontaneously as by-products or routine copying errors. All new genetic variants must arise as exaptations. Exaptations are features that have arisen and are available to be used in some new function. Before we explore this question further it is a good time to review the basic structure, growth, and function of brain cells.

BASIC STRUCTURE OF THE BRAIN

1. Neurons.
2. Dendrites.
3. Axons.

The basic cell or building block of the brain is the neuron. There are at best estimates anywhere from 30 to 100 billion neurons in the human brain (Leisman, 1976a). By comparison, our nearest relatives, the chimpanzees and gorillas have about 7–8 billion neurons (Nieuwenhuys et al., 1998). There are a few different categories of neurons, but all areas of the brain are the same. The neuron has two basic jobs; to receive and to transmit information. It does these jobs through what is known as a nerve impulse as outlined in Fig. 2.1. A single neuron can handle as many as 50,000 messages per minute at least (Leisman, 1976a; McClurkin and Optican, 1996). The neuron consists of a cell body, which has one nucleus, long fibers at one end called the axons, which transmit impulses, and short spine-like branches called dendrites, which receive information. The neuron receives signals in the form of electrochemical impulses and then transmits the impulse through the neuron by way of the axon ending at branched synaptic terminals as outlined in Fig. 2.2. The impulse can travel along the length of axons at speeds exceeding 100 miles/hr. The axon comes close, but does not touch the dendrites of neighboring neurons. There is a small gap or space between the axon and the dendrite known as a synapse. The neuron has one cell body, but can have many branches as well as many dendrites to receive signals. A random section of the human cortex contains an enormous number of synapses somewhere in the order of 600 millions/mm³ (DeFelipe, 2002). There are, therefore, estimated to be approximately 10^{10} synapses in the human cortex. When one

22 NEUROBEHAVIORAL DISORDERS OF CHILDHOOD

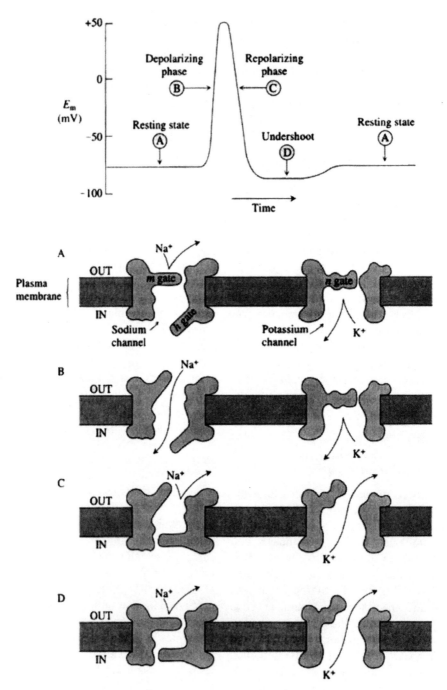

Fig. 2.1. The action potential.

learns something new, one creates new lines of connections between neurons (Ornstein, 1991; Squire, 1991; Koch and Leisman 2000; Leisman and Koch 2000). As one increases the number of connections, one increases the number of *glial* brain cells. These cells act like glue to hold brain cells together and they are thought to act like "housekeeping" cells. They help supply neurons with more energy to do more work and send more signals.

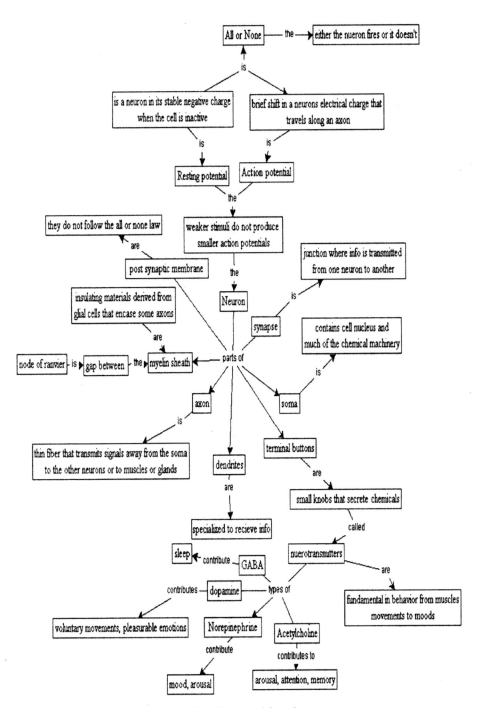

Fig. 2.2. Neuronal function.

Neuron Theory

The ability of a neuron to transmit a message is due to the electrical and chemical properties of the nerve cell. This is mostly because there is a difference between the electrochemical charges of the extracellular fluid with reference to the intracellular fluid of the cell. The outside of a neuron carries a positive charge relative to the inside. This charge is the product of certain ions that have a relatively high concentration outside the cell. These are mostly sodium, Na^{++} and chloride, Cl^-. Sodium is the most abundant ion present in the extracellular fluid, potassium is also present but to a much lesser degree. Sodium carries a positive charge; therefore, the outside of the cell is relatively positive. The inside of the neuron also contains electrically charged ions primarily potassium and calcium and to a much lesser degree sodium. The electrical charges of these ions are primarily positive, but the majority of the inside of the cell consists of protein, which carries a negative charge.

Protein is the main constituent of the internal organelles as well as the cell membrane and neurotransmitters. Therefore, the outside of a cell is positive with its main ion being sodium while the inside is negative due to protein, but the main ion concentration in the intracellular fluid is potassium. The separation of negative and positive charges across the membrane produces an electrical potential. This potential in central cells is generally around -65 mV. There is approximately a $35:1$ concentration of sodium outside the cell to inside the cell. Potassium has an approximate $10:1$ concentration on the inside relative to the outside. When the difference between the charge of the inside and the outside becomes less this is known as depolarization. When this becomes low enough it reaches a threshold, an action potential takes place, and a current is propagated down the nerve cell. Whether a nerve cell fires a signal or not, is a product of incoming signals that cause changes in the relative flow of sodium and potassium ions across the membrane.

When a cell is stimulated at its terminal end or dendrites, a local change in the permeability of the cell occurs. If the permeability of the cell favors sodium rushing into the cell, the cell will move closer to its firing threshold, which is approximately -45 mV. If the stimulus changes the permeability of the neuron and favors potassium rushing out of the cell, the cell becomes hyperpolarized or more negative and it moves the cell further away from its firing threshold. There are thousands of dendrites on the end of each cell and they contact thousands of other cells. At any given time in the area of the dendrite, there are thousands of signals some of which cause sodium to rush in, depolarizing the cell, bringing it close to the threshold and there are also signals that hyperpolarize an area of the cell making that area move further away from the threshold. The integration of all these different signals occurs in the cell body of the neuron. It adds all the positive and negative impulses and assures either that a cell will not fire or eventually if there is enough sodium rushing in, a wave of this change flows down the cell body until it reaches an area where the cell body meets the axon. This area is known as the axon hillock and it is significant because it is much more permeable to sodium than any other area. In fact, it is approximately seven times more permeable to sodium than to any other ion. Therefore, when this wave of depolarization makes its way down the cell body to the axon hillock, there is a sudden rush of sodium in the axon hillock, an action potential takes place, and the signal is propagated down the axon to the terminal end.

When an impulse reaches the end of the axon, spherical bodies called vesicles are released and fused with the axon membrane. The vesicles then open and release chemicals called neurotransmitters, which cross the space or synapse and cause the neighboring cell to then send a signal along its axon. To end the signal, the axon reabsorbs some neurotransmitters and enzymes in the synapse.

The neurotransmitter specifically contacts receptors on the post-synaptic neuron to take one of the two actions: either it will cause sodium to rush in, or potassium to rush out. Either it will produce a local excitation of the cell, or it will produce local inhibition of the cell. No single neuronal input triggers an action potential; there is not enough power for the flow of electrochemical signals to continue. It is the combination of excitation and inhibition of thousands of synapses that give the neuron a virtually limitless possibility of combinations that may affect it. So just as the simple binary system of computing in 0s and 1s gives computers tremendous processing capacity, the brain has a related process of excitation and inhibition.

Plasticity

The brain has approximately 100 billion neurons with trillions of possible connections. It has been estimated that there are more synaptic connections in one human brain than there are stars in the known universe. This gives the brain virtually limitless processing power. There is tremendous variability in this system. There are thousands of different neurotransmitters, the number of dendrites and connectivities change with activity (Eccles, 1987; Koch and Leisman, 1990). The type and the density of receptors can be regulated up or down. Hormones can influence the sensitivity of the cells and/or neurotransmitters and the number of synaptic connections can vary. The human brain is extremely plastic, which means it is modifiable based on activity and experiences. It can change its shape, size, number of branches, and number of connections as well as the strength of its connections.

There is increasing evidence that the strengths of individual synapses in the different regions of the human brain are enhanced by repeated activation (Eccles, 1987; Bliss and Collingridge, 1993; Benke et al., 1998). This provides strong support of the original hypothesis by Hebb (1949) that generation of memory, for example, results from such a process, as do other neuropsychological functions. The *hippocampus* is known to be both the seat of short-term memory and the initiator of long-term memory through its connections to the higher centers of the brain. While the means through which long-term memory is implanted is not well settled, it is highly likely that recollection of memories, once they are implanted, is a collective process of integration of information distributively stored in different regions of the neo-cortex (Zola-Morgan and Squire, 1993). This idea is reinforced by the direct connections of specialized hippocampal cells (as seen in recognition tasks in humans (Fried et al., 1997), "place cells" in rats (Muller, 1996), and "spatial view" cells in primates (Rolls et al., 1998)) to specific cortical areas (Eccles, 1987; Zola-Morgan and Squire, 1993).

Neurons in the brain are interconnected with many other neurons and, in turn, each neuron receives input from many synapses of its dendrites and cell body. The resulting neuronal loops according to Hebb (1980) contain neurons whose output signal may be either excitatory or inhibitory. Although the neuronal loops are usually drawn as though they were in the cortex, many of the loops probably run from the cortex to the *thalamus* or other subcortical structures and back to the cortex. Because each neuron is thought to both send and receive thousands of outputs and inputs, the number of possible neuronal loops is truly immense.

In Hebb's (1949, 1980) theory, each psychologically important event, whether a sensation, percept, memory, thought, or emotion, is conceived to be the result of activity flow in a given neuronal loop. Hebb proposed that the synapses in a particular path become functionally connected to form a cell assembly. In Hebb's view, the most probable way in which one cell could become more capable of firing another is that synaptic knobs grow or become more functional, increasing the area of contact between the afferent axon and efferent cell body and dendrites.

Hebb assumed that if two neurons are excited together, they become linked functionally. In Hebb's view, the cell assembly is a system that is initially organized by a particular sensory event but is capable of continuing its activity after the stimulation has ceased. Hebb proposed that to produce functional changes in synaptic transmission, the cell assembly must be repeatedly activated. After the initial sensory input, the assembly would therefore reverberate. Repeated reverberations could then produce the structural changes. Clearly, this conception of information storage could explain the phenomenon of short- and long-term memory: short-term memory is reverberation of the closed loops of cell assemblies; long-term memory is more structural, a lasting change in synaptic connections.

In Hebb's theory, there is yet another factor in neuropsychological function. For the structural synaptic changes to occur there must be a period in which the cell assembly is left relatively undisturbed. Hebb referred to this process of structural change as consolidation, a period thought to require 15 minutes to an hr. Its existence was supported by observations that retention failed when brain function was disrupted soon after learning, as, for example, in the amnesia for events just before a concussion. The case of patient H. M. (Milner, 1968) invited a logical extrapolation of Hebb's theory: the *hippocampus* was assumed to be especially important to the process of consolidation, although just how it is involved could not be specified. New material is not remembered because it is not consolidated; old material is remembered because it was consolidated before the hippocampal damage. Finally, Hebb assumed that any cell assembly could be excited by others. This idea provided the basis for thought or ideation. The essence of an idea is that it occurs in the absence of the original environmental event to which it corresponds.

In a thoughtful review, Goddard (1980) discussed the cell assembly and found that with few modifications, it was then a sound metaphor for psychological behavior. Goddard based his argument on the work of Eccles (cf. 1987) and others. Bliss and Gardner-Medwin (1973) demonstrated unequivocally that electrical stimulation of a neuron could produce either brief or long-lasting changes in synaptic transmission, according to the characteristics of brain stimulation. Brief pulses of current are delivered to an axon over a few seconds and the magnitude of the response is recorded from areas known to receive projections from the stimulated axon. After a stable baseline of response to the stimulation has been established, the stimulation is changed to one of high frequency, driving the system very hard. This high frequency stimulation is then discontinued and the brief test pulses are resumed.

The magnitude of the post-synaptic response can be compared with the original baseline and the time course of the decay of changes in response magnitude can be measured as well. Two significant findings emerged from this study. First, response magnitude markedly increased immediately after the high frequency stimulation. This increase declined over time and returned to baseline. The rate of decline depends on the details of stimulation. This short-term increase is called *post-tetanic potentiation*. Second, the changes in response magnitude may not decline to baseline but instead remain elevated, possibly for days or as long as is practical to measure it. This has been called *Long-Term Potentiation* (LTP). In some cases, LTP may be present after two months and Barnes (1998) has shown that LTP is prolonged by occasional repetition of the high-frequency stimulation. The original studies of LTP were performed on the *hippocampus*, but these phenomena can be demonstrated elsewhere in the brain.

Goddard (1980) emphasized the similarity between the phenomena of short-term memory and post-tetanic potentiation and between long-term memory and LTP, supporting Hebb's (1949) original theory. As attractive as the physiological work is as a model for short- and long-term memory, it is still a substantial theoretical leap to understanding the effects of lesions on memory, such as the differential effects of temporal and parietal lobe lesions on short- and long-term memory, respectively.

The demonstration of LTP is important but it still leaves open the question of what change in the brain allows such physiological phenomena. The nature of the changes that occur at the synapse in information storage is still uncertain. Greenough and colleagues (1985) have shown that when organisms are trained in specific tasks or are exposed to specific environments, there are changes in the dendrites of neurons. If there is an increase in the number of dendrites of particular neurons, then it follows that there might be an increase in the number of synapses on these neurons. Greenough and his colleagues have shown this to be the case.

In addition, they have shown that there is a qualitative change in the synapses, presumably including not only new ones but also existing ones that have been changed by experience.

These include changes in the size of various synaptic components, in vesicle numbers, in the size of post-synaptic thickenings, and in the size of the dendritic spines. (For a more complete review see Greenough and Chang, 1988.) Similar changes have also been found in neurons exhibiting LTP, thus adding evidence that LTP may be an analogue of normal learning.

The cause of these changes in the synapse is unknown at present. Various hypotheses have been advanced to suggest that alterations in protein synthesis in neurons might be responsible, possibly because of some sort of change in gene expression in the neurons, which may be expressed through changes in RNA (Black et al., 1987). It follows that blocking protein synthesis ought to block both LTP and new learning, and this appears to be so. Another hypothesis is that the use of neurons leads to changes in pre-synaptic calcium permeability, which, in turn, leads to a series of biochemical changes (Lynch and Baudry, 1984). At present, although all of these hypotheses remain speculative, it seems highly likely that long-lasting behavioral change stems from a morphological change in neurons.

From an evolutionary perspective, the human brain's large size is partly due to genetic mutations, which created a larger pool of neurons. The process of synaptic stabilization, however, where some cells live and others atrophy is related to the activity level of those cells. In addition, it is not the number of neurons that is unique to the human brain alone, but the plastic changes of the neurons, the density of the connections, and the thickness of the cells themselves. The number of *glial* cells is also significant. *Glial* cells have a "housekeeper" function, but they also are the cells that insulate and improve the speed of conduction of neurons. They also help provide nutrition and support and it is estimated that the human brain consists of 60 percent fat contributing to its extremely large size, mostly constituted by *glia*. There is some controversy as to whether the human brain has the ability to create new nerve cells. There is some recent evidence to suggest that neurogenesis in adult brains is possible and may also be activity dependent (Rakic and Goldman-Rakic, 1982; Rakic et al., 1994; Chen et al., 2002; Gage, 2002; Kato et al., 2002; Kim and Diamond, 2002; Munte et al., 2002). There is consensus however that *glial* cells can readily reproduce in adult brains and this is dependent on the activity and metabolic needs of the neurons. It is well documented that plasticity also occurs within the neuron itself, not just chemical plasticity, but also structural plasticity. This is thought to occur because of environmental activity through either a depolarizing or hyperpolarizing stimuli.

With any stimulus to the cell, there is the activation of cellular immediate-early-gene responses. These genetic responses regulate the production of protooncogenes that are the precursors to the production of proteins within the cell. When the cell is stimulated, it will produce more protein for production of membrane, neurotransmitters, and organelles, like mitochondria. The increased protein makes the intracellular fluid more negative and maintains a polarized state. This is important because neurons work best when there is a clear delineation when the cell is "on" or "off." This is also known as signal-to-noise ratio; brain cells should have a high signal-to-noise ratio. Cells that do not have a clear separation between activation and rest are thought to be "jumpy." Cells that are "jumpy" are more likely to fire spontaneously as in epileptiform activity. They also may fatigue or reach oxidative stress levels more quickly, where they may cease to work, or they may produce free radicals, which can permanently damage or kill the cell. The mitochondria provide more energy in the form of ATP, which gives the cell the ability to work for longer periods before fatigue. The proteins that are formed are also used to increase the density of dendritic spines, axonal branches, as well as receptors on the membrane of the cell. These changes can increase the size and density of the cell and the brain as a whole.

The most significant factors in plasticity are the frequency and duration of activation of the cell. Integration at the neuronal level occurs because of a process known as summation. *Temporal summation* occurs when the same input is triggered repeatedly so that stimulus threshold is lowered to permit the cell to fire. *Spatial summation* results when sufficiently different dendritic spines are stimulated simultaneously, so that firing threshold is reached. In *spatial summation*, the more contacts that activate the cell simultaneously, the greater likelihood the cell will depolarize enough to produce an action potential. *Temporal summation* concerns the speed at which the contacts fire. Each time a wave of polarization passes through the cell, it will degrade in approximately 15 msec. However, if we activate that cell before it degrades, we can then super add that wave on top of previous waves.

Evolutionary Growth of the Human Brain

The human brain is extremely consistent in terms of its cellular makeup. The fact that the brain cells are made up of the same basic cell structure in all areas is one of its most striking features. Therefore, what is true for one brain cell is true for all brain cells in all areas of the brain.

In order to function and grow, brain cells need two things. One is fuel in the form of oxygen and glucose; and the second is stimulation. However, increasing fuel alone does not cause brain cells to grow; only stimulation does. As the brain cells grow in size, they require more fuel to have the ability to do more work. If we supply our brain cells with an abundance of fuel in the form of oxygen and glucose, but fail to stimulate it, the brain cells degenerate and die. Therefore, the most important element in growth of brain cells, and the growth of the brain itself, is based on increased stimulation. The most important aspect of stimulation is frequency, which means that the more often a brain cell is stimulated the more it will grow (Snyder et al., 2001; Gemmell and O'Mara, 2002; Sommer et al., 2002).

The human brain is larger than that of any other land mammal relative to body size. What makes our brain truly unique is not just its size, but its structure. The anatomy of our brain is similar to that of primates, with the exception of our cerebral cortex and *cerebellum*. Although we have an increased number of neurons in the human brain, the increase in neurons is due to the brain's greater size, not greater density. Humans have only about 1.25 as many neurons per cubic centimeter as chimpanzees (Nieuwenhuys, 1998).

There are approximately 146,000 neurons per mm^2 of cortical surface. The human brain has an area about 2,200 cm^2 and about 30 billion neurons. Chimpanzees have brains that measure about 500 cm^2 with about 6 billion neurons. Gross anatomical examination shows that the cerebral cortex is significantly larger and more complex in man than in the rest of the primates. This is accompanied with major changes in connectivities, especially the high development of cortical projections (Nieuwenhuys, 1998). However, in contrast, subcortical structures do not appear to achieve the same level of specialization in man, although there are concurrent increases in areas of the *cerebellum, thalamus,* and *basal ganglia*. How this came to be and how this relates to bipedalism is yet to be determined.

It has been proposed that random genetic changes may be responsible for the larger growth of the human brain (Williams, 2002). As of yet, no genes have been identified that are responsible for this process. It has been suggested that nerve cells establish their connections by way of specific chemicals that allow them to recognize their targets (Kandel and Schwartz, 1995). If this were correct, then neural connectivity would be genetically determined. Although, this appears to be true in some invertebrates, there is evidence that a more plastic mechanism may be involved in higher vertebrates. It has been noted that in this case, the nervous system initially develops an aggressive pattern of connectivity. However, later in the perinatal period, many of these connections become synaptically stabilized against others because not all of them are able to generate functional synapses. The development of functional synapses is thought to be dependent on activity, which is a product of both spatial and temporal frequencies within the neuronal network to which it is thought that it is connected. It is thought, therefore, that there is a natural selection process of synapses based on the function of the whole network (Carlen et al., 2002; Chen et al., 2002; Ding et al., 2002).

The primary increase in brain size from rodents to primates and to humans is seen by the relative increase in the size of the association cortices (the temporal, parietal, and frontal cortices). The increase in cortical volume is attributed to an enlargement of the number of cells and the overall structure. It is thought that most neurogenesis is restricted to the perinatal period in primates. Therefore, there are a limited number of neurons during postnatal life in primates. Neuroblasts are thought to stop dividing before they begin to migrate through radial glia to different areas of the cerebral cortex where they then differentiate into neurons.

It is therefore thought that changes in the regulation of neuroblast proliferation within specific areas of the ependymal layer produces the size and number of cells of their associated cortical areas. This may have been the way in which the cerebral cortex initially increased its size in hominids. This increase in cell number in the cortex, especially in the association areas must have resulted in dramatic consequences in the development of connectivity. These areas may have then offered good targets for several types of axons as well. Their efferents may have overcrowded their subcortical targets since they did not grow to the same degree. However, there are subcortical areas that seem to have expanded in parallel with the association areas, particularly the lateral *cerebellum* and the ventrolateral *thalamus* as well as the areas of *basal ganglia* (*neostriatum*).

Due to the process of synaptic stabilization, an increased frequency of this network would have allowed for the maintenance of a larger population of neurons to survive. This increased activity would have been provided by bipedalism itself. The change in posture from quadruped to biped position would place greater gravitational stresses on postural muscle receptors and joint receptors. This would be so especially since the greatest density of these receptors is in the spinal musculature and joints and closer to the head where there would be the greatest gravitational stress. This increase in afference would find a good place to make synapsis in the *cerebellum, thalamus,* and cortex to which it connects. The ability to stand upright would require greater *cerebellar* and vestibular cortical control mechanisms which would have resulted in expansion of these areas allowing for more connection sites, maintenance of a larger amount of *cerebellar* cells, and through the principle of synaptic stabilization. Direct feedback of spinal joints and muscle afferents back to the *cerebellum* would increase exponentially due to the dramatic increase in transduction of gravity by these receptors. This may explain why the *cerebellum* actually has a larger numbers of cells than the cerebral cortex. That may also be why the *cortico-ponto-cerebellar* tract, which consists of an estimated 40 million fibers, is so much larger than any other tract in the human brain.

The neurons in the cortex would have found the *cerebellum* and its connecting areas in the pons to have the most available subcortical neurons to which they could connect. The *cerebellum* also has connections to all areas of the neocortex, especially the parietal areas, which would receive incoming proprioceptive input, and the frontal cortex, which would control voluntary motor activity, and coordination of motor activity. Therefore, increased temporal and spatial summation within networks associated with postural motor activity would ensure a greater survival of cells within this network, which includes the association areas of the cerebral cortex. A second possibility, however, was still available once these pathways were filled to capacity, the development of *cortico–cortico* connections. Here, any access to efferents would find another place on which to make synapsis. It is thought that in this way, an increase in *cortico–cortico* projections was facilitated. There is good evidence that rearrangements in activity do in fact occur in the developing cortex as witnessed by cell death or terminal retraction suggesting that selective stabilization of synapsis does exist in this system (Elbert et al., 1994, 1995).

This may also explain why the human brain became so large before we developed any of our more advanced human abilities such as language and social behaviors.

In this scenario, however, bipedalism along with genetic mutations were the only factors needed to create and maintain the larger number of cells seen in the human brain. Once this was established and pressures for bi-symmetry were released, humans then could develop asymmetric functions in their brains that were not directly tied to motor or autonomic control. Hemispheric specialization then could develop different control centers consistent with the previous function of that hemisphere, creating most of the unique human characteristics. The other demand that bipedalism would place on the brain would be the need to be more precise and complex in the synchronization of muscles to be able to walk, run, and jump. This increased synchronization would require greater frequency of oscillation of control centers within the *inferior olive* and *cerebellum* and their feedback to the *intralaminar nucleus* of the *thalamus* and its reciprocal *thalamo-cortical* projections. This increase in oscillation into the 40 Hertz range is thought to be required to achieve binding within various cortical sites into one continuous conscious percept of the world. This appears to be the foundation of human consciousness, which is thought to be unique in humans and due to unique connectivities in the human brain.

Therefore, a proposal of an increase in neuroblast proliferation in the human brain is consistent with the concept of *neoteny* in the human evolution. This concept states that certain characters are delayed in their development with respect to others (*Paedeogenesis*) (Bjorklund, 1997; Penin et al., 2002). This resulted in changes in adult morphology during evolution. This is thought to be the process in the human skull in which infantile dimensions are comparable to other primates. This first factor explains the increase in cell size that concurs with a minimal genetic change. However, the maintenance of these cells would not continue without the appropriate activity, presynaptically, and postsynaptically.

In essence, they require a power source as well as they would in turn require connections to expanded areas subcortically. Bipedalism would provide both by increasing exponentially, the amount of temporal and spatial summation within sensory motor networks, especially *cerebellum, thalamus*, and cortex. This would require expanded areas of *cerebellum* and *thalamus* that would evolve in parallel with the expanded areas of cortex and could provide a site for connection to these increased numbers of neurons. This would take place because although the genetic change would increase cell number, it would do so with a nondirectional force, which would not specify any specific shape. Posterior epigenetic reorganization (synaptic stabilization) would determine the shape and configuration of the networks within the brain itself. Therefore, genetic factors would produce the density of cells required but environmental factors would trim and shape it in a specific fashion.

Therefore, it can be speculated that there are no genes specifying particular types of neuronal networks involved in higher cognitive function. The human brain is about four times larger than those of primates, because the brain cells or neurons are spread about (Blinkov and Glezer, 1968; Nieuwenhuys, et al., 1998). The thicker cortices of the large mammals and humans as well, seem to be primarily a function of larger nerve cell bodies, and more extensive dendritic and axonal systems, and more numerous *glial* cells. Although neurons cannot reproduce after birth, *glial* cells can. They reproduce based on increased metabolic demand of the neurons or increased stimulation. This increase in growth of *glial* cells allows the neurons to make more connections, which increases the ability and speed of the cell to transmit signals. The increase in size and strength of connections allows both to happen more efficiently. The growth, size, and complexity of the human brain comes from the number of supporting cells which in essence, feed the neurons with more fuel and encourage growth of new connections. It is not the increased number of neurons, but the increase in connections between the cells and the increase in separating and supporting cells that accounts for the large growth of the human brain. This is the very definition of "plasticity."

Plasticity is the ability of the brain to grow and whether it is growing on a short-term basis or on a long-term basis in the case of evolution, the facts of plasticity are consistent. This can only mean that there was some increase in the frequency, duration, and intensity of stimulation of the human brain over time for it to have evolved as uniquely as it has. There are two things that make humans unique among other organisms: (1) we have a larger cortex and, (2) we stand upright (bipedal). Assuming what we have thus far indicated is so, we must then determine the source of the hypothesized increase in stimulation that may have created the human brain as it now is and how it relates to bipedalism.

Although the brain does provide itself with certain intense stimuli (e.g., dreams), when it comes to functioning in the real world, the brain is a stimulus-based system. It is dependent on outside sources for stimulation. Outside sources are the natural environmental stimuli that are available: light, sound or vibration, odor or smells, food and liquids for taste, heat and cold for temperature, and pressure or gravity for touch. These are known as sensory inputs. We call the ability to use sensory inputs "senses," of which there are many.

One of these senses, proprioception is the ability to be aware of one's body position in space, body movements in relation to gravity, and to be aware of body movement in relationship to itself. This information is collected from the environment in specialized structures known as mechanoreceptors. They are called receptors because they receive information. Receptors are specific (for the most part) for different forms of environmental stimuli. The retina has light receptors, rods and cones; our ears possess sound receptors, hair cells; and our joints and muscles possess receptors for movement and gravity.

The more densely populated the receptor in an area, the more information or stimulation can be collected from that area. These receptors are like buttons that are pushed by environmental stimuli. Once the buttons are pushed, they send information through the nerves to the spinal cord and up to the *brainstem, cerebellum, thalamus*, and ultimately the cortex itself. The brain cells are then stimulated and the signal processed to decide the best response to the environmental stimulus. The more a brain cell is stimulated the more it will increase its size and strengthen its connections and the more supporting cells will be produced (Carlen et al., 2002; Chen et al., 2002; Ding et al., 2002).

Three factors cause brain growth. The first is frequency of stimulation or how often a cell is stimulated. The next is duration of stimulation, and the last is intensity or how much stimulation the cells receive. Therefore, the impulses that stimulate the cell most frequently, for the longest period of time, and the greatest degree of intensity will have the greatest effect on the growth of the cells and the brain.

Most environmental stimuli are not constant, and not always available to stimulate the brain in the same way. Light, sound, taste, and temperature are all of variable frequency, duration, and intensity. For example, sunlight is only present during the day. The only constant source of stimulation from the environment is gravity. We live in what is known as IG environment. We are compelled to resist gravity by using our muscles and joints continuously because that is where the receptors for gravity and/or proprioception are located. Because gravity steadily exerts force on us, and because we are perpetually forced to resist it by using muscles and joints, the amount of time it stimulates our brain based on frequency and duration is much greater than that of any other stimulus. Every movement one make stimulates the brain. Simply standing upright involves responses that not only allows us to see further, but also requires us to constantly resist gravity, even when we are standing still. This simple condition steadfastly increases stimulation to our brain therefore, accelerating its growth.

Receptors are not evenly distributed in all muscles and joints. Certain muscles and joints have a greater density of receptors and therefore, supply a greater amount of stimulation to the brain. Standing upright may well have been the single most important act responsible for increasing our brain size. The greatest concentration of receptors in human muscles and joints is found along the spine; the closer to the top of the spine, or head, the greater the density of receptors in our muscles and joints. Imagine how suddenly moving from a quadrupedal stance to standing erect would shift the degree of gravitational force on our muscle and joint receptors up to the top of the spine where the most receptors are located. In fact, Sperry (1962) has stated that it has been shown that "... 90 percent of the stimulation and nutrition to the brain is generated by the movement of the spine. Kind of like a windmill generating electricity."

Envision how that enables us to harness gravitational forces to an even greater degree and constantly increase the stimulation to our brain cells. We have not only increased the frequency and duration of stimulus, but we have also significantly increased the intensity of stimulus. This is a significant and dramatic increase beyond creatures that walk on all fours most of the time.

3

Why the Brain Works the Way it Does: Evolution and Cognition from Movement

The answer to why and how the brain works the way it does and why it may dysfunction, is based on its evolutionary development. There are two main characteristics that, above all, make human beings unique among all organisms: (1) humans are bipedal, and (2) we have a much larger brain in relation to body size, especially the cortex. The simple act of standing and walking upright is the most sophisticated and complex movement that any organism has yet achieved. Because of the intricacy of that movement, the brain developed to a greater size and complexity. The human brain has grown due to its dependency on the increased environmental stimuli that walking upright creates.

The one pre-adaptation that clearly occurred before the growth spurt of the brain was the ability to stand upright. Human beings are bipedal, which means we walk on two feet instead of four. We are the only organisms who can do this all the time. Monkeys can stand up at times, but when they want to go anywhere, they typically do it on all fours. The first ancestor who walked with two legs is thought to be Lucy, an *Australopithecine* discovered by Dan Johansen and who is thought to have died in what is now known as Ethiopia 3.5 million years ago. The impetus to stand erect and walk on two feet has been universally recognized as the one possible characteristic that made Lucy different from the creatures around her. At that time, no other major physical change occurred. Our vision was the same as it is now, as was our hearing, and the other bodily functions. The only thing that changed was that we stood straighter and our brain grew larger. From the point in time of this tripling effect over a mere few million years and when our ancestors achieved a fully upright position, the brain stopped growing in size but continued to grow in complexity and intricacy. The two are obviously tied together but how this understanding relates to our brain's function now needs to be examined.

One theory is that standing upright increases heat to the head and therefore, the need to decrease heat production arose. The increased need to dissipate body heat resulted in increased blood supply to the head, which stimulated increased brain growth. Although increased blood may have played some role, when we understand how the brain cells grow, we know that just supplying them with increased blood and oxygen will not increase their size unless there is an increased stimulus demand first, which causes an increased use of oxygen and not the other way around. Brain cells that are supplied with plenty of oxygen, but are not stimulated, will die. How did it change its configuration as it evolved its huge size? In general, large primate brains have relatively expanded volumes of neocortex. In a chimpanzee, for example, the neocortex forms a larger proportion of total brain volume than it does in the smaller brain of a monkey. The general trend for primates predicts that a brain as massive as ours should have an exceptionally large volume of neocortex, yet the human neocortex turns out to be more voluminous than expected. In addition, the regions devoted primarily to thinking are especially large. These regions, known as the association cortex, are the ones that are not dedicated to sensory or motor functions. It appears that the evolutionary expansion of the human brain automatically produced a disproportionately large volume of tissue devoted to thinking. Simply put, brain size greatly outstripped body size, and the excess brain tissue, or the parts not required to control the body, was available for higher functions.

Recent theories on how higher mental functions arose in this excess tissue have been proposed by Jerison (1985) which he calls the principle of *proper mass*. According to this principle, the amount of neural tissue devoted to a particular function is appropriate to the amount of information processing that the function entails. In effect, as it grows, the brain organizes itself according to this principle. In both monkeys and humans, the phase of fetal brain growth or neurogenesis begins about 40 days after conception. This phase lasts for about a 100 days in monkeys and about 25 days longer in humans. Neurogenesis occurs deep within the brain, and the neurons assume specific positions in the neocortex by migrating to locations that are specified by genes. Through their migration, the neurons build the six layers that make up the neocortex, starting with the innermost layer and ending with the outermost layer. The human neocortex is identifiable about 2 months after conception, and cell migration ends by the end of the fifth month.

Another aspect of maturation is myelination, a process in which fatty sheaths enclose neurons, insulating them and improving their ability to conduct electrical signals. To some degree, the connections that neurons make with one another are genetically programmed, but the genetic controls are imperfect and feedback from the body and its sensations influence both the production and the elimination of specific connections. Cells that form synaptic connections between neurons receive more nutrition and stimulation than those that do not, and those whose synapses fire off the most frequent messages are particularly well supplied. This is the process of synaptic stabilization, much of this natural selection at the cellular level takes place prenatally, but the process continues well into the postnatal period. A human baby enters the world with perhaps a trillion synapses connecting its cortical neurons, but a large fraction of these disappear during the first decade after birth. Predictably, the maturation of brain tissue parallels the maturation of brain functions. The development of brain and the body involves reciprocal feedback systems. Maturation of a particular sector of the brain stimulates activity in a corresponding area of the body or in a connected area of the brain. The stimulated function then matures more rapidly through use, but this use stimulates development of the area of the brain that controls it. As a human baby grows into a toddler and then a school age child, the brain permits interactions with the environment. This relationship also molds the developing brain, however, favoring the fixation of beneficial neural maps and allowing pruning of useless neural connections. The result is that although genes specify some traits of the developing brain, particular neuronal maps are created through interaction with the environment, especially in the later stages of development. The body automatically sends fibers to the brain proportional to the body's size. This means that when the human brain, in growing so large, outstrips the body, there is no way the body can recruit the excess neurons and neuronal connections that develop within the brain especially when the brain suddenly evolves to a much larger size without a corresponding expansion of the body. The sensory and motor regions of the brain and also the auditory and visual regions are at a disadvantage in competition for connections with new brain cells.

Neurons outside the brain are probably not more numerous than those of our immediate hominid ancestors (except for certain areas such as the neo-*cerebellum* and *neostriatum* and areas of *thalamus*). Areas of the brain such as the *prefrontal* cortex are connected not to neurons of the rest of the body but only to neurons within other parts of the brain itself. When the human brain originally expanded, these regions were therefore favored; they had the right wiring and received nourishment. In this manner, areas of the brain that were devoted to conceptual functions expanded dramatically, while those concerned with bodily functions underwent little change. Therefore, it is no accident that the *prefrontal* cortex of the human brain expanded more than any other area, this is the region that plays the dominant role in thinking. In Deacon's model (1992), these areas differentiated out of preexisting areas with which they retained some similarities. The new functions arose because axons invaded new target areas of the brain, forming new neuronal connections. This may also help to explain the evolution of asymmetry in the human brain. If new areas had developed out of old areas in both hemispheres of the expanding brain, they would have totally displaced preexisting functions. The emergence of a new function in only one hemisphere, however, conserved the displaced function in the other hemisphere. Overall brain expansion, compensated for the disappearance of its original function from one side by expanding the tissue devoted to it on the other side. This "displacement model" implies that heightened intellectual powers were a natural consequence of the evolution of the large brain of *Homo*. Because there was no commensurate increase in body size, the added brain tissue was available for thinking. A pattern observed for monkeys suggests that the displacement model is on target.

A few species of monkeys that have unusually large brains for their bodies also have brains that resemble ours in their general proportions. The way in which *Homo sapiens* grows its large brain allows us to conclude that a population of *Australopithecus* could only have evolved such a structure after abandoning the habit of climbing trees every day. The human pattern of brain growth early in life is unique among primates; the key interval is the first year after we are born. During this brief period, an average infant adds slightly more tissue to its brain than it will add throughout the remainder of its life. The brain makes up slightly more than 10 percent of total body weight at birth. It may seem surprising that the same can be said for a newborn chimpanzee or monkey. In fact, this so-called 10 percent rule for brain weight at birth holds for primates generally. It is only during infancy that humans surpass lower primates. After birth, a monkey or chimpanzee fails to maintain the high rate of fetal brain growth that endowed it with such a large head when it entered the world. It continues in the next phase of growth in which its brain expands much more slowly all the way to its final adult proportion.

Humans differ from lower primates in retaining the first stage of high fetal rate of brain growth through the first year of life after birth. The result is a one-year-old infant who has an extremely large head that encases a brain more than twice as large as that of an adult chimp. Not until an age of about one year do humans settle into the next slower stage of brain growth. Then, between the end of our first year and adulthood, while we grow in total body weight by about 800 percent, our brain grows slowly. In humans, the rate of brain growth beyond birth amounts to a retardation of the brain's development. It is not simply our brain that matures slowly, however, but our entire body. This condition arose because natural selection could not find a way of singling out the brain for delayed maturation. It accomplished the delay by slowing down the overall rate of bodily development immediately after birth. The result is that, although we grow rapidly in physical size after birth, we remain physically helpless while the fetal pattern of brain growth more than doubles our brain size by the end of our first year of life.

Of all mammals, we exhibit the most dramatic increase in brain growth immediately after birth, and we experience the longest interval of infantile helplessness. No other species of mammals more than doubles its brain size by its first birthday, and no other species requires fifteen months to begin to walk without parental support. The resemblance between a human and a juvenile chimp, our relatively flat face, weak brow, and tall forehead are among the juvenile traits of our ancestors that we retain into adulthood. In fact, the proportions of fossil skulls of unknown gender reveal that *Homo rodolfensis* had a much larger brain than this. The adult fossil skulls of this species had an estimated capacity of 760 cm^3 (Ramirez Rozzi, 1998). This is quite large compared to estimates for various skulls of *Australopithecus*, which range from about 430 to 485 cm^3. A portion of another fossil skull of *Homo rodolfensis*, points to an even larger brain size than this hominid. Although this is only a large fragment of skull, we can see that it belonged to a very small child with a very large brain. The skull's cranial capacity, though not precisely measurable, would have approached 900 cm^3 in adulthood. While this adult figure is roughly 30 percent below the modern human average (1,220 cm^3 for women and 1,420 for men), it is nearly twice as large as the average estimate for skulls of *Australopithecus*. A delay of the developmental process was probably the only mechanism by which evolution could have produced the dramatic human encephalization. By using this mechanism, natural selection made use of a pattern of growth—the high rate of brain growth *in utero*—that was already present in the ancestral organism. All that was required was a change in timing.

The driving force behind the evolution of the brain must have been the brain itself. The brain ultimately facilitated an upright posture and it must have done this because it got something in return. There must be some exchange or positive feedback that occurs between standing upright and brain growth.

What is this connection between upright posture and the muscles that maintain that posture and the brain? To understand this completely, we need to explore in more detail how the brain functions.

Biomechanically, it was the change in the development of our spinal muscles and the orientation of our spinal joints, as well as the hips, knees, and feet that allowed us to stand up. In doing this, the amount of gravitational force distributed through the joints and muscles increased dramatically and therefore increased the stimulation to the brain.

If we look at the brain as a greedy master that will do anything to increase its own stimulation and growth and that standing upright will maximally accomplish this, we can see that the brain would augment the muscles it controls so that it can better accommodate the upright position and increase its own stimulation, which in turn gives it the ability for superior movement. This increase in brain function and corresponding bipedalism was so dramatic that it could be accomplished over a relatively short period. As early man achieved a more upright position, the size of the brain increased accordingly. This is exactly how we see our evolution progress. Man stood upright and nothing else changed. There was no increase in any of the other sources of environmental stimulation, which is the only thing capable of causing the brain to grow. Vision did not change, hearing did not change, and taste and temperature did not change. Man stood up. The straighter the spine got, the more the brain grew, and the greater control it exercised over the muscles. This made it easier to stand up for longer periods until an erect posture was eventually achieved with *Homo erectus*. This is consistent with all previous forms of evolution, the more complex the movement, the more complex the nervous system. The more complex the nervous system, the better able the species is to interact with its environment; the more interaction with the environment, the more the brain grows; the more the brain grows, the better the species is able to adapt and flourish.

Biomechanically speaking, the one thing that was sacrificed for the multiple advantages of standing upright was the size of the pelvis. Since the human spine was not originally designed for upright posture, it was modified based on the development of greater muscular control and changes in the orientation of the joints. Standing upright exerts greater gravitational force on the spine and pelvis, which increases the stimulation; however, in order to support the increased rate demand, the human pelvis grew thicker than the apes. This made the birth canal, the opening infants are born through, much smaller. This decrease in pelvic size, forced the human head to be smaller at birth. This resulted in a decrease in infant brain size at birth attributing to the prolonged period of human infant dependency. Relative to other mammals, human babies take much longer to develop. This required early human parents to carry and protect their children for long periods.

Since early human arms were free, the upright posture allowed parents to hold the child. They could now walk long distances carrying a child, which allowed them to migrate more easily. However, this method of locomotion over an extended period requires that the spinal muscles especially, be extremely active. Now not only were early parents fighting gravity, but the muscles had the additional forward rate of an infant pulling them back. This additional weight caused further development of stronger, more coordinated spinal and postural muscles or the muscles to allow one to stand upright. It also put more compression on the joints of the spine, hips, knees, and feet, causing further stimulation of receptors and stimulating the brain.

The fact that children had a longer dependency period also meant that the male, as well as the female, had to share childcare since multiple births, other offspring, and the period of time during which children learn to walk, necessitated more attention than just the mother carrying the child. Two or more children would have to be carried which put the same stresses on the father's spine causing the same increased stimulation and the same growth of the brain. This may have been instrumental in facilitating the development of family units and tighter parental nurturing. This close physical contact also helped to develop the child's brain because the child was now exposed to a "sensory bath" from the parents, which included smell, sounds, and touch. This close contact also cultivated more communication between the parent and child. The longer, closer contact developed deeper bonds between the parents themselves, and between the parents and their children. It also enhanced survivability of the children and the species as a whole.

Walking upright allowed our ancestors to travel and hunt in new, unexplored territories, which further increased interaction with the new, stimulating environment, which further developed the brain. Hunting in broader territories further from home base enhanced Early Man's chances of finding food, but also required carrying the food back home longer distances. This required the development of spinal muscles and joints just as carrying a child did, however, it also nurtured cooperation amongst the hunters.

We can now clearly see that the "prime mover of evolution" is bipedalism and its enhanced ability to transduce musculoskeletal and gravitational forces which cause increased stimulation to brain cells. This is associated with thicker neurons, with more connections and branches, and increased supporting cells to supply the increased fuel required. Increased capacity and increased fuel produced better use and development of the brain.

Certainly, all other factors such as vision changes, tool use, throwing, painting, social organization, nakedness, clothing, increased heat dissipation, played a role in shaping the brain into what it is today. It is vital however, to understand that the brain probably evolved based on its dependency on information and stimulation from the muscles and joints, especially spinal muscles and joints and the ability to maintain an upright posture. Once the brain achieved the size that we saw with *Homo erectus*, its potential increased vastly. The more exciting new function the brain would now have is the capacity for complex and abstract though and eventually language.

Within short periods of time, when a human is not under the constant force of gravity, as he is when in outer space, the body and all of its systems, especially the brain, dysfunction significantly (Casler and Cook, 1999; Baroni et al., 2001; Bock et al., 2001). There is a breakdown of the spinal muscles and the curves of the spine (Whalen, 1993). In reported cases of "space dyslexia," astronauts describe the same symptoms that we see in children with Attention Deficit Hyperactive Disorder and other learning disabilities and cognitive processing problems (Eddy et al., 1998; Casler and Cook, 1999). Humans cannot survive for very long without gravity, whether in space, underwater, or lying in bed. The results and the damage are similar. All of the physiological changes that we see in astronauts, we also see in people who are in extended bed rest.

Organisms that have been sent up in space showed rapid and significant degeneration of brain cells (Kosik, 1998; D'Amelio et al., 1998a; Holstein et al., 1999). The rats from the University of California at Berkeley (D'Amelio et al., 1998b) showed increased plasticity and growth of the brain when they used their muscles and joints in "novel and interesting ways." When the same type of rat goes into space however, they have reverse plasticity and show rapid degeneration of their brain cells. It does not take more than a small leap to understand that a child who is sedentary most of the time or who is sitting in front of the television for several hours a day, is not maximizing his or her use of gravity as a stimulus.

As we evolved, we developed a dependency on the ability to move physically. For most of our history on earth, the human race has been hunter–gatherers, which utilized all of the advantages evolution had given us and has stimulated the growth and complexity of our brain. It is only in our most recent history where we see humans no longer dependent on their physical prowess for survival; actually going out to hunt and walk and run for long distances, just to stay alive.

It is only in the past few decades that we see the development of sedentary entertainment, such as television, computers, and computer games. As we become more sedentary, there results a breakdown of the muscles and joints that allow us to stand upright, the very action that enabled us to climb down from the trees. We see an increase in spinal related disorders such as low back pain, neck pain, headaches, and herniated disks. Low back surgeries because of back injuries have increased dramatically over the past several years, probably due to the latest technology that make such surgery possible. Most of the ailments that we have today are lifestyle diseases that are brought on by sedentary activity.

Most stimulation to the brain comes from muscles and joints, especially from the postural muscles in the form of unconscious touch or proprioception. The source of this stimulation is gravity, which is the only constant source of stimulation in our environment. We discussed how changes in the genetic production of cells increased the pool of cells in the early hominid brain, but required activation and increased connectivities in order to survive. The bipedal posture of early hominids required a larger *cerebellum* and cortex to coordinate and control this movement. This increased size most likely provided new cortical cells with available connection sites. The cortical cells also connected to other cortical cells forming association areas in the brain, which in turn allowed for better cross communication and intra- and inter-hemisphericity.

However, all the cells still required a source to keep them active, that source we hypothesize to be gravity. It was harnessed by the bipedal posture, which would dramatically increase the frequency and duration and constant firing to the brain from muscle and joint receptors in the spine and the postural muscles. This would not only maintain an increased number of cells that would require an increased number of glial cells for support, but would also form a more dense network of connections. This would account for the vast size of the brain before we had any specific need for a larger brain. However, the larger brain would endow its user with greater intelligence which could develop over the next million or so years to be sculpted into the modern human brain. Once the brain was able to evaluate and think, and not just react to the environment, a powerful newly increased source of stimulation unto itself was created. With the combination of increased stimulation that came from increasingly complex thought, the growth of the brain increased exponentially.

We can envision Early Man as a creature on all fours impelled to look at nothing more than the ground and a few feet in front of itself. This organism could not imagine what lay ahead in the near future. We may then envision that creature standing up for the first time, looking up into the sky and seeing a bird flying. Then too, this organism might have imagined what it must have been like to fly like a bird and sowed the seeds of what eventually would cause that same kind of being, with the same basic brain structure to eventually slightly further tilt its head to eventually look to the moon and wonder what it would be like to be able to go there. More than anything else, what standing upright did was create a limitless potential of "brainpower" and an almost infinite number of ways in which to utilize that brain. We have yet to answer the question, "from where did the ability to cognize arise?"

THE EVOLUTION OF COGNITION FROM MOVEMENT

There was unquestionably an increase in average anatomical but not behavioral aspects of brain function. The arrival of the modern cognitive capacity did not simply involve adding just a bit more neural material or adding any major new brain structure. The basic brain design remains remarkably uniform among all higher primates. Instead, an expanded brain equipped with a neglected potential for symbolic thought, was somehow put to use. Further, if at some point, perhaps around 70 to 60,000 years ago, a cultural innovation occurred in one human population or another that activated a potential for symbolic cognitive processes that had resided in the human brain all along, one could readily explain the rapid spread of symbolic behaviors by a simple mechanism of cultural diffusion. Other species certainly possess consciousness in some sense, but as far as we know, they live in the world as simply as it presents itself to them.

Presumably, for lower organisms, the environment seems very much like a continuum, rather than a place, like ours, that is divided into the huge number of separate elements to which we humans give individual names. Beyond what makes this possible is the ability to form and manipulate mental symbols that correspond to elements we perceive in the world within and beyond ourselves. Rodolfo Llinas (2001) has done a brilliant job of describing how movement and cognition are bound together. He states that, "the generation of movement and the generation of mindness are deeply related; they are in fact different thoughts of the same process. In my view, from its evolutionary inception mindness is the internalization of movement." These are powerful statements and have direct connection to the main theme of this book. We are discussing cognitive and emotional symptoms, yet the ability to think, dream, or consciousness itself is a mystery to even most physicians, therapists, and scientists. These abilities are what Llinas refers to as mindness and since we do not understand how it arose through evolution and how and where it is produced in the brain, how can we propose to restore mindness to its normal state when it is dysfunctioning or fails to develop in childhood. Because it has been such a mystery, science has always treated cognitive ability separately from sensory, motor, or autonomic functions.

For years, it was thought that areas of the brain, which control cognitive function, were completely separate from areas that were responsible for motor function. Areas like the *cerebellum* and *basal ganglia* were always considered purely motor areas of the brain, even the *frontal lobes*, where both cognitive and motor control have been known to reside, were still not considered to be directly connected to each other. More recently, however, it has been recognized that these areas have non-motor functions, every bit as important as their motor functions that allow them to produce and control cognitive and emotional behaviors as well as motor behaviors. This is important to note because it is these very motor areas along with the *frontal lobe* of the cortex that seem to be malfunctioning or developmentally delayed in many neurobehavioral disorders of childhood.

This concept of interaction between motor and cognitive function has great clinical implications for the diagnosis and treatment of these disorders. Motor dyscoordination, especially of gait and posture, is the most consistent comorbid condition associated with this spectrum of disorders. It tends to reason then that if we can administer therapeutic intervention strategies to improve motor function and consequently improve the function of motor areas of the brain we should thereby expect improvement in the cognitive and/or behavioral abilities of the child as well.

In the late 19th and early 20th centuries, there were two different opinions with regard to execution of movements. The first proposed by William James was that movement was reflexive dependent solely on the inputs from the sensory system. Movement was thought to be driven and in response to sensory cues. Graham Brown (1911, 1914, 1915) thought that movement was intrinsically generated, even in the absence of sensory input, and that sensory input helped to modulate movement, but was not necessary for the act of movement. He proposed that there were areas in the spinal cord that acted as "pattern generators" that produced the ability to walk. Therefore, these areas would initiate walking and sensory input was needed to modify walking so that we did not fall. More recently, research has confirmed his initial observations that areas within the *brainstem* and spinal cord are the basis of breathing and locomotion in vertebrates (Kandel and Schwartz, 1995).

Llinas (2001) has proposed that in his conception, sensory input provides the specifics of ongoing cognitive states or the context rather than the content. Therefore, the importance of sensory information depends on the preexisting environment of the brain itself. Organisms must move in an intelligent way for movement to be beneficial to them. The nervous system evolved to provide organisms with a goal-oriented plan that would make short-term prediction based on ongoing sensory input possible. This allows the organism to move in a specific direction based on a sensory–motor image. The concepts of corollary discharge, efference copies, spatial maps, contingent negative variation, and expectancy, while beyond the scope of this book, speak of our abilities as humans to predict what is likely to happen next in sensory–motor interaction (Evarts, 1971; Johnstone and Mark, 1973; Robinson and Wurtz, 1976; Leisman, 1976a; 1978; Sommer and Wurtz, 2002). It is precisely through these processes that we are capable of descending a staircase without looking. We are capable of predicting the next likely movement necessary by assuming something about the riser height of each step. When we look at our feet, the motor programs that predict the next step need to be restarted. The same mechanism also holds true in reading a printed page. Our eyes jump from phrase to phrase until they reach the end of the line. The eyes then jump to the left hand margin and overshoot the left-justified text, but only once; from then on, our eyes successfully hit the left hand margin precisely each time.

The next critical point is an understanding of how the brain evolved to perform prediction. In prediction, the organism must anticipate the outcome of the movement depending on sensory cues. A change in the environment must initiate an action or movement (approach or withdrawal) for survival. Llinas states that this ability to predict the outcome of future events, which is critical for successful movement, is the ultimate and most common of all global brain functions. For prediction to be effective, it must be centralized, which is what is thought to be the "self." The nervous system to be able to predict must first quickly compare all sensory input to gain an image of the world and it must then convert this *premotor* image into a specific movement strategy. However, for prediction to be efficient, it cannot compare all sensory input and movement possibilities at once. This would require too much time, processing space, and energy in the brain. To understand how predictions evolved coincidently with more complex movement, we must understand how the control of movement takes place.

Even the most basic movement utilizes most of the body's muscles, which result in an astronomical number of possible simultaneous or sequential muscle contractions and directions. Llinas states that even the simple act of reaching for a carton of milk, requires 10–15 combinations of muscle contractions. Along with this, there must be 10^{18} decisions that need to be made every second. This is obviously not feasible, as this would overload the capacity of our brain. In addition, if the brain is required to be "online" 100 percent of the time during movement this would put a tremendous strain, both energetically and computationally, on the brain. The solution to these problems is to have a system where movements are not continuous, but are made up of a series of smaller movements that are synchronized. Movement would be discontinuous or pulsatile, therefore the system would be online less, and it would turn off and on in specific discrete timeframes. This is in fact the way that our movements take place, with clearly defined rhythmic movement in adults at 8–13 Hz (cf. Condon and Ogston, 1966; Condon and Sander, 1974). This has come to be known as the physiological tremor and it takes place in all adult muscles at a rate of approximately 8–13 times per second. This tremor is present not only during voluntary movement, but also largely in either maintaining posture or at rest.

It is thought that these "rhythmic oscillation of muscles" allows for a mutual synchronization involved in rhythmic movements, which is the basis of all movement. In addition, voluntary movements occur in the direction of the phase of the tremor (Condon and Ogston, 1966), therefore it is simply an exaggeration of the pre-existing movement. This pulsatile movement reduces the work overhead of the nervous system. It brings in line a population of independent muscles, so that they may act in a uniform fashion and it provides an inertial break to overcome frictional forces of the muscle. It also allows for input and output to be bound in time. This allows the sensory input and motor activity to be bound and integrated in the same system. As we can see, evolution provided a brilliant solution. It was further shown that these pulsatile movements were not based on an inherent property of muscle tissue itself, but rather a reflection of a higher descending command from the brain. This pulsatile system brings neurons and muscles closer to threshold for a particular action. This is thought to compensate for the potential problems of discontinuities because there is a synchronizing effect in all the independent parts and at all levels of the motor system. This pulsatile activity occurs so quickly that it appears to any outside observer that our movement is continuous and smooth even though it is actually discontinuous. Although this physiological tremor is a reflection of a descending control system in the brain, it does not spread out that way. Developmentally, the tremor is an intrinsic property of the muscle itself. This is known as the *myogenic movement of motricity*, which takes place before motor neurons even make contact with the muscles.

Let us examine this further using Llinas's example of *elasmobranchs* (sharks). The shark embryo is in an egg that allows oxygen to pass through. As the oxygen is distributed evenly to all the tissues, there must be continuous movement of the fluid in the egg, so the embryo must continually move rhythmically. However, at this point, movement is not generated by the nervous system, in fact the muscle cells are not yet innervated by motor neurons. At this stage of development, the muscle cells are electrically coupled in a way similar to the nodal tissue of the heart.

In this way, the electrical signal that causes contraction of a single cell spreads easily and quickly from one cell to the other producing rhythmic movements within the organism, the *myogenic stage of motricity*. In the next stage of development, the spinal cord begins sending axons from the motor neurons to innervate muscle cells. At this point, the motor neurons in the spinal cord become electrotonically coupled. As they innervate the muscle cells, the muscle cells cease being electrotonically coupled and now movement is purely under control of motor neurons in the spinal cord. At this stage, the movement of the muscle mass has been embedded into the spinal cord known as the stage of *neurogenic motricity*. Therefore, the capacity of the organism to interact with its external environment has begun to be internalized in the nervous system. The spinal cord motor neurons stay electrotonically coupled until the *brainstem*, which now has become electrotonically coupled, and makes its connection with spinal cord motor neurons. At this point, the motor neurons in the spinal cord cease to be coupled as they start to be innervated by other inputs of the nervous system that do not relate to specific activities of muscle groups. These inputs relate to the total movement of the organism as a whole and involve the vestibular system. These other inputs allow the organism to relate its movement to a frame of reference outside its own body, such as gravity, so the organism starts to "think" left and right, and up and down.

The next stage of development is *encephalization*. When the brain matures, the organism starts to move forward along its long axis. In the front are developed telereceptive sensory systems such as vision, hearing, and olfaction. At the front end, it develops jaws and its head with a brain, and skull for protection. The excretions of the organism exit the rear in the opposite direction of the organism's forward movement. Through these stages, we see that the organisms take the initial properties of the muscle and the outside world and internalize them, eventually projecting them up to the brain. Most importantly, as Llinas states, this process gives us "the ability to think, which arises from the internalization of movement." "Thinking was the central event born out of an increasing number of successful possible motor strategies. The issue is that thinking ultimately represents movement, not just of body parts or of objects in the external world, but of perceptions and complex ideas as well." The next solution that evolution developed to reduce energy and computational demands on the nervous system was to use muscle collectives. The muscle collective is a group of muscles that are activated simultaneously as when we pick up a carton of milk. If the brain moves groups of muscles rather than individual muscles, the demand on the brain is significantly reduced. The central control system during complex movement must then use muscle collectives as necessary, transiently, and rapidly. Therefore, the brain must have an area that can choose from a list of various muscle collectives that is efficient and successful so that it does not waste time and energy on useless movements. We will later examine the *basal ganglia*, the area of the brain that aids in controlling muscle collectives or fixed action patterns. We will see that these properties of the brain help to more efficiently control movement and especially prediction, so that all these properties can be understood as a single construct using cognitive binding or conscious thought.

Specific areas of the brain, such as the *inferior olive*, have evolved as controller systems. In regard to controlling timing mechanisms or pacemaker activity, these areas must possess oscillatory neuronal activities of individual cells as well as oscillatory ensemble activity where a large group of neurons oscillate simultaneously and they need to be functions associated closely with movement. Many types of neurons in the nervous system are imparted with the ability to oscillate. Oscillations are made up of ordinary volleys of nerve impulses that have been previously discussed. However, there is a different element that regularly repeats these impulses in the form of bursts and the bursts repeat as well. These bursts are superimposed on an oscillating generating system or "pacemaker" that causes a slow fluctuation in the membrane potential of the neuron (Changeux, 1980, 1981, 1983; Eccles, 1987; Koch and Leisman, 1990, 1996, 2001).

The membrane potential oscillates between two extreme values on either side of the threshold for the nerve impulse. When the potential reaches the threshold an impulse is generated and then repeats as long as the potential remains above threshold. If it falls below threshold, the impulse stops, as does the burst. The pacemaker itself is a product of two different molecular channels (Kandel and Schwartz, 1995). These channels open slowly over several seconds and do so differently than channels involved in the propagation of a typical nerve impulse, which acts over milliseconds. Additionally, their selective permeability to ions is different from other type of channels. One is permeable to potassium and the other to calcium. At the start of an oscillation, the electrical potential decreases and consequently, the calcium channel, which is slow, opens. Calcium then enters the cell, but before being pumped out again, it produces and opens the other slow channel, which is specific for potassium. At this point, potassium leaves the cell and in leaving the cell, it causes an increase in potential (Kandel and Schwartz, 1995). At this point, we are back to the beginning of the oscillation. In this case, the membrane potential oscillates slowly. If the oscillation is of sufficient amplitude for the membrane potential to reach threshold for a nerve impulse, a burst will form on the crest of each slow oscillation. Neuronal oscillators are regulated as much by the membrane potential as by the internal calcium concentration (Changeux, 1985).

According to Llinas (2001), the peaks and valleys of the electrical oscillations of neurons can dictate the waxing and waning of the cell's responsiveness to incoming synaptic signals. It may determine at any moment in time whether the cell chooses to "hear" and respond to incoming electrical signal or "ignore it." Another property that is close in relationship to neuronal oscillation is coherent rhythmicity and resonance. Neurons that produce rhythmic oscillatory activity may entrain to each other through action potentials. This then can produce neuronal groups that oscillate in phase or coherently which help to support simultaneity of activity. There are several areas and groups of central nuclei, such as the *inferior olivary nucleus (IO)*, that plays an important role in the coordination of movement.

In the case of the *inferior olive*, its neurons connect to the *cerebellum*. The fibers from the *inferior olivary nucleus* branch onto the main neurons of the *Purkinje* cells of the *cerebellar cortex*. The connections are made through climbing fibers and connect to the dendrites of the *Purkinje* cells. Most movement control processing occurs in the *cerebellum* and the climbing fibers, which are some of the most powerful synaptic inputs in the vertebrate central nervous system, playing an important role in that motor control (Eccles, 1987; Eccles et al., 1967). Damage to the *inferior olive* or to the climbing fibers causes immediate severe and irreversible termination of many aspects of motor coordination, especially timing of movement and in movements through three-dimensional space.

It has been shown that the *inferior olive* plays such an important role in timing that organisms with damage to these nuclei have problems learning new motor behaviors (Welsh et al., 1995; Welsh 1998). Intracellular recordings from cells in the IO have shown that these cells oscillate spontaneously at 8–13 Hz. The IO cells fire their action potential in a rhythmic fashion and it is thought that through its connection to the *cerebellum* the IO is responsible for the timing signal that helps to control all movements. It is thought that the oscillation of the inferior olive results in a slight tremor of 10 Hz and occurs even when one is not moving (Llinas et al., 1975). This movement, as previously described, is known as physiological tremor, allowing us to time movements as a metronome, when we learn to play the piano. It also has been demonstrated that with the experimental destruction of IO, behavioral tremor is abolished (Llinas et al., 1975).

A similar type of timing mechanism is found in the cerebral cortex to help generate conscious thought. We require a mechanism with which we will be able to bind information from different sensory sources, so that the essential result will be an internal representation or sensory motor image that can associate memories or thoughts with this internal construct such as imagining or remembering. As Llinas (2001) states, the task of cognition is to create an experience, which brings together elements that are truly ours with elements that are truly foreign. This same oscillatory function occurs in the brain and produces temporal coherence (Leisman, 1976a; 2002). Temporal coherence according to Llinas is thought to be the neurological mechanism that underlies perceptual unity, the binding together of independently derived sensory information, or cognitive binding. This is a mechanism similar to that produced in motor binding where through the inferior olive motricity precise temporal activation of muscles is required in order to implement even the simplest movement correctly. Synchronous activation of neurons that are spatially distant is most likely the mechanism that improves the efficiency of the brain. Fixed action patterns set well-defined motor patterns. This has been described as motor tapes or engrams that produce well-defined and coordinated movements such as walking and swallowing. These patterns are called fixed because they are stereotyped and relatively unchanged not only from individual to individual, but within the species. These patterns, however, can be seen as simple or complex motor patterns.

The fixed action patterns are seen as more elaborate reflexes that seem to group lower reflexes together to achieve a more complex goal-oriented behavior (Leisman and Koch, 2003). This allows the brain freedom in efficiency and diminishes processing capacity as the brain does not need to focus time and attention on each aspect of the specific movement, only when it needs to modulate that movement due to change in repetition. In other words, fixed action patterns allow the brain time to do and "think" about other things rather than concentrate on a specific stereotyped movement. Fixed action patterns are more sophisticated than simply the control mechanisms for walking, which can be controlled by the *brainstem* and the spinal cord. Therefore, fixed action patterns are thought to reside in the higher centers of the brain.

In the case of more complex fixed action patterns like playing an instrument, throwing a ball, swinging a bat, it is thought that these are generated centrally by the *basal ganglia* (Saint-Cyr et al., 1995; Hikosaka, 1998). It is thought that the *basal ganglia* acts as a storehouse of motor programs, but how it actually works is not understood. As within the *cerebellum*, the majority of connections within the *basal ganglia* are inhibitory and have many reciprocal contacts. Neural pathology of these nuclei may be due to either producing an excess of fixed action patterns thought to be seen in *Tourette's* syndrome or defects associated with loss of them as in *Parkinson's* syndrome. Very importantly, fixed action patterns have evolved to improve the survivability of organisms. A correct choice needs to be made and made quickly and repetitively for an organism to move through the world successfully. Natural selection has fine-tuned this process and given us a mechanism which can reduce the possible alternatives. The underlying basis of movement is built around conflicting alternatives such as approach–avoidance or approach–approach behaviors. These potential conflicts require too much time to decide whether to approach or especially avoid a predator. Significant decision time will not only prove inefficient, but deadly. Therefore, natural selection has chosen a system providing for a reduction of choice and decision time through *fixed action patterns* (FAP).

FAP's require synchronous and coordinated activations of a number of different and very specific muscle synergies, driving this motor event in a synchronous and coordinative firing of very specific motor neurons with functionally specific firing patterns, frequencies, and durations. However, the cerebral cortex has the ability to override a FAP at any given time, which still allows us an enormous degree of possibilities. Even language as well as emotions is considered a FAP. Activities may not start out as FAP as learning how to play an instrument, but through repetition they can become fixed action patterns and thereby free the cortex from the responsibility of control and it can focus on other things. These differences are most clearly seen in the difference between letter and word-habits in learning out to type or, in fact, the learning curve associated with any sensory motor skill (Fitts, 1954; Fitts and Peterson, 1964; Leisman, 1989a, 1989b, 1989c; Leisman and Vitori, 1990). Therefore, *fixed action patterns* are subject to modification, they can be learned, remembered, and perfected.

Llinas (2001) considers that emotions are elements of fixed action patterns, but the actions are not motor but rather *premotor*, similar to the way that muscle tone serves as a basic platform for the execution of movements. Emotions represent the *premotor* platform as either drives or deterrents to most of our actions. Therefore, emotions provide motivation for our actions. The relationship between emotional state and the motivation for action is extremely important because under usual conditions, it is some emotional state that provides the trigger and the internal context for any specific action. However, the *premotor FAP* does not only trigger an action as an *FAP*, it is also expressed in other forms of an accompanying type of motor *FAP* such as facial expression which allows others to understand our motivation for the action that we are taking. For example, when one touches a hot stove, pulling the hand back is a reflexive fixed action pattern, but it is usually accompanied by a facial expression such as a grimace, which is also a fixed action pattern and possibly even a "scream," which is yet another fixed action pattern. With other fixed action patterns, emotions can be learned and we can, in turn, learn how to suppress them as well. Therefore, emotional states give context to motor behavior. It is interesting to note that children with *autistic spectrum disorder*s and even with attention deficit hyperactive disorder are known to have stereotypical behaviors. We can look at these behaviors as being a release of fixed action patterns, which can be associated with other *hyperkinetic* types of activities. Emotions clearly relate to areas of the brain that are distinct and separate from the *basal ganglia*, but are nonetheless closely associated with them (Saper, 1987; Heilman and Gillmore, 1998). Emotions are linked to the motor aspects of fixed action patterns through their access to the *amygdala* and *hypothalamus* and their associated connections with the *brainstem*.

In summary, movement needs to be accomplished in an intelligent and coordinated fashion to not overload the brain and nervous system as an information processor. The brain seems to have evolved two main strategies. The first was to develop an internal clock or timing mechanism that would turn all of the muscles on and off thereby reducing demand. The perceived temporal continuity of both sensory and motor behavior, exemplified by the apparent smooth and coordinated fashion in which muscles move, belies the fact that neither sensory nor motor behavior function continuously in actuality. This perceived continuity allows all muscles, which are not directly connected to one another to be connected in time. Therefore, functionally connected but spatially distant muscle groups could be coordinated into a purposeful movement. This is thought to be the beginning of abstract thought. An abstraction is something that does not occur in reality. Organisms coordinate their motor systems as one when they are, in fact, made up of separate independent muscles that are not directly connected, is by definition an abstraction. We have also shown how the external properties of muscles eventually become imbedded in internal areas of the nervous system and eventually the brain. This is then integrated with other sensory input to obtain a larger picture of the organism (or self) and the surrounding world. This is then used to form a sensory motor image of that world which is critical for the nervous system to predict the most important function to be performed.

We can see that cognitive functions developed as ways to improve purposeful movement for either approach or withdrawal behaviors. The properties of muscles were imbedded deeper and deeper into the nervous system so that the nervous system would be able to compare movement to other properties of the world and generate the most accurate prediction of the proper response. These control mechanisms involved in sensory–motor interaction are the largest and unique in humans and reside in the *frontal* and *prefrontal* areas of the cerebral cortex. These areas perform executive functions and it is this region of the brain that is primarily affected in function and efficiency in neurobehavioral disorders of childhood. The timing mechanism strategies that developed to make motor activity more efficient were used to eventually allow us to make sense of the world cognitively. The pacemaker for muscles resides in the *inferior olive* and *cerebellum*. The oscillator or pacemaker in the cognitive realm is the *thalamus*. Just as muscles have no direct connection to one another, sensory information is never fused together in the cortex (Koch and Leisman, 2001a, 2001b). There is no one area in the brain to which all sensory input converges that allows for thinking and emotional responsivity. Yet, to make sense of the world we need to combine sensations and body movement to provide a temporally and spatially resolved reality.

The brain employed the same evolutionary strategy that it used for motor binding of separate muscles. Sensory input is never bound spatially but is connected in time synchronized by an internal clock generated by the *thalamus*. Just as movements appear to us to be continuous, but are not, our perception of the world likewise appears continuous although it is not. Like a movie that seems to show a smooth continuous image of the world but is really made up of a series of frames of pictures moving so fast that we do not perceive discontinuity, oscillations in our brain occur so quickly, at 40 Hz or 40 times per second, that we likewise do not notice the discontinuity of reality (Leisman, 1973, 1976a, 1976b; Gaarder, 1975). These oscillations allow us to connect sensory input so that it appears as a single continuous precept of the world. This is so to permit an accurate *premotor* image of the world so that we may accurately predict events and the consequences of our actions—moment to moment. The second major strategy that our nervous systems developed was the use of muscle collectives or fixed action patterns. FAPs link together lower reflexive movements into increasingly more complex movements and behaviors until they become automated. These can be simple or complex, they can be innate or learned, or a combination of both.

4

The *Cerebellum* and *Basal Ganglia*

ANATOMY AND FUNCTION OF THE *CEREBELLUM* IN THE CONTEXT OF NEUROBEHAVIORAL DISORDERS OF CHILDHOOD

Structure of the *Cerebellum*

What is the *cerebellum* and what does it do in the human brain? In the past, the answer would be that it only contributes to motor performance and skill. Still we recognize its contribution to motor control, however to limit its function to only motor coordination is a great understatement to say the least. Recent evidence has shown that the *cerebellum*'s contribution to control of all brain functions especially cognitive and behavioral controls may be just as great as its control over motor functions. In fact, we will see how the *cerebellum* may in fact be the key to normal cognitive and emotional development of the brain and is in fact the key to learning anything, whether it is motor or cognitive learning.

The *cerebellum* is a baseball size structure that lies under and behind the cerebral cortex. On a microscopic level, the *cerebellum* consists of three distinct cellular layers. The outermost layer is called the molecula layer, the innermost is known as the granular layer and the middle layer is the *Purkinje* layer. These layers make up the *cerebellar* cortex, which is an outer mantle of gray matter. The anatomy of the cerebellum and its associated tracted are presented in Figs. 4.1 and 4.2.

Molecular Layer

The molecular layer contains axons of *granule* cells known as parallel fibers. These parallel fibers are excitatory in their output. The molecular layer consists of two types of cells; *basket* cells, and *stellate* cells. These cells; when excited, have an inhibitory output. The molecular layer finally also contains dendrites of the *Purkinje* neurons the cell bodies of which are found in the *Purkinje* layer.

The Purkinje Layer

The *Purkinje* layer is found deep to the molecular layer in the Cortex. The *Purkinje* layer as its name would imply houses *Purkinje* neurons whose axons project to the *cerebellar* white matter, which consists of three *cerebellar* output nuclei. The *fastigial* nucleus, the interposed nuclei, which is subdivided into the *globos* and *emboliform* nuclei, and the most lateral nucleus known as the *dentate* nucleus. The neurotransmitter that is released by the *Purkinje* cell is, *Gamma amino buteric acid (GABA)*, and the most common

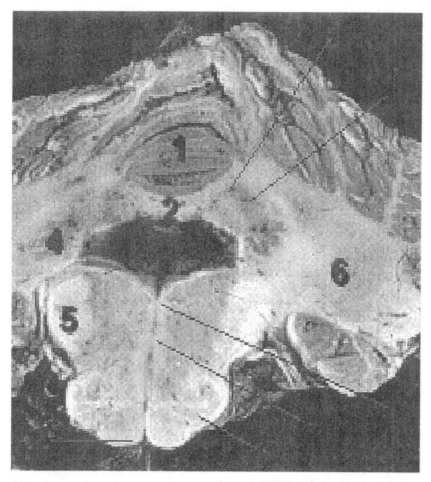

Fig. 4.1. The deep cerebellar nuclei and associated fiber tracts. 1. Cerebellar vermis. 2. Superior medullary velum. 3. Fourth ventricle. 4. Dentate nucleus. 5. Restiform body, inferior cerebellar peduncle. 6. Middle cerebellar peduncle. 7. Corticospinalis tract (pyramidal tract). 8. Medial longitudinal fasciculus. 9. Medial lemniscus. 10. Inferior olive. 11. Emboliform nucleus. 12. Globose nucleus.

inhibitory transmitter substance in the CNS. *Purkinje* cells therefore exert inhibitory control over the deep *cerebellar* nuclei and therefore modulate their output.

The Granular Layer

The innermost layer of the *cerebellar* cortex is the *granular* layer. This layer of cells is packed with small neurons, the number of which exceeds the number of neurons in the cerebral cortex. The density and population of cells belies the importance of the *cerebellum* in the overall function of the brain. The granular layer also consists of two types of cells; *granule* cells, and *Golgi* cells. *Granule* cells, when stimulated, excite *Purkinje, basket,* and *stellate* cells in the molecular layer. The excitatory impulse is carried by way of axon fibers known as parallel fibers. The parallel fiber will also stimulate the dendrites of the *Golgi* cells in the granular layer. The *Golgi* cells will in turn inhibit the *granule* cell, which, therefore, inhibits the *granule* cell excitation of *Purkinje* cells thereby indirectly inhibiting the output of the *Purkinje* cells.

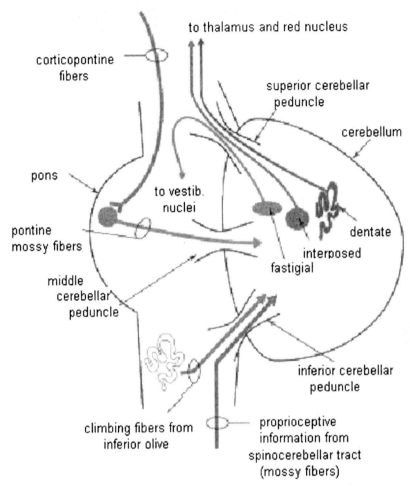

Fig. 4.2. The cerebellum operates with three pathways leading in and out of the cerebellum. There exist three main inputs and three principal outputs from three deep nuclei. The three pathways are called *peduncles*, or "stalks" of which there exist three pairs: the *inferior, middle,* and *superior peduncles*. The three inputs include: *Mossy* fibers from the *spinocerebellar* pathways, *climbing* fibers from the *inferior olive*, and more mossy fibers from the *pons*, which carry information from cerebral cortex. The *mossy* fibers from the spinal cord arise ipsilaterally, so they do not need to cross. The fibers from cerebral cortex, however, do need to cross. These fibers synapse in the *pons* (hence the huge block of fibers in the cerebral peduncles labeled *corticopontine*), cross, and enter the cerebellum as *mossy* fibers. The three deep nuclei are the *fastigial, interposed,* and *dentate* nuclei. The *fastigial* nucleus is primarily concerned with balance, and sends information mainly to *vestibular* and *reticular* nuclei. The *dentate* and *interposed* nuclei are concerned more with voluntary movement, and send axons mainly to the thalamus and the *red* nucleus.

The output of the *cerebellum* runs through the deep *cerebellar* nuclei and their projections. One of their projections runs to the vestibular nuclei, which lie immediately in front of the cerebellum. There are four vestibular nuclei on either side, superior, inferior, middle, and lateral. The only output of the *cerebellar* cortex is through the *Purkinje* cell, which is inhibitory to the deep *cerebellar* nuclei. Their primary output is modulated or gated by the inhibitory action of the *Purkinje* cells in the *cerebellar* cortex. The firing rate of the *Purkinje* cells is the product of two incoming or afferent fibers; *mossy* fibers, and *climbing* fibers.

Mossy fibers and *climbing* fibers also have collateral projections to the deep *cerebellar* nuclei directly. This is known as the primary *cerebellar* circuit. The primary circuit is modified by the action of the *Purkinje* cells of the *cerebellar* cortex.

Mossy Fibers

Mossy fibers constitute the majority of afferent input into the *cerebellum*. Afferent input that is transmitted by the mossy fibers originates primarily from various *brainstem* nuclei as well as input from the spinal cord tracts especially the *spino-cerebellar* tracts. *Mossy* fibers synapse on the *granule* cells at the *cerebellar glomerulus*. The *granule* cells excite the *Purkinje* cells in the *Purkinje* layer. *Purkinje* cells are also excited by parallel fibers in the molecular layer. Each *Purkinje* cell will receive as much as 200,000 connections from parallel fibers. *Mossy* fibers input to *granule* cells and their connections to *Purkinje* cells result in a small excitatory post-synaptic potential or (EPSP's). Spatial and temporal summations of the EPSP's are required for *Purkinje* cells to produce a single action potential. That is to say, for *Purkinje* cells to fire a certain threshold must be met, this threshold is the product of the number of connecting fibers that are being fired (spatial) and the speed or frequency of firing by those same connections (temporal). When enough connecting cells fire at a high enough frequency, *Purkinje* cells will reach a high enough threshold and an EPSP will cause a firing of the *Purkinje* cells. This single action potential is known as a simple spike.

Climbing Fibers

The other major fibers that transmit afferent input into the *cerebellum* are the *climbing* fibers. The afferent input that is transmitted by the *climbing* fibers originates in the inferior *olivary* nucleus in the *medulla*. The inferior olivary nucleus receives its excitation from the red nucleus by way of the central *tegmental* tract. *Climbing* fibers enter the *cerebellar* cortex and synapse on the soma of the *Purkinje* cells but the majority of the connections are to the dendrites of the same *Purkinje* cell. The *climbing* fibers are largely excitatory and each fiber connects between 1–10 *Purkinje* cells. Each *Purkinje* cell however receives only one *climbing* fiber connection. These synaptic connections are considered the most powerful in the CNS. Each synapse results in large EPSP's followed by a small burst of smaller action potentials. This type of characteristic grouping of action potentials is known as a complex spike, which is associated with a large influx of calcium into the cell.

When *granule* cells are excited by afferent *mossy* fibers, they synapse on the dendrites of the *Purkinje* cells, stimulating the *Purkinje* cell by way of the parallel fibers. Parallel fibers run perpendicular to the dendrite tree of the *Purkinje* cells. Once the *Purkinje* cells reach threshold by spatial and temporal summation they are stimulated to release *GABA*, which inhibits the deep *cerebellar* nuclei to which it connects. *Basket* and *stellate* cells are also stimulated in the molecular layer by the same input and they inhibit the action of other *Purkinje* cells, this is known as *surround inhibition*. The importance of this factor for both motor and cognitive functions will be discussed later.

Mossy fibers and their connecting *granule* cells fire spontaneously at high rates (50–100 spikes/s), on the other hand, *climbing* fibers fire at a low rate (1 spike/s). Low frequency (Tonic) *climbing* fiber input does not cause a significant direct afferent on *Purkinje* cells. However it is thought that the function of the *climbing* fibers which come from the *interior olivary* nucleus, form a type of pace-maker function for the *Purkinje* cell influencing its excitability to *mossy* fiber and *granule* cell input. This pace-maker activity discovered by Llinas (2001) is thought to occur in 10 Mhz bursts. The pace-maker effect on the *Purkinje* cells and its influence on the excitability of *mossy* fiber connections is known as *heterosynaptic action*.

There are two groups of *brainstem* nuclei that project *amineregic* fibers to the *cerebellar* cortex. The medial and dorsal *Raphe nuclei* project fibers that use *serotonin* as their transmitter substance. These fibers terminate in the granular and molecular layers. The *locus coeruleus* sends fibers that use *norepinephrine* as their transmitter and terminate in a plexus in all three layers: *molecular, Purkinje,* and *granule*.

Functions of the *Cerebellum*

The *cerebellum* is anatomically subdivided into three functional areas known as the *vestibulocerebellum, spinocerebellum,* and *cerebrocerebellum,* and its cranial nerve attachments are presented in Fig. 4.3(A) and Pathways in Fig. 4.3(B).

Vestibulocerebellum

Among the lobes of the *cerebellum,* include the *flocculonodular* lobe *(FL)* corresponding to the *vestibulocerebellum.* The FL has afferent and efferent connections to the vestibular nuclei. The *FL* controls eye movement and equilibrium of stance and gait. The dominant afferent input to the *vestibulocerebellum* arises from the semicircular canals and from the inner ear. The semicircular canals are concerned with changes in head position; there are three canals on either side, anterior, posterior, and lateral (or horizontal). Each canal, based on its orientation in space, is sensitive to different directional movements of the head. Anterior or forward movement stimulates the anterior canal, posterior or backward movement stimulates the influence on specific eye muscles by driving the eyes in the opposite direction and to maintain a level orientation of the eyes as the head is moved, also known as the *righting reflex.* The eyes move in a coupled fashion known as conjugate movement. Therefore, eye muscles are stimulated on both sides to allow for smooth coordinated movement that stimulates the posterior canal, and lateral or rotational movements of the head toward the same side stimulate the lateral canal. The canals work together on both sides of the head. When one canal is stimulated, it exerts an inhibitory effect on the contralateral canal. The canals also exert what is known as the *tracking reflex.* The midbrain control regions (Fig. 4.4) and the ocular-motor pathways (Fig. 4.5) are presented below.

Several types of eye movement exist with control centers in the *brainstem* and cortex. Anterior canal stimulation results in activation of the ipsilateral superior *rectii* muscles and the contralateral inferior *oblique* muscles, it also results in inhibition of ipsilateral inferior rectus muscle and contralateral superior oblique muscle, as noted in Fig. 4.5. Posterior canal activation results in activation of ipsilateral superior *oblique* and contralateral inferior *rectii* extraocular muscles while causing inhibition of the ipsilateral inferior *oblique* and contralateral superior *rectii.* Lateral canal activation causes contraction of ipsilateral medial *rectii* and contralateral lateral *rectii.* Lateral canal activation will simultaneously cause inhibition of the ipsilateral lateral *rectii* and contralateral medial *rectii* muscles. Coincident with the vestibular apparatus initiating effects on the eye muscles, muscles of the cervice are activated so that head and eye movement may be coordinated.

The fact that natural shifts of gaze involve movement of the head is well recognized. Robinson and colleagues (1994) state, "coordinated activation requires that the immediate *premotor* machinery for the neck and eye motor systems have ready access to the movement structures of the other. One possibility is that the structures that synapse directly on neck and extraocular motor neurons be near each other so that they can efficiently receive common input." Afferents from the semicircular canals and otolith organs are the only primary projections that reach the *cerebellar* cortex without an intervening relay. At the same time as input from the vestibular apparatus and the neck is influencing coordination there is also visual information from the striate cortex, *superior colliculus* and *lateral geniculate nucleus* of the *thalamus* that project to the vestibulo*cerebellum* by way of the *pontine* nuclei. *Vestibular nuclei* themselves also are a major source of direct input to the *vestibulocerebellum.*

Afferent input from the vestibular apparatus synapses in the medial vestibular nucleus, the *fastigial nucleus* of the vermis, and the *flocculonodular* cortex of the vestibulo*cerebellum*. The *vestibular nucleus* also project to the *flocculonodulus*. The *flocculonodular* lobe projects to vestibular nuclei and to the central group of reticular nuclei. The vestibular nucleus gives rise to the *descending medial longitudinal fasciculus (DMLF)* and the lateral *vestibulospinal* tract. The group of central reticular neurons gives rise to the *reticulospinal* tract. The contralateral inferior olivary nucleus also synapses on the fastigial nucleus and the *vestibulocerebellar* cortex. The *vestibulocerebellum* exerts its primary

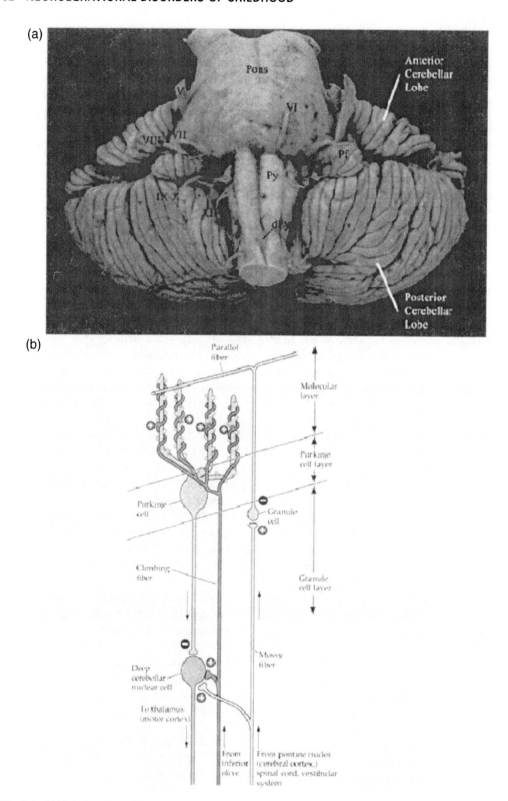

Fig. 4.3. (A) Inferior view of the cerebellum and brainstem, showing cranial nerve attachments and some internal features of the medulla oblongata. The *pyramidal tract* (Py) is represented as are the *trigeminal* (V), *abducens* (VI), *facial* (VII), *cochleovestibulaar* (VIII), *glossopharyngeal* (IX), and *Vaugs* nerves (X). (B) Cerebellar pathways.

THE *CEREBELLUM* AND *BASAL GANGLIA* 53

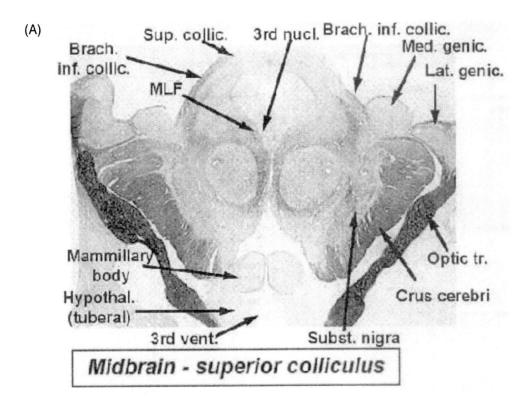

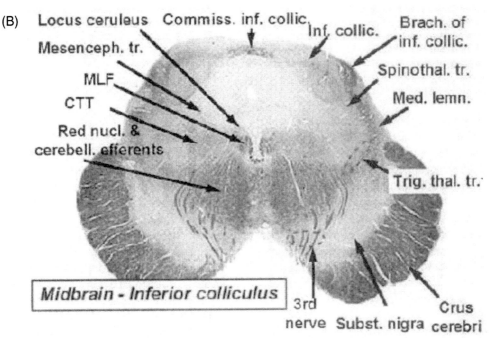

Fig. 4.4. Midbrain; (A) superior and (B) inferior colliculus.

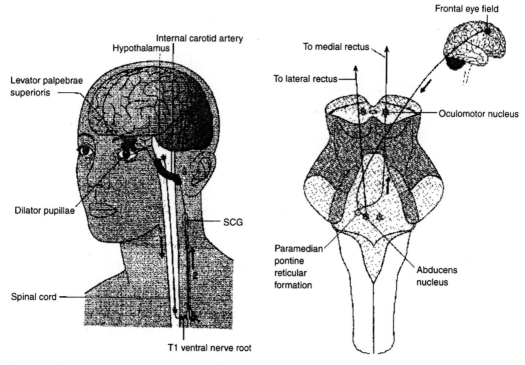

Fig. 4.5. Ocular–motor pathways. (A) Three neuron pathways from the hypothalamus to the eye. Arrows indicate direction of impulse conduction; *SCG*, superior cervical ganglion; (1) central fibers descending from the hypothalamus cross to the other side in the midbrain are also joined by ipsilateral fibers descending from the reticular formation; (2) pre-ganglionic fibers emerge in the first thoracic ventral nerve root and run up in the sympathetic chain to the superior cervical ganglion; (3) post-ganglionic fibers run along the external and internal carotid arteries and their branches. (B) Principal pathways involved in a voluntary ocular saccade to the left (after Fitzgerald and Folan-Curran, 2002).

influence on axial spine musculature and proximal limb musculature. The actions of the bilateral areas of the *vestibulocerebellum* need to be coordinated, therefore there is a need to share information. This is done through a bundle of *fastigobulbar* fibers that cross the midline through the contralateral *fastigial* nucleus to join efferent fibers from the opposite *vestibulocerebellum* that enter through the *restiform* body (Inferior *Peduncle*) of the *cerebellum*. This bundle of fibers is known as the *uncinate fasciculus* of Russell.

The *cerebellum* is also subdivided into three zones that correlate to areas of cortex and their underlying nuclei. The area above and including the fastigial nucleus is known as the medial zone. There are also the intermediate and lateral zones that correspond to the nucleus interpositus and the dentate nucleus respectively.

Spinocerebellum

It is thought that the main function of the spinocerebellum is to adjust the execution of ongoing movements and indirectly regulate muscle tone. It receives *mossy* fiber input that terminates in the vermis (medial zone) and intermediate zone of the *cerebellum*. Four major pathways carry information to the *cerebellar* cortex. Each pathway carries with it information that arises primarily from muscle spindle receptors or joint mechanoreceptors from specific areas of the body: (1) the dorsal spinocerebellar tract transmits information primarily from the trunk and legs, (2) the ventral spinocerebellar tract transmits information primarily from the legs, (3) the *cuneocerebellum* tract carries information primarily from the arms, (4) the rostral *spinocerebellar* tract sends information primarily from the neck. The dorsal spino*cerebellar* tract reflects sensory events from the periphery, informing the *cerebellum* about evolving or expected movements. Ventral *spinocerebellar* information reflects the status of the segmental interneurons that integrate descending and peripheral input allowing the *cerebellum* to monitor the operation of spinal circuits.

Medial Zone

The vermis in the medial zone sends efferent information though the *fastigial nucleus* to the *vestibulospinal* and *reticulospinal* pathways to control the axial (spine) and proximal musculature. It accomplishes this not only be connection to the *brainstem* nuclei but also through higher cerebral central integration. The *fastigial nuclei* have crossed ascending projections to the ventral lateral nucleus of the *thalamus* that relays this information to sensory motor areas cerebral cortex.

Intermediate Zone

This area of the *cerebellar* cortex overlays and connects to the interposed nuclei. The interposed nuclei, consisting of the *globose* and *emboliform* nuclei, function to modulate control commands for movement through their connections with the *brainstem* and central components of the lateral descending system. The lateral descending system consists of the *rubrospinal* tract and the lateral *corticospinal* tract. The interposed nuclei project to the contralateral *magnocellular* layer of the *red nucleus*. These projections exit the *cerebellum* through the superior *cerebellar peduncle*. The *nucleus interpositus* also receives input from the *red* nucleus. The intermediate zone sends afferent projections to the ventral lateral nucleus of the *thalamus*, which relays that information in turn to the sensory motor regions of the cerebral cortex. The interposed nuclei mainly control distal limb muscular action. The interposed nuclei also stimulate the central group of reticular nuclei and eventually control the gamma motor output through reticulospinal projections thereby influencing ipsilateral cortical control over all muscle tone.

Also integral to this system are pre*cerebellum brainstem* nuclei that project to the *spinocerebellum*. The lateral and paramedian reticular nuclei receive information from cutaneous receptors via the spinoreticular tract as well as from motor areas of the cortex. The *pontine reticulotegmental* nucleus also receives afferent input from the cerebral cortex and the vestibular nuclei. The *spinocerebellum* likewise receives *trigeminal* afferents from the *trigeminal mescenephalic* nucleus as well as *climbing* fiber input from the inferior *olivary* nucleus.

Cerebrocerebellum (Neocerebellum) (Lateral Zone)

The cerebro*cerebellum* or lateral zone coordinates the planning of limb movements. It receives the majority of its afferent input from the sensory motor cortices as well as *premotor* and posterior parietal cortices by means of the pontine nuclei. This *cortico*-ponto-*cerebellar* pathway consists of nearly twenty million fibers (Tomasch, 1969) and constitutes the largest pathway in the entire CNS. In comparison, the *cortico*-spinal pathway carries one million fibers. The *cortico*-ponto-*cerebellar* pathway enters the *cerebellum* through the middle *cerebellar peduncle* through the *pontis*. The lateral zone overlies and projects to the *dentate* nucleus; the *dentate* sends efferent projections that exit the *cerebellum* through the superior *cerebellar peduncle* to the ventral lateral nucleus of the *thalamus*. By means of the dentate connection to the central lateral nucleus, it influences motor and *premotor* regions of the cerebral cortex. The *dentate* nucleus also projects to regions of the cerebral cortex and to the *parvocellular* area of the *red* nucleus. This portion of the *red* nucleus does not contribute to the *rubrospinal* motor tract, but it is part of a complex feedback circuit that sends information back to the *cerebellum* via the ipsilateral inferior *olivary* nucleus. The lateral parts of the *cerebellar* hemispheres are largely associated with achieving precision in the control of rapid limb movements and in tasks requiring fine dexterity. Lesions of the cerebrocerebellum or *dentate* nucleus produce delays in excitation or initiation of movement of all muscles except the extraocular muscles. The *cerebellum* controls termination of movement in regard to extraocular motor activity. Therefore, the lateral *cerebellum* is responsible for initiating and terminating movements. Terminal tremor at the end of movement also known as *intention tremor* is involved with lateral lesions. Other symptoms include disorders of the temporal coordination of complex movements involving multiple joints and disorders in spatial coordination of hand and finger muscles.

The *cerebrocerebellum* contributes to the mechanisms for the preparation for movement (feed forward and expectancy) activities. In contrast, the *spinocerebellum* is more concerned with movement execution or (feedback) adjustments. The intermediate zone is fed a copy of the motor program that is being sent by the motor cortex to the muscles, this is known as efferent copy activity. The *cerebellum*, especially the lateral *cerebellum* is the initiator of all motor learning. In regard to motor learning, the *cerebellum* responds primarily to novel activities, primarily in the *dentate*, more so than continued already learned activities. It also appears to play a role in the stimulation and memory storage of learned behavior.

The Cerebellum and Motor Learning

Understanding the way the *cerebellum* responds to novel situations to produce motor learning is important since it has been recently shown that it is also involved in higher cognitive and behavior learning in much the same fashion (Seitz and Roland, 1992; Contreras-Vidal et al., 1997; Attwell et al., 2002; Hazeltine and Ivry, 2002; Small et al., 2002). This is an extremely important point since this book focuses on learning disorders, either cognitive-academic or social-behavioral learning. As we will see all human learning of behavior and movement seems to involve the *cerebellum*. The *cerebellum* responds to novel movements that are complex rather than simple in a continuous single plane (Leisman, 1987; Leisman and Vitori, 1990) due in part to the interaction of *granule* and *Purkinje* cells with *stellate* and *basket* cells. If an individual muscle is stretched or contracted causing stimulation or stretching of the muscle spindle receptors, these receptors send fibers that fire back to a specific area of the *cerebellum*, which has a somatotopic representation of body schema. Therefore, specific body areas and specific muscles will fire to specific discreet areas of the *cerebellum*. For instance, leg muscles will fire back to specific leg areas of the *cerebellum*. Therefore, if an arm movement is produced in a unitary and linear plane, specific *granule* cells will fire to *Purkinje* cells and nuclei in a specific area associated with that arm (Leisman, 1989a). These same *granule* cells associated with that specific arm movement also activate *basket* and *stellate* cells that spread out and provide a surround inhibition of *Purkinje* cells around the area that is being activated (Leisman, 1989a). This inhibition of *Purkinje* cells outside of the area responsible for prime movement produces disinhibition of the nuclei that are involved with the initiation of movement of other muscles not associated with the exemplified arm motion (Leisman, 1989a). This has the potential to bring contiguous areas of the *cerebellum* not directly responsible for the specific arm motion described, closer to threshold making them better able to react to a lesser stimulus (Leisman, 1989a). Such a situation would allow the creation of a smoother coordinated movement that is characteristic of normal *cerebellar* function.

Dysdiadchokinesia described originally by Gordon Holmes early in the 20th century is a breakdown of coordinated movement, and is recognized as a classic symptom of *cerebellar* dysfunction. On the other hand, continuous simple movements fire back into a specific area of *granule* cells in the *cerebellum*. As the *granule* cells fire to the *Purkinje* cells, the *Purkinje* cells in that area fire to inhibit the underlying output nuclei, decreasing the reactivity to incoming stimuli driving the neurons further away from threshold, while at the same time causing the surrounding areas to be closer to threshold and thereby prepared to kindle the next new movement. Movements resulting from changing muscle activity increase the *cerebellar* output of neurons associated with other muscles. The *cerebellar* cortex does not just affect its output neurons but also sends projections that exit from the somatotopic area to higher centers of the *brainstem* and cortex. The somatotopic information is then specifically relayed to areas of the *thalamus* as well as to the *red* nucleus, mescenephalic reticular formation, *nigro striatal* system, and areas of the somatosensory cortex. Novel movements then, stimulate the *cerebellar* area associated with that movement as well as all associated areas to which the *cerebellum* projects. As a movement becomes repetitive in the same plane, the increased activation of *granule* cells and of *Purkinje* cell inhibition causes increased inhibition of output of that area of the *cerebellum*; its associated areas are fired at a lesser rate in the *thalamus*, midbrain, and cerebral cortex. However, due to surround inhibition by *basket* and *stellate* cells, local regions connected with other somatotopic areas are closer to threshold than they would have been without the previous movement and are ready to fire with the next novel activity, firing the other central areas at a faster rate. This not only helps provide good smooth coordinated activity of muscles but also results in smooth coordinated activation of other areas of the CNS.

This process may be one way that the *cerebellum* promotes motor-cognitive as well as emotional learning. Since similar pathways and areas are involved in cognitive and behavior learning the same principles may apply using the *cerebellum* as a way to promote novel learning of all types. Therefore, any dysfunction or lesion within the *cerebellum* that disrupts or affects the function of *Purkinje* inhibition may affect smooth coordinated movements and the ability to learn new activities. Likewise, anything that affects projections to the *cerebellum* or areas of the brain with projections from the *cerebellum* such as the *thalamus, motor* cortex, *premotor* cortex, or *basal ganglia* may result in a learning disability.

There are specific types of symptoms that are associated with *cerebellar* dysfunction outlined in Table 4.1 such as an inability to stop the limb rapidly (as a result of not being able to initiate agonist or antagonist muscles) resulting in excessive rebound, delays in initiating responses with an affected limb, *dysmetria* errors (judgments of distance) in the range and force of movement, *dysdiadchokinesia* (clumsiness in performing rapidly alternating movements) in the rate and regularity of movements, *dysnergia* or errors in the relative timing of the components of complex multi-joint movement, *intention tremor*, a tremor that becomes most marked at the end of movement as the individual attempts to achieve fine motor precision, *titubation* (unsteadiness) manifest as spontaneous tremor activity that affects *axial* and *trunkal* muscles especially of the head and neck, *dysarthria* which are disorders in articulating coordinated muscles of speech, *hypotonia*, global decrease in muscle tone, and *ataxic* gait that result in a wide stance and unsteady balance. Ataxia involving the legs and gait are most common with *vermal* dysfunction, whereas *ataxia* in a nongravitational or lying position is mostly associated with a *flocculonodular* lesion or dysfunction. Intermediate or interposed nuclear dysfunction most commonly produces *ataxia* of limb movements. Dysfunction of the lateral *cerebellum* consists primarily of delays in initiating movements or decomposition of multi-joint movements especially distal joints and higher cognitive dysfunction that will be discussed in greater detail later.

Ocular motor dysfunction is one of the most common symptoms of *cerebellar* disorder.

TABLE 4.1 Common Symptoms of Cerebellar Disorder

Cerebellar Disorder	Common Symptoms
Excessive rebound	Inability to stop the limb rapidly
Delayed motor response	Delay in initiating responses with an affected limb
Dysmetria	Judgment errors in the range and force of movement
Dysdiadchokinesia	Clumsiness in performing rapidly alternating movements
Dysnergia	Errors in timing complex multi-joint movement
Intention tremor	Tremor with fine motor precision
Titubation	Tremor of head and neck muscles
Dysarthria	Disorder of muscles of speech
Hypotonia	Decrease in muscle tone
Ataxia	Gait with wide stance and unsteady balance

Wessel and colleagues (1998) recently studied the effects of various *cerebellar* disorders and the presence of ocular motor abnormalities. Their patients consisted of: 7 patients with *Freidrick's ataxia* (spino*cerebellar* disease), 9 with *cerebellar* atrophy, and 10 with *olivopontocerebellar* atrophy. They noted that previous attempts had been made to determine the association between oculomotor abnormalities and degenerative ataxic disorders with mixed results (Murphy et al., 1975; Maki et al., 1994). The researchers tested various types of ocular motor movements including *saccades*, pathological *nystagmus*, fixation instability, *optokinetic nystagmus, smooth pursuit,* and *vestibular nystagmus*. They stated that, "it can be assumed that most oculomotor disorders observed frequently in our patients are mainly related to a degenerative lesion of the *cerebellum*."

With regard to saccadic *dysmetria*, according to electrophysiological studies in monkeys and clinical observations in humans, accuracy of *saccades* is primarily maintained by a neuronal circuit involving *Purkinje* cells in lobule VI and VII of the dorsal *vermis* and their inhibitory projections to neurons in the underlying *fastigial* nucleus (Collins and De Luca, 1993). We learn that saccadic dysmetria, impairment of smooth pursuit, optokinetic *nystagmus*, deficient suppression of the visual ocular reflex either by visual or otolith input (Mauritz et al., 1981), and pathological *nystagmus* can all be attributed to degenerative lesions in various parts of the *cerebellum* (Wessel et al., 1998). Therefore, in individuals who present with oculomotor defects especially saccadic *dysmetria*, the oculomotor deficits may be directly related to a primary dysfunction of the *cerebellum*.

Non-Motor Functions of the Cerebellum: Evolutionary Implications for Cognitive Function

Recent studies and clinical evidence have shown that the contribution of the *cerebellum* to human behavior is greater than simply the control of motor function. In fact it appears that the *cerebellum* is involved in almost all functions of the brain including the control of motor, sensory, autonomic, cognitive, emotional, and behavioral responses (Sanford and Andy, 1969; Passingham, 1975).

As the human brain enlarged in its course of evolution along the phylogenetic scale, it appears that the *cerebellum* enlarged more dramatically than any other part of the brain except the cerebral cortex (Braitenberg and Atwood, 1958; Noback and Demarest, 1981). However, within the enlarged *cerebellum* the number of nerve cells apparently exceed the population in the cerebral cortex (Leiner et al., 1987). The phylogenetically new parts of the *cerebellum* seem to have developed in parallel not with the cerebral cortex as a whole, but with specific areas of the cerebral cortex known as associated areas. The *cerebellum* is a hindbrain structure containing billions of nerve cells (Ito, 1984; Eccles, 1987). The *cerebellum* is connected by millions of nerve fibers to many parts of the *brainstem* and forebrain including all lobes of the cerebral cortex (Larsell and Jansen, 1972; Brodal, 1981). It is therefore thought, based on the significant and widespread connections of the *cerebellum*, that the enlarged *cerebellum* must serve a variety of functions in the human brain beyond primarily motor control (Brodal, 1981). The question is why would the *cerebellum*, during the course of evolution, enlarge more than the cerebral cortex as well as actually containing more cells. This would suggest that natural selection would favor organisms that maintain a larger *cerebellar* function. The human *cerebellum* is uniquely large relative to other organisms, phylogenetically. Therefore, there must be a unique human function that is related to *cerebellar* functions and that quality may have either resulted in or been a part of an adaptive advantage that favored a larger *cerebellum* and subsequent unique cerebral cortical function.

There are two reasons that the *cerebellum* would maintain a larger size and greater amount of cells than other areas of the brain: (1) Genetic mutation and natural selection have favored organisms with large *cerebellums*, (2) during the development of a child, through the activity of synaptic stabilization the *cerebellum* loses less neurons due to high synaptic activity so that when the smoke clears the child is left with more cells in the *cerebellum* than anywhere else. This would mean that the synaptic activity, afferent, efferent, or both would exceed all areas of the CNS especially during the development process of synaptic stabilization. Most likely it is a combination of the two factors, that an enlarged *cerebellum* would impart an adaptive advantage to humans either with enhanced motor or cognitive control or both and that the importance of this function places extremely high demands on the *cerebellum* which requires cells greater than any area of the CNS.

Enhanced motor control was required of the *cerebellum* to permit bipedalism. Most postural control occurs reflexively through the *cerebellum* without direct control of the cortex, the same is true for the fine motor coordination of our hands, although it appears that enlargement of the brain occurred before advanced tool use, but after bipedalism. It is also thought that the non-motor functions of the *cerebellum* could be expected to confer a considerable adaptive advantage on humans because with stringent constraints that cranial size imposes on brain enlargement (Armstrong, 1980), natural selection would favor those enlargements that especially subserve advantageous functions (Leiner et al., 1991). Cognitive abilities such as language may be one of these advantageous human functions. Language capabilities in humans are dependent on a phylogenetically new area of the cortex and neocortex (Benson and Geschwind, 1983; Ojemann and Creutzfeldt 1987). Along with the evolution of the newer areas of the neocortex, association areas in the lateral or neo*cerebellum* respond to linguistic signals received from the posterior lobes of the cerebral cortex (Leiner et al., 1991). These responses are then transmitted to Broca's area in the *frontal lobe* of the neocortex. It is thought therefore that new areas of the *cerebellum* serve as a link between the posterior and frontal language areas of the cerebral neocortex, thereby providing an additional association cortex (Leiner et al., 1991). These new areas therefore, do not only provide motor control of speech but also verbal cognitive thought. Advanced techniques for scanning and imaging the brain have recently confirmed this (Murphy et al., 1997; Ackermann et al., 1998; Papathanassiou et al., 2000; De Nil et al., 2001; Marien et al., 2001). PET studies show activation has been greater in the *cerebellum* when subjects generate a semantic association to a word (e.g., EAT to APPLE) in comparison to a control condition where subjects simply repeat the target word continuously. This ability of the *cerebellum* to respond to novel cognitive abilities is similar to the *cerebellum*'s motor control responses described earlier.

The *Cerebellum's* Role in Information Processing

In regard to information processing, it was predicted that phylogenetically new parts of the human *cerebellum* can affect language and mental skills in the same way that older parts of the *cerebellum* influence motor skills (Leiner et al., 1986, 1987, 1989). It is also thought that the older parts of the *cerebellum* through their output to the cerebral motor cortex coordinate skilled manipulation of muscles. Similarly, newer parts of the *cerebellum*, through their connection to newer association areas of the cerebral cortex can influence the skilled manipulation of symbols. Therefore, in regard to both motor and nonmotor functions it is thought that the *cerebellum* can function at the "subconscious" level as an adaptive mechanism in all vertebrates. It can improve different skills in different species on the phylogenetic scale depending on the different connections of afferent and efferent information that evolved between the *cerebellum* and other parts of the nervous system.

In species where the *cerebellar* output is connected to the autonomic nervous system, the *cerebellum* can subserve a vegetative or visceral function (Haines et al., 1984, 1990; Haines and Dietrichs, 1987). In species where the *cerebellar* output is connected to motor areas, the *cerebellum* subserves a motor function (Ito, 1984, 1991). *Cerebellar* output to the sensory system can modulate sensory input (Snider, 1967; Bell et al., 1981). *Cerebellar* output to the *limbic system* can allow it to exert control over emotional behavior (Watson, 1978; Berntson and Torello, 1982; Lalonde and Botez, 1990; Schmahmann, 1997; Schmahmann and Sherman, 1998). In human species where the *cerebellar* output is connected to areas of the brain that are involved in mentation and language, the *cerebellum* can subserve those functions as well (Peterson et al., 1989; Roland et al., 1989; Decety et al., 1990; Ryding et al., 1993). By sending signals to other parts of the brain, the *cerebellum* can enable those receiving parts to improve incrementally the performance of learned skills, which can then be carried out optimally and rapidly (Leiner et al., 1991).

Now that we understand the evidence behind the variety of functions of the *cerebellum* and its widespread connections to other areas of the nervous system, we wish to further review research related to non-motor *cerebellar* functions.

Ideational Tasks

The term ideography was first introduced by Ingvar (1977) to describe investigations in which an individual performs purely mental tasks, which are influenced neither by ongoing perception of sensory signals or by ongoing control of movements, speech, or behavior (Ingvar, 1977, 1991). With improved imaging techniques, it has been demonstrated that the lateral *cerebellar* areas are activated during mental tasks (Martin and Raichle, 1983; Fox et al., 1985). Normal subjects were asked to do various cognitive tasks while the *cerebellum* was being imaged. These tasks included mental arithmetic (Silent Counting) (Decety et al., 1990; Ingvar, 1994), mental simulation of tennis playing, and other motor-ideational activities (unaccompanied by any motor activity) (Decety and Ingvar, 1990; Decety et al., 1990; Lassen and Ingvar, 1990; Ingvar, 1993, 1994), mental association of a word with its use (Peterson et al., 1989) and learning to recognize complicated geometric objects (Roland et al., 1989). In all of these investigations, activation of the *cerebellum* was observed.

Mental Imagery

Initial studies of *cerebellar* activity were focused on cognitive function during motor ideation (Decety et al., 1989; Decety and Ingvar, 1990). However more recent findings of *cerebellar* involvement in cognitive processes (Decety et al., 1988) have led to additional studies of *cerebellar* involvement with mental activity (Decety et al., 1990; Ingvar, 1991, 1994).

Tactile Learning and Sensory Processing

Investigations of tactile learning of complicated geometric objects provided evidence of high activation in the lateral *cerebellum* (Roland et al., 1989). Involvement of the *cerebellum* in sensory perception has been demonstrated by chronometric tests (Posner, 1986) of timing functions in *cerebellar* patients (Ivry et al., 1988; Inhoff et al., 1989; Ivry and Keele, 1989). *Cerebellar* patients were found to be impaired in their perception of time intervals and in their judgment of velocity of a moving stimulus. In other species, *cerebellar* involvement has been noted in sensory function (Dow and Moruzzi, 1958; Bell et al., 1981; Finger et al., 1981; Cristino and Bullock, 1984; Leaton and Supple, 1986; Bower and Kassel, 1990). Other studies have proposed that the base role of the *cerebellum* is to adjust motor performance and, to obtain the highest quality sensory input while the organism is exploring peripheral objects (Bower and Kassel, 1990). Therefore, by monitoring the acquisition of sensory input and by adjusting motor performance accordingly, *cerebellar* circuits could substantially improve the efficiency of sensory processing by the rest of the nervous system. Such improvement could extend to different sensory processes in different species (Leiner et al., 1991). Another way that the *cerebellum* may modulate sensory perception is through its connection to the central lateral nucleus of the *thalamus* (Asanuma et al., 1983; Jones, 1985). The central lateral nuclei have a diffuse projection to the cortex and may provide a substrate through which the *cerebellum* can contribute to cortical arousal, alertness, and attention (Saper, 1987). It has been shown that lesions involving the intralaminar nuclei of the *thalamus* produce different sensory perceptual decreases than lesions of discrete or specific *thalamic* nuclei. Lesions of *specific thalamic* nuclei like the medial *geniculate* nucleus cause specific *sensory* defects. However, lesions of the *diffuse thalamic intralaminar* nuclei result in *global* sensory perception loss (Llinas, 2001).

Arousal and attention deficits have also been thought to result in widespread perceptual, cognitive, and behavioral dysfunction (Leisman, 1976a; Davidson and Hugdahl, 1995). The other major connection to the *thalamus*, the ventral lateral nucleus, provides a more specific route for converging information to particular columns of the cerebral cortex (Steriade et al., 1990). This route projects primarily to the *frontal lobe* in humans, it has also been shown to project to the posterior parietal cortex in monkeys (Schmahmann and Pandya, 1990). The ventral lateral nucleus may particularly affect specific sensory acquisition therefore; it may also impact specific sensory perception as well as other functions of the frontal and parietal lobes and their connections.

Cerebellar Control of Autonomic Function, Emotions, and Motivation

The *cerebellum* has been shown to have both direct and indirect control over autonomic and emotional function. Two direct routes from the *cerebellum* to the cerebral cortex have been confirmed and traced. One route connects the *cerebellum* to the *thalamus* and to the neocortex. The second route connects the *cerebellum* to older structures of the brain like the *hypothalamus* (Dietrichs and Haines, 1989). The *cerebellum* is also connected to reticular structures in the *brainstem* where they provide a less direct route to the *limbic system*, which is concerned with autonomic, emotional, and motivational behavior (Berntson and Micco, 1976; Berntson and Torello, 1982; Haines and Dietrichs, 1987; Supple and Leaton, 1990). In previous neurophysiological studies in cats (Harper and Heath, 1973) the anatomic connections of the *fastigial* nucleus with the *hypothalamus* had been identified, the central nuclei of the *thalamus*, the *septal* nuclei, the nucleus *accumbens septi*, the *diagonal band* nucleus, the *cingulum*, and orbital *gyrus*. Cerebellar projections from *fastigial* nucleus to the (*noradrenergic*) *locus coeruleus* and to the (*serotonergic*) *Raphe nucleus* have also been identified (Snider, 1975; Marcinkievicz et al., 1989). Studies have also identified reciprocal connections between the *hippocampus* and the *cerebellum*. Efferent projections from the *cerebellum* to the *hippocampus* and *amygdaloid* complex (Fig. 4.6) originate mainly from

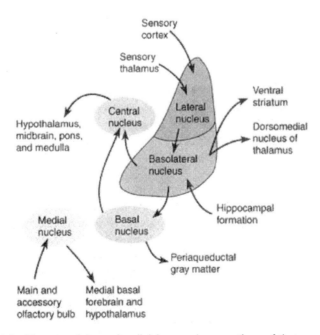

Fig. 4.6. Diagram of the major divisions and connections of the amygdala.

fastigial nuclei, which are both mono- and polysynaptic and are bilaterally arranged (Heath and Harper 1975; Newman and Rezea 1979). The *cerebellum* is connected to the ventral *tegmental* and *mesolimbic* areas of the midbrain that reach the *limbic system* through polysynaptic projections (Snider, 1975; Snider et al., 1976). Studies have revealed that the deep *cerebellar* nuclei of the intranuclear collateral neurons, probably use glutamate as a transmitter and mediate excitation of midbrain *dopamin*ergic neurons, (Snider, 1975; Snider et al., 1976; Audinat et al., 1992) thereby increas-ing meso*limbic* and mesocortical activity. The paeleo*cerebellar* and *limbic* projections from the anterior vermis and *fastigial* nuclei that connect to various *limbic system* structures are thought to modulate sensory input to the *hippocampus* (Newman and Rezea, 1979). These projections have also been shown to shorten or stop seizure activity produced by electrical stimulation of the *amygdala* and *hippocampus* (Maiti and Snider, 1975). Efferent hippocampal-*cerebellar* descending projections end synaptically mainly in the *vermis* (Newman and Rezea, 1979). Organism studies have shown that the *cerebellum* is involved in the learning and production of classically conditioned responses (McCormick and Thompson, 1984; Yeo, 1991), in long-term habitation of acoustic startle responses (Leaton and Supple, 1986), and in behavioral abnormalities similar to rodents with *hippocampal caudate* and *prefrontal* lesions. All abnormalities are thought to be a deficiency of spatial learning (LaLonde et al., 1988). Impaired discriminative learning and recognition learning have been shown in organisms following lesions of *cerebellar* hemispheres and *vermis* (Monjan and Peters, 1970).

Stimulation of the *fastigial* nucleus has been shown to cause *sham rage*, first noted by Cannon in the 1930s, and vegetative reactions in decorticated cats (Zanchetti and Zoccolini, 1954). These reactions are seen in hypothalamic and other *limbic* areas. Other behaviors such as grooming, feeding, and attacking or biting prey accompanied by visceral responses are also produced by stimulation of *fastigial* nucleus in non-decorticated cats (Reis et al., 1973). Lesions of the *vermis* in monkeys mitigate or abolish aggressive behavior (Berman et al., 1978). PET scans measuring regional blood flow changes during lactate-induced anxiety attacks have demonstrated that the left anterior *vermis* and other cerebral structures are involved in panic attacks (Reiman et al., 1989). There has been speculation that the older *cerebellar* regions including the *vermis* and *fastigial* nucleus could be considered as the equivalent of the *limbic cerebellum* and that the lateral (neo) *cerebellum* may be concerned with higher cognitive functions (Courchesne, 1991; Leiner et al., 1991). Therefore the *vermis* and its excitatory and inhibitory *Purkinje* cells modulate the output of the *fastigial* nucleus and its output to *limbic* and *hypothalamic* structures, while the lateral *cerebellum* and its connections to higher cortical regions, especially *prefrontal*, may also modulate the output of *limbic* and *hypothalamic* structures. These connections can affect many autonomic, emotional, and behavioral responses as well as arousal, vigilance, motivational, and attentional functions of the subcortical and cortical structures.

The *cerebellum* also has direct connections through the *fastigial* nucleus to the pontine and medullary reticular nuclei especially the nucleus *tractus solitarius* (NTS), which controls vagus nerve (cranial nerve X) function. The NTS, through its *vagus* nerve, can modulate a wide variety of cardiovascular and visceral functions including blood pressure as well as the strength and speed of cardiac contractions. The autonomic control properties of the *cerebellum* improve fuel delivery of nutrients to all areas of the body and brain.

The vestibular apparatus and its connection to vestibular nuclei also have connection to NTS as well as *fastigial* nuclei. Therefore, with lack of cortical *cerebellar* stimulation of inhibitory *Purkinje* cells, the *fastigial* nucleus along with vestibular stimulation can cause increased firing to the NTS. This may result in a sudden drop in blood pressure, syncope, and visceral responses such as nausea and vomiting as well as bladder contractions. Because of decreased *Purkinje* inhibition, these reactions can occur with subtle stimulation of the *vestibular apparatus* such as turning or bending the head, movement of eyes, or movement in an automobile or other moving vehicle. These responses are commonly seen as symptoms of motion sickness and are modulated primarily by the *cerebellum*. These effects may be triggered with thought alone.

The term non-motor function of the *cerebellum* implies that there exists a distinction between motor function and other cerebral cortical functions such as perception, autonomic, emotion, and cognitive. These distinctions may in fact be artificially imposed. The same neuronal substrates involved in motor control may well be the same as in cognition and emotion (Martin and Albers, 1995).

Sensory Functions

The *cerebellum* has not been considered a sensory organ because *cerebellar* lesions have not been noted to cause gross sensory defects (Thach et al., 1992). However, the *cerebellum* receives input from virtually every sensory system (Brodal, 1981) including tactile, independent of movement (Fox et al., 1985). It does connect to the *intralaminar* nucleus of the *thalamus*, lesions of which have been noted to be associated with gross global sensory defects (Llinas, 1995). It is also known that motor behavior is guided by ongoing sensory acquisition of object information and motor efficiency (the accuracy, coordination, and smoothness of behavior), which depends on continuously updated sensory data (Gao et al., 1996) as evidenced by magnetic resonance imaging of the lateral *cerebellar (dentate)* output nucleus during passive and active sensory tasks (Gao et al., 1996). Gao and colleagues tested the hypothesis that the lateral *cerebellum* is not activated by control of movement per se, but is strongly engaged during acquisition and discrimination of sensory information. The conclusion of the study was that the *cerebellum*, based on MRI activation, might be active during motor, perceptual, and cognitive performance specifically because of the requirement to process sensory data. This would be consistent with the fact that the lateral neo*cerebellum* is connected to all neocortical association areas (Leiner et al.,

1991), which are known to be involved in higher processing of sensory information. This is performed through reciprocal connections to the *thalamus* and other primary cortical and association areas (Llinas, 1995).

The Cerebellum, Cerebral Cortex, and its Frontal Lobes

Each output nucleus of the *cerebellum* sends fibers to ventral lateral nucleus and central lateral nucleus (Kyuhou et al., 1997; Middleton and Strick, 2001; Sakai et al., 2002). The central lateral nucleus sends diffuse projections to the cerebral cortex including the *frontal lobes*. The other connection is through the ventral lateral nucleus, which provides a more direct route for conveying information to particular columns of the cerebral cortex (Llinas, 1990; Steriade, 1998, 2000, 2001; Steriade et al., 1998). This route projects primarily to the *frontal lobe* in humans (Schmahmann and Pandya, 1990). These are specific areas in the *prefrontal* cortex to which the ventral lateral *thalamus* projects. For years, it has been thought based on primate studies that the ventral lateral *thalamus* serves as a motor substrate and transmits information solely to the motor cortex (Jones, 1985). However, it has been shown that through the course of hominid evolution the ventral lateral *thalamus* enlarged concomitantly with enlargement of the *cerebellar dentate* nucleus (Armstrong, 1980). It is thought that these newly evolved parts of the *cerebellum* and *thalamus* may provide a neural substrate for the evolution of cognitive and language skills (Jerison, 1973; Galaburda, 1984) through its connection to the *prefrontal* cortex.

It has been shown that the human ventrolateral *thalamus* is involved in cognitive and language as well as motor functions (Ojemann, 1977; Armstrong, 1980: Ojemann and Creutzfeldt, 1987). It has also been shown that the ventral lateral *thalamus* can send output projections to the *prefrontal* and motor cortex (Stern, 1942; Norman, 1945; Freeman and Watts, 1947, 1948; Hassler, 1964; Van Buren and Borke, 1972). The lateral *prefrontal* cortex is an association area that includes cortical areas of Brodmann 8–12 and 44 through 47

64 NEUROBEHAVIORAL DISORDERS OF CHILDHOOD

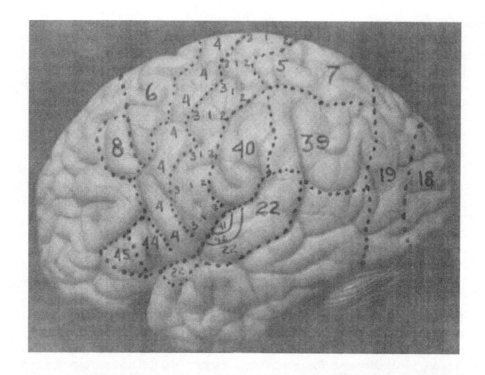

(B)

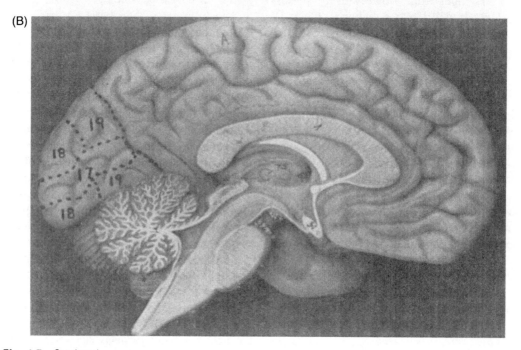

Fig. 4.7. *Continued*

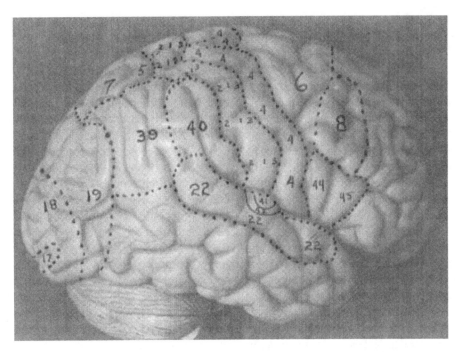

Fig. 4.7. Areas of Brodmann. (A) Left lateral surface, (B) left medial surface, and (C) right lateral surface.

(Fig. 4.7). This excludes the motor areas 4 and 6 (Leiner et al., 1991). Immediately anterior to these motor areas are the *prefrontal* areas that can receive *cerebellar* projections from the ventral lateral *thalamus* (Stern, 1942; Norman, 1945; Freeman and Watts, 1948a, 1948b; Hassler, 1964; Van Buren and Bork, 1972).

These areas include area 8, the frontal eye field, as well as areas 44–45 which are part of Broca's speech area. These areas have been regarded as motor areas (Zilles, 1990) and they have been shown to possess cognitive functions as well (Penfield, 1975; Grossman, 1980; Tonkonogy and Goodglass, 1981; Risse, et al., 1984). Experiments that are more recent involve the injection of *Herpes simplex* virus 1 (HSv-1) into the dorsolateral *prefrontal* cortex of monkeys tracing its axonal spread (Middleton and Strick, 1994). Two of the areas that were labeled were the *cerebellum* and the *globus pallidus* of the *basal ganglia*, which suggests that these two areas may be part of a cognitive neural network.

To further examine if the deep *cerebellar* nuclei participate in cognitive processing, MRI was used to study functional activation during a task that included a cognitive component (Kim et al., 1994). During imaging, participants were asked to do a visually guided pegboard task, and the cognitive component of an activity known as an insanity task. These results suggest that the cognitive processing associated with the solving of the insanity task lead to activation of the *dentate* nucleus that was noted by MRI. They further thought—based on their results that the dramatic increase in *dentate* activation during a visually guided motor task—that the regions in the *dentate* involved in cognitive activities are distinct from the ones involved in control of the eyes and limb movement. Other neuroanatomic and electrophysiological studies indicate that the *cerebellum* is part of a neural system that includes the *thalamus, basal ganglia,* and front lobes (Thach, 1980). PET studies have also shown a correlation between the metabolic rates of the *cerebellum* and frontal cortices (Junck et al., 1988). The *cerebellum* may interact with other subcortical and cortical structures that are known to be concerned with cognitive as well as motor functions. It is, therefore, likely that the *cerebellum* has a role in cognitive operations especially with those that have been identified with *prefrontal* and *basal ganglia* functions such as planning (Grafman et al., 1992).

PET scan was used to image regional blood flow during memory tasks. The study was designed so that recall of well practiced (previously learned) versus novel material could be compared. Their conclusion was that based on the results of this and other studies on memory, there is a suggestion that the human brain may contain a distributed multinodal general memory system. Nodes on this network include the frontal, parietal, and temporal cortices, the *thalamus*, the anterior and posterior *congulate*, the *precuneus*, and the *cerebellum*. They concluded that there appears to be a commonality of components across tasks (e.g., retrieval and encoding) that are not dependent on content, as well as differentiation of some components that may be context-specific or task-specific. In addition, these results suggest a significant role for the *cerebellum* in cognitive functions such as memory (Andreasen et al., 1995).

In another study examining lobular patterns of *cerebellar* activation in verbal working memory and finger tapping tasks using functional MRI (Desmond et al., 1997), for both tasks bilateral regions of the superior *cerebellar* hemispheres (left superior H VII A and right H VI) and portions of *vermis* (VI and superior VII A) exhibited increased activation during relatively high to low load conditions. The right inferior *cerebellar* hemisphere (H VII B) exhibited this load effect only during working memory tasks. Desmond and his colleagues concluded from this study that the (H VI) and superior H VII A activation represents input from the articulatory control system of working memory from the *frontal lobes* and that H VII B activation is derived from the phonological store in the temporal and parietal regions. They also thought that from these inputs the *cerebellum* could compute the discrepancy between actual and intended phonological rehearsal and use this information to update a feed-forward command to *frontal lobes* thereby facilitating the phonological loop. This study is important in that it not only confirms that the *cerebellum* connects to the *frontal lobe* for both motor and cognitive functions, but also shows that in the course of this interaction the *cerebellum* actually initiates function in the *frontal lobes* by firing before the *frontal lobe*. It also relays connections to parietal and *temporal lobes*, which subserve important non-motor functions.

Another recent study suggests that biochemical changes in *cerebellum* may affect cognitive functions other than language. Individuals with *Williams* syndrome (Galaburda and Bellugi, 2000) are late in reaching motor and cognitive milestones and display a characteristic physical and behavioral phenotype. The syndrome is of interest to cognitive neuroscientists because there is a relative spacing of aspects of language and fine processing accompanied by serious defects in non-verbal tools such as muscular, spatial, cognition, planning, and problem solving. The study suggests that a decreased neuronal marker *N-acetylaspartate* in the *cerebellum* is in some way involved in the condition. Significant correlations were found between the *cerebellar* ratios cho/na and cre/na and the ability of all subjects at various neuropsychological tests including verbal and performance IQ, British picture vocabulary scale, Ravens Progressive Matrices and inspection time. In the *William's syndrome* subjects, enlarged neo-*cerebellar* lobules and reduced paeleo*cerebellar* lobules had been noted. These investigators conclude that evidence exists to support the notion that the *cerebellum* is instrumentally involved in cognitive function, either on the developmental time scale (i.e. intact *cerebellar* functioning is required for normal cerebral functional or anatomical development) or on the immediate real time scale. In other words, proper development of the *cerebellum* may be a prerequisite for proper development of cerebrum both anatomically and functionally in both motor and cognitive abilities.

The *cerebellum* is part of a neuronal system that includes the *thalamus, basal ganglia*, and *frontal lobe* (Thach, 1980). *Frontal lobe* areas 8 and areas 44–45 are regarded as highly differentiated control areas (Pandya and

Yeterian, 1990). These areas in the cerebral cortex have enlarged in the human brain through evolution and receive projections from the lateral *cerebellum* through the ventral lateral *thalamus* (Freeman and Watts, 1947). These projections provide an anatomical substrate whereby the lateral portions of the *cerebellum* can contribute to cognitive functions including language as well as motor functions (Leiner et al., 1991). It has also been associated with other functions of the *prefrontal* cortex such as planning (Grafman et al., 1992) and verbal working memory (Desmond et al., 1997). The *cerebellum* has also been demonstrated to be part of a network with other areas of the cerebral cortex associated with memory including the frontal, parietal and *temporal lobe*s, the *thalamus*, the anterior and posterior *cingulate gyrus*, and the *precuneus* (Andreasen et al., 1995).

The *paeleocerebellum* including the *vermis* and *fastigial* nucleus could be considered the equivalent of the *limbic cerebellum* (Schmahmann, 1991) and therefore can affect more primitive autonomic and emotional functions as the lateral *cerebellum* can influence higher cognitive functions. Cerebellar connections to and from the *locus coeruleus* (*noradrenergic*) and *Raphe nucleus* (*seratonin*) have also been identified (Snider, 1975) and these neurotransmitters are thought to play a role in regulating emotional and affective behavior as well as arousal, motivation, and global behaviors such as sleeping and walking (Saper, 1987). The *cerebellum* is also connected to reticular structures in the *brainstem* and is concerned with autonomic, emotional, and motivational behavior (Berntson and Micco, 1976; Berntson and Torelloi, 1982; Haines and Dietrichs, 1987; MacLean, 1990; Supple and Leaton, 1990). The *cerebellum* also connects to the central lateral nucleus of the *thalamus*, and has diffuse projections to the cortex and may provide a substrate whereby the *cerebellum* may contribute to cortical arousal, alertness, or attention (Saper, 1987). The *cerebellum* can communicate with higher centers in the frontal, temporal, and parietal lobes and in some instances may provide a feed-forward command to the *frontal lobe*s (Desmond et al., 1997). Newer areas of the *frontal lobe* appear to have evolved in parallel with newer areas of the *thalamus* and neocortex, particularly the association areas of the neocortex (Leiner et al., 1991). Association areas are the areas involved with higher processing of information. The population of nerve cells in the *cerebellum* exceeds any other part of the nervous system and its input fibers far exceed the quantity that the cerebral cortex sends to any other part of the nervous system (Leiner et al., 1991). This would appear to make the *cerebellum* the most functionally active area in the central nervous system. Lastly, intact *cerebellar* functioning may be required for normal cerebral function or anatomical development in both higher cognitive and motor functions (Galaburda and Bellugi, 2000). Any lesion or dysfunction of the *cerebellum* should therefore have significant input or output connections based purely on the shear volume and frequency of stimulation that it receives and transmits, especially its input from gravitational receptors from the vestibular system and proprioceptive muscle and joint receptors. The *cerebellum* also seems to act as a link between posterior and anterior areas of the cortex providing the cerebral cortex with an additional association area (Leiner et al., 1991).

5

The *Basal Ganglia*

The *basal ganglia* is part of a neuronal system that includes the *thalamus*, the *cerebellum*, and the *frontal lobe*s (Thach, 1980). Like the *cerebellum*, the *basal ganglia* was previously thought to be primarily involved in motor control. However, recently there has been much written about the role of the *basal ganglia* in motor and cognitive functions (Alexander et al., 1986, 1990; De Long, 1990; Graybiel, 1995; Albin et al., 1995; Brown et al., 1997; Feger, 1997; Levitan et al., 1998).

The *basal ganglia* is located in the *diencephalon* and is made up of five subcortical nuclei: *globus pallidus*, *caudate*, *putamen*, *substantia nigra*, and the *subthalamic* nucleus of *Luys*. The *basal ganglia* is thought to have expanded during the course of evolution as well and is therefore divided into the neo and *paeleostriatum*. The *paeleostriatum* consists primarily of the *globus pallidus*, which is derived embryologically from the *diencephalon*. During the course of its development it further divides into two distinct areas; the external and internal segments of the *globus pallidus*. The *neostriatum* is made up of two nuclei, the *caudate* and *putamen*. These two nuclei are fused anteriorly and are collectively known as the *striatum*. They are the input nuclei of the *basal ganglia* and they are derived embryologically from the *telencephalon*. The sub*thalamic nucleus of Luys* lies inferior to the *thalamus* at the junction of the *diencephalon* and the *mescencephalon* or midbrain. The *substantia nigra* lies inferior to the *thalamus* and has two zones similar to the *globus pallidus*. A ventral pole zone called *pars reticulata* and a dorsal darkly pigmented zone called the *pars compacta*. The *pars compacta* contains *dopaminergic* neurons that contain *melatonin*. The *globus pallidus internum* and the *pars reticulata* of the *substantia nigra* are the major output nuclei of the *basal ganglia*. The *globus pallidus internum* and the *pars reticulata* of the *substantia nigra* are similar in cytology, connectivity, and function. These two nuclei can be considered a single structure divided by the *internal capsule*. Their relationship is similar to that of the *caudate* and *putamen*. The *basal ganglia* is part of the *extrapyramidal* motor system as opposed to the *pyramidal* motor system that originates from the sensory-motor cerebral cortex. The *pyramidal* motor system is responsible for all voluntary motor activity except for eye movement. The extrapyramidal system modifies motor control and is thought to be involved with higher-order cognitive aspects of motor control, the planning and

70 NEUROBEHAVIORAL DISORDERS OF CHILDHOOD

execution of complex motor strategies, as well as voluntary control of eye movements. There are two major pathways in the *basal ganglia*, the direct pathways, which promote movement, and the indirect pathways, which inhibit movement. Figures 5.1(A–F) illustrate the relation between *basal ganglia* and *cerebellum*.

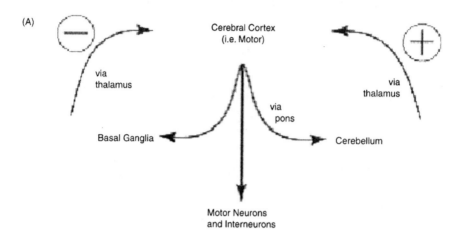

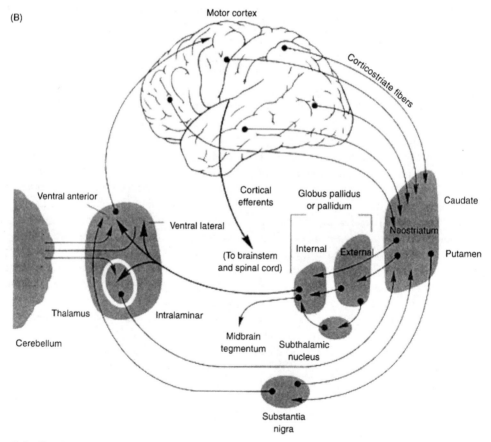

Fig. 5.1. *Continued*

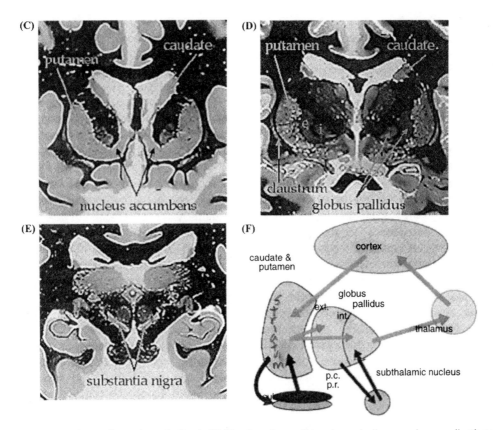

Fig. 5.1. (A) Basal ganglia and cerebellum. (B) The *basal ganglia* and *cerebellum* are large collections of nuclei that modify movement on a minute-to-minute basis. The motor cortex sends information to both, and both structures send information back to cortex through the *thalamus*. The output of the cerebellum is excitatory, while that of the *basal ganglia* are inhibitory. The balance between these two systems allows for smooth, coordinated movement, and a disturbance in either system will show up as movement disorders. The *basal ganglia* are a collection of nuclei deep in the white matter of cerebral cortex and include the *caudate, putamen, nucleus accumbens, globus pallidus, substantia nigra, subthalamic nucleus, claustrum*, and the *amygdala*. As to the *claustrum* and the *amygdala*; neither essentially deal with movement, nor are they interconnected with the rest of the *basal ganglia*, and they are not included. The figures; (C) rostral, (D) middle, and (E) caudal sections present the basal ganglia as they appear when stained for myelin. The *caudate* and *putamen* receive most of the input from cerebral cortex; in this sense they are the doorway into the *basal ganglia*. There are some regional differences: for example, medial caudate and nucleus accumbens receive their input from frontal cortex and limbic areas, and are implicated more in thinking and schizophrenia than in moving and motion disorders. (F) The *caudate* and *putamen* are reciprocally interconnected with the *substantia nigra*, but send most of their output to the *globus pallidus*. The *substantia nigra* can be divided into two parts: the *substantia nigra pars compacta* (*SNpc*) and the *substantia nigra pars reticulata* (*SNpr*). The *SNpc* receives input from the *caudate* and *putamen*, and sends information right back. The *SNpr* also receives input from the *caudate* and *putamen*, but sends it outside the basal ganglia to control head and eye movements. The *SNpc* produces dopamine, which is critical for normal movement. The *globus pallidus* can also be divided into two parts: the *globus pallidus externa* (*GPe*) and the *globus pallidus interna* (*GPi*). Both receive input from the *caudate* and *putamen*, and both are in communication with the *subthalamic nucleus*. It is the *GPi*, however, that sends the major inhibitory output from the *basal ganglia* back to *thalamus*. The *GPi* also sends a few projections to an area of midbrain (the *PPPA*), presumably to assist in postural control. This schematic (F) summarizes the connections of the basal ganglia as described above. Although there are many different neurotransmitters employed within the *basal ganglia* (principally *ACh*, *GABA*, and *dopamine*), the overall effect on *thalamus* is inhibitory. The function of the *basal ganglia* is often described in terms of a "brake hypothesis." To sit still, one must put the brakes on all movements except those reflexes that maintain an upright posture. To move, one must apply a brake to some postural reflexes, and release the brake on voluntary movement. In such a complicated system, it is apparent that small disturbances can throw the whole system out of balance, often in unpredictable ways. The deficits tend to fall into one of two categories: the presence of extraneous unwanted movements or an absence or difficulty with intended movements.

The *basal ganglia* receives afferent input from the entire cerebral cortex but especially from the *frontal lobes*. Almost all afferent connections to the *basal ganglia* terminate in the *neostriatum* (*caudate* and *putamen*). The *neostriatum* receives afferent input from two major sources outside of the *basal ganglia*, the cerebral cortex (*cortico-striatal* projections), and the *intralaminar* nucleus of the *thalamus*. The *cortico-striatal* projections contain topographically organized fibers originating from the entire cerebral cortex. An important component of that input comes from the centromedian nucleus and terminates in the *putamen*. Because the motor cortex of the *frontal lobes* projects to the centromedian nucleus, this may be an additional pathway by which the motor cortex can influence the *basal ganglia*. The *putamen* appears to be primarily concerned with motor control whereas the *caudate* appears to be involved in the control of eye movements and certain cognitive functions. The ventral *striatum* is related to *limbic* function, and therefore may affect autonomic and emotional functions.

The major output of the *basal ganglia* arises from the internal segment of the *globus pallidus* and the *pars reticulata* of the *substantia nigra*. The nuclei in turn project to three nuclei in the *thalamus* (cf. Figs. 4.1 and 4.6), the ventral lateral nuclei, the ventral anterior nuclei, and the mesio-dorsal nuclei. Internal segments of the *globus pallidus* project to the centromedian nucleus of the *thalamus*. *Striatal* neurons may be involved with gating incoming sensory input to higher motor areas such as the intralaminar *thalamic* nuclei and *premotor* cortex that arise from several modalities to coordinate behavioral responses. These different modalities may contribute to the perception of sensory input (Chudler and Dong, 1995) leading to motor response. The *basal ganglia* is directed, in a way similar to the *cerebellum*, to *premotor* and motor cortices as well as the *prefrontal* cortex of the *frontal lobes*. Recent experiments where *Herpes simplex virus 1* (*HSV-1*) was administered into the dorsal lateral *prefrontal* cortex of monkeys to determine its axonal spread or connection, labeled the ipsilateral neurons in the internal segments of the *globus pallidus* and the contralateral *dentate* nucleus of the *cerebellum* (Middleton and Struck, 1994). It is therefore thought that this may show the role of both the *cerebellum* and *basal ganglia* in higher cognitive functions in association with the *prefrontal* cortex. This would also substantiate a *cortico*-striato-*thalamo-cortical* loop, which would have a cognitive rather than motor function. The *substantia nigra* is also thought to connect to the superior *colliculus* through non-*dopaminergic* axons that form an essential link in voluntary eye movement.

It is thought that normal *basal ganglia* function results from a balance of the direct and indirect *striatal* output pathway and different involvement of these pathways account *for hyperkinesia* or *hypokinesia* observed in disorders of the *basal ganglia* (Alexander et al., 1990; De Long, 1990; Albin et al., 1995). *Hypokinesia* is a disinhibition or increase in spontaneous movement (tics, tremors). It is thought that *hypokinesia* and *hyperkinesia* may relate to *hypoactive* behavior and *hyperactive* behavior associated with subcortical hypo-stimulation or hyper-stimulation of medial and orbito-frontal cortical circuits (Litvan et al., 1998). It is important to review these connections further to understand the role of *basal ganglia* in control of cognitive and emotional behavior.

Five *fronto*-subcortical circuits unite regions of the *frontal lobe* (the supplementary motor area; frontal eye fields; dorsolateral, *prefrontal*, orbito-frontal, and anterior *cingulate* cortices) with the *striatum, globus pallidus*, and *thalamus* in functional systems that mediate volitional motor activity, saccadic eye movements, executive functions, social behavior, and motivation (Alexander et al., 1986; Alexander et al., 1990; Mega and Cummings, 1994; Cummings, 1995; Litvan et al., 1998).

DIRECT AND INDIRECT PATHWAYS

Five major cortical to subcortical loops exist that make up *cortico-striatal* pathways. All cortical pathways initiate the direct and indirect pathways with the *basal ganglia* through excitatory *glutamatergic corticostriatal* fibers. The direct pathway from the striatum sends *GABA* fibers (associated with *dopamine* receptors) from the striatum to the *globus pallidus* and *substantia nigra*. The indirect pathway sends inhibitory *GABA*/encephalin fibers (associated with D2 *dopamine* receptors) from the striatum to the *globus pallidus*. Indirect pathways then continue with inhibitory *GABA* fibers from the *globus pallidus* to the sub*thalamic* nucleus of Lluys. Indirect excitatory *glutamatergic* fibers then connect from the sub*thalamic* nucleus to the *globus pallidus* and *substantia nigra*. The *basal ganglia* then sends inhibitory outflow by *GABA* fibers from the *globus pallidus* and *substantia nigra* to specific *thalamic* nuclei. The *thalamus* has excitatory fibers that return to the cortex (Litvan et al., 1998). Abnormalities of direct and indirect pathways result in different pathological functions.

Hyperkinetic disorders (increased movement) are thought to be a selective loss of *GABA*/enkepholinergic intrinsic *striatal* neurons projecting to the lateral *globus pallidus* and *substantia nigra*. This results in decreased inhibitory stimulation to the *thalamus* leading to increased activity of the excitatory *glutamatergic* thalamocortical pathways and in turn greater neuronal activity in the *premotor*-motor and supplementary motor cortices (Albin et al., 1995). The result is over-facilitation of motor programs resulting in increased motor activity. *Hypokinetic* disorders (decreased movement) are associated with decreased *dopamine*rgic nigro*striatal* stimulation from the *substantia nigra* to the *striatum*. This results in both excess outflow of the indirect *striatal* pathway and an inhibited direct *striatal* pathway. Both of these pathways increase *thalamic* inhibition and therefore decrease thalamocortical stimulation of motor cortical areas resulting in *hypokinesia* or decreased output of the frontal cortex (Litvan et al., 1998). It is possible that the difference between *hypokinetic* and *hyperkinetic* syndromes may be different only in the timing and or the severity of the dysfunction. In this model, decreased *thalamic* excitation of the frontal cortex results in decreased excitation of the *cortico-striatal* fibers of the *neostriatum*. The *neostriatum* therefore decreases its inhibition of the *globus pallidus*. There is then increased inhibition of the thalamocortical pathways leading to progressive *hypokinesia*. Eventually the lack of *striatal* inhibition of the *globus pallidus* results in its metabolic dysfunction and the rapid loss of *GABA* neurons. This can then result in decreased inhibition of thalamocortical pathways causing a sudden onset of *hyperkinesia* (increased movement) with the increased *thalamic* firing of the frontal cortex. There also appear to exist cognitive and emotional symptoms that parallel the motor effects. Previous studies have shown that patients with *hyperkinetic, hypokinetic, Tourette's*, and *OCD* disorders may exhibit neuropsychiatric disturbances such as apathy, depression, agitation, or excitability (Starkstein et al., 1992; Chiu, 1995; Commings and Commings, 1998; Litvan et al., 1998).

It has been hypothesized that neuropsychiatric symptoms exhibited by patients with *basal ganglia* disorders are a consequence of an involvement of *fronto-striatal* connections. In addition to expressing contrasting motor dysfunction patterns, these disorders would also differ in the presenting psychiatric symptoms (Litvan et al., 1998). In this study patients with *Huntington's* disease (*hyperkinetic*) and *Parkinson's* disorder (*hypokinetic*) were observed to determine if they would present with hyperactive behavior (agitation, isolation, euphoria, or anxiety) and *hypokinetic* behavior (apathy) respectively. The results of this study demonstrated that patients with *Huntington's* (*hyperkinetic*) more frequently exhibited hyperactive behaviors such as agitation, irritability, euphoria, and anxiety whereas patients with *Parkinsons'* (*hypokinetic*) frequently displayed hypoactive behavior (high levels of apathy). The investigators thought that in *Huntington's*, these behaviors result from excitatory subcortical output through the medial and orbito-frontal circuits to the *pallidum, thalamus,* and cortex as well as *premotor* and *motor* cortex. In contrast, patients with *Parkinsons'* (hypoactive) in whom apathy is present were thought to demonstrate these behaviors as a consequence of hypo-stimulation of frontal subcortical circuits resulting from damage to several integrated nuclei (*substantia nigra, striatum,* and *globus pallidus*) (Litvan et al., 1991). It had been previously noted that patients with *Huntington's* and other *hyperkinetic* disorders like *Tourette's* exhibit mania, *OCD*, and intermittent explosive disorder (Cummings and Cunningham, 1992; Cummings, 1993). PET studies of *Huntington's* patients without hyperactive behavior have shown frontal metabolism to be normal but with decreased *caudate* and *putamen* metabolism (Kuhl et al., 1982; Young et al., 1986). However it is thought that normal frontal metabolism in *Huntington's* may result from a coexistent neurological degeneration and the resultant thalamo-frontal hyper-stimulation. This may result in normal appearing frontal-cortical regional blood flow even when overt *prefrontal* type cognitive defects are manifested. This suggests that in this case, a dysfunctional *prefrontal* cortex may appear to be at baseline levels that appear normal when in fact the *prefrontal* cortices may be over stimulated by the *thalamus*. (Gomez-Tortosa et al., 1996; Weinberger et al., 1998). In fact, it was noted that with further atrophy of the *caudate* there was increased *fronto-cortical* metabolism while the patient performed cognitive tasks (set-shifting) and a greater increase in cerebral metabolism over baseline. The poorer the subject performed on cognitive tasks, the greater was the cortical activation. (Weinberger et al., 1988; Gomex-Tortosa et al., 1996). It has been speculated that in early Huntington's when there are no *frontal lobe* lesions, a relative balance between frontal and increased *thalamic* functions may explain behavioral symptoms (Litvan et al., 1998).

PET scans of patients with *Parkinsons'* have also provided support that frontal-subcortical connections are disrupted by subcortical dysfunction showing decreased glucose consumption in frontal cortex, and decreased nigro*striatal* D2 receptor uptake ratios (D'Antona et al., 1985; Brooks et al., 1990, 1992; Blin et al., 1990, 1992; Brandel et al., 1991; Brooks et al., 1992; Brooks, 1994). Researchers at Stanford University may have observed similar results in children with *ADHD* also known as childhood *hyperkinetic* disorder. The Stanford study used functional MRI to image the brains of boys between the ages of 8 and 13 while playing a mental game. Ten of the boys were diagnosed with *ADHD* and six were considered normal. When the boys were tested there appeared to be a clear difference in the activity of the *basal ganglia* with the boys with *ADHD* having less activity in that area than the control subjects. After taking Ritalin, the subjects were scanned again and it was found that boys with *ADHD* had increased activity in the *basal ganglia* whereas the normal boys had decreased activity in the *basal ganglia*. Interestingly, the drug improved the performance of both groups to the same extent (*Newsday*, 1998). This may be a similar finding as the PET scans on patients with hyperactivity disorder, where normal appearing frontal metabolism existed with decreased *caudate* and *putamen* metabolism (Kuhl et al., 1982; Young et al., 1986). Ritalin, which is thought to be a *dopamine* uptake inhibitor, may increase function in a previously dysfunctional *basal ganglia* whereas raising *dopamine* levels in normal individuals would most likely result in decreased activity of the *basal ganglia* to prevent overproduction of *dopamine*. The previously dysfunctional *basal ganglia* would have most likely resulted in decreased frontal metabolism with increased *thalamo-cortical* firing; this would result in decreased cognitive function with increased *hyperkinetic* (hyperactive) behavior.

Increasing *dopamine* levels may increase frontal metabolism due to increased activity of the striatum with decreased firing of the *globus pallidus*; inhibiting *thalamo-cortical* firing decreases *hyperkinetic* behavior. This would make sense based on the findings of MRI before and after, and the fact that both groups showed equal improvement in performance.

Along with the *basal ganglia*, there are five basic projections of the *dopamine* system. Four of these projections are significant in regard to behavior. These four pathways are the: meso*cortical*, meso*limbic, tuberinfundibular,* and *nigrostriatal* systems. The *nigrostriatal system* is concerned with excitation and maintenance of motor behavior. It sends projections from the *substantia nigra* to the *caudate* and *putamen*.

THE DOPAMINE SYSTEM

The meso*limbic* and mesocortical systems arise from the ventral *tegmental* area (VTA). This cell group lies extremely close and is almost inseparable from the *substantia nigra*. As its name suggests, the meso*limbic* region projects from the *mescencephalon* to various *limbic* structures including the *amygdala, hippocampus, nucleus accumbens*, the *septal* area, and olfactory tubercle. These areas are sometimes referred to as the *limbic striatum* due to their proximity to the *striatum* (*caudate* and *putamen*). The mesocortical system projects from the VTA in the *mescencephalon* to the frontal cortex; primarily the *prefrontal cingulate* and *entorhinal* areas. Due to separate feedback loops, there oftentimes exists cross activation. The mesocortical and meso*limbic systems* are often activated during stressful situations. Both pathways seem to be important in the excitation and maintenance of goal directed and reward mediated behavior and in the maintenance of cognitive sets and is likely due to their connections with the *frontal lobes*. Dysfunction can cause breakdown of *frontal lobe* function that results in the inability to screen non-meaningful stimuli resulting in perseverative or repetitive behavior. It can also cause a child to attend to non-critical stimuli or overly-attend to sensory input resulting in non-selective attention (Leisman, 1976a). This may also lead to other symptoms such as delusions, flight of ideas, and loosening of associations. These cognitive or behavioral symptoms are often associated with other disorders of decreased *dopamine* concentration such as *Parkinsons'* or increased *dopamine* concentration such as *Huntington's* resulting in *hypokinetic* or *hyperkinetic* behavior. The neurobiologic processes of reward and reinforcement affect motivational behavior.

It appears that the *limbic system* and its *hypothalamus* support these behaviors more commonly known as self-stimulating behavior. Pathways involved in this type of behavior seem to involve the medial *forebrain bundle*, carrying *noradrenergic* fibers from the *locus coeruleus* to the anterior *limbic system*, which includes the *septum* and nucleus *accumbens*. This pathway also carries some of the mesocortical *limbic dopamine*rgic fibers from the ventral *tegmental* area. The balance and degree of *dopamine*rgic innervation from the SN and VTA may play a role in different affective disorders such as mania and depression. *Caudate* hypofunction has been associated with both mania and depression and may relate to hemisphere laterality of the lesion (Castellanos et al., 1994). Anything that disrupts the normal frontal-*basal ganglia* loop may result in abnormal *thalamo-cortical* excitation which can cause "abnormal reverberatory" activity that maintains the fixed cognitive emotional set of depression (Drevets and Raichle, 1992). This abnormal reverberatory effect may be the basis of all *hyperkinetic*/hyperactive behavior whether it is manifested as repetitive motor activity as in *Tourette's*, inability to filter stimuli or attend to minimal stimuli, or perseverative or repetitive as seen in *autism*, PPD, *OCD*, which is associated with *frontal lobe* dysfunction.

Much of the afferent-initiating input to this loop may arise from postural antigravity muscles to the *cerebellum, thalamus, basal ganglia*, and *frontal lobe*. Studies have demonstrated that the *neostriatum* is probably involved in posture regulation (Johnels and Speg, 1980). It has also been suggested that the *nigral dopamine*rgic/*GABA*/cholinergic balance is involved in positive regulation (De Montis et al., 1979). De Montis and associates analyzed components of conditional postural adjustments on 15 adult dogs. They concluded that activation of *nigrastriatal dopamine*rgic pathways can diminish muscle tone, acting through the *basal ganglia*, while activation of the *substantia nigra* and *globus pallidus* pathways can increase muscle tone. They concluded that their results suggest that the various components of postural adjustment and the tonic and phasic components of learned movements are probably controlled by different transmitter systems of the neo-*striatum*. They thought this was essential for understanding how to connect disturbances of posture and muscle tone that commonly occurs with astronauts in hypogravity (Shapovalova, 1991).

The *cerebellum* is connected to the VTA of the midbrain (Snider, 1975; Snider et al., 1976). Therefore anything that causes decreased muscle tone or dysfunction of postural antigravity muscles not only will decrease feedback to *cerebellum, thalamus*, and frontal cortex, but may also result in decreased feedback to the *basal ganglia*, meso-*limbic* and/or meso-cortical system. Each of these components has motor as well as cognitive, emotional, and autonomic effects. The *cerebellum* can influence the *basal ganglia* loop through several connections: through direct connections to VTA, through connections to the *thalamus*, and through its connection to frontal and *prefrontal* cortex, each of which have reciprocal connections to the *basal ganglia*. The *basal ganglia* can also influence the *cerebellum* by affecting *frontal lobe, thalamus*, as well as motor tone, all of which are controlled by the *cerebellum*.

Abnormal muscle tone and motor performance is the most commonly observed symptom in relationship to *basal ganglia* dysfunction. Brain lesions may cause *dystonia*, with the responsible sites being thought to include the *basal ganglia, thalamus*, and *brainstem* (Hallett, 1998). *Dopamine* is thought to be involved in some disorders characterized by dystonia as in *Parkinsons'* disease. *Dystonia* can develop either spontaneously or because of treatment, and can occur with high or low *dopamine* levels (Hallett, 1998). *Dystonia* can apparently also be produced behaviorally by excessive repetitive activity. An organism model of *dystonia* was created in non-human primates using synchronous widespread stimulation to the hand during repetitive motor tasks (Byl et al., 1996). Over a period of months, the organism's motor performance decreased eventually developing a movement disorder (*dystonia*). The primary somatosensory cortex was mapped and receptive fields in the area 3b were increased 10–20 fold after extending across the surface of two or more digits.

The conclusion of the study suggested that synchronous sensory input over a large area of the hand could lead to re-mapping of the receptive fields and subsequently to a movement disorder. These tasks also involve repetitive movements, which can lead to a re-mapping of the motor system. It is important to note that repetitive movement along the same plane may reduce the control effectiveness of the *cerebellum* in the areas associated with that movement thereby decreasing the firing rate from the *cerebellum* to higher cortices. This may also affect that state of readiness of other muscles, both proximal and distal, to muscles being used. *Dystonic* movements are characterized by abnormal patterns of activity on EMG with co-contraction of antagonist muscles and overflow into extraneous muscles (Berardelli et al., 1998). It has been reasoned that the problem of excessive co-contraction of antagonist muscles could be caused by deficient reciprocal inhibition. That this process is represented at multiple levels of the CNS, and that it produces inhibition of a muscle when its antagonist is activated (Rothwell et al., 1983), led to studies of deficient reciprocal inhibition in the spinal cord reflexes of individuals with *basal ganglia* disorders. Other reflexes such as eye blink reflexes in individuals with blepharo-spasm were examined leading to the conclusion that inhibition was deficient at the *brainstem* and spinal cord levels. Even though decreased spinal cord and *brainstem* inhibition may be an important mechanism in *dystonia*, it is thought that the fundamental dysfunction may be at higher supraspinal centers (Hallett, 1998). In rigid patients with *Parkinsons'*, spinal mechanisms have been studied. Regional inhibition of inhibitory interneurons has been shown to be increased (Conce and Delwaide, 1984). Inhibition of Ib fibers could also theoretically explain rigidity. Ib fibers from *golgi* tendon organs are large diameter fibers that project onto interneurons located at *Rexes lamina V and VI*, which in turn inhibits homonymous motor neurons. Other afferents converge on the Ib and Ia interneurons and cutaneous and joint afferents. Ib interneurons are facilitated by both the *cortico-spinal* and *rubro-spinal* tracts, and are inhibited by dorsal *noradrenergic reticulo-spinal* tracts (Delwaide et al., 1991). If inhibition by Ib interneurons is reduced, the mechanism of maintaining a linear relationship between length and tension (Nichols and Houk, 1976) would be altered and muscle stiffness may result.

In a study conducted on *Parkinsons'* patients, it has been shown that they did in fact exhibit less active Ib inhibitory interneurons and more active Ia inhibitory interneurons which was thought to produce rigidity (Delwaide et al., 1991). The investigators hypothesized that the nucleus *gigantocellularis* may be responsible since it is thought to receive projections from the *substantia nigra*. Delwaide and colleagues thought that the nucleus gigantocellulan's might be more active and therefore responsible for the observed findings at the spinal cord interneurons. In normal subjects, the *pars dorsalis* of the *reticularis gigantocellularis* nucleus exert inhibition of the spinal Ib inhibitory interneurons and motorneurons by the pathway of the dorsal longitudinal *fasciculus*. The pars medialis and ventralis of the same nucleus produce respectively, facilitation and inhibition of proprio-spinal interneurons, by the pathway of the medial longitudinal *fasciculous* which arises from vestibular nuclei. The proprio-spinal interneurons activate Ia inhibitory interneurons and motorneurons. Hyperactivity within the *gigantocellularis* might explain rigidity in *Parkinson's* patients. Hyperactivity of a *pars dorsalis* and medialis would lead to increased excitability of Ia interneurons by propriospinal interneurons and a more pronounced inhibition of Ib interneuron.

Various studies also indicate that there is some type of deficiency in the cortical motor system. PET and tracer H_2O, which were used to measure regional cerebral blood flow in a study of voluntary movement in patients with dystonia, found lower activity in supplementary motor area and primary sensory motor cortex bilaterally. There were also some areas of increased activity in the contralateral lateral *premotor* cortex, rostral supplementary motor area, *lentiform* nucleus, anterior *cingulate*, and ipsilateral dorsolateral *prefrontal* cortex (Mizuno et al., 1989). Another PET scan investigation examining patients with *dystonia* of the hand while they performed neurological tasks found decreased blood flow in the sensori-motor cortex contralaterally, *premotor* cortices bilaterally, *cingulate* cortex, and supplementary motor area. Still other studies using transcranial magnetic stimulation (Ikoma et al., 1996) show increased excitability of the motor cortex. Some think that all of these results can be explained due to lack of inhibition of cortex leading to a hyperexcitable cortex and *hyperkinetic* activity (Hallett, 1998). The basis of this theory is that thalamocortical influences on the cortex are thought to exert excitatory and inhibitory influences. The *basal ganglia's* major role may be to balance the excitation and inhibition. It has been postulated that the direct and indirect pathways may create a type of surround inhibition to focus motor commands (similar to what we previously described in the *cerebellum*). The loss of this surround inhibition might well explain the essence of *dystonia* leading to increased motor commands with overflow into inappropriate muscles (Hallett, 1998). In this model of *dystonia*, decreased cortical inhibition could result from dysfunction of the *basal ganglia* anywhere in the pathway including the *thalamus*. Increase or decrease of *dopamine* depending on its effect on indirect or direct pathways would cause it. *Dystonia* may also be caused by repetitive use, which may lead to larger cortical representation, which may be caused by lack of inhibition (or increased stimulation) (Hallett, 1998).

The loss of surround inhibition from the *cerebellum* or altered cerebello-thalmocortical firing could also result in many of the findings of the aforementioned studies. A decrease in *cerebellar* cortical stimulation could be a result of decreased *corticopontine-cerebellar* stimulation, or decreased muscle spindle and/or joint mechanoreceptor stimulation to the *cerebellum*. PET studies in patients with *dystonia* show reduced activation to vibration stimuli (Hallett, 1998). This may be a result of decreased peripheral large afferent firing, decreased central summation of the dorsal column, or spino-*cerebellar* pathways which would alter thalamocortical firing to the somatosensory cortex. This could therefore decrease subsequent frontal *striatal* firing to the *basal ganglia*. It is thought that the decreased firing of the *basal ganglia* pathways may result in metabolic damage to the *globus pallidus* which then increases its firing to a previously decreased area of the frontal cortex. This can result in *dystonia* or even epileptiform activity in the cortex.

One of the most interesting and fairly recent discoveries with regard to *dystonia* and similar motor disturbances that have been previously thought to be related to *basal ganglia* or *thalamic* dysfunction is that they may be *cerebellar* in origin. *Parkinson's* disease and *essential tremor* are the two most common movement disorders. *Parkinson's* tremor is slower and happens mostly at rest where *essential tremor* is more rapid and occurs with movement (Lance and Schwab, 1963; Koller et al., 1989; Palmer and Hutton, 1995; Henderson et al., 1995, 1994; Deuschl and Krack, 1998; Ondo et al., 1998). The ventral *thalamus* is thought to play a role in both types of tremor (Deiber et al., 1993; Ondo et al., 1998). The pathophysiology of Parkinson's Disease is fairly well established although not completely understood (Wichman and Delong, 1996). It is thought that loss of *dopamin*ergic cells results in increased *globus pallidus* activity in both direct and indirect pathways. Increased activity of the *globus pallidus* inhibits ventrolateral *thalamic* output to result in a decrease of motor output (Ondo et al., 1998). *Essential tremor* is considered a central nervous system disorder that is partially modified by body mechanics and peripheral influences (Elble et al., 1994). However, $C_{15}O_2$ PET studies of patients with *essential tremor* have demonstrated inferior olive hyperactivity (Dubinsky and Hallett, 1987) and bilateral *cerebellar* cortex hyperactivity that is inhibited by alcohol at doses sufficient to suppress tremor (Hallett and Dubinsky, 1993; Jenkins et al., 1993; Bowcker et al., 1996). These could both signify increased *Purkinje* inhibition of deep nuclei output since alcohol is known to be toxic to *Purkinje* cells and therefore may reduce firing.

CNS tremor generators have been hypothesized to reside in the *cerebellum*, the inferior olivary nucleus, the VIM *thalamus*, or a combination of these centers (Higgins, 1997; Gulcher et al., 1997). It has been determined that the VIM is the ventral portion of the *cerebellar* territory in the *thalamus* (Caparros-Lefebvre, 1994). Surgical ablation of the Ventralis lateralis (VL) has been used to treat tremor of *Parkinson's* and *essential tremor* for decades (Ondo et al., 1998). More recently, electrical *thalamic* stimulation for tremor has been used. Although the mechanisms of action are not well understood, high frequency deep brain stimulation appears to mimic the beneficial effects of ablation (Benwabid et al., 1987). It was previously thought that the electrical deep brain stimulation was effective because it destroyed cells in the VL *thalamus* causing an ablative lesion similar to surgery. It now appears that some think that it relieves tremor because it electrically stimulates the *thalamus* or pre*thalamic* fibers. This would lead us to think that previous inhibition or lack of firing of the *thalamus* by pre*thalamic* fibers may well be the cause of *dystonia*. The most likely site of dysfunction seems to be the *cerebellum*. In a reported case of a *Parkinson's* patient who experienced successful reduction of tremor following the placement of a neuropacemaker or deep brain stimulator (Caparros-Lefebvre et al., 1994), it was postulated that because the inferior placement of the stimulator site was very close to the upper part of the superior *cerebellar peduncle*, the favorable effects could be linked to the excitation of afferent *cerebellar* axonal endings more than that of the *thalamic* neurons themselves. Caparros-Lefebvre and colleagues also found that their PET studies showed regional blood flow decrease after stimulation was started, in post- and pre-cortical *cerebellar* nuclei and in the contralateral *vermis*. They thought therefore that this was an argument in favor of the role of the *cerebellum* in the genesis of *Parkinson's* tremor. It has also been noted that the site of dysfunction for the manifestation of resting tremor in *Parkinson's* does not correlate pathologically or physiologically with other cortical regions thought to be responsible for *Parkinson's*, and tremor improvement by VIM deep brain stimulation reduces blood flow PET activity of *cerebellar* cortical regions (Ondo et al., 1998).

What can be interpreted from these studies is that much of the *thalamic*-cortical dysfunction that has been previously related to *basal ganglia* dysfunction and decreased *dopamine* levels may also be attributed to *cerebellar* dysfunction, which in turn may be a result of dysfunction of afferent firing to the *cerebellum*, either from the cerebral cortex, *inferior olive* or muscle spindle afferents from postural muscles. It has been noted that in some cases of *thalamic* tremor, interruption of the *cerebellar* outflow tract to the *thalamus* may induce severe postural tremor (Holmes, 1904; Denny-Brown, 1958; Miwa et al., 1996). *Dystonia*, which has been classically thought of as a primary lesion in the *basal ganglia*, may in fact have a primary source in the *cerebellum* and therefore any structural or functional damage observed in the *basal ganglia* may in fact be secondary to direct or indirect disruption of cerebello-thalamo-*cortico-striatal* pathways as well as a primary disorder of the *basal ganglia*. In addition, in any case of decreased *dopamine* production in the *basal ganglia*; the *cerebellum* should also be considered as a potential primary cause. This can be applied to motor, cognitive, or emotional symptoms for which both the *basal ganglia* and the *cerebellum* may be responsible.

As in the *cerebellum*, the *basal ganglia* has previously been viewed as solely involved with motor control. Recently, much has been written about the role of the *basal ganglia* in cognitive as well as motor control (Alexander et al., 1986, 1990; De Long, 1990; Albin et al., 1995; Graybiel, 1995; Brown et al., 1997; Feger, 1997; Levitan et al., 1998). It has also been shown that the *basal ganglia* is part of a neuronal system that includes the *thalamus*, the *cerebellum*, and the *frontal lobes* (Thach, 1980). The *basal ganglia* has also undergone an expansion in humans, with enlargement and development of the *neostriatum*; the *paeleostriatum* is primarily the *globus pallidus*. The motor functions of the *basal ganglia* include the modification of motor control and higher cognitive aspects of motor control such as planning and execution of complex motor strategies, as well as some control of eye movement. The balance of motor activity is between two major pathways in the *basal ganglia*; the direct and indirect pathways. The *basal ganglia* is connected to the entire cerebral cortex, especially the *frontal lobes*. Afferent input proceeds to the *neostriatum* through both direct and indirect pathways whereby the output of the *basal ganglia* is modulated. The output arises from the internal portion of the *globus pallidus* and the *substantia nigra*. Their output inhibits the *thalamus* thereby modifying thalamocortical activity. Through this connection to the *thalamus* the *basal ganglia* can modify motor and sensory function (Chudler and Dong, 1995). Abnormal balance of the indirect and direct pathways can result in *hyperkinetic* (increased movement) or *hypokinetic* (decreased movement) disorders (Alexander et al., 1990; De Long, 1990; Albin et al., 1995). It is thought that *hyperkinetic* or hyperactive motor activity may relate to hyperactive behavior (agitation, incitation, euphoria, anxiety, mania, or hyperactivity) or *hypokinetic* behavior (apathy, depression).

Various *hyperkinetic/hypokinetic* disorders associated with the *basal ganglia* have characteristic motor and cognitive symptoms. Disorders such as *Tourette's* and *OCD* have associated *hyperkinetic* motor and emotional symptoms and *hypokinetic* disorders such as *Parkinson's* exhibit *hypokinetic* behavior such as depression and apathy (Albert et al., 1974; Starkstein et al., 1992; Chiu, 1995; Comings and Comings, 1998; Litvan et al., 1998). It is thought that neuropsychiatric symptoms exhibited by patients with *basal ganglia* lesions are a consequence of disruptions of frontal-*striatal* connections (Litvan et al., 1998).

There are five pathways that make up *frontal-striatal* connections, the *mesolimbic* and *mesocortical* being the most important in regulating emotional behavior. Both of these pathways use *dopamine* as the neurotransmitter and therefore are considered part of the *dopamine* system. The mesol*imbic system* affects several *limbic* structures, which are known to affect emotional behavior as well as autonomic function. The mesocortical pathway projects to the *prefrontal* cortex and both pathways seem to regulate goal directed behavior, reward mediated behavior, and the ability to screen out non-meaningful stimuli. Breakdown of these functions is commonly seen in children with *ADHD* also known as *childhood hyperkinetic syndrome* and *autism*. Perseverative or repetitive behavior may also be due to dysfunction of these systems. Dysfunction may produce abnormal reverberatory activity between the *thalamus* and cortex (Drevets and Raichle, 1992). Much of the afferent and efferent connections to this loop appear to come from postural antigravity muscles and are involved in regulation of muscle tone and posture (Martin, 1967; Jonnels and Speg, 1980). Abnormal function of posture and muscle tone associated with *basal ganglia* dysfunction may be related to these symptoms, which are commonly seen in astronauts working in hypogravity (Shapovalova, 1991).

Abnormal muscle tone and motor performance is the most commonly observed symptom in relationship to *basal ganglia* dysfunction. Brain lesions or dysfunction can cause *dystonia* with the responsible sites being thought to include *basal ganglia, thalamus,* and *brainstem* (Hallett, 1998). Dysfunction of normal production of *dopamine* is thought to be associated with disorders characterized by *dystonia* and can occur with high or low levels of *dopamine* (Hallett, 1998). *Dystonia* can apparently be produced behaviorally through abnormally excessive voluntary motor activity and some types of *dystonia* are genetically transmitted.

The *basal ganglia*'s major role may be to balance excitation and inhibition of the cortex through direct and indirect *basal ganglia* pathways. This form of surround inhibition might explain increased motor commands and overflow into inappropriate muscles, which is the essence of *dystonia* (Hallett, 1998). A similar type of surround inhibition is found in the *cerebellum* and a similar model of *dystonia* may be the result of *cerebellar* dysfunction. In *Parkinson's* patients, *dystonia* is thought to also be a product of various abnormal inhibitions or excitations of Ia or Ib inhibitory neurons at the spinal cord level (Delawaide et al., 1991), or abnormal *brainstem* reflexes involving Ia and Ib inhibitory neurons (Rothwell et al., 1983). It is thought that the fundamental underlying dysfunction in *dystonia* may be dysfunction at the higher supraspinal levels (Hallett, 1998). Increased firing of the nucleus *gigantocellularis* of the *brainstem* reticular formation and descending reticulospinal pathways may cause abnormal reciprocal inhibition. The *basal ganglia* and *cerebellum* may both affect the firing of these two pathways. Decreased and increased areas within the cortex have been noted in patients with *dystonia* (Mizuno et al., 1989; Ikoma et al., 1996). Some think that all of these results can be explained by decreased inhibition leading to a hyperexcitable cortex (Hallett, 1998). Therefore, we see that *basal ganglia* dysfunction as in *cerebellar* dysfunction can cause characteristic motor, cognitive, and emotional symptoms that are commonly seen in children with cognitive or emotional/affective disorders.

Forty percent of patients with Huntington's disease manifest an associated major depression, which often precedes the onset of involuntary movements and dementia. If cognitive or emotional dysfunction occurs before obvious motor tics and tremor, vocal tics, or incoordination, it may not be recognized as *basal ganglia* dysfunction, or the symptoms may appear to be unrelated to the untrained physician. *Basal ganglia* dysfunction can be caused by abnormal voluntary muscle use, which may be related to underlying genetic factors. Abnormal muscle tone especially of postural antigravity muscles characteristic of *basal ganglia* dysfunction has been commonly seen in situations that decrease gravitational influences as in space flight.

6

The *Thalamus*

ANATOMY AND FUNCTION

In the preceding chapter, we noted that there appears to be a neuronal circuit that includes the *thalamus, basal ganglia, cerebellum*, and *prefrontal* cortex (Thach, 1980). The *thalamus* has strong connections to the cortex as a whole, and is the "relay station" for sensory information to the cortex except olfactory sensation. Some of the early neuroanatomists such as Von Manakow and Missl discovered that extensive parts of the *thalamus* undergo atrophy after removal of the cerebral cortex. They were also able to indicate specific nuclei in the *thalamus* that were connected to different parts of the cortex. At the beginning of the 20th century, the major ascending afferent connections to the *thalamus* had been identified and it was understood that this nuclear complex represents the final link in most afferent fiber systems transmitting to the cerebrum.

Today the *thalamus* can be viewed as being composed of functionally subdivided nuclei. The *thalamus* has two major functions with regard to the cortex. There are specific relay nuclei for specific sensory input. The relay nuclei project to specific primary association areas in the cortex. There is also a group of non-specific nuclei that possess widespread connections to the entire cerebral cortex. It is now thought that these nuclei may possess pacemaker-type activity and provide a baseline level of arousal or "consciousness" to the brain.

The *thalamus* is located in the dorsal *diencephalon* and the third ventricle separates the two egg-shaped *thalamic* nuclei. *Thalamus* is the Greek term for "inner chamber" and it lies at the most rostral end of the *brainstem* as illustrated in Fig. 6.1. The *internal capsule* borders it laterally, the fibers project from the motor cortex to the *brainstem* and spinal cord. The *internal capsule* separates the *thalamus* from the *basal ganglia*. The two halves of the *thalamus* are connected by a bridge of gray matter called the *massa intermedia*. The *thalamus* can be subdivided into three main nuclei, the anterior, the medial, and the lateral *thalamus* (which extends posterior to include the *pulvinar*). These three main nuclei are separated by white matter, which is shaped like a "Y" and called the internal *medullary lamina*. Microscopically each of the three main nuclei can be seen to have various subgroups. Several small cell groups are found within the internal medullary lamina, as a group these nuclei are referred to as the *intralaminar* nuclei. Found to extend along almost the entire lateral

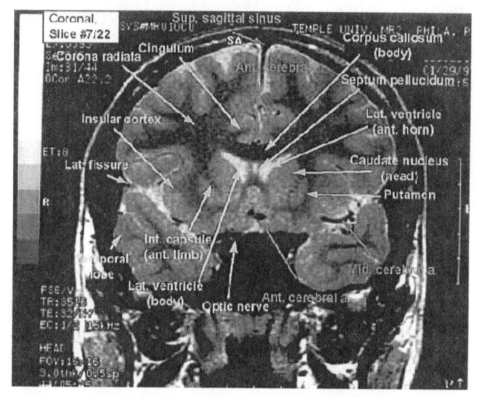

Fig. 6.1. Coronal section of the *caudate nucleus* and *putamen*, and *corpus callosum*.

surface of the *thalamus* is a thin layer of cells known as the *reticular nucleus* of the *thalamus*. This is also separated from the main body of the *thalamic* nuclei by a thin white matter lamina called the external *medullary lamina*. The basic pattern of the *thalamus* appears to be the same in higher organisms along the phylogenetic scale. There are, however, some differences especially in humans. The *pulvinar* and the centromedian nuclei appeared when hominids evolved the ventrolateral *thalamus* and enlarged concomitantly with the enlargement of the *cerebellar* dentate nucleus (Armstrong, 1980). The ventral lateral *thalamus* is a phylogenetically new part of the *thalamus*. The new parts may possibly provide a neuronal substrate for the evolution of cognitive and language skill (Jerison, 1973; Galabuda, 1984) through their connection to the *prefrontal* cortex.

It has been demonstrated that the human ventrolateral *thalamus* is involved in cognitive and language function as well as motor functions (Armstrong, 1980; Ojemann and Crutzfeldt, 1987). Many of the *thalamic* nuclei project fibers to the cerebral cortex (thalamocortical fibers). These fibers are considered cortically dependent because they show retrograde cellular changes following ablation of specific parts of the cerebral cortex. It appears that most of these thalamocortical projections have a geographic relationship between a specific *thalamic* nucleus and a corresponding cortical area. One of these is the *anterior thalamic nucleus*, of which there are three subdivisions and all send their connections to the medial surface of the cerebral cortex. In the medial group, the largest subdivision, the *nucleus medialis dorsalis* sends fibers to regions of the *frontal lobe* especially the orbital cortex. In the lateral portion, the ventral or inferior part of the nucleus can be subdivided into three units; the *ventralis anterior* (VA), the *ventralis lateralis* (VL), and the *ventralis posterior* (VP). The VP is further subdivided into three parts: the *ventralis posterior lateralis* (VPL), the *ventralis posterior medialis* (VPM), and the *ventralis posterior inferior* (VPI). It is thought that with the exception of parts of the VPI and parts of the VA, these three nuclei all project to the cortex. The VL projects to the precentral cortex in the *frontal lobe*, and the VPL and VPM project to the post-central cortex. The VPL is the main final relay station in the *dorsal column-medial leminiscal* system, and the VPM receives fibers from the *trigeminal* sensory nuclei. These three areas are ventral parts of the lateral mass. The dorsal part of the lateral mass is subdivided into two areas; the *nuclei lateralis* (LD) and the *lateralis posterior* (LP). The LD appears to have cortical projections primarily to the posterior part of the *cingulate gyrus* and *parietal* cortex. The LP projects to the parietal cortex alone. The *pulvinar* sends fibers to the parieto-temporal cortex. The *pulvinar* has several subdivisions that make different connections with the cortex and other subcortical areas. Some think that the pulvinar is important in attentional processing because of its heavy reciprocal connection with regions of the cortex thought to be involved with attentional control (Leisman, 1976a). Lastly, there are the lateral and medial *geniculate bodies* which are relay stations in the pathway for vision and hearing respectively. They too have specific cortical projection sites. The *lateral geniculate nucleus* receives information from ganglion cells of the retina and sends axons to the primary visual cortex. The *medial geniculate nucleus* receives information from the inner ear via other *brainstem* nuclei in the ascending auditory pathway and sends axons to the primary *auditory* cortex (AI). The posterior area of the parietal lobe and the visual cortex are where feature analysis and object recognition are accomplished.

The structures of the posterior *thalamus* receive fibers that originate in subcortical structures such as the *mescencephalic reticular formation*, the deep *cerebellar* nuclei, the *locus coeruleus*, the dorsal *Raphe nucleus*, and cholinergic neurons of the *peduncle-pontine* nucleus (Shute and Lewis, 1967; Graybiel, 1974; Mackay-Sim et al., 1983). Most of these areas are thought to be involved in arousal, attention, motivation, and vigilance due to their effect on the cerebral cortex. All of these areas may be involved in the *ascending reticular activating system*, thought to be responsible for arousal and attention. It is thought that the main *ascending activation system* arises from the non-specific *thalamic* nuclei, about which more will be said later. The *mescencephalic reticular formation* seems to have most of its effect on the cerebral cortex through its interaction with the non-specific *thalamic* nuclei. There are direct projections from the *bulbar reticular formation* to intralaminar *thalamus* that have been identified anatomically (Bowsher, 1967) and physiologically (Steriade et al., 1984). There exists five principal connections between the *mescencephalic reticular formation* and the cortex, three of which pass through the *thalamus*. It appears that the primary brain areas involved in arousal, attention, and alertness include the intralaminar *thalamus* or non-specific nuclei which appear to receive projections from a number of subcortical areas especially the *cerebellum, basal ganglia*, and *reticular formation* as well as reciprocal connections from the cortex.

The area of importance for the *thalamic* relay of somatic sensory information is the posterior complex. The medial part of the posterior nucleus projects to the retro-insular field lying posterior to the second *somatosensory area* (SII). The lateral part projects to the post-auditory cortical area. The *thalamic* nuclei that project to particular areas of the cerebral cortex for the most part also receive afferent feedback from the same cortical region (*cervico-thalamic* fibers), so that intimate reciprocal connections are established. It appears that all *cortico-thalamic* fibers arise from cells located in the internal layers of the cerebral cortex. However, the thalamocortical fibers end in different cortical layers according to the *thalamic* nucleus of origin.

The *intralaminar nuclei* of the *thalamus* project diffusely upon the cerebral cortex and possess numerous connections to the *basal ganglia*. Most of the *thalamic* nuclei that send massive reciprocal projections to specific parts of the cortex receive their afferent input, primarily from the somatic sensory, *cerebellar, pallidal*, and *hypothalamic* afferents. For some time it has been recognized that there are two types of *thalamic* nuclei; specific and non-specific. This description is based on electrophysiological and functional observation. Nonspecific *thalamic* nuclei on electrical stimulation have been shown to evoke potentials over wide cortical areas of both hemispheres. They appear after a long latency (delay) and demonstrate a characteristic known as "recruitment" (see Leisman and Koch, 2003). The specific nuclei in contrast show, discretely localized and rapidly occurring potentials. The reticular *thalamic* nucleus is also generally considered a nonspecific nucleus. All of the non-specific *thalamic* nuclei are considered a major functional part of the "ascending activating system," which is thought to be responsible for arousal and attention of the cortex. Many of the descending *cortico-thalamic* projections end in the *thalamic* reticular nucleus. Neurons in the reticular nucleus are thought to form a lateral inhibitory network of cells that may act in the modulation of *thalamo-cortical* outputs, possibly to fine-tune sensory transmission or partially gate the flow of information to the cortex.

Closed vs. Open Systems and the Thalamus: The Binding Problem

Systems neurophysiologists regard the central nervous system as a closed vs. an open system. An open system would be one that accepts input from the environment, processes it, and returns it to the external environment. However, recent studies appear to imply that the central nervous system functions more as a closed system. This means that the basic organization of the central nervous system is geared toward the generation of intrinsic images (thoughts or predictions), and is primarily self-activated and capable of generating a cognitive representation of the outside environment even without incoming sensory input, as it does in dreams (Llinas, 1995). This intrinsic order may represent the core or baseline activity of the brain, which can be modified through sensory experience and through the effects of motor activity. Motor activity may occur either in response to external environmental cues or stimuli, or in response to internal images or concepts. For example, emotions generated by the *limbic system* are the result of internally created stimuli, which are thought to be *premotor* templates in primitive form. In other words, we have two simultaneous systems functioning at the same time: a baseline or constant level of stimulation which may be in part an intrinsic property of neuronal and other systems. The second system is then superimposed on the baseline activity, which allows us to perceive specificity of stimulation from both external and internal sources, which allows us to appropriately respond. This closed system of the central nervous system is thought to have developed over time, as a product of evolution and initially was responsible for basic reflexes (Leisman and Koch, 2003).

The functions of the central nervous system require a series of networks that can interact with external stimuli. This needs to occur with enough speed and accuracy to allow effective interaction with the environment. The *thalamo-cortical* system, which has reciprocal connections to the cortex, can create a resonant function state and is thought to be responsible for cognition. This coordination of activities occurs at several levels. The reticular *thalamic* nuclei which use *GABA* neurons are at the interface of the *thalamus* and neocortex (Llinas, 1995, 2001). This nucleus receives collaterals of *thalamo-cortical* and *cortico-thalamic* axons that project to the dorsal *thalamus* (Llinas, 1975, 1990, 1995, 2001). In addition, the second in parallel with the specific *thalamo-cortical* sensory input is a non-specific system made up of the *intralaminar* and posterior *thalamic* nuclei which have widespread reciprocal connections to the cortex (Llinas, 2001). Third, the main targets of the reticular *thalamic* neurons are the *intralaminar* nuclei (Llinas, 1995).

We can understand the function of these networks better by observing what happens following lesions of *thalamic* nuclei. In a lesion of one of the specific sensory nuclei such as the *lateral geniculate body*, a specific sensory modality is lost (e.g., blindness), but the lesion does not necessarily result in a loss of function of other specific nuclei such as hearing. However, if there exists a lesion of the *nonspecific intralaminar nuclei*, patients are unaware of any input by the specific intact nuclei even though those pathways still exist, they cannot perceive or respond to them. In essence, the person does not exist from a cognitive viewpoint. These results suggest that non-specific nuclei are required to achieve binding or summation of specific sensory input into the context of ongoing activities (Llinas, 2001). In other words, there are two systems, a constant system that sets a baseline level of stimulation of the brain that appears to arise from non-specific nuclei, as well as a non-constant or variable information that is sent by modality-specific nuclei. The perception of variable input is dependent on constant input and if the constant input is decreased or absent, specific stimuli will be perceived at a lower level than may exist in the environment or may not be perceived at all. This ongoing constant activity has been observed to occur in bursts or oscillations. These oscillations at 40 Hz are thought, therefore, to be centrally related to cognition (Leisman, 1973, 1974, 1976a, 1976b; Koch and Leisman, 1990, 1996, 2001a, 2001b). These continuous 40 Hz oscillations can be recorded over large areas of the surface of the heads of alert subjects. These oscillations are not in phase and, there is a 12–13 ms phase shift between the rostral and caudal parts of the brain that appear as continuous phenomena. When sensory stimuli are presented, these oscillations show a phase locking, which is related to cognitive processing and temporal binding of sensory stimuli (Mayevsky et al., 1996).

We can use face recognition to better understand this process. How do we recognize a familiar face? It had been previously thought that there was one area in the brain or a specific group of neurons devoted to face recognition, called *grandmother cells*. However, if this is the case, there are not enough cells in the brain to compute and store all of the bits of information we need to retain all the information that we acquire (Leisman and Koch, 2000, 2003; Koch and Leisman, 2001a). We now know that there is not a single part of the brain where all arriving information is processed. While we know there are brain regions with a high degree of localized function and control, parts of the brain that do one thing, and other parts that do something else, we also know that there is no single region responsible for the integration of perceptions and memories. How the brain integrates its perceptions has been referred to as the *binding problem*.

A team of scientists in France at the *National Center for Scientific Research* have studied this problem recently. Varela and his associates (Rodriguez et al., 1999) think that for the first time they have been able to measure binding in a momentary firing in the human brain. "What we are measuring is the integration of the brain," said Francisco J. Varela who led the research. "We were for the first time able to calculate synchrony between emissions of brain cells widely distributed in the brain." This is thought to be a step toward how the human brain produces "consciousness." It is thought that there cannot be consciousness or self-awareness without perception, therefore consciousness must involve different parts of the brain working in concert. In their research, Varela and his colleagues asked 11 subjects to glance at images known as "*mooney faces*." These black-and-white images, seen right side up can appear to be faces; upside down, they look like abstract blotches. The subjects were asked to look at a series of *mooney faces* shown upside down and right side up, pushing a button with their right hand if they saw a face and the left hand if they did not. Electrical activity in their brains was measured. When the subjects recognized a face the electrical activity of different regions of the brain became synchronized for a moment. The synchrony appeared to be established by brain regions firing in a similar manner, known as "gamma oscillations." In an interview Varela said that,

Different regions of the brain get active when you do anything, look at a face, move your hand, have a memory, any cognitive act implies the working together of very different neurons in the brain that are widely distributed. The hypothesis is that the gamma oscillation is the medium through which the neurons act together by being synchronous. They time their oscillations together. It is like a transitory glue, a transitory pattern.

During their experiment, the gamma oscillation synchrony would last about a fourth of a second, disappear, and then reappear in a different pattern using different parts of the brain. Varela who reported his findings in the February 4, 1999 issue of the journal *Nature* said,

> it increases quite substantially during the recognition of a face, then synching goes down, and everything gets unglued before it becomes synchronized again for the pushing of the button. Thought and deed followed each other punctuated by an active blank. These transient patterns of synching are certainly something that relates to the moment of consciousness.

The fact that neurons are linked in space and time allows the brain to have virtually unlimited possible combinations of neurons with a resultant limitless capacity for retaining information. When we view a movie which we perceive to be a continuous event we are in actuality viewing that movie as a series of pictures put together at such speed that it seems continuous. Each frame presented quickly in succession allows for perceptual linkage by our brains (Gaarder, 1975; Leisman, 1976a), at each frame everything in that picture is frozen in time and *we* link those frames together.

This may be how our brain works, a series of moments where sensory input arising from different regions is bound together by a process which phase-locks all of the pieces, freezing the units together thereby providing perceived temporal continuity. It all happens so fast that it appears that our perception of the world is continuous. In this scenario, if we lose one sensory modality or if we lose the nonspecific input (constant), we no longer take a complete picture.

Studies suggest that 40 Hz oscillations are therefore not only involved in primary sensory processing, but form a time-conjunction or binding property that brings together sensory events into a single experience (Llinas, 2001). 40 Hz oscillations are prevalent in the mammalian central nervous system (Llinas et al., 1991; Metherate et al., 1992; Hari and Salmelin, 1997; Jones et al., 2002) and are seen at both a single cell (Sukov and Barth, 2001) and multicellular levels (Rols et al., 2001).

The specific *thalamic* nuclei establish cortical resonance through activation of pyramidal cells and feed-forward inhibition through activation of 40 Hz inhibitory interneurons in layer III–IV of the cerebral cortex. These oscillations re-enter the *thalamus* by way of layer VI pyramidal cell axon *collateris* (Llinas, 2001), providing *thalamic* feedback inhibition by the *reticular nucleus* (Llinas, 1995). The *intralaminar nonspecific nuclei* project to cortical layers I and V and to the *reticular nucleus* (Llinas et al., 1991). Layer V pyramidal cells return oscillations to the *reticular* and *intralaminar nuclei* (Llinas, 1995). It is also clear that neither of these two circuits alone can generate cognition. Damage to non-specific *thalamic* pathways produces deep disturbances of consciousness; damage to the specific system produces loss of a specific modality (e.g., vision, hearing). In essence, the specific system would provide content whereas the non-specific system would provide the context or alertness (Llinas, 2001). It has been demonstrated that *thalamo-cortical* oscillations mediate both physiological and pathophysiological behaviors including sleep and generalized absence of epilepsy (Huguenard and Prince, 1994). EEG findings can be used to measure the state of consciousness. It has been shown that the alpha rhythm in EEG (waves of relatively high voltage and frequency ranging between 8–13 Hz) changes to low voltage fast activity (18–22 Hz) when the subject passes from a state of relaxation or drowsiness to an alert, attentive state. The former pattern is referred to as synchronization or hypoarousal. In contrast, arousal or alertness is represented by desynchronization of these wave patterns. This may reflect the caudal-ventral oscillations referred to earlier. Low frequency stimulation of non-specific *thalamic* nucleus produces inattention, drowsiness, and sleep accompanied by slow-wave synchronous activity and so-called spindle bursts. High frequency stimulation arouses a sleeping subject or alerts a waking organism and there exists an associated desynchronization of electrocortical activity. Electrical stimulation with a 3/s stimulus was shown to produce regular waves and spioke complexes as seen in petit mal (absence) epilepsy (Jasper and Droogleever-Fortuyn, 1947).

Is there a source for these *thalamic* oscillations? Some think that it is an intrinsic property of neuronal cells that may be genetically programmed, but recent studies have been conducted to determine if there is a main source for these oscillations. Intracellular recordings were taken from relay neurons of the dorsal *thalamus* in rats under urethane and anesthesia (Pinault and Dechev, 1992). Pinault and Dechev found that neurons of the ventro-postero-lateral and ventral-lateral nucleus produced spontaneous potentials with a pattern similar to synaptic potentials triggered by somatosensory stimulation and could not be recorded after transection of the dorsal column. These investigators also noted that afferents deep within the *cerebellar* nuclei discharged spontaneously in a rhythm within the same frequency band as that of the synaptic potentials recorded in ventral-lateral cells. They concluded that the rhythmic depolarization observed in *thalamic* neurons (40 Hz oscillations) are not generated intrinsically but rather represent excitatory postsynaptic potentials of pre-*thalamic* origin. The neurons of the dorsal column and deep *cerebellar* nuclei are capable of encoding their output rhythmically. These potentials are not affected by lesions of the *internal capsule* or cortex, revealing that they are not of cortical origin. The VPL responses were affected by limb position or stimulation by slowly adapting receptors. Slowly adapting receptors are characteristically found in slow twitch muscle fibers, which are most densely populated in postural or antigravity muscles, which fire continuously against gravity. In other words, the oscillatory 40 Hz activity which has been noted in the *thalamus*, appears not to be intrinsic, but appears to arise from the *cerebellum* and from muscle and joint afferents which fire to the *thalamus* either directly or through the *cerebellum*.

It is possible that decreased firing of dorsal column fibers can decrease or stop *thalamic* activity. Since it appears the *thalamus* may act as a pacemaker to the brain through the regular bursts of 40 Hz oscillations and that this is thought to be the source of cognition and possibly consciousness itself, interruptions of this input from the *cerebellum* or dorsal column could cause gross deficits in cognition and consciousness. A recent case study was reported of a female with a *thalamic* stroke involving the mammillo-*thalamic* tract, *intralaminar* nuclei, parts of the dorsomedial, and ventral-lateral nuclei bilaterally (Chatterjee et al., 1997). SPECT-scan showed decreased *thalamic* and *basal ganglia* blood flow. General decreases of blood flow were noted in right frontal, left temporal, and left temporal-parietal regions. Chatterjee and associates noted "the most striking clinical feature was a bizarre disconnected and at times incoherent speech output. Analysis of the patient's speech revealed relatively preserved lexical and morpho-syntactic linguistic production. We interpret her disordered speech output as representing the surface manifestation of a thought disorder (rather than a language disorder per se) characterized by the inability to maintain and appropriately shift themes that normally guide discourse." In conclusion, the study notes that median and *intralaminar* nuclei appear to be critical for neurophysiological regulation of *thalamo-cortical* and striato-cortical circuits, which in turn may be critical for the functional regulation of continuously appropriate transitions of thought.

Lesions of the *thalamus, cerebellum,* or *basal ganglia* can result in various cognitive deficiencies. Lesions of the *basal ganglia* and *cerebellum* result in characteristic motor deficits. Damage to the *thalamus* also results in characteristic motor disturbances even though the *thalamus* has primarily a sensory function. In a study conducted on two patients with unilateral resting and postural tremor because of *thalamic* stroke (Miwa et al., 1996), neuroradiological examination evidenced a lesion in the postero-lateral *thalamic* region in both patients with no involvement of *brainstem, cerebellum,* or *cerebellar* outflow tract to VL nucleus. Other studies have indicated that interruptions of the *cerebellar* outflow tract to the *thalamus* may induce severe postural tremor (Holmes, 1904; Denny-Brown, 1958; Narabayashi, 1986). In contrast, the *nucleus intermedius* (VIM), the most caudal portion of the VL (the *cerebellar* portion), is the target for ablative surgery (Hallett, 1986), and is expected to abolish tremor and contracture. In addition, deep brain stimulation in the same area has a similar result (Ondo et al., 1998). Other regions of the postero-lateral *thalamus* are related to tremor. The VPL and VPM contribute to the signal processing of somatosensory information and project to somatosensory cortices (Jones, 1985). It is thought that reverberating circuits between the *thalamus* and cerebral cortex are implicated in the generation of tremor and contracture in a system that is also implicated in the process of attention. Just as in the *basal ganglia* and especially the *cerebellum*, the *thalamus*, when damaged, can produce similar motor and cognitive defects that appear to be related to cortical function in general and *prefrontal* cortex function in particular. It is worth noting that in patients with *thalamic* stroke, postural tremor becomes more severe on goal directed movement or emotional stress (Miwa, 1996). Goal directed motor activity is a well-known function of the *frontal lobe*. If goal directed motor activity which is *frontal lobe* in nature can be disrupted by damage to the *thalamus* and especially the *cerebellum*, then it is possible to conceive of goal directed emotional or cognitive behavior (which is also *frontal lobe* in origin) being severely disrupted by damage or dysfunction to the *thalamus* and *cerebellum*.

Subcortical Neurotransmitter Systems Facilitating Interregional Communication

The last subcortical structures that are important to review are areas that utilize important neurotransmitters that play an important role in emotion and behavior especially attention and motivation. Of these the monoamine systems seem to be the most important, including *dopamine, norepinephrine,* and *serotonin*. Having previously discussed the *dopamine* system, we will focus our attention here on *norepinephrine* and *serotonin*. Among other places, *norepinephrine* is produced in the *brainstem locus coeruleus* and lateral *tegmental neurons*.

Locus Coeruleus

The *locus coeruleus* is found in the *pons* at the floor of the fourth ventricle and is represented in Fig. 6.2. A6 neurons containing *norepinephrine* correspond approximately to the nucleus *locus coeruleus* (LC). There are several roles that the LC plays with regard to memory, learning, and behavior. This system is thought to be involved with the orientation of the brain to stimuli in the environment and viscera. This system is thought to be activated by a variety of sensory stimuli and it appears to be related to vigilance. These orientation and vigilance responses are needed to adequately explore the environment. It is a necessary part of central control of autonomic function. This system is also involved in sleep–wake cycle as well as in reward and reinforcement. The efferent connections to the *locus coeruleus* are widespread. Two main pathways have been identified, a dorsal pathway innervates the entire cerebral cortex, especially the frontal cortex (Shimizu et al., 1974), the *hippocampus,* and the *amygdala* as well as projecting numerous collaterals to *thalamic* relay nuclei. A ventral or intermediate pathway supplies the *hypothalamus*. Other fibers pass through the superior *cerebellar peduncle* to the *cerebellum*. A caudal projection passes to the lower *brainstem* including the *reticular formation*.

Afferent projections *to* the *locus coeruleus* also arise from the *hypothalamus* (Mizuno and Nakamura, 1970), *cingulate gyrus* (Domesick, 1969), *cerebellum* (Kobayashi et al., 1975), *Raphe nuclei, substantia nigra* (Sakai, 1977), and *amygdala* (Hopkins and Holstege, 1978).

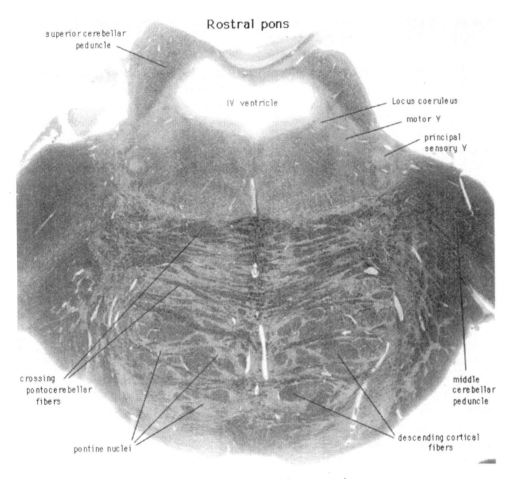

Fig. 6.2. Rostral pons and *locus coeruleus*.

The lateral *tegmental* neurons send fibers caudally into the spinal cord and anteriorly into the *diencephalon* and *basal forebrain* region. The basal forebrain region is the region inferior to the anterior *corpus callosum* in the cerebral cortex. This area is thought to be extremely important in cognition and motivated behavior. It includes the *septal nuclei*, which are particularly important in reward and reinforcement. The LC is thought to be an essential part of ascending activating system and is therefore involved in cortical arousal (Jouvet, 1972).

The Raphe Nuclei

The *Raphe nuclei* appear to be the principal part of the brain that contains *serotonin (5 H-T)* neurons (Jouvet et al., 1966, 1967). These nuclei are thought to facilitate information processing and slow wave sleep (Buhusi and Schmajuk, 1996; Jouvet et al., 1966, 1967). These nuclei have a role in gating incoming sensory stimuli and directing attention (Buhusi and Schmajuk, 1996). Along with the *locus coeruleus, the Raphe nuclei* are important in arousal and vigilance (Bronzino et al., 1976; Balaban, 2002). Vigilance is the state of increased arousal and is a requirement for focused or directed attention, but it is not in itself sufficiently able to control the process. Focused attention requires that incoming sensory stimuli be given a priority according to importance. A lack of directed attention can appear as impaired concentration. Habituation can occur if stimuli are not reinforced. Disruption of normal *serotonin* level in organisms has been shown to affect exploratory behavior (Dray et al., 1978). Organisms seem to treat meaningless stimuli with relative importance when *serotonin* levels are depleted (Blokland et al., 2002). This leads to repetitive or perseverative activity in *serotonin* deficient states. Hyperactive states seem to result from excess *serotonin*.

Efferent connections of the *Raphe nuclei* have been studied extensively in the cat (Brodal et al., 1960). Clear-cut retrograde cellular changes were found in the *Raphe nuclei* following transection of the spinal cord, the *brainstem* at the mescencephalic level, or in lesions of the *cerebellum*. Ascending efferent connections have been studied in the rat (Conrad et al., 1974) and in the cat (Taber et al., 1976). Based on these studies it is thought that fibers supply the *mescencephalon*, nuclei in hypo*thalamus*, intralaminar and other *thalamic* nuclei, the *cerebellum*, parts of the *amygdala* and *hippocampal formation*, the *septum*, the *caudate* and *putamen* (*neostriatum*), and the cerebral cortex especially the *frontal lobes*. The descending efferents are not as abundant. They connect to the *pontine* and *medullary reticular formation*, cranial nerve nuclei, *cerebellum*, LC, and spinal cord. Spinal cord projections appear to descend in the dorsolateral *funiculus* and end in Laminae I, II, and V (Basbaum et al., 1976; Anderson et al., 1977). It appears that descending *Raphe nuclei* fibers end in the IML and exert a monosynaptic inhibitory action on preganglionic *sympathetic* fibers (Cabot et al., 1979). Afferent connections of the *Raphe* complex appear to arise from the spinal cord and the *cerebellar* and cerebral cortex (Brodal et al., 1960). Other connections have been described including the *septum*, the lateral preoptic region, the lateral hypo*thalamus*, and the *prefrontal* cortex (Aghajanian and Wang, 1977; Mosko et al., 1977). Lesions of ascending *serotonergic* system can produce various disturbances, such as increased motor activity, insomnia, hyperactivity, and aggression. *Serotonin* is also thought to play an important role in pain inhibition and sleep (Jouvet et al., 1966, 1967). The role of *serotonin* containing cells with regard to sleep appears to be involved with the dampening of certain sensory stimuli and inhibiting motor output in sleep onset (Jones, 1993). Lesions of the *Raphe nuclei* may produce total insomnia (Jones, 1993).

The *thalamus* is part of a neuronal system that includes the *basal ganglia, cerebellum*, and *frontal lobes*, which includes the *prefrontal* cortex (Thach, 1980). Functionally the *thalamus* is the last major relay station for all sensory input to the cortex with the exception of olfaction. There exist *specific* nuclei in the *thalamus* that transmit signals through specific sensory channels to the control centers for vision, hearing, touch, and somatosensation. These areas have strong reciprocal connections to corresponding projection sites in the cortex. There also exist *non-specific* nuclei that have diffuse projections to the cerebral cortex. It is thought that these *non-specific* nuclei provide a pacemaker function for the cortex through gamma oscillations. These 40 Hz bursts of electrochemical activity originating in the *thalamus* appear in turn to be associated with desynchronous activity from the caudal to the ventral portions of cortex. These 40 Hz gamma oscillations form a baseline level of brain activity, which may be the foundation of all perception and possibly consciousness itself. The *non-specific* and *specific* nuclei together, and the information they relay are necessary for all perceptual or cognitive experience. Vision, hearing, somatosensation, or any sensory perception is based on the proper functioning of these pathways. The source of these gamma oscillations do not appear to be intrinsically generated but have been traced to dorsal column pathways and *cerebellar* nuclei that transmit signals from somatosensory receptors that are found in skin, muscle, and joints. The relay of most of the signals originates from slowly adapting receptors that are most prevalent in high-endurance slow-twitch muscle fibers. The majority of these fibers are found in postural antigravity muscles especially of the spine, most specifically in and around the cervical spine muscles. These gamma oscillations do not appear to emanate from the cortex. Therefore, any decrease in activity or function of joints and muscles of the spine and neck can decrease firing of non-specific and specific *thalamic* nuclei to the cortex through *thalamocortical* projections.

The *cerebellum* projects to intralaminar and VL *thalamic* nuclei. The intralaminar nuclei are non-specific nuclei, which diffusely project to the cortex. The VL nucleus projects especially to *prefrontal* cortex, which is responsible for goal directed motor, cognitive, and emotional behavior. Damage to these regions of the *thalamus* has been shown to negatively affect all of these functions.

The *Limbic System* and the Evolution of the Cerebral Cortex

The cerebral cortex has undergone significant changes during the course of its evolution. As we ascend the phylogenetic scale, we note the development of newer areas of the cortex that have annexed the function of older areas including the improvability of species. From the study of comparative anatomy, three main regions can be distinguished within the *pallium* of amphibians that correspond to areas of higher vertebrates. The medial part of these is thought to represent the *archicortex*, which corresponds to the *hippocampus*. The lateral area is considered the *paleocortex* or the *piriform* area. Between these two areas is located a dorsal area which corresponds to the neocortex of mammals. In reptiles, these three divisions seem to become more distinct although their structure is still considered primitive. It is only in mammals that the dorsal cortex undergoes a dramatic development and increases progressively along the phylogenetic scale to reach its peak in humans, where it forms the majority of the entire *pallium*.

The *paeleo-* and *archicortex* do not appear to undergo further evolutionary development in higher mammals. The *archicortex* becomes folded and, by the development of the *hippocampal* fissure, will expand into the medial wall of the lateral ventricle as the *hippocampus*. With the development of the neocortex, the two more primitive areas are eventually pushed medially, where in humans they are found on the medial aspect of the cerebral hemispheres (Fig. 7.1). The expansion of the neocortex results in a change of shape of the other areas of the brain. When the occipital pole of a hemisphere is developed, the *archi-* and *paleocortex* are pushed posteriorly, and with the development of the *temporal lobe* are again pushed anteriorly and ventrally. As a result of this process the *archi-* and *paleocortices* in humans are found as almost circular structures that extend posteriorly from the front of the interventricular foramen and curve downward and forward to eventually reach the base of the brain below their starting point. This appearance is thought to have inspired the name, *limbic system* (Fig. 7.2). The concept of a *limbic system* apparently came from the term *limbic* lobe. The name is often credited to Paul Broca who in 1878 wrote about *le grand lobe limbique*. It is thought he termed it as he did because of its appearance as a ring of gray matter bordering the intra-ventricular foramen in the central part of the brain.

94   NEUROBEHAVIORAL DISORDERS OF CHILDHOOD

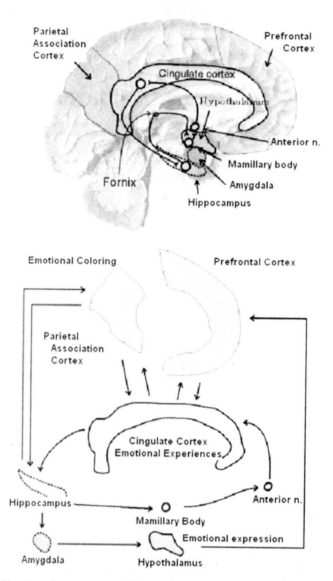

Fig. 7.1. Components of *limbic system* and cerebral hemispheres.

The *archicortex* is represented mainly by the *hippocampus* and the *dentate gyrus*. In humans, with the development of the *corpus callosum*, the *archicortex* becomes relatively reduced. The *paleocortex* in mammals develops posteriorly to form the *piriform* lobe, in humans this is thought to represent most of the *hippocampal gyrus* with the rest representing the *cingulate gyrus* and *retrospenial* cortex, connecting the *cingulate* with the *hippocampal gyrus*. The paleo- and archicortices in lower species are thought to be primarily concerned with olfactory functions. In mammals and humans however, they have taken on other important functions. The term *rhinencephalon* or nose brain is sometimes used to describe the archio- and paleo-cortices, parts of the basal areas of the *telencephalon* such as the septal area, the olfactory tubercle, tract, and bulb. However, in mammals, these other parts are not related exclusively to smell as the name implies. A distinction between the *rhinencephalon* and the *limbic* lobe is not clear amongst most authors. The *limbic* lobe

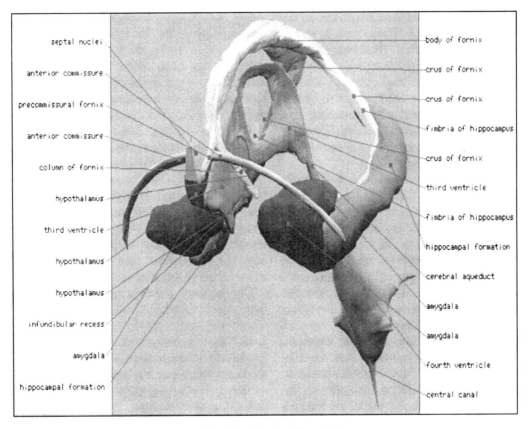

Fig. 7.2. The *limbic system*.

is thought to consist of the *cingulate gyrus* and the induseum griseum, the *hippocampus* and *dentate gyrus*, the *subiculum*, the *presubiculum, parasubiculum*, the *entorhinal* area, the *prepiriform* cortex, the *septum*, the *olfactory tubercle*, and the medial and cortical *amygdaloid* nucleus.

The *limbic system* is typically the term used collectively for the functions of the *limbic lobe*. Functionally the *limbic system* is thought to be concerned with visceral functions, especially those associated with the emotional status of the individual. Therefore, the term *visceral brain* and *emotional brain* have been used synonymously to describe the region. The *limbic system* has been described as a set of interconnected structures, which have intimate, multisynaptic connections with the hypo*thalamus* and mescencephalon (Mesulam et al., 1977). Others think that the *limbic system* is considered a system only because each of its components have relatively direct connections to the hypo*thalamus*. On this basis some think that the *Raphe nucleus, locus coeruleus*, the nucleus of the solitary tract, the dorsal motor nucleus of the *vagus* nerve, and the intermediolateral cell column could be considered part of this system as well because they all connect, some of them directly to the *hypothalamus* (Isaacson, 1975b).

The Anatomy of the *Amygdaloid* Nucleus (*Amygdala*)

The name *amygdala* was given to this nucleus due to its resemblance to an almond. In humans, it is found under the uncus, and it can be subdivided into a number of subnuclei. The main divisions are the *cortico*-medial and basolateral groups of nuclei. Through the course of mammalian evolution, the baso-lateral group, consisting of the basal and lateral nuclei, has progressively increased in size being largest in humans (Crosby and Humphrey, 1941; Humphrey, 1972). In contrast, the relative size of the medial, central, and cortical nuclei are small in humans. As would be expected, the difference in size is related to the connections and functions of the two main divisions. The afferent connections are numerous and come from many different brain regions.

The afferent connections are specifically placed however within the *amygdaloid* complex. In general, one afferent pathway supplies one or a very few *amygdaloid* nuclei. The afferent connections have a similar distribution, which is thought to confirm that the *amygdala* is a complex of structurally and functionally specific and different subunits. The afferent connections to the *amygdala* arise from the olfactory bulb and anterior olfactory nucleus, the anterior *cingulate gyrus*, the *prefrontal* granular cortex, the temporal neocortex, the dorso-medial *thalamic* nuclei, and other regions of the *thalamus*, the hypo*thalamus*, the *Raphe* nuclei, the nucleus of the solitary tract, and other regions of the *brainstem*. The olfactory bulb and the anterior olfactory nucleus as expected connect to the older *cortico*-medial nucleus. The primary olfactory cortex projects to the baso-lateral group and this group is thought to be indirectly influenced by olfactory stimuli.

In monkeys, the anterior *cingulate gyrus* has been seen to project to the basal nucleus (Pandya et al., 1973). From the more recently evolved parts of the *amygdala*, the baso-lateral group is the only terminal projection to areas of the neocortex. In addition, fibers in monkeys have been traced primarily to the basal nucleus from the granular *prefrontal* cortex (Leichnete and Astrucm, 1976) as well as from the temporal neocortex (Whitlock and Nauta, 1956; Jones and Powell, 1970a). These temporal projections maintain a topographical pattern on the *amygdala*, especially those that go to the basal and lateral nuclei. The *thalamus* has multiple connections to the *amygdala*.

As with the cortex, some of the *thalamic* afferents have a diffuse distribution, whereas others have a specific connection to specific *amygdaloid* nuclei. The dorso-medial *thalamic* nuclei project to the baso-lateral nucleus (Krettek and Price, 1977a). It is thought that the lateral *amygdaloid* nucleus receives afferents from several non-specific midline and intralaminar nuclei of the *thalamus* (Venning, 1978; Ottersen and Ben-Ari, 1978a). The *parafascicular thalamic* nucleus project to the central *amygdaloid nucleus* (Ottersen and Ben-Ari, 1979). The ventro-medial (taste) *thalamic* nuclei are connected to the central and medial *amygdaloid* nuclei. It has also been reported that connections from the medial geniculate body of the *thalamus* to the centro-*cortico*-medial part of the *amygdala* exist (Ottersen and Ben-Ari, 1979).

The *pulvinar thalamic* nucleus projects to the lateral *amygdaloid* nucleus in monkeys (Jones and Burton, 1976). From the *brainstem* area, the dorsal *Raphe nucleus, locus coeruleus, parabrachial nucleus* of the *pars*, the substantia nigra, and connections from the *mescencephalon* project to the central *amygdaloid nucleus* (Veening, 1978a; Ottersen and Ben-Ari, 1978b). This nucleus also receives afferents from the nucleus of the solitary tract (Ricardo and Koh, 1978) and from the *hypothalamus* (Conrad and Pfaff, 1976a; Veening, 1978a). The *hypothalamus* is also thought to connect to other areas and appears to arise primarily from the ventro-medial *hypothalamic* nucleus.

The efferent connections of the *amygdala* are almost as diffuse as its afferent connections, and many of these connections are reciprocal. There are two main efferent pathways; the dorsal *amygdolofugal* pathway, which passes in the stria terminalis, and the second pathway is the ventral *amygdalofugal* pathway. It is important to note that these pathways also carry afferent fibers.

The *stria terminalis* passes along the medial surface of the *caudate* and continues to send fibers, some of which descend anteriorly and others posteriorly to the anterior *commisure*. These are connected to the *hypothalamus* and preoptic areas as well as to other smaller nuclei. It is thought that most of these fibers arise from the *cortico*-medial *amygdaloid* nucleus. The fibers to the hypo*thalamus* end mostly in the ventro-medial *hypothalamic* nucleus. The large ventral *amygdalofugal* pathway connects to the *olfactory tubercle*, the *thalamus*, caudally to the *mescencephalon* and anteriorly to association bundles of the cerebral hemispheres. Most efferent connections to the *thalamus* appear to arise from the *amygdaloid* nuclei except the central and lateral (Krettek and Price, 1977b) and they are primarily connected to the dorso-medial *thalamic* nucleus (Pribram et al., 1953; Nauta, 1967). The neocortex of the *frontal lobes*, especially the *prefrontal* granular cortex, is thought to be connected to the *amygdala* through the dorso-medial *thalamic* nucleus and it is thought to influence the *orbitofrontal* cortex. It has also been shown that the *amygdala* has direct monosynaptic connections to the *prefrontal* cortex (Krettek and Price, 1974). These fibers appear to be ipsilateral to the medial and convex surfaces of the *frontal lobe*. Connections arise from the basal nucleus and project to the motor and *premotor* regions of the frontal cortex. These may explain the motor and effects elicited from the *amygdala* (Jacobson and Trojanowski, 1975).

Other connections to the *cingulate gyrus* and the *caudate* arise from the baso-lateral *amygdaloid* nucleus (Royce, 1978a). Descending fibers project from the *amygdala* to the central nucleus of the *substantia nigra*, *periaquiductal* gray matter, the reticular formation in the *mescencephalon, pons* and *medulla*, the *parabrachial nucleus*, the nucleus of the *solitary tract*, and dorsal motor nucleus of the *vagus*.

Functions of the *Amygdala*

The *amygdala* is thought to be involved in autonomic, hormonal, and motor activities that relate to emotions and motivation. The *amygdala* receives sensory information from different sources. Somatosensory, auditory, and visual sensory input has been shown to affect the baso-lateral *amygdaloid nucleus* (Machne and Segundo, 1956). Visual stimuli are thought to affect the *amygdala* through its connection to the inferior *temporal lobe*.

Olfactory stimulation projects to the *amygdala* by way of the *olfactory bulb*. It is thought that there exists a large degree of integration of sensory input from varied sources. This is thought to be why electrical stimulation of the *amygdaloid complex* produces a variety of motor and autonomic reactions termed *arrest reaction*. The reactions are exemplified by the arrest of spontaneous ongoing movements, inhibition or facilitation of spinal reflexes and cortically evoked movements, contraversive turning movements of head and eyes, complex rhythmic movements related to eating, like swallowing, chewing and licking, changes in cardiovascular and respiratory functions, inhibition or activation of gastric motility and secretion, micturition and defecation, uterine contractions, pupillary dilation, and piloerection among others.

It is thought that most of these responses are components of a more complex behavioral reaction, which can be elicited in organisms by stimulation in or close to the *amygdala* (Kaada, 1951; Kaada et al., 1954). In cats, there have also been observed searching movements to the contralateral side with facial expressions of attention with some bewilderment, anxiety, and sometimes fear or anger. This has been referred to as the *attention* or the *anxiety response*. Interestingly, it appears to be accompanied by EEG desynchronization (Ursin and Kaada, 1960). It has been shown that the attention response always precedes certain behavior patterns, which are indicative of emotional change resulting from stimulation of the *amygdala*. During a *fear response*, searching movements become more rapid, coupled with anxious glancing movements, and the organism becomes restless, runs away, and hides. The *anger response* is characterized by growling, hissing, and piloerection. The rage or anger appears to be directed at something imaginary. The *fear response* is also known as the *flight response* whereas the *anger response* is also known as the *defensive reaction* (Kaada, 1967, 1972). Therefore, the classic *fight or flight* reaction is constituted by the two classic reactions elicited by the *amygdala*. Stimulation of the *cortico*-medial nuclei which, receive olfactory stimulation, elicits autonomic responses, chewing, sniffing, and licking, as well as *tonic* and *clonic* (*dystonic*) movements also seen in seizure disorders. The baso-lateral areas of the *amygdala*, being evolutionarily more recently developed, seem to produce affective and attention response (Ursin and Kaada, 1960).

Particular regions of the *amygdaloid complex* appear to be more related to fear (flight) and others to anger (fight). Bilateral ablation of the *amygdala* most commonly produces tameness. However, with lesions of *amygdala*, aggressive behavior has also been reported and is thought to be caused by removal of areas of inhibition of aggression, known since the work of Walter Cannon in the 1930s. Stimulation and destruction of the *amygdala* seems to also affect the endocrine system and can affect the secretion of various hormones. Hormones such as *gonadotrophic* hormone, *adrenocorticotrophin, threotrophin*, and *vasopressin*, are some of the hormones affected by the *amygdala*. *Cholinergic* and *catacholaminergic* neurotransmitters (*dopamine, norepinepherine*) are active in the *amygdala* (Kaada, 1972).

In summary, functions of the *amygdala* are related to emotional experience and behavioral responses, but it is not the only region of the brain involved in these reactions. In fact, all of the reactions of the *amygdala* can be elicited from other areas of the brain. However, not all of them can be elicited from one area as they can in the *amygdala*. Some other areas include the *hypothalamus, septum,* and area 24 of Brodmann (Fig. 4.7), which may be associated with aggression; the anterior *cingulate gyrus* produces tameness and decreased aggression. *Mesencephalic gray* matter areas, when stimulated, produce rage reactions (Gloor, 1960; Delgado, 1964; Kaada, 1967, 1972; Clemente and Chase, 1973). The *amygdala* is thought to be the area where all of these reactions are integrated.

Anatomy and Function of the Septal Nuclei

Another major area of the *limbic* lobe is the *septal region* or *septal nuclei*. The septal region embryologically arose from the *telencephalon*. It is a layer of gray matter that can be found in the medial wall of the *anterior horn* of the lateral ventricle anterior to the anterior *commissure*. The *septal nuclei* are well-developed in lower mammals and there seems to be a relationship between the size of the *septal nuclei* and the *hippocampus*. In humans, it is thought that the upper region of the septal nuclei forms a thin *septum pellucidum*, the lower area is subdivided into the lateral and medial *septal nuclei*. Dorsal to the *septum* is the *fornix* (Stephan, 1975). Also related to the septal nuclei are *the septo-hippocampal nucleus*, the *septofimbrial nucleus*, and the *nucleus accumbens* (Nauta and Haymaker, 1969; Stephan, 1975). The bed nuclei of the *stria terminalis* and the bed nucleus of the *anterior commisure* are also related to the *septal nuclei*.

The *septal nuclei* seems to possess reciprocal connections with the *hippocampus*, lateral septal nuclei gives off efferents and the medial region receives most of the afferents. The afferents from the *hippocampus* that end in the lateral *septal nucleus* passes through the *fornix* and produce monosynaptic excitation of its cells (Defrance et al., 1973). The *septal nucleus* also receives fibers from the *subiculum* (Chronister et al., 1976; Swanson and Cowan, 1977). There also exist afferent projections from the *periaqueductal gray matter, locus coeruleus, Raphe nuclei*, and ventral *tegmental* areas. The latter three neurons are *noradrenergic*, serotonergic, and *dopamin*ergic respectively. Afferents also arise from *the fastigial nucleus* of the cerebellum (Harper and Heath, 1973). The *amygdala* appears to send afferents only to the bed nucleus of the *stria terminalis*. The lateral *hypothalamus* is thought to project afferents, as well as the *hippocampus, mammillary body*, and *substantia nigra*. The *cingulate gyrus* is the only area of the cortex that sends afferents (Robinson, 1975).

As in the *amygdala*, much of the *septal* efferents are reciprocal with afferents. It is thought that the most important of these efferent connections is with the *hippocampus* (Meibach and Siegal, 1977a). The fibers appear to arise mostly from the medial *septal nucleus* and seem to supply the entire *hippocampus* and appear to be cholinergic (Lewis and Shute, 1967; Mellgren et al., 1977). Efferents have also been traced to the *hypothalamus* (Meibach and Siegel, 1977). Connections that are more extensive have been recorded in the anteriomedial, anterioventral, and dorsomedial nucleus of the *thalamus* (Powell, 1973; Meibach and Siegel, 1977), the *mammillary bodies* (Swanson and Cowan, 1976), *cingulate gyrus* (Kemper et al., 1972), and to the central and medial portions of the *amygdala* (Swanson and Cowan, 1977).

The functions of the *septal nuclei* are not very well known, however, the term *septal syndrome* applies to a set of symptoms related to damage to the *septum*. It involves behavioral overreaction to most environmental stimuli. Sexual and reproductive, feeding and drinking behavioral changes as well as rage reactions have all been noted. Reduction of aggressive behavior in particular has been noted with *septal* lesions. The septal changes seem to be accompanied by normal changes as well. In general, it appears that the majority of the functions of the *septum* are primarily related to connections to the *hippocampus* and secondarily to the *amygdala* and *hypothalamus*.

Another major structure of the *limbic* lobe and the *limbic system* is the *cingulate gyrus*. It is located above the *corpus colliculus* and is related primarily to Brodmann's area 23 and 24. Afferents to the *cingulate gyrus* are varied, with a major source arising from the *thalamus* and the *anterior thalamic nucleus* in particular in turn, subdivided into antero-medial (AM), antero-dorsal (AD), and antero-ventral (AV). In humans, the anterio-ventral appears to have expanded beyond that found in lower organisms and occupies most of the space of the nucleus. This nucleus projects to the *cingulate gyrus* in a topographical pattern. In lower organisms, the AV appears to project to all areas of the *cingulate gyrus* and *presubiculum*. The *thalamus* also indirectly connects other *cingulate* areas such as the *mammillary bodies*. The *cingulate gyrus* receives afferents both directly and indirectly though the *hippocampal-cingulate* pathway by way of the *thalamus*. The anterior *cingulate* is also connected to the posterior *cingulate*, receiving afferents from the lateral *septal nucleus* (Kemper et al., 1972). One of the most significant connections to the *cingulate gyrus* arises from the cortex. The parietal, temporal, and *prefrontal* cortices all send fibers to the *cingulate gyrus*. There appear to be efferent reciprocal connections to neocortical areas as well. Efferent connections from the cortex of the *cingulate gyrus* appear to project to the *hippocampus, amygdala, septum*, the *thalamus*, and cortex, particularly the *prefrontal* and association cortex in the parietal lobe. In addition, *brainstem* efferent projections from the *superior colliculus, pretectal* area, *periaqueductal gray matter*, midbrain *tegmentum, locus coeruleus*, and *amygdala*, and efferents from the anterior *cingulate gyrus* terminate in the basal *amygdala* (Pandya et al., 1973).

The *cingulate gyrus* has been extensively studied in relation to its behavioral function. Many behavioral effects seem to be similar to the effects of the *amygdala* and *septum*. Bilateral lesions of anterior *cingulate* increase tameness and what is described as *social indifference* (Ward, 1948). However, some studies have reported increased aggressiveness (Mirsky et al., 1957).

Anatomy and Function of the Mammillary Bodies

The mammillary body is the most posterior part of the *hypothalamus*; there exist medially large and small cell divisions. In humans the medial is the largest. The afferent connections to the *mammillary body* that appear to be the most important arise from the *hippocampal* formation and project mostly to the medial portion of the *mammillary nucleus* (Swanson and Cowan, 1977). Other important afferent fibers arise from the *septum* and from the medial *septal nucleus* and project to the medial *mammillary nucleus* (Swanson and Cowan, 1977), other afferents project to the *mammillary peduncle*, which arise from the ventral and dorsal *tegmental nucleus* (Cowan et al., 1964). There exist very large efferent connections by way of the *mammillo-thalamic bundle* from the mammillary body particularly to the *anterior thalamic nucleus* which projects to the *cingulate gyrus*. Efferents also descend into the *brainstem* through the mammillary tract. They end primarily in the ventral and dorsal *tegmental nucleus, pontine gray matter*, and in the *nucleus reticularis tegmenti pontis*. Some authors think that the mammillary bodies may be related to certain memory functions (Rosenstock et al., 1977; Leisman and Koch, 2000), and lesions seem to decrease the ability of spatial discrimination.

The descending *mammillo-tegmental* tract and its reciprocal connections to *brainstem* are thought to form a *limbic*-midbrain circuit. It also appears to regulate the subicular (*hippocampal complex*) effects on the *cingulate gyrus* by way of the anterior *thalamus*. The mammillary body is a relay in the *circuit of Papez* (Papez, 1937) and is thought to be an important part of the structural basis of emotions in general and aggression in particular. The circuit is thought to include the *hippocampus, mammillary bodies, thalamus*, and *cingulate gyrus*. It has been suggested that the *hippocampal formation* combines information of internal and external origin into affective feelings that find further elaboration and expression through connections with the *amygdala, septum, striatum*, and *hypothalamus*, as well as through re-entry paths to the *limbic* lobe via the *mammillo-thalamic tract* and *thalamic* projections to the *cingulate gyrus* (the so-called *Papez circuit*) (MacLean, 1975).

Anatomy of the *Hippocampus*

The *hippocampus* was given its name because it resembles a sea horse. In humans, it is located along the floor of the temporal horn of the lateral ventricle. The surface facing the ventricle is the deepest layer and consists of myelinated fibers, which collect on its surface as the *alveus*. Most of these fibers are efferent and unite to form the *fornix*. The *hippocampus* is often referred to as the *hippocampal formation* and includes the *hippocampus, dentate gyrus* that accompanies the *hippocampus* as a band of cortex, and the *subiculum*. The *subiculum* includes the *hippocampal* cortical areas medial to the *dentate gyrus*. This complex is actually made up of the *presubiculum, subiculum*, and *parasubiculum*. All are collectively referred to as the *parahippocampal gyrus*. The *entorhinal* area, the posterior part of the *piriform* cortex, is also part of the *hippocampal region*. The *hippocampus* and *dentate gyrus* are composed of a three-layer cortex, whereas the *entorhinal cortex* and *parahippocampal gyrus* is composed of a six-layer cortex in humans. The two separate fornix areas are connected by fibers that pass underneath the *corpus callosum*, these fibers are called the *hippocampal commissure*.

The internal structure of the *hippocampus* is similar to the *cerebellum*. The efferent fibers that exit the *hippocampus* in the *fornix* are axons of *pyramidal* cells. Afferent fibers that enter the *hippocampus* arise from the *entorhinal* area. In the *dentate gyrus* of the *hippocampus* a dense layer of *granule* cells is present. The dendrites of these *granule* cells extend into the molecular layer where afferent impulses also enter. The axons of the *granule* cells of the *dentate gyrus* are called *mossy fibers*. These fibers project along the *hippocampal pyramidal* cells and have synaptic control of the dendrites of these cells.

The *hippocampus* is subdivided into different areas along its length, which are referred to as CA1, CA2, CA3, and CA4. CA4 is the closest and is partly fused to the *dentate gyrus*. CA1 is adjacent to the *subiculum*. The afferent projection from the *entorhinal* area is the largest and possibly the most important. The *entorhinal* area has greatly expanded in humans compared to lower organisms. The *entorhinal* area consists of two main parts: medial and lateral. The lateral part receives afferent information mostly from the *olfactory bulb*, the *pre-piriform* and *peri-amygdaloid cortex* (Van Hoesen and Pandya, 1975; Krettick and Price, 1977). Afferent fibers also project from neocortical *temporal lobe* regions (Van Hoesen and Pandya, 1975; Leichnetz and Astruc, 1976). The *entorhinal* area also receives afferents from the *baso-lateral nuclei* of the *amygdala* (Beckstead, 1978; Veening, 1978b) as well as from other parts of the *hippocampal formation*. Still other afferents are thought to arise from the medial *septal nucleus*, the dorsal *Raphe* nucleus, and from the *locus coeruleus* (Segal, 1977; Beckstead, 1978). The medial part of the *hippocampus* has afferents arising from the *presubiculum* (Segal, 1977; Beckstead, 1978; Kohler et al., 1978), *hippocampal field* CA3 (Hjorth-Simonsen, 1971), *locus coeruleus*, *Raphe* nuclei, *septum*, and the *thalamus* (Segal, 1977).

Many afferent projections maintain a topographical arrangement as well as a laminar distribution (Hjorth-Simonsen, 1972; Steward, 1976). Other afferents of interest are known as *septo-hippocampal connections*, although these afferents from the *septal nuclei* are smaller in number than those from the *entorhinal area*. Most of these fibers arise from the medial *septal nucleus*, are diffusely connected to the *hippocampus*, and are cholinergic. The *hypothalamus*, from the vicinity of the *mammillary body* also sends afferent fibers to the *hippocampus* (Segal and Landis, 1974; Pasquier and Reinosa-Swarey, 1976). The *hypothalamus* has powerful inhibitory action on the *hippocampus* (Segal, 1979). The anterior *thalamic nucleus* sends afferents to the *hippocampus* as well (Swanson, 1978). Significant influences exist from afferent connections of the *Raphe* nuclei and *locus coeruleus*. The *locus coeruleus* primarily sends *noradrenergic* fibers to the *hippocampus*. *Dopaminergic* fibers from the *entorhinal* area most likely arise from the ventral *tegmental* area. Serotonergic fibers, on the other hand, likely arise from the dorsal and median *Raphe* nuclei projecting to the *hippocampus* (Blackstad, 1977; Swanson, 1978).

Most efferents appear to travel in the *fornix*, which, in turn, travels beneath the *corpus callosum*. Some fibers descend in front of the *anterior commisure* (*pre-commissural fornix*) and some descend posterior to it (*post-commissural fornix*). Fibers of the *post-commissural fornix* are thought to connect to the *mammillary body* of the *hypothalamus*, anterior *thalamic nucleus, peri-aqueductal gray matter* of the *mesencephalon*, and the *pontine nuclei* (Cragg and Hamilyn, 1959). The *pre-commissural fornix* fibers also project to the *mammillary body* and the anterior *thalamic nucleus* as well as to the preoptic region, lateral *hypothalamus* and nuclei of the *septum*. Most of the fibers of the fornix are thought to arise from the *subiculum* especially all fibers to the *Hypothalamus* and many of the fibers to the *septum* traveling in both the *pre-* and *post-commissural fornix*. Fibers in the fornix arising from the *hippocampus* proper (CA1–3) all pass through the *pre-commissural fornix* and project to the lateral nucleus of the *septum*. Other fibers from the *hippocampus* proper project to the *cingulate gyrus*, the *entorhinal* area, and to the *subiculum* (Hjorth-Simonsen, 1971, 1973; Swanson and Cowan, 1977). The antero-ventral nucleus of the anterior *thalamus* (Sikes et al., 1977; Swanson and Cowan, 1977) receives efferents from the *subiculum*. The *subiculum* also appears to supply fibers to the *hypothalamus*, primarily the ventro-medial *hypothalamic nucleus* (Swanson and Cowan, 1977).

Functions of the *Hippocampus*

Although the *hippocampus* is part of the *rhinecephalon* or *nose brain* it seems to have very little to do with olfaction (Brodal 1947; Allison, 1953). The *hippocampus* has shown electrophysiologic reactions to stimulation of visual, acoustic, gustatory, somatosensory, as well as other types of peripheral stimuli. In addition, responses have been recorded after stimulation of certain areas of the cortex and subcortical regions and some of these reactions are reciprocal. Behavior responses that have been associated with the *hippocampus* include sexual and reproductive as well as visceral and endocrine responses (Kurtz, 1975; Hecht et al., 1976; Micco et al., 1979). The *hippocampus* is thought to be involved with attention and alertness (Hendrickson et al., 1969; Wall and Messier, 2001). *Hippocampal* stimulation has been reported to be associated with reactions such as quick glancing or searching movements to the contralateral side (Hendrickson et al., 1969; Wall and Messier, 2001), and facial expressions of attention, bewilderment, and anxiety (Williams et al., 2001). Other behaviors such as orientation, hallucinations, and defensive reactions have been noted and are thought to be related to the function of the *hippocampus* (Rajarethinam et al., 2001). There are different opinions as to the effect of *hippocampal* function on motor performance. Some think that the *hippocampus* plays a major role in the modulation of motor control directly (Bingel et al., 2002). Others think that the *hippocampus* is involved with non-motor behavioral and psychological processes (Rajarethinam et al., 2001; Williams et al., 2001). Most authors agree that the *hippocampus* is thought to have a role in attention and arousal. However, when the neocortex becomes desynchronized or alert, the *hippocampus* becomes synchronized with a coincident production of EEG *theta* waves (Buzsaki, 2002). When the cortex demonstrates synchronization, the *hippocampus* manifests a concomitant desynchronization (Buzsaki, 2002). These activities may reflect connections from the reticular formation or *thalamus*. It is thought that the *septum* may serve a pacemaker function for the *hippocampus*.

Other activities ascribed to the functioning of the *hippocampus* include: processing of information (Burgess, 2002), general motivational changes (Tracy et al., 2001), reactions of frustration to non-reward and punishment (Nagaratnam et al., 2002), learning and memory (Leisman and Koch, 2000; Burgess, 2002), motivational responses in approach behavior (Tracy et al., 2001), voluntary phasic skeletal activity (Muley et al., 2001), and motor processes initiated by *brainstem* reticular formation (Black, 1975). The *hippocampus* is also thought to play a role in spatial discrimination and in the formation of cognitive maps (O'Keefe and Nadel, 1978).

Lesions of *hippocampus* and *temporal lobes* are known to be involved in epileptic seizures and memory disturbances (Leisman and Koch, 2000; Martin et al., 2002). The memory changes that appear to be associated with the *hippocampus* appear to be primarily involved with *recent memory*. This involves the ability to store new information or retrieve stored information. Overall, the main function of the *hippocampus* appears to be *integrative*. In patients with seizure disorders marked cell loss in the *hippocampus* is frequently seen (Kotloski et al., 2002).

Most recently it has been noted that virtually all sensory input that the *hippocampus* receives arises from higher order, multi-modal cortical regions (O'Reilly and Rudy, 2001). This would imply that whatever processing the *hippocampus* does is in the service of forming long-term memories and is accomplished with abstract representations of experience. It is now apparent that in addition to its subcortical connections, the *hippocampus* has massive reciprocal efferent connections to the *neocortex*. This is consistent with the theory that the *hippocampus* does not store the final long-term memory. The site and nature of long-term memory storage is at present unknown (Leisman and Koch, 2000, 2003).

The *hippocampus*, however, seems to be designed for forming associations. It appears that with its extensive associational connections, *hippocampal* neurons form a massive network to create representations of perceived experiences. Connections originating in one part of the *hippocampal* formation, project to almost half of the region of the next processing step. Studies of monkeys have demonstrated that the largest input of sensory information to the *hippocampal formation* arises from the *peri-rhinal* and *para-hippocampal* cortices adjacent to the *hippocampal* formation in the *temporal lobe*. It appears that these two areas not only play an important part in contributing sensory information to the *hippocampal formation* but also play a role in some forms of memory function on their own. This has been noted in studies where selective lesions were created in the *peri-rhinal, ento-rhinal*, and *para-hippocampal* regions (Meguro et al., 1999). This was shown to decrease memory performance in delayed non-matching to sample tests. Subsequent studies have shown that lesions of only the *para-hippocampal* and *peri-rhinal* cortices also cause significant memory deficits (Zola-Morgan et al., 1993; Malkova and Mishkin, 2003). It has also been noted that the *hippocampus* will be activated when explicit (conscious) retrieval of memory includes significant frequency of exposure and not one-time retrieval of information. It is thought that the *hippocampus* encodes new information and retrieves recent information when explicit recollection is involved.

The *Hypothalamus*

The one thing that seems to tie the various structures together into the *limbic system* is their common connection to the *hypothalamus*. It is therefore important to understand the basic structure and function of the *hypothalamus* even though it is not anatomically part of the cortex (cf. Figs. 4.6 and 7.2). The *hypothalamus* is one of the major areas that are involved in the control of autonomic or visceral function. The *hypothalamus* has been called the head or chief nuclei in the autonomic nervous system. We have already reviewed most of the afferent and efferent projections of the *hypothalamus*. Many regions of the brain and *brainstem* are involved with autonomic changes, areas such as cerebral cortex, *hippocampus*, *entorhinal* area, *thalamus, basal ganglia, reticular formation*, and the *cerebellum*. It is thought that the effects of these regions on autonomic function is in part due to their connections with the *hypothalamus*. The *hypothalamus* is considered the highest level of the brain concerned in the integration of autonomic function. The *hypothalamus* lies beneath the *thalamus* and on the floor of the third ventricle. The *hypothalamus* is a small collection of nuclei including the *mammillary bodies, the substantia grisea centralis*, the *supraoptic nucleus, para-ventricular nucleus, nucleus tuberis*, and the dorso-medial, ventro-medial, and lateral *hypothalamic nuclei*. The lateral nuclei are especially well developed in humans. The *hypothalamus* appears to be the region of the brain with among the highest concentrations of *norepinepherine*, it also contains *acetylcholine, serotonin, dopamine*, and *histamine*. Afferent connections include retinal fibers, olfactory fibers, spinal cord, reticular formation, *locus coeruleus, Raphe* nuclei, *peri-aqueductal gray matter, cerebellum, hippocampal formation, piriform cortex, cingulate gyrus*, and the *orbito-frontal cortex*.

Efferent connections include the anterior *thalamic nuclei, amygdala, septum, hippocampus*, pre-ganglionic sympatheic and *parasympathetic* neurons, *pretectal* area, *superior colliculus*, dorsal and ventral *tegmental nuclei of Gudden*, the *Raphe* nuclei, *nucleus lovus coeruleus*, midbrain *reticular formation*, dorsal motor nucleus of *vagus*, nuclei of *solitary tract, nucleus ambiguous*, and the intermedio-lateral cell column. There are also significant fibers from the *hypothalamus* to the posterior lobe of the *hypophysis* (*pituitary* gland) and a vascular link to its anterior lobe. It has been noted that lesions of the central nervous system and *hypothalamus* result in stress induced increases in adrenal hormone release. This is thought to be caused by the effect of the *hypothalamus* on the anterior lobe of the *pituitary* gland, and other endocrine organs such as gonads, adrenal cortex, and thyroid. Supra-optic and para-ventricular nuclei also produce *vasopressin* and *oxytocin*. Other functions that are related to the *hypothalamus* are cardiovascular with increase in blood pressure, acceleration of heart rate with stimulation of posterior-lateral portions and vasodilation, decreased blood pressure, and slowing of heart rate with stimulation of the anterior hypo*thalamus*. The hypo*thalamus* appears to play a role in regulating body temperature as well as piloerection, regulation of water and food intake, and satisfaction of appetite, growth, regulation of digestive organs, sexual function, waking, and sleeping centers. Lesions of the *hypothalamus* in humans have been followed by both manic and depressive states (Martin and Riskind, 1992; Benabarre et al., 2002). Other behavioral changes associated with lesions of the *hypothalamus* include anxiety and depression (Balaban and Thayer, 2001), flight of ideas (Maes et al., 1989), and motor hyperactivity (Peled et al., 1997; Weissenberger et al., 2001). The *hypothalamus* has been famously the site of Delgado's demonstration of the control of aggression (Delgado, 1967), Old's experiments on pleasure centers in rats (cf. Olds, 1967), and has long been implicated as the control site for reinforcement and operant behavior (Olds, 1969).

In summary, the *limbic system* is considered to be constituted by the structures of the *limbic lobe*. The component regions function as a system by way of their connections to the hypo*thalamus. The* more primitive *archi-* and *paleocortex* constitute the *limbic system*. It therefore seems to control and maintain the more primitive survival driven characteristics in organisms: fear and aggression, autonomic regulation, feeding, sexual behavior, motivation, and attention. These are all the most basic behavioral characteristics of any organism.

8

The Cerebral Cortex

The cerebral cortex (neocortex) and the *limbic system* generate all sensory and cognitive perception, as well as voluntary motor activity; the *limbic system* generates most of our basic emotional and autonomic regulations and it is modified through its connections to the neocortex. We have discovered most of the subcortical areas, their connections to the cerebral cortex, and their role in the development; anatomically and functionally. The neocortex and the *limbic system* cannot develop or function properly without the input of intact subcortical connections. We need to understand more fully the evolution and development of the cerebral cortex before we can completely understand its function and dysfunction. There are also important cerebral interconnectivities between different areas of the cortex that determine normal function as well. We cannot completely understand the function of the cortex or *limbic system* and their subcortical connections without exploring the concepts of cerebral lateralization and specialization of function. As we will see this is probably the single most important aspect of the human cerebral cortex; without proper interaction of the different functional areas of the cortex and specialized areas of function we cannot understand or even begin to consider treatment interventions.

CEREBRAL CORTEX: FUNCTION AND DEVELOPMENT

The cerebral cortex is developed from the *telencephalon* and becomes the most rostral part of the *prosencephalon*. The phylogenetically older parts of the cortex are known as the *rhinencephalon*. In lower vertebrates like amphibians and reptiles, these parts make up the majority of the *telencephalon*. The neocortex has significantly developed through evolution in mammals, and in humans it has become the largest in size. Its surface area has become greater than its volume due to its convolutions and fissures. Six layers of cells characterize the neocortex in humans. The earliest event in the development of the cortex is the generation of cells in the ventricular zone that are destined for different layers of the cortex. These layers are generated in an inside-out fashion with the cells of the deepest layer born first and cells from the other layers born progressively later. The cells all grew out of the same *parent cells* and maintain a columnar type of arrangement. It appears that certain cells are destined for a certain layer before birth although cells born at a specific location in the ventricular zone may end up in widely separate parts of the cortex (Goodman and Shatz, 1993).

Targeting and innervation of the cortex by thalamocortical axons proceeds as the cortical layers form. This is thought to be the key to the development of the cortical regions devoted to specific functions. Somehow, the *thalamic* axons recognize specific cues at the appropriate cortical location. It is also thought some interaction exists between ingrown *thalamic* axons and the developing cortex. It has been shown that lesions that reduce the size of the *thalamic* nuclei also reduce the size of their target cortical areas. Initially this appears to be driven chemically or genetically.

It has been demonstrated that specific membrane-bound molecules regulate the termination of *thalamic* axons in appropriate layers and do not depend on electrical activity for growth and development (Jones and Powell, 1969a, 1969b; Molnar and Blakemore, 1995, 1999; Aumann et al., 1998; Rouiller et al., 1998; Aboitiz, 1999; Darian-Smith et al., 1999). It is only later in development that cortical tissue becomes activity-dependent (Jones, 1990; Brenneman et al., 1998; Ben-Ari, 2002; Desai et al., 2002; Laurent et al., 2002). It has been shown that the arborization of thalamocortical axons measured *in vitro* is influenced significantly by activity (Jensen and Killackey, 1987; Molnar and Blakemore, 1995, 1999; Kakei et al., 1996; Arnold et al., 2002). This process has been studied most extensively in the visual system. It has been reported that abnormal activity from either brain damage or reduced environmental stimulation can disrupt normal development of the visual cortex during critical periods (Singer, 1983; Henderson and Blakemore, 1986; Carlson et al., 1987; Cramer and Sur, 1995; Mataga et al., 2001; Desai et al., 2002; Schweigart and Eysel, 2002).

Hubel and Wiesel (Carlson et al., 1986; Hubel and Wiesel, 1979) were the first to demonstrate that monocular lid suture in kittens during a *critical period* caused cortical cells to be dominated by the open eye. Most recent authors are mainly surprised at how rapidly the effects of lid suture occur and can include reduced size of axon connections to the deprived eye (Pizzorusso et al., 2000). It appears also that pre- and post-synaptic activity together influence the *thalamo-cortical* connections; blocking post-synaptic activity causes *thalamic* afferents to spread and activate inappropriate neurons. Some studies suggest that it is not only the level of activity but also the temporal pattern of the activity that is most important (Tieman et al., 1983; Lund et al., 1991; Todd and Perotti, 1999; Crewther and Crewther, 2002). This means that any condition like *strabismus* or monocular deprivation prevents the normal synchronization or phase of cell discharges, not just their rate. This synchronized activity binds and reinforces the structural and functional connections of certain groups of cells. In this example, the functional role of the two eyes will be united, and the deprived eye will have fewer cortical connections because it is unable to maintain correlated cortical activity. The cortical activity associated with the deprived eye is not only weak but also disorganized. It is thought that the pattern and amount of afferent activity is a key determinant of synaptic efficacy during cortical development. This is therefore another form of activity-dependent strengthening of connections. Temporal connections between pre- and post-synaptic cells lead to strengthening of the synapses between them; lack of correlation leads to a weakening of synapses.

It appears that intra-cortical connections occur through similar mechanisms, initially genetically driven and then activity-dependent. When one understands the principle of plasticity and activity-dependence the cortex becomes an extremely dynamic organ that can be molded or shaped by activity in either development or adulthood. The developmental patterns of *thalamo-cortical* and intra-cortical projections, when examined at the level of a single axon and dendrite branches allow for an amazing lack of precision. A single cortical cell receives 5,000–10,000 synapses in its dendrites arising from more than one pre-synaptic axon. In addition, single *thalamo-cortical* axon arbors spread widely and any one axon might potentially contact several hundred post-synaptic cortical cells. Looking at the spread of *thalamo-cortical* arbors and the spread of intra-cortical horizontal axons, the potential convergence grows even more. This makes it virtually impossible to think that all synapses are preprogrammed or specified.

It would be more reasonable to think that an appropriate set of synapses is selected physiologically from the potentially large number present anatomically. In this case, it should be possible to change the selected set of synapses by altering the relative strength of a group of neurons. The demonstration of rapid plasticity and dynamic alteration in receptive field properties even in adults indicate that this is the case. A single neuron in any area like the primary visual cortex receives and integrates inputs not only from the *thalamus* but also from other cortical neurons, which can be located in distant areas of the brain regions concerned with vision. It has been described that inputs to a neuron resemble an iceberg, cells have a large sub-threshold zone of activation, and although a spike response might be elicited by stimulating only a small central portion of this space, the remaining portion could modulate the spike output in significant ways (van Vreeswijk, 2000; Volgushev et al., 2000). For example, it has been shown that decreasing the level of central activation with an *artificial scotoma* can cause outer portions of the visual field to drive the cell, enlarging the cell's *receptive field*. Therefore, a cell's response to a stimulus within the *receptive field* can be altered by other distinct stimuli. This constitutes the first level of complexity and conditionality with respect to neuronal integration in the cortex. Responses in higher sensory areas such as association areas are influenced not only by specific sensory stimuli but also by attention and motivation, and other aspects of behavior. We will examine some of these factors in cognitive perception.

In other words, any cell that perceives a primary stimulus is affected not only by the cells related to that modality, but also by other neurons that are connected in space and time (Leisman and Koch, 2003). In fact, the sub-threshold stimulation of other neurons may be the determining factor of whether the cell fires at all. Other stimuli may affect this sub-threshold level and in fact may be more important to the function of that cell than is the primary stimulus. For instance, visual neurons may be more dependent on *somatosensory* input that is connected to those neurons, than it is on input from visual pathways. Therefore, the perception of visual input or the loss of vision, or visual processing, may be affected more by losing *somatosensory* input to visual neurons than by visual input. This may be why when suturing closed the eye of an adult cat, vision was not lost and may return even though that pathway was not fired. Cells remained viable because of other sub-threshold stimuli. In this case, we see that baseline oscillations from *nonspecific thalamic* neurons may be one of the major factors that establish a certain threshold level of activation of neurons. Therefore, this may explain why levels of attention or arousal may affect the level of perception. The greater firing or arousal due to the constant firing of diffuse *thalamo-cortical* synapses will bring cortical neurons either closer or further from threshold depending on the frequency and temporal patterning of firing. Perception through any sensory modality or cognitive processing through that sensory modality by the *association cortices* is dependent on constant, continuous, or nonspecific factors or specific input from many different areas of the *thalamus* and cortex.

Constant input to the cortex seems to consist primarily of *somatosensory* input through the *dorsal column* and *cerebellum* to the ventral, lateral, and intra-laminar nuclei of *thalamus*. This may be represented by 40-Hz oscillations, which may impact this baseline sub-threshold cortical arousal as well as affect the temporal patterning so neurons are connected not only in space but in time as well. Short term, rapid changes in responses based on stimulators content can be explained by rapid alterations in the excitability of neurons and circuits. In fact, long-term changes in input activity can lead to major changes even in the adult cortex (cf. Fig. 8.1). It has been shown that cortical changes can accompany or even be the basis of skill acquisition and learning (Niemann et al., 1991; Grafton et al., 1992; Salmon and Butters 1995; Pascual-Leone et al., 1995a, 1995b; Karni and Bertini,

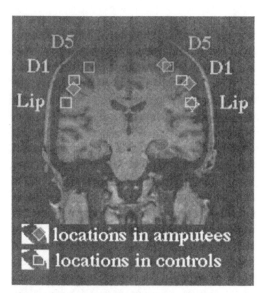

Fig. 8.1. An example of cortical representational plasticity in amputees. The central nervous system and the cerebral cortex in particular have the potential to reorganize itself after alterations of its input from peripheral neural structures. Using MEG/MSI in humans, amputation of an upper extremity results in alterations of the homuncular organization, whereby the representational zone of the face shifts toward the zones formerly represented by the hand and digits. The focus of cortical activation elicited by facial stimulation is shifted up to several centimeters toward the receptive field, which would normally receive input from the now amputated hand and fingers (Elbert et al., 1994).

1997). It seems that the pattern of input activity appears to be critical for selecting and strengthening specific sets of cortical or spinal synapses, and this strengthening might form the basis for learning related plasticity.

The cerebral cortex from a gross anatomical standpoint has two mostly symmetric hemispheres. These hemispheres have four main divisions or lobes. These lobes have different functions and are distinguished from one another by various anatomical landmarks. The four lobes are the frontal, parietal, temporal, and occipital. The *central sulcus* divides the frontal from the parietal lobes and the *lateral fissure* separates the temporal from the parietal lobes. The occipital lobe is separated from the parietal and *temporal lobes* by the *parieto-occipital sulcus* on the brain's dorsal surface and the *pre-occipital notch* located on the ventral-lateral surface. The *interhemispheric* fissure that runs from the rostral to the caudal end of the forebrain separates the left and right hemispheres. Interconnections between the hemispheres are accomplished by axons that travel from one side to the other in the *corpus callosum*. In 1909, Brodmann identified approximately 50 distinct regions of the cerebral cortex. He categorized these areas based on cellular morphology and organization and numbered these areas on this basis (cf. Figs. 4.7 A,B,C). The numbers are unsystematic and have more to do with the order in which Brodmann sampled than any true functional relationship between regions. Each lobe has basic functions with which it is primarily involved.

Frontal Lobes

The *frontal lobe* plays a major role in motor activities like planning and in the execution of movements. The primary motor area proximal to the *precentral gyrus* and is known as the *motor strip* (Brodmann's area *4*). This is located just anterior to the *central sulcus*. The *primary motor area* is also referred to as motor area 1 or MI. Anterior to this area are two additional primary motor areas (Brodmann's *4*, *5*, and *6*). This supplementary motor cortex lies anterior to the motor strip and extends around to the hemisphere's medial surface. The *premotor cortex* lies anterior to the supplementary motor cortex and on the lateral surface of the hemispheres. These motor areas contain motor neurons whose axons extend to the spinal cord, *brainstem*, and synapse on motor neurons in the spinal cord. The motor neurons are located in layer 5, the output layer of the motor cortex. This layer contains large *pyramidal* cells; they are the largest neurons in the cerebral cortex. The most anterior region of the *frontal lobe*, the *prefrontal cortex* is responsible for higher aspects of motor control, planning, and in the execution of behavior; these tasks require integration of information over time. The *prefrontal cortex* has two main areas, the *dorsolateral prefrontal cortex*, which is found on the lateral surface of the *frontal lobe* anterior to the *premotor* regions, and the *orbitofrontal cortex*. The *orbitofrontal cortex* is located on the *frontal lobe*'s anterior-ventral surface and is more medial. The *orbitofrontal cortex* includes *limbic* lobe structures and is connected to them. The *frontal lobe* is the largest lobe in humans and the *prefrontal cortex* constitutes approximately 50 percent of the size of the *frontal lobes*. The *prefrontal cortex* is included in a neuronal system that includes *the basal ganglia*, the *thalamus*, and the *cerebellum*. Most of the higher and more complex motor, cognitive, and emotional behavioral functions are thought to be found primarily in the *frontal lobes*. This area of the neocortex has expanded more than any other in the human brain.

The Parietal Lobes

The *somatosensory* cortex is found in the *postcentral gyrus* and adjacent areas (Brodmann's areas *1, 2*, and *3*). These cortical areas receive afferent input from the somatosensory relays of the *thalamus* and detect information about touch, pain, temperature, and proprioception (body and limb position). The *primary somatosensory cortex (SI)* is immediately caudal to the *central sulcus*. The *secondary somatosensory cortex (SII)* receives information from projections primarily from *SI* and is located ventral to *SI*. Somatosensory inputs projecting to the posterior parietal cortex arise from *SI* and *SII*. Somatosensory information to the *thalamus* follows two main pathways: the *anterolateral system* for pain and temperature and the *dorsal column-medial lemniscal system* for information about touch, proprioception, and movement. Receptors in the periphery transduce physical stimuli into neuronal impulses transmitted to the spinal cord and up to the brain making synaptic connections at various relay sites along the ascending pathway.

The Occipital Lobes

The primary visual cortex, also known as the *striate cortex* or *VI* (Brodmann's area *17*), receives visual input from *the lateral geniculate nucleus* of the *thalamus*. In humans, the *primary visual cortex* is found mostly on the medial surface of the cerebral hemispheres, and extends slightly into the posterior hemisphere pole. The cortex in this area has six layers and is primarily responsible for coding visual features like color, luminance, spatial frequency, orientation, and movement. These properties are organized into two main projection streams of visual processing. Visual information from the outside environment is processed by retinal cells and transmitted by the optic nerve to the *lateral geniculate nucleus* of the *thalamus* and then to *VI*. This pathway is called the *retino-geniculo-striate* or *primary visual pathway*. Visual projections from the retina also reach other subcortical brain regions through a secondary system. The *superior colliculus* of the midbrain is the main target of the secondary pathway and controls visual-motor functions. Surrounding the *striate cortex* is a large visual cortical region known as the *extrastriate visual cortex*. The *extrastriate* cortex includes Brodmann's areas *18* and *19*. It has been shown in lower primates that there are more than 3-dozen distinct visual areas in the *extrastriate* cortex (Blatt et al., 1990). These visual areas contain partially redundant maps of the visual world, each specialized to analyze specific aspects of a scene including color, motion, location, and form. Two pathways from *the striate cortex* convey primary streams of information. One pathway flows from *VI* to the *temporal lobes* and conveys analysis of stimulus features. Ultimately the information is used to carry out form description and object identification. The other pathway projects from *VI* to the parietal lobe and carries information about stimulus motion and localization within visual space. Each of the *extrastriate* areas maintains strong reciprocal interconnectivity with areas prior to it in the visual hierarchy and with other areas in the same processing stream. Many interconnections exist between the ventral and dorsal visual processing streams.

The Temporal Lobes

The auditory cortex lies in the superior part of the *temporal lobe* and is found deep within the *sylvian fissure*. The projection from the *cochlea*, through the subcortical relays to the *medial geniculate nucleus* of the *thalamus*, then proceeds to the *supratemporal cortex* in a region known as *Heschl's gyri*. The *auditory association areas (AII)* surrounding the *primary auditory cortex (AI)* is referred to, in Brodmann's scheme, as areas *41* and *42*. Area *22*, which surrounds the auditory cortex, aids in the perception of auditory input. When this area is stimulated, sensations of sound are produced in the human brain. One can represent the sensory inputs to the auditory cortex using a *tonotopic map*. The orderly representation of sound frequency within the auditory cortex can be determined with several *tonotopic maps*. The *temporal lobe* is also involved in memory (Leisman and Koch, 2000).

The majority of neocortex that is neither motor nor sensory has been called the *association cortex*, which is made up of regions that receive inputs from one or more modalities. One example is *the visual association cortex* in the temporal and parietal lobes, which, in addition to the primary visual cortex, are necessary for the conscious sensation of vision. Parts of the *limbic lobe* take part in emotional processing and the frontal *association* cortex plans and calculates the long-term outcome or goals of certain acts. The *association* cortex determines what emotions mean and what the outcome or reaction should be. The association areas of the *parietal-temporal-occipital* junction have a significant role in language processing. Therefore, higher mental or cognitive processes are the responsibility of the association cortex in interaction with sensory and motor areas of the cortex. The association areas constitute the majority of the neocortex. These are thought to represent the highest evolutionary expansion of the human brain and they have evolved in parallel with areas of the *thalamus* and especially the newer areas of the *cerebellum*. The *cerebellum* is thought to serve as a link between anterior and posterior areas of neocortex and provides the neocortex with an additional association area (Leiner et al., 1991).

The *Prefrontal* Cortex

The *frontal lobes* comprise one third of the neocortex and the *prefrontal* cortex constitutes 50 percent of the *frontal lobes*. The *prefrontal* cortex is unique to humans; the reference to highbrow, for example, is a reference to the structural changes of one's forehead that humans underwent to provide more space for our *prefrontal* cortices. It is thought that most of the unique qualities that humans possess are found or connected in some way with the expansion of the *prefrontal* cortex. This brain region is also important because the *frontal lobes* include areas of motor control as well. Proceeding anteriorly in the *frontal lobes* from the motor strip to the supplementary motor areas and the *premotor* cortices, we see the control of motor activity becoming more sophisticated. We also see that as the brain expanded and evolved anteriorly, the *frontal lobes* became more concerned with the cognitive control, timing, and duration of movement whereas the motor strip was an evolutionary advance giving humans greater gross voluntary motor control. The newer areas of the *frontal lobe* provide more precision and direction to the movement. Eventually we see that the *prefrontal* cortex has little to do with the movement per se, but is largely concerned with the control of direction of the movement and the behavior that drives that movement.

It is well established that humans need normal *frontal lobes* to accomplish goals, make decisions, express creativity, and navigate through complex social situations (Chatterjee, 1998a). The *frontal lobes* regulate goal directed behavior, a hierarchy of reflexive movements, cross-temporal contingencies, approach and avoidance behavior, response inhibition, and perseveration. As we will see, all of the activities that the *prefrontal* cortex controls revolve around improvement of goal-directed behavior. We hypothesize that the development of the human *prefrontal* cortices was a natural expansion of the evolutionarily earlier developed areas of the *frontal lobe* and that goal-directed movements and behavior provided for an expansion of those areas. The same regions of the human central nervous system that were already employed for better control, coordination, and timing of movements, expanded in parallel with the frontal cortex. The lateral portions of the *cerebellum*, for example, are more involved with the cognitive coordination and control of motor activity than with the control of the actual movement of muscles. The ventral lateral *thalamus*, linking the lateral *cerebellum* to the *prefrontal* cortex, is witness to the fact that these two areas evolved together. There must have developed a partnership of sorts between the *cerebellum* and the *prefrontal* cortex; the initial focus of the *frontal lobe*s was the control of motor activity as it was for the *cerebellum*, but as the movements became more goal directed, greater cognitive control over movement was needed.

The *prefrontal* cortex was required in higher organisms and in humans not just for more speed, precision, and coordination, but also for the provision of a control mechanism for memory of previous motor actions, projection of future movements, facilitation and inhibition of movement, and the reaction or inhibition of reaction to stimuli; to know when to move toward prey or away from a predator. All of these involve higher cognitive control that are linked to the emotion-controlling *limbic system* which provides motivation and autonomic regulation. Early in the evolutionary scale, the sense of smell through the *rhinencephalon* or "nose brain" provided the primitive brain with basic information about the environment. The sense of smell allowed the organism to detect food or danger. This was naturally connected with emotional centers like the *amygdala* that provide motivation for *fight or flight* behavior, and was linked to the autonomic nervous system to increase blood, respiration, and adrenaline to respond to stress. The links of frontal and *prefrontal* regions of the cortex to the *limbic system* had as its main function the provision of a mechanism to either allow the organism to catch the prey, to run from a predator, or to seek a mate. The *frontal lobe*s then, in coordination with the *cerebellum* and *basal ganglia*, have expanded beyond their control of movements and have evolved to control the behaviors that guide movement and most of our basic actions.

The *prefrontal* cortices regulate movement directed at accomplishing goals (Luria, 1966; Kolb and Whishaw, 2001) along with the *basal ganglia* and *cerebellum*. These structures are known to control primary motor output. However, damage to or dysfunction in the *cerebellum* and *basal ganglia* results typically in obvious motor defects of tone and coordination than with damage to the *prefrontal* cortex. Goal-directed movements controlled and regulated by the *prefrontal* cortex are complex and have multiple sequences. The *basal ganglia* seems to be involved with the sequencing of movements, together with the *prefrontal* cortex. These complex movements are assembled using combining sets of more simple reflexive movements (Leisman and Koch, 2003). These reflexive movements are simple movements involving individual muscle contractions, simple automatic movements, as well as more complex reflexive movements. Both Pavlov and Luria thought that following *frontal lobe* damage, subordinate reflexes are no longer integrated seamlessly into goal-directed movements (Luria, 1966).

The complex cognitive coordination of goal-directed movements must involve temporal organization. It has been shown that information that is not sequenced properly will disrupt plastic changes along the pathway (Merzenich et al., 1993). More complex time considerations involve *cross-temporal contingencies*, a behavior whose control is thought to reside in the *prefrontal* cortex (Fuster, 1989). *Cross-temporal contingencies* are movements that take into account events or information from experience or the predicted events of information in the future. Behaviors in which there is a time delay between the stimulus and the response particularly activate the dorsolateral *prefrontal* cortex. Goal-directed control of movement in relation to *cross-temporal contingencies* allows an individual to initiate movement in the absence of external stimuli and inhibit movements in the presence of external stimuli that might otherwise be stimulating. In dysfunctional states of the *frontal lobes* or *prefrontal* cortices, we would expect therefore that individuals would have difficulty initiating movements without guidance or explicit stimuli. These individuals would possibly be apathetic or passive and possibly depressed. We would also expect that individuals with *frontal lobe* dysfunction might have difficulty inhibiting responses to stimuli. They would be easily distracted, have poor attention, and react randomly to environmental stimuli, behaviors typical of those with *hyperkinetic* or *hyperactive* disorders (Heilman and Watson, 1991). In other words, individuals with *frontal lobe* dysfunction do not know when it is appropriate to move and when it is not. Knowing when it is important to proceed toward a goal or move away from danger is critical to survival. These behaviors are known as approach and avoidance behavior.

The *prefrontal* cortex receives information from the parietal somatosensory areas as well as from visual and auditory stimuli. These are important to provide attention to sound or movement that may require a response by the organism—either to approach prey or a mate, or avoid a predator or enemy. This information is integrated between different areas of the cortex. It is thought, based on monkey ablation studies (Denny-Brown, 1958), that there are inhibiting interactions between the parietal and *frontal lobes* that help to mediate approach and avoidance behavior. It appears that the parietal lobes may mediate the approach to and the *frontal lobes* mediate avoidance of certain stimuli. However, this may be differently controlled by each cerebral hemisphere. The important aspect to understand is that the frontal and parietal lobes appear to work together in the regulation of certain behaviors.

Clinically, individuals with frontal or parietal lobe dysfunction may present with abnormal or exaggerated approach and avoidance behavior. Abnormal approach activities can be as simple as a grasp reflex, or complex like using whatever tools or implements within view. Avoidance behavior can be simple or complex such as extension of finger to touch, spatial neglect of a body part or area, withdrawal, or severe shyness behavior. Also controlled by the *frontal lobes* are movements over time and, therefore *perseveration*, are a result of the dysfunctioning of the region. *Perseverations* are the inappropriate repetitions of movement over time. *Perseveration* may be simple or complex. Simple activities may be impaired such as being unable to stop shaking someone's hand, or being unable to stop drawing circles or loops, once started. Some individuals will continue to write their name after they have been told to stop. They cannot stop one action to begin another or vice versa. These behaviors may be viewed because of an inability to terminate normally voluntary activity in one muscle or group or initiate coordinated voluntary responses in the next muscle or group. With damage to the *cerebellum*, postural or intentional tremors result as does disruption of goal-directed movement. These motor symptoms may be similar to *perseverative* behaviors, tics, obsessive thoughts or actions that must be repeated, as in *obsessive-compulsive disorder* or mania, all of which can also be called *hyperkinetic* behaviors typical of basal ganglia dysfunction. What is important to understand is that it had been previously thought that the functions of the *prefrontal* cortex and its role in cognitive or emotional activities were separate from other motor areas of the *frontal lobe* (Luria, 1966). Therefore, motor dysfunction of the *frontal lobe* did not necessarily relate to non-motor function or dysfunction of the *frontal lobe*. However, it can be alternatively viewed that goal-directed behaviors are merely evolutionary expansions of goal-directed movements, and that all of the activities of the *frontal lobe* are variations and refinements of the same function. Achieving a goal is provided by the stimulation of the *limbic system*. All functions of the *frontal lobe* are natural expansion and development that were selected to facilitate better survival.

The most basic function of goal-directed movements are no different from cognitive functions of approach and avoidance, or the development of communication or verbal skills. Being able to sequence motor patterns is just a more elementary form of sequencing used in calculation or higher mathematics. Therefore, these same pathways that control motor activities, arising from muscle receptors and projecting to the spinal cord, *cerebellum*, *thalamus*, frontal cortex, and downwards to the *basal ganglia* are part of the same functional system. These pathways expanded in humans as natural selection favored those individual members of the species that could achieve goal-directed movement. We see that expansion in the *neocerebellum*, the ventral lateral *thalamus*, the *neostriatum*, and *prefrontal* cortex were all involved not so much in the initiation or cessation of movement but rather in the cognitive control, planning, timing, and coordination of actions and concepts. Therefore, we would expect that any disruption in goal-directed movements would result in abnormal goal-directed behavior.

A recent study may confirm this relationship. In this study, the authors use the presence or absence of *paratonia* to predict frontal cognitive impairment (Beversdorf and Heilman, 1998). Paratonia is defined as an "alteration of tone to passive movement, and divided into *oppositional paratonia* (*gigenhalten*, or *paratonic rigidity*) and *facilitory paratonia*." The authors also note that although *paratonia* has been thought to be related to *frontal lobe* dysfunction, previous studies did not note a correlation between *paratonia* and cognitive impairment. To test this, Beversdorf and Heilman flexed and extended their patients' arms three times to assess *paratonia*. They scored the patients' performance 1 if there was no movement, to 4 if the patient flexed his or her arm completely or repeated the flexion-extension cycle. Defects of these simple movements reflect several kinds of *frontal lobe* dysfunction. The patients who cannot relax and allow their limbs to be passively moved show dysfunction of response inhibition. Being touched by the examiner represents environmental stimuli that the patients cannot avoid. The continued or repetitive movements or the inability of the patients to stop the movements represents *perseverative* activity. These activities were then correlated with tests more cognitive in nature. The investigators assessed the presence or absence of *echopraxia*, inhibition of eye movements, verbal fluency, and administered to the patients a mini-mental status examination.

Beversdorf and Heilman found that abilities on closely related activities correlated well in patients with diffuse *frontal lobe* damage. The level of *facilitatory paratonia* correlated most with *echopraxia* and the ability to inhibit eye movements. These findings reflect dysfunctioning inhibition of reflexive movements to external stimuli. *Facilitatory paratonia* also correlated well with semantic and phonemic verbal fluency, a measure of *frontal lobe* abilities in cross-temporal contingencies. *Paratonia* requires inhibiting a response in the presence of stimuli. Based on these studies we conclude that the general principles underlying motor control may also apply to cognitive operations of the highest order (Chatterjee, 1998b). In practical terms, we see that the amount of motor dysfunction seems to relate to the degree of cognitive or emotional dysfunction. It has been shown that by examining the motor functions of the *frontal lobe* the non-motor functions of the *frontal lobe* can likewise be assessed. We should also understand that motor and non-motor *frontal lobe* areas are systems connected to the *basal ganglia*, *thalamus*, and *cerebellum*. Therefore, any disruption of this pathway, dysfunction of these structures, or impairment in the function of their presynaptic pathways (*dorsal column*, *spinocerebellar*, peripheral nerve, muscle or joint receptor) would be expected to result in a combination of motor, cognitive, and behavioral impairment.

On the other hand, overlooked by most researchers and clinicians is that improvement in motor function should result in improvement in non-motor function of the *frontal lobe*. Therefore activities that improve muscle tone, coordination, timing, and the control of skeletal muscle, should also improve cognitive activities of the *frontal lobe* like verbal fluency, *echopraxia*, *perseverative* or obsessive activity, attention, inhibition of eye movements, and more. Damasio (1994) thinks the *prefrontal* cortex helps us navigate through complex social situations. These behaviors are guided by somatic markers, which are the perception and awareness of our own affective and autonomic reactions. He thinks that these somatic markers help guide us through social behavior. Somatic afferent input also helps create affective and autonomic reactions.

Therefore, adult patients with *frontal lobe* dysfunction may make poor personal decisions or exhibit social disinhibition. Children may be hyperactive, depressed, have poor attention, academic learning disabilities, language difficulties, and impulsive or impetuous behavior. The use of motor activities as a treatment intervention will improve motor function of the *frontal lobe*. We also have to consider that since the *cerebellum* and *thalamus* may both be important to the anatomical and functional development of the frontal cortex, if a child does not have normal or proper motor development, we would expect that the higher *frontal lobe* functions of cognition and behavior would be delayed in their development. Likewise, helping a child to develop their motor skills should also help develop their non-motor skills.

BRAIN ASYMMETRY

Of fundamental importance to the understanding of brain organization for cognition and behavior is knowledge of the nature of the interaction of the two cerebral hemispheres. The development and function of the cerebral hemispheres, their interaction and their balance of activity, as well as their interaction with subcortical structures is the key to understanding brain function and its dysfunction. Any attempt at diagnosis and treatment that does not take these interactions and differences into account may not only fail but also may increase the dysfunction.

Much of our present understanding of laterality of brain function initially came from split-brain research started approximately 50 years ago by Sperry and Bogen. However, the earliest concepts of laterality of brain function actually go back to ancient times (a historical review by Lauren Harris (1980) is worth reading in this regard). It is generally accepted that cerebral localization as a formal clinical concept started at the beginning of the 19th century with the work of Franz Joseph Gall. Gall became known for his system of phrenology. This system was based on three basic principles: (1) the brain is the organ of the mind; (2) the brain is a composite of parts, each serving a specific mental function; (3) the size of the different parts of the brain may be assessed by examining the bumps on the skull that relate to the relative strength of the different functions served (Spurzheim, 1818). Gall proposed that the brain was composed of 30 self-driven organs that together added up to the totality of human mind and personality. Gall was one of the first to predict that behavior and function of the brain had a relationship to its structure. This is thought to be the foundation of Neuropsychology. Gall also thought of regions of the brain as possessing unequal worth, meaning that some areas were of greater or lesser importance when viewing man in comparative physiological terms with other organisms. He thought that "lower organism behaviors" were located in the lower posterior lobe and *cerebellum*, for example. The higher functions like intellect, sentiments, verbal memory, calculation, and spatial recognition were localized in the *frontal lobe*. In the 1860s, a French neurologist Jean-Baptiste Bouillard accelerated the idea of cerebral localization and the concept of laterality of brain function. Bouillard thought that there was some basis for Gall's theories especially that verbal memory could be found in the *frontal lobe*. In a timespan of over 40 years, he collected and presented more than 100 cases of *frontal lobe* damage resulting in loss of speech. His work was still met with skepticism until 1861 when a neuroanatomist and anthropologist Paul Broca decided to test Bouillard's theory.

Broca is considered the founder of modern localization theory. Broca also thought strongly that the highest human intellectual functions would be localized in the *frontal lobes*. Broca's first patient had an expressive language loss, and after autopsy, a lesion was noted in what ultimately became known as Broca's area in the frontal as well as the parietal-*temporal lobes*. Broca thought that this patient's language loss due to the damage of his *frontal lobe*s was a form of memory loss of the movements needed to pronounce words (Broca, 1861a). This first case was an important step and was soon followed by a number of corroborating cases from various sources (Broca, 1861b). Broca's work combined with the social and scientific environment at the time turned the scientific community toward a localization model. This model took an even more dramatic turn when it was realized that most of Broca's cases of speech loss were not only localized to the *frontal lobe*, but actually to the left *frontal lobe*. This discovery of asymmetry for language would truly change the way neurologists and the scientific community viewed the function of the human brain.

Broca also thought that brain asymmetry was not inherited but was rather due to environmental influences of education and civilization. Eventually attention was drawn away solely from the left *frontal lobe* for speech and the concept of global asymmetry of function was examined. In 1972, Arthur Benton noted the importance of the right hemisphere (cf. Figs. 8.2 and 8.3) for visuo-spatial function. Sensory-motor asymmetries were noted as early as the 1870s. The Viennese physiologist Sigmund Exner noted that motor-tactile function was represented more in the left hemisphere, whereas sensory function was represented more in the right. In 1881 Jules Bernard Luys, a neuroanatomist who was known for his work on the *thalamus*, thought that the right hemisphere possessed an emotional center, that worked in a complementary fashion with the intellectual center thought to be centered in the left hemisphere. He observed that patients with left hemisphere damage presented with more apathy or passive symptoms, while patients with right hemisphere dysfunction presented with emotional volatility, paranoid, and manic-like symptoms. Charles Edward Brown-Sequard as well as Charcot in the late 1800s thought that the right hemisphere was more important in the nutritional needs of various body parts. It was

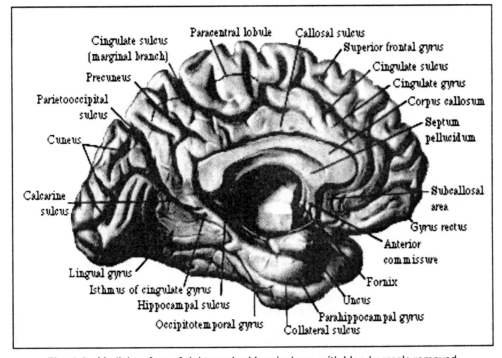

Fig. 8.2. Medial surface of right cerebral hemisphere, with blood vessels removed.

116 NEUROBEHAVIORAL DISORDERS OF CHILDHOOD

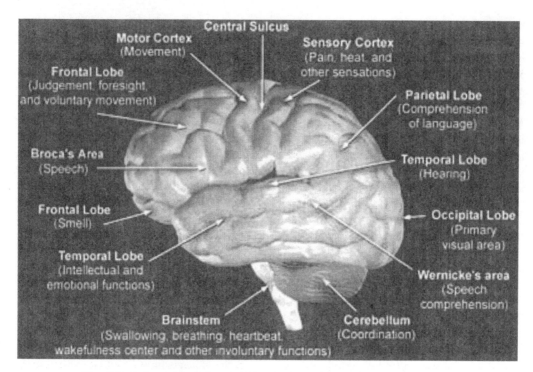

Fig. 8.3. Lobes of the left cortical hemisphere and their localized function.

also thought that the right brain harbored a more emotionally primitive or organismistic tendency that caused people to go mad. These early writers thought that madness revealed that all humans harbored a more organismistic nature.

At the beginning of the 20th century, the 19th century laterality and asymmetry research was thought to be obsolete. In the 1960s, a new generation of researchers mainly in the United States rediscovered brain asymmetry and laterality. This renewed interest was fueled mainly by split-brain research. This involved patients with severe epilepsy who underwent surgery severing their *corpus callosa* in the treatment of intractable epilepsy. The surgery was performed by neurosurgeons Joseph Bogan and Phillip Vogel. Roger Sperry and his students subsequently studied the patients. This research led to the modern concepts of the left hemisphere being the site of analytic and verbal cognitive function and the right brain the control site of more affective and perceptual functions. Today, 50 years later, a new interest in laterality and asymmetry has been fueled by modern technology for functional brain imaging. However, new studies are not only focused on the functional significance of the right-left axis, but also along the frontal-caudal and cortical-subcortical axes.

EVOLUTION AND DEVELOPMENT OF CEREBRAL LATERALIZATION: A FOCUS ON THE ADAPTATION FOR MOVEMENT

Lateralization is loosely defined as a deviation from complete bisymmetry in the specialized nervous systems of bisymmetric organisms. Bisymmetry is thought to be an evolutionary adaptation for movement. To move forward an organism needs to have a symmetric body: two fins, two arms or legs, and requires the control of a symmetric nervous system. It also helps if the brain or head, with specialized sensory adaptation to explore the environment (like olfaction), is located at the front. Forward movement then would allow the head to explore the environment and a symmetric body and nervous system would allow for directed movement based on instructions from the head or brain. Bisymmetry is primarily important to the locomotor system, bones, joints, muscles, and the part of the nervous system that controls their actions namely, the somatosensory and motor systems.

The relationship between the side of the body and neurological control is ipsilateral or same side in all species except for *chordata*, in which it becomes contralateral or crossed. It is thought that a deviation away from bisymmetry toward asymmetry of brain function is due to the development of higher mental functions. This appears to be true because the lateralization of specific higher mental functions was first seen in humans. Lateralization and hemispheric specialization in humans is thought to be an advanced evolutionary step in phylogenetic and antogenetic progression from simple and symmetric to complex and asymmetric nervous systems. Luria (1973) thought that the higher or more abstract a function, the more asymmetric is its cerebral control. It is further thought that hemispheric specialization originated because of bipedal locomotion that allowed the forelimbs to become specialized for complementary functions, (e.g., one for grasping and the other for manipulating) (Bruner, 1968; Lovejoy, 1981). The right hand became specialized for manipulating and the left for holding. Left hemisphere speech specialization is thought to have developed from this manual asymmetry, possibly because the left hemisphere is better for controlling sequences of rapid movements (Kimura and Archibald, 1974).

To understand the phylogenetic development of human lateralization we can look for similar differences in lower organism forms. Recently the relevance of the asymmetries of other organisms to humans has received more attention. In fish, amphibia, and reptiles the *habenula nuclei* in the dorsal diencephalons of these organisms have been shown to favor left side in some species and right in others (Braitenberg and Kamali, 1970; Galaburda et al., 1985). The functional significance of these findings is at present unknown. In male passerine songbirds, such as canaries, their brains have been found to be lateralized for control of song (Nottebohm, 1977). The left *hyostriatum ventrale* and *pars caudate* have primary control over singing even though the same structure in the right hemisphere is anatomically similar. When these pathways are present, a left *hypoglossal* nerve lesion disrupts singing automatically and permanently, whereas cutting the right does not (Nottebohm, 1977). This is an interesting finding because it is one of the first examples of a functional asymmetry without cytoarchitectonic differences. Interestingly, if the *hypoglossal nerve* is cut early in development, the right side takes over and compensates for the loss. This capacity continues during the period of motor learning and ends once the behavior is established. However, in parrots, sectioning the right or left *hypoglossal nerve* has little effect and vocalization returns to normal within a few days. Various asymmetries have been noted in domestic chicks. Visual discrimination learning could be disrupted by glutamate injection into the left but not right *striatal* hemisphere (Nottebohm et al., 1980; Rogers, 1980). Other studies seem to confirm a left-brain advantage for visual discrimination learning (Andrew et al., 1982). Some authors think that the asymmetries in chicks depend on the position of the chick embryo within the egg. They think position effects how much light each eye receives during the end of the incubation period (Rogers, 1982; Zappia and Rogers, 1983). The domestic chick also displays a trait known as imprinting which is a tendency to follow a moving object to which the chick is exposed soon after hatching. This is thought to be related to functional as well as structural brain asymmetry (Arnold and Bittjer, 1985). Researchers identified an asymmetric structural change that could be related to the length of time they were exposed to an imprinting stimulus (Bradley et al., 1981). It is thought that the right brain in chicks is more involved with initial imprinting and the left plays an inverse role the longer the stimulus is exposed (Cipolla-Neto et al., 1982). This type of shift of control with increased familiarity or practice is similar to previous theories of the formulation of differential hemisphere control in humans.

The right hemisphere appears to be more important for initial responses to novel situation and the left hemisphere for subsequent response as behavior develops (Goldberg and Costa, 1981).

In humans, it has been thought that gonadal steroids influence the asymmetric pattern of the cerebral cortex. Very recent evidence supports this view where developing bodies can in fact determine right from left. Levin and colleagues (2002) report that a completely novel mechanism exists that functions in left-right asymmetry. This allows the embryo to figure out which side is left and right. Studying frog embryos, Levin and associates found that an ion pump begins operating earlier than previously expected. The researchers observed a voltage gradient at the four-cell stage of *Xenopus* within approximately one and a half hours after fertilization. It had been previously thought that asymmetry begins at *gastrulation*; the developmental stage in which the embryo begins to form layers two cells thick. This radically revises the current knowledge of when the embryo knows left from right.

There is some evidence in rats as well that gonadal steroids may affect the development of male versus female characteristics. Studies have shown that in male Long-Evans rats, the right cerebral cortex is significantly thicker than the left in most regions. The cortex of female rats is more symmetric, with a tendency of the left being thicker than right (Diamond et al., 1979). It has been shown that sex steroid hormones play some role in determining asymmetry in the cortex. This was demonstrated in studies where the gonads were removed (Diamond, 1984). Differences also exist in the left–right, male–female distribution of estrogen receptors in rat cortex.

In rats, the cerebral cortex is sexually dimorphic. In male Long-Evans rats, the right cerebral cortex is thicker than the left (Diamond et al., 1979, 1980; Diamond, 1979, 1983, 1987, 1988). This asymmetry is apparently present at birth (Diamond, 1983) and continues throughout the rat's lifetime, with differences disappearing in very old age (Diamond et al., 1979, 1983). The Long-Evans female rats have more symmetric hemispheres, with the left being greater that the right (Diamond et al., 1979, 1980, 1983; Diamond, 1983, 1985a, 1985b, 1987, 1988).

Differences in cortical thickness and volume reflect difference in cell size or number or both (McShane et al., 1988). In female mice, focal cortical lesions on the right side have different immunologic effects than left side lesions (Renoux et al., 1983; Barneoud et al., 1987). Asymmetries in male rats are seen in cortical *aminopeptidase* activity (Alba et al., 1985) and in *serotonergic* responses to focal cortical lesions (Mayberg et al., 1990). Males *gonadectomized* at birth lose their usual right-greater-than-left asymmetry and are left-greater-than-right asymmetric in some cortical regions. In addition, females *ovariectomized* at birth show an opposite shift in cortical asymmetry to the right (Diamond et al., 1979). Interestingly, exposing females to stressful conditions during pregnancy lowers fetal serum *testosterone* in male offspring (Ward and Weisig, 1980) and is not seen in females (Ward and Weisig, 1980). Prenatal maternal stress causes male offspring to shift from right-greater-than-left to left-greater-than-right in somatosensory cortical asymmetry (Anderson et al., 1986). It appears that the critical period for establishing asymmetry begins prior to delivery and continues into early postnatal life.

Interventions that alter cortical asymmetry alter gonadal steroid levels also. Prenatal stress lowers fetal serum *testosterone* (Ward and Weisig, 1980), but not in females (Ward and Weisig, 1984). Like prenatal stress, early *gonadectomy* reduces testosterone levels in male pups and has similar effects on male cerebral asymmetry (Diamond, 1985a, 1985b), hypo*thalamic* structure (Jacobson et al., 1981), and behavior (Corbier et al., 1983; Heinsbroek et al., 1987). *Naltrexone* prevents the effect of prenatal maternal stress on male behavior (Ward et al., 1986) and postnatal *naltrexone* increases brain weight and somatosensory cortical thickness in rats (Zagon and McLaughlin 1983a, 1983b). In *naltrexone* treated organisms enlarged brain size is accompanied by an increase in the number of neurons and *glia*, particularly those arising postnatally (Zagon and McLaughlin, 1983a, 1983b).

Behavior

There are significant findings of behavioral asymmetry in the rat. Ablation of the right hemisphere has been shown to cause greater change in activity level, however, this was only seen in organisms that had been handled during the first 20 days of life. The direction of the change in activity level, up or down, depended on whether the organism had been reared in a standard cage or in an enriched environment (Denenberg et al., 1978). Studies have also shown that in rats, representation of fear is more a right hemisphere function than left, but expression of fear may be inhibited by an intact left hemisphere (Denenberg et al., 1980; Ledoux, 1987). It also appeared, in split brain rats, that cutting the corpus callosum blocked the inhibiting influence of the left hemisphere (Denenberg et al., 1986).

Researchers have speculated that hemisphere-specific lesions on open field activity in rats results from aberrations of both spatial behavior as well as emotionality. Lateralized processing of spatial information could explain why the leftward turning tendency in non-handled male rats with intact right hemispheres exceeds the rightward turning tendency of rats with intact left hemispheres (Sherman et al., 1980). Rats as well as other mammals show spontaneous circling under certain circumstances (Glick and Shapiro, 1985) with some moving leftward and others to the right. A few neural systems exist whose stimulation can result in circling (Buckenham and Yeomans, 1993). However, the most aplastic area is the *nigro-striatal* system consisting of the *substantia nigra, nigrostriatal bundle*, and the *corpus striatum*. Unilateral lesions of this system cause rats to circle at a high rate in a direction ipsilateral to the lesion (Ungerstedt, 1971). Amphetamine and other *dopamine*rgic drugs reverse this surgically-induced lateralized behavior (Ungerstedt and Arbuthnott, 1970). This is thought to suggest that an imbalance in *dopamine* activity in the left and right *nigrostriatal* systems causes the organism to turn in a direction opposite to the more activated side (Arbuthnott and Crow, 1971).

Both D1 and D2 *dopamine* receptors appear to contribute to the turning activity (Pazo et al., 1993). It is thought that since systemic amphetamine induces circling in rats without lesions (Jerussi and Glick, 1974) spontaneous circling in normal rats can be attributed to an endogenous asymmetry in *nigrostriatal dopamine* levels (Glick et al., 1974; Jerussi and Glick, 1976). There is a higher concentration of *dopamine* in the striatum contralateral to the rats preferred direction of rotation (Zimmerberg et al., 1974) and the direction in which the same organism rotates spontaneously (Glick and Cox, 1978). However, research that is more recent shows that asymmetric *dopamine* uptake can result either in contralateral or ipsilateral turning (Shapiro et al., 1986). This seems to suggest that, "the net influence on *dopamine*rgic input could be either excitatory or inhibitory in different rats" (Glick and Shapiro, 1985). Based on the results of other studies, researchers have concluded that systemic injection of *dextroamphetamine* produces perseveration rather than asymmetric circling tendency. Therefore, the *amphetamine* induced circling may reflect perseverative running rather than the potentiation of preexisting *dopamine*rgic motor asymmetry. Right hemisphere lesions produced by the ligation of the middle cerebral artery (Robinson et al., 1975), suctioning cortical tissue (Pearlson and Robinson, 1981; Moran et al., 1984), or cutting transcortical connections (Kubos and Robinson, 1984) results in as much as a 50 percent increase in running wheel activity. Comparable lesions in the left hemisphere do not increase activity level. Asymmetric lesion effects on *catecholamine* concentration in the brain accompany this asymmetric behavior. Right hemisphere lesions cause a bilateral depletion of *dopamine* and *norepinephrine* (Robinson, 1979) as well as a bilateral increase in *serotonin* S2 receptor binding (Mayberg et al., 1990). These changes are not noticed with corresponding lesions of the left hemisphere. The more anterior the lesions are placed, the greater the effect on activity level and *norepinephrine* concentration (Pearlson et al., 1984).

Studies in rats confirm a right hemisphere role in regulating activity level (Robinson and Bloom, 1977; Robinson and Stitt, 1981). When researchers experimentally induce *norepineperine* depletion in the right hemisphere the rat becomes hyperactive. The same procedure applied to the left hemisphere does not produce the same result. However, lesions of the right hemisphere that do not deplete *catecholamines* may also produce hyperactivity (Kubos et al., 1982; Kubos and Robinson, 1985). Robinson (1985) thought that depletion of right hemisphere *norepinephrine* is sufficient but not necessary to elicit hyperactivity.

Lateralization of auditory processing has been noted in rodents. It has been reported that mice have a right ear advantage for artificial pup calls (Ehret, 1987). It has been further noted that pup calls are preferred over neutral stimuli only in binaural and left-ear plugged conditions. This is thought to suggest a left hemisphere advantage for recognizing pup calls. In subsequent studies it was concluded that left hemisphere superiority is manifested only when the auditory stimuli has intrahemispheric communicative significance. This is similar to a left hemisphere advantage for words that is seen in human brains.

Non-Human Primates

Obvious anatomic differences exist between the brains of human and non-human primates in the type rather than in the degree of brain asymmetry. The *planum temporale* is not fully evident in a non-human primate. In chimpanzee and orangutan brains, the *gyrus of Heschl* is apparent enough to suggest a posterior *planum temporale*.

The orangutan and the papions show frontal asymmetries that favor the left side not seen in human brains, where more commonly the right *frontal lobe* is more prominent. However, baboons show frontal asymmetry that favors the right (LeMay and Geschwind, 1975; Falk, 1978; Cheverad et al., 1990; Falk et al., 1990).

Beyond structural asymmetries, non-human primates show some behavioral asymmetries although the interpretation of the data is difficult. In regard to handedness, the data suggests a left hand preference for reaching, the strength of which increases from simple to more complex tasks, and a right hand preference for object manipulation (MacNeilage et al., 1987). Studies of old world monkeys seem to also show a left hand preference for picking up their babies in an emergency situation (Hatta and Koike, 1991), and that monkeys usually use the left hand when reaching for food under circumstances that require postural adjustments (Ward, 1991). Recent studies indicate Rhesus monkeys and chimpanzees may exhibit a right hand preference for tasks that require fine motor skill (Hopkins et al., 1989; Morris et al., 1993). These results may be interpreted that the left hand and right hemisphere are more adapted for lower frequency–higher stability activities whereas the right hand and left hemisphere are more adapted to higher frequency–more coordinated activities. This may be an adaptation to arboreal life where monkeys are seen to stabilize or hang more with the left hand while the right is used more for targeting or grasping.

Studies of visual processing in split-brain monkeys have mixed findings demonstrating that in general, performance between left and right hemispheres appear equal (Downer, 1962; Hamilton and Gazzaniga, 1964; Butters and Rosvold, 1968). Split-brain rhesus monkeys have been shown to have left hemisphere superiority for discriminatory line orientations (Hamilton, 1983; Hamilton and Vermeire, 1988). This finding is different from the right hemisphere superiority for line orientation normally seen in humans (Benton et al., 1978). Other studies of spatial discrimination (Hamilton and Lund, 1970; Hamilton et al., 1974; Hamilton, 1983; Jason et al., 1984) show a left superiority, which is also opposite to findings in humans. Baboons show human-like performance on mental rotation tasks with minor image stimuli, but only when the rotated stimuli are presented to the right visual field represented in the left hemisphere (Vauclain et al., 1993). As for facial processing, humans show a right hemisphere advantage focusing on the right half of a picture of a face. Split-brain monkeys show a similar preference, and are more adept than humans at distinguishing individual organisms and facial expressions with their right than left hemisphere (Hamilton and Vermeire, 1983, 1988). Additional evidence of human-like hemisphere specialization in non-human primates derives from studies on auditory processing in monkeys. One example is a study that reports that lesions in the left superior temporal *gyrus* disrupt auditory-visual matching; the same lesions on the right have no such effect (Dawson, 1972).

The results in dichotic listening tasks in primates are similar to the findings for mice but are also similar to results found in humans. Right ear advantage seems to be most related to sounds that have communicative significance to the listener. In addition, vocalizations are disrupted by removal of the left and not the right superior *temporal gyrus*. This is similar to what would be expected in humans (Kimura, 1961a). With regard to attention in monkeys, there appears to be an attentional bias favoring the hemisphere that controls the dominant hand (Kinsbourne, 1970) that is also similar to that found in humans.

Ontogeny of Human Lateralization

To understand fully the function of the adult human brain we need to examine not only its phylogenetic but also its embryologic and childhood development. There are specific lateralized functions that have ontogenetic significance. With regard to language development, excluding left handers who sometimes have reversed laterality of language most recent studies show that only about 5 percent of childhood aphasias are associated with right-sided damage (Woods and Teuber, 1978). It has been further estimated that 84 percent of right-handed children below the age of 5 years have left hemisphere speech and language and the other 16 percent have bilateral speech and language. However, the 16 percent estimate for bilateral speech does not differ significantly from zero, the null hypothesis that no child has bilateral speech could not be rejected (Carter et al., 1982). Language development depends on laterality of function even though in children the right hemisphere can compensate for damage to the left hemisphere. However, based on several studies (Dennis and Kohn, 1975; Dennis and Whiteher, 1976; Dennis, 1980; Dennis et al., 1981) it has been argued that the intact and isolated right hemisphere in children cannot support as complete a development of language as can the isolated left hemisphere. Further studies have shown that children with early left hemisphere damage are specifically impaired on syntactic production (Aram et al., 1986). Children with right-sided lesions have spatial impairments (Vargha-Khaden et al., 1985). With regard to non-verbal function, it has been shown that left-lateralized lesions result in more mathematical difficulties than do right-sided lesions (Ashcraft et al., 1992) as well as temperamental problems, involving mood and rhythmicity all being noted after right and not left *hemispherectomy* (Nass and Koch, 1987). These findings all argue for early lateralization of hemispheric specialization.

Neuroanatomic asymmetries have been noted in infants (Tesiner et al., 1972; Witelson and Pallie, 1973; Wada et al., 1975; Chi et al., 1977; Weinberger et al., 1982) that seem similar to those seen in adults (Geschwind and Levitsky, 1968; Lemay and Calebras, 1972; Wada et al., 1975). Asymmetries in normal children's brains have been studied and provide even greater support for early lateralization of language.

Asymmetric postures and orienting suggest that the left hemisphere is more dominant in infancy. It has been noted that infants consistently orient or turn toward the right side suggesting that the left side is more activated (Kinsbourne, 1972, 1974a, 1974b). The asymmetric tonic neck reflex, in which the head is turned sideward with ipsilateral limbs extended and contralateral limbs flexed can be observed during the first 4 weeks of life. Most infants turn to the right most of the time (Gesell, 1938; Gesell and Ames, 1950). This rightward bias predisposes the infant to coordinate their gaze with pointing or reaching with the right hand when these behaviors emerge and, generally to orient to the right side of space. Both spontaneously and in response to external stimuli, infants appear to favor rightward over leftward turning (Turkewitz et al., 1965; Siqueland and Lipsitt, 1966; Turkewitz et al., 1967; Coryell and Michel, 1978; Liederman and Kinsbourne, 1980; Michel, 1981; Harris and Fitzgerald, 1983). These turning tendencies appear to reflect parental handedness (Liederman and Kinsbourne, 1980) and predict the child's eventual hand preference (Coryell and Michel, 1978; Viviani et al., 1978; Goodwin and Michel, 1981; Coryell, 1985). Infant orienting behavior has been linked to speech perception (Lempert and Kinsbourne, 1985) which appears to involve the left hemisphere in both.

It has been shown that 6-month-old infants could detect the correspondence between the acoustic and visual components of an adult's speech articulation, but only when they were looking to the right (MacKay-Sim et al., 1983). This is thought to suggest a tendency of the left hemisphere to favor orienting to the right side. There are also speech specific phasic activities of the left hemisphere that potentiate rightward turning within a linguistic context.

Manual asymmetries can be demonstrated at a very early age. Infants as young as 17 days can grasp objects for a longer time with the right hand than with the left (Caplan and Kinsbourne, 1976; Petrie and Peters, 1980; Hawn and Harris, 1983) and the right arm tends to be more active that the left during the first 3 months of life (Coryell and Henderson, 1979; VonHofsten, 1982; Liederman, 1983). During the first 4 months of life the right hand is preferred for "directed, target-related" acts (Young et al., 1983). There are those that think that there is a temporal linkage between preference and language milestones, periods of slowing in development cause manual asymmetry to coincide with transitions between stages of language development (Ramsay, 1985). These "cycles in the development of handedness" have been attributed to fluctuations in the degree to which speech interferes with the use of the dominant hand (Bates et al., 1986). Such interference is thought to be due to the proximity of these two control processes in *functional cerebral space* (Kinsbourne and Hicks, 1978) within the same hemisphere. Another study concluded that speech interferes most with right hand activity while children are mastering a new problem in language development (Bates et al., 1986). It is interesting to consider the possibility that if there is interference in movement during the course of manual development, that this may be expected to impact on language development as well.

Perceptual asymmetries have been studied in infants. Consonant shifts were more likely to be detected when they occurred at the right ear, but shifts from one musical sound to another were more likely to be detected at the left ear. This study concluded that hemispheric asymmetry is manifest at 3 weeks of age (Wada and Davis, 1977). It was also found that newborns turn more often to the right than left when they hear speech sounds (Young and Gagnon, 1990). Perceptual asymmetries have been noted even in premature infants. One group was repeatedly exposed to speech while the other was repeatedly exposed to music. The neonates exposed to speech showed a disproportionate reduction of right-sided movement, whereas those exposed to music showed the opposite asymmetry (Segalowitz and Chapman, 1980).

Numerous findings in infants strongly suggest that the human brain is functionally asymmetric long before language and other higher cognitive skills have developed. It has been proposed that these asymmetries are not indicative of the lateralized neural processors that manifest later in development, but instead indicate precursors of these lateralized processors, especially given the immaturity of the newborns' cerebrum (Dobbing and Sands, 1973). These precursors are thought to be at the level of the *basal ganglia* and *thalamus*, structures that are now known to share in the complementary specialization that is familiar in the case of cerebral hemispheres. The concept of precursors, which is an important explanation for early behavior (Kinsbourne and Hiscock, 1977), allows us to appreciate that lateralization of behavior or the lateralization of certain early developing components of a behavior may preclude the expression of the behavior itself. Therefore, development of the cortical areas that direct behaviors is dependent on the proper development of subcortical structures. If these precursory structures do not develop appropriately, then the subsequent cortical areas would resultantly not develop appropriately. In addition, the behavior for which these inadequately developed cortical areas are responsible will likewise not develop adequately.

The asymmetries that are seen at the neonatal level are to some degree regarded as precursory to a more mature form of hemispheric specialization for perpetual, motor, and cognitive functions (Kinsbourne and Hiscock, 1977). As lateralized precursors develop and the degree of asymmetry increases (Moscovitch, 1977) these precursors evolve to become the controllers of increasingly mature behaviors. For example, it has been suggested that development of the lateralization of linguistic processing depends on the level of processing required (Porter and Berlin, 1975). It is thought that the lower level acoustic and phonological processes are fully lateralized early in development, whereas the semantic processes would come later. Therefore, development of the more mature function can be delayed if the precursor areas do not develop properly. There are numerous studies that show other kinds of asymmetry in normal children including visual perception, tactile (haptic) perception, and interference between two tasks performed concurrently.

Anatomic studies show that increases in absolute brain weight occur almost entirely before birth and in the first 2 years of life. Over 90 percent of human brain growth is completed by the age of 6 years (Coppoletta and Wolbach, 1933; Blinkov and Glezer, 1968). However, head circumference continues to increase beyond that age. *Planum temporale* asymmetry is already present in the fetal brain (Wada et al., 1975; Chi et al., 1977). Increased size of right anterior and left posterior brain is also seen in the fetus (Lemay, 1984). Therefore, we see that precursors of adult asymmetries are present at a very early stage of development.

What is the adaptive significance of asymmetric lateralization of the brain? Corballis (1983) indicates that bilateral symmetry is itself an evolutionary adaptation and that, "for bilateral symmetry to have emerged in so precise and comprehensive a fashion there must have been adaptive advantages associated with symmetry sufficiently strong to overcome a natural predisposition to asymmetry." One advantage may have come from the neural network characteristic of the forebrain and from the hemispheric representation of contralateral turning. Activation of each vertebrate half-brain occasions contralateral turning (Kinsbourne, 1974b). The processing of information relative to a target is best accomplished in the hemisphere opposite its location because that is where processing and orienting functions are congruent. Processing information ipsilateral to the target would incur cross-talk interference between hemispheres (Kinsbourne, 1970). It is thought that mental operations not targeted to specific points in space (e.g., language, emotions, problem solving) do not need to be bilaterally represented and therefore the relaxation of the need for bisymmetry may be a sufficient condition for lateralization to evolve (Kinsbourne, 1978).

Specialization can also occur in orientations other than along a lateral left–right plane. There is also evidence for a dorsal–ventral specialization of the brain. It has been postulated that there may be a phylogenetic shift from dorsal–ventral to right–left complimentarity in a hominid ancestor (Kinsbourne and Duffy, 1990).

In summary, we can see that structural and functional asymmetries are seen in non-human vertebrates. Therefore, we cannot say that brain lateralization is an exclusively human characteristic. It appears that natural selection favors bisymmetry in motile organisms. Only when selection pressures are relaxed do asymmetries appear in species. Human brain asymmetries exist much earlier in development than previously thought. The fact that asymmetries appear so early implies that a subcortical mechanism is the foundation of infantile lateralization. The more commonly recognized cortical asymmetries might therefore be an expression of the corresponding subcortical asymmetries.

Some well-reorganized anatomic asymmetries are commonly found in the human brain. However, the question exists, do these anatomic asymmetries have any functional significance? Initially research into lateralization started with the belief that researchers would discover fundamental differences in the ways the left and right hemispheres are organized structurally and that these differences may lead to a better understanding of functional differences of the hemispheres. However, for the most part, researchers have only found side differences in the amount of brain substrate devoted to a particular architectonic area or a particular gross anatomical landmark. That is to say, that even though the left and right hemispheres have very different functions there does not appear to be any significant structural or chemical constituent that is present in one hemisphere and not the other. All architectonic areas are present on both sides. There are no cell types found in one hemisphere but not the other. There is no known pattern of connections that appear to be specific to the dominant hemisphere. There has not been found any physiologic properties in one neuron of one hemisphere that are not present in the other. The only differences therefore, are quantitative or in the size of certain areas on one side in comparison to the other. This is not to say quantitative differences in and of themselves may or may not be important. However, it is also likely that quantitative differences lead to qualitative differences by permitting the arrival at thresholds and emergent properties (Galaburda, 1994).

It is also possible that qualitative or functional differences lead to quantitative changes in the size of an area on one side versus the other. There are three issues important in any discussion of brain asymmetry. It is important to note whether a brain area is larger on the left or the right. Some areas would be expected to be larger and functionally dominant on one side or the other based on the function. For instance, in humans, larger areas are normally found in the left hemisphere for most linguistic capacities and in the right hemisphere for most visuo-spatial functions. Theoretically, the side in which the brain area is larger would be expected to correspond to the hemisphere in which a particular function is dominant. It is possible, therefore, to assume that the side with the larger area could be dominant because most of the circuitry related to a particular function is located in that hemisphere (Leisman and Ashkenazi, 1980). It is also possible that quantitative cortical differences could be a reflection of different sizes and activity in subcortical precursor areas such as *cerebellum, basal ganglia*, and *thalamus*. In this case the increased size of cortical areas could be the result of increased plasticity due to increased frequency of firing from asymmetric subcortical areas. We would also expect that if there was damage in a hemisphere, such damage would injure the larger portion of the bilateral structure leading to a more substantial deficit. For instance, after examining 100 autopsied human brains, it was found that 65 percent of the sample showed a larger left *planum temporale*, the right side was larger by 11 percent, and equal in 24 percent of cases (Geschwind and Levitsky, 1968). Their results have been replicated in subsequent studies (Witelson and Pallie, 1973; Wada et al., 1975; Campain and Minkler, 1976).

The *planum temporale*, which contains several auditory association cortices, is thought to be an important part of the language network of the left hemisphere. Lesions affecting significant portions of the *planum temporale* usually on the left often lead to *Wernicke's aphasia*. *Heschl's gyrus*, which represents the anterior portion of the *planum temporale*, and is thought to contain the primary auditory cortex, has been reported to be asymmetric also, with the left being more oblique than the right (Galaburda, 1994). Gross anatomic asymmetries are present in the cerebral cortex before birth, even though cerebral dominance can be modified after birth. The anatomic asymmetries that underlie functional lateralization are fixed at least in their gross designs before birth. However, this does not mean that functional expression of the asymmetries could not be changed by later modulation of detailed connectivity, synaptic architecture, and other lower-level structural changes that do not change the gross appearance (Galaburda, 1994). This is more consistent with researchers who think of a functional brain organization in which the beginning state of asymmetry in the nervous system can be modified by learning (modification of synaptic weights) so that either more or less functional asymmetry results from that which would have been predicted from the original, innate structural pattern (Kosslyn, 1987; Kosslyn et al., 1992; Koch and Leisman, 1996).

Studies on the effects of learning on anatomy (Greenough et al., 1987) suggest that changes take place at lower levels of structure and would not be reflected in the gross anatomy or even the cytoarchitecture and gross connectivity. Therefore, gross inspection of anatomy may reveal very little about the actual function of that area. An area may appear grossly intact and be severely functionally impaired (Koch and Leisman, 2001; Leisman, 2002; Leisman and Melillo, 2004). Other anatomical differences have been noted in the *frontal lobe*. Cytoarchitectonics refers to the regional cortical differentiation, which consists of fluctuations in thickness (and sometimes number) of layers, sizes, and shapes of neurons in the layers, the packing density of these neurons, and local arrangements in clusters and columns. The frontal *operculum* contains mostly the cytoarchitectonic areas 44 and 45 of Brodmann. Both of these regions are part of *Broca's area*. Both of these cortices belong to the motor association cortex, with area 44 being an inferior *premotor* cortex and area 45, an inferior *prefrontal* cortex. The *prefrontal* cortex is considered more multi-nodal (cognitive) than the *premotor* cortex, although lesions that result in *Broca's aphasia* affect area 44 alone more often than area 45 alone. Area 44 was found to be of greater volume in six out of ten brains, symmetric in three, and greater volume on the right in one (Galaburda, 1980).

Studies have also been performed on the inferior *parietal lobe* in ten normal brains and it was found to be larger in the left area (Brodmann's area *39*) in 8 out of 10 cases. Brodmann's area *39* lies mainly on the *angular gyrus* and is considered prototypical high order association cortex. It is highly multi-nodal and found between cortices dealing with somesthetic, auditory, and visual functions. Lesions of the *angular gyrus*, which appear to cause anatomic aphasia, acquired reading and writing disorders, and *Gerstmann's syndrome* are likely to impair function in this area P.G. (or area *39*).

The right *frontal lobe* has been noted quite frequently to protrude ahead of the left. This is known as *right frontal petalia*. However, the left occipital lobe tends to protrude further toward the back than the right. This is known as *left occipital petalia*. This *left occipital petalia* is seen in 78 percent of right-handers whereas left-handers are more symmetric. As a group, left-handers with a family history of left-handedness are particularly more symmetric than left-handers without a family history. This suggests an influence of environment, either through injury or through learning, to the expression of left-handedness in the latter group. Studies that concentrate only on *petalia* have been found to be difficult to interpret. This may relate to the fact that these gross anatomic asymmetries may reflect sub-regions of the brain that are asymmetric in opposite directions and that cannot be predicted by gross anatomic asymmetry alone. For example, the right occipital pole may protrude because of growth of visual association cortices, while the left may push out because of growth of temporal-parietal cortices just anterior to it (Galaburda, 1994).

An important question to consider is to what degree size of region relates to the brain function. The phrenological approach to neurology implies that the basis for cognitive capacity is large size. An alternative to the phrenological concept is that although size may be an important factor in some cognitive functions (e.g., memory) other highly specialized functions may depend on the specificity of organization (e.g., rule learning) (Galaburda, 1994; Koch and Leisman, 1996, 2001). In this case, building a phonologically competent brain area may require the specificity of the brain substrate by customized reduction rather than by enlargement. In this sense, an asymmetric language area would result from the pruning of one side instead of increasing the size of the other. This is the principle of *synaptic stabilization*. It is known that the cerebral cortex during development creates more neurons, axons, synapses, and receptors than it eventually keeps. Some of those structures are removed by what appears to be a process of pruning, while others are removed because of environmental stimulation (e.g., axons' input to neurons, electrical or chemical stimulation). It is therefore thought that the intended purpose of overproduction and pruning is to create an optimal anatomy to match environmental requirements, which may be unique and cannot be predicted in advance.

It is thought that hemispheric differences in neuron numbers account for the size difference between areas. It is important to understand how the number of neurons is determined in cortical development. In the human brain between the 16th and 24th weeks of gestation most neurons destined for the cerebral cortex are produced from neuroblasts (mother cells) residing in germinal zones. Before that, it is the number of mother cells that is determined. It is known that after neuronal production and migration, some of the neurons are eliminated (*synaptic stabilization*) but it is not known whether mother cells are also overproduced and eliminated. So it appears that neuronal numbers in the cerebral cortex depend on the early production and possible eventual elimination of mother cells and the production and subsequent elimination of neurons during development. Neuronal numbers in a particular architectonic area may also rely on assignment of neurons to it rather than to an adjacent architectonic area, by shifting borders between the two areas. In this scenario, while one area loses neurons and becomes smaller, its adjacent area would gain neurons and becomes larger. These principles are thought to play a role in determining side differences in the number of neurons, which leads to structural brain asymmetry.

It is possible that at the time when the midline of the neuraxis is established, in some cases this line is not exactly in the middle, leading to asymmetry in the distribution or assignment of neuroblasts to either side of the line. If we assume that the mother cells then divide the same number of times on both sides, some cases will end up symmetric and others asymmetric. This scenario however, does not explain the fact that the asymmetric areas end up with fewer neurons. There are a number of possible explanations. After the establishment of the midline, neuroblasts may divide the same or a different number of times on either side to produce a final pool of neuroblasts. After the final pool of the mother cells is established bilaterally, neuroblasts divide either symmetrically or asymmetrically to produce either symmetric or asymmetric number of young neurons. A similar hypothesis can be applied to the elimination of young neurons after they have migrated to their destined location. Either there exists a symmetric elimination after symmetric or asymmetric production or a symmetric reduction. Asymmetries in the birth of neurons may reflect side differences in the rates at which other cells could generate young neurons or in the length of time during which they can divide. In addition, asymmetries in the death of neurons may reflect side differences in the rates or in the length of time in which young neurons die (Galaburda, 1994). At this point studies seem to indicate that the generation of asymmetry is associated with the asymmetric production of neurons. This in itself reflects the asymmetric generation or death of neuroblasts early in histogenesis, before neuronal migration and the establishment of cortical connectivity (Levin et al., 2002). The fact that this occurs so easily is thought to be initially consistent with a genetically controlled process.

The last issue to consider in anatomic asymmetry and its effect on function is to examine inter- versus the intra-hemispheric connections. This has been studied in rats with severed *corpus callosa*, and degenerating axons (Rosen et al., 1989). It appears that the more asymmetric areas have relatively fewer colossal projections and a larger portion of colossal connections reach under the cortical plate but only some of these establish connectivity with neurons in the cortex (Innocenti and Frost, 1980; Innocenti, 1981; Ivy and Killackey, 1981, 1982). Some of the neurons that do not penetrate are withdrawn and disappear, whereas others project to other areas. It appears that in symmetric areas, fewer neurons are withdrawn while in asymmetric areas more are withdrawn. In the asymmetric scenario, it is not understood whether more neurons are rerouted or just disappear. If rerouting occurs in asymmetric areas, we could infer that inter-region connectivities would be more effective within the same hemisphere rather than between hemispheres. Therefore, asymmetric areas would be more connected intra-hemispherically whereas symmetric areas would be better inter-hemispherically connected, resulting in important behavioral, clinical, and cognitive implications.

Motor Asymmetries

In this section, we wish to look at the relationship between specific asymmetric motor activity and its relationship to other specialized brain functions. Handedness is probably the most easily observed expression of cerebral lateralization. As we stated most motor functions are bisymmetrical to improve motility. Handedness, however, is the only motor activity that is unilateral; therefore, it is of particular interest. There is still much discrepancy as to the classification of left-handed and right-handed people. There are many ways to classify handedness including writing and throwing preference, etc. Quite often, these preferences are mixed. The easiest way of defining handedness is a preference of one hand over the other if a choice is possible. There are also significant cultural variables in right-handedness. For instance, it is thought that prevalence of left-handed writers is as low as 1 percent in countries like Korea (Kang and Harris, 1993) to 12 percent in North America (Gilbert and Wisocki, 1992). In the United States, population estimates of people writing with the left hand are reported as 10.5 percent for females and 13 percent for males.

Differences also exist in the styles of left-handed writers. Left-handers who write with an inverted hand position are theorized to have ipsilateral control of their hand as opposed to left-handers who write much like right-handers. This theory does not meet with great acceptance (Levy and Reid, 1976; Weber and Bradshaw, 1981). However, in one study it has been demonstrated that a relationship exists between hand movement and visual half-field in inverted writers and it is thought that this is influenced by their habit of crossing hands during writing (Guiard and Millerat, 1984). This is important because it implies that the way in which the visual half-fields relate to fine motor control is not because of underlying neurological differences but because of differences in hand movement relevant to a specific visual field. It demonstrates that hand motor activity influences neurologic organization. There are many factors considered in left-handedness including: familial versus non-familial, hormonal, and factors associated with classification confusion such as a reported incidence of approximately 12 percent left-handed writers in the U.S. population and 17 percent of the U.S. population who do a number of important activities with the left hand. Reports indicate that 30 percent of all left-handed individuals throw a ball with their right hands and write with their left (Gilbert and Wisocki, 1992). This may explain why some of the results reported in the literature show that left-handers as a group show greater ambidextrousness than do right-handed individuals. When hand strength is compared for left and right hands for left-handers, no significant group differences are found. Right-handers, however, show significant right hand superiority in strength. However, left-handers who throw with the left and write with the left have a stronger left hand, whereas left-handers who throw with the right and write with the left have a stronger right hand (Gilbert and Wisocki, 1992).

Next to hand preference, foot preference is the next most common behavioral asymmetry.

There appears to be a relationship between handedness and footedness. Right-handers as a group show a marked preference for the right foot in kicking (over 90 percent of right-handed writers). Left-handers as a group do not seem to prefer the left foot for kicking, whereas 78 percent of left-handers who prefer the right arm for throwing preferred the right foot for kicking. Studies have also been performed that explore the relationship between eyedness and handedness (Porac and Coren, 1981) and ear dominance and handedness. It appears that right-sided eye and hand preference is common. Apparently ear and hand association is more difficult to calculate because the majority of left-handers have left hemispheric language and speech dominance.

The least studied asymmetry is the tendency of pyramidal tract fibers from the left hemisphere to cross over at a higher level than the fibers from the right side (Yakoviev and Rakic, 1966). The uncrossed anterior *corticospinal* tract is larger on the right side in the majority of cases (Nathan et al., 1990). These studies suggest that the dorsal *corticospinal* tract favors the left hemisphere while the anterior or ventral *corticospinal* tract appears to favor the right. The dorsal *corticospinal* tract controls the musculature of the fingers and is associated with skilled manual movements. The ventral *corticospinal* tract controls more the proximal portion of limbs and the trunk. It appears from the point of view of structure that there exists an asymmetry favoring the left hemisphere for fine manual movement and an asymmetry favoring the right hemisphere for limb and trunk movements.

Regarding the *cerebellum* and handedness, it appears that the lateral portions of the *cerebellar* hemispheres are implicated in movements of the digits. The evolution of the lateral portions of the *cerebellum* parallels the degree to which organisms can make use of the distal portions of their limbs. In lower organisms like birds, there is almost no lateral development of the *cerebellum*. In organisms that do not have independent function of digits, the intermediate portion of the *cerebellum* is well developed, but not the lateral portion. The fact that the lateral *cerebellum* is related to higher order aspects of motor control is suggested by the fact that there is a significant amount of information that reaches the *cerebellum* through the *pons* that originates from the cortex. This information not only originates from motor and sensory areas but also from other regions of the cortex, especially the *association areas*. The efferent information from the intermediate *cerebellar* areas is directed to motor machinery involved with distal portion of the limbs, whereas the efferent information from the lateral *cerebellar* regions is directed to those regions of the *prefrontal cortex* and *basal ganglia* that are involved with the planning and organization of goal directed movements. There exists an especially close relationship between the outflow nuclei of the lateral *cerebellum*, the *dentate nuclei*, and the *premotor* and supplementary cortices, which are directly related to the organization, planning, and modification of motor activities (Allen et al., 1978; Halsband et al., 1993). Midline areas of the *cerebellum* play a more important role in the control of trunk and proximal muscles through their connections to the *vestibular nuclei*. There is a powerful reciprocal connection between these muscles and the *cerebellum*.

It has been noted that the lateral neo-*cerebellum* is likely to play a role in the control of the organization, planning, and learning, as well as execution of skilled movements of the hands. Neuropsychological reports on the effects of *cerebellar* lesions is especially important as when neo-cortical lesions are examined with conventional neurologic methods, no or minimal impairments are noted (Keller et al., 1937; Russell, 1970; Peters and Filter, 1973). The lack of clear signs suggests that higher-order motor mechanisms are involved and their study demands sophisticated behavioral approaches such as those developed in Neuropsychology (Peters, 1995). It has also been pointed out that the focus on *cerebellar* involvement in hand movement is particularly pertinent in view of the fact that there has been an evolutionary shift between primate and non-primate mammalian hand control mechanisms. This implies the existence of a highly developed neo-*cerebellar* role in the control of the distal musculature in primates. Original studies by Holmes (1917) clearly point out this relationship when he states, "The slowness, awkwardness, and irregularity of finger movements in handling objects and difficulty of bringing each finger of the affected hand separately and accurately to the tips of the thumb have been described alone, but these effects are even more apparent when the patient attempts to use simple familiar tools" (p. 490). These observed effects are not only due to problems in execution but also due to those aspects of movement that require timing and assembly of the component acts into a recognized whole (Peters, 1995). Holmes comments specifically that, "In writing, too, these disturbances are very obvious when the wound involves the right half of the *cerebellum*" (p. 490).

The right half of the *cerebellum* would be related to the left hemisphere in the cerebral cortex. The point is that *cerebellar* involvement in hand movement will interact with hand preference at levels close to motor execution and also at levels that are close to the planning and organization of movement. For example, in fine-skilled movements the timing and component movements that are assembled in the course of voluntary movement is crucial in execution (Ivry et al., 1988; Ivry and Keele, 1989). To the extent that the timing involves proper sequencing, such a function involves both the actual precision of timing as the movement process progresses and the syntax of timing specific elements, which involves, at a very minimum, the *premotor* cortex, and the supplementary motor areas (Halsband et al., 1993). It is thought unthinkable that hand preference which largely involves the development of expertise for certain movements, does not involve *cerebellar* function (Peters, 1995). It is also interesting to consider the relationship between handedness and the *cerebellum*, and motor and cognitive control of speech. The lateral *cerebellum* has been implicated in cognitive control of language, whereas the more medial areas of the *cerebellum* are involved in motor control of speech. The right *cerebellum*, and left cerebral hemisphere, especially the right lateral *cerebellum* and left *prefrontal cortex*

have been shown by PET scan to be active during certain verbal skills (Marien et al., 1996, 2001; Ackermann et al., 1998; Oki et al., 1999; Papathanassiou et al., 2000; Cabeza and Nyberg et al., 2000; Harris et al., 2001). A right-sided *cerebellum* would be involved in right-handed fine motor control.

It has been hypothesized that a relationship exists between the development of language abilities of the left hemisphere and the same hemispheric control of fine motor activity of the right hand. It is possible that the development of fine motor skills in the right hand influences the development of fine motor coordination of muscles required to produce speech. In fact, it is thought that the cognitive organization of speech or the *syntax* in the way humans organize speech is similar to the organizational pattern of the use of muscles. It may be that the way we organize our thoughts is modeled after how we organize our muscles, this would explain the relationship between fine motor coordination of motor activity in the right hand for tool use and the development of speech in the left hemisphere, as well as the cognitive expression of language in the left hemisphere. The *syntax* or timing of muscles and thoughts would involve the right *cerebellum*, left *basal ganglia*, and the left *frontal cortex*. From an evolutionary perspective, it appears that the development of fine motor activity and tool use with primarily the right hand preceded the development of language. It is thought that the flexible rapid speech of *Homo sapiens* did not evolve until 150,000 to 200,000 years ago (Corballis, 1989).

Corballis thinks that modern human language and the construction of complex tools mark the emergence of *generativity* or the ability to combine elements according to rules to form "novel assemblages by words, sentences, or multipart tools" (Corballis, 1989, p. 499). He also thinks that *generativity* depends largely on the left hemisphere. We also noted that the development of speech and manual dexterity appear to be interrelated to one another. There also appears to be a temporal linkage between hand preference and language milestones (Leask and Crow,

2001). The left hemisphere of the cerebral cortex and the right *cerebellum* appear to be responsible for the *syntax* of movement (Halsband et al., 1993) and the *syntax* of speech. It has been shown that children with early left hemisphere damage are particularly impaired in syntactic production (Aram et al., 1986). It has also been indicated that children with learning disabilities or behavioral difficulties, may have language problems, in both motor production (e.g., lisping, stuttering) and in the cognitive organization of language concepts. It has also been noted that many have difficulty with handwriting and fine motor coordination. It is possible that all of these symptoms are related to a similar underlying problem. In fact, the relationship between language and movement is even more interrelated in adults that previously thought. A recent study suggests that we gesture when we speak, not necessarily for the benefit of the person with whom we are speaking, but more to help ourselves think.

Iverson and Goldin-Meadow wanted to know if we gesture while talking because we learn by observing others. In one study, they videotaped 29 children responding to a series of reasoning tasks known to elicit hand signals. Surprisingly the children all gestured in remarkably similar ways despite the fact that 12 had been blind since birth (Iverson and Goldin-Meadow, 1998). Goldin-Meadow was quoted in *Psychology Today* (Howe, 1999) as saying that, "the gestures looked just the same.... It really shows that gestures are not unique to the sighted." In this article, a common assumption is discussed that we gesture to convey information to others.

This idea was apparently proved to be incorrect when 4 additional blind children were asked to run through the same tasks for a researcher they had been told was also blind. Again, the children made the same gestures even though they apparently knew their movements could not be seen. The researchers suggest that people who talk with their hands think with them as well, an idea supported by the preliminary results of studies that she and Iverson have recently conducted (Iverson and Goldin-Meadow, 1998; Goldin-Meadow et al., 2001). Both indicate that children remember events more clearly when they are allowed to gesture while explaining what happened. Says Goldin-Meadow "gesturing may make thinking a little easier by easing the burden on verbal communication" (Howe, 1999). It is also possible that this somehow allows greater access to the *cerebellum*, which may improve sensory feedback or feed-forward mechanisms, which in turn may help access information in the *prefrontal cortex* such as language and working memory. It is also possible then that dysfunction of the *cerebellum* which develops before the cortex and which seems to act as one of the subcortical precursors to asymmetric cortical development, may affect the development of fine motor control and motor and cognitive language development. Therefore, poor fine motor coordination in children may later reveal language-based problems such as speech or reading difficulties known to be inter-hemispheric communication inefficiencies. Likewise, therapy that is directed at improving fine motor coordination may also have an effect on cognitive and motor language-based problems.

Handedness has also been shown to have a relationship with the *corpus callosum*. It has been reported that the *corpus callosum* of ambidextrous individuals is significantly larger than that of *consistent right-handers*. This statement is true for males only (Witelson, 1985). Male mice with a congenitally absent *corpus callosum* show weaker paw preferences (Gruber et al., 1991). It has also been shown that organisms with malformations of the *corpus callosum* show a greater asymmetry favoring the left than do organisms of the same strain without these malformations (Schmidt et al., 1991). What this all means is difficult to determine, however, it is thought to represent a conceptual link between handedness and colossal function. It has been intimated that the corpus callosum is in some way involved in the direction of attention (Kinsbourne, 1973; Levy, 1985; Peters, 1998) suggesting multiple colossal roles. This would suggest that hand preference is more directly related to asymmetries in the allocation of attention than to the structural asymmetries at the executing level of the motor system (Peters, 1995).

It has been speculated that there are different levels of the neurophysiological mechanisms that are the foundations of handedness. In voluntary bimanual movement, it is thought that there are three conceptually separate levels (Peters, 1995). Level one is thought to involve formulation of the goal and determination of the functional contribution of the two hands, for what function each hand performs and where attention is directed in the control of hand movement. It is thought, therefore, that the most important functional aspect of hand use is that the hands can act in a complementary way. In naturally occurring activities, complementarity implies specialization (Peters, 1985). In this case, this specialization would involve one hand's manipulative activity on some object while the other hand stabilizes or positions it. It has been noted (Peters, 1991, 1992, 1995) that because attention cannot be divided in the guidance of two different motor activities at the same time, focal attention can only be directed at one hand at a time. Therefore, when movements of both hands are simultaneously of the same difficulty and complexity, but are different in terms of temporal demand, attention is asymmetrically distributed in right-handers, favoring the right hand when both hands compete for attention (Peters, 1985, 1991). In this case, it is thought that asymmetric attentional focus is a prerequisite for skilled bimanual activity.

Attentional direction may also affect unimanual activities as the hand that is more directly related to the movement's goal requires most directly the proper function of the left hemisphere (Oldfield, 1969; Corballis, 1989) and will be the hand chosen for skilled voluntary unimanual activities. However, focusing of attention must still be flexible enough to be switched back and forth between hands. Understanding these factors brings up the question why the left hemisphere in most individuals is associated with generativity in action and with selective focusing of attention to the right hand (Peters, 1995).

It has been argued that consistent lateralization develops through motor speech control, manipulation of food and objects with lips, tongue, and jaw movements (Peters, 1988c) in addition to bimanual hand control, which all require unilateral control. By far the most efficient way of ensuring congruence is to have lateral specialization of generative mechanisms that is one order removed from the actual machinery that operates the jaw, larynx, intra-thoracic musculature, and the hand, rather than having a separate mechanism that performs the same sort of task for each domain (Corballis, 1989). Anything that would draw attention from the left hemisphere to the right either because of decrease in left-sided function or decreased right hemisphere function, causing attention to be potentially drawn to unimportant stimuli, can disrupt these left hemisphere functions. In this instance a child may develop delays in speech and manual abilities because of a failure to be able to take focus off the right hemisphere and focus it on the left. Inability to stop one function affects the initiation of the next. These with regard to motor difficulties are reflective of classic *cerebellar* defects leading to the breakdown of goal-directed movements. It is also of interest to note that each of these functions is primarily *frontal* cortex in origin.

In level two, the precisely timed commands for the initiation and termination of movement trajectories of the two hands are controlled. Once an attentional bias is lateralized, the lateral specialization of control at level two is also determined. It is possible that the timing of onset and termination of the complementary movements of two hands would be inefficient if it had to be arranged by back-and-forth communication between the hemispheres (Peters, 1995). It is reasonable to expect that the initiation and termination commands for the two hands may come from one side, and in right-handers, this would be the left hemisphere. This speculation has been confirmed by the predominance of left hemisphere lesions in the production of aphasia in right-handers (Liepmann, 1905), the disruption of the timing of component manual sequences especially after *premotor* cortex lesions of the left hemisphere, and the absence of problems in movement execution (Halsband et al., 1993).

Finally, in level three there is the governing of the final outflow of control for the particular hand that allows the hand to perform the movement as required. This, however, does not appear to have a specific cerebral asymmetric localization. The reason may be that in human skilled bimanual activities, both hands have to perform complex activities. It would be limiting if structural concerns were to make one hand less capable of performing certain movements (Peters, 1995). It is interesting to note that deficits in the termination of a skilled movement is most evidenced with *cerebellar* dysfunction and is known as *intention tremor*.

Cerebellar Components of Speech and Language Asymmetries

It has been argued that the underlying variable in hand preference is attentional in nature and that the direction of lateral bias is a more fundamental variable than skill differences (Peters, 1995). If this is true for handedness, than it may also be true for other functions like speech or reading. The use of the term *speech* refers to the specialized motor mechanisms that translate language into vocalization. On the other hand, the term *language* refers to all linguistic processes that occur before the provision of motor output specifications. While there may be a need for lateral specialization in language processes, this is likely to be flexible and allow for inter-hemispheric interactions in the integration of syntax, semantics, and the general context of communication. At the level of cognition, there is likely great flexibility in where and how hemispheric processes interact. However, in translating thought into motor instructions for speech production, the necessity to produce a finely tuned and modulated movement of the bilateral speech apparatus does not allow for great flexibility. Motor instructions have to be precisely sequenced and formulated. Therefore, it is thought that speech production may be more sharply localized to only one hemisphere at a time whereas language processes may require more flexibility of lateralization and specialization (Vallar, 1990).

Coordinated control of bilateral muscles is seen elsewhere in the central nervous system including the eyes and the spinal column. Both of these have significant *cerebellar* input control. It may be possible that the *cerebellum* therefore would be involved in the bilateral coordination of speech muscles as well. With the *basal ganglia*, the *cerebellum* would also be expected to be involved in the timing and sequencing of muscles associated with speech production. Likewise, the lateral *cerebellum* has been implicated in the coordination or organization of the cognitive processes of language (Leiner et al., 1991). There is little disagreement that the overwhelming majority of right-handers have left hemisphere lateralization for speech and language. However, a not-insignificant number of left-handers also show left hemisphere specialization for language. There have also been claims that left-handers show a greater prevalence of bilateral involvement in language functions (Segalowitz and Bryden, 1983). Dichotic listening studies also suggest that the majority of left-handers are left hemisphere dominant for language. It should also be noted that there is a strong attentional component to the dichotic listening task (Bryden et al., 1983). The conclusion is that the underlying asymmetry especially for motor speech control favors the same side in right and left-handers.

It is interesting to consider the medial portions of the *cerebellum* for bilateral motor control and the right lateral *cerebellum* in specific asymmetric control of higher cognitive processes of language. PET scans have confirmed this relationship where medial areas are activated during motor activities involved in speech production and the lateral *cerebellum* is activated during the cognitive component of language (Leiner et al., 1991).

Handedness and Abilities

There has been interesting literature that attempts to relate verbal and spatial abilities to handedness as well as other abilities. For instance, there are claims that there is a high prevalence of left-handed architects (Peterson and Lansky, 1977). There are still questions about the validity of these assessments. At this time, the most interesting correlation between handedness and abilities is that strong right-handers are generally at a disadvantage across the entire range of the intellectual spectrum (Annett and Manning, 1989). This may imply that habitual unilateral motor use may in some way interface with interhemispheric access to information and that a balance of activity is better. Since overall intellectual ability would require skills of both hemispheres, one hemisphere being too dominant or active may affect crosstalk or may disrupt the temporal sequencing of activities of both hemispheres. It may also be indicative of too much attentional bias to one hemisphere, which is also not optimal. This means that any process that favors one side over the other-whether it be due to over, activity of one hemisphere or a dysfunction of one hemisphere that is activated less than the other, may have a negative impact on intellectual abilities.

Visual Processing

Vision is thought to be like all other high level abilities and therefore does not involve a single process. Vision has a number of different subsystems, which compute information about, spatial properties of objects, movement, shape, color, etc. One type of visual processing is accomplished by the so-called *ventral system* because computations take place in the more ventral *occipital-temporal* and *inferior-temporal* cortices. This system has also been characterized as the *what system* (Ungerleider and Mishkin, 1982) as opposed to the *where system* functioning in the *parietal* lobe which focuses on object recognition. The two hemispheres are known to act differently in the way they encode shapes (Leisman, 1976a, 1976b).

It has been argued that many different functions of vision could be achieved effectively if the system could encode information at multiple levels of scale (Marr, 1982). In other words, one way to distinguish whether you were seeing an edge or just a change in texture is to determine whether changes in intensity are present at multiple scales. For instance, if they are noticeable with only high resolution, they are most likely texture variations; if they are present at multiple levels, they are probably edges (DeValois and DeValois, 1988). The evidence from research at this time suggests that the two hemispheres focus on different types of features of visual input when forming object representations. The left hemisphere is thought to focus on smaller parts, higher spatial frequencies, or details. The right hemisphere is thought to focus on the global form, lower spatial frequencies or course patterns. To explain this asymmetry there are two theories that have been proposed (Brown and Kosslyn, 1995). One is called *structural theory* that proposes that one or more processing subsystems have become specialized in the hemispheres. The *allocation theory* states that the hemispheres tend to employ different strategies that often produce these results but there are no specific structural differences between hemispheres.

Visual processing can be divided into three phases of *low* level, *intermediate*, and *high* level (Marr, 1982) and the hemispheres can differ structurally or in terms of allocation at any of these levels. There is an assumption also, as in most lateralized functions that each type of processing is found in both hemispheres, to different degrees. Therefore, the hemispheres are different in the relative efficiency of the individual subsystems for a particular type of processing (*structural theory*) or in their predominance for using certain strategies (*allocation theory*) (Brown and Kosslyn, 1995).

Low Level

At the lowest level, subsystems organize the input so that distinct figures are separated from the ground. This processing takes place in a structure known as the *visual buffer* (Kosslyn et al., 1990). Computations in the *visual buffer* specify edges, regions of common color, and texture, and other characteristics that distinguish one object from its background. It is thought that not all the information in the *visual buffer* can be considered in detail, therefore, some information is chosen for additional processing. This has been referred to as an *attention window* that can be focused to a specific size, shape, and location to select a specific area of the *visual buffer* for more processing (Treisman and Gelade, 1980).

According to structural theories, the right hemisphere may detect more effectively large variations in light intensity over space. The left hemisphere detects more efficiently small variations in light intensity over space. This would suggest that the hemispheres differ in their sensitivity to different spatial frequencies (Jonsson and Hellige, 1986; Sergent and Hellige, 1986; Sergent, 1987b).

Intermediate Level

At this level visual processing organizes the stimuli into perceptual groups that will be useful for later object recognition. In this model the contents of the attention window are sent to a *preprocessing* subsystem in the *occipital-temporal* regions. Features such as texture gradients and color are found at this level (Biederman, 1987; Kosslyn et al., 1990). According to the structural theories, the subsystem is focused to detect different kinds of information in the two hemispheres. For instance, an individual may wish to look at an overall pattern like a face, whereas in other instances an individual may want to look at one of its components, like an eye. The two functions are not compatible with one another. The global function needs to incorporate into a whole the same things that need to be separated out by the local process. Therefore, it may be more efficient to have separate processes operate in parallel at the two levels.

High Level

High-level vision is concerned with matching input to representations in stored memory. It is thought that an object is reorganized when a match is made. In this same model, the output from the preprocessing subsystem serves as the input to the *pattern activation* subsystem found in the inferior *temporal lobe*s. It is here that the perceptual input is compared to the stored visual information, and recognition is achieved if a match is made. If the input does not match a previously stored representation well enough, then the new pattern is stored. It is thought that size per se is not likely to be represented at the level of object recognition. It is thought that neurons in the inferior *temporal lobe* that are sensitive to high level visual properties are insensitive to changes in visual angle (Leisman, 1976b; Plaut and Farah, 1990). The hemispheres may function differently in their ability for encoding parts in the whole or at different levels of hierarchy in a structural description. A structural description specifies how components are organized to compose a whole. The shape of a person is one example given to illustrate how parts are organized to compose a whole. In this example, a person is represented as a tree diagram, with the *body* at the top, *head, trunk, arms*, and *legs* as branches; *upper arms, forearm*, and *hand* as branches from the arm (Palmer, 1977; Marr, 1982).

It is possible according to one theory that one hemisphere could store the (larger) wholes and the other could store the (smaller) parts. Another theory suggests that it is not size that is different but that the hemispheres store by preferred level of hierarchy. The left hemisphere may compute input farther down in a structural hierarchy whereas the right hemisphere may compute parts or wholes on higher or lower levels in hierarchy. Several experiments have been used to verify the functional differences between the two hemispheres. These are useful to review because they emphasize the functional differences in a practical way. In addition, the same techniques that are used to identify functions can be used to diagnose dysfunction, or if one hemisphere is decreased in activation as compared to the other. Additionally, if we know what each hemisphere responds to we can later use this information to concentrate rehabilitation on the performance of one hemisphere.

One of the most common experiments (Heinke and Humphreys, 2003) involves use of letter stimuli that are most commonly used in global precedence studies; the features of the global-level object (e.g., two vertical lines and one horizontal line forming an H) are determined by the positioning of the local elements. Therefore, there is a confounding between size and level of hierarchy; the larger letter is made up of smaller letters. In this case, the term hierarchy refers to objects that are made up only of their constituent parts. For example, a dog's body is hierarchically structured because it is made out of head, trunk, legs, and so on; if these parts are removed, nothing remains. In contrast, patterns on a shirt are not hierarchically related to the shirt, if one removes them, the shirt would remain intact.

In other experiments (Pacquet and Merikle, 1988), investigators removed this confounding for letter stimuli. In addition to letter stimuli, picture stimuli can be employed that are not hierarchically arranged which removes the possibility of processes that are specialized for reading. In most real world objects, global features provide general information about object identity, whereas local feature can be used to identify specific information. In one experiment (Martin, 1979), stimuli were pictures of garments with smaller pictures on them, the larger pictures were not composed of the smaller ones, and therefore there was not a hierarchic relation between the two. The smaller pictures were also garments, providing the same types of objects at the local and global level without a hierarchic arrangement. In one divided visual feed study of global precedence (Martin, 1979) it was found that the global (larger) was processed faster than the local (smaller) level in both hemispheres when the global and local level letters were different. However, the global level was processed faster following right hemisphere presentation, than following left hemisphere presentation when subjects attended to the global level. In addition, there was a greater interference from the global level letter when stimuli were presented originally to the right hemisphere than when the subjects selectively attended to the local level. In this case, the subjects evaluated the stimuli more quickly when they were presented initially to the left hemisphere than to the right hemisphere.

To test whether local–global hemisphere specialization is different for hierarchic letter or non-hierarchic pictures (Martin, 1979; Lamb et al., 1989, 1990), which would suggest high-level visual processing, subjects were shown two types of stimuli: letters composed of smaller letters and articles of clothing with patterns of smaller articles of clothing printed on them. Results showed that for pictures, subjects did detect targets at the local level faster when stimuli were shown initially to the left hemisphere than the right hemisphere, however, they evaluated targets at global level equally well with both hemispheres. This confirms a local precedence and a trend toward the expected pattern of hemispheric specialization. In comparison, although a global precedence was found for the letter stimuli, the effect was the same when stimuli were presented initially to the right hemisphere or to the left. For letters, subjects responded faster and made fewer errors when the targets were at the global level than when they were at the local level. Overall, study results are consistent with findings of studies of patients with damage to the left hemisphere superior *temporal gyrus* (LSTG) or right hemisphere superior *temporal gyrus* (RSTG). Patients with RSTG lesions show local preference in similar tasks (Lamb et al., 1989, 1990). These patients showed a local preference across a range of sizes, except the very smallest where global preference was demonstrated. Patients with LSTG lesions showed global preference across all stimulus sizes. It is thought that the best way to interpret these results is that the hemispheres are different in how they preferentially set the attention window. It is thought that attentional factors must affect the encoding process.

In an attempt to explain the attention allocation hypothesis, Kosslyn and colleagues (1992) propose a specific mechanism where they suggest that the right hemisphere preferentially monitors outputs from neurons that have relatively large receptor fields. Also the hemispheres may differ in their ability to monitor the outputs from different size receptive fields even if the same outputs are available in both hemispheres (Kosslyn et al., 1992). Kosslyn and colleagues (1992) suggest that the bias to encode outputs from neurons with different size receptive fields allows the *ventral* (object) and *dorsal* (spatial) systems to be coordinated. It is thought that an individual during movement not only needs to know the precise metric distances of objects (dorsal) but also the specific shapes of objects (ventral). The two types of processing need to be linked and it is thought that this is best achieved if they are both in the same hemisphere—the right (Kosslyn et al., 1992; Marsolek et al., 1992). In addition, when an individual attempts to identify objects, they may need to ignore variations in shape among specific examples and may need only to know the type of spatial relations among parts, not the specific positions of parts of a given object. For example, to reorganize the shape of a dog, one would ignore the type of dog and the exact position of limbs. The left hemisphere is thought to have a special role both in generalizing over shapes (Marsolek, 1992) and in categorizing spatial relations (Kosslyn et al., 1989). Therefore, differences in receptive field size may act to coordinate the encoding of shapes and spatial relations, and could therefore cause the right hemisphere to specialize in computing metric spatial relations and specific shapes and the left hemisphere to specialize in computing categorical spatial relations and categories of shapes.

So far we have suggested that hemispheres function differently in their effectiveness of focusing attention at scales of different size, and that the underlying mechanism involves sampling outputs from neurons with different size receptive fields. This is similar to a spatial-frequency hypothesis. In fact, the receptive field and spatial-frequency theories predict similar results. The two concepts are closely related. The smaller its receptive field, the higher the spatial frequency a cell will respond to, on the other hand, the larger the receptive field the lower the spatial frequency. In this case it is thought therefore that a large receptive field, or a lower spatial frequency is more of a right hemisphere function, whereas small receptive fields and higher spatial frequency is more of a left hemisphere function with the effects modulated by attentional variables. Normal processing of global aspects is thought to depend on the posterior superior *temporal lobe* of the right hemisphere. Normal processing of local elements depends on the posterior superior *temporal lobe* of the left hemisphere (Lamb et al., 1990).

Another important quality of visual information processing is the ability to localize a visual image in space. The observer is required to detect whether one object touches another object, this would be an *on* and *off* quality. Another requirement is the discernment of near or far and *above* or *below*. There are hemispheric performance differences for all of these tasks. Studies have shown that there is a left hemisphere advantage for the *on-off* tasks and a right hemisphere advantage for distance judgment tasks (Kosslyn et al., 1989). Hellige and Michimata (1989) tested individuals by having observers indicate whether a dot was within 2 cm of the line (a coordinate or distance task called *near-far* task). For the *above-below* task, there is a left hemisphere advantage whereas the right hemisphere shows an advantage for the *near-far* task. Researchers have examined how these different asymmetries arise during the course of ontogenetic development. Compared to adults, the visual sensory system of newborns is especially limited in its transformation of information carried by high spatial frequencies (Banks and Danner-Miller, 1987; De Schonen and Mathivet, 1989). It has been suggested that the development of various brain areas is more advanced in the right hemisphere than in the left hemisphere at the time of birth and possibly for a short time after. Hellige (1993) postulates a certain critical period for incoming visual input modification, which occurs earlier for the right hemisphere than for the left hemisphere. Once modified by highly degraded visual input, the right hemisphere is not only predisposed to become dominant for processing low spatial frequencies but is also less able than the left to take full advantage of higher frequencies when they finally do appear. If this notion is accurate, then it would also follow that the resulting hemispheric differences in visual processing would influence asymmetry for any task that depends on the relevant aspects of visual information whether the activity requires stimulus identification or stimulus location.

Hemispheric asymmetry for auditory stimuli has been found to be related to temporal frequency. For instance, in a pitch discrimination task, Ivry and Lebby (1993) found that subjects were faster and more accurate in judging relatively low frequency sounds when presented to the left ear (right hemisphere). However, they were found to be faster and more accurate in judging relatively high frequency sounds when they were presented to the right ear (left hemisphere). It has been found that evoked potentials recorded over temporal (not occipital) leads show opposite hemispheric asymmetries depending on whether a sine wave grating reversed its phase at low (e.g., 4 Hz) or high (e.g., 12 Hz) temporal frequencies. The right hemisphere was more responsive than the left at low temporal frequencies, but the left hemisphere was more responsive than the right at high temporal frequencies (Rebai et al., 1986).

Another interesting comparison is seen between visual processing and motor activity of the hands (Winstein and Pohl, 1995; Classen et al., 1998; Flanders et al., 1999). It is postulated that the two hands work together, with the non-dominant hand performing movements of lower temporal and spatial frequency while the dominant hand performs activities that consist of high temporal and spatial frequencies. In most people who are right handed, the left hemisphere (right hand) performs high temporal and spatial frequency movements and the right hemisphere (left hand) performs low temporal and spatial frequency movements. One possibility of how this asymmetry arises is that certain motor areas of the right hemisphere are more fully developed than homologous areas of the left hemisphere at a time in development when only gross movements (and sensory feedback from them) are available to the brain. This would encourage the right hemisphere to become dominant for making these movements. This may eventually be followed by a time when the left hemisphere develops enough to be shaped by more precise movements (and with the attendant sensory feedback).

Auditory Processing

Dichotic listening (DL) is described as a behavioral technique used to study a wide range of cognitive and emotional processes related to brain laterality and hemispheric asymmetry (Kimura, 1961b; Bryden, 1998a, 1998b; Hugdahl, 1992a), attention (Broadbent, 1954; Nataan, 1982; Hillyard et al., 1973), conditioning and learning (Corteen and Wood, 1972; Hugdahl and Brobeck, 1986), memory (Christianson et al., 1986; Hugdahl et al., 1993), and psycholinguistics (Repp, 1977; Lauter, 1982). It is recognized that in neuropsychological practice DL is used as an index of language function (Studdert-Kennedy and Shankweiler, 1981; Tartter, 1984). Dichotic listening is thought to be a measure of *temporal lobe* function (Spreen and Strauss, 1991), attention, and stimulus processing speed, as well as a way of measuring hemispheric language asymmetry. It is further thought that auditory lateralization is probably not related to a single mechanism (Jancke et al., 1992) but is related to several mechanisms involving both perceptual and other cognitive components. It is known that the right and left hemispheres each participate in multiple functions that do not necessarily correlate with one another. However, measures of general activation of a hemisphere and tasks that use specific functions within that hemisphere should correlate.

Davidson (1984) showed that the magnitude of the right ear advantage (REA) in the DL task significantly correlated with resting electroencephalographic (EEG) asymmetry. Individuals with larger-left-than-right EEG resting activation also had better recall from the right as compared to the left ear in DL. The basic principle of DL is to provide to the brain more information than the brain can process consciously at one time. The question then is to which elements of the stimulus input will be recognized or attended (Holender, 1986). A typical task involves monitoring and shadowing the message in one ear while ignoring the message presented to the other ear. One of the uses of DL is in examining the function of selective attention. In these tests, the subject is told to focus their attention to one ear, and then discriminate between two tones varying in pitch, while at the same time ignoring a similar input presented to the unattended ear. It has been shown that the typical dichotic procedure is sensitive to attentional influences, and that a hemispheric-specific advantage can be shifted if attention is focused to the ipsilateral ear (Hugdahl and Anderson, 1986; Hugdahl, 1992b). DL is thought to be particularly sensitive to *fronto-temporal* cortex functional integrity (Hugdahl, 1988). One of the most common findings in DL is known as REA or left hemisphere advantage.

According to Kimura (1967), the REA is a consequence of the anatomy of the auditory projections from the *cochlear nucleus* in the ear to the *primary auditory cortex* in the *temporal lobe*, and of left hemisphere superiority for processing language-related material. It is thought that an auditory stimulus activates neurons in the *cochlear nucleus* at the level of the *vestibulo-cochlear* nerve. One of the subdivisions of the *cochlear nucleus* is the *ventral acoustic striata*, which enters the second level at the *superior olivary complex*. From there, inhibitory and excitatory impulses proceed to the lateral *lemniscus*. From the nuclei of the lateral *lemniscus*, the projections are mainly contralateral projecting to the inferior *colliculus* in the *tectum*. The contralateral fibers then innervate the *meniculate body* in the *pulvinar thalamus*, which sends its axons into the auditory cortex in the posterior superior *temporal gyrus* (Brigge and Reale, 1985; Price et al., 1992). Even though auditory stimuli from one ear reach both hemispheres, the contralateral connection is stronger. Therefore, REA is the result of stronger auditory input to the contralateral brain. In addition, the left hemisphere for right-handers is specialized for language processing.

An auditory stimulus that is sent to the ipsilateral hemisphere is inhibited by contralateral information and auditory information that reaches the ipsilateral right hemisphere has to be transferred across the *corpus callosum* to be processed in the language-processing left hemisphere. The REA is reflective of a left hemisphere language dominant hemisphere. A left ear advantage would normally indicate a right hemisphere language processing dominance, and no ear advantage would indicate a bilateral or mixed processing dominance. It is thought that people with crossed laterality (e.g., crossed hand and eye preference) do not show a significant REA in DL as opposed to people with non-crossed eye–hand laterality (Hugdahl, 1995). REA subjects show enhancement of the P3 component over the left auditory cortex, whereas the LEA subjects show the reverse.

Studies have demonstrated that children show letter recall from the right compared with left ear input with single ear presentation (Bakker, 1969, Bakker et al., 1978). There are two primary models that explain REA in DL, one model refers to the anatomy and physiology of the auditory system, and the other refers to attentional and other cognitive or emotional influences (Hugdahl, 1995). In one test, subjects are asked to focus their attention either to the right ear (FR) or the left ear (FL) and then compare their performance with a divided or non-focused (NF) task. Results indicate that REA is enhanced during the FR condition and decreased or even reversed to a LEA during FL condition. Children, however, cannot perform this task. Children still demonstrate a REA during the FL attentional condition—especially the boys. However, when the data is split for 8 and 9 year olds a clear difference is evidenced. The 8 year olds cannot modify their REA, whereas 9 year olds do (Hugdahl and Anderson, 1986). This difference may reflect an increase in the development of reading skills. It has been shown that the ability of the child to shift attention away from the right ear increases as the child advances in reading ability with no difference evidenced between boys and girls (Hugdahl and Anderson, 1987). This suggests a link between the development of literacy and attentional capacity. It appears clear that attention may modify the ear advantage observed in the typical DL task. However, attention usually affects the magnitude rather than the direction of ear advantage.

It has been found that normal children differ from reading-delayed children on a DL task, when tested in the morning but not in the afternoon. The normal children are more lateralized compared to the reading delayed children when instructed to focus attention to the right ear stimulus. This suggests a lack of control of attentional resources in the morning for reading-delayed children. These findings also demonstrate temporary and dynamic modulations of a supposed structural laterality pattern. Therefore, ear advantage in DL reflects a structural hard-wiring of lateralized cognitive functions that are put under modulatory influences by dynamic processes, being switched on and off depending on the general level of activation, attentional focus, and emotional significance (Hugdahl, 1995).

Another effect that may alter DL is level of motivational effort or arousal. Level of arousal has been shown to have a significant effect on complex cognitive tasks (Hockey, 1979). A LEA has been reported for the identification of emotional stimuli (Haggard and Parkinson 1971; Carmon and Nachshon, 1973; Ley and Bryden, 1982; Bryden and Mac Rae, 1988). In these studies, subjects were asked to identify emotional sounds or words. These studies attempted to determine if the processing of affect may be different from the processing of content, so the emotional value of the stimulus might activate the right hemisphere. Additionally, the presentation of an emotional stimulus coupled with the instruction to attend to and identify the emotional properties of the stimulus may make the task an emotional one leading to increased arousal and generalized activation of the right hemisphere (Kinsbourne, 1975; Tucker, 1981). It has also been found that extremely anxious individuals show a universal REA for positive words (Wexler and Halwes, 1983). This suggests a left hemisphere advantage for positive emotions. It has also been noted that a highly negative condition abolishes the REA (Ley and Bryden, 1982). It has been argued that aversive, but not positive arousal results in suppression of the REA pointing to a unique role of the right hemisphere in aversive or negative affect.

There is also the thought that spatial orientation through either head or eye movement may affect the ear advantage (Asbjornsen et al., 1990). In their study, Asbjornsen and colleagues instructed subjects to actively attend to either left or right side by either turning their head or their eyes to left and right space during DL. The study indicates that REA is affected by spatial orientation. Results show that the ear of stimulus application seems to be a significantly more powerful determinant of ear advantage in DL than the perceived sound source in space (Hugdahl, 1995).

Hemispheric asymmetry for auditory stimuli has also been found to be related to frequency of sound. Low frequency sounds have a LEA which projects to right hemisphere. High frequency sounds have a REA or left hemisphere advantage (Ivry and Lebby, 1993).

Face Processing

Psychological and cognitive functions are no longer viewed as constituted by organized sets of independent processes (e.g., reading, writing, object recognition) but instead are now viewed as being composed of several sub-processes (e.g., feature analysis, structural encoding, activation of biographic memories) that are organized in specific ways (e.g., in parallel or in succession, independently or interactively). In addition, the cognitive structure of mental functions is now thought of as a compartmental organization of interactive component operations (Leisman and Koch, 2003). Therefore, by breaking down any given function into its smaller operations we can specify the nature, goal, order, and the interrelated individual functions that we are trying to understand. The understanding of what is known as brain-behavior relationships can be thought of as an endeavor that tries to map a suspected set of interrelated mental operations and their underlying cognitive functions onto their associated interconnected cerebral structures (Sergent, 1995). Although it is most common to think of left and right hemispheres as separate entities each with unique processing abilities, there is likely more similarity in processing ability between two homotopic areas of the two hemispheres than there is between two areas of the same hemispheres (Sergent, 1995).

One process that has been extensively studied is the process underlying the perception and recognition of faces. This has led to a description of the several operations that underlie the ability for a face to become recognized and be identified. Faces convey various bits of information about an individual person. Some of that information is visually derived, based on that it can be accessed on the sole basis of physical attributes of the person even if the person is not familiar. Other information is semantically derived in that it can be accessed only after the perceived image of the face *makes contact* with a corresponding stored image from which biographic information about the individual can then be reactivated (Bruce and Young, 1986; and Sergent, 1989a). It is thought that with face recognition too, there exists a hierarchic organization of the underlying processes. The first step requires a structural encoding of the physical facial information to obtain the best description possible of the perceived face. Information can include gender, age, or identity of the face (Sergent, 1989a). To interpret the identity of the face also requires that physical characteristics that specifically describe a face be recovered from the structural description such as the appearance of faces, their emotions, viewpoint, and lighting conditions which are always changing and can complicate the perceptual operations necessary to discover what is unique to a face and makes it belong to a specific person (Sergent, 1995). The process may trigger pertinent memories related to the bearer of the face, which includes different aspects of memory like episodic, semantic, emotional, and other biographic information. From this information, the name of the person can be accessed; although the name may not be directly "connected" to the representation of the face, for we know that a face may be recognized even when the name cannot be recalled (Sergent, 1995). The face, however, is only one of the several means available where knowledge that is stored about an individual can be accessed.

The breakdown or inability to perform normal face processing, results in a condition known as *prosopagnosia*. The symptoms associated with *prosopagnosia* can vary greatly amongst patients. Even though the location of the lesion in the cerebrum can also vary it has been suggested that *prosopagnosia* is the result of a bilateral posterior damage involving mainly the mesial *occipital-temporal junction* (Damasio, 1985). Although the pattern of injury does not conform to this generalization, in fact there are cases in which the lesion is unilateral (e.g., Michel et al., 1989) or does not involve the posterior cerebrum (Sergent and Poncet, 1990). In an attempt to further locate an anatomical control function for face processing, various case studies have been analyzed. One such study was performed on a long-standing *prosopagnosia* patient (Sergent and Signoret, 1992b) with a lesion in the white matter surrounding the right *lingual* and *fusiform gyri*. MRIs of this patient's brain revealed no apparent damage to the cortex of the right *fusiform* and *parahippocampal gyri*, however, PET studies revealed that metabolically the posterior ventro-medial region of the right hemisphere including the *fusiform* and the *parahippocampal gyri* were inactive (Michel et al., 1989) suggesting that functionally the cerebral damage was much more extensive than one would have expected based solely on the structural (MRI) radiologic data.

It can be concluded that both cerebral hemispheres are equipped with the necessary structures to carry out the discrimination and recognition of faces as well as the processing of visually derived semantic property of faces (e.g., age, race, gender). However, there are several studies showing better left hemisphere ability to identify famous or well known faces (Marzi and Beslucchi, 1977). This is thought to suggest a special role for the right hemisphere in the initial storage of facial information, but equal ability to access stored information. Most experiments of accuracy in face processing have found right hemisphere superiority. It has also been shown that a lesion in the posterior region of the right hemisphere disrupts the discrimination and recognition of faces, where similar lesions in left hemisphere do not affect this process (DeRenzi et al., 1968; Hécaen, 1972; Benton et al., 1983). It also appears that the posterior right hemisphere plays a critical role in the processing of facial features regardless of the types of operations that are performed on faces. However, lesions in different regions of the posterior right hemisphere result in qualitatively different deficits (Warrington and James, 1967). Functional imaging as MRI has revealed a large number of cases of *prosopagnosia* with cortical injury restricted to the right hemisphere (Landis et al., 1986; Michel et al., 1989). The fact that *prosopagnosia* can result from the unilateral right hemisphere damage strongly suggests a crucial role for the right hemisphere in the processing of faces and may indicate that the right hemisphere is necessary to face recognition abilities.

Whereas unilateral brain damage produces face-processing impairment almost exclusively following right hemisphere lesions, evidence from split-brain or *hemispherectomized* patients suggests that there may be functional equivalence of the two hemispheres with respect to the processing of faces or that destruction of areas in both hemispheres is necessary to produce a complete inability to process faces. PET studies have been performed and have included three main tasks: (1) a face gender categorization (Is the face male or female?), (2) a professional categorization (Is it the face of an actor or a politician?), and (3) object semantics categorization (Is object natural or man-made?). The results show (Sergent et al., 1992a): (1) the gender task activates the right *lingual gyrus* and the posterior part of the *fusiform gyrus* along with the lateral *occipital* cortex of the left hemisphere, (2) the face recognition task produces activation of both *fusiform gyri* of the right *parahippocampal gyrus* and the anterior cortex of the *temporal lobe* of both hemispheres, and (3) the object recognition task causes activation of the left *fusiform gyrus* and the lateral *occipital-temporal* cortex as well as the left middle *temporal gyrus* (Sergent et al., 1992a). Overall findings appear to indicate a bilateral yet asymmetric involvement of the cerebral hemisphere in the normal processing of faces.

The major role of the right hemisphere in face processing appears to be in perceptual aspects associated with reactivating the biographic information necessary for a face to become memorized and for identification. The left hemisphere, although not completely silent, does not appear to be involved with processing of faces (Sergent, 1995). However, the left hemisphere contributes to object processing as observed in PET studies (Kertesz, 1979; Sergent and Signoret, 1992a). Right *lingual* and *fusiform gyri* are specific to perceptual operations of encoding and in extracting a configuration that is unique to a face. The right *parahippocampal gyrus* forms a link between facial representation and related memories. In right *temporal lobe* injury, patients' access to biographic information, accessed either through face or name, is disrupted. However, proper name finding is related to function of the left anterior *temporal lobe*.

It is thought that subcortical systems cannot efficiently mediate the required synchronization. It has been speculated that the ventral visual system, which processes identity information, may depend more on inter-hemispheric synchronization than the portions of the visual system processing non-identity information (e.g., location). In order to remember a stimulus in its full content, the dimensions of the stimulus must be bound into an object perception, which needs to be linked with emotional and temporal cues to place it in autobiographic form. This spatio-temporal pattern of activation requires multi-regional synchronization of activation within the hemisphere that receives the stimulus. This must then be integrated with the opposite hemisphere. It is thought that this type of integration is specifically dependent on the synchronizing effects of the cerebral cortex (Liederman, 1998). On the other hand, stimulus location, which is computed primarily by the dorsal system, may be effectively integrated inter-hemispherically by asynchronous subcortical mechanisms and may be able to effectively control behavior even though the identity of the object being localized is not conscious.

Autonomic Asymmetry

Although it has been known for some time that the brain controls most of the body's physiologic functions largely without the necessity for an individual's conscious intervention, it was not realized how important the brain is in the control of specific regulatory functions of the cardiovascular, immune, and hormonal functions. That neurotransmitter distribution is asymmetrically lateralized in the brain was not recognized either. The understanding of how these functions are localized changes a clinician's approach to the diagnosis and testing of symptoms related to autonomic, hormonal, and immune dysfunction. We had previously viewed the function of these systems as peripheral to a fundamental understanding of the brain and cognition. However, recent work elucidating central neurological, cognitive, and emotional mechanisms as a principal source of autonomic symptoms has become well recognized. Approaching the treatment of autonomic or immune dysfunction through treatment directed at the central nervous system makes much more sense than other approaches. In this context, it is also useful to note that autonomic, immune, or behavioral breakdown may not be the primary problem but rather a signal that central neurological control mechanisms are at the root. Autonomic function diagnoses may be made more accurate by examination of the effectiveness of lateralized function and dysfunction of cognitive or emotional behavior.

Much of the research concerning autonomic asymmetry comes from studying cardiovascular effects in adults. Much can be learned from how the cardiovascular system is regulated by the brain since other autonomic functions would be expected to be controlled in a similar fashion (Fig. 8.4). There is an important reciprocal relationship between the brain and autonomic system that can be demonstrated quite well through the cardiovascular system. The brain controls the regulation of heart rate, respiration, and blood pressure, which then regulates fuel delivery of oxygen and glucose back to the brain. Which system however, takes the lead role has been in question for some time. Many have thought that emotional and cognitive brain disorders were mostly secondary to reduced fuel delivery capacities of the heart and blood vessels and hormonal dysfunction. Now it appears that the brain or central dysfunction may be the initiator of cardiac dysfunction and blood pressure dysregulation, and associated emotional or cognitive symptoms are not secondary but are primarily related to the neurobiological dysfunction. One of the primary reasons this has been extensively studied of late is due to the large number of unexplained cardiac deaths each year and the realization that emotional stress can kill people suddenly (Lane and Jennings, 1995).

Experiments have demonstrated that emotional stress can significantly lower the threshold for ventricular fibrillation, the most common terminal event in sudden cardiac death (SCD) (Lown et al., 1977; Lown, 1979; Lown et al., 1980). Specific central nervous system (CNS) (Skinner and Reed, 1981) and peripheral nervous system interventions (Schwartz, 1984; Schwartz and Priori, 1990) also successfully affect fibrillation and ischemia. Therefore, it is reasonable to conclude that brain functions that mediate or regulate emotion can be contributing factors to SCD. If we understand how the brain is related to SCD, then we will also understand its role in other autonomic functions or dysfunctions that may be associated with emotional or cognitive control mechanisms.

Since most children with learning or affective disorders also present with a variety of autonomic symptoms which can include abnormal regulation of pulse rate, cardiac system, bowel and bladder function, digestive disorders, and respiratory dysfunction, this relationship is extremely important to

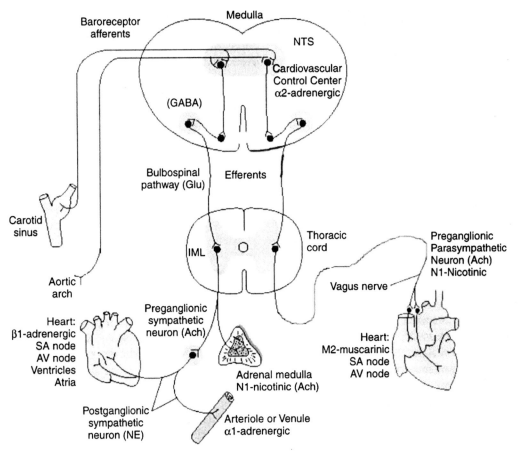

Fig. 8.4. Autonomic control of the heart.

understand. With regard to SCD, a role for both *sympathetic* and *parasympathetic* mechanisms and their interactions is well documented (Verrier, 1990). Of special interest has been the observation that stimulation of the left *sympathetic* nerves to the heart lower ventricular fibrillation threshold more than stimulating the right side does (Verrier et al., 1974; Lown et al., 1977). Being aware of the fact that cerebral hemispheric asymmetries are well recognized in the mediation of emotion (Silberman and Weingartner, 1986), we can hypothesize that emotional aberrations related to brain function may induce a lateralized difference in *sympathetic* input to the heart which may increase the possibility of ventricular fibrillation and sudden death (Lane and Schwartz, 1987). It has been suggested that activity in each cerebral hemisphere may be experienced in a characteristic pattern of autonomic activity based on the fact of the existence of autonomic asymmetries (Galin, 1974). Research has suggested that the right hemisphere (RH) has a preferential influence in regulating autonomic function (Zoccolotti et al., 1981, 1986). Verrier and Dickerson (1991) have shown that dogs angered by food withdrawal showed significant vasoconstriction of the coronary arteries after transient reactions to anger showed in heart rate (HR) and blood pressure (BP). Cessation of *noradrenergic* input by *stellectomy* was proven to eliminate the vasoconstriction. These studies demonstrate that psychological factors can increase work demand of the heart by increasing HR and BP and can indirectly cause *ischemia*. Studies in lower organisms and humans show that psychological stress relates to spontaneous ventricular *arrhythmias* and decreases the threshold for triggering ventricular *arrhythmia* (Lane and Jennings, 1995). It is thought that the psychological effects that may result in acute cardiac symptoms are produced by activation of the *sympathetic* nervous system and its subsequent influence on *adrenal medullary* hormones (Manuck and Krantz, 1986). *Sympathetic* activation can produce changes directly through release of *norepinephrine* and through action on the *adrenal medulla*, which releases *epinephrine*. These two functions will cause an increase in HR, BP, and peripheral vascular resistance. This can result in decreased fuel and oxygen delivery. Although *techolomine* levels increase, it has been shown that their increase alone cannot account for stress-induced *arrhythmia* (Taggert et al., 1973).

In regard to *parasympathetic* control mechanisms, activation of the *vagus* nerve generally results in *cholinergically* mediated slowing of HR and decrease in cardiac contractility, mostly due to inhibition of *sympathetic* effects in lower organisms (Cerati and Schwartz, 1991; DeFerrari et al., 1991). Proper autonomic function at any level depends primarily on an appropriate balance of *sympathetic* and *parasympathetic* activity, which is mainly regulated by the brain. It has been shown that *sympathetic* and vagus interactions can create more *arrhythmias* than *sympathetic* influences alone (Verrier, 1990). The *sympathetic* system is activated most rostrally by the *hypothalamus* which is intimately related to all *limbic* structures. Therefore, more primitive emotional or survival driven reactions will tend to activate areas of the *hypothalamus*, which will stimulate the *sympathetic* nervous system directly and indirectly through the *pituitary* and *adrenal medulla*.

On the other hand, vagal activation is primarily controlled by the *nucleus tractus solitarius* (NTS) in the *medullary reticular formation*. The higher central regulation of NTS is primarily from descending neocortical projections especially from the *frontal lobe*. Neocortical activation produces increased vagal inhibition of the *sympathetic* nervous system directly and produces inhibitory control over *limbic–hypothalamic* reactions. In humans this is thought to give greater control of emotional responses, which also results in a decrease of resting HR, BP, and circulation stress hormones extending life expectancy over organisms that do not have as large a neocortex as ours. However, decreased activation of the neocortex, especially the *prefrontal cortex*, or a lateralization of brain activity in the neocortex or *prefrontal cortices* can result in ipsilateral decreased inhibition of the *sympathetic* nervous system. This could cause widespread autonomic dysfunction and an apparent increased *stress response* that may not be a reaction to stress but rather a result of decreased stimulation of the brain, especially the *prefrontal cortex*. This may result in associated cognitive and emotional dysfunction based on the side of the brain whose activation is decreased.

Another factor that has been examined is the importance of timing or phase in cardiac control. The heart is seen as a cyclically controlled organ, which is reset to different periods by external events such as neural stimulation. It has been suggested that resetting at certain points in the cycle can result in instability and even the termination of the heart beat (Winfree, 1983, 1987). It is well documented clinically that a premature wave of ventricular depolarization occurring during repolarization, the *R* on *T* phenomenon, can induce malignant ventricular *arrhythmias* (Glass and MacKay, 1988). This again is primarily a function of balance between *parasympathetic* and *sympathetic* factors. A number of clinical and experimental studies in humans now indicate the important role of anger, hostility, and impatience in predicting SCDs (Williams et al., 1980). These are similar to manic symptoms, which are thought to be lateralized to the RH. Other studies also indicate that depression or depressed effect is a significant predictor of subsequent cardiac events including SCD (Ahem et al., 1990; Almada et al., 1991; Frasure-Smith, 1991; Frasure-Smith et al., 1993). Cortical activation associated with depressive symptoms are thought to be localized primarily to the left hemisphere. These findings are thought to suggest that the way the brain mediates emotion may influence susceptibility to SCD.

It would also seem to highlight how the brain mediates autonomic dysfunction in general, especially since pathophysiological effects on the heart and coronary vessels seem to be primarily related to *sympathetic-adrenal* influences held in check by vagal restraints (Lane and Jennings, 1995).

The cerebral cortex also plays a significant role in the regulation of autonomic function (Cechetto and Saper, 1990). The existence of direct projections to subcortical structures regulating autonomic functions, as well as neurophysiological studies showing autonomic changes with stimulation or inhibition of these structures, substantiates their contribution to the control of autonomic function (Fig. 8.4). These areas include the medial *prefrontal* cortex, the *cingulate gyrus*, the *insula*, the *temporal lobe*, and the primary *sensory* and *motor* cortices (Mesulam, 1985; Cachetto and Saper, 1990). The first three areas directly connect to the *amygdala, hypothalamus, brainstem,* and spinal cord areas involved in autonomic control. These higher centers are thought to coordinate autonomic output with mental or cognitive function, emotional reactions, and homeostatic needs of the body. With the exception of the midline nuclei in the *brainstem* such as the *locus coeruleus* and *Raphe* nuclei regulating *sympathetic* function, the *amygdala, insula,* and *medial prefrontal* cortex are bilateral structures. It is thought therefore that relative activation lateralized to one side may occur to end-organs such as the heart. The most important factor may be the balance of activation and inhibitory influences acting on the *midbrain* and *brainstem* centers controlling autonomic function on each side.

Using the heart as an example of autonomic control over organs, we can see that lateralized functions as well as a balance between *parasympathetic* and *sympathetic* innervation is a key factor in maintaining autonomic regulation. Evidence suggests that the right postganglionic *sympathetic* system innervates the *atria*, while the left *sympathetic* system mostly supplies the *AV node* and *left ventricle* (Levy et al., 1966; Randall and Ardell, 1990). It is generally thought the right and left *vagus* nerves follow a similar distribution, as do the right and left *sympathetic* systems. In contrast, the *vagus* nerves have greater influence over the *sympathetic* neurons in the *atria*, having more of an effect over (SA) *sinus node*-generated HR, than do *sympathetic* nerves (Levy and Martin, 1984). Sectioning the right-sided *vagus* fibers results in greater acceleration of heart rate than does sectioning left-sided *vagus* fibers (Hamlin and Smith, 1968). Studies on dogs (Levy et al., 1966) showed that right *stellate* (*sympathetic*) stimulation has primarily *chronotrophic* effects resulting in significant increases in HR, whereas left *stellate* stimulation has primarily *inotrophic* (contractility) effects causing a major increase in systolic BP. Yanowitz and colleagues (1966) reported that right *stellate* stimulation increased HR, lowered T-wave amplitude, and did not change the Q–T interval. On the other hand they found that left *stellate* stimulation did not change HR but did increase T-wave amplitude and prolonged the Q–T interval, which is thought to reflect repolarization phenomena that are associated with life threatening *arrhythmias* and SCD (Schwartz et al., 1998; Zipes, 1991).

It is thought that if neurogenic *sympathetic* input to the ventricle is imbalanced, the areas that receive *sympathetic* input repolarize more quickly than adjacent areas that were not stimulated. Therefore, adjacent areas that are still depolarized can prematurely depolarize the adjacent ventricular areas that have repolarized. This phenomenon is known as *reentry*. This mechanism can result in disruption of the normal depolarization–repolarization cycle and deteriorate into *ventricular fibrillation* (Lane and Jennings, 1995).

Studies have also observed that right *stellate* block results in *bradycardia*, while left *stellate* block does not affect heart rate (Rogers et al., 1978). What all of these studies reveal is that physiologically important autonomic asymmetries do exist (Lane and Jennings, 1995). These asymmetries are also reflective of other brain asymmetries and that the deactivation of inhibiting structures may result in a net ipsilateral activation effect.

In addition, given the fact that both *callosal* fibers (Cook, 1984; Trevarthen, 1984) and frontal-*hypothalamic* fibers (Brutkowski, 1964) are primarily inhibitory to *sympathetic* activation, it is possible that activation of one hemisphere will lead to contralateral autonomic activation. Therefore, decreased neocortical *frontal lobe* activation can result in primarily ipsilateral decreased inhibition of the *sympathetic* nervous system by at least two pathways. One is loss of direct inhibition of the *hypothalamus* and the second is loss of stimulation by *brainstem* vagal centers that inhibit the *sympathetic* activity. Loss of the inhibitory control ipsilaterally can result in HR or rhythm changes and depending on the side of decrease, can cause ipsilateral increase in BP, and can result in a wide variety of *sympathetic* and *parasympathetic* right and left imbalances. Studies have confirmed a model of ipsilateral transmission of lateralized hemispheric activation through the *sympathetic* chain to the heart. Decreased activation of the brain and neocortex that inhibits *sympathetic* activation can be caused by decreased sensory or afferent input. One study showed that evoked potentials in the right hemisphere varied significantly as a function of HR while the left hemisphere did not show the same effects (Katkin et al., 1991; Aftanas et al., 2002). In addition, P_1 and P_2 amplitudes were larger during slow HRs than during fast HRs. This is thought to be consistent with the restriction of sensory input with higher HR stimulation (Walker and Sandman, 1979, 1982).

Autonomic changes during increased attention are extremely relevant because the right hemisphere appears to be superior in the control of attention. Attentive observation of the environment, particularly in anticipation of an important event has been shown to decrease *sympathetic* activation and lower HR, through vagal mechanisms (Lacey and Lacey, 1974). It is further thought that the cerebral organization of behavior, primarily affective-emotional behavior, in the right hemisphere extends to the organization of the autonomic changes supporting those behaviors (Lane and Jennings, 1995). Recent studies suggest that stable individual differences exist in the degree of asymmetry of frontal activation as measured by EEG that can be used to predict the intensity of emotional responses to films (Tomarken et al., 1992; Wheeler et al., 1993). In addition, *hemisphericity* research suggests that stable individual differences exist in cognitive style (Yeap, 1989). It has also been shown that mathematically gifted children appear to have stable patterns of hemispheric arousal asymmetry that are more lateralized to the right hemisphere then in children with average mathematical ability (O'Boyle et al., 1991). If accurate, these results suggest that specific predictions could be made about the patterns of autonomic activity based on these stable individual differences in hemispheric arousal asymmetry (Lane and Jennings, 1995). In other words, individuals have been shown to have differences in activation and arousal of one hemisphere in relationship to the other. This could reflect a decrease or increase in activation of one side of the brain that creates a *hemisphericity* or imbalance of activity. This has been documented in individuals and has been correlated with predicting the response of individuals to emotional stimuli and with their cognitive abilities.

This stable imbalance can also reflect specific predictable changes in the *autonomic nervous system* based on difference of hemispheric activation. On the other hand, autonomic symptoms should be useful in predicting asymmetric hemispheric activation and associated behavioral or emotional activities in specific individuals. Children with learning disabilities or affective disorders often are thought to show autonomic or neuroendocrine changes that are consistent with increased *sympathetic* activity. This would most likely be due to a hemispheric decrease or imbalance of neocortical activation. This could be reflective of decreased higher cognitive and/or emotional abilities with decreased inhibition on *limbic–hypothalamic* mediated *sympathetic* activation. This could very well be a result of decreased sensory afferent input from subcortical or peripheral structures. The *cerebellum* and *thalamus* are two likely candidates as the source for decreased sensory afferent input since decreasing sensory input from these structures has been shown to affect autonomic function and both project to *hypothalamus* and *limbic* structures.

Kubitz and Landers (1993) have noted that an 8-week aerobic exercise-training program decreases EEG *alpha* laterality and increases vagal tone. It has also been shown that regular exercise decreases resting HR and BP. These changes reflect decreased *sympathetic* and increased vagal tone. The effects of exercise on autonomic tone may therefore actually be through direct control over brain activation, which causes a subsequent decrease in *sympathetic* tone. This would most likely be a result of increased afferent input from muscle spindle and joint afferent activity transmitted through *cerebello-thalamic-frontal* pathways. Therefore, decreased use of muscle and joint activity would be expected to have an opposite effect on autonomic function due to decreased activation of the same pathways.

It is thought that an imbalance or asymmetric activation of the brain can result in sudden death at an early age (Lane and Jennings, 1995). Schwartz and colleagues (1982) have reported evidence of *sympathetic* imbalance to the left in *sudden infant death syndrome*. From an evolutionary standpoint, it can be postulated that natural selection acted to embrace survival by promoting autonomic and cerebral laterality. This same mechanism could also contribute to man's demise or dysfunction in such a way that natural selection could not anticipate (Lane and Jennings, 1995).

NEUROTRANSMITTERS AND NEUROENDOCRINE FUNCTION

Alteration in levels of neurotransmitters, especially *dopamine, serotonin,* and *norepinephrine* has been implicated as major factors in learning disabilities and affective disorders in both children and adults. Recent research shows that there is asymmetric distribution of these neurotransmitters within the cerebrum. How this is involved with specific cognitive and/or emotional processes is important to understand. It is now well established that the hemispheres are clearly different in their basic mode of cognitive processing (Trevarthan, 1990). It is also thought that emotionally related processes are lateralized in the brain (Gainotti, 1972, 1983). It was previously thought that processes that are mostly under subcortical control could not be regulated in an asymmetric fashion; that is now thought to be an unjustified assumption (Wittling, 1995). Of all of the neurotransmitters in the CNS, the brain monoamines including *norepinephrine, serotonin,* and *dopamine* are thought to be of critical importance for the regulation of a wide spectrum of human and organism behavior, including cognitive and emotional responses, as well as arousal and autonomic processes (Whittling, 1995). *Noradrenergic* neurons have their cell bodies primarily in the *locus ceruleus* in the *pons*. Efferent projections include areas of the *mesencephalon, thalamus,* and *limbic system* including the *amygdala, hippocampus, entorhinal cortex,* and *septum*. The bulk of neocortical innervations arise from fibers that enter the cortex at the *frontal lobe*, the majority arising from the ipsilateral *locus ceruleus*. There is an equal distribution to all cortical areas with an especially dense distribution of *noradrenergic* fibers in the *cingulate gyrus* and *prefrontal orbital* cortex (Fallon and Loughlin, 1987). It is thought that the *noradrenergic* system is extremely important for the cerebral regulation of arousal, attention-related functions, and adaptive response of the organism to stress and environmental stimulation. Studies have shown that activity of neurons in the *locus ceruleus* are significantly increased in situations that have a potential for physiologic challenge to the organism (Morailak et al., 1987), evoke orienting behavior, or require selective attention to novel or relevant environmental information (Clark et al., 1987). *Norepinephrine*, along with its arousing effects, is thought to be involved in the modulation of affective behavior. It has been shown that *norepinephrine* is decreased in some types of depression. In addition, drugs, which are thought to increase *norepinephrine* levels, have been shown to reduce depressive symptomatology (Schildkraut, 1965; Garver and Zemlan, 1986). It is also thought that *norepinephrine* plays a role in immuno-regulation and its primary effect seems to be immunosuppressive, although it has been noted that some aspects of the immune response may be increased by central *noradrenergic* activity (Besedovsky et al., 1985; Dunn, 1989). *Norepinephrine* levels have been examined at subcortical levels. The *thalamus* of five human brains, after routine autopsy, was studied for *norepinephrine* concentrations. It was found that the whole *thalamus* showed a significantly higher concentration on the right side. The *pulvinar* was the only area that had a higher concentration on the left (Oke et al., 1978). Similar findings were found in the rat *thalamus* with higher left-sided values in some anterior regions and a right-sided *norepinephrine* lateralization extending over the middle and posterior *thalamic* regions (Oke et al., 1980). Higher *norepinephrine* values were also found in the right anterior *hypothalamus* of the mouse brain (Boineoud et al., 1990). Right hemisphere infarction leads to a significant reduction of *norepinephrine* in the right cortex and left un-lesioned posterior cortex, as well as the right *locus ceruleus*. *Norepinephrine* concentrations are also significantly reduced in the *frontal–temporal* and posterior cortex of the contralateral hemisphere. On the other hand, left hemisphere infarction does not result in any significant effects on *norepinephrine* concentration in either the cortex or *locus ceruleus* (Robinson, 1979). These studies and others appear to clearly show that *noradrenergic* innervation, the biochemical substrate of arousal (Tucker and Williamson, 1984), shows a clear right hemisphere asymmetry.

Serotonin

Serotonin or *5-hydroxytryptamine (5-HT)* innervation in the brain mainly arises from neurons located in the dorsal and to a lesser degree the median *Raphe nuclei* in the *mesencephalon*. The projections of the *Raphe nuclei* are primarily ipsilateral, traveling through the *median forebrain bundle* (Fallon and Loughlin, 1987). The *serotonergic* connections diffuse to all regions of *hypothalamus, thalamus, amygdala,* and *hippocampus*, as well as the *mesencephalon,* and *cerebellum*. In the cortex, there are dense generalized innervations with a higher concentration in the *frontal* and *visual* cortex and lower in the *primary motor* cortex (Fallon and Loughlin, 1987). *Serotonin* is thought to be an inhibitory neurotransmitter that acts to decrease firing of cerebral neurons, decreasing arousal and inhibiting the activity of the *noradrenergic* system (Birnbaumer and Schmidt, 1991). With regard to the immune system, *serotonin* is thought to have an immuno-suppressive effect (Hall and Goldstein, 1985). *Serotonin* has been extensively studied with regard to its role in affective states, behavior, and psychopathologic disorders (Spoont, 1992). It is generally thought that *serotonin* dysfunction may result in depressive disorders and suicidal behavior (Asberg et al., 1987; Meltzer and Lowy, 1987; Arora and Meltzer, 1989a, 1989b). Measurement of cortical S_2 *serotonin* binding by PET scan was performed in normal control subjects and patients with either right or left hemisphere stroke. There did not appear to be any left–right asymmetry in receptor binding in any cortical region in controls or left hemisphere stroke patients. Right hemisphere stroke patients showed significantly higher S_2 receptors in the injured hemisphere thought to result from a significantly greater extent of *serotonin* depletion by right hemisphere stroke patients. This is most likely due to a greater right hemisphere involvement in *serotonergic* activity. Based on the majority of studies, there is a tendency toward right hemisphere *serotonin* lateralization in normal humans and a reversed lateralization in suicide victims (Wittling, 1995). Right lateralization of *serotonin* has been demonstrated in neonates (Frecska et al., 1990), therefore, it appears to be an inborn trait in the human brain. Therefore, it can be concluded that the two neurotransmitters that are mostly involved in the increase or decrease of arousal, *norepinephrine*, and *serotonin* respectively, are reciprocally inhibitory to each other (Flor-Henry, 1986) and are also both asymmetrically distributed to the right side of the brain. It is also thought that changes in their innervation are correlated with affective and behavioral abnormalities.

Dopamine

Dopamine brain innervation originates from neurons in the ventral *mesencephalon*, which forms two subsystems. In the cerebrum, there are dense *dopamine* connections to the *prefrontal, premotor*, and *motor* areas and a minor distribution to the posterior cerebrum. There appears to be a preferential innervation of motor as opposed to sensory cortices, as well as association cortices over primary sensory areas (Fallon and Loughlin, 1987). Therefore, it is thought that *dopamine* is concerned with higher integrative cortical functions and in the regulation of cortical output activity, especially in motor control. Studies show that the normal function of *dopaminergic* pathways is extremely important for the selection and function of appropriate motor responses and coordination of motor output with sensory output. It would seem to make sense that *dopamine* would favor the hemisphere thought to be primarily responsible for motor control. Therefore, in humans it has been found that there is a higher *dopamine* concentration in the left hemisphere. In postmortem studies of 14 normal human subjects, the results indicate that *dopamine* content is significantly higher in the left *globus pallidus* as opposed to the right (Rossor et al., 1980). In addition, PET imaging of *dopamine* receptors in live brains reveals higher concentrations of *dopamine* terminals in the left than in the right *basal ganglia* in humans (Wagner et al., 1983). Organism studies have also revealed asymmetries that have been found in the striatum and cortex of rat and mice brains with regard to *dopamine* content (Zimmerberg et al., 1974; Glick et al., 1980), *dopamine* metabolism (Yamamoto and Freed, 1984), *dopamine* stimulated *adenylate cyclase* activity (Glick et al., 1983), and *striatal dopamine* receptors (Schneider et al., 1982). It appears these asymmetries in *dopaminergic* innervation are related to various kinds of motor-behavioral asymmetries (Castellano et al., 1989). The vast majority of studies appear to show that *dopamine* concentration is much higher in the *neostriatum* contralateral to the organism's side of paw preference (Zimmerberg et al., 1974).

In conclusion, the lateralization of brain neurotransmitters seems to be distributed based on various functions with which they are involved. *Norepinephrine* and *serotonin* are associated with an increase and decrease in autonomic and psychological arousal, and are located in the right hemisphere, which is known to control cerebral arousal and autonomic regulation. On the other hand, *dopamine* is more concerned with control of motor behavior and higher integrated functions, which is known to be a function primarily of the left hemisphere.

NEUROENDOCRINE SYSTEM

After the nervous system, the hormonal system is the second most important control system of the body. As opposed to the specific pathways of the nervous system, the hormonal system utilizes the circulatory system as its mode of transport to target organs. There is thought to be significant reciprocal effects between the two systems (Wittling, 1995). Cerebral control of hormonal activity is mostly provided by the *hypothalamus*, which is primarily under the control of the *midbrain, limbic system, amygdala*, and *frontal* cortex (Feldman, 1989; Schmidt and Thews, 1989). There has been research to identify lateralization of neuroendocrine control. Organism experiments that involve unilateral lesions of the medial *preoptic* area, lower *brainstem*, or medial basal *hypothalamus* are found to have a two to three times higher morality rate if the lesions are performed on the right brain, which is not the case with left-sided lesions (Gerendai, 1984). It was concluded from these results that the right brain is more important in vital functions than the left.

Cortisol is an important hormone that has been associated with the stress response. There appears to be asymmetry in cerebral regulation of *cortisol* secretions in humans (Wittling and Pfluger, 1990). In previous studies, it has been found that cerebral asymmetry in the control of various physiologic functions, such as *cortisol* levels (Roschmann et al., 1992), BP (Wittling, 1990), or HR (Wittling and Schweiger, 1992) is significantly enhanced if lateralized visual input of a film presentation that is either aversive or neutral, is combined with lateralized motor responses to the films content. The results show that *cortisol* secretion is significantly affected by the film's emotion-related content. The aversive or negative film resulted in significantly higher *cortisol* increases than the neutral film did. Right hemisphere presentation of the aversive film correlated with significantly higher increases in *cortisol* secretion than the left hemisphere presentation of the same film. It appears that only the right hemisphere is able to respond neuroendocrinologically in a different way to emotional (aversive film) or non-emotional (neutral film) stimuli, showing a clearly higher *cortisol* response to the emotional film. Therefore, it would appear that *cortisol* regulation in emotionally related situations is under the primary control of the right hemisphere (Whittling, 1995). However, subsequent results have shown that brain asymmetry in *cortisol* regulation cannot necessarily be attributed to hemispheric differences in subjective-emotional arousal (Wittling and Roschmann, 1993).

There has been much attention of late to the role that *cortisol* plays in physical health and well-being. In addition, studies have been performed examining how changes in brain asymmetry of *cortisol* may affect health. It is thought that chronically elevated *cortisol* levels are associated with the development of several diseases like *atherosclerosis* (Troxler et al., 1977) and neoplastic disease (Sklar and Anisman, 1981). However, an appropriate level of *cortisol* is apparently necessary for properly functioning mechanisms of resistance to stress (Munck et al., 1984), immunoreactivity (Cupps and Faud, 1982; Dunn, 1989), and in preventing disease (Munck et al., 1984). Therefore, secretion of *cortisol* is increased in a variety of stress related situations whether they are physical or psychological such as physical or emotional danger, threats of violence, work stress, noxious stimuli, antigen challenge, and illness. The function of the increase in *corticosteroid* levels is primarily to protect the organism against its own normal defense reactions that are activated by stress, and, if not controlled, may damage the organism (Munck et al., 1984). *Cortisol* therefore, prevents defense reactions from getting out of hand and, thereby, threatening homeostasis. Therefore, either a chronically high or low *cortisol* level can be associated with abnormal physiologic function and health problems. Too much can reduce the body's immune response to stress or pathogens, too little and the body may suffer from autoimmune-type diseases. Too much *cortisol* may result in a child who is chronically infected by viral or bacterial agents, chance colds, ear infections, etc. A child with too little *cortisol* may be prone to chronic allergies, food allergies, asthma, etc. Both of these may be associated with right hemisphere dysfunction.

During the last decade, evidence of a close bi-directional communication between the immune and nervous systems has increased, indicating that the brain plays the most important role in the control of immune functions (Calabrese et al., 1987; Dunn, 1989; Daruna and Morgan, 1990). The brain controls the immune system by two mechanisms. First, it has a direct effect on the immune system through connections to the *sympathetic* and *parasympathetic* innervation of lymphoid tissue such as *thymus, spleen*, bone marrow, and *lymph nodes*. Second, brain control of the immune system can be indirect by neuroendocrine connections under *hypothalamic* and cortical control with reciprocal feedback by the immune system. The *corticosteroids* of the *hypothalamic-pituitary-adrenal* axis appear to play a significant role in the modulation of immune responses, which appear to be mainly immunosuppressive (Tsokos and Balow, 1986). It also seems apparent that as with emotional arousal, *sympathetic* innervation, neurotransmitter concentration, and neuroendocrine activity, immune responsiveness is also asymmetrically lateralized in the human brain. Most of the studies on this topic have been performed on organisms utilizing lesion techniques. In one study, it was found that the *natural killer (NK)* cell activity of the mouse spleen cells, which is thought to indicate the amount of spontaneous resistance to tumors, was significantly decreased after left cortical lesions, and no effect on *NK* cells was noted with right-sided lesions (Bardos et al., 1981). Unilateral neocortical ablation of the dorsal and lateral aspects of frontal, parietal, and *occipital* regions decreased the development of lymphoid organs significantly, especially the *thymus* and *spleen* in left-lesioned mice. Additionally, right-lesioned mice had a significantly increased weight of the *thymus* as compared to sham-operated organisms. In addition, the number of *splenic T (THY-1$^+$)* cells was reduced in left-lesioned mice to approximately 50 percent of sham-operated mice; it was not affected in right-lesioned mice. The number of *IgG-PFC*'s, a *T* cell-dependent response, was significantly depressed after left cortical ablation and significantly increased with right cortical ablation (Renoux et al., 1983; Bizier et al., 1985; Renoux and Biziere, 1986). These and other results suggest that the left hemisphere increases responsiveness of several *T*-cell-dependent immune factors; right hemisphere activity seems to be mostly immunosuppressive (LaHoste et al., 1989; Neveu, 1992).

Another important fact, with regard to altered brain asymmetry and immune function is that studies have shown that changes in hemispheric balance in neuroimmuno modulation, whether by neocortical ablation of one hemisphere or by a relatively higher activation of the other hemisphere (Kang et al., 1991), have been found to result in altered immune responses. In other words, a functional imbalance can present with the same change in function with which a structural lesion can present.

In summary, we can see that brain asymmetry is not just observed with cognitive or emotional functions, but also with cerebral regulation of autonomic-physiologic processes. Lateralized neural control has been documented at virtually all levels of the nervous system including peripheral pathways to lower *brainstem, midbrain, hypothalamus, thalamus, basal ganglia, amygdala,* as well as the *prefrontal* and *orbitofrontal* cortices. The functions that are asymmetrically controlled include neurotransmitter activity, neuroendocrine activity, immune control, and cardiovascular activity. *Dopamine* activity, control of *AV* node conduction, and myocardial contractility all appear to be left hemisphere in nature; whereas, *noradrenergic* and *serotonergic* activity as well as most neuroendocrine activity, *SA* node chronotropic cardiac activity, and BP appear to be primarily under right hemisphere control. There appears to be a differential specialization of the two sides of the brain in the control of different functional systems or single function within a functional system (Wittling, 1995). The left-brain also appears to be most important for regulation of an individual's defense responses against invading agents. The right-brain, however, is primarily involved in the control of vital functions that are key to survival and which allow the individual to cope actively with stress and external challenges. The right-brain also controls the up-regulation and down-regulation of autonomic-physiologic arousal. It is important to understand that although there is preferential control of brain function, there is not exclusive control of any function by one hemisphere. With most functions, both sides of the brain appear to be different only in the strength or efficiency of their regulatory activity. Other functions like the immune regulatory are reciprocally inhibitory by both sides of the brain. One side causes activating effects, while the other is inhibitory. In addition, it is important to note that brain asymmetries in the control of autonomic-physiologic functions are thought to be independent of emotionally related brain asymmetry. It has also been noted that descending fibers involved in the control of autonomic functions mostly take an ipsilateral path. Ipsilateral paths are taken by virtually all autonomic functions controlled by the brain and to connections at every brain level including amygdalo-fugal and amygdalo-pedal connections with cortical areas (Iwai and Yuckie, 1987), *Amygdaloid* connections with *hypothalamus, thalamus, midbrain, pons,* and *medulla* (Price and Amaral, 1981; Cechetto et al., 1983; Sakanaka et al., 1986), and direct *hypothalamo-autonomic* pathways connecting *hypothalamic* regions with pre-ganglionic nuclei of *sympathetic* and *parasympathetic* systems (Saper et al., 1976). Abnormalities in brain functioning by a structural or anatomic brain damage, cerebral inactivation, decreased cortical activity, or by altered neurotransmitter concentrations have been shown to be associated with a variety of immunologic, cardiovascular, and related physiologic disorders (Wittling, 1995).

ASYMMETRY AND EMOTIONS

It is clear that in complex organisms, and especially in humans, the cerebral cortex plays an important role in aspects of emotional behavior and experience (Kolb and Taylor, 1990), especially anterior or frontal cortical regions, which have extensive anatomic reciprocity with both subcortical centers and with posterior control circuits, all extremely important in emotional behavior. The *frontal lobes* or anterior cortical zones are the brain areas, which have shown the most dramatic growth in relative size over the course of phylogeny in comparison to other brain regions (Jerison, 1973; Luria, 1973). Asymmetries of *frontal lobe* function have been implicated in different forms of emotional behavior. Some of the first observations of asymmetries and their role in emotional behavior were with patients with unilateral cortical lesions (Jackson, 1878). Most of these reports seemed to show that injury to the left hemisphere was more likely to result in what has been called *catastrophic-depressive reaction*. This was seen with similar injury to the right hemisphere (Goldstein, 1939). Recent studies show that damage specifically to the left *frontal lobe* results in depression. In addition, the closer the injury is to the frontal pole, the more severe the depression. However, patients with right *frontal lobe* injury are more likely to develop *mania* (Robinson et al., 1984). It has been stated that a fundamental asymmetry in the control of functions related to emotion should not be surprising based on speculation with regard to the evolutionary advantage of cerebral asymmetry (Levy, 1972). Most researchers agree that *approach and withdrawal* are fundamental motivational behaviors, which are found at all levels of phylogeny. It has been postulated that the *frontal lobes* or anterior regions of the left and right hemispheres are specialized for approach and withdrawal behavior respectively (Davidson, 1984, 1987, 1988; Davidson and Tomarken, 1989; Davidson et al., 1990b). This is thought partly because the left frontal region has been noted as an important center for intention, self-regulation, and planning (Luria, 1973). This area is thought to be the region that produces behaviors that have been described as the "will." Also, during development, the child will approach and reach out to objects that it is drawn to using the right hand more than the left, which would involve the left frontal motor areas more than the right (Young et al., 1983). Right-handed reaching and positive behavior or affect are thought together to be an expression of a brain circuit controlling approach behavior, and the left frontal region is thought to act as a "convergence zone" for this current circuit (Davidson, 1992a, 1992b). It has also been noted that injury to the left frontal region results in a deficit in approach behavior. Patients with damage in this area of the brain are reported to present with apathetic behavior, experience a loss of interest and pleasure in objects and people, and have difficulty initiating voluntary action. Therefore, it is thought that hypoactivation in this brain region would be expected to be associated with a lowered threshold for sadness and depression (Davidson and Sutton, 1995). On the other hand, it is thought that the right anterior region is specialized for withdrawal behavior. Some of the most informative studies have been those on normal human subjects that involve measurement of regional hemispheric activation based on electrophysiologic measurement. These studies show that activation of withdrawal-related emotional states (e.g., fear and disgust) occur when the right frontal and anterior temporal regions are specifically stimulated; also, baseline tonic activation on these same areas show greater likelihood of response with increased withdrawal-related negative behavior to emotional stimuli. In addition, individuals with chronically increased baseline activation of these areas are reported to have greater negative disposition in general (Davidson and Sutton, 1995).

Recently, a study of a patient with right *temporal lobectomy* was studied after being exposed to positive and negative emotional stimuli (Morris et al., 1991). Surgical ablation included the right *temporal lobe* including the anterior part and the complete right *amygdala*. The investigators recorded skin conductance responses to stimuli. They found that skin response to positive stimuli was normal whereas skin conductive response to negative stimuli was markedly decreased. Interestingly, PET studies have reported increased activation of a resting baseline activity in a right hemisphere subcortical area which projects to the *amygdala* in panic prone patients (Reiman et al., 1984). These results are thought to suggest anterior cortical and subcortical right hemisphere regions for specializing in the control of withdrawal-related negative affect (Davidson and Sutton, 1995). Therefore, it can be stated that activation of the left anterior region of the cerebral cortex is associated with approach-related emotions. Decreased activation in this same area is associated with approach-related deficit behaviors, such as sadness, apathy, and depression. Conversely, activation in the right anterior region is associated with withdrawal-related emotions such as fear and disgust, and withdrawal-related psychopathology like anxiety or mania. However, it is important to understand that the areas of the hemisphere that perceive emotional information are thought to be different from those that experience the actual emotion (Davidson and Sutton, 1995). Research appears to suggest that the right posterior cortical region is specialized for the perception of emotional information not depending on the type of emotion.

Another important consideration is that although an individual may have a baseline anterior asymmetry, which can be used to predict reactivity to emotional stimuli, without the actual stimuli they would not present with symptoms (Davidson and Fox, 1989; Tomarken et al., 1990; Wheeler et al., 1993). For example, left anterior damage is not by itself enough for the production of depressive symptoms. However, the threshold and reactivity to emotional stimuli will be more likely to be accentuated. It is important to understand that a patient with an emotional or affective disorder may not present with symptoms even though that person may have dysfunction of the frontal region, and this dysfunction may not be apparent unless the individual is emotionally tested. Therefore, the absence of symptoms does not rule out the fact that significant dysfunction may exist.

Although emotional stimuli may not be present, it may be important to cross check for other functions that may be affected by dysfunction in the same area, especially motor, autonomic, or cognitive change; although it is possible that these areas may not present with active symptoms without the appropriate stressors being applied. For example, a child may not be depressed, yet in school being stressed with cognitive challenges, the left-brain deficit becomes apparent. It has been shown that periods of peak emotional intensity are unpredictable, and they can happen at different points in time for different subjects in response to the same emotional stimulus (Davidson and Sutton, 1995). In a study where subjects who were exposed to film clips to induce either approach-related positive emotion (happiness) and withdrawal-related negative emotion (disgust) (Davidson et al., 1990b), results measured by EEG activity, showed greater right-sided frontal activation during disgust emotions than during left-frontal happy conditions. The difference between happy and disgust conditions appear to be due to the fact that these emotions were superimposed upon the subjects basal level of asymmetry. This would confirm the fact that individuals possess wide variations in baseline levels of activity that can differ inter- or intra-hemispherically and that this can play a significant role in how individuals will react to environmental or emotional stresses. The baseline may be set high or low which may account for the unpredictability for the periods of peak emotional intensity. The time differences may reflect these high or low baselines, which would alter the time to threshold, making it shorter or longer. In another study, EEG activity was again measured during reward or punishment trials (Sobotka et al., 1992). Results showed that there was significantly greater frontal activation with punishment compared with reward trials, on the other hand there was a greater left-frontal activation with reward trials. Baseline activity appears to be an important factor that must be considered. It has been shown that subjects' overall (across-task) EEG asymmetry is highly connected with his or her asymmetry during resting baseline (Davidson et al., 1979) and that resting baselines are stable over time (Tomarken et al., 1992). This implies that either high or low baseline levels may be a chronic condition in individuals. In a group of individuals who were both low and stable scorers on resting frontal asymmetry, results showed that depressed individuals had less frontal activation compared with non-depressed subjects (Schaffer et al., 1983). These same results have been demonstrated with a group of clinically depressed subjects (Henriques and Davidson, 1991). Additionally, with EEGs recorded during resting baseline conditions, it was shown that remitted depressives, similar to acutely depressed subjects, had significantly less frontal activation as compared to normal control subjects in the frontal region (Spitzer et al., 1978). In conclusion, the results indicate that decreased left anterior activation, thought to be characteristic of depression, appears to remain stable when depression is remitted. It could therefore be predicted that relative to a comparison group, a higher percentage of subjects who are shown to have decreased left anterior activation would develop subsequent psychopathologic disorders (Davidson and Sutton, 1995). It should also stand to reason that we would expect similar results with chronically low right-sided frontal activation. In addition, if other functions of the left or right frontal cortical areas, such as cognitive, motor, and autonomic functions, should also be predicted by baseline levels. Studies have shown that subjects who were measured with greater right-sided frontal activation at rest reported more intense negative affect in response to films designed to elicit fear or disgust (Tomarken et al., 1990). It has also been shown that patients with decreased right frontal activation are immuno-suppressed relative to left-frontal counterparts. Additionally, during baseline state, right-frontal subjects show significantly less *NK cell* activity compared with left-frontal subjects (Kang et al., 1991).

Results of baseline asymmetry and affective style in adults have proven to be the same as in children. It has been noted that among 10-month-old infants, there can be extreme differences in response to maternal separation. Some infants become distressed right away and cry as soon as their mother leaves. Other infants show a much different pattern of response and show almost no negative emotions when separated from their mothers. A study was conducted where 10-month-old infants were separated into two groups based on whether they cried or not after being separated from their mother for approximately 60 s (Davidson and Fox, 1989). It was found that about half the group cried and half did not. Baseline measures of *frontal* and *parietal* activation from both hemispheres were taken 30 min prior to separating the infants from their mothers. It was found that there was a large difference in *frontal* asymmetry that could predict which infants would cry and which would not. The infants that cried had greater right-sided and less left-sided *frontal* activation during the baseline period as compared to those who did not cry. There did not appear to be the same asymmetry in the *parietal* lobes. This is thought to be the first study that demonstrated that in infants individual differences in *frontal* asymmetry can predict emotional reactivity. In addition, this relationship is the same as seen with adults.

Another study examining behavior and *frontal* asymmetry in children specifically examined behavioral inhibition (Kagan and Zajonc, 1988). Behavioral inhibition is a young child's tendency to withdraw or freeze in novel or unfamiliar situations. In these new or unfamiliar situations, behaviorally inhibited children will stay close to their mothers without playing or interacting with other children. Three hundred and eighty six children aged 31 months were tested in a peer play session. Brain electrical activity was taken at rest and in response to several tasks. Results showed that inhibited children show right *frontal* activation whereas uninhibited children show left *frontal* activation. The question raised after examining the results was whether the behavioral inhibition is due to a decreased left-sided approach behavior, or due to increased right-sided withdrawal activity. Further information confirmed that in these children, the reason for inhibited behavior appeared to be due to a decrease in left *frontal* activation rather than due to an increase in right *frontal* activation. The pattern of decreased left *frontal* activation found in inhibited children was practically the same that has been reported in depressed adults. It would be expected with these children, that they would be more likely to experience sadness and depression-like reactions to emotionally stressful situations. However, it has been speculated that although only a small percentage of the vulnerable children would be expected to actually develop an affective disorder, more of them would be expected to have sub-clinical characteristics like *dysthymic* mood, shyness, and decreased dispositional affect (Davidson and Sutton, 1995). It is possible that children with the opposite finding of decreased right-sided withdrawal behavior would show the same results. Nevertheless, it would be expected that these uninhibited children would display impulsive behavior and show an overreaction to reward. We see that these asymmetries can be acute or chronic over time and can predict the affective behavior or threshold to appropriate stressors.

We can now explore in more detail the cortical and *limbic* interactions that are involved. The most basic division of *cortico-limbic* structures is not left/right, but rather dorsal/ventral. It is thought that evolution of the neocortex from *para-limbic* cortices occurred by two paths of network differentiation: The first, the dorsal archi-cortical network, concentrated on the *hippocampus*; the second, the ventral paleo-cortical network, focused on the *olfactory* cortex with important interconnections with the *amygdala* (Pandya et al., 1988). Studies have been conducted to explore the cognitive and perceptual differences between these two systems. These studies suggest that the dorsal pathways are specialized for spatial memory and that the ventral pathways are specialized for object memory (Ungerleider and Mishkin, 1982). It is also thought that the functional differences between these two networks are significant in the motivational realm as they are in the cognitive realm. It is also thought that there has not been an equal distribution of dorsal and ventral networks within the left and right hemisphere. It is thought that the left hemisphere has specialized to express the so-called "cybernetic" characteristics of the dorsal cortical network. It is further thought that asymmetries of dorsal and ventral expansion lead to hemispheric asymmetries in the *limbic* control of cognition (Liotti and Tucker, 1995). Some of the research has shown that lesions in the dominant (right-handers) or the left hemisphere typically result in what has been termed "catastrophic reaction," for example, tears, despair, and anger. Damage to the right hemisphere, which in most people is the minor hemisphere, is accompanied by indifference reactions such as unawareness, euphoria, or lack of concern (Goldstein, 1952; Gainotte, 1970, 1972). Therefore, the left and right hemispheres, based on this early research, were thought to result in opposite emotional tone with the left normally being more oriented toward a positive mood, whereas the right was oriented toward a more negative mood. Other studies have confirmed that pathologic crying occurs with left hemispheric lesions (Poeck, 1969; Rinn, 1984). It has been reported by Sackheim and associates (1982) that changes in affect following unilateral injury is due to disinhibition of contralateral cortical regions and not because of release of ipsilateral subcortical areas. A significant finding has been the relation between arousal and hemisphericity. Right hemisphere lesioned patients have been described as "hypo-aroused" (Heilman et al., 1978) and show reduced cortical and autonomic responsivity, in skin conductance or HR, especially when exposed to emotionally charged stimuli (Morrow et al., 1981; Zoccolotti et al., 1981). It has been shown that normal individuals express emotions more noticeably on the left side of the face, especially the lower half, which is almost exclusively innervated by the right hemisphere (e.g., Borod and Caron, 1980). Patients with right hemisphere damage are less expressive facially than individuals with left hemisphere lesions (Borod et al., 1985, 1992). It is thought that this type of decrease in facial expression is especially seen with right-sided pre-*rolandic* lesions (Ross, 1981) and mediofrontal lesions that include the *supplementary motor area* and *cingulate gyrus* (Tucker and Frederick, 1989).

The ability to recognize facial expression has also been considered. Studies of normal individuals suggest that the right hemisphere is faster and more accurate than the left in recognizing facial expressions of emotion (Ley and Bryden, 1979; Strauss and Moscovitch, 1981). Additionally, patients with right hemisphere injury show a greater decrease in recognizing facial expression of emotion (Kolb and Taylor, 1981; Bowers et al., 1985). It has also been found that recognition of emotional tone in speech is specifically impaired by right hemisphere damage (Ross, 1981). These results can be explained by the right hemisphere's increased ability to process the larger holistic perspective of visual or auditory stimuli (global vs. local).

The right hemisphere is better for complex visuo-spatial configurations such as facial processing, especially when global organization is required. This is an example of low spatial frequency as opposed to focusing on small details, which requires high spatial frequency processing (DeRenzi and Spinnler, 1966; Rizzolatti et al., 1971; Kitterle et al., 1992). It is also thought that the same process exists for the interpretation of auditory stimuli, such as speech prosody. The right hemisphere is specialized in processing auditory stimuli in which the emotional message is embedded.

Again, low temporal frequency stimuli are processed more by the right hemisphere; high temporal frequency by left hemisphere (Sidtis, 1980; Ivry, 1991) suggesting that low temporal frequency is the primary auditory parameter characterizing emotional prosody. The right hemisphere is also superior in processing emotional words (Graves et al., 1981; Ley and Bryden, 1983). *Aphasic* patients who are thought to have dysfunction of the left hemisphere have been found to read preferentially, emotional compared to non-emotional words (Landis et al., 1982). It has also been reported that right hemisphere patients have more difficulty in remembering emotional as opposed to non-emotional stories (Wechsler, 1973). This is supported by evidence that the right hemisphere is important in the ability to appreciate verbal or pictorial humor, metaphors, and connotative meaning. Individuals with decreased function of right hemisphere are more likely to have literal interpretation and inappropriate emotional responses (Gardner et al., 1975, 1983). Patients with right hemisphere deficits have difficulty in picking out similarity of facial expression from photographs, whereas they do not have difficulty when similarity judgments involve emotional words, suggestive of a more perceptual than a symbolic or verbal defect in emotional processing (Etcoff, 1984). It has been shown that the right hemisphere is dominant in both identifying faces as well as recognition of emotional expression and that the ability to identify one factor is not dependent on the other (Strauss and Moscovitch, 1981).

Prosopagnostic patients have been shown to have normal autonomic responses when presented with familiar faces that they cannot identify. However, patients with *orbito-frontal* lesions, especially on the right, are able to recognize familiar faces, but have dampened emotional responses to them (Tranel and Damasio, 1985). These results are thought to suggest parallel processing of both cognitive and emotional aspects of emotional stimuli (Liotti and Tucker, 1995). It is thought that the right hemisphere may utilize nonverbal or emotional communications. This would focus on the use of facial, prosodic (intonation), and gestural aspects of nonverbal signs involved in social communication (Blonder et al., 1991). In fact, individuals with right hemisphere dysfunction have difficulty not only with facial expression and prosody but also with facial, prosodic, and gestural expression using verbal language (Blonder et al., 1991). Right hemisphere lesions decrease an individual's ability to generate and use visual images of emotional expressions (Cancelliere and Kertesz, 1990).

Anterior and posterior cortical regions appear to express different levels of control over the vertical hierarchy of subcortical centers (Robinson et al., 1984; Starkstein and Robinson, 1988). It is further thought that the right hemisphere global conceptual skills may be critical to combining external and internal environmental information to achieve the integration of emotional experience (Safer and Leventhal, 1977). In other words, it is thought that the right hemisphere is able to access internal feelings that monitor individuals internal state (Buck, 1985). It appears that the right hemisphere has greater interconnectivities between areas of the right hemisphere than the left hemisphere (Leisman and Ashkenazi, 1980; Tucker, 1991). This suggests that the specialized psychological abilities are due to its more diffuse interconnections, which provide it with a more dynamic and holistic integration across different sensory modalities (Semmes, 1968). EEG studies appear to confirm this belief, showing in both children (Thatcher et al., 1986) and adults (Tucker et al., 1986) with greater coherence among right hemisphere regions than left (Leisman and Ashkenazi, 1980; Leisman, 2002) (cf. Fig. 9.1). Dense interconnections between regions are characteristic of the *paralimbic* cortex. In fact, the greatest density of interregional connections is achieved by the *paralimbic* cortex (Pandya et al., 1988). If connection density reflects the level of functional integration, it is thought that the brain's functional integration would be more likely to occur in the denser *paralimbic* areas than in higher "association" areas (Tucker, 1991). It is also thought that greater connection density in the right hemisphere would indicate that its representation would be formed with greater interaction with *paralimbic* influences. Recent research has shown that the antero-posterior location is an important factor in predicting the occurrence of emotional changes that result from unilateral lesions. If the location of a lesion is more anterior, it is more likely to cause depression in the left, and the severity of the depression depends on how far from the frontal pole the lesion is (Robinson and Szetela, 1981). It has also been noted that in right hemispheric lesions, anterior ones are associated with pathologic cheerfulness and apathy whereas posterior lesions in the *parietal lobe* are more likely to be associated with depression (Robinson et al., 1984).

Besides anterior/posterior relationships, other important relationship studies are cortical/subcortical. Several metabolic studies in unipolar depression have shown changes in blood flow and glucose utilization in subcortical nuclei such as the *caudate nucleus* (Baxter et al., 1985; Buchsbaum et al., 1986). Left anterior subcortical lesions (body and head) of the *caudate* and anterior limb of *internal capsule* have been shown to be significantly associated with major depression. Interestingly, the severity of the depression correlates with the distance from the frontal pole, in the same way as it does with cortical lesions (Startstein et al., 1987). Also, it has been shown in patients with unilateral *Parkinson's* disease, that patients with right-sided (left *basal ganglia*) as opposed to the left-sided symptoms reported more depression. This is thought to be consistent with left *caudate* lesions and greater depression (Starkstein et al., 1990). In addition, post stroke patients with major depression have been shown to have more subcortical atrophy than nondepressed patients (Startstein et al., 1998).

It is fairly well documented that subcortical structures like the *basal ganglia* are associated with the development of affective disorder. A possibility is that lateralized cerebral changes may result from the release or disinhibition of ipsilateral subcortical centers that may have a vertical hierarchy of emotional control (Tucker, 1981; Tucker and Frederick, 1989). Poeck (1969) and Rinn (1984) in their research noted involvement of subcortical structures mainly the *basal ganglia* and the internal capsule in practically all the cases they studied of patients who suffered from pathologic emotional outbursts in the absence of *brainstem* lesions. Subcortical centers may be able to affect emotional expressions independent of the cerebrum; an example of this may be seen in *anencephalic* newborns that show normal facial expression of emotion (Buck, 1988). It has been postulated that with lesions that exclude the frontal convexity, both dorsolateral and dorsomedial *prefrontal* cortices are associated with slowness, indifference, apathy, and lack of initiative. On the other hand, lesions of the *orbito-frontal* cortex appear to lead to disinhibition, lack of social constraint, hyperactivity, grandiose thinking, and euphoria (Kleist, 1931; Benson and Blumer, 1975). Others have shown a relationship between secondary mania being associated with *orbito-frontal* and basal-temporal lesions, especially in the right hemisphere. In addition, mania is more frequent in right subcortical lesions of the *thalamus* and *basal ganglia* as opposed to left (Starkstein et al., 1987; Starkstein and Robinson, 1988).

In an attempt to explain the findings, it has been noted that many *manic* symptoms, such as euphoria, hyperactivity, and insomnia can be explained by an interruption of the control normally thought to be exerted by the *orbito-frontal* cortex over the *septal, hypothalamic,* and *mesencephalic* regions (Nauta, 1971). Therefore, *mania* may be a result of disinhibition of normal subcortical centers because of cortical dysfunction. This has been observed in rats where right hemisphere lesions, especially in the *frontal lobe,* result in hyperactivity (Robinson, 1979; Pearlson and Robinson, 1981). Similar effects have been reported with *orbito-frontal* or anterior *cingulate* ablations (Tucker and Derryberry, 1992). Studies utilizing PET scans (Drevets et al., 1992) reveal that depressed patients have increased blood flow in the *orbito-frontal* cortex and parts of the *cingulate gyrus,* primarily on the left side. This is seen in association with reduced metabolic activity in *temporal* and *parieto-occipital* regions. A similar pattern of activation of the left ventro-lateral *prefrontal* cortex is found in PET studies of normal subjects told to think sad thoughts or memories (Pardo et al., 1991). It has also been noted that there seems to be a metabolic relationship between anterior and posterior cortical areas. There also appears to be an inverse relationship in some cases between these hemispheric areas in depressed patients. Additionally, an inverse relationship has been reported between frontal and posterior activation in normal EEG studies (Tucker et al., 1981; Davidson, 1984; Davidson et al., 1987) and are possibly related to the posterior right hemispheric decreases noted in depressed patients (Tucker and Liotti, 1989). Also noted is an imbalance of activity with increased activity of *para-limbic* and *limbic* areas in relation to neocortex during periods of intensive emotional states and is thought to show a release or disinhibition of the *limbic system* involved in emotional processing (Nauta, 1971).

It is common to find patients with *Parkinson's* disease also suffer from depression. It is thought that this is due to disruption of frontal cortical function that causes the depression. The *caudate* has direct and indirect connections to the *frontal lobe* through the dorso-medial *thalamus*. There are five functional parallel *cortico-striatal-thalamic* loops; two of them are purely motor (Alexander et al., 1986). The three non-motor loops have different levels of *limbic* connections. It is thought that each loop represents a functional unit, which includes as the primary target, the *prefrontal* region. One of these non-motor loops, a dorso-lateral *prefrontal* network is thought to support temporary storage in working memory of spatial locations (Goldman-Rakic, 1987a, 1987b, 1987c), or rules for stimulus response contingencies (Fuster, 1992). There is also a lateral orbito-frontal circuit, projecting from the *orbito-frontal* cortex and connecting to different parts of the *caudate* and *globus pallidus,* which project to the *thalamus* and back to the *orbito-frontal* cortex. This circuit is thought to be involved in the control of inhibitory responses during learning and recognition tasks requiring frequent shifts of set. This may explain *perseveration* or repetitive compulsive behavior seen with damage to the *orbito-frontal* cortex. Another circuit, the anterior *cingulate* circuit, includes the ventral *striatum, nucleus accumbens*, and medio-dorsal nucleus of the *thalamus*. The *hippocampus* and *entorhinal* cortex is thought to send inputs in this circuit, which integrates information from the *para-limbic association* cortex.

The *nucleus accumbens* is also the target of *dopaminergic* terminals from the *mesencephalic* ventral *tegmental* area. This loop's function is thought to include the integration of emotional cues and the individual's response to the hedonic quality of a stimulus. It is thought that a disconnection of these loops may be related to depression and other affective and cognitive disorders. PET studies were conducted on two different groups of patients that had *basal ganglia* strokes (Mayberg et al., 1988). One group included patients that suffered lesions of the motor loops (*putamen* alone, or *putamen* and posterior *internal capsule*). The second group included patients with lesions to the non-motor loops (head of the *caudate* alone, *caudate* and anterior limb of *internal capsule*, and anterior and dorsomedial *thalamus*). Patients with *putamen* lesions and motor symptoms were found to have widespread ipsilateral and *thalamic* hypo-metabolism. Those patients with *caudate* strokes with cognitive symptoms had focal ipsilateral hypo-metabolism involving localized areas in the *frontal, temporal,* or *cingulate* cortex. In a different study, patients with stroke in the head of the *caudate* were divided into euthymic, depressive, and manic patients. Compared to euthymic patients, all the patients with mood change showed hypometabolism in the *orbital-frontal* cortex, anterior *temporal* cortex, and *cingulate* cortices (Mayberg, 1992).

Some of the ascending neurotransmitter symptoms, especially the *norepinephrine (NE), 5-hydroxytryptamine (5-HT),* and *serotonin,* thought to be dysfunctional in depression, are inherently lateralized to the right hemisphere (Tucker and Liotti, 1989). Studies in rats show right-sided ischemic lesions and not left, result in a depletion of brain and *locus ceruleus NE* bilaterally, and the more anterior the lesion the greater the depletion (Robinson, 1985). In addition, a greater concentration of plasma *NE* and resulting BP increases are noted (Hachinsky et al., 1992). Lesions in the *frontal* cortex or *basal ganglia* may possibly interrupt the connecting fibers since they pass through the *basal ganglia* and ascend to the frontal pole. In another PET study, Mayberg and associates (1988) showed that patients with right hemispheric strokes have significantly higher ratios of ipsilateral to contralateral binding for S_2 (*serotonin*) receptors than patients with left hemispheric strokes. This means that up-regulation of S_2 receptors occurs in the non-injured *temporal* and *parietal* cortex. In addition, in the left hemisphere stroke group the S_2 binding ratio significantly correlates with the severity of depression. This is thought to show greater depression in left *frontal* and *basal ganglia* lesions, yet more ipsilateral S_2 up-regulation for right-sided lesions. The authors explain this finding thinking that a right hemisphere stroke would produce higher compensatory up-regulation in the intact ipsilateral tissue than the left hemisphere lesion would. Therefore, a left lesion would produce a relative S_2 and *NE* depletion that would not be accompanied by an adequate up-regulation of ipsilateral receptors. This may be enough to result in depression (Mayberg et al., 1988). These same findings in S_2 binding have been reproduced in rats (Mayberg et al., 1990). It is also thought that dysfunction of *NE* may be associated with *mania* in humans; right hemispheric lesions involving the *limbic* regions may cause secondary *mania* (Robinson et al., 1988; Starkstein et al., 1991). In rats, induced lesions of the right middle cerebral artery result in hyperactivity significantly more frequently than left-sided lesions do (Robinson, 1995). *Dopamine (DA)* terminals are thought to specifically connect to *cortico-limbic* and *limbic* regions, especially the *prefrontal* and *cingulate* cortex, *nucleus accumbens*, and *amygdala*. Most of these fibers arise in the *mesencephalic ventral tegmental area (VTA)*. A significant loss of *DA* neurons in the *VTA* has been found in a group of patients with

Parkinson's disease and depression (Torack and Morris, 1988). A dysfunction of the *meso-limbic DA* system is thought to play a role in the production of deficits in *"frontal lobe"* tasks within *Parkinson's* disease patients with major depression. It is further thought that emotional states may entail asymmetric activation of *DA* pathways. In studies where stress is induced by electric foot shock, conditional fear, food deprivation, and restraint, it is found to increase *DA* metabolism in the terminal regions of the mesocortical (*prefrontal* cortex), or *meso-limbic* (*nucleus accumbens*) systems in rats (Thierry et al., 1976; Lavielle et al., 1979). It has been shown that stress induced *DA* metabolic changes show hemispheric differences that depend on the amount of time exposed to the stress. Prolonged uncontrollable foot shock stress, or long-lasting restraint lead to the organism equivalent of depression. They also result in greater activation of *prefrontal DA* in the right hemisphere (Fitzgerald et al., 1989; Carlson and Glick, 1991; Carlson et al., 1991). However, brief food deprivation (24 hrs), or brief restraint stress (15 min), results in greater activation of *DA prefrontal* metabolism in the left hemisphere (Carlson and Glick 1991; Carlson et al., 1991). This increased *prefrontal DA* metabolism in an organism model of depression and stress may be relevant to the finding of increased frontal activation of the EEG (alpha desynchronization) in depressed patients (Perris et al., 1979). It may also prove to be an increased frontal inhibitory influence in depression (Tucker et al., 1981).

It has been hypothesized with regard to neural systems for activation and arousal that a "tonic" activation system, supporting motor readiness applies an influence on ongoing information processing or a "redundancy" bias. This influence relying on controls from the *mesolimbic* and possibly *striatal DA* projections maintains not only motor readiness, but also a focal form of attention. The authors further postulate that this control mode is inherent to states of anxiety and hostility. Furthermore, consistent with the greater left lateralization of *DA* projections, this primitive control mode of the "redundancy bias" may underlie focal attention and analytic cognition shown by the left hemisphere (Tucker and Williamson, 1984). A different control bias is thought to emerge from the *NE* projections, which support a "phasic" arousal system. This system applies a "habituation bias" to attention that is thought to be critical to the orienting response to novel stimuli, and to perceptual orienting processes generally. Consistent with this hypothesis, increased *NE* activity in *mania* is thought to lead to a shortened attention span and stimulus seeking, both of which are produced by the habituation of bias of an overactive "phasic" arousal system. This system is thought to have both attentional effects and affective qualities, where high "phasic" arousal is associated with *mania* and low "phasic" arousal associated with depression. The inherent right lateralization of the "phasic" arousal systems is thought to be integral to the right hemisphere's control of elementary attentional orienting as seen in neglect, arousal, and affective functions evidenced by its responsiveness to mood variations (Tucker and Liotti, 1989; Liotti et al., 1991; Liotti and Tucker, 1992; Liotti and Tucker, 1995).

There is always the question of whether symptoms that are produced are due to injury or dysfunction of ipsilateral controls or contralateral release. To explore this question patients with unilateral epilepsy were studied (Tucker, 1981). It was found, by Bear and Fedio (1977) that patients with left *temporal lobe* epilepsy presented with clinical symptoms and a psychometric profile that showed ideational exaggeration in their cognitive function. These subjects became focused on moral ideas or religious concerns. In contrast, individuals with right *temporal lobe* epilepsy had an exaggeration of emotional processes, suggesting dysfunction of affective self-regulation. This suggested that epilepsy caused a hyper-connection of cortical *limbic* structures. In addition, it was shown that left *temporal lobe* epilepsy subjects could emphasize their negative characteristics with self-evaluation, whereas right *temporal lobe* subjects would deny problems and emphasize their positive traits which was thought by Tucker (1981) to suggest ipsilateral release. In analysis of how this could be related to dorsal and ventral cortical pathways, Bear (1983) suggested that right hemisphere specialization for spatial functions of dorsal cortical pathways may be significant to both the attentional deficits (*spatial neglect*) and the emotional deficits (denial and indifference) seen following right hemisphere damage. It is thought that the right hemisphere is responsible for "emotional surveillance" and when there is a dysfunction, this function is decreased; therefore, the individual does not recognize the significance of emotionally important information, hence emotional neglect. In regard to left hemisphere, it has been suggested that affective traits may have evolved in line with its expansion of the motivational and cognitive functions of the ventral *limbic system*. Therefore, the emotional qualities of anxiety and hostility modulated by the *amygdala* and ventral *para-limbic* cortex including *orbital-frontal* regions may be especially important to mediating left hemisphere function, so that hyper-connection of *limbic* and cortical functions, as witnessed in left *temporal lobe* epilepsy, results in a more negative critical affect. Subjects who were tested on visual field effects appear to show different cognitive styles based on hemispheric strengths. Individuals who show a strong right visual field effect on a symbolic identification task, thought to emphasize left hemisphere function, demonstrate a negative more critical bias in judging their left field performance. Those that we thought to show a more right hemispheric cognitive style show a more positive bias judging their left field performance. Those that are thought to show a more right hemispheric cognitive style show a more positive bias in rating their left visual field performance (Levy et al., 1983). In other words, the left more analytical hemisphere is more critical or pessimistic of performance whereas the right hemisphere-holistic individuals are less critical or more optimistic in their self-assessments. This suggests that there are important limits between the two hemisphere's cognitive capacities and different styles of emotional self-regulation (Liotti and Tucker, 1995).

It is thought that right and left hemisphere differences in information processing reflect different elaborations of dorsal and ventral cortical regions. In regard to information processing within the visual system, a model has been proposed based on two parallel visual systems, one being connected ventrally to the *inferotemporal* cortex, and one connected dorsally to the inferior *parietal lobe* (Ungerleider and Mishkin, 1982). In monkeys, lesions of the *inferotemporal* cortex result in severe deficits in performance in a number of visual discrimination learning tasks, but not in visual-spatial tasks. However, lesions of the posterior *parietal* cortex result in significant defects in visual-spatial tasks but do not affect visual discrimination performance (Ungerleider and Mishkin, 1982). The ventral visual pathway terminates in the inferior *temporal lobe*. Neurons in this area have large, bilateral receptive fields thought to be responsible for mediating the perceptual equivalence of objects over translation in retinal position. A small percentage of cells respond specifically to higher-order complex stimuli such as faces or hands. Most of the neurons respond specifically based on object features like color, shape, and texture as opposed to specific objects. It has been shown that lesions of the *inferotemporal* area result in loss of perceptual constancies and in profound loss of visual memories (Mishkin et al., 1984). Memories of visual inputs would be limited to a "habit" learning system involving the *basal ganglia* (Mishkin et al., 1984). Mishkin thinks that the *inferotemporal* area contains "central representation" for visual objects and that damage results in the availability to form new memories as well as the loss of old memories. Limbic lesions especially those specifically of the *hippocampus* have been found to impair memory storage only temporarily. The dorsal visual pathways, which terminate in the inferior *parietal lobe* and in the *intra-parietal sulcus* are thought to provide the basis for perceptual vision, motion perception, stereopsis, perception of three-dimensionality of objects based on perspective and shading, as well as most of the gestalt phenomena of "linking operations" (Livingstone and Hubel, 1987).

In monkeys and humans, the posterior *parietal* cortex is one of the areas involved in control of spatial attention. Lesions of this area decrease the ability to direct attention to the contralateral side when presented with directional cues (Posner and Cohen, 1984; Posner et al., 1987). The ventral system has direct reciprocal projections to the *amygdala* and to the *orbito-frontal* cortex. The dorsal system projects to the *cingulate gyrus*, the *hippocampus*, and surrounding *limbic* areas. Therefore, these regions are thought to participate in the perceptual and mnemonic function of two visual streams, providing the integration of emotional, motivational, and interceptive quality of visual percepts. Developmental studies have shown that neocortical visual areas like inferotemporal cortex evolved from the *olfactory* cortex and the *hippocampus*. This is thought to suggest that the role of *limbic* representations within the dorsal and ventral systems goes beyond anatomic and functional connections to neocortical pathways.

In origin, the two visual pathways may be entirely *limbic* (Liotti and Tucker, 1995). The neutral connections that subserve visual object recognition include connections to two *limbic* structures, the *amygdala*, and the *orbito-frontal* cortex (VanHoesen and Pandya, 1975). Bilateral change to the *amygdaloid complex* produces severe and long-term recognition loss (Mishkin, 1978; Zola-Morgan et al., 1982). Lesions that include the neocortical temporal areas produce disturbances in visual discrimination learning. However, lesions of the temporal pole and *amygdaloid* nuclei cause changes characteristic of *Klüver-Bucy* syndrome (Ledoux, 1987). It is thought that the emotional components are added to visual input via connections between the *temporal* neocortex and *amygdala* (Liotti and Tucker, 1995). It is therefore possible that without these connections it would not be possible to associate reward conditions with visual stimuli. In humans, electrical stimulation of the *amygdala* results in the complex combination of perceptual, mnemonic, and affective features that have been noted in the *temporal lobe* epilepsy (Gloor, 1990). This is thought to indicate that the *amygdala* may play a crucial role not only in combining emotional cues to perceptual stimuli, but also in integrating perceptual and affective memories into a subjective experience (Ledoux, 1987; Gloor, 1990). A direct monosynaptic connection between the *thalamus* and *amygdala* has been noted in rabbits (Ledoux, 1987). This is thought to precede the *amygdala* with rapid primitive representations of the peripheral stimulus, as opposed to the highly processed and more precise neocortical inputs to the *amygdala*. Therefore, it is speculated that the *thalamo-amygdala* projection may exist to prepare the *amygdala* for the subsequent reception of highly processed information from the neocortex (Ledoux, 1987).

The *orbito-frontal* cortex has been described as the neocortical representation of the *limbic system* (Nauta, 1971). This area of cortex direct connections to the temporal pole and the *amygdala*, and is thought to be the direct frontal extension of the ventral visual pathway. Monkeys with *orbito-frontal* lesions have been noted to present with significant difficulty with visual object learning (Thorpe et al., 1983). The orbital frontal region is also thought to play a significant role in visual object memory by creating connections between stimuli and reward, especially when the task calls for frequent making and breaking of the stimulus–reward association (Ledoux, 1987; Derrybery and Tucker, 1992). Without *orbito-frontal* neurons, it has been shown that the individual will continue to respond to stimuli that are not rewarded. This is thought to explain the *perseverative* behavior documented in monkeys and humans with injury to the *orbito-frontal* cortex (Sandson and Albeit, 1984).

Regarding learning and unlearning the emotional import of sensory input, it is thought to require a close relationship between *amygdala* and the orbital regions (Liotti and Tucker, 1995). The *orbito-frontal* cortex modulates septal, *hypothalamic*, and *mesencephalic* areas

(Nauta, 1971). Abnormal social behavior and *manic* behavior may occur due to the disinhibition of subcortical areas resulting from *orbito-frontal* dysfunction, especially right-sided lesions in humans (Starkstein and Robinson, 1988). Dysfunction of the *orbito-frontal* cortex may have a cognitive explanation (Damasio and Tranel, 1988). It has been hypothesized that disinhibited emotional behavior may be thought of as a category-specific memory dysfunction. In this case, the ventro-medial *frontal* cortex would store complex specialized memories about social events. A lesion of this area of the *frontal lobe* therefore would cause an "acquired sociopathy." It is thought then that the orbitofrontal cortex would be one of the highest convergence zones in this region; complex aspects of spatial and temporal features that identify social events would be stored as memory and could be accessed later (Damasio, 1989a, 1989b).

Massive inputs are sent from the *limbic* areas to the posterior parietal lobe. Those *limbic* areas include the *cingulate gyrus*, the *retrosplenial* area, and the *cholinergic nucleus basalis*, which has a nonspecific distribution to other cortical areas. The *cingulate* and *retrosplenial* projections are more selective and may be related to complex and learned aspects of motivation. This connection between *limbic system* and *parietal* cortex is thought to allow parietal neurons to correlate motivational relevance in complex sensory events (Nesulan, 1985). Studies of monkeys with unilateral *cingulotomy* exhibit contralateral inattention in visual and somatosensory modulation (Watson et al., 1973). Infarcts of the right *parietal* or medial *frontal* cortices that also include the *cingulate* cortex result in unilateral neglect (Heilman et al., 1983). It also results in a significant inability to address spatial memory in that half of the environment (Mesulam, 1985). It has been proposed that inattention due to removal of *cingulate* may be due to inability to understand the significance of the stimuli (Pandya and Yeterian, 1985). Therefore, the *cingulate gyrus* may play a role in adaptive attention as opposed to the spatial orientated attention found in the posterior *parietal lobe*.

The creation of spatial memories is thought to occur by two different loops (Pandya and Yeterian, 1985). The first goes from *area 7* and the *cingulate gyrus* to the *hippocampus* by way of the *presubiculum*. This is thought to be a fast-acting memory pathway for significant objects located in space. Damage to this connection bilaterally produces an *akinetic* state and attentional dysfunction, which is thought to imply a role that pertains to immediate survival. The second circuit connects *area 7* and the *cingulate* to the *hippocampus* and *amygdala* by way of the *parahippocampal gyrus*. This is a slower pathway that is involved in complex spatial memories. Bilateral damage in this pathway causes a deficit in recognizing geometric forms and faces (Parkinson and Miskhin, 1982).

Evolutionarily, it is thought that the ventral cortical system that is specialized for object recognition and central vision evolves from the paleo-cortical structures of the temporal pole and *olfactory* cortex. The evolution of the dorsal cortical system for spatial perception and memory in peripheral vision is thought to have developed from the archi-cortical structure of the *hippocampus* (Pandya et al., 1988). This may be the foundation of reciprocal *cortico*-cortical connections. This "backward" feedback from *limbic* areas to higher centers may be how the *limbic system*, which controls motivational and emotional states, exerts regulatory and adaptive functions and helps to determine what input will dominate information processing and behavior of the brain as a whole (Derryberry and Tucker, 1991; Tucker, 1991). Specific lesions of the right posterior cortex, including the *parietal lobe*, have been shown to decrease visual judgments of line orientation (Benton et al., 1978), visually guided maze performance (DeRenzi et al., 1978), as well as the ability to identify objects presented from an unusual visual perspective (Warrington, 1982). Right posterior cerebral injury has also been shown to produce *prosopagnosia* and constructional apraxia (Hécan and Angelergues, 1962; Arrigoni and DeRenzi, 1964). There is also thought to be a right hemisphere/left visual field advantage for depth perception (Kimura, 1961b), spatial localization (DeRenzi, 1982), and for identifying complex geometrical shapes (Umilta et al., 1978), as well as for tasks that require exploratory eye movements (Sava et al., 1988). The left hand has been shown to be more accurate in judging the orientation of a rod by palpation (Benton et al., 1978). The right hemisphere can attend to both sides of extrapersonal space but the left hemisphere can only attend to contralateral space (Mesulam, 1981). Cortical and subcortical areas that mediate visual orienting is thought to consist of areas that are functionally part of the dorsal visual system including posterior *parietal*, the *cingulate cortex*, the dorsolateral *frontal* cortex (Heilman, 1979; Mesulam, 1981), *inferior colliculus*, and *pulvinar*. The right hemisphere is responsible for the ability to explore and navigate in the environment and to position attention to specific locations where specific objects are to be found (Kosslyn, 1987). Injury to the left *occipital-temporal* cortex may be sufficient to result in visual object *agnosia* (McCarthy and Warrington, 1990). The left side is involved in object recognition for objects shown in a visual view, and identification of stimulus as natural or manmade (Sergent et al., 1992). The left hemisphere maintains long-term semantic representations—the lexicon—to be able to identify or categorize a visual object (Benton et al., 1978). The left hemisphere is better at line orientation, when a line is vertical, horizontal, or at 45-degree diagonal orientation. The right hemisphere is better for detecting the oblique orientation of a line (Umilta et al., 1974).

Temporal-occipital lesions of the left hemisphere have been shown to result in decreased ability in the generation of images of visual objects (Farah, 1984; Kosslyn, 1987). The left hemisphere is also thought to play a role in access and long-term storage of categorized, prepositional, or semantic information about an object (Kosslyn, 1987). In addition, sequential information (high spatial frequency) of auditory stimuli is superior in the left hemisphere (Ivry, 1991). Broca's area *44* is part of *prefrontal* area *6,* which is thought to be important for the programming of learned motor sequences such as sign language and *ideomotor praxis* (Heilman and Vallenstein, 1985). The lateral *premotor* region of the *frontal lobe* of the monkey is thought to have evolved phylogenetically from the cortex of the olfactory *paleocortex*, which is the *limbic* core of the ventral visual pathway (Goldberg, 1985). Therefore, this aspect of left hemisphere sequential motor processing may take advantage of the cybernetics of the ventral *cortico-limbic system* (Liotti and Tucker, 1995).

What we see is that hemispheric specialization for cognitive and perceptual processes are a byproduct of different elaborations of the dorsal and ventral cortical networks. The function of the cerebral cortex needs to be viewed in terms of balances between not only left/right, but also between anterior/posterior and dorsal/ventral. *Cortico-limbic* interactions are therefore located at the crossroads of the essential regulatory functions of subcortical structures and the flexible representative functions of the massive cortical networks (Liotti and Tucker, 1995). One study has found left *cerebellar* abnormalities in depressed patients versus controls who were studied by PET during a spatial matching test and an inability of the depressed patients to activate the dorsal lateral *prefrontal* cortex. However, they did activate a wider region of the temporoparietal cortex. The depressed patients were shown to unusually activate the left *cerebellum* (George et al., 1994).

ATTENTIONAL ASYMMETRIES

It is known that the brain is supplied with more simultaneous sensory input than it can possibly process. Therefore, it is thought that the brain must create priorities based on its goals and needs (Leisman, 1976a, 1976b). Based on these priorities the brain chooses a portion of the stimuli to process fully (attend to stimuli) and a portion of the stimuli that it will not fully process (unattended stimuli). The brain also decides when to continue attending to stimuli or space, waiting for stimuli (vigilance), and when to stop attending to stimuli (extinction or habituation) (Heilman, 1995).

Besides attention, there is also intention. In intention, there is thought to be two *premotor* pathways that influence motor acts. The *praxis* system is crucial for programming the turning, trajectory, and sequencing of motor acts; the *praxis* system is also known as the "how system" (Heilman and Rothi, 1993). The intention, or "when" system, however, supplies the goals and needs of the motor acts. In a similar way to the attentional system, there are thought to be four intentional instructions "when" to initiate a motor act, "when" not to initiate a motor act, "when" to continue a movement, or "when" to stop a movement. It is thought that the attentional (Heilman, 1995) and intentional systems are asymmetrically organized. Visual inattention can be confused with *hemianopia* as both can result in loss of a specific visual field. *Hemianopia* can be differentially diagnosed from body-centered inattention by instructing the patient to gaze to the side of the brain lesion. Patients with inattention will not perceive the contralateral stimuli when looking straight ahead or toward contralateral hemispace. However, when their gaze is directed to ipsilateral hemispace, patients may be able to perceive the visual field (Kooistra and Heilman, 1989; Nadeau and Heilman, 1991), although *hemispatial neglect* is more commonly associated and is more severe with right than with left hemisphere lesions (Heilman, 1995).

Tactile Inattention

Right-sided cerebral lesions have been shown to produce contralateral *hemianesthesia* even though there are normal electrical skin responses or evoked potentials (Vallar et al., 1991a, 1991b). An interesting finding is that symptoms associated with *neglect* syndrome may temporarily improve after stimulation of the contralateral ear with cold water (Rubens, 1985). Vallar and coworkers (1990) have reported that their patients who appear to have *hemianesthesia* have it reduced with caloric stimulation. Each of their patients had left *hemianesthesia* resulting from right cerebral lesions. This is interesting because caloric stimulation primarily affects the *vestibular apparatus* and the *vestibulocerebellum* on the ipsilateral side. The *vestibular* stimulator may therefore increase firing to the *thalamus* in order to increase the *thalamic* nonspecific stimuli to increase the attentional system's activation thereby improving the temporal binding of stimuli. Alternatively, this could affect the specific pathways to the *somatosensory* area of the *parietal lobe* to increase perception or stimulation of the *prefrontal* cortex and the cognitive or attentional abilities specific to the right *frontal lobe*. Patients who suffer with what is known as *extinction* are able to perceive contralesional stimuli. However, when presented with bilateral simultaneous stimuli, they cannot perceive the contralesional side. This has been shown to occur much more frequently with the left hand than with the right hand (Schwartz et al., 1979). In addition, it has been shown that contralateral *extinction* is more commonly associated with right hemisphere than with left-sided lesions (Meador et al., 1988). When testing individuals with unilateral lesions they may show specific unilateral neglect. For example, patients may not eat off one side of their plate or when asked to copy a picture they may only copy one half (the ipsilesional) of the picture. Additionally, when asked to bisect lines they may favor the ipsilesional side. The severity of *hemispatial neglect* may be decreased by providing attentional cues (Riddoch and Humpreys, 1983). It is thought that these attentional cues may modify the distribution of attention in a "top down" manner.

As opposed to attentional cues, novel stimulation may affect attention in a "bottom up" fashion. It has been shown that novel stimuli presented on the contralesional side reduce *hemispatial neglect* (Butters et al., 1990). One hypothesis is that some patients with *hemispatial neglect* do not show inattention but are just simply less aware of stimuli on the left versus the right. In other words, they have ipsilesional attentional bias. It is also possible that patients with hemispatial neglect have other objects in ipsilateral space that draw their attention and in this case, the bias is so strong that these patients cannot draw their attention away from these objects.

AROUSAL

Arousal has two definitions, both behavioral and physiologic. An aroused organism is alert; it is prepared to process incoming information. An unaroused organism is thought to be comatose. Such an organism is therefore not prepared to process information and is not even aware of stimuli. From a physiologic perspective, arousal refers to the excitatory state or the propensity of neurons to discharge when appropriately activated (neuronal preparation) (Heilman, 1995). This can also be referred to as being closer to threshold; an aroused organism will have neurons that are closer to threshold at a resting state than an un-aroused or unaware organism. An aroused organism is prepared to process information and therefore may require less stimulation to reach the processing level because their neurons are closer to threshold. Whereas a less aroused organism is not as prepared to process incoming information, their neurons are further from threshold and may require more stimulation to reach the processing level.

Changes in arousal are often reflected equally by changes in the level of activity of the peripheral *autonomic* nervous system. Using *autonomic* response as a measurement of arousal, investigators have shown that patients with right hemisphere lesions have reduced responsivity when compared with the left hemisphere patients or to normal uninjured control patients (Heilman et al., 1978; Schmidt et al., 1989). Using EEG as a measure of control arousal, it has been shown that individuals with right hemisphere lesions have more *delta* and *theta* activity in their non-lesioned left hemisphere than left hemisphere-lesioned patients over their right hemisphere (Heilman, 1979). The ability to sustain attention is known as *vigilance*. *Arousal* and *vigilance* are closely related so that when *arousal* decreases *vigilance* does as well and vice versa. Patients with right hemisphere lesions have been found to have decreased sustained attention as compared to patients with left hemisphere injury (Wilkins et al., 1987). Split-brain patients have shown that the right hemisphere is more vigilant than the left (Dimond, 1979). Blood flow studies have shown a greater increase in blood flow to the right hemisphere versus the left with sustained attention tasks (Deutsch et al., 1987). In general, lesions of the inferior parietal lobe are most often associated with disorders of attention (Critchley, 1966; Heilman et al., 1983). *Temporo-parietal* lesions in monkeys have been shown to result in contralesional attentional disorders especially *extinction* (Heilman et al., 1970; Lynch, 1980). The *parietal lobe* also receives input from other sensory modalities and therefore the coding of spatial location is most likely multimodal. Primary sensory cortices project only to their *association* cortex. Therefore, all of these single modality association areas converge on multimodal association areas; in monkeys, this includes the *frontal* cortex (*periarcuate, prearcuate,* and *orbito-frontal*) and the superior *temporal sulcus* (Pandya and Kuypers, 1969). Multimodal conversion area involvement can produce pathologic laughing occurring with right hemisphere lesions thought to be important for sensory synthesis and cross-modal association. Multimodal convergence areas project to the supra-modal *parietal lobe* (Mesulam et al., 1977).

Deciding when a stimulus is novel is mediated by the sensory association cortex, whereas the significance of a stimulus requires knowledge of the meaning of the stimulus and the motivational needs and goals of the organism (Heilman, 1995). The *limbic system* is important in deciding the stimulus significance and provides information about immediate biological needs. However, the *frontal lobe* provides input to the attentional systems about goals that are not motivated by immediate biological needs. The *frontal lobe* and *cingulate* connections to the *parietal lobe* may provide the structural basis whereby motivational states affect attentional systems. It has been shown that both in monkeys and in humans, unilateral *neglect* can result from lesions to the dorsolateral *frontal* (Welch and Stuteville, 1958; Heilman and Valenstein, 1972), as well as from *cingulate* lesions (Heilman and Valenstein, 1972; Watson et al., 1973). Profound sensory neglect has been shown to result from lesions of the *mesencephalic reticular formation* (Reeves and Hagaman, 1971; Watson et al., 1974).

Arousal is a physiologic state that prepares the organism for sensory processing by increasing neuronal sensitivity and the signal-to-noise ratio. Stimulation of the *mesencephalic reticular formation (MRF)* increases behavioral *arousal* and is associated with desynchronization of the EEG, a measurement of central arousal (Morruzzi and Magoun, 1949). The *MRF* is thought to affect the cortex primarily through its projections to the *thalamus*, either through the *centralia lateralis* and *paracentralis thalamic* nonspecific nuclei or through the *nucleus reticularis*, which projects to the *sensory thalamic* relay nuclei and may inhibit the *thalamic* relay of sensory information to the cortex (Scheibel and Scheibel, 1966).

When a stimulus is important to the organism, *corticofugal* projections may inhibit the *nucleus reticularis* thereby allowing the *thalamus* to send sensory input to the cortex. One hypothesis is that each hemisphere orients attention to its contralateral space and therefore after a right hemisphere lesion, there exists an ipsilesional bias (Kinsbourne, 1970). However, others think that this bias is not caused by the hypoactive hemisphere (Heilman and Watson, 1977). If each hemisphere attends in a contralateral direction and one hemisphere is hypoactive this may also result in an ipsilateral attentional bias. PET studies have shown lower metabolic activity in the hemisphere contralateral to the lesion thought to be inducing the *neglect* (Fiorelli et al., 1991). It is thought that decreased arousal, resulting from right hemisphere lesions is related to the right hemisphere's superiority of communicating with the *reticular activating system*, so that the bilateral arousal defect associated with right hemisphere lesions is related to a decrease of the right hemisphere's afferent influence on the *reticular formation*. *Vigilance* tasks may be heavily dependent on *frontal lobe* input into attentional systems. It has been shown that patients with primarily right frontal lesions are impaired in sustained attention tasks (Wilkens et al., 1987). The *frontal* cortex has been shown in imaging studies to be activated by *vigilance* tasks (Cohen et al., 1988).

In normal subjects, visual attention is biased toward upper visual space or away from the lower visual field (Heilman, 1995). In cancellation tasks, subjects are more likely to make errors in the section of the paper closer to their body, which is the lower visual field (Geldmacher et al., 1991). It has been noted that visual cognitive activation performed by left hemisphere, such as reading, writing, and motor activity, are performed close to the body or in peripersonal space; whereas visual cognitive activities performed by right hemisphere, such as facial and emotional recognition and route finding, take place away from the body in extrapersonal space (Heilman et al., 1993a).

LEARNING

Classical conditioning is a basic form of learning where a weak sensory stimulus gains signaling qualities due to repeated associations with a strong stimulus (Leisman and Koch, 2003). The associative process therefore changes the weak stimulus into a *conditioned stimulus (CS)* causing a similar response to the one elicited by the strong stimulus (*unconditioned stimulus UCS*). The action that results from the presentation of the *UCS* is the unconditioned response (*UCR*) and the response that follows the simultaneous combination of the *CS* and *UCS* is known as the *conditioned response*. Conditioning may act as a prototype for the study of brain functions involved in unconscious acquisition of knowledge (implicit learning) (Lazarus and McCreary, 1951; Leisman and Koch, 2003). It is thought that the two hemispheres have different capacities to form associations in perceptual analysis of sensory stimuli as well as in the formation of an association into memory. The fact that the two hemispheres are different in their capacity to form associations is thought to be of critical importance for a neuropsychological understanding of learning disabilities (Hugdahl, 1995). The importance of modern views of conditioning is that concepts related to attention, expectancy, and memory is the focus, emphasizing an information processing view (Ohman, 1983; Davey, 1987). It is thought that classical conditioning under certain circumstances is a result of the differences in functioning of the cerebral hemispheres and especially within the right hemisphere (Hugdahl, 1995). *Laterality* is a term typically used to describe various cognitive and emotional functions in different hemispheres of the cerebral cortex. However, more recently the term *laterality* has taken on a more global meaning that includes subcortical, peripheral, autonomic, endocrine, and immune functions. It has been noted that recent improvements in psychophysiologic research have documented a number of important physiologic asymmetries with regard to immune function (Barneoud et al., 1987), neuroendocrine responses (Wittling and Pfluger, 1990), and cardiovascular function (Heller et al., 1990; Weisz et al., 1992). An example is a study showing that the hemispheres are significantly different in their capacities to regulate BP during emotional situations. *Diastolic* and *systolic* BPs are both increased after right hemisphere exposure to film clips of emotional content (Wittling, 1990). It is currently thought that activation of the *autonomic* nervous system may be centrally under different influences of the right and left hemisphere (Werntz et al., 1983). This is suggested by a study that showed that cardiovascular arousal is directly related to stimulation of the right hemisphere (Heller et al., 1990). Other studies have shown much more dramatic *sympathetic* effects following right hemisphere than after left hemisphere damage (Hachinsky, 1992).

Implicit memory refers to an experimental situation where information learned and encoded during a specific point of an experiment is revealed at a later point without the subject having any explicit or conscious awareness of the learning scenario (Graf and Schacter, 1985; Schacter, 1987). This implies that a reaction to a current situation can be affected and influenced by a previous event of which even the subject is unaware (Lockhart, 1989). It is thought that behavior adaptations to changes in the environmental perceptual input are largely "set" by post-*implicit memory* processes. This is thought to be the way in which we are able to cope with a constantly overwhelming information load (Hugdahl, 1995). In fact, it has been argued that the way we behave most of the time is through *implicit memories* of past events (Lockhart, 1989). A classic example of an *implicit* conditioning experiment is one in which subjects are conditioned to visual presentations of nonsense syllables with electric shock to the hand as the *UCS*. Later they are tested with quick exposure to the conditioned syllables mixed with other syllables. Subjects show greater skin conductance response to the *CS* syllables as compared to the control stimuli, even though they are completely unaware of the stimuli (Lazarus and McCleary, 1951). This has been called *subliminal implicit conditioning* phenomena. It has been shown experimentally that patients with visual *neglect* after right hemisphere lesions are capable of implicitly processing stimuli that are presented in the neglected visual field, even when they deny ever having seen the stimuli. Studies examining the *implicit* processing during altered states of consciousness show a right hemisphere dominance (Spiegel et al., 1985).

When subjects are compared in normal and hypnotic consciousness, they are found to have a greater right hemisphere involvement in a dichotic listening task (Frumkin et al., 1978). It has also been shown that individuals with multiple personality disorders show constant differences in electrodermal responses when they "change" personality (Coons et al., 1988). Negative emotional films have a more significant effect on heart rate changes when presented to the right as compared to the left hemisphere (Dimond and Farmington, 1977). *Implicit* learning in *classical conditioning* is thought to be mediated through right hemisphere processes. MRI imaging has shown positive correlations between electrochemical responses and size of *prefrontal* cortical areas. In addition, skin conductance responses are decreased in monkeys after lateral *frontal* lesions and are completely absent after removal of the *dorsolateral* cortex (Bagshaw et al., 1965).

It has been argued that the medial *prefrontal* cortex, as well as the *infra-limbic* components may act as an *autonomic* motor cortex (Cachetto and Saper, 1990). The electrodermal pathway is thought to be controlled by two cortical structures. The first is an ipsilateral system including the *hypothalamus*, the second consists of a contralateral system which includes the *prefrontal* cortex and the *basal ganglia* (Schliar and Schiffter, 1979; Boucsein, 1992). *Implicit* repetition priming particularly activates the right *hippocampal* region in the brain (Squire et al., 1992). The same group also found changes in activity in the right *prefrontal* and posterior areas. Learning of emotionally relevant CS are found in the right hemisphere especially for negative aversive stimuli (Hugdahl, 1995). It has been noted that the medial and lateral areas of the *amygdala* appear to play a major role in the mediation of fear conditioning in rats (Ledoux, 1991). Along with the *limbic* structures human conditioning involves both *frontal* and *temporo-parietal* cortical areas. The brain mechanisms involved in the cortical representation of conditioned associations appear clearly to be right hemisphere especially involving *prefrontal* and *posterior* areas (Squire et al., 1992).

THE FUNCTIONS OF THE *CORTICO-CORTICO* FIBERS

It has traditionally been thought that the main function of the *cortico-cortical* association pathways is to "transfer" information and therefore lesions of these pathways cause a disconnection syndrome where information from one area is unavailable to a second. Inter-hemispheric *cortico-cortical* pathways are known as *commissures*. A split-brain patient generally has had a surgical section of at least the *corpus callosum* and the *anterior commissure*, the largest inter-hemispheric pathways. It is thought that disconnection syndrome occurs in split-brain patients and that the only way of integrating the two hemispheres that remains is subcortical. This traditional view has recently been challenged. It has been proposed that even after sectioning of the inter-hemispheric *cortico-cortical* pathways, subcortical pathways still allow one hemisphere to have access to input restricted to the opposite hemisphere, however that access is on an *implicit* (unconscious) but not *explicit* (conscious) level.

Two important points that have been considered is that to a greater degree than had been previously thought, subcortical inter-hemispheric pathways, even in the intact brain, may modulate inter-hemispheric integration during tasks that may be performed on either a categorized or *implicit* level. The second point is that information about the perceptual identity of an item is more vulnerable to disruption in synchronization and equilibration of arousal than categorized information or information that can be processed automatically, and that even in the normal brain, these disruptions happen more frequently between than within hemispheres. It has been argued that the dissociative behavior noted after colossal section is primarily the result of desynchronization of the two regions and the imbalance in activation, which may be secondary to the desynchronization as opposed to the structural limitations of the use of subcortical pathways (Liederman, 1998).

It has been shown through experiments in *commissurotomized* patients that the non-viewing hemisphere has *implicit*, but not *explicit* access to information restricted to the opposite hemisphere. For example, in one study, two numbers were flashed to the subjects, one to each hemisphere and the *commissurotomized* patients were instructed to cross-compare them. This request makes this an *explicit* or conscious task. Subjects are able to properly compare the numbers in terms of single features or dimensions of the stimuli. They can for example indicate: (1) on which side the larger number appears, (2) whether the sum of the numbers is odd or even, (3) whether the sum is greater than or less than 10. However, these subjects are not able to cross-compare the numbers with regard to their physical appearance. When asked whether the two numbers are the same, performance was at chance levels (Sergent, 1990).

Several explanations for these results have been proposed. The first is that only low resolution, coded, or categorized information is able to be transferred inter-hemispherically, and that the physical appearance of the numbers is beyond the resolution of the subcortical pathways. The second is that there is a decontexualization and fragmentation of information when the synchronizing effects of the commissures are taken away. Therefore, the non-viewing hemisphere can infer several aspects of a stimulus, but the stimulus itself cannot be reconstructed into a whole, including the specific markers that can place it in an autobiographical timeline. The third possibility is that the integration dysfunction is a result of the under-active state of the non-viewing hemisphere, which is not able to operate upon the desynchronized signals that arrive to fuse them into a distinct percept (Liederman, 1998). To summarize these theories, the first proposes a poor quality of information that can be transferred subcortically. The second considers the desynchronized process by which information is transferred subcortically due to the loss of the synchronizing effects of the *commissures*, and the third considers the under-activated state of the hemisphere receiving the subcortically transferred information, which results in a *neglect* of that hemisphere to inputs to it from other areas of the brain.

For the first hypothesis, it does not make sense that the "transfer" process itself is of too low resolution. For the *implicit* areas, the right hemisphere information seems to be accessible to the left with considerable fidelity. Details that can be communicated between hemispheres, even on an *implicit* level, is thought to depend on the number of dimensions that need to be computed simultaneously. Therefore, it is thought that there are some dynamic aspects to inter-hemispheric interaction that affect whether information made accessible to the non-viewing hemisphere can be consciously perceived.

This dynamic model takes us to the second and third hypotheses, broken into three subcomponents. The first is that interregional synchronization is crucial for binding the dimensions of an object into a single idiosyncratic event memory. The second is that *commissurotomy* may result in the two hemispheres being desynchronized. The third is that inter-hemispheric desynchronization interferes with stimuli that are experienced by one hemisphere as being accessed by the opposite hemisphere along with contralateral features that bind it into an *explicit* event memory. Temporal synchronization of activity within a distributed neuronal cell assembly allows the binding or integration into a single event memory of various physical dimensions of an object such as location, shape, direction of movement, color, texture, and sound (Leiderman, 1998). Various researchers support this concept (Poppel, 1988; Gray and Singer, 1989; Gray et al., 1989; Poppel et al., 1990; Engel et al., 1991).

It has been argued that intersensory, inter-hemispheric integration happens by way of a synchronization mechanism, which is thought to temporally integrate distributed activity every 20 to 40 ms. Therefore, stimuli from these different senses are grouped within the same time frame, even though their receptors may transduce at different rates. It is further thought that each of the individual population of cell assemblies oscillate at about 40 Hz. The term oscillates means recurrent synchronous bursting of neuronal cell groups; the oscillations allow cell groups, which communicate with each other by reciprocal pathways to fall into synchrony.

Damasio has argued that areas of the brain integrate their activity through "binding" instead of through the direct transfer of information. Therefore, integration of multiple modalities depends on the time-locked co-activation of geographically separate sites of neural activity within sensory and motor cortices and not a neural transfer and integration (Damasio, 1989a, 1989b; Koch and Leisman, 2001; Leisman and Koch, 2003). In this case, no information is actually sent to other areas, or is moved around the system. Instead, for networks of information to be unified into a single conscious percept they need to be simultaneously co-activated and "bound" by a temporal attentional enhancement. With regard to access memory explicitly, one must reactivate regions that processed the physical dimensions of the object as well as the regions that processed the observer's idiosyncratic emotional and cognitive reactions to the stimulus. If an individual is unable to reactivate the entire network including the contextual network, retrieval of the memory is more likely to be fragmented and may be restricted to a single region, coded for a single physical feature, or can be over-generalized.

It has been shown in cats that the *corpus callosum* is necessary for the inter-hemispheric synchronization of two populations of cells in area *17*, each of which shows a local pattern of oscillation at a rate of about 40 Hz (Engel et al., 1991). It has been hypothesized that this synchronization allows temporal integration or "binding" of groups of cells which are focused on different dimensions of a single visual stimulus positioned to overlap in the midline. What is significant about this study is that it demonstrates that after the *callosal* transection, there is no longer inter-hemispheric synchronization of homologous neuronal populations in the two hemispheres in response to a single object overlapping the midline. In a different study, researchers showed synchronization between regions in areas *17, 18,* and *19* at inter-cortical distances as far as 7 mm. The authors think that this "binding" is a *cortico-cortical* phenomenon that may not be secondary to a subcortical pacemaker or conjoint input from the *thalamus* (Engel et al., 1992). Therefore, it can be concluded that the synchronization of neuronal activity between the two hemispheres is a major function of the *corpus callosum* (Liederman, 1998). It is also thought that the *callosal* section disrupts cross-comparison of stimuli that simultaneously appears on opposite sides of space even when neither of them actively cross the midline.

Therefore, what is also proposed is that cross-comparison is also mediated by a similar process of inter-hemispheric synchronization. The role of the *corpus callosum* may then be to provide a parallel wave of activation within each hemisphere. This would result in widespread activity occurring at the time of encoding the stimulus, which may then be integrated into a single event perception. This then, could be fed-forward independently by each hemisphere to many different regions receiving their conjoint output. This would suggest that with the *callosum* cut, information from many different cortical areas could be channeled through subcortical regions to the opposite hemisphere. However, the transfer may be fragmented and asynchronized and cannot allow all of the dimensions to be transmitted as a single event. In this case the available memory will not be linked with specific markers of emotion, timing, history, etc. that allows it to be remembered as a single episode, as an experience "bathed in contextual surround" (Liederman, 1998); this all results in a specific kind of inter-hemispheric transmission loss, or loss of conscious experience of normally available information. Elements, pieces, and even abstractions are relayed, but are only retrievable by the "semantic" or context-independent system and therefore do not allow for conscious retrieval of the initial idiosyncratic and specific event that produced the memory. From an anatomic perspective it should be noted that the majority of fibers in the *corpus callosum* which connect *associational* areas, are small caliber or unmyelinated. These types of fibers have been postulated to be best suited to convey "tonic" impulses between assemblies of neurons located in opposite hemispheres (Lamantia and Rakic, 1990), even though *callosal* structure reveals that fibers vary widely in size and level of myelination (Lamantia and Rakic, 1990). It has been shown that coupled oscillations can be synchronized even if the conduction delays show a wide distribution and are not homogenous across the network (Sompolinsky et al., 1990).

In the third hypothesis, the non-viewing hemisphere's inability to access *explicit* information that has been shown only to the opposite hemisphere is also due to its under-activated state. In this scenario, the *corpus callosum* acts as part of a system that equilibrates the level of activation of the two hemispheres. Therefore, after sectioning of the *callosum*, the non-viewing hemisphere becomes under-activated and therefore functions similarly to the unilateral neglect syndrome. Therefore, in this scenario, lack of *explicit* awareness is a result of an arousal imbalance, which prevents the under-activated hemisphere from further operating on the input that it receives.

Therefore, it is thought the under-activated hemisphere is not able to attend to input available from other areas due to the fact that the less activated hemisphere cannot fuse the different elements made available synchronously from different subcortical areas into a unified percept embedded into a rich episodic contextual surround. In this case, other symptoms of split-brain such as mutism may be explained in terms of an arousal imbalance (Liederman, 1998).

Several researchers think that the primary function of the *commissures* is equilibration of activation between the hemispheres as opposed to transfer of information (Levy, 1985, 1990; Kinsbourne, 1987a, 1987b, 1988). Guiard and Requin (1978) speculate that a single "attention distribution center" located in a subcortical structure (e.g., the *reticular formation*) distributes attention between the two cortices. In this case the commissures serve a stabilizing function. The *corpus callosum* sends mostly excitatory signals demonstrated electrophysiologically (Lassonde, 1986). It is thought that these excitatory signals allow cortical modules in one hemisphere to "alert" similar areas in the other hemisphere. Without this signal, one hemisphere may be more aroused than the other. There is good evidence that the *corpus callosum* is involved in the equilibration of arousal between hemispheres. An increase in arousal in the hemisphere contralateral to a stimulus is usually accompanied by an inhibition of arousal in the opposite hemisphere thought to occur in part by reciprocal inhibitory connections through the *brainstem*. An example of this may be seen in studies that suggest that unilateral sensory stimulation not only produces facilitation of *dopamine* release in the contralateral *substantia nigra* region (and ipsilateral *caudate nucleus*), but also inhibition of *dopamine* release in the ipsilateral *substantia nigra* and contralateral *caudate* (Leviel et al., 1981). It has also been shown in rats that this same type of *striatal dopamine* asymmetry and the orientation biases associated with them are significantly exaggerated after anterior *callosal* sections (Glick et al., 1975). It is also thought that the amount of CNS asymmetry is exaggerated when the modifying influences of the *callosum* are removed.

These new theories affect the way we view conditions such as *hemispatial neglect*. It has previously been thought that *neglect* was due to a lack of intake of information from furthest left side of space. It has been postulated that a right *parietal* lesion disinhibits the left which then is allowed to pull attention to the contralateral side of space (Kinsbourne, 1987b). Therefore, *neglect* is a result of dysequilibrium. However, the more recent view is that the *neglect* is a processing dysfunction after the input is received. In this case, the sensory input arrives in the CNS but after it is received it is under-processed as opposed to being blocked or neglected. To illustrate, one study demonstrated that right *parietal lobe* lesions cause a deficit in integration not detection. The neglected patients do not have difficulty directing their attention to single feature targets. However, they do have specific difficulty looking for conjunctions among distractions consisting of the two elements, which together would constitute a conjunctive stimulus (Parther et al., 1992). In this case, *neglect* does not affect *autonomic* recognition of features, but patients deficient in conjoining or integrating information after attention are focused on the features. They are not able to utilize "selective attention" as the glue to bind these features (Liederman, 1998).

It is argued that after *callosal* section, the under-activated hemisphere is left inattentive in a fashion similar to what is observed following a right *parietal* lesion. Other symptoms can also be explained by this under-arousal hypothesis. Split-brain patients are known to have an extremely sluggish mental set; one hemisphere becomes aroused by some stimulus and it is hard, sometimes impossible, for the inactive hemisphere to be made active. It has been noted that sometimes a split-brain patient may appear to become so focused on a right hemisphere task that speech and left hemisphere functions are temporarily depressed as if consciousness has been shifted completely to the working hemisphere (Sperry, 1962). Mutism, after callosal section, may be the most extreme example of an arousal imbalance that disrupts performance. It has been noted that whether a patient will be mute after *callosal* section is dependent on whether the hemisphere that controls speech is different than the hemisphere that controls the dominant hand (Sullivan and McKeeser, 1985).

What is thought to implicate an arousal imbalance explanation for mutism is that attentional manipulation alleviates the speech problem. It has been reported in one study that when a patient palpated an object with the hand contralateral to the hemisphere that controls speech, speech is improved (Sullivan and McKeeser, 1985). The fact that using the non-dominant hand impairs speech suggests that the inability is fundamentally a result of arousal imbalance (Liederman, 1998). Another symptom that may result from an arousal imbalance is the change that has been observed in left ear extinction of split-brain patients during dichotic listening. In one example, it has been demonstrated that as task difficulty is increased (by increasing the number of digits to be remembered) there is increased suppression of verbal material presented to the left ear (Peschstedt, 1986). A dynamic explanation is that the verbal left hemisphere is even further aroused by the increased verbal demand and that there is reciprocal inhibition of right hemisphere processing at the *brainstem* level (Liederman, 1998). Additionally, it has been shown that intra- and inter-cortical pathways are so similar in origin and structure that it is assumed that they should have similar functions.

The next step is to propose that inter-hemispheric *cortico-cortical* pathways, which are between *association* areas, have the same equilibrating and synchronizing function. In fact, it has recently been argued that all disconnection characteristics can be viewed as secondary to an imbalance in activation as opposed to a lack of transfer of information (Kinsbourne, 1987a). One example of this may be *conduction aphasia*, the primary symptom of which is the inability to repeat what is heard, even though language production is normal and aural comprehension is intact (Benson, 1979). It is possible that the lack of ability to repeat what was said may be because of a situation-specific under-activation of *Broca's area* by *Wernicke's area* (due to the decreased activation along the *arcuate fasciculus*) rather than by a lack of transfer per se. Measures of metabolic activity in *conduction aphasia* suggest this type of situation-specific under-activation of *Broca's area*. However, baseline levels appear normal and this area is physically intact (Demeuirsse and Capon, 1991). In general, it can be assumed that intra- and inter-cortical pathways have the same synchronizing and equilibrating functions. In the normal intact brain where *commissural* and subcortical mechanisms are functioning, inter-hemispheric integration is still deficient in comparison to intra-hemispheric integration. There are some exceptions to this rule, however where inter-hemispheric is equivalent to intra-hemispheric integration with the exception being information that can be accessed without conscious awareness. Studies have shown that when information between the hemispheres is exchanged inadvertently, unconsciously, and implicitly there is no inter-hemispheric integration deficit (Liederman, 1998). The only exception to this appears to be in inter-hemispheric access to *implicit memory* (Banich and Belger, 1991).

There are several reasons postulated to cause inter-hemispheric integration to be decreased relative to intra-hemispheric integration. The deficits are thought to be due to both structural and dynamic factors. The structural factors are primarily that there are fewer inter-cortical connections between—than there are within—hemispheres and this may make inter-hemispheric integration more difficult. As a result of less inter-cortical connection, there are dynamic factors that arise. It is thought that even in the normal brain, the two hemispheres can have more independent arousal levels than adjacent regions within the same hemisphere. This should not normally occur except where stimulus conditions result in the untrained or non-viewing hemisphere being either too busy or too minimally alert to properly allow inter-hemispheric integration or when two hemispheres are activated by different emotions (Liederman, 1998). It has been argued that in the normal state, input from the *colossum* has to compete with direct *thalamo-cortical* input which will be dominant and possibly at times incompatible (Lepore et al., 1982). There is a second situation where inter-hemispheric processing is affected by asymmetric hemispheric arousal, when one hemisphere is favored for processing or maintained at a high level of arousal for an extended time. This results in an activation imbalance, and the underutilized hemisphere becomes refractory. For example, when individuals repeatedly perform a task during a session that selectively utilizes only one hemisphere, normal subjects present with symptoms of disconnection syndrome when attempting to use the under-aroused non-performing hemisphere (Landis et al., 1979, 1981).

Inter-hemispheric integration may be decreased because of asymmetric arousal in a third situation. The *amygdala* does not have any major inter-hemispheric *commissural* connections (Demeter et al., 1990) therefore, the aspects of emotions that are mediated by the *amygdala* may occur in the two hemispheres independently. If the two hemispheres are then differently aroused, this may form the basis of emotions that are stored in somewhat different emotional surrounds. Different memories would therefore result in inter-hemispheric deficits in access to stored material whenever the two hemispheres were momentarily in different emotional states. Three primary mechanisms of integration of information are thought to exist between remote cortical regions including: Direct transfer, third party convergence, and non-convergent temporal integration.

Direct Transfer

Direct transfer is a mechanism in which it is assumed that the products of processing in one area are transferred directly to a second region where they are integrated with that area's processing and then stored. It has been traditionally thought that this is the major function of *cortico-cortical* pathways. One study showed that when input is provided only to one hemisphere the other hemisphere has its own store of that memory (Ringo and O'neil, 1993). It is assumed that this task is shared inter-hemispherically by way of the process of *direct transfer*.

Third Party Convergence

Third party convergence is a mechanism in which two or more areas feed-forward the products of their processing to a third convergence area, however they do not share information directly with one another (Goldman-Rakic, 1988). In this type of integration, the purpose of the *cortico-cortical* pathways is to synchronize the outputs of these two regions to the third convergence area. One example indicates that by means of double-labeling techniques in monkeys, the principal sulcus of the *frontal lobe* and the posterior *parietal* cortex are mutually interconnected to as many as 15 other cortical areas and most of these projections are reciprocal (Goldman-Rakic, 1988). In terms of inter-hemispheric *third party convergence*, it is reasonable to assume that these zones are located in the *brainstem* (subcortical) where both hemispheres could project (Liederman, 1998).

Nonconvergent Temporal Integration

Nonconvergent temporal integration by means of *cortico-cortical* pathways synchronizes the output of two or more remote regions and enables the output to feed-forward independently but synchronously to various other areas. Output from these two areas initiates a synchronized distribution of patterned activation. This forward progression of patterned activation can affect relatively independent systems for the control of behavior in a parallel but synchronized way. In this model of non-convergent synchronization, there is only temporal integration; there is no *grandmother* convergent zone and no region where traces of the two areas of activity are stored as a compound single engram (Liederman, 1998). Two regions that can be used as an example of regions that might operate primarily by the mechanism of *non-convergent temporal integration* are the dorsal and ventral visual systems. There are very few connections between the two systems (Livingstone and Hubel, 1988). It is thought that the two systems may not need to communicate directly. However, it is more likely that they operate in terms of third party convergence. For example, they both project to the *frontal lobes* and it may be possible that they interact by way of local circuits. It is thought that they may frequently operate by means of *non-convergent temporal integration* by influencing relatively independent behaviors. Therefore, the two systems do not need to be integrated at a single point of convergence but they need to be synchronized. This point is expressed by Kinsbourne (1987a)

> The experience of the moment (the event) is represented by a distributed but specifically configured pattern of neuronal excitation across wide areas of cortex. It would be the pattern as a whole rather than by some bystanding area that observes it, that generates the experience and initiates the decisions. The neurons that fire in concert during specifically patterned activation need not be conceived as interconnected while doing so. Instead, they might fire in parallel to yield an output the stochastic characteristics of which exert the required control over some components of behavior (p. 425).

It has been argued that subcortical pathways are adequate for the inter-hemispheric integration of categorized information or context independent, fragmented information which can be accessed only *implicitly* (Liederman, 1998). Guiard (1980) states that there has been too much emphasis on horizontalist (*cortico-cortical*) pathways and that instead most integration is likely to occur "vertically" by way of subcortical pathways (Leisman and Zenhausern, 1982). It has been stated by Karl Pribram (1986), "The hierarchical aspects of visual processing can be readily attributed to systems of cortical-subcortical loops as to the operations of a transcortical mechanism" (p. 534). He concludes that there is a hierarchy of pre-cortical visual mechanisms and that the cortex adds finger grain and enables "reflective awareness of the resultants of the process." The idea that cortical pathways mediate an intra-cortical reflex and are similar to spinal reflexes, with cortical sensory areas inducing movements via projections to motor areas has been criticized by Kornhuber (1974). With regard to *conduction aphasia* which had been thought to result in a lack of transfer of auditory signals from *Wernicke's area* to *Broca's area* because of damage of *cortico-cortical* pathway (the *arcuate fasciculus*), Koenhuler suggests that both *Broca's* and *Wernicke's* areas have direct access to the *basal ganglia* and the *cerebellum*, and that these structures have been implicated in movement and speech disorders. Therefore, he thought that a cortical-subcortical loop exists in which cortical activity is transformed into spatio-temporal motor patterns by subcortical generators such as the *basal ganglia* and *cerebellum*. It appears possible that item-specific information is not transferred from one hemisphere to the other but is integrated by synchronization.

It is thought that subcortical systems cannot efficiently mediate the required synchronization. It has been speculated that the ventral visual system, which processes identity information, may depend more on interhemispheric synchronization than the portions of the visual system processing nonidentity information (e.g., location). In order to remember a stimulus in its full content, the dimensions of the stimulus must be bound into an "object perception" which needs to be linked with emotional and temporal cues to place it in an autobiographic form. This spatio-temporal pattern of activation requires multiregional synchronization of activation within the hemisphere that receives the stimulus. This must then be integrated with the opposite hemisphere. It is thought that this type of integration is specifically dependent on the synchronizing effects of the cerebral cortex (Liederman, 1998). On the other hand, stimulus location, which is computed primarily by the dorsal system, may be effectively integrated inter-hemispherically by asynchronous subcortical mechanisms and may be able to effectively control behavior even though the identity of the object being localized is not conscious.

9

Signs and Symptoms of Neurobehavioral Disorders of Childhood

Traditionally clinicians have looked at various disorders like *attention deficit hyperactive disorder* (ADHD), *Tourette's syndrome* (TS), and *obsessive–compulsive disorder* (OCD) as separate and distinct clinical entities. However, recently many of these disorders have been shown to have significant overlap and many children that have one disorder will often be diagnosed with at least one other. Neurobiological research and functional imaging tests have given science a new way of examining the brain in its functional and dysfunctional states. These new tests have also revealed striking similarities in the brains of children and adults with various cognitive and behavioral disorders. As a result, we are arriving at the view that rather than consisting of separate processes, most of the common cognitive and behavioral disorders of childhood rest along a continuum all with a similar underlying mechanism. At one end of the spectrum is *attention deficit disorder* (ADD), then ADHD, *learning disability*, *pervasive developmental disorders*, *autism*, OCD, TS, and the *schizophrenias*. From an anatomical and functional perspective, we find that many of the same areas of the brain are affected. Some regions are smaller or appear atrophied, other regions appear to be functionally hyperactive, and some are hypoactive.

What seems to disturb parents and teachers alike and confuse many clinicians is that they become overwhelmed by the wide variety of different symptoms that a child may express. Most of these children do not present solely with behavioral problems or learning disabilities although these may be of greatest concern or the most obvious. These children usually exhibit a combination of emotional, behavioral, cognitive, sensory, motor, and autonomic symptoms. The broad range of symptoms causes even more anxiety and confusion to parents and professionals and leads them to think that the problem is overwhelming and insurmountable. Guiding principles in understanding these conditions and in fomenting intervention strategies start with the caveat that the greater the number of symptoms, the more likely the problem is centered in the brain where all bodily and behavioral systems are controlled. If one can isolate specific dysfunctioning regions, one can then better understand the nature of the problem making treatment more specific and effective.

While many of the same pathways are involved in disorders within the continuum, their involvement appears restricted to either right-sided or left-sided hemispheric and subcortical areas. Throughout the book, we have stressed a common theme of how the brain functions including the existence of a baseline arousal level of subcortical and cortical structures that appears to make up the contextual aspect of cognition, emotion, and perception. Superimposed on this baseline arousal level we have described specific pathways that convey content specific information to areas of the brain. Normal functions can be disrupted if the contextual information, which is arousal dependent, or the context specific information is disrupted in its progression through the nervous system.

From a functional perspective, we can examine by brain imaging techniques underactive or overactive brain regions that in turn relate to ineffectiveness of the arousal nonspecific system or the specific content pathways. We have discussed how if neurons are active they will have maintained normal metabolic resting level. If not effectively stimulated they will be underactive, but eventually more active and relatively unstable. The result is that these neurons will fatigue more quickly and may even fire spontaneously as in the case of seizures or *hyperkinetic* disorders.

From an anatomic perspective, we see that certain areas of the brain are physically smaller and different from normal in children with these disorders (Leisman and Ashkenazi, 1980; Singer et al., 1993; Rosenberg et al., 1997; Saitoh and Courchesne, 1998; Aoyama et al., 2000; Hardan et al., 2000, 2001a, 2001b; Blatt et al., 2001; Castellanos et al., 2001). When an area is abnormally smaller, it is either usually due to failure to develop properly or is a result of disease. As with most areas of our body, when we use a structure it becomes larger with use, when we do not use it, we will often see atrophy of that area. Even with lack of development, we see that a region will fail to develop due to lack of exposure to stimuli, which will promote a developmental delay of the growth of the understimulated region.

In most of the cognitive and behavioral disorders that are discussed here, we will see that there is a combination of alteration of function usually underactivity or hypoactive states, as well as atrophy or smaller physical size of neural structures. We also will recognize a common link between hypoactivity and atrophy of many of the same areas in all of these disorders. The only difference is that they are usually restricted to the left or the right side of the subcortical and cortical structures resulting in different symptoms. Many of the symptoms result from lack of inhibition of one area resulting in that region's functional *hyper*activity; this in turn we hypothesize to be due to the *hypo*activity of a region that would normally inhibit that function. In behavioral disorders, we know that the *limbic system*, especially on the right, is responsible for the production of many primitive or survival-driven emotions, such as fear, rage, anger, or approach and avoidance behaviors. These brain regions in humans are normally modulated or inhibited by higher centers in the neocortex, especially the *prefrontal* cortex. Decreased stimulation or hypoactivity of the *prefrontal cortex* will reduce inhibitory control over the *limbic system* which can result in abnormal behavior, uncontrollable emotions especially aggression, increase in either avoidance or withdrawal behavior, perseveration, inability to focus attention and *hypo*- or *hyper*activity in a way not unlike that seen in the early Klüver–Bucy preparations in chimpanzees. The type of symptoms will depend on which hemisphere is dysfunctional. Autonomic systems can be affected or imbalanced due to dysfunction of the cortex and failure to modulate the *limbic system* and the *hypothalamus*. Neurotransmitter systems, especially *serotonin, norepinephrine,* and *dopamine* systems can be dysfunctional. Which system is affected is usually the result of the side of the brain affected as there exists an asymmetric distribution of neurotransmitters in the brain.

Finally, most of the development and normal function of the cerebrum is dependent on subcortical structures especially the *cerebellum, thalamus,* and *basal ganglia*. A failure to develop and/or a dysfunction in these areas can affect both the nonspecific arousal system as well as specific transfer of information in the brain. Dysfunction in these areas will usually result in specific motor and sensory symptoms that are commonly seen in children with cognitive or behavioral disorders. These brain regions are often seen to be underactive or atrophied as well in these children. These cortical loci have been shown to be connected with the *prefrontal cortex*, which have also often been noted to be underactive or atrophied in children with the neurobehavioral developmental disorders. The underactivity and or atrophy is usually either restricted to the right or left side of the sub-cortex and cortex.

An imbalance of activity or arousal of one side of the cortex or the other can result in a functional disconnection syndrome similar to what is seen in split-brain patients, which could be an underlying source of many if not all of the symptoms that we see with children with behavioral and cognitive disorders.

For example, postmortem examinations have indicated structural differences between the brains of good and impaired readers. High concentrations of micro-dysgenesis are noted in the left temporoparietal regions of dyslexic brains. The concentration is most evidenced in the *planum temporale* region (Galaburda et al., 1985; Duane, 1989; Kaufman and Galaburda, 1989). These micro-dysgeneses seriously impair the normal pattern of architecture of dyslexics and remove the asymmetry normally observed between the enlarged language areas of the left temporoparietal region and the smaller homologous areas of the right hemisphere (Leisman and Ashkenazi, 1980; Galaburda et al., 1985). The capacity for language is generally correlated with a significant development in the magnitude of the left temporoparietal region and an attrition of neurons in the right hemisphere. These neuronal casualties may produce the observed asymmetry between corresponding areas in the left and right hemispheres (Geschwind and Levitsky, 1968; Leisman and Ashkenazi, 1980). The relative symmetry in the dyslexics' brains might reflect their impaired linguistic development.

In one study (Leisman, 2002), left parieto-occipital EEG leads recorded a frequency spectrum in dyslexics that is consistently different from the spectrum obtained from normals. It is suggested that these effects represent significant differences in the functional organization of these areas. EEG coherence values indicate that normals have significantly greater sharing between hemispheres at symmetrical locations. Dyslexics demonstrate significantly greater sharing within hemisphere than do normals as evidenced in Fig. 9.1 and Table 9.1. The data

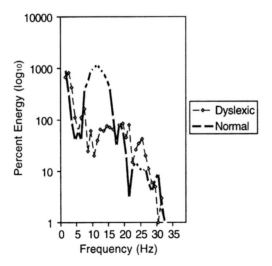

Fig. 9.1. Mean values of EEG autospectral density by frequency recorded from P_3–O_1 electrode placements for normal and dyslexic subjects. (From Leisman, 2002.)

TABLE 9.1 Average Frequency (in Hz), Power (in dB), Left–Right Asymmetry of Power (in dB) Between Hemisphere and Within Hemisphere Coherence Values at P_3–O_1/P_4–O_2 Locations for Dyslexics and Normals

	Dyslexic					Normal				
S	Freq. (Hz)	Power (dB)	L–R (dB)	Bilat. Coher.	W/in Coher.	Freq. (Hz)	Power (dB)	L–R (dB)	Bilat. Coher.	W/in Coher.
1	09.2	12	−03	—	1.1	09.2	28	—	—	0.8
2	10.4	21	−04	—	1.8	10.8	24	—	2.4	—
3	11.7	22	10	—	2.4	12.7	18	—	1.9	—
4	09.8	18	04	—	1.6	10.9	20	−4	1.3	—
5	10.8	17	03	—	1.4	08.6	16	—	1.9	—
6	10.6	24	−01	—	0.8	08.9	08	—	1.8	—
7	10.6	28	−05	—	1.5	11.2	11	—	2.4	—
8	11.2	12	−07	—	2.1	11.7	13	−2	1.5	1.8
9	12.0	19	−04	—	1.9	10.0	12	—	1.3	—
10	09.8	14	—	0.7	0.6	10.7	15	−1	1.3	0.9
11	10.8	25	−02	—	1.0	10.6	11	—	1.2	1.4
12	11.7	22	—	1.0	—	12.0	09	—	0.8	1.1
13	08.7	13	−01	—	0.9	11.7	07	—	1.0	—
14	09.0	27	08	—	2.1	08.9	11	—	1.9	—
15	10.7	13	−04	—	2.4	09.5	10	—	1.7	0.6
16	10.3	08	−06	—	1.8	08.8	11	−2	2.1	—
17	09.5	22	−07	—	2.0	08.6	14	—	1.4	—
18	12.2	20	−07	—	1.9	09.3	09	—	1.8	—
19	11.9	09	−01	—	0.9	12.4	12	—	1.9	—
20	08.4	15	−04	—	1.6	11.6	10	—	0.9	—

supports the notion that developmental dyslexia is a functional hemispheric disconnection syndrome. Other conditions in the spectrum of disorders that we are discussing yield similar results.

This spectrum of childhood disorders that we are discussing generally relates to an increase or decrease in activation of the brain and the balance of activation between brain regions. These conditions result from two primary system effects: (1) primary arousal deficit or imbalance, and (2) a specific activation deficit, imbalance, or asynchrony. The brain is driven by sensory input. We know that the brain receives more simultaneous sensory input than it can possibly consciously process (Broadbent, 1958, 1965a, 1965b; Leisman, 1976a; Heilman, 1995). In general, the more stimulation the brain cells receive, the better their function allowing it to process more information faster, for longer periods of time (Szeligo and Leblond, 1977; Venables, 1989; Pascual and Figueroa, 1996; van Praag et al., 2000; Mohammed et al., 2002). Therefore all sensory input is important although not all of it can be consciously processed and perceived. In fact, without subconscious baseline stimulation, higher conscious processing of sensory stimuli would be difficult if not impossible.

Before higher brain centers can develop, the lesser supportive brain structures must develop. In the cortex, Luria (1973) thought that lateralized cortical functions progressed from primary cortical areas to secondary and tertiary areas as the child matures. Going back even further we see that development of cortical areas and the cortex itself are dependent on the anatomic and functional development of subcortical areas, especially the *cerebellum* and the *thalamus*. Studies suggest that intact *cerebellar* functioning is required for normal cerebral functional and anatomical development (Llinas, 1995; Rae et al., 1998). The same has been seen for the *thalamus*— that intact *thalamic* function is necessary for cortical development and function (Albe-Fessard et al., 1983; Gil et al.,

1999; Kalivas et al., 1999; Scannell et al., 1999, 2000; Young et al., 2000; Amino et al., 2001; Alonzo, 2002; Castro-Alamancos, 2002). Developmental dysfunction of the same brain areas as seen in acquired disorders such as posttraumatic aphasia may be the basis of developmental learning disabilities and neurobehavioral disorders (Dawson et al., 1982, 2001, 2002a, 2002b; Dawson, 1988, 1996; Obrzut, 1991).

As Orton (1937) had indicated, it is generally assumed that persons with learning disabilities have abnormal cerebral organization including atypical or weak patterns of hemisphere specialization (Corballis, 1983; Bryden, 1988; Obrzut, 1988). The developmental lag hypothesis proposed by Lenneberg (1967) suggested that learning-disabled persons are slower to develop basic language skills and demonstrate weak hemispheric specialization for language tasks. In a reformulation of the progressive lateralization hypothesis (Satz et al., 1990), it may be that subcortical and antero-posterior progressions have a differential developmental course with learning-disabled children and adults compared to control subjects or those with acquired syndromes.

Since learning-disabled children exhibit deficient performance on a variety of tests thought to be a measure of perceptual laterality, evidence of weak laterality or failure to develop laterality has been found across various modalities (audio, visual, tactile) (Boliek and Obrzut, 1995). It is thought that these children have abnormal cerebral organization as suggested by Corballis (1983) and Obrzut (1988). The basic assumption is that dysfunction in the central nervous system either prenatally or during early postnatal development, results in abnormal cerebral organization and associated dysfunctional specialization needed for lateralized processing of language function and nonlanguage skills. It is thought that cortical and subcortical dysfunction which results from aberrant patterns of activation or arousal (Obrzut, 1991), inter- and intrahemispheric transmission deficits, inadequate resource allocation (Keshner and Peterson, 1988), or any combination of these may compromise hemispheric specialization in those with cognitive and behavioral deficits (Bolick and Obrzut, 1995).

Development of higher processing areas in the *cerebellar* cortex would develop after other more primary areas. For example, the lateral *cerebellum* would be dependent on proper development of the more midline areas in the inter-medial and medial zones first. Similarly, any region to which lateral *cerebellum* projected would be dependent on the effective development of the lateral *cerebellum* and it in turn would be dependent on the more medial *cerebellar* development. Therefore, if the medial aspects of the *cerebellum* do not develop adequately, then the lateral areas would still grow; however, they may be smaller or atrophic, and dysfunction would be expected.

The *cerebellum* is thought to be part of a neuronal system that includes the *thalamus*, *basal ganglia*, and *prefrontal cortex* (Thatch, 1980). Anatomic and functional development of the nervous system is dependent on sensory input, which is associated with growth of a given brain area and its associated connectivities with other brain regions. Brain area growth and the capacity to make functional connectivities is highly dependent on continued regional stimulation and by global stimulation through connected and coordinated function. If specific regions are inadequately stimulated, then we may see failure of anatomic or functional development in that region with a preservation of basic lower level functionality. Higher functions that depend on greater areas of integrated stimulation may be lost or dysfunctional. If the sensory loss develops after a *critical period*, these areas may still be smaller due to atrophy or reverse plasticity, with either global or specific effects depending on the modality of dysfunction. In children with learning disabilities or affective disorders, there are specific areas of the nervous system that have been noted in imaging studies to be smaller than normal (Larsen et al., 1990; Chiron et al., 1999; Cohen et al., 2000; Frank and Pavlakis, 2001; von Plessen et al., 2002). Most often, these areas involve the *prefrontal cortex, basal ganglia, thalamus*, and *cerebellum*.

Some neurophysiological experts regard the central nervous system as partly a *closed* and partly an *open* system (Llinas, 1995). An open system is one that accepts input from the environment, processes it, and returns it to the external environment. A closed system suggests that the basic organization of the central nervous system is geared toward the generation of intrinsic images and is primarily self-activating and capable of generating a cognitive representation of the outside environment even without incoming sensory stimuli. Although it is possible that a certain level of activation or stimulation will be intrinsic to single neuronal cells and the nervous system as a whole, this stimulation does not seem adequate to sustain a conscious, awake, individual. Behaviorally, arousal is a term used to describe an organism that is prepared to process incoming stimuli. From a physiological standpoint, arousal also refers to the excitatory state or the propensity of neurons to discharge when appropriately activated (neuronal preparation). A non-aroused organism is comatose (Heilman, 1995). Therefore, an aroused, alert individual that is prepared to process information is in a state dependent on sensory input with an attendant intrinsic excitability. Remove stimulation and the individual will eventually lose conscious awareness and become comatose or at least inattentive. The majority of brain activity associated with arousal comes from the ascending *reticular activating system*. The majority of this activity is relayed by the nonspecific *thalamic nuclei* or *intralaminar nuclei*.

All sensory perception is based on the effectiveness of the arousal level of nonspecific, mostly subconscious, activity of the brain. There can be no specific sensory modality perception like vision or hearing without a baseline arousal level. The more the stimulation or greater the frequency of stimulation, the more aroused an individual will be. Low-frequency stimulation of midline *thalamic nonspecific nuclei* produces inattention, drowsiness, and sleep accompanied by slow wave synchronous activity and so-called spindle bursts. High-frequency stimulation on the other hand has been shown to arouse a sleeping subject or alert a waking organism (Tanaka et al., 1975; Arnulf et al., 2000; Halboni, 2000). Specific sensory perception and processing are dependent on specific *thalamic* relays; if one of the specific *thalamic* nuclei is damaged such as the *lateral geniculate body*, that specific sensory modality is lost (e.g., blindness), but it does not result in loss of other specific nuclei input like hearing. However, if lesions of the nonspecific intralaminar nuclei exist, patients cannot perceive or respond to any input by the specific intact nuclei even though those pathways are intact. In essence, the person does not exist from a cognitive standpoint (Llinas, 1995).

Luria postulated that the brain was divided into three functional units: (1) the arousal unit, (2) the sensory receptive and integrative unit, and (3) the planning and organizational unit. He subdivided the last two into three hierarchic zones. The primary zone is responsible for sorting and recording incoming sensory information. The secondary zone organizes and codes information from the primary zone. The tertiary zone is where data are merged from multiple sources of input and collated as the basis for organizing complex behavioral responses (Luria, 1973). Luria's dynamic progression of lateralized function is similar to Hughlings Jackson's Cartesian coordinates with respect to progressive function from *brainstem* to cortical regions (Kinsbourne and Hiscock, 1983).

Satz and colleagues (1990) have suggested that developmental invariance describes the lateral (x-axis) dimension of asymmetry, whereas current formulation of equipotentiality and the *progressive lateralization hypothesis* better describes vertical (subcortical–cortical) and horizontal (antero-posterior) progression during infancy and early childhood. Interestingly it has been noted that most research designed to address laterality issues in developmental disabilities (i.e., learning disabilities) has not dealt systematically with subcortical–cortical development or antero-posterior progression, all based on the concept of arousal unit.

The arousal unit is really the *nonspecific thalamic nuclei*. We know that arousal is dependent on external and internal environmental sensory input. The largest proportion of subconscious sensory input passes between the *thalamus, cerebellum,* and *dorsal column* from slowly adapting receptors found in muscles with a preponderance of slow-twitch fibers—or slowly adapting muscle spindle receptors. The highest percentage of these is found in antigravity postural muscles, especially muscles of the spine and neck (Guyton, 1986). The receptors, which provide the major source of input to the brain, only receive sensory information. These receptors only work when muscles are stretched or contracted with gravity being the most frequent and constant sensory stimulus.

In summary, brain development and the adequacy of its continued functioning is dependent on sensory input. Specific sensory perceptual processes like vision and hearing are dependent on nonspecific sensory input. This, in turn, creates a baseline arousal and synchronization of brain activity (consciousness). This is a form of constant arousal and is dependent on a constant flow of sensory input from receptors that are found in muscles of the spine and neck. These receptors receive the majority of their stimulation from gravity, creating a feedback loop that forms the basis of most if not all of brain function. Sensory input drives the brain, and motor activity drives the sensory system. Without sensory input, the brain cannot perceive or process input. Without motor activity provided by constant action of postural muscles, a large proportion of sensory stimuli are lost to further processing. This loop is the *somatosensory system*.

Higher processing is also dependent on the baseline sensory functions. For example, it has been shown that when performing a complex task, it is likely that transfer of motor commands to produce a final output is preceded to some degree, by transfer of information between association areas, which in turn may precede transfer between sensory regions (Banich, 1995).

BALANCE IS ESSENTIAL

It is clear that the sensory motor system then cannot be separated. A dysfunction of the sensory system results in a motor weakness or coordination deficit and a decrease in motor tone or coordination, which in turn decreases the sensitivity of the muscle receptor and, therefore, the frequency of firing of the sensory pathways. The ability to sense body movement or position in space relative to gravity is known as *proprioception*, or the sense of subconscious touch (Fig. 9.2). *Proprioception* is dependent on the amount of force or functional state of somatosensory receptors of joints, tendons, and especially muscle spindle receptors. These are pressure receptors similar to skin touch receptors for tactile sensation, the only major difference being the frequency of stimulation and the resultant conscious or subconscious perception. *Muscle tone* may be defined as the resistance of stretch of muscles. *Muscle tone* is a product of the output of the brain or its level of activation and arousal. Output of the brain is directly related to input plus arousal level. Effective *muscle tone* is characteristic of an aroused individual. *Proprioception* and *muscle tone* are therefore related. The adequacy of *proprioception* results from the effective sensitivity of muscle spindle receptors. High sensitivity of muscle spindles is due to effective muscle tone, which in turn results from the adequacy of brain output.

Of all the sensory input, *proprioception* constitutes the largest part. Many researchers have noted that in learning-disabled children, there often exists poor proprioception and low muscle tone. Symptoms of poor *proprioception* and low muscle tone include poor balance, posture, clumsiness, and poor coordination (Kaluger and Heil, 1970; Ayres, 1972; Njiokiktjien et al., 1976, Ottenbacher, 1978; Kohen-Raz and Hiriatborde, 1979; Freides et al., 1980; Voronin et al., 1980; Hulme et al., 1982; Conrad et al., 1983; Morrison et al., 1985; Morris et al., 1988; Kohen-Raz, 1991; Cammisa, 1994). Jean Ayers (1972a) recognized over 30 years ago that these children demonstrated a variety of problems that she thought were related to what she called poor *sensory integration*. She noted then that these children possess different degrees of sensory deficits including

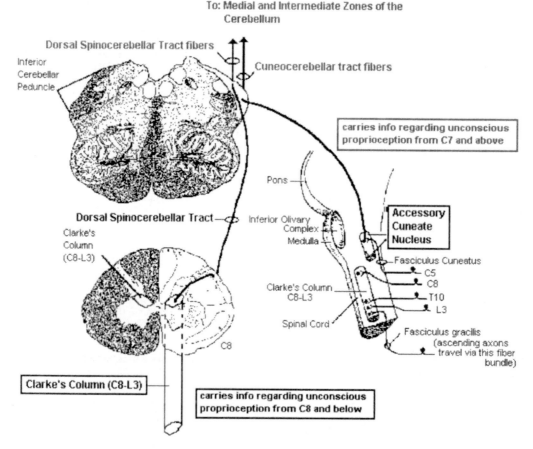

Fig. 9.2. Proprioceptive pathways (cf. Leisman, 1989a, 1989b).

hearing, visual, tactile, and *proprioceptive*. Associated with these symptoms were speech and language deficits, motor dysfunction, learning disabilities, and emotional problems. She thought that the vestibular system was the principal source of the dysfunction (Fig. 9.3). Ayres arrived at this conclusion because she recognized that gravity plays a significant role in *proprioception* and further concluded that the major gravity receptors, which control balance, were primarily concentrated in the vestibular system. While her clinical observations were keen, we now recognize, however, that the majority of proprioceptive and gravity data arises from spinal and neck muscles, joints, and their sensory input to the *cerebellum*.

While vestibular receptors do not change their sensitivity, the *cerebellum* may alter it responsiveness to the vestibular apparatus based on the *cerebellum's* baseline arousal level, which results mostly from spinal and neck motor activity (Bruggencate, 1975; Gdowski et al., 2001; Matsushita and Xiong, 2001). These midline motor structures of the spine are connected to the midline *cerebellar* structures in the medial zone and the *fastigial* nuclei of the *cerebellum*, and send sensory input directly back. These midline muscles are stimulated bilaterally by the *cerebellum*. Signals then project to midline nonspecific nuclei in the *thalamus*, which then project to the entire cortex (Leisman, 1989a, 1989b). This *cerebellar* feedback mechanism also connects through the *vestibular nuclei* and proceeds to primarily affect the proximal trunk muscles. They also ascend to effect the coordination of eye muscles. The intermediate and lateral

SIGNS AND SYMPTOMS 185

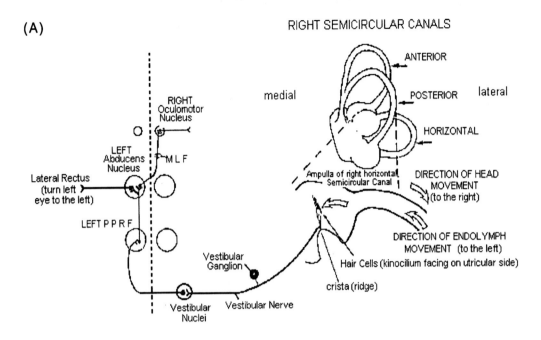

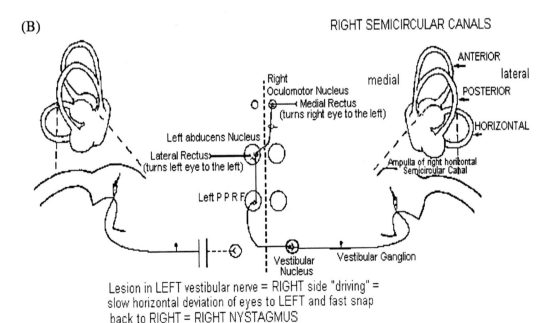

Lesion in LEFT vestibular nerve = RIGHT side "driving" = slow horizontal deviation of eyes to LEFT and fast snap back to RIGHT = RIGHT NYSTAGMUS

Fig. 9.3. *Continued*

186 NEUROBEHAVIORAL DISORDERS OF CHILDHOOD

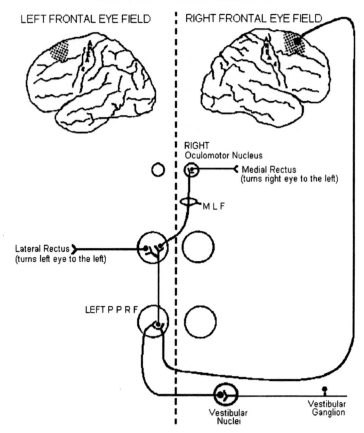

RIGHT frontal eye field turns eyes to the LEFT

Fig. 9.3. Compensatory eye movements induced by the vestibular apparatus, opposing head movements or changes in head position and act to keep the fovea of the retina on an object of interest. (A) Rotatory movement of the head to the right results in an increase in discharge of the right vestibular nerve, an increase in firing of the right vestibular nuclei, an increase in firing of neurons in the left paramedian pontine reticular formation (PPRF), an increase in firing of both small and large neurons in the left abducens nucleus and reflex turning of the left eye to the left (via left lateral rectus; Cranial nerve VI) and the right eye to the left (via ascending MLF input to the right medial rectus; Cranial nerve III). This is called the vesibulo-ocular reflex, critically important for stabilizing visual images in the presence of a continuously moving head. (B) Results of a lesion in the left vestibular nerve on eye movements. Such a lesion puts the right vestibular nerve in "control." This imbalance results in the eyes being pushed slowly to the left (right vestibular nerve turns on right vestibular nuclei, which turns on the left PPRF, which turns on the left abducens which turns both eyes to the left). When the eyes are pushed as far left as possible, they snap back very quickly to the right by mechanisms not fully understood. The eyes then slowly move to the left again, and this vicious cycle continues. This nodding back and forth is called *nystagmus* (to nod). A lesion of the left vestibular nerve will result in a right *nystagmus*. Thus, the right (intact) vestibular nerve is "driving" the left PPRF and left abducens to move the eyes slowly to the left, after which they reflexively snap back to the right. (C) Pathways involved in voluntary turning of both eyes horizontally to the left to see a new object of interest, called a left horizontal saccade. The left lateral and the right medial rectii contract synchronously. The two eyes then move together (conjugately) to the left, using a pathway beginning in the frontal eye fields of the cerebral cortex (area 8). This is a cortical area that lies rostral to the primary motor area (area 4). To voluntarily move eyes to the left, information from the right frontal eye fields is conveyed to the left (contralateral) PPRF. The right frontal eye field directs the left PPRF to turn on both large and small neurons in the left abducens nucleus. The left eye then turns left (laterally) and the right eye turns left (medially). This is a voluntary left horizontal saccade. (From Harting (1997).)

cerebellum send more specific information about extremity and distal muscles, which goes to specific *thalamic* nuclei that are connected to the *somatosensory cortex, prefrontal cortex,* and *association cortices* on the contralateral side of the body. These specific areas of neocortex, especially the *frontal lobe*, then send fibers downward to the ipsilateral *brainstem reticular formation* which then sets global muscle tone of all muscles on that side, both spinal and postural, as well as distal extremity muscles. It also sets the balance of *muscle tone* between anterior and posterior compartment muscles on the ipsilateral side of the body to help promote an upright bipedal posture.

Cerebellar lesions have been recognized, since Gordon Holmes, as resulting in problems with *proprioception* and motor tone. Some of these symptoms affect the bilateral coordination of muscles of the spine and eyes and occur primarily at the level of the *cerebellum*, and always occur bilaterally due to both contralateral and ipsilateral fibers (Ross et al., 1979; Ramnani et al., 2001). However, unilateral dyscoordination, unilateral decreases in muscle tone in both postural and extremity muscles and the balance of vestibular activity occur at the level of the cortex, dependent on contralateral *cerebellar* input plus arousal activity (Leisman, 1989a, 1989b). Levinson (1980), in his study of almost 4,000 learning-disabled subjects showed that 94.1 percent presented with what he termed *cerebellar–vestibular* (C–V) dysfunction. His data suggested that learning disabilities and dyslexia are C–V based and represent one disorder, that this C–V dysfunction continues with increasing age and that the symptomatic overlap within subjects and samples appear to reflect a common C–V basis rather than a group of separate neurophysiological disorders. He went on to state that a variety of so-called "pure" disorders and terms such as *dyslexia, dysgraphia, dyscalculia, dysphasia, dysnomia, dyspraxia, ADD, perceptual-motor disorder,* etc. merely reflect highly selected learning-disabled symptoms and samples. Although he recognized *cerebellar* dysfunction, he related many of these aforementioned problems to a primary source in the inner ear of the vestibular system. He also did not consider asymmetric cerebral lateralization and its development as a significant factor. Therefore, a sensory deficit of proprioception and subsequent abnormality of muscle tone may reflect a global decrease in global *cerebellar, thalamic,* and cortical arousal levels.

Other specific sensory deficits may result from the asymmetric arousal of these areas, which would present with specific localized perceptual dysfunction. Since *proprioception* is a subconscious sense, processed subconsciously through the *cerebellum, thalamus,* and *somatosensory cortex,* when cortical *proprioceptive* input is decreased we are not consciously aware of the loss or dysfunction. However, there are symptoms that we may experience when the *proprioceptive* input becomes too low or imbalanced. Dysequilibrium, or the feeling of unsteadiness of gait, is one primary symptom. Poor balance, dizziness, or vertigo can be manifest if vestibular input is not modulated effectively by the *cerebellum*.

In the past, most of the emphasis for balance dysfunction, such as vertigo, *nystagmus,* or motion sickness, had been placed on the vestibular apparatus of the inner ear. However, research suggests that as humans evolved from quadrupeds to bipeds, the emphasis of control of balance, posture, and the coordination of head and eye movements was shifted from the inner ear to the spinal muscles and joints. Wyke (1979) notes that during the course of the evolution of the erect from the quadrupedal posture, there has been a shift in the relative perceptual and reflexogenic significance of the *mechanoreceptors* in the labyrinth of the internal ear and those located in the cervical spinal joints in favor of the latter. This evolutionary decrease in the functional significance of the vestibular system in man was first suggested by Rudolf Magnus (1924) and is powerfully illustrated by the later clinical observations of Puidon Martin (1967) which show, *inter alia*, that this system in man is of no significance at all in static postural circumstances or in the production of reflex righting reactions. In this connection, it may be relevant to note that the cervical articular *mechanoreceptors* became functionally active in the developing human fetus long before the vestibular *mechanoreceptors* (Wyke, 1975, 1979). Wyke goes on to note that patients experience a feeling of postural instability and unsteadiness of gait in situations of poor lighting after these individuals have been provided with a cervical collar that affects their neck joints and muscles (Wyke, 1979). He also notes that based on neuropsychological study (Wyke, 1965) showing that, "precision of voluntary control of arm movement in the absence of vision is markedly impaired by rotation of the head and neck to right and left of central position." It has also been observed by de Jong and colleagues (1977), and others (Leisman, 1989a, 1989b) that local anesthetic infiltration of cervical spinal and other joints in normal individuals results in a feeling of static body dysequilibrium (often with vertigo). Also noted is *kinesthesia*, upon which the accurate control of voluntary movements including walking depends even in the presence of a normally functioning vestibular system (Wyke, 1979).

In regard to *proprioceptive* input and its relationship to various sources of input, Carpenter and colleagues (2001) states: "by far the most important proprioceptive information needed for the maintenance of equilibrium comes from neck joint receptors and muscles." Guyton (1986) also notes,

> mechanisms governing equilibrium and orientation in three dimensional space *proprioception* are largely reflexive and subconscious in character and depend on input from several sources. The most important of these are as follows: *kinesthetic* sense conveyed *by spino-cerebellar* system from receptors in muscles, tendons, and joints. Sense provided by the vestibular organ of the inner ear and visual input from the retina.

The reason input of spinal muscles and joints are so important is that they are made up of slowly adapting tonic receptors that continuously fire to the *cerebellum* and brain. The sheer volume of stimulation because of continual firing more than any other source of stimulus to the brain may be why the spinal muscles and joint have more neurons than the rest of the nervous system. The spinal muscles and joints process more information than any other region of the body, and are the only continuous source of stimulus to the *cerebellum* and brain, not the inner ear or the eyes, only the muscles that are forced to resist the forces of gravity.

To understand the effects of gravity on the nervous system we will review studies on the effects of micro-gravity on organisms and humans. In one study, the ultrastructure of *somatosensory*, visual, and olfactory cortex of the brain in rats exposed to space flight for 7 days ("Cosmos-1667" biosatellite) and 14 days ("Cosmos-1887 and -2024" biosatellites) as well as rats of ground-based synchronous control and Vivarium control groups has been studied (Dyachkova, 1991). The flight organisms showed similar changes in the ultrastructure of all the brain cortex regions under study. The observed differences were primarily due to the extent of observed changes, the relationship distribution density of changed structures, and specific features of their localization in the corresponding brain areas. These changes were most distinct in the structural components of the *neuropile-pre*synaptic axon terminals, dendrites, synapses, as well as *glial* components. Studies revealed *light* and *dark* degeneration of axon terminals, destructive changes in the *post*synaptic areas of dendrites, as well as enlargement of area occupied by *glial* components in the cortical *neuropile*. These changes of the ultrastructure in the *somatosensory cortex* of the flight organisms were noted to be more pronounced than those in the *visual cortex*. In the *olfactory cortex* of flight organisms, the ultrastructural changes are less expressed than in the neocortical areas under study and are mainly localized at the sites of Axo-spinal contacts. Ultrastructural changes were observed to depend on the duration of space flight. Changes of the ultrastructure in the brain cortex of flight rats reflect the morphological–functional rearrangement of a system of inter-neuronal contacts that develop due to the loss (*dark* degeneration of axon terminals) or decrease (*light* degeneration) of the functional activity of some synapses. The gradually developing *light* degeneration of terminals is the result of deficient *afferentation* in weightlessness.

In another study, cytochemical and morphometrical analyses were performed on motor neurons in the *lumbar enlargement of the spinal cord* (*LESC*) and sensitive neurons in the *lumbar inter-vertebral ganglia* (*LIVG*) in rats after a 19–32-day space flight on board *Cosmos*-605, -782, -936 biosatellites. Results revealed a decreased content of RNA, protein, body and nuclei volume in motor neurons, and the content of RNA and protein in large living LIVO neurons. The investigators also noted that results support a conclusion of hypofunction of LIVO neurons in weightlessness. This conclusion, in turn, is thought to reflect a decline of afferent impulsation in weightlessness arising from *proprioceptors* of muscle joints and tendons of the hindlimbs. In addition, results show that in rats after space flight on *Cosmos*-2044, 6 out of 8 rats demonstrated a 12 percent decrease in the volume of the nucleolus in LESC motor neurons and a 33 percent decline in the number of peri-neuronal *glial* cells, as compared to ground-based synchronous control rats. The investigators conclude that these results provide evidence for the hypofunction of motor neurons in weightlessness and suggest similar data previously obtained for rats in longer space flight (Polyakov et al., 1991).

These studies demonstrate that with loss of gravitational force, the spinal cord, brain, and especially neocortex, show rapid structural changes and atrophy. This is as a direct result of the loss of the constant flow of *proprioceptive* input from muscles, tendons, and joints involved in maintaining posture. In humans in space flight or micro-gravity situations such as underwater, *vertigo*, *nystagmus*, and motion sickness are the most concerning problems. The primary source of these symptoms has been studied. Human data was collected on the peculiarities of sensory interaction under varying experimental conditions including microgravity. Aizikov and colleagues (1991) attempted to determine the most significant afferent input to the cerebral cortex. They state,

> The main role here appears to be the postural reflexes caused by an exposure of musculo-articular and *vestibular* systems to the gravitational field of earth and by forming during the life a subjective vertical. In this case of great importance is still antigravity musculature with its more powerful input of information as compared with that of the vestibular system. Thus owing to an integrative function of the brain cortex there exists gravitational vertical which by its nature is a peculiar balance of vestibular sensitivity, muscular sense and vision.

They continue by indicating that the greatest protective effect against experimental motion sickness is produced by a procedure, which uses activation of basal *proprioceptive*-tactile coordinates of the subjective vertical and consecutive tension of antigravity muscles, followed by tension of the antagonists. They point out that this is associated not only with several hundred percent increase of tolerated duration of swinging (from 4 to 10 mm), but also with a marked (5 points vs. a baseline of 13) decrease of vegetative symptoms (motion sickness). They further note that with the aid of muscular tension, it is possible to eliminate illusions appearing during tilts and rotation as well as at the end of the procedure on a rotating chair.

The effect of electrical muscle stimulation is markedly smaller than that of muscular tension stimulation. Analysis of the protective anti-motion sickness effect of the muscular activity was tested with the administration of the drug *scopolamine* (Gordon et al., 2001). It was revealed that the effectiveness of the muscular tension procedures was enhanced with a simultaneous administration of *scopolamine* and that the drug clearly potentiated the effect of muscular tension. The authors concluded that volitional efforts that accompany a purposeful motor act or muscular tension hinders or blocks completely spatial disorientation illusions and decreases significantly the severity of the *vegetative vestibular response* (VVR). The investigators thought that, "ascending *proprio-tactile* efferent impulses developing from muscle tension of the spine and lower extremities provide information on body position irrespective from vestibular sendings but they are adequate to overcome the negative effect of the latter." Furthermore, the investigators go on to state that they think that an integration of these polysensory inputs is realized in coordinating structures as the *cerebellum* and cortex of the brain. They recommend that based on the results of this study, muscular tension should be used for normalizing vestibular–motor interactions as the means of nondrug prevention of motion sickness.

What we have seen in these studies is that motion sickness, a severe symptom of loss of *proprioceptive* input, is a common finding in micro-gravity situations. That the loss of *proprioceptive* input is mostly the result of postural antigravity of the muscles of the spine and lower extremity and this presents with a more powerful effect than loss of vestibular input and active contraction, and use of these muscles can completely compensate for the loss. We also see that this loss of *proprioceptive* input also results in *nystagmus* of the eyes and an inability to efficiently control eye movement. Loss of *proprioceptive* input can also result in significant abnormal *autonomic* reactions, which again can be alleviated by increasing *proprioceptive* input from those specific muscles. We have also seen that the loss of this input is thought to be mediated centrally through the *cerebellum* and cortex. This loss of input in rats has been shown to cause rapid *light* and *dark* degeneration of neurons in widespread areas of the brain, but especially the neocortex. Areas of neocortex involved include the *somatosensory* and visual areas especially.

Therefore, in children that present with signs and symptoms of *proprioceptive* loss, which includes disorders of balance and gait, abnormal muscle tone—especially *hypotonia, nystagmus, vertigo, dysequilibrium,* nausea, motion-sickness symptoms, etc.—these symptoms are due essentially to a primary dysfunction of *cerebellar-thalamo-cortical* input. The primary source of input to the *cerebellum* of *proprioceptive* input is from muscles, joints, and tendons of spine and lower extremity, and postural antigravity muscles. This input is more significant than vestibular input and can even compensate for vestibular loss. Therefore, anything that may decrease the ongoing *proprioceptive* input of postural antigravity muscles and/or vestibular input as seen in micro-gravity experiments can result in sensory loss, degeneration or reverse plastic changes in the *cerebellum* and neocortex. Such conditions have the possibility of affecting any or all other brain regions but especially neocortical functions.

TACTILE, VISUAL, AND AUDITORY SYMPTOMS

Tactile

Ayers (1989) and others (Kaluger and Heil, 1970; Ayres, 1972a; Njiokiktjien et al., 1976; Ottenbacher, 1978; Kohen-Raz and Hiriatborde, 1979; Freides et al., 1980; Voronin et al., 1980; Hulme et al., 1982; Conrad et al., 1983; Morrison et al., 1985; Morris et al., 1988; Kohen-Raz, 1991; Cammisa, 1994) have noted that children with learning disabilities and behavior disorders often have altered sense of touch or tactile sensations. Most often, there exists a decrease in tactile sensitivity in these children, but others also appear to have a hypersensitivity to touch. Interestingly, tactile sensory input or conscious touch input travels much the same path as *proprioceptive* input. Therefore, the same dysfunctioning mechanisms that affects proprioception can also affect tactile perception. Both senses are mainly activated and are processed in the *somatosensory cortex*; however, for touch and spatial orientation there does appear to be asymmetric lateralization favoring the right hemisphere. The *somatosensory cortex* is proximal to the *postcentral gyrus* and areas adjacent to it (Brodmann areas 1, 2, and 3). These areas receive input from *somatosensory* pathways originating in the *thalamus*, which transfer tactile information of *pain, temperature, and proprioception*. The primary somatosensory cortex (*SI*) is immediately caudal to the *central sulcus*, the *secondary somatosensory cortex* (SII), which obtains most of its afferent input from SI, is located ventral to SI. Somatosensory inputs projecting to the posterior *parietal cortex* arise from *SI* and *SII*. Somatosensory input projecting to the *thalamus* and proceeding to the *primary somatosensory cortex* transverses two main pathways: *the antero-lateral system* for pain and temperature sense, and the *dorsal column-medial lemniscal pathway*, which carries information about tactile *proprioception* and movement sensation. Any process that physically disrupts the flow of information such as a lesion at the receptor, nerve, spinal cord,

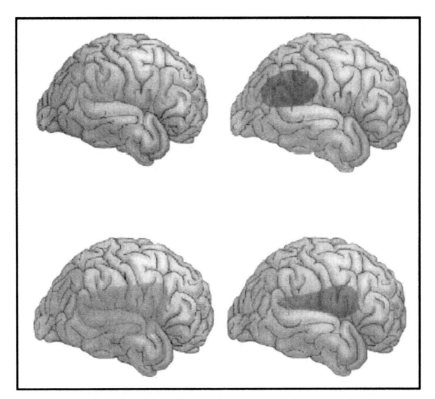

Fig. 9.4. Cortical source of hemispatial neglect syndromes, most often seen after large areas of damage to right areas of parietal lobe. It is a frequent consequence of stroke on the right side of the brain and thus neglect of everything on the left. It is also evidenced in other forms of brain dysfunction especially with a right hemisphere source. Patients ignore everything on the side opposite to the lesion. It is not blindness as patients can recognize and label objects appropriately. Thus, it is a disturbance in the spatial distribution of directed attention. Patients frequently bump into objects, tend not to groom themselves, all on the side opposite the lesion. In the majority of patients, the lesion involves the supermarginal *gyrus* in the inferior parietal lobe at the temporo-parietal junction. Although less frequently, damage or dysfunction to the dorsolateral *premotor* and medial frontal regions can also bring about hemineglect. Lesions confined to the primary motor, somatosensory, and visual cortices anre not associated with neglect. The posterior parietal component provides an internal sensory mechanism, the libic component regulates the spatial distribution, the frontal component coordinates the motor programs for scanning, reaching, and exploring, and the reticular component provides levels of arousal.

brainstem, cerebellum, thalamus, or cortical level can cause abnormalities in the attendant perceptual processes. Abnormal perception, decreased arousal, or imbalance of arousal may result even with a physically intact pathway (Fig. 9.4).

Arousal level or attention is the foundation of all perception and without proper arousal, any or all effective sensory perception can be impaired. These perceptual alterations and resultant inattention syndromes are known as *neglect*, and can occur with or without spatial awareness, tactile, visual, and auditory sensations. In these cases, the physical pathways of the stimulus are intact but the cerebral cortex does not integrate or perceive the information, in essence rendering the stimulus non-existent to the individual. This can occur bilaterally, but more often, the result is a unilateral loss. The asymmetric lateralization of the function, and the localization of the primary dysfunction will determine which sense will be most affected, which helps us to localize the primary problem. There are numerous studies of various asymmetries of tactile (*haptic*) perception, visual perception, and of interference between two tasks performed at the same time (Hiscock and Kinsbourne, 1995). Tactile inattention is also known as *pseudo hemi-anesthesia*.

Vallar and colleagues (1991a) reported on a patient who had apparent contralateral *hemi-anesthesia* from a large right hemisphere cerebral infarction. Although the patient denied being stimulated on the left side, tactile stimulation did produce an electrochemical skin response. Three other patients were reported by the same author (Vallar, 1991b). All three had left *hemi-anesthesia* from right hemisphere lesions; however, all had normal evoked potentials recoded from the left side. This suggests that although the pathways were intact, the tactile sensations were not perceived due to lesions in the right hemisphere. Another study (Meador et al., 1988) using selective hemisphere anesthesia in 18 epileptic patients, found that contralateral tactile inattention was present in 8 patients after right hemisphere injection of anesthesia but only in two subjects after left-sided injection.

Another form of tactile inattention is known as *extinction* to *simultaneous stimulation*. It has been reported that as patients with neglect or hemi-inattention symptoms improve, they may develop extinction to simultaneous stimuli. The difference is that individuals with inattention are not aware of tactile stimulation to the side opposite the impaired hemisphere. However, patients with *extinction* are aware of contralateral stimuli but are unable to perceive simultaneous bilateral stimulation, specifically failing to perceive the stimulus to the contralateral side. In most cases, studied contralateral extinction was more frequently associated with selective anesthesia of the right hemisphere than the left (Meador et al., 1988).

In *spatial neglect*, most commonly *hemispatial neglect*, patients with injury or dysfunction localized to one hemisphere may not be able to perform normally on spatial tasks. This is usually found in the contralateral hemi-space so that the stimuli to the left of the patient's body are neglected more often than those to the right side of the body (Heilman and Valenstein, 1979). *Neglect* of the environment may also occur so that even when the individual lies on his or her side, the left half of the environment is neglected (Ladavas, 1987).

Spatial neglect is complex and can involve: (1) attentional disorders, (2) intentional or motor activation disorders, or (3) representational disorders (Heilman, 1995). It has been hypothesized that *hemispatial neglect* may have four possible attention related causes: (1) contralateral spatial inattention, (2) ipsilateral attentional bias, (3) inability to disengage, and (4) reduced sequential attentional capacity or premature habituation. One theory (Heilman and Watson, 1977) posits that patients with left *hemispatial neglect* are unaware of stimuli presented to the left hemisphere because they may simply be less aware of stimuli on the left than on the right resulting in an ipsilateral bias with attention being drawn to that side and away from the neglected side. Another possibility for explaining *neglect* is that objects besides the critical stimulus in ipsilateral space draw the patient's attention with the bias becoming so strong that the individual may not be able to disengage their attention (Posner et al., 1984). It is also possible, as exemplified in one case, where bias or inattention may determine where stimuli are neglected, a limited sequential capacity or inappropriate habituation may also lead to *hemispatial neglect* (Chatterjee et al., 1992a, 1992b; Heilman, 1995).

Hemispatial neglect is seen more frequently with right hemisphere lesions (Costa et al., 1969; Gainotti et al., 1972). Although several studies note that spatial neglect caused by right hemisphere lesions mostly affects the left side of space or body, it may also affect the right side of space (Albert, 1973; Heilman and Valenstein, 1979; Weintraub and Mesulam, 1987). In humans, the inferior *parietal lobe* is more often associated with disorders of attention (Critchley, 1966; Leisman, 1976a; Heilman et al., 1983).

Temporo-parietal ablation in monkeys also results in contralesional attentional disorders, primarily extinction (Heilman et al., 1970; Lynch, 1980). Therefore, it is suggested by this research that *tactile* as well as *spatial neglect* can be a result of decreased or imbalanced arousal or attention. In the right hemisphere, a lesion or dysfunction can result in *neglect* that affects both sides of the body or space with the left being more frequently affected. The primary brain area reportedly responsible for the majority of cases is the inferior *parietal lobe* or *temporo-parietal* areas.

Functionally, decreased afferent input in the absence of a physical lesion could be another reason for *neglect*. In the *parietal lobe*, which is important in directing attention, it has been noted that the rate of cell firing appears to be associated with the importance of the stimulus to the monkey, so that important stimuli are associated with a higher firing rate than unimportant stimuli (Lynch, 1980; Bushnell et al., 1981; Motter and Mountcastle, 1981). It has been demonstrated that the activity of some parietal attentional neurons are spatially selective. However, the *parietal lobe* also receives input from other sensory modalities and the coding of spatial location is thought to be multimodal (Goldenberg and Robinson, 1977; Robinson et al., 1978). The primary sensory cortices project only to the *association* cortex. These single modality primary sensory areas converge upon polymodal association areas in the monkey including the *frontal cortex (periarcuate, prearcuate,* and *orbitofrontal)* and both banks of the superior *temporal sulcus* (Pandya and Kypres, 1969). Multimodal association areas are important for sensory synthesis and cross-modal association. These multimodal convergence areas project to the inferior *parietal lobe* (Mesulam et al., 1977).

The *cerebellum* has also been noted to have a close connection to all association areas of the neocortex. The *cerebellum* has also been implicated in tactile learning and sensory tactile deficits besides its primary role in the processing of *proprioceptive* input. Leiner and colleagues (1991) have stated that, "concomitant with the evolution of new association areas in the cerebral cortex, new neuronal connections evolved that descend (via enlarged structures in the *brainstem*) to new area 5 in the lateral *cerebellum*." It is further stated, "therefore, the neo-*cerebellum* can serve as a link between the posterior and frontal language areas of the cerebral neocortex in effect the *cerebellum* provides the cerebral cortex with an additional 'association area'."

Blood flow studies during tactile learning of complicated geometrical objects show a high activation in the lateral *cerebellum* (Roland et al., 1989). The activation of the lateral *cerebellum* is thought to be due to cognitive processing during tactile learning, whereas the medial *cerebellar* activity relates more to the actual movement (Fox et al., 1985). The *cerebellum* has been specifically shown to be involved in sensory perception; this has been shown during chronometric tests (Posner, 1986) of timing functions. *Cerebellar* patients demonstrate deficits in perception of time intervals and judgment of velocity of a moving stimulus. It has been postulated that one role of the *cerebellum* is to modify motor performance to improve the efficiency of sensory processing by the rest of the nervous system by means of arousal.

Arousal is a physiological state that prepares an organism for sensory processing by increasing sensitivity and signal-to-noise ratio (Heilman, 1995). This is an extremely important concept because it forms the basis of all perception and perhaps even consciousness itself. The baseline arousal of all cortical areas makes all of these cells more active and therefore closer to threshold or more sensitive to incoming stimuli. The closer to threshold, the better the signal-to-noise ratio resulting in global or specific effects on the brain. The results of such a process in turn could lower the effectiveness of global possessing of the brain or one specific function may be affected more than the others. This can happen with an imbalance of arousal, where one side is too high in relationship to the other.

EMOTIONALLY SIGNIFICANT

The significance of the stimulus requires knowledge of the meaning of the stimulation and the motivational needs and goals of the individual (Heilman, 1995). The *limbic system* provides the more primitive motivation of immediate biological needs, whereas the *frontal lobe* is involved with goal-directed behavior. The difference between the *limbic* and frontal input to attentional or arousal systems, is that the *frontal lobe* may provide motivation and goals that are of a higher level beyond the basic biological needs of the child. It is thought that the frontal and *limbic (cingulate gyrus)* connections with the *frontal lobe* may provide the neuronal basis for the ability of motivational states to affect attentional systems and perception. Unilateral *neglect* in humans and monkeys can be induced by lesions in the dorsolateral *frontal* and *prefrontal* cortex (Welch and Stateville, 1958; Heilman and Valenstein, 1972) as well as with *cingulate gyrus* lesions (Heilman and Valenstein, 1972; Watson et al., 1973). Affective behaviors have also been shown to be influenced by *cerebellar* dysfunction, especially midline *cerebellar* atrophy (CA) (Gutzmann and Kuhl, 1987). The *cerebellum* is also known to be connected to the dorsolateral *frontal* and *prefrontal* cortex. Not only can the *cerebellum* affect global arousal, but it also has specific connections to *association cortices* for multimodal sensory processing. It has strong specific connections to the *somatosensory cortex* in the *parietal lobe*, as well as the *frontal lobe*.

The overall neuronal preparedness and signal-to-noise ratio is dependent on arousal input. The *cerebellum* may be the largest contributor to the arousal system. The *cerebellum* also works to modify motor behavior to maximize all sensory input. Therefore, the *hemispatial neglect* can be seen with all sensory input, not just tactile and spatial, but also visual and auditory. Besides signal-to-noise ratio, neuronal synchrony is essential for sensory perception to exist (Koch and Leisman, 1990, 1996, 2001a, 2001b; Leisman and Koch, 2003). Liederman (1998) describes how synchronization dysfunction may result in a *neglect* syndrome for any sensory modality. She describes how *cortico*-cortical pathways can function to synchronize activity across remote cortical regions. This synchronization then is thought to have the effect of equilibratory activation between different regions. She is quoted,

thus disconnection symptoms such as those seen in split-brain patients can be accounted for mainly by two factors: the desynchronized and fragmented manner by which subcortical pathways permit inter-hemispheric integration and the diminished arousal state of the non-viewing hemisphere without the synchronizing influence of the cerebral commissures. The underactivated hemisphere displays an *internal neglect* that is marked by an abnormality of processing of input. After sensory reception it does not effectively process inputs to it (a) directly from the contralateral side of space, or (b) indirectly from the opposite hemisphere. These under-processed inputs are not consciously perceived though they are reacted to on an implicit level.

Synchronization of a hemisphere is thought to result from 40 Hz oscillations that arise from the *intralaminar thalamic nuclei*. These subcortical oscillations in the *thalamus* have been traced back and are thought to originate from the *cerebellum* and *dorsal column* fibers that are thought to transmit information from slowly adapting receptors. These types of receptors are found most commonly in slow-twitch muscle fibers of postural spinal muscles. It is known that even in the intact brain the two hemispheres can maintain different arousal levels as well if not more than differences within adjacent areas of the same hemisphere. Large differences should not occur in a normally functioning intact brain. However, in situations where stimulus conditions cause the untrained or non-viewing hemisphere to be either busy or too minimally alert to adequately subserve inter-hemispheric integration (Liederman, 1998), inter-hemispheric arousal differences may exist.

It has been demonstrated that right *parietal lobe* lesions cause a deficit of integration and not detection. It is thought that the *neglect* subjects do not have difficulty with directing their attention to a single-feature target. They do appear, however, to have difficulty with searching for conjunctions among distractions consisting of two elements, which together would constitute a conjunctive stimulus and patients present with deficiencies in their ability to integrate information after their attention has been focused on the features. They do not have the ability to use *selective attention* as the glue to bind these features (Prather et al., 1992).

Kinsbourne (1970) thought that each hemisphere directs attention to the contralateral space and that a normal attentional system of one hemisphere inhibits the other hemisphere. Therefore, if one hemisphere is injured or less active, the other becomes attentionally hyperactive, and attention is biased to the side opposite the normal hemisphere. This may be one of the explanations for the increased sensitivity to touch that learning-disabled children exhibit. The right side of the brain is known to direct attention to both sides of the body, therefore if it were functionally attentionally hyperactive, the child may be hypersensitive to stimuli on both sides of the body and space.

Another reason for hypersensitivity of tactile stimulation maybe that *proprioceptive* and tactile input is inhibitory to pain input. Therefore, with decreased *proprioceptive* input, a child would be more sensitive to pain and therefore even touch would be painful. This would be compounded by the fact that loss of *proprioceptive* or arousal activity to the neocortex would cause a decrease in inhibition of *limbic* autonomic output, which in turn would result in increased *sympathetic* activity and increased adrenal catecholamine release. These substances sensitize *alpha 2* receptors found in pain transmitting fibers, bringing the fibers closer to threshold globally. This may result in a child being hypersensitive to touch because of decreased stimulation or arousal.

Occupational therapists for years have used brushing techniques where they lightly brush the child's skin, the child eventually becomes "desensitized" to the stimulus, and the child will not be as defensive to touch. This may not be due to actual desensitization, but rather increased sensory input may result in an increased arousal and reciprocal inhibition of each hemisphere so that one is not hyperactive and one is not underactive. It may increase large afferent fiber input to produce inhibition of pain directly as well as inhibition of *sympathetic* output in the spinal cord and brain.

In summary, sensory processing is a product of multiple factors. The baseline arousal level and signal-to-noise ratio or neuronal preparation form the basis of all sensory perception. This can be globally deficient and/or specifically decreased in a limited area. It can be imbalanced from one side to the other with one side becoming hyperactive and the other hypoactive. Any decrease in arousal will affect at least one sensory perceptual process. The arousal level of a hemisphere is set by the subcortical structures, especially the *thalamus* and *cerebellum*. They receive most of their input from muscle joint and tendon receptors in the postural and antigravity muscles. These receptors, which are continually transducing gravitational forces into the central nervous system, create the arousal level and the *gamma* oscillations that are the basis of all perception. Hillyard (1999) indicated,

What you are asking about is the nature of neuronal codes for sensory perceptual information. There is a general assumption that more activity—greater firing rates etc.—within a nerve cell population means that more information is being represented or processed by those cells. Conversely, a suppressed response reflects diminished processing.... In simple low-level sensory signals, greater physical stimulus energy produces systematic increases in neuronal response amplitudes that are paralleled by elevations in the perceived stimulus magnitude. Also signal detection experiments in humans and monkeys have shown that moment-to-moment fluctuations in neuronal response amplitudes can predict precisely whether the person or organism detects a faint signal.

Vision and Hearing

Children with learning disabilities have also been noted to demonstrate various types of visual and auditory dysfunction. If we look at all sensory perception in light of Luria's theory (Luria, 1973) we find that the brain has three functional units: (1) the arousal unit, (2) the sensory receptive or integrative unit, and (3) the planning and organizational unit, and that the second and third are subdivided into three zones. The primary zone sorts and records incoming sensory stimuli, the secondary zone organizes and codes information from the primary zone, and lastly the tertiary zone is where data is merged from multiple sources of input and collated as the basis for organizing complex behavioral responses. These last two, especially the third or tertiary zones, are what make humans unique. In regard to vision, the primary zone gives us the ability to see, to perceive light at its most basic level. The secondary zone allows us to make basic responses based on immediate biological needs—the *limbic response*. Most organisms have evolved to this level. However, the human neocortex, which fits Luria's description of *tertiary zone* gives us the ability to put a higher meaning to this stimulus, to go beyond basic primitive needs and motivations, and to utilize the stimulus to allow us to obtain higher more complex goals. This higher goal-directed behavior is supported by the *prefrontal cortex*. This region allows us to put higher cognitive and emotional significance to a stimulus like light or sound. The *parietal lobes* allow us to integrate this information with other sensory stimuli before the meaning of it all is processed in the *frontal cortex*. This area of integration seemingly occurs in the *temporal, parietal,* and *occipital* cortices (approximately Brodmann's area 39). The *primary zone* would correspond to the primary *visual cortex*. (See Figs. 9.5–9.7.)

The final projection of the visual pathway is via the *geniculo-cortical* pathway. This group of axons exits the *lateral geniculate nucleus* of the *thalamus* and proceeds to the cerebral cortex terminating in the primary visual area of the *occipital lobe* (Brodmann's area 17). This area is regular and stippled in appearance, hence its name, the *striate cortex*. This area is also known as the primary visual area and it is thought to be the region of the cortex where the first level of visual processing takes place. If the signal-to-noise ratio is not high enough due to a decrease arousal in the *occipital lobe*, a child may appear to have a primary visual deficit, even though the eyes and the visual pathway may be intact. A severe example of this is related by Glen Doman (1974) in his book *What to do About Your Brain Injured Child* where he notes that he would often see children in his clinic who were apparently blind. They would have no apparent reaction even to the brightest light. These brain-injured children were then placed on a program of therapy. Surprisingly they noticed that many of these allegedly blind children improved in their visual function coincident with improved movement ability. Doman assumed that this meant that the pathways associated with vision in these children were always intact but for some reason this information was not summating to the level of perception in the brain.

While Doman's clinical observations were not examined under controlled circumstances, we can postulate that as these children increased their movement or crawling, they increased the tone of postural muscles and their feedback to the *cerebellum* and *thalamus*. This would increase the baseline arousal, signal-to-noise ratio, and neuronal sensitivity so that with greater degrees of motor activity the greater would be the degree of *cerebellar* feedback and the greater the effectiveness of sensory perception. The fact that the primary pathways were intact could also be considered a *visual neglect syndrome* that would be the same as *tactile* or *spatial neglect*, which would mean that the primary information may be effective but under-processed and therefore ignored or neglected. This, however, would be more characteristically seen in a hemi-field of vision. Besides primary visual loss or visual neglect, there is an asymmetric distribution of visual processes. The specificity of symptoms of visual loss, such as object identification vs. spatial localization, or global vs. local effects, can help us to localize the problem to the right or left hemisphere and ventral or dorsal systems.

Ungerleider and Mishkin (1982) proposed a model of cortical visual processing based on

▶ The Human Visual System

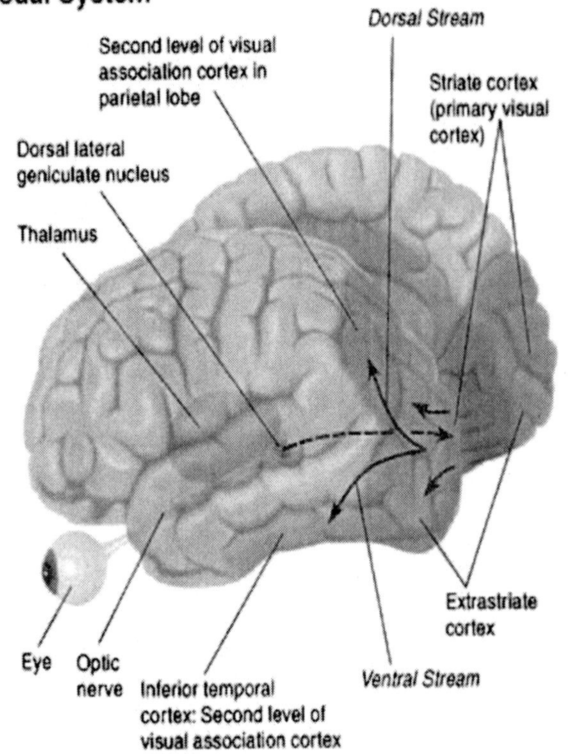

Fig. 9.5. The human visual system. Goodale and Milner (1992) claim: "two cortical visual systems have evolved: a ventral stream for visual perception and a dorsal stream for the visual control of skilled actions." Damage to the ventral pathway: Their case of D.F. demonstrates a complete dissociation between visual form perception and visuomotor abilities. She can accurately reach for and grasp an object, which she cannot identify. D. F. suffered anoxia and subsequent damage to the ventrolateral regions of her cerebral cortex, with sparing of primary visual cortex and the dorsal pathway. Damage to the dorsal pathway: Optic ataxia, difficulty reaching for and grasping objects, often results from damage to superior portions of the parietal cortex. These patients can recognize and describe objects. Thus, there is a double dissociation between the ventral/dorsal pathways and the grasping/perceptual matching tasks. In the dorsal pathway, motion sensitive cells abound, particularly in the areas MT and MST. Cells in posterior parietal cortex show evidence of anticipating the retinal consequences of saccadic eye movements. Egocentric coding of these cells: Ventral pathway studies show that inferotemporal cortex cells are tuned to highly specific forms, for example one IT area is particularly responsive to face stimuli. These cells have extremely large receptive fields. These large cells are thought to underly object-centered representations. In studies of dorsal pathways in brain-damaged monkeys, deficits in visually guided reaching resulted from damage to the posterior parietal cortex. In studies of the ventral pathway, if the inferotemporal cortex is damaged or dysfunctioning bilaterally, the monkeys display severe difficulties in visual recognition and discrimination learning tasks.

two parallel visual systems, one thought to be directed ventrally to the *infero-temporal cortex*, and the other directed dorsally to the inferior *parietal lobe*. It has been shown in monkeys that lesions of inferior *temporal cortex* produce severe deficits in performance in a variety of visual discrimination learning but not in visuo-spatial tasks. However, posterior *parietal* lesions cause severe impairments in visuo-spatial performance, such as in visually guided reaching and in judging which of two objects is closer to a visual landmark; however, these lesions do not affect visual discrimination performance (Ungerleider and Mishkin,

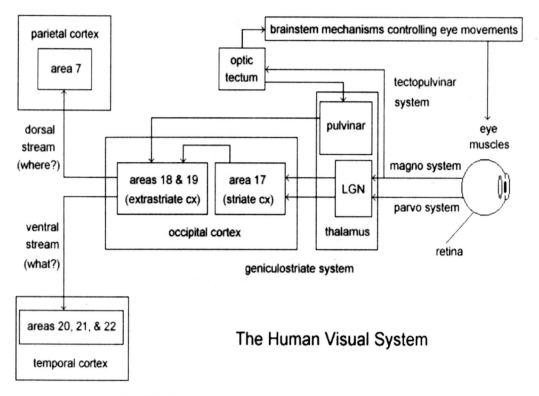

Fig. 9.6. The human visual system's network of pathways.

1982). Some cells in the *parietal* lobe are most active in the monkey when it fixes it gaze on an important object (*fixation neurons*), other cells fire when the organism is visually tracking a moving object that is important (*tracking neurons*), while some others fire before a visual saccadic eye movement to a significant stimulus (*saccadic neurons*). Finally, there are neurons that fire when important stimuli present to peripheral vision (*light-sensitive neurons*) (Lynch, 1980; Bushwell et al., 1981; Motter and Mountcastle, 1981).

The ventral visual pathway terminates in the inferior *temporal lobe* (area *IT* or *TE*). In TE area neurons, a small percentage of cells respond selectively to high-order complex stimuli, such as faces or hands. Most of the neurons, however, respond specifically based on object features such as color, shape, and texture rather than specific objects. *IT* area cells are thought to make up a network of cells that appear to be involved with the representation of general object features. Lesions of the *IT* areas produce loss of perceptual constancies and loss of visual memories (Mishkin et al., 1984). It is thought that memories of visual inputs are limited to a *habit* learning system that involves the *basal ganglia* (Mishkin et al., 1984).

The Dorsal Visual Pathway

This system provides the basis for peripheral vision, motion perception, stereopsis, perception of three-dimensionality based on perspective and shading, and most of the *gestalt* phenomena of *linking operations* (Livingstone and Hubel, 1987). The *dorsal pathway* is thought to terminate in area PG of the inferior *parietal* lobe and *intra-parietal sulcus*. In monkeys it has been shown that a percentage of cells in the PG area fire when the organism fixates on interesting objects in a particular region in space (Mountcastle, 1978). This is known as *ambient vision* (Trevarthen, 1968; Motter and Mouncastle,

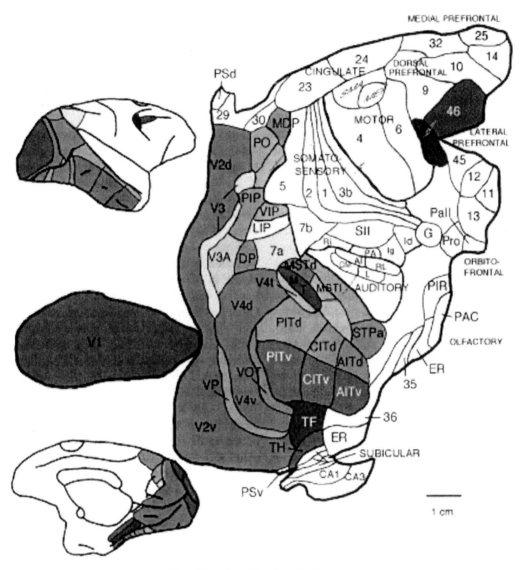

Fig. 9.7. Visual-fugal projections area.

1981). Some cells in area P6 have been shown to increase their firing rate when the organism expects a stimulus in a specific region of the visual field; this is called the *enhancement effect* (Robinson et al., 1978). The posterior *parietal cortex* is involved in the control of spatial attention, lesions or dysfunction in this region impair the ability to orient attention to the contralateral side with presentation of *invalid* directional cues (Posner and Cohen, 1984; Posner et al., 1987). The two visual networks that have been described are parallel hierarchies of sequential neocortical cognitive representations. However, they are not exclusively neocortical and they are dependent on the function of the *limbic system*. The *ventral system* has direct reciprocal connections to the *amygdala* and to the *orbitofrontal cortex*. The *dorsal system* connects to the *cingulate gyrus*, the *hippocampus*, and surrounding *limbic* areas. These connections are thought to provide the integration of emotional, motivational, and interoceptive quality of visual perception (Liotti and Tucker, 1995).

Bilateral damage to both TE and the *amygdaloid complex* produce severe longlasting recognition loss (Mishkin, 1978; Zola-Morgan et al., 1982). The affective components are added to visual stimuli by connections between neocortical aspects of the *temporal lobe* and *amygdala*. Without these connections, it is not possible to associate reward conditions with visual stimuli. The *amygdala* may also play a role in integrating perceptual and affective memories into a subjective experience (Ledoux, 1987; Gloor, 1990).

There exist monosynaptic connections from the *thalamus* to the *amygdala* (Ledoux, 1987). The *thalamo–amygdaloid* system supports premature emotional reactions that are loosely coupled with the stimulus that do not require object recognition. Inputs to the *amygdala* from multimodal association areas can be involved in evaluating object information integrated from different sensory modalities (Ledoux, 1987). The *orbito-frontal cortex* possesses direct reciprocal connections with the *temporal pole* and *amygdala* and is thought of to be the frontal extension of the *ventral visual pathway*. Severe impairments in visual object learning have been noted because of *orbito-frontal* lesions in monkeys (Thorpe et al., 1983). Learning and unlearning the emotional significance of sensory input may require close interaction with the *amygdala* and orbital region.

Limbic System and Dorsal Visual Processing Deficits

The convergence of *limbic* inputs from the *cingulate-retrosplenial* cortex with preprocessed sensory information may allow parietal neurons to recognize the motivational relevance in complex sensory events (Mesulam, 1985). Monkeys with unilateral *cingulotomy* display contralateral inattention in visual and somatosensory modalities (Watson et al., 1973). Right parietal or medial frontal infarcts including the *cingulate cortex* result in unilateral neglect for the left hemisphere (Heilman et al., 1983) and significant inability with spatial memory in half the world (Mesulam, 1985). The *cingulate cortex* may mediate adaptively significant attention as opposed to more spatially oriented attention in the posterior parietal region. Bilateral *cingulate* infarcts in humans produce an akinetic state and attentional dysfunction. Bilateral *para-hippocampal* lesions result in decreased discrimination of complex geometric forms and faces.

The right hemisphere is more connected to the *dorsal system*. The right hemisphere's primary role in humans is visual-spatial behavior (Fig. 9.8). Visual-spatial behavior can be disrupted by right posterior regions including the *parietal lobe*. Visual-spatial behavior that is typically disrupted includes visual judgment of line orientation (Benton et al., 1978a), maze performance (DeRenzi et al., 1977), and the ability to identify objects presented from an unusual visual perspective (Warrington, 1982). Other functions of visual processing associated with right posterior damage include *constructional apraxia, dressing dyspraxia*, and *prosopagnosia* or disorders of face recognition (Hécaen and Angelergues, 1962; Arrigoni and DeRenzi, 1964), depth perception (Kimura, 1961a), spatial localization (DeRenzi, 1982), identification of complex geometric shapes (Unilta et al., 1978), and exploratory eye movements (Sava et al., 1988). The network of cortical and subcortical regions that control visual orienting in monkeys and humans are functionally part of *dorsal visual system* and include the posterior *parietal* region, the *cingulate cortex*, the dorsolateral *frontal cortex*, the *pulvinar*, and the *inferior colliculus* (Heilman, 1979; Mesulam, 1981).

The right hemisphere stores global configurations of visual objects as opposed to local features or details of objects. Line orientations are oblique (Umilta et al., 1974). The left hemisphere is associated with the *ventral visual pathway*. Lesions in the left occipital-temporal regions produce visual *object agnosia* (McArthy and Warrington, 1990). The left hemisphere is also superior for object identification shown from a usual viewpoint. The left hemisphere is better for line orientation when vertical, horizontal, or 45° diagonal orientations are used (Umilta et al., 1974) generation of images of visual objects (Farah, 1984; Kosslyn, 1987). Patients with left-hemisphere lesions are unable to assemble

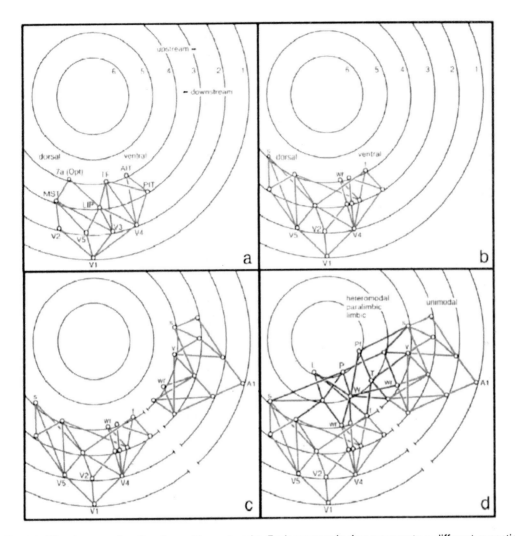

Fig. 9.8. Visual system functional cognitive networks. Each concentric ring represents a different synaptic level. Any two consecutive levels are separated by at least one unit of synaptic distance. Level 1 is occupied by the primary sensory cortex. Small empty circles represent macroscopic cortical areas or "nodes," one to several centimeters in diameter. Nodes at the same synaptic level are reciprocally interconnected by the black arcs of the concentric rings. Lines represent reciprocal monosynaptic connections from one synaptic level to another. (a) Visual pathways as demonstrated by experimental neuroanatomical methods in the macaque brain. (b) The inferred organization of the homologous visual pathways in the human brain. (c) Visual and auditory pathways in the human brain. (d) Visual, auditory, and transmodal pathways in the human brain. In b–d, the anatomical details of individual pathways are inferred from experimental work in the monkey. The anatomical identity of many of the nodes is not specified because their exact anatomical location is not critical. This review is guided by the hypothesis that these types of anatomical interconnections and functionally specialized nodes exist in the human brain even though their exact location has not yet been determined. The terms 'dorsal' and 'ventral' in (a) and (b) refer to the separation of visuo-fugal pathways, especially at the fourth synaptic level, into dorsal and ventral streams of processing. The gaps in the circles at the first four levels indicate the absence of monosynaptic connections between modality-specific components of auditory and visual pathways. Abbreviations: A1 5 primary auditory cortex; AIT 5 anterior inferotemporal cortex; f 5 area specialized for face encoding; L 5 hippocampal–entorhinal or *amygdaloid* components of the *limbic system*; LIP 5 lateral intraparietal cortex; MST 5 medial superior temporal cortex; P 5 heteromodal posterior parietal cortex; Pf 5 lateral *prefrontal* cortex; s 5 area specialized for encoding spatial location; PIT 5 posterior inferotemporal cortex; T 5 heteromodal lateral temporal cortex; TF 5 part of medial inferotemporal cortex; v 5 area specialized for identifying individual voice patterns; V1 5 primary visual cortex; V2, V3, V4, V5 5 additional visual areas; W 5 Wernicke's area; wr 5 area specialized for encoding word-forms; 7a(Opt) 5 part of dorsal parieto-occipital cortex. (From Mesulam (1998).)

parts of visual image into a unified ensemble according to spatial and temporal relation such as above–below right–left (Kosslyn, 1987), long-term storage of categorical, prepositional, or semantic knowledge concerning an object (Kosslyn, 1987), and analysis of local levels or detection of visual scene, sequential analysis of local information, or high spatial frequency.

Auditory Processing Symptoms

Many children with learning disabilities and behavioral disorders are also diagnosed with auditory processing problems. It is most commonly described as a *central auditory processing disorder*. In this case, many children have been diagnosed as having hypersensitivity to various auditory stimuli (Geffen et al., 1992; Geffner et al., 1996). Others have been thought to have decreased perception to sound. In most of these cases, it appears that the peripheral pathway is intact and that the problem lies in the central nervous system.

In 1895, Daniel David Palmer was studying the effects of spinal manipulation and health. In the building where Palmer was conducting research, a janitor named Harvey Lillard was working. Harvey Lillard was deaf and D. D. Palmer inquired of the man how this came to be. Lillard explained that it happened all of a sudden while lifting something, he heard a pop in his spine and from that time, he could not hear. As the account goes, D. D. Palmer went to the area of the spine pointed out by Mr. Lillard and he performed the first chiropractic adjustment manipulating the man's spine. Lillard's hearing reportedly returned explained by Palmer on the basis a hypothesized connection between spinal function and brain and nervous system function. We now understand that spinal joints and muscles have the greatest concentration of receptors that respond to gravity and movement with the largest degree of activity arising from postural, antigravity muscles. We hypothesize that by increasing the movement of spinal joints and allowing those muscles to stretch and contract against the forces of gravity on a continuous basis, we would expect to increase the arousal level of the brain through the *cerebellar, thalamic*, and spinal pathways. The result should be an increase in signal-to-noise ratio and the same level of sound that could not be perceived previously, now is better able to summate and be processed. This is one simple explanation.

The *cerebellum* and *thalamus* can influence the perception of sound in the same manner that they are thought to influence vision. The *cerebellum* and *thalamus* not only affect the arousal zone, but also the *primary, secondary*, and *tertiary* zones, through connection to *association cortices, prefrontal lobe, limbic system*, and the effect on the ability to integrate and temporally bind auditory input with other sensory and emotional events to perceive and remember these events. Besides these global effects, we must consider the fact that as with visual and spatial functions, certain auditory functions are asymmetrically localized in the brain. These specific pathways can also be affected, and *hemi-neglect* symptoms may arise.

One method that is commonly used to test these asymmetric functions is the dichotic listening task. Dichotic listening (DL) is a measure of *temporal lobe* function (Spreen and Strauss, 1991), attention, and stimulus processing speed as well as a way to measure hemispheric language asymmetry (Hugdahl, 1992a). It is thought that auditory lateralization is probably not related to a single mechanism (Jancke et al., 1992), but is thought to be related to several functions involving both perceptual and other cognitive factors (Hugdahl, 1995). Hemispheric perceptors of sound are dependent on the level of activation and attention. Hugdahl has indicated that the level of right ear advantage (REA) in the DL task is significantly correlated with resting EEG asymmetry. Individuals with larger left-than-right EEG resting activation also had better recall from the right as compared to the left ear in DL (Hugdahl, 1995). DL is thought to be particularly effective at testing *frontal–temporal* functional integrity (Hugdahl, 1988).

DL has revealed an REA for consonant vowel (CV) syllables in about 80 percent of subjects and in 65 percent of left-handers.

This REA has been found to be significant in children down to the age of 5 years. The REA for CV syllables is thought to correlate with the left hemisphere advantage for language skills. REA is thought to be related, therefore, to the integrity of the auditory system as well as to the structural organization of the two cerebral hemispheres. However, it is clear that dynamic cognitive and emotional factors can affect the level of the ear advantage. Also attention, transient changes in activation, and arousal have also been shown to influence the ear advantage. Emotional states especially the threat of negative events have been shown to affect the ear advantage; this is thought to be due to the right hemisphere advantage for negative emotions which changes the balance between the hemispheres in activation level for processing of dichotic stimuli.

As with vision, an auditory bias to one side or the other may result in an auditory *neglect*. Kinsbourne (1970) postulated that language activation of the left hemisphere might result in making the *neglect* associated with right hemisphere lesions more severe because it may increase the ipsilateral bias. On the other hand, it is thought that language-induced left hemisphere activation may make neglect from left hemisphere lesions less significant because it may decrease the ipsilateral bias. In contrast, emotional words elicit an REA (Ely et al., 1989). Stimuli such as pure tones and musical chords yield a left ear advantage (LEA) or right hemisphere advantage (Hiscock and Kinsbourne, 1995).

The REA/left hemisphere specialization for language processing has been postulated by Kimura (1961b, 1967) to be due to four factors: (1) the advantage of the contralateral over ipsilateral pathways in the auditory system, (2) the left hemisphere is preprogrammed to selectively attend to the right side of space, (3) ipsilateral perception is suppressed by contralateral input, and (4) input from the nondominant hemisphere must travel across the corpus callosum to be processed by the dominant hemisphere (Bradshaw and Nettleton, 1988). This concept has significant relevance for examining cerebral organization in learning-disabled populations. Issues of attention and arousal are the most intriguing factors known to interact with dichotic asymmetries in both normal and learning-disabled populations (Boliek and Obrzut, 1995). Boliek and Obrzut (1995) state that,

Evidence derived from neurobiological studies as well as data derived from our current investigations using behavioral paradigms, have led to the conclusion that for a large subset of learning-disabled children atypical cerebral organization and unusual patterns of laterality, attention and arousal may underlie deficits in auditory and visual, language and non language information processing abilities.

Directed attention tests have been widely used with learning-disabled children and have shown a lateralized performance difference between these children and control children" (Obrzut et al., 1983; Hugdahl and Anderson, 1987; Boliek et al., 1988; Kershner and Micallef, 1992). Using some form of behavioral task like DL, visual half-field, and tactual *dihaptic* tasks, Bryden (1988) reviewed 51 studies and found 30 of those studies indicate that learning-disabled children (poor readers) were less lateralized than normal readers. In one study of 51 children aged 6–12, there were 34 normal control children and 17 learning disabled, all of whom had above average IQs (Obrzut et al., 1993). The learning-disabled children all demonstrated disability in reading and related language skills. A tone was presented to either one ear or the other prior to every dichotic CV trial. In the control children, the results proved that the REA for verbal functions that have been found in DL experiments is dependent on attentional processes. A subgroup of mixed learning-disabled and control children had difficulty performing the dichotic task above chance levels because they could not orient attention to the cue. It is thought that their inability to orient attention is related to a possible underlying atypically organized functional (or structural) system or lack of motivation. An interesting finding in the learning-disabled children who could perform the task identified as the *to-be-attended* and *to-be-unattended* items equally often, which is thought to be indicative of children who exhibit selective attention deficits, but whose performance tends to vary depending on what they are doing. The authors of this study concluded that this data seems to support the notion that auditory perceptual asymmetries in children are the result of the interaction of hemispheric ability and attentional factors.

Furthermore, it has been postulated (Obrzut et al., 1993) that learning-disabled children may be less lateralized for both linguistic and non-linguistic tasks. The results show that the learning-disabled children demonstrate a significant disability in the auditory-linguistic domain causing a significant dysfunction of reading and related language skills. *DL* tasks were used, and the results suggest that control children possess the normal hemisphere shift in laterality (REA for CVs, LEA for tones). In contrast, learning-disabled children show a processing bias to the same hemisphere regardless of the stimulus characteristics. It was speculated based on these results that performing both verbal and nonverbal processing with the same hemisphere is problematic, and that children who do so are likely to show cognitive processing deficits and are more likely to be labeled as learning disabled even though they may have above average IQs.

Another study examined children with reading disabilities compared with a clinical group of children with left *temporal lobe* brain tumors. DL performance showed significant laterality differences among dyslexic subgroups and between children with acquired left *temporal lobe* lesions (Cohen et al., 1992). Hemisphere asymmetry for auditory stimuli has been found to be related to temporal frequency. In a pitch discrimination task Ivry and Lebby (1993) report that individuals are faster and more accurate in judging relatively low frequency sounds with the left ear/right hemisphere. However, individuals are faster and more accurate in judging relatively high frequency sounds with the right ear/left hemisphere.

Hypersensitivity

One of the symptoms of children with central auditory processing disorders is hypersensitivity to certain frequencies. This may be explained by an arousal imbalance where there may be an attentional bias to one ear. It may also be explained by a lack of inhibition of one side by the other due to a hemispheric imbalance of activation. A decrease of activation of one side may also result from chronic ear infections, which may interfere with development and produce a developmental lag due to decreased sound stimulus.

MOTOR SYMPTOMS

A large percentage of children with learning disabilities demonstrate motor symptoms. These can range from poor tone, poor posture, slowness or unsteady gait, clumsiness, high tone, hyperactive motor behavior, weakness, poor coordination, imbalance of muscle tone, poor eye muscle control, tracking difficulty, poor handwriting, etc. These motor symptoms involve both voluntary and involuntary motor control. The sensory pathways of the nervous system that control motor activity involve the receptors in muscles, tendons, and joints that transmit information about various aspects of movements to the *cerebellum, thalamus, basal ganglia, somatosensory cortex,* and *frontal lobe*. The majority of efferent flow of information from the cerebral cortex is directed to the motor system, but the motor system can only respond appropriately based on sensory feedback; therefore, it makes sense that the majority of subconscious input to the brain originates from the same motor system. Hence, any interruption to the flow of this loop that sends sensory input from the motor system to the brain and then from the brain back to the motor system will have significant negative effects, not just on the *sensory–motor* system but to motor, sensory, emotional, cognitive, and autonomic functions as well. The majority of this sensory input and motor output is subconscious, and is processed in the *cerebellum, thalamus,* and *basal ganglia*. Many functions of the motor system are reflexive, and are therefore a direct reflection of the adequacy of brain activity. For example, resting muscle tone is reflexive and subconscious and while we may be able to consciously modulate it, the baseline muscle tone is a reflection of base output in a structurally intact system.

High/normal tone is a reflection of a highly active, aroused brain that is prepared to process all the sensory input it can get. An individual with low/normal tone has a brain that has a low arousal, is not as active and not as prepared to process information. It would not be unusual to expect therefore that children with learning disabilities often have low muscle tone or motor dysfunction, especially when we consider that the majority of all incoming stimuli and antigravity motor information is processed in the *cerebellum* and this flow of input is essential to the baseline arousal of the *cerebellum, thalamus,* and *cerebral cortex*. If there is any disruption, dysfunction, or lack of development of this loop, we would expect to see *cerebellar* motor symptoms usually demonstrated in most learning-disabled children.

The *basal ganglia, thalamus,* and *frontal lobe* are intimately connected to one another processing mostly *somatosensory* and motor input. If there exists a dysfunction of this loop, we would expect to see typical *basal ganglion* motor symptoms of *hypo*kinesia or *hyper*kinesia, as well as dysfunction of motor and non-motor functions of the *frontal lobe*. Although the global arousal and activity level is important, what is likewise important and more often at fault is an imbalance of brain and motor activity. It is essential that both sides of the brain and both sides of the motor system work together in a balanced coordinated way. This balanced function is a product of evolution, the survival of species, and related to the organism's ability to develop goal-directed movement. To have goal-directed controlled movement, it is a prerequisite for an organism to be bisymmetric.

Bisymmetry appears to be an adaptation for motility and therefore applies especially to the locomotor system, including bones, joints, and muscles, as well as those parts of the nervous system that controls their actions (Hiscock and Kinsbourne, 1995). Therefore, balanced activity is of singular selective advantage for the motor system. To have a balanced bisymmetric organism and motor system, one must have a balanced and bisymmetric nervous system.

Likewise, to have effective goal-directed movement, the brain must integrate the sensory–motor loop with other brain functions that help motivate and direct movement as well as provide fuel and oxygen for that movement. The more goal directed the movement, the more integration is required, and the greater the requirement for stimulus processing to support the increasing size of the nervous system which is devoted primarily to sensory–motor functions. Any deviation from bisymmetry of the motor system results in an imbalance that may result in improper integration of brain function and over time creates an ever-increasing imbalance.

It has been shown that an increase in arousal or activation of one hemisphere (usually contributed by the stimulus) is typically accompanied by an inhibition of arousal in the opposite hemisphere, mediated at least partly by reciprocal inhibitory connections through the *brainstem* (Liederman, 1998). Recent literature also states that when one hemisphere is maintained at a high level of arousal for an extended period, the result is an activation imbalance and the underutilized hemisphere becomes refractory (Liederman, 1998). At the spinal cord level, especially in midline spinal muscles, increased activation, or tone, over one side will be associated with the inhibition of contralateral muscles, which will alter feedback through the *cerebellum* and *thalamus* to the cerebrum resulting in an imbalance of activation of these structures. In the periphery, the balance of muscle tone can be imbalanced not only from left to right but also in the anterior and posterior direction. In the motor system, the midline postural muscles of the spine and the *ocular–motor* muscles are primarily controlled reflexively by the medial zone of the *cerebellum* or *spino-cerebellum*. This process is mediated by the *vestibular nuclei* and the *fastigial nuclei*. These muscles have both decussating and non-decussating controlling efferent pathways so that there exists bilateral innervation or control. In normal situations, therefore, these muscles should be equally balanced. However, activation of muscles on one side is usually coupled with inhibition of antagonist muscles on the other side. Decreased activation of the *spino-cerebellum* will normally result in decreased activation and coordination of the spinal and *ocular–motor* musculature.

Learning-disabled children quite often are seen to have poor postural tone, and ocular-deficiencies especially saccadic *hyper-* or *hypo*-metria (Leisman, 1976a). In infants, delays in the ability to roll over, head lift, and crawl may be a result of midline postural muscle tone deficiencies or *hypotonia*. In older children, the spinal muscles must be able to react as shunt stabilizers that are needed to create a stable base before spurt muscle activity of the extremities; therefore, apparent weakness of extremity muscles may be due to failure of the spinal muscles to provide stability for extremity movement. Persistent tonic neck reflexes that have been noted in hyperactive and learning-disabled children may be due to decreased activation of the *cerebellum*. Poor spinal muscle tone may lead to early fatigue of these muscles and may result in muscle pain, neck pain, and headaches in children. These children are also known to often sit in unusual postures, shift position, which may be the cause of fidgety behavior. In an effort to compensate for their instability, they may hook their legs in their chairs.

In the extremities, muscles are innervated in an anterior and posterior distribution. This distribution of anterior compartment and posterior compartment control becomes reversed in the lower extremities during embryologic development so that in the upper extremities, the anterior compartment muscles are neurologically the same as the posterior compartment muscles in the lower extremity. Voluntary motor activity is continually controlled contralaterally by the *corticospinal* and *rubrospinal tracts*. Muscle tone is regulated ipsilaterally by the *reticulospinal tract* and this regulation involves all muscles ipsilaterally, which includes midline spinal or postural muscles as well as anterior and posterior compartment muscles. However, the *reticulospinal tract* also controls the normal balance of muscles so that anterior muscles of the upper extremity and posterior muscles of lower extremity are slightly less activated or more inhibited in relationship to their antagonist muscles. This ipsilateral increase of muscle tone as well as balance of activation of anterior and posterior compartment balance is thought to be a phylogenetic development in humans, which promotes an upright bipedal posture.

Reticulospinal output is under the direct control of descending neocortical activation especially *frontal lobe* output. Therefore, asymmetric activation of the neocortex may result in an imbalance of global motor tone as well as an imbalance of anterior and posterior compartment muscle activation and inhibition. In the spinal or postural muscles and ocular–motor muscles, this may result in an imbalance, and in severe cases, in spinal muscles, this may result in a *scoliosis* or curvature of the spine. It may also result in a chronic head tilt to one side. In *ocular–motor* muscles, it may promote a *stabismus* or lazy eye that can be aggravated by a chronic head tilt. Subjective clinical reports often indicate that both *strabismus* and *scoliosis* are coincident, possibly a result of asymmetric hemispheric or neocortical activation or arousal. Both conditions are often associated with a history of learning disability (Geschwind and Galaburda, 1985; Herman et al., 1985; Rosner and Rosner, 1987). In the extremities, this may result in a fixed arm or leg on the side of decreased cortical activation or arousal. This results in antagonist muscle weakness as well as decreased arm or leg swing on the side of decreased cortical activation. This condition may also resemble or can be a form of *hemi-motor neglect* ipsilaterally.

Motor distribution and development also appear to be asymmetrically distributed in the cortex as observed by handedness. The left hemisphere appears to favor fine manual movements, and favors control of distal muscles, whereas the right appears to be superior for control of proximal limb muscles and trunk musculature (Peters, 1995). The outflow of the intermediate regions of the *cerebellum* is directed to muscles involved with the distal portion of the limbs. The lateral region of the *cerebellum* is directed at areas of the cortex and *basal ganglia* that are involved with planning and animation of movement. There is thought to be an especially close relationship between the output nuclei of the lateral *cerebellum*, the *dentate nuclei*, and the *premotor* and *supplementary* cortices, which are directly implicated in the organization, planning, and modification of motor activities (Allen et al., 1978; Halsband et al., 1993).

Basal ganglia dysfunction results in movement disorders of which we have described two types, *hyperkinetic* and *hypokinetic*. *Hyperkinetic* disorders present with increased motor activity and tone. Hyperactivity in children is often referred to as *childhood hyperkinetic disorder*. Hypokinetic dysfunction is often seen as slowness of movement or *hypotonia*. Frontal lobe motor *dyspraxia* dysfunction is seen in *paratonia*, and neurological examination of hyperactive children suggests that they experience more difficulties with coordination and more *overflow* movements than in control children (Gillberg et al., 1982, 1993; Shaywitz and Shaywitz, 1984; Denkla et al., 1985). Studies that used various tests of motor coordination such as balancing, gesturing, copying, writing, maze tracing, and pursuit tracking reported more difficulties on the part hyperactive children (Shaywitz and Shaywitz, 1984; Korkman and Pesonen, 1994; Mariani and Barkley, 1997). Joe Sergeant and his associates performed studies in which a range of task variables was manipulated to create differential demands at different processing stages. They found that hyperactive children were slower in responding and more variable in their response time (Sergeant and van der Meere, 1990). They also found an event rate effect (van der Meere et al., 1989). In a low event rate condition, the hyperactive children had a significantly longer reaction time. This was thought to demonstrate a deficit at the motor adjustment stage, which controlled the motor preparatory processes (Sanders, 1983).

It has also been noted that a conceptual model linking performance and hyperactivity requires the consideration of factors related to energy supply. It has been suggested that motor adjustment is optimized by preparatory processes. However, preparatory processes could only be effectively maintained for short period requiring energy resources of alertness or activation (Sanders, 1983; Meulenbroek and Van Galen 1988). One hypothesis for this is that a low event rate, which causes a longer foreperiod, taxes activation and therefore hyperactive children cannot maintain the activation and the necessary preparatory processes. In contrast, it is postulated that a high event rate, which has a shorter foreperiod, puts less demand on activation and therefore the preparatory processes remain optimized (Leung and Connolly, 1998). This may explain why in a high event rate condition, hyperactive children perform as well as controlled children (van der Meere et al., 1992). It has been suggested, therefore, that hyperactive children are primarily defective in energy resources, and a motor deficit is a secondary consequence (Leung and Connolly, 1998).

These results may be interpreted in the context of cerebral asymmetry. The right hemisphere is favored for low temporal frequency movement and therefore decreased activation or arousal of the right hemisphere would result in a low event rate deficit and fatigability, whereas the left hemisphere, which favors high temporal frequency movements, would perform at normal levels. Barkley (1997) interpreted performance impairments in complex motor coordination tasks such as copying designs, writing, maze tracing, and pursuit tracking as indicative of a motor-control deficit, which, he suggested, was a link to a more central deficit in behavioral inhibition. Behavioral inhibition has been recognized as a core deficit associated with hyperactivity (Pennington and Ozonoff, 1996; Barkley, 1997).

Much has been written about the role of the *basal ganglia* in motor and cognitive functions (Alexander et al., 1986, 1990; DeLong, 1990; Albin et al., 1995; Graybiel, 1995; Brown et al., 1997; Feger, 1997). Five *frontal-subcortical* pathways connect areas of the *frontal lobe* (*supplementary motor area, frontal eye fields*, and the *dorsolateral prefrontal, orbito-frontal*, and *anterior cingulate cortices*) with the *striatum, globus pallidus*, and *thalamus* in functional systems, that mediate volitional motor activity, saccadic eye movements, executive functions, social behavior, and motivation (Alexander et al., 1986; Alexander and Crutcher, 1990; Mega and Cummings, 1994; Cummings, 1995). It has been suggested that normal *basal ganglia* function comes from a proper balance between direct and indirect *striatal* output pathways, and therefore abnormal imbalance or differential involvement of these pathways can result in *hyperkinesia* or *hypokinesia* of the type seen in *basal ganglia* disorders (Alexander and Crutcher, 1990; DeLong, 1990; Albin et al., 1995).

Litvan and colleagues (1998) found that patients with *hyperkinetic* disorders exhibited predominantly *hyper*active behaviors (e.g., Tourette's agitation, *ADHD, OCD*, irritation, euphoria, or anxiety), and those patients with *hypokinetic* motor activity manifested *hypo*active behaviors like apathy. Injury of the *caudate* is thought to be responsible for neuropsychiatric symptoms that include depression, mania, apathy, disinhibition, *OCD*, defects in planning and sequencing, attention, and free recall (Salloway and Cummings, 1994). Lesions of the *globus pallidus* can cause *OCD, TS*, apathy, irritability, *mania*, and *amnesia*. Lesions in the medial *caudate* (and projecting to the *orbito-frontal cortex*) can produce apathy and depression. Inferior lesions can produce disinhibited behavior, with dorsolateral lesions of the *caudate* (and projecting to the dorsolateral *prefrontal cortex*) being associated with decreased ability in executing executive functions. Bilateral *caudate* dysfunction is thought to be involved in most patients with *mania* (Alexander et al., 1994). Therefore, with *basal ganglia* disorders, we can see an increase in motor activity, such as tremors, motor and vocal tics, uncontrolled movement, or motor overflow, collectively known as *hyperkinetic* disorders. These appear to correlate with hyperactive behaviors. In contrast, *hyperkinetic* disorders characterized by slowness or decreased motor output as in *Parkinson's* disease appear to correlate with *hypo*active behaviors.

Regarding motor activity and hemispheric asymmetry, it has been shown that the right hemisphere has a somatotopic representation for sensory input for both right and left sides of the body, as well as an attention control function for both sides of space. The left hemisphere, however, is thought to have a similar pattern for the control of motor activity and intention, Whereas the right hemisphere controls motor activity and intention for only the contralateral side of the body, the left hemisphere appears to have control over the right (contralateral) and left (ipsilateral) side of the body for motor activity and intention. This was seen as early as 1905 when Liepmann reviewed results of 89 patients tested for signs of *apraxia* with unilateral hemiplegia. He found that right brain-injured patients had normal right-hand function while their left hand was paralyzed and approximately half of the left brain-damaged patients showed signs of severe *apraxia* or *dyspraxia* when trying to function with their left hands while their right hands were paralyzed. The conclusion was that an intact left hemisphere was not only necessary for normal function of right-sided movements, but it was also important for the normal function of intentional actions on the left side. It appeared that the left hemisphere played an executive role in voluntary motor function for the entire body both left and right side (Liepmann, 1900; Harrington and Haaland, 1991).

Interestingly *dopamine*, which is thought to regulate movement and is intimately involved with mood and motivation, is asymmetrically distributed favoring the left hemisphere. Decreased *dopamine* activity has been implicated in *ADHD* as well as a number of *basal ganglia* disorders including both *hyperkinetic* and *hypokinetic*. *Dopaminergic* brain innervation arises from neurons located in the *mesencephalon* forming two subsystems. In the cortex, there is extensive *dopamine* innervation in the *prefrontal, premotor*, and *motor* areas and a relatively sparse innervation in posterior regions. It appears that in general there is a preferential innervation in *motor* vs. *sensory* cortices and *association* areas vs. *primary sensory areas* (Fallon and Loughlin, 1987). Therefore, it is assumed that *dopamine* is involved in higher integrative cerebral functions and in the regulation of cortical output activity especially in motor control (Wittling, 1995) with conclusions based on anatomic studies and confirmed by findings from lesion and pharmacological studies (Clark et al., 1987). These studies show that the integrity of *dopaminergic* pathways is crucial for the selection and operation of appropriate motor responses and for the coordination of motor with sensory input.

Volkow and colleagues (2000) studied brain scans of 30 healthy males and females spanning many adult ages. The brain scans measure *dopamine*, which they think regulates human behavior including movement, working memory, and the experience of pleasure and reward. They found a direct relationship between depleted stores of *dopamine* and a decline on tests that measure motor and cognitive abilities. In another study, Glick et al. (1982) examined neurotransmitter concentrations in the left and right sides of various brain regions collected by Rossor and colleagues (1980) in postmortem studies of 14 normal human subjects. Results showed that *dopamine* content is significantly higher in the left *globus pallidus* than in the right *globus pallidus*. Wittling (1995) states, "Neurotransmitters such as *dopamine*, being more intimately involved in the control of motor behavior and higher integrative functions, obviously favor the left side of the human brain, whose leading role in the control of these functions is undisputed."

Recent studies have described the first direct evidence that *dopamine* triggers the major symptoms of *schizophrenia*, including psychosis. Nora Volkow (cf. Vastag, 2001) was recently quoted as saying "*schizophrenia* is caused by a very significant disruption of the *dopamine* system." The area of the brain thought to be targeted by the disease is the *striatum*, which is rich in *dopamine*. In *schizophrenic* patients, findings in visual half-field tasks seem to provide evidence of a left hemisphere dysfunction (Bruder, 1995). In studies of identical twins where one of the twins develops *schizophrenia*, it has been noted that in comparing the movements of the two as children that the twin who develops *schizophrenia* demonstrates motor deficiencies or jerky movements as compared to the unaffected twin (Torrey et al., 1994). Abnormalities in *dopaminergic* neurotransmission have been implicated in a number of neurological and psychiatric disorders including *schizophrenia*, *ADHD*, TS, *autism*, alcoholism, drug addiction, and *Parkinson's* disease (Young and Penny, 1984; Bunzow et al., 1988; Cummings, 1990; Cummings et al., 1991). In each of the different conditions, it has been noted that a large percentage of individuals have motor difficulty.

Ayers (1972) noted that learning disabled, *autistic*, and children with *ADHD* had motor problems as a prominent feature. She thought that this was primarily a result of poor sensory integration especially from the inner ear *vestibular system*; this is probably because the majority of motor symptoms could best be described as *cerebellar* symptoms. Ayers referred to this problem as *developmental dyspraxia* and she is quoted as saying, "*Developmental dyspraxia* is one of the most common manifestations of sensory integrative dysfunction in children with learning disorders or minimal brain dysfunction." Ayers discussed five aspects of movements and movement disorders that were disrupted in learning-disabled children: (1) smooth control of movement such as picking up a pin, (2) postural reactions such as rolling over or balancing on one foot, (3) patterns of movement that are programmed into the central nervous system (e.g., crawling or walking), (4) specific motor skills (e.g., writing the alphabet), and (5) motor planning. Ayers felt that *developmental dyspraxia* is a brain dysfunction that decreases the organization of tactile, *vestibular*, and *proprioceptive* sensations and interferes with motor planning. Motor planning is primarily performed by the *cerebellum* in connection with the *frontal lobe*. The *cerebellum* is also the brain region that processes *tactile*, *vestibular*, and *proprioceptive* sensations. Children with *autism*, *ADHD*, and patients with *schizophrenia* have all been shown to have structural changes and atrophy of the *cerebellum* (Harris et al., 1999; Fatemi et al., 2002a, 2002b; Keller et al., 2003). Some of these patients have been shown to have smaller areas of the *cerebellum, basal ganglia*, and *frontal lobes* than those not affected.

A subset of *autism* is *pervasive developmental disorder (PDD)*. *Autism* is characterized by an inability of relating to other people. The *autistic* child is often described as being in its own world. Speech or communications is often limited and, if they learn to speak, it has been noted that their speech lacks *prosody*, or intonation. In addition, *autistic* children may have emotional problems; sometimes they may show too much emotion and at other times they appear to show no emotion at all. It is estimated that *autism* is a disorder of the brain and behavior that affects 5 out of 10,000 children (Wing and Potter, 2002). *Autistic* children appear healthy but may stare into space for hours, throw tantrums, show no interest in people, and perform perseverative or repetitive activities, like head banging; they may also have their attention drawn to small details and they may have difficulty inhibiting or stopping their actions.

Recently it has been reported that *autistic* children have subtle abnormalities in body movements or motor activity that can allow diagnosis as early as 3 months of age. Teitelbaum and colleagues (1998) described their findings obtained by examining videotapes of babies that were later diagnosed with *autism*. The infants apparently showed specific motor deficiencies, which included difficulty rolling over, sitting up, crawling, and walking. It is thought that examining motor symptoms may also help diagnose other developmental disorders like *schizophrenia* or ADD. Anne Donnellon, of the University of Wisconsin at Madison, was quoted by the *NY Times* (*New York Times*, January 19, 1999) as saying, "Teitebaum's work is important because it reflects a reality about *autism* that has been missed. We tend to think that it is a problem with the mind. Now that we are really beginning to see how the brain works we know that the mind is embodied. Body is part of the mind and there's no way to separate." Teitebaum, in the same article, commented that he got the idea of looking at *autism* as a movement disorder partly because of his work with brain-damaged organisms.

As they recover, he said, they go through predictable stages that reflect fundamental aspects of brain organization. Because human babies also pass through predictable stages of development, he theorized that deficits in the brain might show up in early movements. He notes that none of the *autistic* babies learned to roll over as normal children did. In addition, unlike normal infants who usually learn to sit up at 6 months, *autistic* infants fall over easily falling to one side "like a log" and failing to break their fall with their hands. Abnormalities are also seen in crawling and every *autistic* child shows some degree of asymmetry in walking.

Glen Doman (Thomas, 1969; Doman, 1974) has noted in his subjective clinical observations, that brain-injured children, often are unable to crawl or walk. He thought similarly that children go through predictable stages of recovery from injury. He thought that if a child missed a stage of recovery that child would have difficulty progressing beyond that point. He felt there were different stages of crawling that a child needed to go through before they could walk. If children did not progress from crawling to walking, their brains and nervous systems would not develop normally. Left unanswered by this approach is whether the problem arises first in the *cerebellum* or whether it starts as a motor problem. In addition, we do not know whether the motor problem causes the difficulty in development and with the associated symptoms or whether it is a dysfunction of the *cerebellum* or subcortical motor regions that causes the problem.

In infants, it is known that the cortex is not completely developed, especially the *frontal* and *prefrontal lobes*, which are thought to assist in the control of motor planning. In infancy, the control of movement is centered at the subcortical level, especially in the *cerebellum* (Trevarthen and Aitken, 2001). The *cerebellum* receives information from the motor system, the *spino-cerebellum*, or *vermis*. The lateral *cerebellum* is thought to be involved with higher cognitive function and is especially connected to the *prefrontal cortex*. The lateral *cerebellum*, or *cerebro-cerebellum*, receives most of its input from the cerebrum. Therefore, examining the *cerebellum* may give us an idea as to which problem is the primary. If the *vermis* shows an abnormality, lack of development, or atrophy, it is more likely that the motor system is at fault and the developmental lag starts with failure of development of the *vermis* or its presynaptic connections. If the *vermis* alone is smaller, then we can think that the problem arose principally after the development of the *lateral cerebellum*, but initiated by a dysfunction of the motor system. If the lateral *cerebellum* is affected, then it can be the result of either an intrinsic problem or an inadequacy of stimulation from the *frontal lobe* to the lateral *cerebellum*.

In regard to *ADHD*, imaging studies over the past decade have indicated which brain regions might malfunction in patients with the condition and that may account for the symptoms. This work suggests that involvement includes the *prefrontal cortex*, part of the *cerebellum*, and areas of the *basal ganglia*. In a study in 1996, Castellanos and his colleagues found that the right *prefrontal cortex*, the *caudate nucleus*, and *globus pallidus* are significantly smaller in children with *ADHD*. Castellanos' group also found that the *vermis* region of the *cerebellum* is smaller in *ADHD* children. Patients with *schizophrenia* and *autism* have also shown structural changes of the *cerebellum* (Cahn et al., 2002; Deshmukh et al., 2002; Ojeda et al., 2002). It has been noted that several arguments support the *cerebellum* as a site of pathological or abnormal function in *schizophrenia*. Traditional neuroscience however, still considers the *cerebellum* exclusively associated with control of motor and *ocular–motor* functions. This is thought to arise from research on human *cerebellar* function on patients with *cerebellar* lesions where motor and *ocular–motor* signs are the most pronounced clinical finding (Martin and Albers, 1995). Previously, motor and *ocular–motor* abnormalities known to be present in *schizophrenia* have been ascribed to *basal ganglia, limbic,* and neocortical dysfunction (Martin and Albers, 1995). Blueler (1911) noted gait abnormalities in *schizophrenic* patients. He especially noted the irregular space and timing of their steps. Another unpublished study observed 16 *schizophrenic* patients at the Psychiatric University Hospital Zurich who appeared to show movement abnormalities. It was postulated that the clinical pattern of these patients may correlate to Blueler's observations and that these clinical findings may be blamed on a dysfunction of the anterior *vermis* of the *cerebellum* (Victor et al., 1959; Dichgens and Diemer, 1984).

Various visual–motor abnormalities especially of *saccadic* and *smooth pursuit* eye movements have been noted in *schizophrenia*, and an increase of *dysmetric*, mostly *hypometric, saccades* are a consistent finding in *schizophrenia* (Levin et al., 1981; Schmid-Burgk et al., 1982). *Saccadic dysmetria* is thought to be characteristic of lesions in the dorsal *cerebellar vermis*; *hypermetric saccades* are thought to result when lesions involve larger areas of the dorsal *vermis* and *fastigial nucleus* (Zee, 1984). A postmortem study (Weinberger et al., 1980) showed that an abnormally small anterior *cerebellar vermis* was significantly more frequent in brains of *schizophrenics* as compared to normal controls. Furthermore, the *cerebellar* cortices showed *Purkinje* cell losses of different degrees and thinning of the *granular* and *molecular* layers of the *cerebellar vermis*. It has also been reported in another study, which reported a loss of *Purkinje* cells and *granule* cells in schizophrenia, mainly in patients with *catatonic schizophrenia*, as well as patients with epilepsy and psychosis (Stevens, 1982). Reyes and Gordon (1981) used light microscopy to analyze the brains of eight *schizophrenic* patients. They also found a decreased number of *Purkinje* cells per unit line length of *Purkinje* cell layer in chronic *schizophrenia* as well as an over development (surface density) of the *Purkinje* cell layer.

Computerized tomography (CT) imaging studies also showed structural changes in the *cerebellum* in *schizophrenic* patients. CT scans in 9 of 60 *schizophrenic* patients showed *CA* (Weinberger et al., 1979). Another CT study suggested pathology, mostly atrophy of the *cerebellar vermis* in a large percentage of schizophrenic patients (Heath et al., 1979). Lippmann and associates (1982) showed abnormal *cerebellar* dimensions on CT scans in 30 percent (17 percent abnormal *vermis*) of *schizophrenic* patients. Another study showed enlarged fourth *ventricles* and decreased *vermis* width in chronic *schizophrenic* patients (Dewan et al., 1983). Sandyk and colleagues (1991), in a CT study, found *vermal CA* in 10 of 23 chronic *schizophrenic* patients. Data published by Rossi and associates (1993) shows a gender difference in the *vermis*. MRI imaging shows that male *schizophrenics* have significantly smaller anterior *vermal* areas than females as compared to normal controls.

It is interesting to note that male children are three times as likely to develop *ADHD* as girls (Barkley, 1997). PET scans of the *frontal lobe* in a group of medicated chronic *schizophrenics* (Volkow et al., 1992) show decreased *cerebellar* metabolism that was independent of metabolic activity in the *basal ganglia*. Psychiatric disorders have been noted in various cases of *cerebellar* pathology (Heath et al., 1979; Kutty and Prendes 1981; Hamilton et al., 1983). Neuropsychiatric disorders especially *dementia* has been seen with *cerebellar* degeneration (Schmahmann, 1991). *Autistic* symptoms have been shown in *Joubert syndrome*, an autosomal recessive disorder that has partial or complete agenesis of the *cerebellar vermis* (Joubert et al., 1969; Holroyd et al., 1991). Neuropathological studies have shown loss of *Purkinje* cells and less frequently *granule* cells in the neo-cerebellar cortex. Cell loss was also found in the deep *cerebellar nuclei* in *autistic* individuals. The anterior *cerebellum* and *vermis* were also involved to a lesser extent (Bauman, 1991; Courchesne, 1991) in affective psychosis but are still associated with CA. One study of patients with *bipolar disorders* showed *vermis* atrophy (Lippmann et al., 1982) and Nasrallah and colleagues (1981, 1982) found CA in a large percentage of patients with *mania*.

Alcohol abuse shows typical CA, *cerebellar* lesions due to alcohol toxicity are seen with anterior *vermis* atrophy (Victor et al., 1959). It is known that the *cerebellum* is anatomically connected with areas of the brain that are not primarily involved in motor functions (Leiner et al., 1991). Although for the past hundred years, the *cerebellum* has been regarded as a motor organ (Brooks, 1984), the *cerebellum* receives input from every sensory system (Brodal, 1978, 1981) and is known to be activated by tactile stimulation without the requirement for movement (Fox et al., 1985). The *cerebellum* is not considered a sensory organ as deciphering whether nervous structures have a sensory or motor function is difficult since motor behavior is guided by ongoing sensory acquisition of object information and the accuracy, coordination, and smoothness of motor behavior depends on continuous updated sensory data (Gao et al., 1996).

In regard to non-motor control, the same dependence on sensory input probably exists. Neuroanatomic and electrophysiological studies show that the *cerebellum* is part of a neural system that includes the *thalamus, basal ganglia*, and *frontal lobes* (Thatch, 1980). In addition, PET studies have shown a strong correlation between metabolic rates of the *cerebellum* and *frontal cortex* (Jancke et al., 1998). Ascending projections from the *cerebellum* to the *hippocampus–amygdaloid* complex arise primarily from the *fastigial nucleus*, are bilaterally arranged, and are mono- and polysynaptic (Heath and Harper, 1974; Newman and Rezea 1979). It is thought that it is through the polysynaptic connections that the projections reach the *limbic system* via the ventral *tegmental area* (AlO) of the *mesencephalon* (Snider et al., 1975, 1976). The *paleo-cerebellum-limbic* projections, which are the pathways from the anterior vermis and fastigial nucleus to several areas of the *limbic system*, have been shown to modulate sensory input to the *hippocampus* (Newman and Rezea, 1979) and also shorten or stop seizure discharge produced by electrical stimulation of the *amygdala* and *hippocampus* (Maiti and Snider, 1975). It has been shown that this is a bilateral descending *hippocampal–cerebellar* projection system ending mostly in the *vermian* portions of the *cerebellum* (Newman and Rezea, 1979). *Purkinje* cells project from the *cerebellar* cortex by *GABA* and inhibit neurons of the deep *cerebellar* nuclei of the *cerebellum* (Ito, 1976). According to studies on intra-nuclear collateral neurons of the *deep nuclei*, these neurons most likely use *glutamate* as a transmitter and mediate excitation of *mesencephalic dopaminergic* neurons (Snider, 1975; Snider et al., 1976; Audinot et al., 1992), thereby increasing *mesolimbic* and *mesocortical* activity. In addition, *noradrenergic* and *serotonergic* projections as well as *cholinergic* pathways from the *fastigial nucleus* to the *septal* region (Paul et al., 1973) may mediate non-motor functions of the *cerebellum* (Martin and Albers, 1995).

It has been shown that the *dopamine* D_3 receptor is expressed not only in *limbic* structures but also has been noted in *Purkinje* cells of the *cerebellar lobules* IX and X (*archicerebellum*) in rat brain. Therefore, it is thought that the same neuronal substrates may be involved in motor control as well as in cognition and emotion. Structural changes in other areas as the medial *temporal lobe* in *schizophrenic* patients are called *neurodevelopmental* disorders (Roberts, 1991). It has been postulated that the same may be the case in *vermal* pathology in *schizophrenia*. The embryologic development of the *neocerebellum* and the *paleo-cerebellum*, that is, essentially the anterior *vermis*, differs in regard to neuronal tissue timing and migration (Larsell and Jansen, 1972; Altman and Bayer, 1985). This developmental difference may offer a partial explanation of the preferential involvement of the anterior *vermis* and it may also reflect the different input that arrives in that area which promotes development. Findings suggest that a decrease in anterior *vermal* areas of *schizophrenics* is specific to males (Rossi et al., 1993). Increased activity of the *fastigial nucleus* because of decreased inhibition by *vermian GABAminergic Purkinje* cells is in agreement with the *dopamine hypothesis* of *schizophrenia* (Snider et al., 1976). The excitatory, possibly *glutaminergic* projections of the *fastigial nucleus* to *dopaminergic neurons* in the *mesencephalon* are disinhibited and, therefore, lead to elevated *dopamine* concentrations in the *forebrain* (Snider and Snider, 1979). In addition, the *fastigial nucleus* has been shown to modulate and project through *serotonergic, noradrenergic,* and *cholinergic* pathways, which are thought to be involved in the neurochemistry of *schizophrenia* (Tandon and Greden, 1989; Lieberman and Koreen 1993) as well as depression and other psychiatric symptoms. It has further been reported that electrical stimulation of the *cerebellum* with the use of *cerebellar* surface pacemaker stimulation for intractable behavioral disorders results in at least minimal improvement in the majority of their chronic *schizophrenics* and also in the smaller groups of patients with depression (Heath et al., 1980).

The hypothesis that dyslexia has a C-V foundation was first reported by Frank and Levinson (1973). In their study of 115 dyslexic children, 97 percent showed what the investigators thought to be clear-cut neurological signs of C-V dysfunction. The reported C-V signs included positive *Rhomberg*, difficulty in tandem walking, articulatory speech disorders, *dysdiadokokensis, hypotonia,* and various dysmetric or *dyspraxic* symptoms, pass pointing during finger to nose, heel to toe, writing, drawing as well as ocular fixation, and tests for scanning (Dow and Moruzzi, 1958). In addition, it was reported that 90 percent of the dyslexic sample tested with *electronystagmography* evidenced vestibular abnormalities (Frank and Levinson, 1973). It was further reported that C-V related mechanisms thought to be responsible for each of the various symptoms were examined neuropsychologically (Levinson, 1980, 1984). These findings were seen as proof that dyslexia and learning disabilities reflected a single disorder and were thought to share a common group of C-V determining mechanisms and symptom combinations. This was further thought to be consistent with the current definition of learning disabilities, which include dyslexia as a subcategory, according to Public Law 94-142, *Education for all Handicapped Children Act* (Levinson, 1984). Levinson (1980), and de Quiros and Schaeger (1978) all described C-V dysfunctioning mechanisms and neurological and *electronystagmographic* diagnostic parameters in learning-disabled persons and correlations between reading ability and postural control in elementary school children tested in the United States and in Israel and France by Kohen-Raz and Heriatborde (1979). It has further been reported that *space dyslexia* was experienced by astronauts who began mirror reading at zero gravity during a combined French–Russian space mission (Levinson, 1984).

Ayers (1972) has reported a beneficial result of sensory integration or vestibular stimulation training of learning-disabled persons. Also, positive responses in reading, writing, and concentration were reported for learning-disabled individuals subjected to ocular–motor fixation, scanning, and perceptual motor exercises thought to cause C-V stimulation (Halliwell and Solan, 1972; Pierce, 1977). In addition, a reduction in academic failure has been noted for a large sample of culturally disadvantaged first graders who were given physical education emphasizing exercises requiring C-V control (Kohen-Raz, 1986). It has also been noted that vestibular-determined postural training in deaf toddlers who exhibited gross motor and cognitive retardation became normal when their deficient or reduced *labyrinthine* function was stimulated (Kaga et al., 1981). Furthermore, Levinson (1980) performed neurological and optokinetic measures of *cerebellar* and *vestibular* function on learning-disabled subjects and controls. He reported that a substantial majority of learning disabled (82–90 percent) showed *ADD*-like symptoms. Levinson thought that since learning-disabled sub-samples with or without *ADD* showed similar coexisting symptoms and C-V signs, it was probable that learning disabilities and *ADD* were a reflection of the same underlying C-V determinants. In studies using PET scans of 25 hyperactive adults compared to a control group Zametkin and Ernst (1999) found that there was less overall activity in the brains of hyperactive adults, especially in the *premotor cortex* and *superior prefrontal cortex*. The *frontal lobes* regulate goal-directed behaviors, a hierarchy of reflexive movements, cross-temporal contingencies, approach and avoidance behaviors, response inhibition, and *perseverations* (Chatterjee, 1998a, 1998b). Furthermore, the *prefrontal cortices* regulate goal-accomplishing movements (Luria, 1966). Along with the *basal ganglia* and *cerebellum*, these structures regulate primary motor output (Chatterjee, 1998a, 1998b).

The goal-directed movements regulated by the *prefrontal* cortex are complex and have multiple sequences. Luria suggested that following *frontal lobe* damage, subordinate reflexes are no longer integrated seamlessly into goal-directed movements (Luria, 1966). Frontal damage or dysfunction produces two characteristic defects. First patients with frontal damage may have difficulty initiating movements without guidance of explicit stimuli. Clinically these patients are passive and apathetic, reminiscent of children with *autism* or depression. Second, these patients may have difficulties inhibiting responses to stimuli. Clinically they are distractible and react randomly to environmental stimuli. This is more typical with children with *ADHD*. Many patients with frontal damage show features of both kinds of deficits (Heilman and Watson, 1991). They do not know when to move or when not to move (Chatterjee, 1998a, 1998b). The difference is that the primary symptom probably lies in an asymmetric dysfunction of the *prefrontal* region. *Autistic* children, for example, are thought to have left brain deficits, which normally produce approach behavior, therefore, with decreased activation avoidance behaviors are exhibited. This would be seen as a lack of ability to appropriately perform intentioned movement, disability in initiating movement, or of approach movement. Children with right brain deficits show decrease in avoidance behavior, which is typical of right *frontal lobe* involvement. This decrease in effective avoidance movement translates to increased approach movements, typical of a hyperactive child.

The right brain also governs attention, and with deficits in response inhibition, the child's ability to focus attention will likewise be deficient with an oftentimes seen consequence of *perseverative* or repetitive movements, all typical of right brain deficits. *Right-brain* involved children will demonstrate increased levels of *approach* movements and repetition that they find difficult to stop, whereas with the intention-governing *left-brain*, those children with primary activation impairment in this hemisphere were observed to show lack of initiative, motivation, or effective *intentional* movements. Depression, shyness, and lack of initiative or motivation are typical of left-brain decrease in activation or arousal. Increase activation or arousal on one side or the other produces the exact opposite reaction in each hemisphere.

What seems to be demonstrated is an imbalance in the complementarity of the right and left hemispheres as well as between the *frontal* and *parietal lobes* of the same hemisphere and attendant subcortical structures. For instance, increased firing of the *basal ganglia* and *thalamus* activating one hemisphere will produce *hyperkinetic* movements and increased activation of that hemisphere. Decreased *basal ganglia* and *thalamic* firing may produce *hypokinetic* movement and decreased firing of frontal areas on that side. Depending on whether the decrease or increase is right or left frontal it will be determined whether the child exhibits increased avoidance or approach movements. Again, the *frontal lobe* controls goal-directed movement and behavior.

An individual's approach to environmental stimuli is an important aspect of goal-directed behaviors. Dense interconnections between *prefrontal cortices* and *somatosensory, visual,* and *auditory* cortices provide the neural substrate for the integration of the sensations (Barbas and Pandya, 1989). Based on monkey studies, it has been postulated that inhibitory interactions between *parietal* and *frontal* cortices mediate the approach to and avoidance of stimuli (Denny-Brown, 1958). Therefore, it is thought that individuals can choose to approach or avoid environmental stimulation by modulating *parietal* and *frontal lobe* activity (Chatterjee, 1998a, 1998b). A recent study of patients with *frontal lobe* disease (Beversdorf and Heilman, 1998) compared motor dysfunction to cognitive dysfunction using standard tests of *frontal lobe* cognitive function. This study compared the motor and non-motor functions of the *frontal lobe* to see if there was a correlation between the two. In the past the motor and non-motor function of the *frontal lobe* were considered separate and therefore did not relate directly to one another. This study suggests that there is a direct and equal relationship between motor and non-motor deficits of the *frontal lobe*. Therefore, the degree of cognitive or behavioral dysfunction correlates directly to an individual's motor dysfunction. Therefore, it is postulated that in patients with diffuse *prefrontal* dysfunction, the inability to inhibit simple movements to simple external stimuli may correlate with the inability to inhibit complex social behaviors in response to complex internal stimuli. If this correlation is accurate, then a motor dysfunction could result in cognitive and emotional dysfunction and vice versa. This also has powerful therapeutic implications because it should follow that improved sensory–motor function should result in improved cognitive and behavioral function of the *frontal lobe*.

COGNITIVE DISABILITIES

Although sensory and motor symptoms may be consistent with findings in children with all types of emotional and learning problems, some children will express more cognitive or academic problems. These children demonstrate more typically, what has come to be known as a learning disability and most of them have difficulty in left hemisphere functions of verbal skill, reading, math, and writing. As we start to understand the asymmetric distribution of specific functions in the brain, we realize that traditional education paradigms and instruction mostly emphasizes left-brain activities. This is not to say that the right-brain is not needed; however, the three Rs, the foundation of education, are left-brain activities. Therefore, standard IQ tests are also biased toward testing left-brain function, whereas EQ or emotional intelligence involves mostly right-brain functions. It has been said that to be successful in school, a child needs good left-brain function, but to be successful in life one needs the right. The left-brain focuses on small details, facts, and local stimuli, whereas the right-brain examines the big picture—global stimuli. In reality, the goal is a bilaterally, active, integrative function between the two hemispheres. We want to aspire to a brain that functions like Leonardo DaVinci's, a great scientist, inventor (left-brain), and artist (right-brain). Balance of activity is key and in these children, we see symptoms similar to a disconnection syndrome in split-brain patients. Although learning disabilities may arise secondary to right-brain, attentional, behavioral, or *hyperkinetic* activity, the more severe learning disabilities that affect academic performance are typically lateralized to the left. We have reviewed their function and location earlier and will now examine dysfunction in language, auditory processing, *dyslexia, dyscalculia, dyspraxia* while writing (*dysgraphia*), *autism*, or *autistic* tendencies which tend to constitute the majority of the population of learning-disabled children.

There exists a specific group of children who exhibit noticeable deficiencies performing tasks that require *semantic-linguistic* or *visuo-spatial* skills or both. Characteristics of children with developmental learning disabilities have been well documented over the last few decades and have often been compared to adult patients with known brain lesions (Boliek and Obrzut, 1995). Observation of these children has been summarized by Hynd and Willis (1988) to include a congenital form of learning disability, which appears to affect more males than females. It has also been noted that in general these learning disabilities do not respond to classroom education remediation. It has been further hypothesized that these disabilities may be related to a neuro development process primarily affecting the left hemisphere (Baliek and Obrzut, 1995). Laterality hypotheses have been promoted mostly based on studies of infants, children, and adults with known lesion sites. Developmental dysfunction of the same brain areas as seen in acquired disorders, may be the basis of learning disabilities (Dawson, 1988; Obrzut, 1991). It is generally accepted that individuals with learning disabilities demonstrate abnormal cerebral organization, which includes abnormal or weak patterns of hemispheric specialization (Corballis, 1983, 1991; Bryden, 1988; Obrzut, 1988). It has been shown that learning-disabled children exhibit decreased performance on a number of tests which were thought to measure perceptual laterality, and it has been shown that weak laterality has been found across several modalities including auditory, visual, and tactile as well as two combined modalities such as verbal-manual tasks. It has been proposed that these children exhibit abnormal cerebral organization (Corballis, 1983; Obrzut, 1988). The basic assumption is that abnormal structural development of the central nervous system, developed prenatally or during early postnatal development, causes abnormal cerebral organization and associated functional specialization that is necessary for lateralized processing of language and nonlanguage information. It is thought that cortical and subcortical dysfunction develops from abnormal patterns of activation or arousal (Obrzut, 1991), inter-and-intrahemispheric transmission deficits, or inadequate resource allocation (Keshner and Peterson, 1988). Decreased or abnormal patterns of activation or arousal will result in a delay or dysfunction of cerebral structure and function that will result in a number of functional problems that contribute to learning disability.

We have discussed previously in detail what produces the activation and arousal, how it is balanced, and how it may become imbalanced. It is thought that if there exists a specific injury or dysfunction to the left hemisphere before ages 6–8 and appropriate intervention and or recovery processes have occurred, there may be little or no evidence of speech, language, or reading deficit. However, if damage or dysfunction occurs or persists after this time, the result may be a speech-language type of neuropsychological syndrome (Satz, 1990). Results of studies appear to show that neurodevelopmental abnormalities exist in the brains of developmentally *dyslexic* adults (Leisman and Ashkenazi, 1980). Postmortem studies show that certain structures are symmetric (e.g., *temporal plana*) but abnormal cell migration and organization are seen in both left and right areas of the anterior cortex and left *temporal* cortex (Galaburda et al., 1985) as well as in regions of the right *temporal lobe* (Duane, 1991). Research reviews have linked postmortem findings with specifically impaired behaviors that have been noted in learning-disabled individuals. Examples are language related impairments of *prosody* and emotionality involving the right *frontal* and central *peri-sylvian* cortex. Recent studies using MRI found symmetry and reversed asymmetry of the left–right *peri-sylvian* areas in a small sample of language disabled children as well as in several of their siblings and biological parents (Plante et al., 1989, 1991; Plante, 1991). Another MRI study examined three groups of children (*dyslexic, ADHD,* normal control children), and found the *dyslexic* group to have symmetric or reversed asymmetry of the *plana* length. Of the *dyslexic* group, 90 percent showed this pattern, whereas only 30 percent of the *ADHD* group and normals showed this same pattern. The authors found that reversed or symmetric frontal regions were correlated with behavioral deficits in word attack skills, but not passage comprehension (Hynd et al., 1990). It has also been reported in another study that 70 percent of the sample of adolescents with *dyslexia* compared to 30 percent of normals showed asymmetric left/right *plana temporale* (Leisman and Ashkenazi, 1980; Larson et al., 1989; Leisman, 2002). Although traditional thinking is that the left hemisphere does one thing and the right hemisphere does another, it has been learned from the brains of individuals with developmental *dyslexia* (Heir et al., 1978; Haslam et al., 1981; Galaburda et al., 1985; Humpherys et al., 1990; Hynd et al., 1990; Leonard et al., 1993), that it may also be the same for other developmental disorders (Jennigan et al., 1952; Luchins and Meltzer, 1983; Tsai et al., 1983; Brown et al., 1985; Crow et al., 1989; Gur et al., 1991; Falkai and Boggerts, 1992; Rossi et al., 1992).

In considering degrees of asymmetry (Galaburda, 1994), one characteristic that differs in brains of *dyslexics* is the degree to which the language area in the brain is asymmetric. It seems that the magnitude of asymmetry may be as important as other issues like specialization of the brain for language. With change in the size of the asymmetry, there is significant change in the circuitry so that at some point the functional abilities of the system change. Therefore, in developmental *dyslexia*, lack or reduction of asymmetry may be a crucial factor in explaining the differences in linguistic capacity compared with normals.

In human brains, it must be remembered that the larger areas are normally in the left hemisphere for most linguistic abilities and in the right hemisphere for visual-spatial skills. The *planum temporale*, which contains several auditory association cortices on the left side, is thought to be an important part of the language network of the left hemisphere. A lesion affecting a significant percentage of the *planum temporale*, usually on the left side in right-handed individuals, leads to Wernicke's *aphasia* (Galaburda, 1994). *Heschl's gyrus*, which is the anterior portion of the *planum temporale* and which contains the primary *auditory cortex*, is reported to be asymmetric as well. The left *auditory cortex* is usually more oblique less transverse than the right, also the right *Heschl's gyrus* is doubled more often than the left (Rademacher et al., 1993). Absence of asymmetry therefore is thought to reveal a brain that appears to have two left brains and no right (Witelson, 1977; Leisman and Ashkenazi, 1980, Leisman and Zenhausern, 1982). It is further thought that this finding suggests that asymmetry is produced by developmental curtailment rather than enhancement. Therefore, lack of appropriate development would be more likely to be because of decreased stimulus on one side and not an increase in stimulation on the other.

In the *frontal lobe*, the *frontal operculum* contains mostly areas 44 and 45 of Brodmann; both these cortices belong to the category of motor association cortex. Area 44 is an inferior *premotor cortex* and area 45, an inferior *prefrontal cortex*. The *prefrontal cortex* is considered more multimodal (cognitive) than the *premotor cortex*. However, lesions that produce Broca's *aphasia* affect area 44 in isolation more times than area 45 (Galaburda, 1994). Studies of the inferior *parietal lobe* in normal humans show a larger left area PG in 80 percent of the brains examined. Area PG which is the same as Brodmann's area 39 is found mainly on the *angular gyrus* and is a prototypical high-order association cortex. Area PG (Brodmann's area 39) is extremely multimodal and is found in between cortices dealing with somesthetic, auditory, and visual functions.

Lesions of the *angular gyrus* are known to result in *anomic aphasia*, acquired reading and writing disorders, and *Gerstmann's syndrome*, which most likely result from dysfunction in the majority of area PG. The same brains that were noted to have a left-sided predominance of area PG also show a larger *planum temporale* and left area 44 (Eidelberg and Galaburda, 1984). This finding is thought to suggest that asymmetries in one area may be correlated with asymmetries in another area as long as the areas are functionally related (Galaburda, 1994). Functional imaging, including PET and regional cerebral blood flow (rCBF), has been used to examine brain organization in learning-disabled individuals. One study using PET and a radioactive glucose tracer found that during reading, *dyslexic* subjects show active bilateral participation of the *insular cortex* (Gross-Glenn et al., 1991). Studies using rCBF have shown that recruitment of both left and right, and central and posterior cortical regions were seen in *dyslexic* subjects during reading of narrative text (Hynd et al., 1987; Huettmer et al., 1989).

Sally and Bennet Shaywitz, the co-directors of the Yale Center for Learning and Attention (Shaywitz et al., 1998), asked volunteers to perform a hierarchy of tasks while they imaged the brain to see which areas were active. They discovered that poor readers, people with *dyslexia* have a "glitch" in the wiring of the brains pathway that is used for reading. They note that although there is the popular belief that people with *dyslexia* reverse letters, that is not the essence of the disorder. In fact, *dyslexia* is defined as a great difficulty in reading that is not explained by a lack of intelligence. Reid Lyn, Chief of Child Development and Behavioral Branch at the National Institute of Child Health and Human Development stated that as many as one in five American children have great difficulty reading. A study by the U.S. Department of Education (March/99) stated that 70 percent of non-minority and 90 percent of minority school children are not reading at grade level. The ability to read is thought to require the ability to match letters with sounds that represent them. It also requires phonological awareness or the ability to break words into their component sounds; therefore, it is thought that people who are unable or have difficulty reading cannot recover the sounds from speech.

Using a spectrum of tasks that involve going from letters to sounds to words, Shaywitz and colleagues (1998) designed a hierarchal series of reading tasks and asked 29 adults with dyslexia and 32 good readers to perform these tasks. While the subjects performed these tasks, the researchers imaged their brains with functional MRI. They found that in normal readers, the involved pathway starts with the primary *visual cortex*, and then the *angular gyrus* takes over. The final area involved is the superior *temporal gyrus* or *Wernicke's* area, where it is thought that the sounds of language are converted to words. Subjects with *dyslexia* are found to barely use this reading pathway, instead, another area of the brain lights up, the inferior *frontal gyrus* or *Broca's* area, which is thought to pair words with units of sound.

Electrophysiological studies have been conducted as well on learning-disabled individuals. One study showed that bilateral temporal indices in spontaneous EEG were indicated for a group of learning-disabled children (Morris et al., 1989). Studies using electrical activity mapping (*BEAM*) (Duffy et al., 1980, 1988) suggest that learning-disabled children have dysfunction in cortical language zones, and patterns of disturbances may be related to specific subgroups of disabled readers. Another study reported different visual evoked potentials to language stimuli by two groups of reading disabled children. The results of this study suggest that one group has increased *occipital-parietal* involvement and the other has increased *occipital-temporal* involvement (Harter, 1991). Another study used *event-related potentials* (*ERPs*) to look at hemisphere asymmetry for language, signal processing efficiency, hemisphericity, and frontally based attentional control in good and poor adolescent readers. The results suggest that hemisphericity differences account for reading skill level only in good readers, but that frontally generated attentional *ERP* account for reading skills among poor readers; whereas good readers show the expected *ERP* asymmetries, the poor readers do not. The authors therefore suggested that below some "crucial" reading threshold, frontal attentional skill might be a better predictor of reading disability, whereas above this threshold good reading is predicted by hemisphericity (Segalowitz et al., 1992). What we see is that functional abnormalities that may result in reading and learning disabilities are dependent on baseline arousal and activation, and hemispheric dominance. Therefore, it is thought that these results support a neurobiological basis of developmental learning disorders. It also leads to the conclusion, as stated by Bolick and Obrzut (1995), "for a large subset of learning disabled children, atypical cerebral organization and unusual patterns of laterality, attention and arousal may underlie deficits in auditory and visual, language and non language information processing abilities."

Attentional factors have been studied with DL tasks. In one study, it has been shown that all groups of children demonstrate an REA except the younger poor readers (Obrzut et al., 1993). The results also show that there is an interaction among handedness, age and reading ability, and auditory verbal dichotic tasks. The lack of attentional shifting for left- and right-handed good readers is thought to support a structural theory of lateralization whereas degree and direction of attention shifting for left- and right-handed poor readers is suggestive of inter-hemispheric transfer difficulties and atypical cerebral organization (Obrzut, 1991). Regardless of the particular model one subscribes to, the results support the view that, in contrast to a fixed laterality deficit, learning-disabled children experience an attentional dysfunction which interferes with left hemisphere language processing by over-engaging either hemisphere (Boliek and Obrzut, 1995).

More than 100 years ago, the first reports of children with learning disabilities appeared in the literature (Kussmaul, 1877; Morgan, 1896; Bastian, 1898; Hinshelwood 1900). Most of the reports at the time tried to explain why an estimated 3–6 percent of the school age population could not learn consistent with what would be expected considering their intellectual ability and many attempts at instruction (Gaddes, 1985). After reviewing the early reports, Hynd and Willis (1988) concluded that by 1905 the literature supported: (1) reading disability (cognitive word blindness) could manifest in children with normal ability, (2) male children seemed to be more commonly affected than female children, (3) although children may exhibit a number of different symptoms, they all suffer a core deficit in reading acquisition, (4) normal classroom instruction does not improve reading ability, (5) some reading problems seem to be genetic, and (6) the core symptoms appear similar to those that are seen in adults with left *temporal–parietal* lesions. It is now recognized that learning disabilities may be expressed in many different areas including arithmetic, writing, spelling, etc. However, it is well accepted that reading disabilities or *dyslexia* has been the area most researched (Hynd et al., 1995). Research for years seemed to support the theory that there is a neurological basis for *dyslexia*. For example, research suggests that reading disabled children have an increased incidence of electrophysiological abnormalities (Duffy et al., 1980). Soft signs are also more often found in reading disabled children (Peters et al., 1975) as well as a greater frequency of left or mixed handedness (Bryden and Steenhuis, 1991). Studies also found that children with learning disabilities performed more poorly than normal children on any given task, cognitive or perceptual, however, they did better than children with brain damage (e.g., Reitan and Boll,

1973). Therefore, these children were referred to as suffering from *minimal brain dysfunction* because they appeared to function between normal and known brain-damaged levels.

Language

There is evidence in the literature that suggests that very young children as well as infants are lateralized for language processing (Molfese, 1989). No one would argue with the fact that in the majority, language is lateralized to the left hemisphere. Although language abilities appear to develop over the course of human ontogeny, language lateralization remains stable at least from middle childhood if not earlier in infant development. Based on the work of Dejerine and Gauckler (1911), who thought that there was a left lateralized "word center" in the area of the *angular gyrus*, and contributions of Broca, Wernicke, and others, a complete neuro-linguistic model of language and reading evolved. This model suggests that first, visual stimuli are registered in the *occipital cortex*, with associations made in the *secondary visual cortex*. This input is then compared and contrasted with other sensory input from other modalities in the area of the *angular gyrus* in the left hemisphere. Associations of linguistic-semantic comprehension with input from the area of the *angular gyrus* involve the cortical region of the left posterior-superior *temporal* region, including the *planum temporale*. This process is thought to be completed when inter-hemispheric fibers connect these regions with *Broca's area* in the left inferior *frontal* region. Research has examined anatomic or structural differences in *dyslexia* or individuals with other language problems. Rosenberger and Hier (1980) suggested that there was limited but interesting evidence that symmetry or reversed symmetry may be associated with poor verbal-linguistic ability that is commonly found in children with *dyslexia*. In four consecutive autopsy cases, it was found that focal *dysplasias* clustered specifically in the left superior *temporal* region by a ratio of 11:1 (Galaburda, 1985). Another study (Larsen et al., 1990) found that when symmetry of the *planum temporale* presents in *dyslexia*, individuals exhibit phonological deficits. They concluded that a relationship appears to exist between anatomic brain patterns and neuro-linguistic processes.

The balance of activity or arousal between the two hemispheres as well as intrahemispheric transfer is essential for effective language. One study on split-brain patients, for example, noted, "occasionally the *commissurotomized* patient may become so absorbed in a right hemisphere task that speech and left hemisphere functions are temporally depressed to the extent that one questions whether consciousness may not be shifted entirely to the working hemisphere" (Sperry, 1962). In fact, mutism after *callosal* section may be one extreme example of an arousal imbalance that interrupts performance (Suillivan and McKeever, 1985). In another study, *akinesia* and *mutism* were found to be correlated with mixed hand dominance (Sussman et al., 1983). In another study of two-stage surgery, *mutism* occurred only after the second stage, which involved transection of the posterior part of the *callosum*. What especially points to an arousal explanation for the *mutism* is that manipulating attention alleviated the speech problem. Interestingly, when the patient palpated an object with the hand contralateral to the hemisphere that controls speech, speech was improved (Sullivan and McKeever). That using the nondominant hand improves speech suggests that the description is a result of arousal imbalance (Liederman, 1998).

It has been shown that inter-hemispheric connections are crucial for equilibrating activation levels between two regions or synchronizing activity. The same process is thought to occur within hemispheric *cortico*-cortical pathways, which also equilibrate and synchronize function (Leisman and Ashkenazi, 1980; Leisman and Zenhausern, 1982; Koch and Leisman, 1996, 2001; Leisman, 2002). One within hemisphere syndrome, which is caused by damage of *cortico*-cortical fibers, is *conduction aphasia*. This syndrome is thought to be a result of damage to the *arcuate fascicules* a *cortico-cortical* pathway from *Wernicke's* area to *Broca's* area. It has been shown that a small lesion that is confined to the *arcuate fascicules* is enough to produce *conduction aphasia* (Tanabe et al., 1987). The primary symptom of *conduction aphasia* is the inability to repeat what is heard, even though language production is relatively fluent and without *dysarthria* and aural comprehension is fundamentally intact (Benson, 1979). The inability to repeat what has just been said may be due to a situation-specific under-activation of *Broca's* area by *Wernicke's* area (due to the lack of activation along the *arcuate fascicules*) rather than to a lack of transfer itself. Under-activation of *Broca's* area in this type of situation has been confirmed by measures of metabolic activity. One study found that during a naming task, patients with *conduction aphasia* had less of an increase in *Broca's* area than normal individuals (Demeurisse and Capon, 1991). Under conditions when subjects initiate their own speech, there is enough activation of *Broca's* area to subserve fluent output. However, in conditions when the patient's speech is an attempt to echo that which was just heard, *Broca's* area is not sufficiently activated by *Wernicke's* area to allow word-by-word repetition. In this case, it could be concluded that these patients had lost explicit access to information but retained access to it on both categorical and implicit levels. In this case, arousal level is enough for more lower level perception; however, the activation that would normally be superimposed on *Broca's* area that is obviously necessary for explicit access, is not enough to raise signal-to-noise ratio to reach the level of activation needed.

Activation or arousal levels may be related to motor activity. There is evidence of a temporal linkage between hand performance and language milestones; disruptions in the course of development of manual asymmetry coincide with transitions between stages of language development (Ramsay, 1985). "Cycles" in the development of handedness have been blamed on fluctuations in the degree to which speech interferes with the use of the dominant hand (Bates et al., 1986). This interference is thought to be because of the close proximity of the two control areas in *functional control space* (Kinsbourne and Hicks, 1978) within the same hemisphere. We see these effects early in development, but early childhood is too early to reflect lateralized differences in the cerebrum, but is reflective of the precursors of these lateralized processors. It has been noted that due to the relative immaturity of the newborn's cerebrum (Dobbing and Sands, 1973), these precursors are most likely represented at the level of the *basal ganglia* and *thalamus*.

Recent research implicates the *cerebellum* as playing a major role in the development and maintenance of speech as well as other higher cognitive and behavioral functions. It is generally accepted that language abilities in humans are probably dependent on some phylogenetically new areas of the cerebral cortex (Benson and Geschwind, 1968, 1970; Benson et al., 1973; Ojemann and Creutzfeldt, 1987). However, it is not generally accepted that new areas of the *cerebellum* may be important as well. Along with the evolution of the new association areas in the cerebral cortex, connections to new areas in the lateral *cerebellum* can respond to linguistic signals received from the posterior lobes of the cerebral cortex. These responses are then thought to be able to be transmitted from the neo-*cerebellum* to *Broca's* language areas in the *frontal lobe* of the cerebral cortex. It is thought, therefore, that the neo-*cerebellum* can act as a link between the posterior and frontal language areas of the cerebral neocortex. The *cerebellum* seems to be able to contribute not only to the motor processing of speech but also to the cognitive processes that think of the words to be expressed (Leiner et al., 1989). It is interesting to note that it was this very type of transfer of posterior to anterior cerebral areas by the *arcuate fascicules* that was at fault in *conduction aphasia*.

It has also been noted that there are no motor deficits but rather a cognitive deficiency possibly related to an arousal imbalance. One study designed to distinguish the motor from the cognitive function of speech, was constituted by a subtraction technique. Each individual was asked to perform a sequence of tasks, ranging from simple to complex, where each successive task necessitates an additional word-processing requirement. By subtracting the motor activation measured in the simple task from the activation in the complex task it is possible to distinguish the areas that are activated during word association (Peterson et al., 1989). Results show that the area of the *cerebellum* activated during the cognitive process of word association is structurally separate from the area activated during the motor process. In the motor task of speaking a word activation occurs in the superior *anterior lobe*, near the areas that are activated by movements of *extra-ocular* muscles and movements of digits. However, in cognitive tasks, word association activation occurs in an inferior-lateral area in the right hemisphere of the *cerebellum*. It has been noted that the lateral *cerebellar* area can send its output to *Broca's* area in the *frontal lobe* of the neocortex (Leiner et al., 1991). Brain scans confirm that the medial part of the *cerebellum* is activated during finger and motor speech movements (Peterson et al., 1989; Bellugi et al., 1990), while the cognitive language tasks activate a more lateral part (Peterson et al., 1989; Roland et al., 1989). These results suggest that in normal human brains, the lateral areas of the neo-cerebellum participate in the cognitive aspects of language and learning, while the more midline or medial areas are involved in motor aspects of speech and gesture.

The *cerebellum* is involved in all human learning, motor and cognitive, as well as in linguistic, and sensory and emotional aspects of behavior. It has been shown that when the *cerebellum* is damaged, something of the fundamental learning process is lost, the ability to improve performance with practice. In a study by Fiery and colleagues (1990), a lawyer who had a stroke that impaired a large area of the posterior *cerebellum* on the right was tested by being asked to associate a word to its use. Initially, 60 percent of his responses were incorrect compared to 2 percent for normal subjects. In addition, it was noted that normally when individuals are given successive nouns in successive trials, they become faster at finding verbs to associate with nouns. However, in this case the lawyer did not show this kind of improvement. In the following trials, while normal subjects decreased their reaction time by 27 percent, his reaction time decreased by only 8 percent. It was further noted that lack of improvement was also seen in his performance on other cognitive tasks involving two types of memory. The first was *declarative memory*, which is connected to learning facts and events and the second was *procedural memory*, which involves the learning of skills also known as habit learning. The lawyer was impaired on *procedural memory* tasks but not *declarative memory*. It is well documented that the *cerebellum* is involved in motor learning (Lou and Bloedel, 1988; Solomon et al., 1989). It has also been postulated among those that think that long-term memory is localized in the brain, that learning may be stored within the *cerebellum* or in some other structure that is connected to the *cerebellum* (Ito, 1990; Lalonde and Botez, 1990).

The *hippocampus* is thought to be involved in long-term memory and is connected to the *cerebellum*. Pathways from the anterior *vermis* and *fastigial nucleus* to several areas of the *limbic system* have been shown to modify sensory input to the *hippocampus* (Newman and Rezea, 1979). It seems that involvement of the *cerebellum* in associative learning, not simply motor but cognitive as well, now seems unequivocal (Bracke-Tolkmitt et al., 1989; Roland et al., 1989; Fiez, 1996). To review, the *cerebellum's* role in practice improvement as well as new learning especially of procedural learning is probably a result of the effects of surround inhibition. In the *cerebellum*, a stimulus enters the granular layer where *granule* cells then excite *Purkinje* cells and at the same time, the *granule* cells excite the output neuron immediately below the area of activation. Therefore, the more this area becomes excited or activated, the more the *Purkinje* cells inhibit that specific area associated with the movement or activity. However, due to surround inhibition of *basket* and *stellate* cells, *Purkinje* cells surrounding the area of activation are inhibited, decreasing the inhibition of the output nuclei below that area, increasing signal-to-noise ratio and neuronal excitation, and priming them to respond faster to the next stimulus. Repetitive movements or cognitive tasks that activate the same neuronal areas of *cerebellum* and neocortex will become more inhibited, but other areas are brought closer to threshold ready to respond to the new cognitive thought or motor activity. Therefore, all new learning, which includes relearning an activity after a stroke, is dependent on the *cerebellum*.

We would expect all children with learning disabilities to have a deficit of the *cerebellum*. The right *cerebellar* hemisphere would be associated with left cerebrum and would affect speech, language, arithmetic, reading, and other left-brain activities. Left *cerebellar* hemisphere dysfunction would be associated with right cerebral dysfunction of social learning and behavior. This description would be consistent with the findings of Ayers (1972) and Levinson (1980, 1984) who both found *cerebellar* symptoms in the majority of children with learning disabilities and behavioral disorders. The cerebellum has also been shown to be involved in other cognitive functions as well as disorders such as *William's syndrome, autism*, and *fragile X* syndrome.

In children with *William's syndrome* who have good verbal skills, it is thought that the *neo-cerebellum* is normal in size; however, in others who have difficulty with language, the *neo-cerebellum* is smaller (Jernigan and Bellugi, 1990). *William's syndrome* is a unique form of retardation that affects cognitive abilities while retaining normal linguistic abilities both semantic and syntactic (Rae et al., 1998). MRI studies show that the neo-cerebellum of these individuals appears normal, whereas the *paleo-cerebellum* is reduced and the forebrain is significantly reduced (Jernigan and Bellugi, 1990). MRI studies of *William's* and *Down's syndrome* brains report cerebral *hypoplasia*, alterations in the size of the *paleo-cerebellum* and midline *neo-cerebellum* (*vermis*), as well as decrease in the size of the posterior compared with the frontal cerebrum (Jernigan et al., 1993). In *Down's syndrome*, the reduction of volume is more dramatic in the *cerebellum* and *brainstem* than in the cerebrum (Jernigan and Bellugi, 1990). The difference between these two groups of patients is the fact that children with *William's syndrome* retain their complex language abilities even with severe cognitive deficits. These MRI results also raise a possibility that the *midline cerebellum* or *vermis* may be more connected to the *posterior* cerebrum or dorsal cortex and the *lateral cerebellum* may be more connected to the *frontal* cerebral or ventral cortex.

Phylogenetically the older *paleo-cortex* may be more connected to the evolutionary older dorsal cortex, whereas the *neo-cerebellum* may have evolved in line with the ventral cortex or neocortex, which is the latest evolutionary development. The ventral cortex is more associated with left-brain activities such as speech and language functions. *Autistic* children on the other hand usually have significantly impaired language as well as cognitive and behavioral development. MRI studies of these children in contrast show that the *neo-cerebellum* is decreased in size (Courchesne et al., 1987, 1988; Murakami et al., 1989). Other studies have reported structural abnormalities in the *cerebellum* and *brainstem* of *autistic* patients (Bauman and Kemper, 1985, 1986; Gaffney et al., 1987, 1988). Studies have also shown specific loss of *cerebellar* nerve cells in *autistic* patients (Ritvo, 1986).

Studies have also been performed on patients with *Fragile X Syndrome*, the most common form of mental retardation. These individuals have been shown to have a significant decrease in the size of the *neo-cerebellum* similar to that found in *autistic* children (Reiss and Freund, 1990; Reiss et al., 1991). It has been noted that males with *Fragile X Syndrome* express some *autistic* behavior, such as deficits in social interaction with peers, abnormalities in verbal and nonverbal communication, stereotypical motor behavior, and unusual responses to sensory stimuli (Harvey and Kennedy, 2002; Havlovicova et al., 2002; Jones and Szatmari, 2002).

Verbal IQ is primarily an assessment of the left brain cognitive function: patients with *cerebellar* lesions were tested on a wide range of intellectual and learning abilities to see what effect the *cerebellum* had on intelligence (Neau et al., 2000; Cotterill, 2001; Scott et al., 2001). In contrast to normal subjects similar in age and education, the *cerebellar* patients were significantly impaired on all IQ measures and these deficits were not attributable to motor impairments per se. The researchers found that the decrease in IQ was not expected, that apparently the deficits had not been anticipated even though after testing the deficits were large enough that they should have been apparent (Neau et al., 2000; Cotterill, 2001; Scott et al., 2001). Previously it was not thought that *cerebellar* deficits affected IQ, and that *cerebellar* deficits are not usually associated with gross intellectual impairments, tending instead to be subtle, and therefore requiring subtle and detailed testing for detection. This may be the case with motor symptoms as well; they may be subtle or specific to only one side, therefore requiring experience, skill, and attention to subtle effects. It may be the case that *cerebellar* and motor signs although significant, may often be overlooked or missed.

Studies have been performed using *ideography* and mental imagery. *Ideography* refers to the performance of purely mental tasks, which are not influenced by ongoing perception of sensory signals or by the control of movements, speech, or behavior (Ingvar, 1991). Blood flow and metabolic change studies have shown activation of the *cerebellum* during various mental tasks in normal individuals. These tasks include mental arithmetic (silent counting) (Decety et al., 1990; Ingvar, 1991), mental simulation of tennis playing and other motor ideation (without any actual motor activity) (Decety et al., 1988; Decety and Ingvar, 1990), mental association of word with its use (Peterson et al., 1989), as well as learning to recognize complex geometrical objects (Roland et al., 1989). In another study, 12 patients with CA and 12 normal controls matched for age and education were required to solve the *Tower of Hanoi* puzzle, a nine-problem task that requires cognitive planning (Grafman et al., 1992). CA patients were shown to have difficulty solving the *Tower of Hanoi* task that could not be accounted for by motor impairment, age, education level, level of dementia, depression, visuomotor procedural learning, verbal memory, or verbal fluency. The results suggest that failure of this task is based on the specific demands of cognitive planning. In addition, pure *cerebellar* cortical atrophy (CCA) patients took significantly longer in pre-movement planning time (with no increase in between move pause time) relative to the control group. The authors note that in addition to CA patients, there is evidence that patients with *frontal lobe* lesions or dysfunction due to subcortical disease also fail on cognitive planning tasks similar to the *Tower of Hanoi* (Shallice, 1982). An explanation for failure on the tower-type tasks is that assembling a sequential series of events or actions into a coherent behavioral unit is difficult for patients with *frontal lobe* injury (Grafman, 1989), subcortical lesions (Grafman et al., 1990), and *cerebellar* lesions (Daum et al., 1993). The authors speculate that cognitive planning may be seen as an analogue of cognitive representation to complex motor procedures that require a series of individual movements to function as a unitary sequence (Ito, 1990). Therefore, it may be possible that the *cerebellar-frontal* axis functions in different ways depending on the requirements of the specific task. For instance, initiating motor activity may depend on *frontal* cortex and motor sequences may depend on the *cerebellum* for the timing of the movements. However, cognitive sequences may depend on the *prefrontal* cortex for the initiation of plans or sequential actions and on the *cerebellum* for the temporal integration of events that make up the cognitive plan or action (Decety et al., 1990). The conclusion of the authors is that deficits in cognitive planning suggest a functional link between the *cerebellum, basal ganglia*, and *frontal lobe* concerning specific cognitive processes (Grafman et al., 1992).

Memory

Memory is another important cognitive function that may be affected in learning-disabled individuals. One of the tasks that parents and teachers appear most concerned about is reading comprehension. Reading comprehension is thought to rely heavily on *working memory*, which is thought to be a function of the *frontal lobe* especially the left *frontal lobe*. There are two major types of memory, *explicit* (*declarative*) *memory*, and *implicit* (non-declarative or *procedural*) memory (Leisman and Koch, 2000). Explicit memory is also known as *conscious memory*, which is asymmetrically localized in the left hemisphere, whereas *implicit memory* or *subconscious* (*subliminal*) *memory* is lateralized to the right hemisphere (Hugdahl, 1995). The learning and, therefore, retention of new information about an individual's autobiographical history is supported by medial *temporal lobe* structures and the midline *diencephalon*. It appears that damage to these areas interferes with the ability to form new *explicit memories*. It has been noted that damage to these regions also leads to problems in remembering events in the years immediately prior to the injury, but apparently leaves intact the majority of previous episode and semantic memories acquired during one's life. These areas are not storage sets of information in long-term memory. However, damage to areas of the *temporal lobe* outside the *hippocampus* are thought to cause a loss of *episodic memories* while the ability to acquire new ones may remain intact. It has been indicated by Gazzaniga (1998) that brain injury leading to deficits of *implicit* learning and memory affects some patients while *explicit* memory remains intact. This is thought to affect the medial *temporal lobe* and midline *diencephalon*. It is further thought that the left *supramarginal gyrus* and left *premotor cortex* relate to the phonological loop of the *working memory* system. Studies involving PET scans, and transcranial magnetic stimulation in humans and lower organisms show that the *motor* cortex is critical for *implicit procedural* learning of movement patterns. Activation of the *basal ganglia* and the *putamen* would be anticipated because patients with *Huntington's disease* have been shown to have deficits in sequence-learning tasks. Because the *supplementary motor cortex* is also activated during *implicit learning*, it is thought to be part of a network of a cortical–subcortical motor loop, which regulates voluntary movement.

In contrast, *explicit learning* and awareness of the sequences require greater activation in the *right premotor cortex*, the dorsolateral *prefrontal cortex* associated with *working memory*, the *anterior cingulate* areas in the *parietal cortex* controlling voluntary attention, and the lateral *temporal* cortical area thought to store *explicit memories*. The dorsolateral *prefrontal cortex* (DLPFC) is associated with *working memory*; the DLPFC also has been shown to have strong connections from the *basal ganglia* and *cerebellum* via projections to the *thalamus* (Middleton and Strick, 1994).

EMOTIONAL AND AFFECTIVE SYMPTOMS

Whereas most of the traditional cognitive functions previously reviewed are related to a superiority of the left hemisphere, we recognize that a right hemisphere dysfunction may also result in a learning disability or cognitive dysfunction. In general, when it comes to emotional, behavioral, or affective disorders, the right hemisphere is generally responsible. However, the interaction with the left hemisphere as well as the interaction with *limbic* structures may make emotional problems less clear-cut than cognitive problems. Davidson and Sutton (1995) has stated,

> emotion is a class of behavior that has invited the consideration of its underlying biological substrates since the time it was first studied. Probably more than any other class of behavior, emotion often involves frank biological changes that are frequently perceptible to the person in whom the emotion arises as well as to the observer (e.g. facial blood flow changes as in "white with fear").

Because of the type of observations, it has been noted that most of the early research on the biological substrates of emotion has been focused on the autonomic reactions and the subcortical and *limbic system* structures thought to be responsible for those changes. However, more recently it has become apparent that in humans especially, the cerebral cortex plays an important role in aspects of emotional behavior and experience (Kolb and Taylor, 1990). In this regard, the anterior cortical regions especially appear to play a major role, through their dense anatomic reciprocal connections with subcortical centers, *limbic* structures, and posterior cortical circuits in the control of emotional behavior.

In humans, the anterior cortical zones are the areas of the brain, which have shown the most dramatic growth in relative size over the course of phylogenetic development compared with other brain regions (Luria, 1973) and are thought therefore to be critical to unique human behavior and emotion. Initial research on cerebral laterality in affective disorder concentrated on exploring the hypothesis of right hemisphere dysfunction in these types of disorders (Gruzelier and Venables, 1974; Flor-Henry, 1976; Kronfol et al., 1978; Yozawitz, 1979). These and other research reports are thought to show some support for this theory; however, it has become clear that to completely consider the full range of laterality findings in affective disorders, we must consider right and left, anterior–posterior, and cortical–subcortical interactions (Bruder, 1995).

Much of the research has focused on interaction of these areas with the anterior cortical region. Asymmetries in anterior cortical function have been implicated in different forms of emotional behavior. Even early on, research seemed to suggest that injury of the left hemisphere would more likely produce a catastrophic-depressive reaction compared to similar damage to the right hemisphere (Goldstein, 1939). Research by Robinson and colleagues (1984) reported that damage specifically to the left *frontal lobe* appears to produce depression. In addition, the closer the damage is to the frontal pole, the more severe the depression. In contrast, patients that develop *mania* following brain injury are more likely to have damage to the right hemisphere than the left. This dichotomy will form the foundation of much of the symptoms that we will explore.

There are children with and without learning disabilities who exhibit depressive or *manic* type symptoms such as being withdrawn, sad, irritable, slow moving (*hypokinetic* type motor activity) or aggressive, hyperactive, angry, impulsive, fidgety, lacking concentration, repetitive or persistent, frustrated, anxious, or violent. Most of these symptoms and others will be explained by anterior activation asymmetries and their interaction with other cortical, subcortical, and *limbic* processes. The asymmetric control of different emotional behaviors relates to evolutionary advantages that have developed due to these asymmetries (Levy, 1972).

It is generally agreed that approach and withdrawal are basic motivations that are found at any level of phylogeny (Davidson and Sutton, 1995). In several papers, Davidson and colleagues (Davidson, 1984, 1987, 1988; Davidson and Tomarken, 1989; Davidson et al., 1990a, 1990b) have suggested that the anterior regions of the left and right hemisphere are specialized for approach and withdrawal behaviors, respectively. The left *frontal* region has been described by Luria (1973) as being an important center for intention, self-regulation, and planning. These characteristics have also been described as "will" which is important to approach behavior. It is further noted that during child development, a child will approach or reach out to objects with the right hand more than the left (Young et al., 1983). It is thought that right-handed reaching and positive emotions together are manifestations of a brain region, which controls approach behavior, and the left *frontal* region is thought to act as a "convergence zone" for this circuit (Damasio, 1989; Davidson, 1992a, 1992b). Therefore, as expected damage to the left frontal region produces behavior, that can be recognized as a decrease in approach. These individuals have been described as apathetic, exhibit loss of interest or pleasure in people and things, and have difficulty initiating voluntary action or movement (intention). Hypo-activation in this same area is expected to be associated with a decreased threshold for the experience of sadness and depression and associated behavior.

In contrast, it is thought that the right anterior region is specialized for withdrawal. Electrophysiological measurement of regional hemispheric activation suggests that the right *frontal* and anterior *temporal* regions are specifically activated during withdrawal-related behaviors such as fear and disgust. In addition, it has been shown that baseline tonic activation in these regions shows a superiority to respond with exaggerated withdrawal-related negative affect to appropriate emotional activators. These individuals are said to be generally more "negative" individuals (Davidson and Sutton, 1995). We can again see that the level of arousal or activation is critical to the function. Although activation of emotions is often transient in response to stimuli, individuals also have chronic baseline, arousal, or activation of areas of the brain that can be measured. Chronic imbalances of arousal, with one side activated higher or lower than the opposite hemisphere, can create different consistent affects from one individual to the other. A child with low right hemisphere baseline activity may have a different dispositional affect than a child with a low left frontal baseline arousal level. Not only will their affect reflect the asymmetric arousal imbalance but also their emotional response threshold to stimuli will be lower than normal. The greater the imbalance, the more dramatic the response may be. This imbalance or altered level of arousal can be measured subjectively by testing perceptual asymmetries and thresholds. Dichotic testing and visual field-testing are two ways of examining perceptual levels, which reflect arousal, or activation levels. There is evidence to suggest that differences in right-handed individuals in *dichotic ear advantage* (Hellige and Wong, 1983; Kim and Levine, 1992) and visual field advantages (Levy et al., 1983; Kim et al., 1990; Luh et al., 1991) are reliable and are thought real individual differences in the relative activation or utilization of the right and left hemisphere (Bruder, 1995). Also, it has been suggested that between subjects, variation in perceptual asymmetry in right-handed adults are more related to task independent differences in characteristic arousal asymmetry than to hemispheric specialization (Levy et al., 1983). Using a free field *chimeric* faces test to measure characteristic perceptual asymmetry, it was shown that about half of the variation of asymmetry scores on this and other visual laterality tests could be attributed to a common factor called *characteristic perceptual asymmetry* (Kim et al., 1990; Luh et al., 1991). In another study it was found that about half of the between subjects variation in DL asymmetry could be attributed to *characteristic perceptual asymmetry* and that both modality-specific and modality-general components contribute to the observed asymmetries (Kim and Levine, 1992). This may reflect alteration of specific and non-specific *thalamic* pathways together that create the characteristic perceptual asymmetry.

Expressing and perceiving emotion are thought to be different in their areas of control. In this matter, some researchers think that the right hemisphere plays a more general role in all emotion (Borad et al., 1986). Most of the research in this area is related to the perception of emotional information where evidence suggests that the right posterior cortical region is the primary area for the perception of all emotion, positive or negative. It has also been suggested that anterior activation asymmetry acts to predispose a child to respond most of the time to positive or negative affect in the presence of an actual emotional stimuli. It has been noted that while baseline anterior asymmetry predicts the threshold of reaction to an emotional challenge, it is not always related to an individual's unprovoked state (Davidson and Fox, 1989; Tomarken et al., 1990; Wheeler et al., 1993). The diagnosis of children with affective disorders appears to be increasing.

Depression and anxiety in children can be very similar to what adults experience. Depression can result in deep sadness, loss of interest in friends, tiredness, and thoughts of hopelessness. One important sign of depression is when a child stops participating in social activities, his grades fall, or he does not feel like going to school. These signs can now be recognized as being attributable to lack of approach or expression of withdrawal symptoms, which are most often related to decreased activation of the left hemisphere. The left hemisphere also is superior for most cognitive or academic activities, so we would also expect a decrease in grades. In addition, the left hemisphere is important for a child's self-motivation or "will." If a child has a decrease in left hemisphere baseline activity, especially *frontal* activity, we can expect these accompanying symptoms. It was only a relatively short time ago when depression was thought to be a disorder of middle-aged women. However, the literature currently supports the view that for many, depression may actually have begun in childhood or adolescence. Now it is becoming quite common to hear about children on antidepressants; however, scientific information about the safety and efficiency is only now being studied.

It has been noted that long-term psychotherapy does not appear to benefit teenagers as well as with adults and the older antidepressants are not as effective as they are for adults (Bedi et al., 2000; Martin et al., 2000). Now it appears that there is a greater emphasis on *serotonin selective reuptake inhibitors* (SSRIs). *Serotonin, norepinephrine,* and *dopamine* have all been implicated in a variety of effective disorders. Although *serotonin* and *norepinephrine* are asymmetrically distributed in the right brain, they appear to be involved in left-brain disorders and depression.

There is a considerable body of research examining the potential role of *serotonin* in the regulation of affective states, behavior, and psychopathological disorders. It has been postulated that *serotonergic* dysfunction may contribute to depressive disorder and suicidal behavior (Asberg et al., 1987; Meltzer and Lowy, 1987; Arora and Meltzer, 1989a, 1989b). In postmortem studies, lower concentrations of 5-HT and 5-HIAA (5-*hydroxyendoliacetic acid*), the major metabolite of 5-HT, have been reported in midbrain *Raphe nuclei* of suicide victims as opposed to normal controls. *Norepinephrine* has also been shown to favor the right hemisphere. Both neurotransmitters have been reported to be mostly involved in up-regulation (*norepinephrine*) and down-regulation (*serotonin*) of arousal, and also apparently having reciprocally inhibitory effects (Flor-Henry, 1986) and are asymmetrically distributed favoring the right side of the brain. In addition, it is possible that changes of this normal pattern are correlated with affective and behavioral abnormalities. Even though *serotonin* and *norepinephrine* favor the right hemisphere, when there is a decrease there may be a more substantial negative effect on the opposite hemisphere. This seems to be because the right half of the brain has better compensatory mechanisms to counterbalance a deficit (up-regulation of receptors). Therefore, the left hemisphere cannot compensate as well as the right, and as a result may be more affected as a result.

Previously, alterations of neurotransmitters were looked on as the primary cause of depression and other affective disorders. New research has gone way beyond a purely chemical evaluation and more to a functional neurological examination of dysfunction. A recent article in *Psychology Today* (Marano, 1999, March/April) comments on these changes in focus. Marano writes that recent findings are

radically different from the conventional wisdom on several counts. First, it overturns the widespread belief that depression is "just" a chemical imbalance. Yes, neurotransmitters like *serotonin* function abnormally in depression-but so do many other things. Second it challenges the view that this disorder is "merely" from the neck up depression affects the heart and the bones too, and the body's stress system.

He goes on to state

The evidence that the adult brain is much more plastic than anyone recognized, that experience and environmental changes to the brain circuitry underlying emotion is still, as Rockefeller University neurobiologist Bruce McEwen says taking scientists by surprise. The surprise is not just kicking off a revolution in our understanding of depression. It will force a revised view of all human behavior and the capacity for change.

Davidson is quoted as stating that based on new imaging techniques such as PET and fMRI, "the idea that there is global derangement of the *serotonin* or *norepinephrine* system is not sustainable in light of recent brain imaging data." He further states, "what distinguishes depressed from non-depressed individuals are patterns of regional brain function, differences in specific circuits" ((Marano, 1999, March/April); Davidson et al., 1999). We can imagine that if these newer ideas apply to adults, that it applies even more to children whose brains are much more plastic and changeable. These new concepts must also affect our approach to treatment, moving away from trying to manipulate chemicals and more toward trying to improve the function of the brain. As Davidson states ((Marano, 1999, March/April); Davidson et al., 1999), "non pharmacological treatments may exert quite specific biological effects in being able to affect certain select brain regions." He points out "the deficits in activation of the *prefrontal* cortex that we and others have identified in depression may be something that can be changed with cognitive therapy." The balance of activity between hemispheres is also a strong focus of attention.

The New York Times commented on a rekindling of interest in how the left and right sides of the human brain interact (Pettigrew and Miller, 1998; *NY Times* January 19, 1999). At a meeting of the Society for Neuroscience in November 1998, Jack Pettigrew, a neuroscientist at the University in Brisbane, Australia proposed that people with manic depression have a "sticky switch" somewhere in their brain. He thinks that normally this switch allows either the left or the right hemisphere to be dominant during different mental tasks, with the two sides constantly taking turns. He proposes that in people with manic depression, one hemisphere becomes locked into a dominant position in periods of depression while the other hemisphere is locked at all times of mania. Another theory being proposed is that of Frederic Schiffer, a psychiatrist at Harvard Medical School, who thinks that one hemisphere can be more immature than the other and that this imbalance leads to different mental disorders (Schiffer, 1996; Schiffer et al., 1998). Schiffer has used special goggles to help individuals "talk" to each half of the brain separately to help learn which half is less mature and to bring the two hemispheres into harmony. It has been shown through experimentation that it is possible to stimulate one hemisphere and inhibit the other so that the individual looks at the world using only half of their brain at a time. Schiffer explains that when individuals gaze to the far right and therefore activate the left-brain, they do better on verbal memory tasks, and when they look to the far left to engage the right brain, they feel more inertia and fatigue. In general, Schiffer found when patients look through the goggles they report very specific feelings depending on which side of the brain was being activated. In general, he found that depressed patients felt worse when the right side was stimulated and patients *with posttraumatic stress syndrome* felt worse when the left side was more active. This again points to an arousal or activation imbalance, which would therefore involve subcortical areas that provide arousal.

Pettigrew and Miller (1998) found a treatment that may imply that the *cerebellum* is involved. He reported that placement of ice water in one ear seems to "unstick" the switch. Ice water in the ear is a part of a traditional neurological test known as *caloric* testing. This test activates the *vestibular* apparatus and can test the balance of activation of the *cerebellum* by activity one side more than the other. This test has also been used on astronauts to understand the mechanism of space sickness. Pettigrew indicated that, if one places ice water into one ear, the opposite brain hemisphere will become activated. He further noted that cold water in the left ear activating the right hemisphere might temporarily reduce the symptoms of *mania*. Depression may be reduced by placing water in the right ear. This would clearly imply that stimulation of the ipsilateral *cerebellum* and contralateral cerebrum has an effect on alleviating both *mania* and depression. This indicates the importance of the subcortical structures in the role of depression and other emotional disorders.

Watson and Heilman (1982) hypothesized that a left subcortical lesion would not induce depression because it would not interrupt hemispheric processing of positive emotions and therefore would not "release" the more negative right hemisphere. They felt that a left subcortical lesion would affect the patients arousal level, resulting in unawareness but not depression. However, several studies of *unipolar* depression have indicated abnormal changes in blood flow and glucose utilization in subcortical structures like the caudate nucleus (Baxter et al., 1985; Buchsbaum et al., 1986). *Basal ganglia* dysfunction is thought to berelated to depression in *Parkinson's disease* and *Huntington's* chorea. Also left anterior subcortical lesions including the body and head of the *caudate* and anterior limb of the *internal capsule* have been noted to be significantly associated with major depression. The severity of the depression relates directly to the distance of the lesion from the frontal pole (Starkstein et al., 1987). Also in *Parkinson's* patients with unilateral symptoms, depression was more common in those with primarily right-sided as opposed to left-sided symptoms which is thought to be consistent with left *caudate* lesions and greater depression (Starkstein et al., 1990).

It has also been shown that stroke patients who also exhibit major depression have been shown to have more subcortical atrophy on CT scan than nondepressed stroke patients (Starkstein and Robinson, 1988). In another study it was demonstrated that secondary *mania* patients had lesions confined to right subcortical structures such as the *thalamus* and *anterior caudate*. Contrary to the prediction in 1982 by Watson and Heilman, it appears that the subcortical structures play an important role in depression and *mania*, and that arousal or activation levels are more critical to the processing of emotions in the *frontal cortex* than previously thought. It is also thought that changes in emotional behavior are not satisfactorily explained by release of contralateral cortical regions (Sackheim et al., 1982). Instead, it is now thought that lateralized changes in emotional behavior can be better explained as a result from release of ipsilateral subcortical centers with possibly a hierarchy of emotional control (Tucker, 1981; Tucker and Frederick, 1989).

Studies that have carefully reviewed pathological emotional outbursts in the absence of *brainstem* lesions found that subcortical structures such as the *basal ganglia* and *internal capsule* are involved in almost all cases studied (Poeck, 1969; Rinn, 1984). Examining the effects of the location of the lesion on emotional behavior, it has been proposed that lesions that include the *frontal convexity*, both dorso-lateral and dorso-medial *prefrontal* areas, can be associated with slowness, indifference, apathy, and lack of initiative. On the other hand, lesions of the *orbito-frontal* cortex appear to result in disinhibition, lack of social constraint, hyperactivity, grandiose thinking, and euphoria (Kleist, 1931; Blamer and Benson 1975). It has also been noted that secondary *mania* is significantly associated with cortical *para-limbic* involvement, which includes both *orbitofrontal* and basal *temporal* dysfunction, especially in the right hemisphere. *Mania* is apparently more frequent for right as opposed to left subcortical lesions, which includes the *thalamus* and *basal ganglia* (Starkstein et al., 1987; Starkstein and Robinson, 1988). It has also been found that *manic* symptoms like euphoria, hyperactivity, and insomnia could possibly be a result of interruption of the control exerted by the *orbito-frontal* cortex over *septal, hypothalamic*, and *mesencephalic* regions (Nauta, 1971; Starkstein and Robinson, 1989). This suggests that *mania* may be a result if disinhibition of intact subcortical centers due to cortical dysfunction or decrease activation. These results would be consistent with lesion effects seen in rats with right but not left hemisphere lesions, particularly lesions located in the *frontal lobe* that produced hyperactivity (Robinson, 1979; Pearlson and Robinson, 1981).

Similar effects have been noted with *orbitofrontal* or anterior *cingulate* ablations (Tucker and Derryberry, 1991). This would lead us to think that a decrease in right *prefrontal* cortex increases approach behavior possibly because of decreased inhibition of *orbito-frontal, cingulate*, and *limbic* structures, collectively observed as mania, hyperactivity, and perseverative behavior. PET studies (Drevets et al., 1992) have revealed increased blood flow in *orbito-frontal* cortex and parts of *cingulate gyrus* along with decreased metabolism in temporal and *parieto-occipital* regions in primary depressed patients. It has been postulated that in major depression, the neurological systems responsible for the processing of sensory *exteroceptive* information are inhibited in favor of pathways responsible for processing of internal information, emotion, and negative thoughts, relying on *limbic* and *para-limbic* representations (Nauta, 1971). Therefore, it is thought that this imbalance of activity, with increased activation of *para-limbic, limbic, and* neocortical areas during the experience of intense emotional states, can be seen as a functional "release" or disinhibition of the *limbic* areas responsible for emotional processing (Liotti and Tucker, 1995).

ATTENTION DEFICIT HYPERACTIVITY DISORDER

Since the 1940s, clinicians have applied various labels to children who are hyperactive and inordinately inattentive and impulsive. These children have been thought to have *minimal brain dysfunction, brain-injured child syndrome, hyperkinetic reaction of childhood, hyperactive child syndrome*, and most recently ADD. Russell Barkley thinks that *ADHD* is not a disorder of attention per se, but rather a result of developmental failure in the brain circuitry that underlies inhibition and self-control. He thinks this loss of self-control impairs other important brain functions crucial for maintaining attention, including the ability to defer immediate rewards for later, greater gain (Barkley, *Scientific American*, September 1998). Daniel Goleman (1995), the author of *Emotional Intelligence*, thinks that this ability to delay gratification is the best predictor of future success of a child. Studies estimate that between 2 and 9.5 percent of all school age children worldwide suffer from *ADHD* (Gottlieb, 2002; Pary et al., 2002; Yeargin-Allsopp and Boyle, 2002). Barclay also notes that Joe Sergeant (2000; Sergeant et al., 2002) of the University of Amsterdam has shown that children with *ADHD* cannot inhibit their impulsive motor responses to sensory input. He also states that other researchers have found that children with *ADHD* are less capable of preparing motor responses in anticipation of events and are not sensitive to feedback about errors made in those responses.

Imaging studies suggest involvement of the *prefrontal* cortex, *cerebellum*, and *basal ganglia*. Castellanos and colleagues (1996) at the National Institute of Mental Health have shown that particularly the right *prefrontal* cortex, the *caudate*, and *globus pallidus* are significantly smaller than normal in children with *ADHD*. Castellanos' group (Giedd et al., 1999a, 1999b) found that the *vermis* region of the *cerebellum* is also smaller in *ADHD* children. The *prefrontal* cortex regulates goal-directed behavior, a hierarchy of reflexive movements, cross-temporal contingencies, approach and avoidance behaviors, response inhibition, and perseveration (Chatterjee, 1998). In addition, the *prefrontal* cortex regulates movements directed at accomplishing goals (Luria, 1966), and together with the *basal ganglia* and *cerebellum*, they regulate primary motor output (Chatterjee, 1998).

The goal-directed movements regulated by the *prefrontal* cortex are complex and have multiple sequences. Luria thought that with *frontal lobe* injury or dysfunction, these reflexes are no longer smooth and integrated goal-directed movements (Luria, 1966). Instead, he saw these movements as fragments of the internal action. Coordinating movements to achieve an intended goal requires their organization over time; therefore, the *prefrontal* cortex is important in *cross-temporal contingencies* (Foster, 1989). Patients with *frontal lobe* dysfunction may present with different symptoms depending on whether dysfunctions include the right or left *prefrontal* cortex. Left *prefrontal* cortex dysfunction produces difficulty in initiating movements without guidance of explicit stimuli. Clinically patients are passive, apathetic, and depressed, and symptoms include lack of approach with increased avoidance behavior. Increased avoidance behavior with withdrawal to touch as seen in some *autistic* children is also a result of decreased activation of the left *prefrontal* cortex. Right *prefrontal* cortex dysfunction can result in difficulty inhibiting responses to stimuli. These children are distractible and react randomly to environmental stimuli, all typical of the ADHD child, and can be viewed as a decrease in avoidance and an increase in approach behavior, normal for right *prefrontal* cortex function. Simple approach behaviors exhibited may include simple movements approximating stimuli such as the group or smart reflexes, or complex approach movements (Lhermitte, 1986) where they inappropriately use whatever implements are within view. In addition, with the right *prefrontal* cortex, the child may exhibit *perseverative* behavior, or the inappropriate repetition of movements over time. *Perseverations* may be simple or complex (Sandson and Albert, 1984; Hotz and Helm-Estabrooks, 1995). The simplest form of *perseveration* is the inability to stop a single movement or muscle contraction after it has been completed. For instance, the child may not be able to relax his or her grip after shaking someone's hand. Another type of *perseveration* is the continuous repetition of a coordinated movement. For instance, a child may be unable to stop drawing loops after beginning a circular motion. A third type of *perseveration* is the repetition of a self-contained act even when required to switch to a different movement. For example, a child may continue to write her name even when told to write something else. Therefore, we see that typically right *prefrontal* cortex dysfunction can cause the inappropriate release of reflexive movements that vary in complexity, from a sustained single contraction, to the repetition of coordinated movement. This can also be considered *hyperkinetic* motor activity like *dystonia*, tremor, or tics typical of *ADHD*, or TS. It can also involve the repetition of a coordinated movement or repetition of a complex act typical of OCD and sometimes seen with *autism* or *pervasive developmental disorder*.

In children, these impulsive repetitive behaviors may in part be due to decreased ability to inhibit *limbic* structures like the *amygdala*. This could arise because of decreased activation or imbalance of arousal, or activation of the *prefrontal* areas, which processes cognitive information and inhibits the *limbic system*. It may also be because of delay in development of the *prefrontal* cortex. In some children, a lack of stimulation or arousal may delay the development of the *prefrontal* cortex, which is the latest part of the neocortex to develop.

Researchers at McLean Hospital and Harvard Medical School have conducted imaging studies on children between the ages of 11 and 17. The study found that the younger children process information using the *amygdala*, which controls aggression, fear, anger, and other primitive emotions. However, older children process the same information using the *frontal* cortex (Kilgore and Yurgelun-Todd, 2001; Yurgelun-Todd et al., 2002). Deborah Yurgelun-Todd (WGBH, 2002) of McLean Hospital's brain imaging center is quoted as stating, "If the physiology isn't complete, parents can't expect their children to have responses that are the equivalent to adults." She went on to comment that "we probably need to assume that they don't always understand what they are learning and may not respond accordingly." (WGBH, 2002) The researchers think that the *frontal* cortex is responsible for inhibiting "gut" or emotional responses or emotions. This is thought to explain why children and adolescents seem so impetuous or impulsive. Without activation of the *frontal lobe*, children cannot have fully developed emotional responses. The researchers in Boston tested the developing brain in 16 healthy children and teenagers. They had the children perform two cognitive neuropsychological tasks. One test presented the child with a number of faces expressing fear. They were also asked to determine what emotion the face was expressing. The second task involved adolescents who were asked to complete a number of word-production procedures including counting and word fluency.

The brains of younger children show strong *amygdala* activation when they respond to the faces. The magnitude of reaction, which is shown as an increase in blood flow, is larger on the left. This is consistent with research that face recognition is a right-brain activity; however, it is thought that the right *amygdala* is inhibited by the left frontal cortex. As children get older, there is more *frontal* cortex activity and less *amygdaloid* activity. In language-based activities, which always originate in the *frontal lobe*, the activity in that same area increases with age. In the study children younger than 13 all had difficulty determining what emotion they were looking at, "they are not correctly discriminating facial affect, which could explain why their response to individuals may seem inappropriate." (WGBH, 2002). Recognition of faces and recognition of emotions are both right hemisphere activities. This study suggests a natural delay in development of the right brain to emotional stimuli, which is processed in the emotional *amygdala*, instead of the cognitive *prefrontal* cortex.

If a child has a below normal activation of the right brain, we would expect this process of development to be slow and the *amygdala* to process more information than normal for longer periods of time. This would explain the impulsive behavior of children with *ADHD*; this may also explain why males are more prone than females, because the male right *frontal lobe* grows larger than females. Therefore, the male brain may be more susceptible to delays in right *frontal* cortex development. This may also explain the lack of social development of children with *ADHD*. It has been noted that we have become increasingly aware of the relationship between learning disabilities and social skills development. It is thought likely that children with learning and social deficits make up a group of children with *nonverbal learning disabilities* (Semrud-Clikeman and Hynd, 1990). This disorder was first described as the inability to process environmental cues including those used in communication (e.g., gesture, facial expression, and vocal intonation). These children often have difficulties in processing other (non-communicative) information such as visual-spatial and certain types of arithmetic problems. Hynd and Semrud-Clikeman (1989a, 1989b) have suggested *thalamic* dysfunction with regard to attention resources, which is also consistent with others who suggest the role of the *thalamic nuclei* in *visual-spatial* input and emotional expression (Kelly, 1985). It has been noted that this subtype of *nonverbal learning-disabled* children warrants further study not only for research value but also for advancing more affective differential interaction strategies (Boliek and Obrzut, 1995).

CORTICAL–SUBCORTICAL CIRCUITS AND AFFECTIVE DISORDERS

In considering how *cerebellar*, *thalamic*, or *basal ganglionic* dysfunction may lead to depression or *mania*, it has been proposed that it is the result of the disruption of *frontal* cortical function. The *caudate* is thought to share common pathways with *frontal* cortical structures either directly or through the *dorsomedial nucleus* of the *thalamus*. Alexander and colleagues (1986) described five functionally segregated parallel frontal *cortical-striatal-thalamic* loops, both motor and non-motor with variable degrees of *limbic* input. Part of the non-motor loop, a dorsolateral *prefrontal* network would be used for storage in *working memory* of spatial locations (Goldman-Rakic, 1987). A lateral *orbito-frontal* circuit, proceeding from the *orbito-frontal* cortex to a specific area of the *caudate* and *globus pallidus* would then project to the *thalamus* and the *orbito-frontal* cortex again. This loop is thought to be responsible for learning and recognition tasks requiring frequent shifts of set. This interpretation is thought to explain the pronounced perseveration seen in *orbito-frontal* damage. In addition, there is an anterior *cingulate* circuit, which includes the ventral *striatum*, *nucleus accumbens*, and medio-dorsal nucleus of the *thalamus*. The *hippocampus* and *entorhinal cortex* also send information in this loop integrating information from the *para-limbic association cortex*. It is thought that the *nucleus accumbens* is a target of *dopaminergic* terminals from the *mesencephalic ventro-tegmental* area (VTA).

Mayberry (1992) conducted PET studies of *basal ganglia* stroke patients. They found that compared to *euthymic* patients, all of the patients with mood change showed hypometabolism in the orbital-inferior *frontal* cortex, anterior *temporal* cortex, and *cingulate* cortex. Of course the *cerebellum* could be considered a primary source of many of these findings as well. Recent reports on human emotional behavior show that disorders in control of affect can follow atrophy of the midline *cerebellum* (Gutzmann and Kuhl, 1987). Lack of behavioral self-control, including episodic rage, is sometimes seen as an early symptom of tumor of the *paleo-cerebellum* (Elliot, 1982). It has been reported that after the surgical removal of a benign tumor, several weeks of explosive behavior follow. The *cerebellum* is connected to the neocortex and the *limbic system*. One route connects the *cerebellum* to the *thalamus* and the cerebral neo-cortex. The other route connects the *cerebellum* to the *hypothalamus* and therefore to older structures of the brain (Haines et al., 1984, 1990; Haines and Dietricks, 1987; Dietricks and Haines, 1989). It is thought that the midline *paleo-cerebellum* is connected primarily to the *limbic system* and *limbic* neo-cortex, whereas the lateral *cerebellum* is more connected to the neo-cortex especially the *prefrontal* cortex. It has been noted that a large percentage of children with learning disabilities and affective disorders like *ADHD* present with motor deficits that are consistent with *cerebellar* and or *basal ganglia* dysfunction.

FEAR, ANGER, AND VIOLENT BEHAVIOR

In extreme cases, some children with learning disabilities or emotional disorders may exhibit extreme emotions of fear, anger, and violent behavior. Violent behavior has seemed to increase significantly, especially among adolescents and young teens. These behaviors may reflect extreme cases of approach and avoidance, and failure of the *frontal* cortex to develop to the degree necessary to inhibit primitive emotions of the *limbic system*. Research has shown that interestingly, individuals with history of violent behavior show similar decreased activity in the same area of the brain as children with depression, or *manic* behaviors like *ADHD, OCD*, and TS. Many children and teens who are committing violent crimes have been previously diagnosed with affective disorders (Loeber et al., 2002; Phillips, 2002).

Fear and Anxiety

The *amygdala* appears to be the main area in control of fear and anger as well as other emotions. This primitive fight or flight instinct is the extreme end of approach or avoidance behavior. In childhood anxiety disorders, *posttraumatic stress disorder* (PTSD), and *phobias*, as well as hyperactivity, OCD, and depression may all suffer from a failure to regulate the *amygdala* response (Benarroch, 1992; Castellanos et al., 1996; Filipek et al., 1997; Heim and Nemeroff, 1999; Durston et al., 2001; Kulisevsky et al., 2002). A recent Harvard study has found that in normal individuals, the *amygdala* is not only involved in fear responses, but also with rapidly assessing the emotional importance of a fearful stimulus. For example, if we see a bear in the woods, the *amygdala* will trigger memories as well as physiological responses. If the bear is in the zoo, the neo-cortex will consciously process information and inhibit the *amygdala* overriding the responses. The *amygdala* appears to be able to work on an implicit level to process potentially dangerous stimuli. Paul Whalen, Scott Rauch, and colleagues at Massachusetts General Hospital recently showed that fear responses can be activated implicitly. The researchers used an approach suggestive of subliminal advertising while using functional MRI to image the brain. Subjects were shown photographs of fearful faces for 33 ms followed by a longer, masking exposure to expressionless faces for 167 ms. The subjects had no conscious memory of seeing the fearful faces; however, imaging showed that the *amygdala* lit up even during the brief flash of a fearful face, but not afterward and not during the similarly brief flash of a happy face. "So it's a very fast and preferential way to get information," says Whalen, "anxiety is about hyper-vigilance and there is a vigilance system" (Whalen et al., 1998a, 1998b; Bush et al., 1999; *NY Times Magazine*, February 28, 1999; Rauch et al., 2000; Shin et al., 2001).

There is also growing evidence that the *amygdala* is important for social situations. Patients with Klüver–Bucy syndrome-type behavior, suffer *temporal lobe* lesions that often include the *amygdala*. These patients often act inappropriately in social situations. They are not emotionally expressive and tend to be hypersexual. They may also ingest items such as tea bags and cigarette butts, not being able to recognize the difference between edible and nonedible items. In addition, primates with lesions to this area seem to be shunned by the rest of the group.

Nancy Etcoff (Aharon et al., 2001) has indicated that if the *amygdala* is disrupted, individuals may find it difficult to make judgments about the environment. Other researchers have focused on the brains reward system. Hans Breiter and his colleagues (Breiter et al., 1997; Breiter and Rosen, 1999) used cocaine as a reinforcer in drug dependent patients. They found what they think to be the human reward circuit. Breiter's team found that access of the *nucleus accumbens*, the area of the brain thought to regulate euphoria, and the *amygdala*, rich in *dopamine*, were both activated. *Dopamine* appears to be involved in regulating pleasure and reward. It was found that the *nucleus accumbens* is activated not during the actual dose of cocaine, but during the state of craving or wanting the drug that follows. It is thought that these circuits may be involved in *schizophrenia* which is characterized by lack of motivation, and interest in rewards or pleasure or lack of "will." Because of these circuits, it is thought that we can experience, learn, and unconsciously commit to emotional memory many fearful situations without ever being aware of what has triggered the racing heart and quick pulse. In small children who are exposed to abuse or violence, the *limbic* connections may become extremely strong and reinforced before the *prefrontal* cortex has developed mechanisms to inhibit these responses. These connections may become so powerful that they may subconsciously promote reactions in a child or later in adulthood. Panic attacks, for example, may be a result of this type of early experience.

The implication of fear conditioning experiments in organisms is that we have a separate memory of fearful stimulus, harbored in the *amygdala*, probably informed by things we have heard or seen but do not consciously remember. In a PET study, Roy Dolan (1999, 2002; Dolan and Fletcher, 1999; Dolan et al., 2000) of the Wellcome Department of Cognitive Neurology at the Institute of Neurology in London has shown that in humans, a fear conditioning route bypasses the cognitive part of the brain. Researchers think that anxiety disorders appear to be a uniquely human phenomenon, because of the involvement of the *prefrontal* cortex. It has been stated by one researcher (*New York Times*, February 1999) "bang on the *amygdala* and you're going to get panic attacks. Bang on the *hippocampus* and you are going to get *posttraumatic stress syndrome*. You mess up the medial *prefrontal* cortex and you are going to get much worrying, they are all within the same system."

Bergmann (1995) reviewed a PET study examining the neuroanatomy of *PTSD* symptoms (Rauch et al., 1996). In this study, subjects were exposed to recordings of scripts describing the subject's past personal experiences, including the traumatic experiences thought to have caused the *PTSD*. The script-driven scans showed findings of increased activity in the right-sided *limbic, para-limbic*, and *visual* cortex. No significant changes were found in the *hippocampus* or *thalamus*. Decreases in activity were found in the left *frontal* and middle *temporal* cortex. Bergmann thinks that these results suggest that emotions associated with *PTSD* are mediated by the *limbic* and *para-limbic* areas of the right hemisphere. The right-sided brain activity is apparently consistent with literature supporting the preferential role of the right hemisphere in anxiety, *panic*, and *phobic* disorders. Similar areas of the brain may be involved in aggressive, socially inappropriate behavior.

One of the most disturbing trends in the United States is the increase of youth violence over the past few years. Statistics show (Gore, 2000) that in 1995 more than 862,000 violent crimes in schools were reported by students in nations public and private schools, according to a survey by the *U.S. Department of Justice* and the *U.S. Department of Education* referenced by Gore. Juvenile arrests for murder and non-negligent manslaughter increased 90 percent from 1986 to 1995, according to statistics compiled by the *U.S Justice Department's Office of Juvenile Justice and Delinquency Prevention* (2000). Arrests for aggravated assault increased 78 percent. Access to firearms is thought to have exacerbated the situation according to the Justice Department. From 1984 to 1994, for example, juvenile homicides involving firearms nearly tripled, while homicides by other methods remained the same. It was noted that guns were the weapons of choice in every school attack in the previous 18 months. Violent crimes committed by girls including murder increased 34 percent between 1991 and 1995. Aggravated assaults by girls increased 39 percent, 6 percent for boys. It has been noted that warning signals are depression and a lack of empathy or feelings for others. Youth violence has increased significantly in the past decade (Eron et al., 1994). The homicide rate for young men in the United States is seven times that of Canada and eighteen times that of the United Kingdom (Tedschi and Felson, 1994).

According to Berkowitz (1993, 1994), feelings of frustration and other negative affects give rise to aggression. We know that negative emotions are generally lateralized to the right hemisphere. Frustration does not immediately lead to aggression, but leads to hostile feelings. Frustration is common among children with learning disabilities, especially since many have above average intelligence, know they should be able to learn in a formal educational setting, but cannot. Children with behavioral problems experience the same types of frustrations; their inability to perceive the emotions of others may lead to inappropriate behavior.

Hostility and anxiety are part of the fight or flight approach, or avoidance responses triggered by the *amygdala* in response to a perceived threat. Research indicates that hostile feelings can be modified by environmental means. Adults who have grown up in caring homes with a great deal of nurturance apparently do not show the same level of aggression as those who experience little nurturance. In contrast, children who were abused tend to be more aggressive (Widom, 1989a, 1989b; Dodge, 1993).

It is no coincidence that in the United States as we see a decrease in cognitive skills as witnessed by the decline in standardized test scores, we have also seen an increase in behavioral and emotional problems among our children. The use of Ritalin has increased 700 percent since 1990 (Connor, 2002), and 90 percent of the worldwide use is in the United States. Researchers at Case Western Reserve University recently (Song et al., 1998) set out to find out if an adolescent psyche is influenced by violent behavior in his environment. After studying over 3,700 children ages 14–19, they report that the link is both "sticking and obvious," teens exposed to physical or violent trauma at home are much more likely to be chronically angry and act violently themselves. Violence is rising dramatically among teenagers: 17 percent of teenage boys in this study report having tried to shoot someone during the past year.

Is there a neurophysiological explanation for this behavior? A report in *New York Newsday* (April 14, 1998) may give some answers. In the first study of its kind, neuroscientists used the latest imaging technology to look inside the brains of killers to determine if their brains differ in some way. Adrian Raine (Raine et al., 1997, 1998a, 1998b), a clinical neuroscientist at the University of

Southern California, Los Angeles who led the study, identified 38 murderers, and determined whether they suffered physical or sexual abuse, severe family conflict or parental divorce. Of the murderers, 12 had suffered significant abuse, while the remaining 26 experienced minimal abuse or none at all. The researchers used PET scans to compare those who had suffered trauma as a child with those who had not, and with a group that had not committed violence. Compared with the subjects who had suffered abuse and with nonviolent individuals, the 26 murderers from relatively benign backgrounds averaged 5.7 percent less activity in the medial *prefrontal* cortex. More significantly, they showed an average of 14.2 percent less activity in a specific area of the medial *prefrontal* cortex and the *orbito-frontal* cortex on the right hemisphere.

The medial *prefrontal* cortex has been shown to inhibit the *limbic system*. Raine was quoted as commenting (*New York Newsday*, April 14, 1998), "the *prefrontal* cortex is a bit like a emergency brake on the deeper areas of the brain that are involved in aggressive feelings." Research has shown that the right *orbito-frontal* cortex is involved in fear conditioning, the subconscious association between antisocial behavior and punishment that is thought to be the key to developing a "conscience." The deficit shown in the study may leave an individual with "an emotionally blunted personality lacking in conscience development." Raine and colleagues (1998a) wrote in reporting their findings, "the fact that there is an identifiable biologic disposition suggests its not how the child was raised, it's that they had a biological dysfunction combined with a situation that led to the violence." This may also explain how children with good family backgrounds and good education may still suffer from a learning disability or affective disorder.

The *orbito-frontal* cortex especially on the right is also described as the neo-cortical representation of the *limbic system* (Nauta, 1971). Damage or dysfunction has been noted to lead to impairments in visual object learning in monkeys (Thorpe et al., 1983), *perseverative* behavior (Ledoux, 1987; Dayderry and Tucker, 1992), *mania*, and inappropriate social behavior (Starkstein et al., 1998). These types of symptoms are characteristic of impulsive disorders, which include *borderline personality disorder* and *antisocial personality disorder*, and have been grouped into a larger group of "*obsessive-compulsive spectrum disorders*." Some of these include *dissociative* disorders, tic disorders, *personality disorders*, "*schizo-obsessive spectrum disorders*," and a wide range of neurological disorders including epilepsy, *autism*, and several *basal ganglia* disorders like TS.

OVERLAP OF SYMPTOMS

Asperger's syndrome, a condition that is generally considered a form of *autism*, has shown a dramatic increase over the past few years. *Asperger's* is often confused by physicians with *ADHD, nonverbal learning disorder*, and OCD. Hans Asperger, the Viennese pediatrician who first described the condition in 1944 called his patients "little professors" who use words as their lifeline to the world. Most *Asperger's* patients have average intelligence or above, 80 percent of *autistic* people by contrast suffer some degree of mental retardation. The most striking characteristic of the syndrome is strong interest in arcane subjects. It has been noted that some patients have shown obsessions with clocks, the Titanic, deep fat fryers, lists on Congressional members, etc. The key point to the diagnosis is that their obsessive behavior significantly impairs their social functioning. It is thought that children with the condition are not able to grasp the nonverbal cues that underlie most interactions with others but intelligent enough to understand and regret their deficits. Even though it was first described by Asperger in 1944, it was not until 1994 that the syndrome was included in the *Diagnostic and Statistical Manual of Mental Disorders* of the American Psychiatric Association (*DSM-IV*), the latest version of the mental health professionals guidebook. More recently, researchers have talked of an *autistic spectrum* with *Asperger's* being described as the spans "smart end." *Asperger's* has been described as *autism's* mirror image.

In cases of *autism*, the difficulties are primarily handled by the left hemisphere. However, in *Asperger's*, it has been noted that the deficits are primarily in the right hemisphere, because difficulty is with nonverbal skills, leading some to speculate the syndrome may be some day known as *right-brained autism*. Autism is usually diagnosed around age 3, but children with *Asperger's* are not usually identified until they start school and their lack of social skills stand out. These children have been noted to have problems with motor skills and auditory processing. Their strengths are in intelligence and verbal ability. Other symptoms include: a marked lack of interest in other children, or a consistently inappropriate style of engaging others, difficulty understanding other children's feelings and expressions, inability to understand teasing or jokes, few facial or bodily gestures, speech that is pedantic in tone or vocabulary, overreactions to minor changes in routine or environment, and precocious verbal skills, and marked self-absorption in subjects unusual for the child's age.

It should now be apparent that there is a tremendous overlap of symptoms and similarity of the areas that produce these symptoms. It becomes understandable why these disorders are now seen as being on a continuum. In fact, we see that most children that are diagnosed with one problem often have a second or third diagnosis, probably because the same areas are involved, the symptoms produced are almost identical, and the neurotransmitters involved are the same. This leads us to think they may all have a similar underlying cause, but slight genetic variations may predispose one child to more of one type of symptom vs. another. Genetics combined with developmental plasticity and environmental influences may be the only major difference between disorders like *ADHD, OCD, TS, autism*, and depression. The other factor, and perhaps the most important one, is that most of these disorders can be simply considered as right-brain or left-brain disorders, and can be isolated to a deficit that is either ventral or dorsal. To illustrate this more fully, let us examine some of the major syndromes and compare the symptoms and how they overlap.

A recent paper that examines the neurobiology of OCD (Stein, 1996) describes the condition as being characterized by recurrent intrusive thoughts and repetitive ritualistic actions. The author notes that although symptoms have long been considered symbolic expressions of unconscious issues, recent research has supported a neurophysiological cause. Obsessions are defined as recurrent and persistent thoughts, impulses, or images, which the individual regards as intrusive and inappropriate (*American Psychiatric Association*, 1994). Compulsions are repetitive behaviors or mental acts (also known as *perseverations*), that the individual feels compelled to perform in response to an obsession or according to rules that must be followed rigidly (*American Psychiatric Association*, 1994). The individual with *OCD* is usually aware that symptoms are excessive but notes that they are unable to resist these repetitive thoughts and actions. *Perseverations* are characteristic of right *prefrontal* cortex dysfunction and can be considered hyperactive behavior or *hyperkinetic* behavior. This is similar to hyperactivity, repetitive movements, or vocal tics. These disorders appear to respond to treatment with *serotonergic* uptake inhibitors but appear not to respond to the *noradrenergic tricyclic* antidepressants (Zohar and Insel, 1987). This differentiates *OCD* from depression, which is thought to be due to a left *prefrontal* cortex deficit. It has been suggested that the *dopamine* system may also mediate *OCD* symptoms (Stein, 1996). It has also been noted that tics are common in *OCD* (Pitman et al., 1987) and that *OCD* symptoms are frequent in patients with TS syndrome (Hollander et al., 1989). Recent evidence suggests an important role for the *basal ganglia* in *OCD*. Patients with TS (Hollander et al., 1989), *Sydenham's chorea* (Swedo et al., 1989), and *Huntington's disease* (Cummings and Cunningham, 1992) may have comorbid *OCD* (Breiter et al., 1996). In addition, patients with *OCD* may

TABLE 9.2 Summary Comparison of Physiological Abnormalities in Autism, Attention Deficit Disorder, and other Developmental Disorders

Biochemistry	Low sulfate levels (Alberti et al., 1999) Low levels of glutathione; decreased ability of liver to detoxify xenobiotics; abnormal glutathione peroxidase activity in erythrocytes (Golse et al., 1978; Edelson and Cantor, 1998a, 1998b) Purine and pyrimidine metabolism errors lead to autistic features (Gillberg, 1992; Page et al., 1997; Page and Moseley, 2002) Mitochondrial dysfunction, especially in brain (Lombard, 1998; Chugani et al., 1999)
Immune System	More likely to have allergies and asthma; familial presence of autoimmune diseases, especially rheumatoid arthritis; IgA deficiencies (Warren et al., 1986; Plioplys et al., 1989; Gupta et al., 1996; Comi et al., 1999) Ongoing immune response in CNS; brain/MBP autoantibodies present (Singh et al., 1993; Connolly et al., 1999) Skewed immune-cell subset in the Th2 direction; decreased responses to T-cell mitogens; reduced NK T-cell function; increased IFNg & IL-12 (Warren et al., 1986, 1987; Plioplys, 1989; Gupta et al., 1996; Singh, 1996; Gupta et al., 1998; Messahel et al., 1998)
CNS Structure	Specific areas of brain pathology; many functions spared (Dawson, 1996) Pathology in *amygdala, hippocampus, basal ganglia*, cerebral cortex; damage to *Purkinje* and *granule* cells in cerebellum; *brainstem* defects in some cases (Ritvo et al., 1986; Piven et al., 1990; Hoon and Riess, 1992; Courchesne et al., 1994a, 1994b, 1994c; Hashimoto et al., 1995; Dawson, 1996; Abell et al., 1999; Otsuka et al., 1999; Sears et al., 1999) Neuronal disorganization; increased neuronal cell replication, increased glial cells; depressed expression of NCAMs (Plioplys et al., 1990; Bailey et al., 1996; Minshew, 1996) Progressive microcephaly and macrocephaly (Fombonne et al., 1999)
Neuro-chemistry	Decreased *serotonin* synthesis in children; abnormal calcium metabolism (Plioplys, 1989; Chugani et al., 1999; Leboyer et al., 1999) Either high or low dopamine levels; positive response to peroxidine, which lowers dopamine levels (Gillberg and Svennerholm, 1987; Gillberg, 1992; Ernst et al., 1997) Elevated norepinephrine and epinephrine (Gillberg, 1992) Elevated glutamate and aspartate (Moreno et al., 1992; Carlsson, 1998) Cortical *acetylcholine* deficiency; reduced muscarinic receptor binding in *hippocampus* (Perry et al., 1999) Demyelination in brain (Singh et al., 1993)
Neurophysiology	Abnormal EEGs, epileptiform activity, variable patterns including subtle, low amplitude seizure activities (Gillberg, 1992; Bailey et al., 1996; Nass et al., 1998; Lewine, 1999) Abnormal vestibular *nystagmus* responses; loss of sense of position in space (Ornitz et al., 1991; Goldberg et al., 2000) Autonomic disturbance: unusual sweating, poor circulation, elevated heart rate (Ornitz et al., 1991)

Adapted from: Bernard S., Enayati, A., Redwood, L., Roger, H., Binstock, T., and Bernard, S. ARC Research, Cranford NJ, USA www.biometricdiagnostics.com/content/amp/content.php3.

have comorbid tics (Pitman et al., 1987) or increased neurological soft signs (Hollander et al., 1990) suggestive of *basal ganglia* pathology. Finally it has been reported that both structural and functional brain imaging studies have shown *basal ganglia* pathology especially decreased *caudate* volume in *OCD* patients (Luxenberg et al., 1988; Robinson and Munne, 1995). Patients with *frontal lobe* lesions sometimes present with *OCD* symptoms and *OCD* patients may show evidence of *frontal lobe* impairment in electrophysiological studies (Khanna, 1988). Also reported has been an MRI study that found abnormalities in right *frontal lobe* (Garber et al., 1989). These results together suggest that

TABLE 9.3 Summary of Behavioral Traits

Psychiatric Disturbances
Social deficits, shyness, social withdrawal (Gillberg, 1992; American Psychiatric Association, 1994; Capps et al., 1998; Tonge et al., 1999)
Repetitive, perseverative, stereotypic behaviors; obsessive-compulsive tendencies (Cesaroni & Garber, 1991; Gillberg, 1992; American Psychiatric Association, 1994; Roux et al., 1997; Howlin, 2000)
Depression/depressive traits, mood swings, flat affect; impaired face recognition (Plioplys, 1989; DeLong, 1999; Klin et al., 1999; Piven and Palmer, 1999)
Anxiety; schizoid tendencies; irrational fears (Gillberg, 1992; Muris et al., 1998)
Irritability, aggression, temper tantrums (McDougle et al., 1995; Jaselskis et al., 1996; Tsai, 1996)
Lacks eye contact; impaired visual fixation. (HgP)/problems in joint attention (ASD) (Baron-Cohen et al., 1996; Dawson, 1996; Filipek et al., 1999)

Speech and Language Deficits
Loss of speech, delayed language, failure to develop speech (Gillberg, 1992; American Psychiatric Association, 1994; Prizant, 1996; Filipek et al., 1999)
Dysarthria; articulation problems (Filipek et al., 1999)
Speech comprehension deficits (Bailey et al., 1996; Filipek et al., 1999)
Verbalizing and word retrieval problems (HgP); echolalia, word use and pragmatic errors (ASD) (American Psychiatric Association, 1994; Dawson, 1996; Filipek et al., 1999)

Sensory Abnormalities
Abnormal sensation in mouth and extremities (Gillberg, 1992; Baranek, 1999)
Sound sensitivity; mild to profound hearing loss (Gillberg, 1992; Roux et al., 1997; Rosenhall et al., 1999)
Abnormal touch sensations; touch aversion (Gillberg and Coleman, 1992; Baranek, 1999)
Over-sensitivity to light; blurred vision (Gillberg, 1992; O'Neill & Jones, 1997)

Motor Disorders
Flapping, myoclonal jerks, choreiform movements, circling, rocking, toe walking, unusual postures (Cesaroni and Garber, 1991; Gillberg, 1992; Tsai, 1996; Filipek et al., 1999)
Deficits in eye–hand coordination; limb apraxia; intention tremors (HgP)/problems with intentional movement or imitation (ASD) (Gillberg, 1992; Dawson, 1996; Filipek et al., 1999)
Abnormal gait and posture, clumsiness and incoordination; difficulties sitting, lying, crawling, and walking; problem on one side of body (Bailey et al., 1996; Teitelbaum et al., 1998; Gillberg, 1999)

Cognitive Impairments
Borderline intelligence, mental retardation—some cases reversible (Gillberg, 1992; Filipek et al., 1999; Edelson et al., 1998a, 1998b)
Poor concentration, attention, response inhibition (HgP)/shifting attention (ASD) (Rumsey, 1985; Bailey et al., 1996; Dawson, 1996)
Uneven performance on IQ subtests; verbal IQ higher than performance IQ (Dawson, 1996)
Poor short-term, verbal, and auditory memory (Dawson, 1996) Poor visual and perceptual motor skills; impairment in simple reaction time (HgP)/ lower performance on timed tests (ASD) (Grandin, 1995; Schuler, 1995; Bailey et al., 1996)
Deficits in understanding abstract ideas and symbolism; degeneration of higher mental powers (HgP)/sequencing, planning, and organizing (ASD); difficulty carrying out complex commands (Rumsey, 1985; Bailey et al., 1996; Dawson, 1996; Filipek et al., 1999)

Unusual Behaviors
Self-injurious behavior, e.g., head banging (Gedye, 1992; Filipek et al., 1999)
ADHD traits (Gillberg, 1992; Dawson, 1996; Kim et al., 2000)
Agitation, unprovoked crying, grimacing, staring spells (Gedye, 1992; Filipek et al., 1999)
Sleep difficulties (Gillberg, 1992; Wiggs and Stores, 1998; Richdale, 1999)

Physical Disturbances
Hyper- or hypotonia; abnormal reflexes; decreased muscle strength, especially upper body; incontinence; problems chewing, swallowing (Church and Coplan, 1995; Schuler, 1995; Teitelbaum et al., 1998; Filipek et al., 1999)
Diarrhea; abdominal pain/discomfort, constipation, "colitis" (Deufemia et al., 1996; Wakefield et al., 1998; Horvath and Perman, 2002)
Anorexia; nausea (HgP)/vomiting (ASD); poor appetite (HgP)/restricted diet (ASD) (Gilberg, 1992; Kanner, 1943)

Adapted from: Bernard S., Enayati, A., Redwood, L., Roger, H., Binstock, T., and Bernard, S. ARC Research, Cranford NJ, USA www.biometricdiagnostics.com/content/amp/content.php3.

cortical–*basal ganglia*–*thalamic*–cortical circuits may play a significant role in *OCD* symptoms (Stein, 1996).

In another review paper, the authors (Spencer et al., 1998) looked at the apparent overlap between TS and *ADHD*. They note that TS is a chronic neuropsychiatric condition commonly associated with social, occupational, and academic dysfunctions (Erenberg et al., 1986; Stokes et al., 1991). They also note that children with TS have often been found to have high levels of comorbidity with *OCD* and ADHD. They conclude that their findings confirm previously noted associations between TS and *OCD*, but also suggest that disruptive behavioral, mood, and anxiety disorders as well as cognitive dysfunctions may be accounted for by comorbidity with *ADHD*. However, they state that TS with *ADHD* is apparently more severe than *ADHD* alone. Another study by Faraone and colleagues (2001) notes that there is comorbidity between *ADHD* and *conduct disorder* (*CD*) that has consistently been reported in both clinical and epidemiological studies (Bierman et al., 1991). These studies are just an example of the significant overlap of many childhood learning disabilities and affective disorders that have been noted in the scientific literature.

Depressive disorders, manic disorders, generalized anxiety disorders, and psychotic disorders are all more common following damage or dysfunction of one hemisphere than the other. Left *frontal* and left *basal ganglionic* lesions lead to an increased incidence of major depression, whereas right *orbito-frontal*, basal *temporal, basal ganglia*, and *thalamic* lesions are associated with manic symptoms. Anxiety disorders without depression and psychotic disorders have also been shown to be associated with right hemisphere lesions, especially posterior temporal and parietal lesions. In addition, there are several studies in both humans and lower organisms that have shown that right and left hemisphere lesions result in different affects on the *biogenic amines*, especially *serotonin* and *norepinephrine*. It is thought that these biochemical changes induced by specific brain lesions may mediate through asymmetric brain pathways the clinical manifestation of these disorders (Robinson and Downhill, 1995).

Bernard and colleagues provide effective summaries of the behavioral traits and physiological characteristics of those with *autistic spectrum disorder* culled from the literature of the past thirty years and presented in Tables 9.2 and 9.3.

10

Causation from an Evolutionary Perspective

In the beginning of this book we have focused on answering the question: what normally stimulates the brain, especially the *frontal lobe*? We have traced the path of environmental stimulation of the brain through specific and nonspecific pathways. We have noted that all perception, information processing, as well as motor and autonomic activity is based on the specific pathways for arousal and nonspecific pathways for information. It has also been shown how a decrease in arousal or activation has been documented to produce cognitive, emotional, perceptual, sensory, motor, and autonomic symptoms. We have demonstrated that baseline arousal and synchronization appear to arise primarily from *thalamic* nuclei that are superimposed on possible intrinsic cellular activation and oscillations. The *thalamus* receives its activation primarily from the *cerebellum*, which is activated primarily from slowly adapting somatosensory receptors that send input to the *cerebellum* and *thalamus* through *dorsal column* pathways. The slowly adapting somatosensory receptors are found mostly in antigravity postural, slow twitch muscle fibers. The greatest concentrations of these receptors are found in the small postural muscles of the spine, especially in the upper cervical spine. It appears that these receptors primarily transduce gravitational forces, which is the most frequent and only constant environmental stimulus available. Gravity, which provides the force that postural muscles are called upon to continually resist, provide the power supply to the brain. Stimulation and "nutrition" to the brain is generated by the movements of the spine as a windmill generates electricity.

We have also seen symptoms of depression related to a type of power shortage of the left *prefrontal* cortex. This so-called power shortage can be traced through neuronal pathways that include the *prefrontal* cortex, the *thalamus, basal ganglia*, and *cerebellum*. This type of decreased stimulation can be seen as a primary cause of many forms of brain dysfunction, the type of symptoms depending on the primary area or location of the cerebral cortex affected on a right/left, anterior/posterior, or dorsal/ventral location. This decrease in stimulation to these various pathways and how they actually produce symptoms have been described. Most of the childhood, even adult, syndromes appear to have a significant degree of overlap and all are considered related to similar neurophysiological mechanisms and deficits. This would lead us to think that they all may have a similar underlying cause that is superimposed on a child's unique combination of genetic makeup and environmental exposure.

We know that genetics alone cannot explain the dramatic increase in symptoms that we have witnessed over the past decade or so. The increase is too fast and is too specific for us to explain a 700 percent increase in the need for the use of *Ritalin* in just the last 10 years, 90 percent of which is used in the United States alone (Connor, 2002). The ultimate cause for these disorders must be viewed in the context of societal change that has occurred primarily in the United States over the past 10–15 years that has negatively impacted our children. This environmental or lifestyle condition appears to be interacting with genetic predisposition to cause an epidemic of learning disabilities, behavioral disorders, and violence. Even though these problems are not new, being first identified over 100 years ago, the rapid increase in real symptoms is relatively new. It is therefore imperative to isolate the primary cause or causes in our society that are producing the neurophysiological effects that we have described in previous chapters. We must identify the cause so that we can design aggressive, safe intervention techniques, and most importantly, to reduce the frequency of this spectrum of problems by effective education of parents, teachers, and clinicians.

Although the causes that we have identified may seem simple and obvious, the understanding of how they combine to produce specific neurophysiological deficits we think is unique. It should not surprise anyone that the underlying cause of these apparently complex syndromes is related to lifestyle factors in children. For some reason we think that children do not respond the same way as adults do, but this is not correct. It is commonly known that the major cause of illness and death in adults are lifestyle disorders such as heart disease, cancer, arthritis, and depression. Why then would we not expect that these same factors would not also have a negative impact on our children's physical and mental health and development? In fact, the rise in learning disabilities and behavioral problems in children has increased at roughly the same rate as depression in the adult population and we can see that they both are due to similar mechanisms. As we have witnessed a rise in the use of *Ritalin* and other medications in children, we have seen a similar rise in the use of antidepressants in adults. In fact, antidepressant use is now becoming increasingly common in children. In a recent article in *USA Today* (May 3, 1999) they point out that children and adolescents are being widely prescribed new antidepressants by family physicians and pediatricians even though as they point out there is "little scientific evidence that they are safe and effective for people under 18." The antidepressants are mostly *serotonin selective reuptake inhibitors* (*SSRI*s) like *Prozac*. These drugs are approved by the FDA only for adults because they have not been properly tested in children and adolescents. In a recent survey of 600 family physicians and pediatricians about their prescribing behavior (Rushton et al., 2000), 72 percent of family physicians and pediatricians acknowledged prescribing an *SSRI* to patients under 18, 67 percent said they prescribed drugs for mild to moderate childhood depression, 40 percent said they prescribed *SSRI*s for *Attention Deficit Hyperactive Disorder* (*ADHD*), a condition for which no randomized chemical trial has been conducted, 8 percent said they had received appropriate training for treating childhood depression, and 16 percent said they felt comfortable treating the disorder in children and adolescents. The survey also found that doctors were prescribing *SSRI*s for aggressive-conduct disorder in children and adolescents.

Ray Woosley, now at the University of Arizona was interviewed about the situation (*USA Today*, May 3, 1999). He stated that nothing is known about the long-term effects of the drugs, on the developing central nervous system of the young. Another pediatrician interviewed thought that the use of drugs for *ADHD* without proper studies should raise concerns. Woosley was quoted, "we presume that children will respond the same as adults but we don't know that. Physicians are having to learn on the fly and every child is a new experiment." One thing that using medications

ing in slowed development, growth, and function; resulting in imbalances of activity level of areas of the brain. This inter- or intrahemispheric imbalance and lack of development are, we think, the sources of the main problems found in all children with functioning neurobehavioral disorders, and these need to be the focus of treatment and prevention.

We have identified five primary causes of development neurobehavioral disorders, which result in a deprivation of stimulation, arousal, and which delay development that we will consider here in detail:

1. voluntary reduction in physical activity (sedentary lifestyle),
2. involuntary lack of physical activity,
3. injury and illness,
4. negative parent modeling and lack of communication, and
5. high calorie, low nutrition diets.

VOLUNTARY REDUCTION IN PHYSICAL ACTIVITY

What is observed frequently by teachers and reported anecdotally is that our children are the most sedentary generation that exists in the world today, or has ever existed. This is reflected by the fact that children in the United States have the highest obesity levels in the world (Troiano and Flegal, 1998). It is no coincidence that as we have witnessed a dramatic increase in both sedentary activity and obesity levels along with a parallel increase in learning disabilities and behavioral problems. There are three main reasons for the decrease in physical activity that has occurred over the past 10–15 years, we think, that include the development and availability of technology, such as television, VCRs, computer games which are used extensively and all promote sedentary activity; parental fears about allowing children to stay far from home unsupervised, and economic factors that have forced both parents to work outside the home, so children are left in day-care or with a care-giver and quite often the television or computer becomes the baby-sitter.

Watching television in itself is probably the single biggest problem, and there are many reasons why television may affect the cognitive development of a child's brain. Based on what we now know about how the brain functions and dysfunctions, sitting in front of a television for hours, promotes atrophy of muscles and primarily the postural antigravity muscles. This decreased muscle use and tone, decreases the amount of continual firing of receptors to the *cerebellum, thalamus, basal ganglia,* and cerebrum, especially the *frontal lobe*. We have previously described this relationship and how lack of this input can result in a multitude of symptoms. What we now want to do is explore some of the research that examines the relationship between television viewing and obesity, obesity and behavior, and obesity and its relationship to muscle development and function.

Recent research shows that Americans are just too fat, with 54 percent of all adults heavier than is healthy. It is thought that if the trend continues, within a few generations virtually every U.S. adult will be overweight. The percentage of overweight Americans has increased by about 30 percent in the past 20 years. More than 25 percent of today's children are overweight or obese according to James Hill (Hill and Peters, 1998; Astrup et al., 2001), an obesity researcher. "The trend will continue ... the predictions are that it is increasing at such a rate that we'll all be overweight at some point," said Hill. Hill blames the environment, he thinks Americans have too much food available, social situations encourage overeating, and technology has made it possible to avoid exercise. "Becoming obese is a normal response to the American environment," said Hill. "If the environment continues to encourage high (food) intake and low activity, then we'll all be overweight." Obesity has been shown to be associated with increased risk of diabetes, cancer, heart disease, and other disorders. Some studies have shown an increase of up to 60 percent in risk of death from all causes for obese people. What is not commonly relayed is that obesity is primarily associated with decreased stimulation and activation of the brain. The reason is that the problem of obesity is primarily related to inactivity.

Inactivity leads secondarily to overeating but it is the inactivity and its effect on the brain that is the primary problem, obesity and other diseases are just a reflection of the underlying problem, a neurophysiological one. The reflections in adults are diseases like hypertension, diabetes, arthritis, depression, and anxiety disorders. In children, the same problems are reflected in developmental problems of the body and nervous system, which manifest as learning disabilities and behavioral disorders. The difference in symptoms may be explained by whether the problem is more of a decrease or delay in the right brain development or left brain development. It has been shown by the *National Center for Health Statistics* that the percentage of overweight children ages 6–17 has almost doubled since 1980. However, awareness of the problem by parents has not increased. Prentice and colleagues (Prentice et al., 1996a, 1996b; Prentice and Paul, 2000) think that while genetics may influence an individual's metabolism, obesity is based on a basic principle: more calories are consumed than expended. They thought they have identified two main factors that have resulted in our current situation: a generalized increased consumption of fatty foods, and a marked decrease of energy expenditures. As an example, in 1960, 85 percent of British children walked or rode a bicycle to school. Now that number is 6 percent (Prentice and Jebb, 1995).

Interestingly it has been shown recently that these same two factors are primarily responsible for a decrease in the production of *dopamine (DA)*. DA decrease has been associated with a number of cognitive and emotional problems in children like *ADHD*. Nora Volkow and her colleagues (1992) at the *Brookhaven National Laboratory*, using brain scans showed a direct relationship between depleted stores of *DA* and a decline on tests that measure motor and cognitive abilities, including finger-tapping tasks and tasks that involve *working memory*. *Working memory* is thought to be critical for reading comprehension and is primarily a product of the *prefrontal* cortex. Volkow found that there were two main factors associated with a decrease in *DA*, lack of physical exercise and high caloric consumption. Research on lower organisms has shown that the more the exercise, the more the *DA* receptors, and the less the caloric consumption. She thinks that it is a lack of stimulation of *DA* that leads to the decline of *DA* receptors. What appears clear is that obesity is not a result of genetic factors but is primarily related to decreased physical activity and this leads to decreased brain activation which ultimately leads to decreased production of neurotransmitters especially *DA* which is most closely related to the motor centers and the left brain.

The left brain is associated with learning disabilities, *dyslexia*, and depression. A recent article published in the *Journal of the American Medical Association* examined the relationship of physical activity and television watching with body weight and the muscle-fat ratio among children (Andersen et al., 1998). The study took place between 1988 and 1994 during which time, 4,063 children aged 8 through 16 years were examined as a part of the *National Health and Nutrition Examination Survey III*. The main outcome measures included episodes of weekly vigorous activity and daily hours of television watched and their relationship to body mass index (BMI) and body fatness. In their study, Andersen and colleagues note that the prevalence of overweight adults from 1976 to 1980, 1988 to 1991, has increased from 25 percent to 33 percent. Others also note that the prevalence of overweight children has increased by a similar magnitude among all sex and age groups of children and adolescents (Troiano et al., 1995). There appears to be a parallel relationship between obesity in adults and children. Obesity in children has been associated with subsequent morbidity and mortality in adulthood (Nieto et al., 1992). These trends have persisted even with the intense preoccupation with weight loss in the United States (Brownell and Wadden, 1992).

It has also been shown that the prevalence of obesity in England has doubled in the past decade even though daily caloric energy intake and fat consumption have actually been reduced during this same time period (Prentice and Jebb, 1995). The authors note that a change in the amount of daily physical activity may account for this discrepancy. They think that increasingly, leisure time activities are more sedentary, with television watching, video games, and personal computing being the most popular activities. In addition, people in industrialized countries are expending less energy in activities of daily living and at work (Prentice and Jebb, 1995; *U.S. Department of Health and Human Services*, 1996).

Several studies suggest that an active lifestyle during childhood and adolescence can play an important role in optimizing growth and development (*U.S. Department of Health and Human Services*, 1996). In a study by Cooper (1994), participating children were asked how many times per week they "played or exercised enough to make them sweat or breathe hard." Results show that overall 80 percent reported participating in play or exercise that made them sweat or breathe hard three or more times per week with the rate higher in boys (85 percent) than in girls (74 percent). Among non-Hispanic White boys, 88 percent reported exercising vigorously 3 or more times per week whereas 78 percent of non-Hispanic Black boys and 80 percent of Mexican-American boys met this criteria, 72.6 percent of Mexican-American girls and 69 percent of non-Hispanic Black girls report performing three or more bouts of vigorous activity each week, whereas only 12.2 percent of non-Hispanic White boys report fewer than three bouts of vigorous activity per week. Overall, 26 percent of American children report watching 4 or more hours of television per day; the rate was lower in girls (23 percent) than in boys (29 percent). Forty-three percent of non-Hispanic Black boys and girls reported watching television for more than 4 hr per day. In contrast, non-Hispanic White boys and girls had the lowest prevalence of watching television more than 4 hr per day (25 percent of boys and 18 percent of girls). Boys and girls report similar patterns of television watching across age groups. The highest prevalence of watching 4 or more hours of television per day occurs in 11–13-year-old children. Both boys and girls in this age group have the lowest prevalence of watching 1 or less hours of television per day.

Boys and girls who watch 4 or more hours of television per day have the highest skin fold thickness and highest *BMI*s; conversely, children who watch less than 1 hr of television per day have the lowest *BMI*s. It has been observed that television watching is more closely related to skin folds and *BMI* than was vigorous activity. The same study notes that there are several worrisome trends among adolescent females and ethnic minority groups. The authors point out that of most concern is that 26 percent of all girls and 31 percent of non-Hispanic Black girls report fewer than two bouts of vigorous activity per week. Their data also confirms that vigorous activity among ethnic minority children is lower than non-Hispanic White children. It has also been found that non-Hispanic White children are the most active, with 77 percent of girls and 88 percent of boys reporting three or more bouts of vigorous activity per week. However, only 69 percent of non-Hispanic Black girls and 73 percent of Mexican-American girls reach this level.

In a survey of parents, 46 percent of U.S. adults think that their neighborhoods are unsafe (*Princeton Survey Research Associates*, 1994). Parents in minority populations are twice as likely as non-Hispanic White parents to report that their neighborhoods are unsafe. It is pointed out that this data may partially explain the lower bouts of vigorous activity and higher prevalence of television watching reported in non-Hispanic Black and Mexican-American children. It is known that physical activity is universally related to body weight, body composition, and waist to hip ratio in adults (Andersen et al., 1998). Sedentary leisure time activities such as television watching, playing video games, and personal computing have contributed to the universal prevalence of overweight in America (Dietz and Gortmaker, 1985; Dietz and Strasburger, 1991; Andersen, 1995). The researchers note that their report shows that television watching is associated with universal skin fold thickness and BMI among U.S. youth (Andersen et al., 1998). In general high rates of television' watching have been reported with 26 percent of U.S. children (and 43 percent of non-Hispanic Black children) watching 4 or more hours per day. Strasburger (1992) has calculated that the average high school graduate will likely spend 15,000–18,000 hr in front of a television but only 12,000 hr in school. Next to sleeping, television watching occupies the greatest amount of leisure time during childhood (Dietz and Strasburger, 1991). It has been found that skin fold thickness increases in both boys and girls as the amount of television watching increases (Andersen et al., 1998). This finding is consistent with an earlier study that found a significant relationship between television watching and obesity in children (Dietz and Gortmaker, 1985). There was also found to be a relationship between television watching, physical activity, and body composition (Abraham et al., 1971; Brown and Cramond, 1974; Jeffery et al., 1982; Williams and Handford, 1986; Serdula et al., 1993; Dipietro et al., 1994; Guo et al., 1994; Epstein et al., 1997; Whitaker et al., 1997).

What is known is that increased obesity is usually associated with decreased muscle mass and muscle tone, especially of postural antigravity muscles. Decreased muscle tone is directly related to decreased activation of the brain, therefore, many of the diseases associated with obesity, may not only be a direct result of obesity, overeating or poor nutrition, especially in young children, but instead may likely be a direct result of decreased brain function resulting from decreased motor activity. The lack of physical activity has largely been examined in the literature from its effects from the neck down, but usually not from the neck up. When the neocortex, which requires the most stimulation, fails the decreased inhibition of the *hypothalamus* and *limbic system*, would cause autonomic problems such as cardiac irregularities and hypertension as well as diabetes, increased stress hormones like *cortisol*, etc. When the *amygdala* is not inhibited because the *prefrontal cortex* is not adequately developed, behavioral problems can develop. The neocortex is responsible for high level information processing like calculating, auditory and visual processing, language production, etc., which may dysfunction and therefore result in learning disabilities. The significance of obesity is that it reflects a lack of physical activity, which results in lack of motor activity and failure to provide enough "power" to the brain, especially the neocortex and the *frontal lobe*.

Two scientists at Toronto University have noted that too much fat during childhood and adolescence impaired memory and concentration by preventing the brain from taking up the glucose needed for healthy performance (Greenwood and Winocur, 1990, 1996, 2001; Winocur and Greenwood, 1999). Although the study was carried out on rats, the two scientists said it had implications for adolescent humans. Winocur and Greenwood have found that teenagers who ate fatty foods might suffer permanent damage to their developing brains. After feeding 1-month-old rats a diet rich in either organism or vegetable fat until the rats were 4 months old, the equivalent of late adolescence in humans, 40 percent of the rats' intake of calories came from fat, but otherwise their diet was nutritionally complete. A group of controlled rats was fed the standard laboratory diet, in which only 10 percent of the calories came from fat. Once they reached 4 months, the rats were taught a task in which they had to learn that they would get a food pellet only if they pressed a lever every second time they were shown it. Some rats quickly learned that pressing every time was pointless, but others were unable to remember if they had pressed the lever the last time it had appeared. The delay could be up to 80 s, a long time for a hungry rat. Rats on either type of high-fat diet performed much worse than the rats fed a lean diet. High-fat diets, and by extension obesity, impair performance on virtually all our measures. When high-fat rats were injected with glucose, their cognitive function improved. The researchers think the fat prevents the brain from taking up glucose, possibly by interfering with the action of insulin, which helps to regulate blood sugar levels. High-fat diets often cause insulin resistance and some people with signs of adult-onset diabetes, often caused by obesity, are known to have memory problems. Our brain needs glucose—essentially energy—in order to function. When glucose metabolism in impeded by saturated fatty acids, it is like clogging the brain and starving it of energy. The concern is that developing neural pathways may be permanently damaged at that age as the developing brain is much more susceptible than the older brain.

Jay Giedd examined the brains of 145 normal children by scanning them with MRI at 2-year intervals (Giedd et al., 1994, 1996, 1999a, 1999b; Giedd, 1999). What Giedd and his colleagues have found has shed light on how the brain grows and when it grows. It was thought at one time that the foundation of the brain's architecture was laid down by the time a child is five or six. Indeed, 95 percent of the structure of the brain has been formed by then. These researchers, however, have discovered changes in the structure of the brain that appear relatively late in child development.

In another study of growth patterns of the developing brain, Paul Thompson of the University of California at Los Angeles, along with Jay Giedd and colleagues from McGill University (Thompson et al., 2001), found waves of growth in the *corpus callosum*. Of particular interest to educators and parents is their finding that the fiber systems influencing language learning and associative thinking grows more rapidly than surrounding regions before and during puberty (a similar period to the growth of the *frontal cortex*), but falls off shortly thereafter. These findings reinforce studies on language acquisition that show that the ability to learn new languages declines after the age of 12.

Studies of the environmental effects of *corpus callosum* and *cerebellum* development are part of a large multicentered research study on twins. Previous studies have shown that the *corpus callosa* of twins are so similar, and Giedd and colleagues (Giedd et al., 1996a, 2001; Berquin et al., 1998) hypothesize that this part of the brain is largely controlled by genes. However, in twins, the *cerebellum* is not very similar, leading Giedd and associates to conclude that the *cerebellum* is not genetically controlled and is thus susceptible to the environment.

Interestingly, the *cerebellum* is a part of the brain that changes well into adolescence. We think that the *cerebellum* helps in physical coordination. But looking at functional imaging studies of the brain, we can now also see activity in the *cerebellum* when the brain is processing mental tasks. Giedd thinks that the *cerebellum* functions as a math co-processor and that while not essential for any activity, it makes any activity better. Anything we can think of as higher thought, mathematics, music, philosophy, decision-making, social skill, draws upon the *cerebellum*, allowing us to navigate the complicated social life of the teen and to get through these things instead of lurching seems to be a function of the *cerebellum*. Inactivity and sedentary behavior as well as obesity in development will likely impact the development of this brain region negatively.

Neuromuscular Adaptations to Inactivity

In a study that examines human neuromuscular adaptations that accompany changes in activity, McComas (1994) notes that changes not only occur in the muscle but also in the central nervous system. In fact, he states, "altered physical activity is likely to involve functional and perhaps structural changes at all levels in the motor pathway." In this review, he examines effects on both strength and endurance. McComas notes that the most significant differences between the results of strength and endurance training are that, structurally, the endurance muscles are generally smaller, while in function they show less strength, but much greater resistance to fatigue. These muscle fibers are known as slow twitch muscle fibers. He notes that resistance to fatigue is achieved through a combination of factors. Two of the most important he thinks are increases in *capillarization* and in the *mitochondrial* content of the muscle fiber.

A study which examined seven young females trained for 24 weeks in cross country running (cf. McComas, 1994) found that the number of capillaries increased significantly around all histochemical fiber types, whereas the increase in *mitochondrial* content was greatest in the Type I (slow twitch) fibers. It is therefore thought that because of increased capillary supply, the exercising fibers would be kept better supplied with oxygen and circulating energy sources (glucose and free fatty acids), however, the waste products of muscle metabolism especially HT K^+ and *lactate* would be removed more rapidly. Also, the more dense *mitochondria* would make sure that the production of *ATP* was better maintained during exercise by aerobic metabolism (McComas, 1994). Some of these changes appear to be even more dramatic in a developing child, especially in the Type I slow twitch fibers. In one study of adults, in which the *adductor pollicis* was stimulated supramaximally for 3 hr each day for 6 weeks, the twitch half-relaxation time did not alter (Rutherford and Jones, 1988). However, in infants aged 6–12 months at a time when the incidence of Type I fibers is known to increase (Collins-Saltin, 1978) the twitches in the ankle *plantar flexors* become significantly larger, whereas those in the *dorsi-flexors* remain the same. It has been suggested that these changes in the *plantar flexor* muscles are due to weight bearing, as the child begins to stand. If this is so, then the dynamic nature of the changes would indicate that the infantile muscle fibers are particularly plastic and therefore highly susceptible to exposed forces and or stretches.

Postural muscles are also called *tonic* muscles because they fire continually, whereas non-postural muscles are known as *phasic*. There is a significant difference in the function and structure of *tonic* vs. *phasic* muscles in their fiber type, receptor morphology, and density. *Tonic* muscles tend to consist primarily of slow twitch fibers; also, there is evidence that *tonic* muscles are equipped with longer spindles than *phasic* muscles (Cooper, 1960; Swett and Eldren, 1960). Fitz-Ritson (1982) notes that Barker (1974) quotes Giegor (1904) who demonstrated that muscle spindle density is highest in the hand, foot, and neck muscles. Generally, Fitz-Ritson notes that high spindle density is characteristic of muscle that initiates fine movement (e.g., *lumbricals, extra-ocular* muscles, and small vertebral muscles) or muscles that maintain posture (Fitz-Ritson, 1982). These muscles seem to primarily have a neurological *proprioceptive* role and are largely reflexive in nature. It has been noted that ultra-structural studies are thought to show that in addition to somatic innervation, *intra-fusal* muscle fibers are innervated by *post-ganglionic sympathetic* nerve fibers (Banker and Girvin, 1971; Santini and Ibata, 1971).

In regard to the role of the mammalian muscle spindle, it is thought that its input to the central nervous system, signaling muscle length and rate of change of length, plays a key role in the reflex regulation of muscle movement, at the spinal and supra-spinal level (Fitz-Ritson, 1982). Furthermore, the mammalian spindle fibers are subject to efficient control, primarily by neurons other than those innervating *extra-fusal* muscle fibers. Therefore, the reflex regulation of this spindle efferent system is a critical aspect of the control of motor behavior by the central nervous system (Hunt, 1974). Muscle spindles are responsible for enabling the central nervous system to alter or control the activity of the receptors they contain. This control is achieved by the *gamma fibers* that innervate the *intra-fusal* fibers in the spindle. The *nuclear bag* fibers are made up of a central non-contractile area, which lies between two contractile areas, both of which have an independent source of motor innervation. When the contractile areas on the ends of the fibers are stretched, the central area, which is the area of the primary receptor, fires. It is important to note that both stretch of the muscle or contraction of the muscle produces stretch of the primary receptor, resulting in firing of the receptor. It has been noted that the input from the two primary types of sensory endings, when the muscle is held at one specific length, can be preset by the central nervous system by altering the fusi-motor output frequency (Guyton, 1986). Furthermore, the sensitivity of that muscle spindle is under central control, which is not the case for any other receptor (Fitz-Ritson, 1982).

The ability of a muscle to resist extension is known as the *stretch reflex*. Altering the sensitivity of this *stretch reflex* by excitatory or inhibitory influences from the brain synapsing alpha and/or gamma *motorneurons* in the spinal cord is thought to be the basis of voluntary movement and reflex postural adjustment (Hunt and Ottoson, 1976). It is clear that all of the adaptations of strength and endurance training, as well as those of everyday living, are reversible if the muscles are not used to the same extent (McComas, 1994). It has been shown that these changes are most noticeable in the postural or antigravity muscles since the muscles will have been subjected to great weight bearing previously. An example given is that atrophy of the *quadriceps* can be detected as early as 3 days after immobilization (Lindboe and Platou, 1984) and can amount to a 30 percent reduction in cross-sectional area within 1 month (Halkjaer-Kristensen and Ingemann-Hansen, 1985) of onset of the disease (McComas, 1994).

The absence of weight bearing, as with astronauts in space, may result in weakness and atrophy (Oganov and Potapov, 1976). When a muscle is immobilized as when in a cast, there may be resorption of *sarcomeres* if the muscle has been set at a shortened length (Tabary et al., 1972). Haincut and Duchateau (1989) noted that fatigability in the *adductor pollicis* during intermittent *tetanic* stimulation was most significant after a period of disuse. In addition, it has been shown that the effect on twitch tension is most obvious in antigravity muscles (White and Davies, 1984). An example given is that there was a 50 percent reduction in the twitch tension in a leg that had been restricted from weight bearing and immobilized at the ankle for 3 months. It has been shown that in an organism's hindlimb, this type of restriction would result in a preferential atrophy of *Type I* fibers and even conversion of some *Type I* fibers to *Type II* fibers with prolongation of the twitch (Corley and McComas, 1984).

Therefore, disuse, which is associated with obesity, results in significant and rapid atrophy of muscles. The muscles that atrophy the fastest and the most are postural antigravity muscles or tonic muscles that are primarily made up of slow twitch fibers. These fibers have primarily a neurological control or *proprioceptive* feedback function and have a reciprocal relationship with the brain, which can alter their length and sensitivity to stretch and contraction.

Neurological Adaptations

Experiments with lower organisms have shown that sensory deprivation can rapidly induce alterations (within 3 min) in topographic maps within the *thalamus* (Nicolelis et al., 1993). McComas (1994) summarizes several points. The adaptations seen in muscle ultimately involve changes in translation or transcription of genetic messages that appear to occur because of stretch of the muscle fibers. Structural changes at synapses in the nervous system, on the other hand, must also be controlled through DNA and RNA; there are also rapidly occurring functional adaptations that are dependent only on alterations in the balance of synaptic excitation and inhibition. It is thought that the impulsive activity and *trophic* factors trigger structural adaptations like the number of synapses or the length of the apical dendrites in the motor cortex (Greenough et al., 1985). Although neuromuscular adaptations can occur at any age, they are most prominent in early childhood. These adaptations can result in increases or decreases of neuromuscular structures and function. Rather than regarding adaptations as necessarily following major bouts of exercise, it would be better to envisage them also as smaller events which take place naturally and repeatedly during the course of everyday activities (McComas, 1994). Therefore, although many of the studies and guidelines emphasize the amount of strenuous bouts of vigorous activity, the most significant decrease comes about from the amount of time spent in a sedentary position. In other words, if a child participates in vigorous sports activities several times a week, but spends the bulk of the rest of the time sitting watching television, this would be worse than a child who did not compete in sports, but was generally more active and spent less time sitting in front of the television.

Sedentary Activity and the Stress Response

Many of the physiological effects and diseases that are seen in sedentary or obese children and adults can appear to be stress related. It has been shown that by the time they are teens, many obese children have the beginnings of heart disease or diabetes. Depression, anxiety, stress, and unconscious conflicts have also been linked to obesity. Many obese or overweight individuals have increased circulating stress hormones like *cortisol*, high blood pressure, and poor lung capacity. For many, these affects lead them to the conclusion that stress of work, school, or family life is the main contributor to obesity and all of the diseases, mental and physical that come with it. However, we can view this is a different and probably more accurate way—decreased activation of the neocortex results in decreased inhibition of *limbic system, hypothalamus,* and *sympathetic* nervous system. This produces a number of physiological changes including hypertension, increased *cortisol*, increased fat deposition, decreased blood flow to brain, decreased lung capacity, decreased immune response, etc. Therefore, the body suffers a stress response, but not necessarily in response to stress. This is an important consideration as many disorders are attributed to stress with relaxation as the recommended remedy, when relaxation itself is causal.

Lack of motor activity decreases stimulation to the neocortex which cannot perform its higher cognitive and emotional and information processing duties, as well as not adequately inhibiting the more primitive *limbic system*. This does not require as much ongoing stimulation and takes over as a back-up system to the neocortex when the neocortex suffers a "power shortage." Therefore, a sedentary lifestyle increases the body's stress response; physical and mental activities decrease it. In most cases, children need to be more active and mentally challenged, not less.

In a study published in the journal *Stroke*, Strassburger and associates (1997) found that individuals with hypertension also were shown to show atrophy in the *temporal* and *occipital lobes*, areas especially involved with memory, language, and intellect. The effects were more pronounced in older people, but it did not appear to have been caused by aging. Only individuals with hypertension exhibited the atrophy. It was also noted that degeneration or atrophy of the brain was seen despite drug therapy to control hypertension. This suggests that the hypertension is not the primary problem, and it is the brain atrophy that produces the hypertension.

Bruce McEwen at Rockefeller University commented, "Several brain disorders are associated with high levels of *cortisol*, including depression and memory loss, over time elevated stress hormones may be linked to the acceleration of senility." He also stated it could lead to a fat belly (McEwen, 1997a, 1997b, 1999a, 1999b, 2000a, 2000b, 2001, 2002; Geraci, 1998). The right hemisphere appears to be lateralized asymmetrically for the control of *cortisol* as well as for spatial orientation tasks such as maze or map skills.

Studies have shown that young children learn faster while at school and tend to stagnate intellectually during summer vacation, which is a finding that researchers argue points to nature, not nurture, as the key determinants of achievement. "The study shows a substantial connection between the environment and intellectual growth in ways that have not been revealed by other studies," commented the study's principal author Janellen Huttenlocher, a psychologist at the University of Chicago (Huttenlocher et al., 1998). Children in kindergarten and first grade had been tested four times, 6 months apart in four categories of intelligence. The children came from public, private, and parochial schools from throughout the United States with sample sizes ranging from 1,652 to 2,387 students. The researchers tested in the categories of language, spatial operations such as reading maps and charts, and concepts such as how objects are related and memory. Of particular concern had been to evaluate growth in language and spatial skills to test the view of many child development experts who think these areas reflect innate ability rather than environmental influences.

Findings revealed that no matter what the student's background or school, there is a slowing in the rate of growth in learning abilities during the 6 months between April and October, when children are generally out of school. In the area of language and development of syntax or sentence structure, the child's growth slows markedly during the vacation period compared with the school year. Only in the area of memory is there little difference noted between the school year and the summer vacation period, though overall growth in memory skills slows as age increases. Boys and girls perform at about the same level. It appears that being mentally active is better than being inactive from a physical perspective.

Sedentary activities can produce psychological symptoms as well as physical. According to recent studies of families with Internet access (Kraut et al., 1998; Sanders et al., 2000; Subrahmanyam et al., 2000), spending just a few hours online can leave someone feeling more socially isolated, lonely, and depressed than they were before. Kraut and colleagues at Carnegie Mellon University found that the more time subjects spend on the Internet, the higher the individuals score in measures of loneliness and depression. Teenagers appear the most vulnerable to the darker aspects of the Internet. Kraut has been quoted as saying,

> We were surprised to find that what is a social technology has such antisocial consequences. It is interesting to note that many of the teenagers who were responsible for school violence like witnessed at Columbine High in Colorado were known to spend hours on the Internet. These children are usually socially unpopular, loners, teased, and who already feel alone. Spending hours in sedentary behavior on the computer may be a major contributing factor to their violent behavior, because of what it does or does not do to their brains. This may in fact be as important or even more important factor to their violence, then the material they are accessing on the Internet.

Gravity and it Effects on the Brain and Performance

We have seen the effects of micro-gravity on brain cell development and function. We also have seen that the postural or antigravity muscles, especially those of the spine and neck, are the most susceptible to atrophy following disuse. Even partial gravity loss has been shown to have a significant effect on motor system function. A study conducted at the Central Scientific Institute of Sport, Moscow, examined the effects of bed rest on functional properties of neuromuscular system in humans (Korijak and Koslovskaya, 1991). In this study, the contractile properties of *m. triceps surae* were studied in six men of 31–45 years old 4–8 days before the start and 3 days after the completion of the 4-month *antiorthostatic* bed rest. Results show that all force characteristics decrease significantly after 4 months of bed rest. The values of decrease are not equal in all parameters under study, being close to 37 percent for single twitch, 36 percent for voluntary, and 34 percent for evoked isometric contractions. The 60 percent decreased force efficiency indicates diminished ability of the central nervous system to recruit voluntarily the same amount or the same type of motor units. Another consistent effect of *hyperkinesia* is revealed in this study as an increase in muscle fatigueability in the calf muscles during the

4 months of bed rest by 25 percent. The authors think that this effect can also depend on diminished contractile reserves of the muscle and on alterations of the state of central control mechanisms. They summarized their findings by stating that,

> it is well known that the motor system of all organisms including human beings has been developed under the conditions of Earth's gravity, that effects greatly all mechanisms and components of the system. Thus, complete and even partial loss of gravity has to be followed by alterations of these components and mechanism's functions as well as the function of the system as a whole.

This suggests that sedentary activity reduces gravitational forces on muscles, but also affects negatively the central control of muscles in the brain.

Studies of astronauts have shown a number of widespread physiological effects that are similar to stress-related responses. Youmans and Smith (1991) note that many reports are presently available on the effects of weightlessness during space flight (Decampli, 1987; Grigoriev and Kozlovskaya, 1987; Nicogossian, 1989) and summarize these effects in their paper. There is a decrease in muscle mass, decrease in nitrogen balance, decreased vital capacity (-10 percent), increased resting cardiac output, increased exercise capacity, decreased blood volume, decreased body red blood cell mass, and plasma Na^+ and Cl^- *osmosicity*. In addition, there is an increase plasma K^+ phosphate, plasma *aldosterone*, plasma *cortisol*, plasma *angiotensin*, plasma *adrenocorticotropic hormone* (*ACTH*), and *phosphate* and *calcium* balances. There is also a noted increase in vestibular and vibration sensitivity with a decrease in vibration threshold. The authors also note that there are many similarities between the effects of space and responses to bed rest (Astrino and Rodahl, 1970; Sandler, 1980). These widespread changes suggest that these effects are a result of decreased gravitational forces.

The reported increase in the sedentary lifestyles of U.S. children and the concomitant decrease in standardized academic performance relate as does the results of studies examining the relationship between zero gravity conditions and human performance. The connection is that zero gravity conditions in the context of the evolution of bipedalism affects human perception and performance for the reasons outlined in previous chapters.

Gravity, Posture, Oxygen, and Performance

Brain cells require two primary things to develop and function normally. The most important is stimulation, and the second is fuel, which is supplied through food that is digested and broken down to glucose. The other aspect of fuel is oxygen. Decrease in stimulation or decrease in fuel affect the cell similarly because they both will result in oxidative stress of the cell, although they affect it by different means. Lack of stimulation will result in a failure to trigger genetic responses, especially immediate early gene responses that stimulate the production of protein. Proteins are used by the cell to produce *organelles*, neurotransmitters, and cell membranes. These proteins are part of the process causing cell polarization. This is also maintained by passive and active transport mechanisms of K^+ and Na^+ that maintain a certain equilibrium of these chemicals. Decreasing stimulation causes a decrease in genetic responses, and therefore, decreased production of protein. The energy for the cell is produced by the *mitochondria*; the cell needs energy to produce protein to maintain cell membrane, to actively purge chemicals in and out of the cell. Decreased protein production results in fewer *mitochondria* resulting in *oxidative stress*. Decreased fuel also produces *oxidative stress*, oxygen being a major part of the fuel that *mitochondria* need.

The brain is the main consumer of the body's store of oxygen. Representing only 2 percent of the weight of the human body, the brain requires 20 percent of the body's store of oxygen. Oxygen deficiency is thought to be associated with many pathological processes. As a result, the effects of impaired *mitochondrial* function and the mechanisms of *mitochondrial* protection during oxygen deficiency are of significant importance to both the understanding of cellular homeostasis and to developing improved means to protect against *mitochondrial* failure (Jones, 1995). Several studies have suggested that the *mitochondria* are a primary target of oxidative injury. *Mitochondria* are present inside a brain cell and provide energy as needed. *Mitochondria* are the thousands of threadlike bodies in each cell that contain enzymes, which control the production of energy. It has been said to be the breathing machine of the cell gathering oxygen to use as energy. *Mitochondria* are thought to be the remnants of ancient bacteria that have formed a symbiotic relationship with cells. The cells supply it with protein and fuel and it provides energy to the cell. Because it was a separate organism, it has its own DNA, inherited solely from the mother. This DNA is not the nuclear DNA responsible for the uniqueness of each individual. However, it is thought that the expression of *mitochondrial* DNA is dependant on the expression of nuclear DNA, which is dependent mostly on stimulation. *Mitochondria* also normally contain high levels of minerals, which are degraded by internal processes and never make it back out into the cell, unless of course there is damage to the cell.

Normally 1–2 percent of the oxygen we breathe turns into oxygen-free radicals, highly reactive and potentially toxic. There are proteins in the *mitochondria* that search out these radicals and destroy them before they become toxic. However, when a cell is dysfunctional or damaged due to lack of stimulation or fuel, these free radicals accumulate and eventually destroy brain tissue. This is what is referred to as *oxidative stress*. Minerals can also exacerbate this free radical process (Fig. 10.1). Therefore, decreased stimulation, decreased ability to breathe or oxygenate blood, or increased constriction of blood vessels, as in increased *sympathetic* tone, decreases oxygen to brain cells. In one study, conditions that limit *mitochondrial* function were examined, specifically states of oxygen deficiency that affect the susceptibility of cells to oxidative injury (Jones, 1995).

In the first situation, Jones and colleagues examined whether cells exposed acutely to oxygen deficiency were more vulnerable to oxidative injury than *normoxic* cells (Tribble et al., 1988). The results demonstrate that substantially more cell killing occurs in *hepatocytes* incubated under steady-state *hypoxic* conditions. In a second model, *hepatocytes* were examined that were previously exposed to *anoxia*. Noted was that these cells are more vulnerable to oxidative injury than cells that had not been exposed to *anoxia* (Kowalski et al., 1992). Their results show that previous exposure of 30–60 min of *anoxia* dramatically increase the sensitivity of cells to a constant oxidative challenge. They point out that this finding is important because it shows that even without enhanced generation of oxidants during post-*ischemic* re-oxygenation, cells are more vulnerable to oxidative injury. Therefore, conditions, which would be treated under *normoxia*, can become pathological following transient interruption of blood flow.

Also examined in this paper were the effects of chronic in vivo oxygen deficiency on sensitivity to oxidative injury in a third model with rat *hepatocytes* (Shan et al., 1992). Rats were maintained for 8–10 days under conditions of *normoxia* or moderately severe *hypoxia*. *Hepatocytes* were isolated, and experiments were performed in vitro. Cells from organisms that had previously been *hypoxic* were comparable in viability and several other characteristics but were significantly more sensitive to toxicity induced by *t-butylhydroperoxide*. Therefore, what is demonstrated is that with three different models of oxygen deficiency, cells are more susceptible to injury from oxidants. In addition, because *mitochondria* are a sensitive target of injury due to oxidants and are functionally impaired by oxygen deficiency, these results suggest that *hypoxic* episodes may be particularly important contributors to *mitochondrial* dysfunction. Results also show that impaired circulatory or respiratory supply of oxygen to tissue may substantially exacerbate the oxidative damage (Jones, 1995).

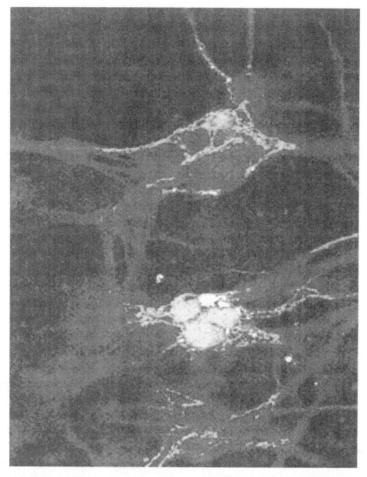

Fig. 10.1. The brighter neuron shows an increase in free radical activity that peaked 3 hr after neurons were deprived of a life-sustaining substance called nerve growth factor. This suggests that free radicals may play a role in the death of neurons and, possibly, neurodegenerative disorders. (Source for photo is Eugene M. Johnson, Jr. cf. Deckwerth et al., 1996.)

Sedentary activity will have multiple effects that exacerbate *oxidative stress*: (1) decreased generalized continuous nonspecific and specific activation of cells, resulting in decreased genetic expression and protein production decreasing production of *mitochondria* and their gene expression, (2) decreased neocortical stimulation resulting in decreased inhibition of *sympathetic* pathways and decreased activation of *parasympathetic* pathways resulting in central vasoconstriction, (3) sedentary activity as well as micro-gravity situations cause decreased lung vital capacity and VO_2 max (17–28 percent) as well as increased blood pressure (Youmans and Smith, 1991) both significantly reducing oxygen to the brain. In addition, sedentary activity and micro-gravity both result in weakness of postural muscles, especially spinal muscles, and loss of *lordosis* of the *lumbar* and *cervical* spine. Loss of *lumbar lordosis* is thought to result in decreasing mechanical efficiency of the diaphragm, also decreasing oxygen consumption and delivery.

It is now a commonly held belief that most Americans are *hypoxic* to which a sedentary lifestyle would be a major contributing factor. It has been shown that severe cerebral *hypoxia* causes cessation of spontaneous cortical activity within 10–30 s in mammals (Mayevsky and Chance, 1975). Within 1–3 min, there is a block of evoked responses (Anderson, 1960; Williams and Grossman, 1970). A study by Lipton and Whittingham (1979) suggests that membrane *depolarization*, not membrane *hyperpolarization*, or neurotransmitter disruption is responsible for the failure of evoked potentials during *hypoxia*. With decreased stimulation and or decreased fuel (oxygen), the loss of protein production results in a shift of the cell toward *depolarization*. This at first may compensate for loss of stimulation by allowing the cell to fire with less stimulation. Eventually, however, the cell may become unstable and start to fire spontaneously as in epileptiform activity. The cell will also fatigue quickly which may produce greater *oxidative stress* and more cell damage and/or death of brain cells. The brain areas that will be affected most by *oxidative stress* are areas that have a high metabolic rate or high blood flow. The *cerebellum*, which has more cells than any other area of the central nervous system, is the most likely to suffer from oxidative stress. In fact, in a state of oxygen deprivation, many of the initial symptoms are *cerebellar* signs.

Lennon and colleagues (1994) state that, "the most significant influences of posture are upon respiration, oxygenation, and *sympathetic* function. Ultimately it appears that homeostasis and *autonomic* regulation are intimately connected with posture." They also point out that the optimal adult human posture includes *lordosis* of the *lumbar* and *cervical* spine. This posture usually develops as infants begin to push themselves up, crawl, sit, and eventually stand. These researchers also note that all mammals are born with the tongue completely in the oral cavity where it remains. This structural lingual oral situation during infancy promotes breathing while nursing enabling both breathing and swallowing at the same time. Humans are apparently the only mammals that lose this innate response sometimes between 6 months and 6 years of age. At birth, the human larynx is situated in front of the first vertebrae of the neck. Because of this, human infants are born nose breathers and are not capable of breathing through the mouth except when crying. As the human structure grows and matures, the tongue and larynx move caudal to their final position anterior to the fourth, fifth, and sixth cervical vertebrae. One third of the tongue forms the anterior wall of the human throat (Crelin, 1987). What this shows is that a bipedal position is essential in developing the ability to speak and produce vocal sounds. In the posterior aspect of the human throat, there is apparently a space that is unique to humans. It is thought possible that standing erect shifted gravity's forces allowing this space to develop as an adaptation to gravity. The sound potential of this space is unique to humans (Crelin, 1987). Therefore, had we remained in a quadrupedal position, it is unlikely that this space would have evolved (Lennon, 1994).

It has also been stated that any learning is enhanced when accompanied by some kind of structured movement (Lennon, 1994). This is probably because all learning occurs through the *cerebellum* and the *cerebellum*, is stimulated mostly by proprioceptive movements of the muscles and joints; the *cerebellum* also requires oxygen.

Plasticity and Activity Dependence

Evidence has been increasing over the last decade that the brain works somewhat like a muscle, the harder you use it, the more it grows. Although researchers have long thought that the brain's circuitry is hard-wired by adolescence and inflexible in adulthood, its ability to change and adapt is much more than was ever thought possible. Until recently, it was generally accepted that during development, no new brain cells could be formed. Recent evidence suggests that this may be incorrect, and that the process of creating new brain cells may be activity dependent as well. The concept of use-dependent plasticity of the nervous system was first proposed by Donald O. Hebb in his 1949 book (Hebb, 1949), *The Organisation of Behaviour*. Rosenzweig and Bennet (1995) in their review of the psychobiology of plasticity note that in the early 1960s, two experimental programs stated that they had research findings demonstrating that the brain can be altered by training or differential experience. They note that the first was the demonstration by their own group at Berkeley that showed both formal training and informal experience in different environments resulted in significant changes in the neurochemistry and neuroanatomy of the rodent brain. Shortly following these results, Hubel and Wiesel (1979) released their report showing that occluding one eye early in development led to reduction in the number of cortical cells responding to that eye.

The first results of tests with rats were focused on neurochemical changes to the brain of rats that were given formal training in a variety of problems. Initially it was shown that there were correlations between levels of the enzyme *acetylcholinesterase* (*ACHE*) in the cerebral cortex and the ability to solve spatial problems (Rosenzweig et al., 1957; Krech et al., 1960). Interestingly it was also shown that the *ACHE* activity was higher in the cerebral cortex of groups that had been trained and tested on more difficult problems than in those that were given easier problems, also all of the tested groups measured higher in cortical *ACHE* activity than groups that received no training and testing. Therefore, to the surprise of the researchers, it appeared that training altered the *ACHE* activity of the cortex.

Even more surprising than the discovery that differential experience resulted in changes in cortical chemistry, was the finding that enriched experience increased the weights of regions of the neocortex (Rosenzweig et al., 1968). This was the first evidence that enrichment of the environment could lead to structural changes in the brain (Bailey and Kandel, 1993). An enriched environment for these rats was constituted by a large cage containing a group of 10–12 organisms and a variety of stimulus objects, which were changed daily and was called the *enriched condition* (*EC*) because it was thought to provide greater opportunities for informal learning than did the other conditions. The *standard colony* (*SC*) situation consisted of three organisms in a standard laboratory cage provided with food and water. In addition, *SC* size cages housed only single organisms, which was called the *impoverished condition* or *isolated condition* (*IC*). Several years and several studies revealed experience-induced increases of the cortex especially in the *occipital* cortex including: increases in cortical thickness (Diamond et al., 1964), sizes of neuronal cell bodies and nuclei (Diamond, 1967), size of synaptic contact areas (West and Greenough, 1972), an increase of 10 percent in numbers of dendritic spines per unit length of basal dendrites (Globus et al., 1973), an increase of 25 percent or greater in extent and branching of dendrites (Greenough and Volkmar, 1973), as well as parallel increases in the number of synapses per neuron (Turner and Greenough, 1985). It was pointed out that these effects are suggestive of a significant increase in processing capacity of the cortical areas involved (Rosenzweig and Bennett, 1995).

It was observed that the training had resulted in changes in specific cortical regions and not in undifferentiated brain growth. Subsequent studies showed that significant cerebral effects of enriched vs. impoverished experience could be caused at any age and with relatively short periods of exposure. It was also noted that although Hubel and Wiesel's experiments altered cortical responses only if the eye was occluded during a critical period, other studies show that modifying sensory experience, particularly in cases of touch or tactile sensation and hearing, could alter both receptive fields of cells and cortical maps (Kaas, 1991; Weinberger, 1995). It was thought that the fact that differential experience could produce changes throughout the life span and relatively quickly, suggested that these effects were due to learning (Rosenzweig and Bennett, 1996). It was also seen that the ability to create plastic changes and for learning are present in older subjects, the cerebral effects of differential environmental experience are produced more rapidly and have a greater overall effect on younger vs. older organisms.

In interpreting this research, it has been somewhat overlooked and underemphasized that the major difference between the environments was the fact that the rats in the enriched environment were given toys like treadmills and the like that they physically interacted with. The cognitive stimulation to these organisms has been mostly emphasized by the researchers but the fact is that the enriched environment was mostly physically stimulating, and therefore the physical activity is what may be mostly responsible for the plastic changes and learning. For example, Rosenzweig and Bennett (1996) note that Hebb (1949) reported that he brought home some of his laboratory rats and allowed them to explore his house for several weeks as pets for his children. When he returned them to his laboratory they showed better problem-solving ability than rats that had remained in the laboratory the whole time. What was interesting was that he noted that they maintained their superiority or even increased it during a series of tests. Hebb is quoted "the richer experience of the pet group during development made them better able to profit by new experience at maturity, one of the characteristics of the 'intelligent' human being" (Hebb, 1949, pp. 298–299). These rats did not have specific training; they were simply allowed to physically interact with their environment more, which apparently made them more "intelligent" than rats that had not interacted.

In a similar study where rats in *IC* cages were placed inside an *EC* cage, in essence a cage within a cage, the rats in the *IC* cage were placed in the middle of all the toys but could not physically interact with them. However, they were still in the middle of an interesting environment and one would expect that if the surroundings were "mentally" stimulating, then the rats should have plastic brain changes even though they did not physically interact with the environment. However, after a period, the rats that were in the *IC* environment were compared with the rats that were in the *EC* environment. It was noted that the rats in the *IC* cages that had been placed in the middle of the EC cage did not have any plastic growth of their brain cells; however, the rats that were more active physically with the toys showed significant growth in their brain.

There is a world of difference between a child who sits and watches a television show about trees and a child who is climbing a tree physically. We cannot fool ourselves into believing that a child is becoming smarter by watching television or playing computer games, because these are sedentary and therefore less stimulating to the brain itself. This concept is something that most parents and teachers instinctively know. In fact, the Montessori method of teaching is based on this concept. At the turn of the century almost 100 years ago, an Italian female physician named Maria Montessori started working with retarded children at the psychiatric clinic at the University of Rome. She noted that most of the so-called "unteachable" children had an advanced capacity for concentration when they were focused on an activity of interest to them. Montessori developed a new kind of classroom where children were allowed to roam at will from one activity to another focusing on whatever caught their attention most. She had equipped the classroom with interactive learning aids that taught children through utilizing all five senses. For instance, the children were taught the alphabet not only by sight but also through tactile sensation, by being asked to identify letters made of sandpaper with their eyes closed. Although this method was initially used for retarded children, it eventually made its way into the mainstream. It is estimated that there are approximately 3,500 Montessori preschools in the United States and about 100 elementary schools that use this method. It has been reported that Montessori children typically learn to read and write well before the age of 5.

The question has been posed by what processes do enriched environmental experience or formal training lead to plastic changes in the cerebral neurochemistry and neuroanatomy? It was found that the enriched experience produces increased rates of protein synthesis and increased amounts of protein in the cortex (Bennett et al., 1964). Formal training increases the rates of precursors into RNA and protein in the *forebrain* of the chick (Haywood et al., 1970), and enriched experience results in increased amounts (Ferchmin et al., 1970; Diamond et al., 1976), and expression of RNA in rat brains (Grouse et al., 1978). In chicks, several researchers have traced a cascade of neurochemical steps from initial stimulation to synthesis of protein and structure (Gibbs and Ng, 1977; Ng et al., 1991; Rose, 1991; Rosenzweig et al., 1992). Sensory stimulation activates receptor organs, which stimulate afferent neurons using various synaptic transmitter chemicals like *acetylcholine* (*ACh*) and *glutamate* (Rosenzweig and Bennett, 1996). One trial training leads to an increase of immediate early gene RNA in the chick forebrain (Anokhin and Rose, 1991) and to increased density of dendritic spines (Lowndes and Stewart, 1994). Many of the effects have been noted to occur only in the left hemisphere, or are more prominent in the left than the right hemisphere. Although plasticity, or the formation of new connections and branches of cells, is now generally accepted, for years neurobiologists have clung to a fundamental truth, that once organisms or humans reach adulthood, they may lose brain cells and can never grow these cells anew. However, in 1997, new research was published showing that monkeys are continually making new brain cells in the *hippocampus*. In addition, it was reported that production of new cells is decreased when the organisms are under stress conditions.

Peter Eriksson and colleagues (1998) published the startling news that the mature human brain does create new neurons routinely in at least one site, the *hippocampus*, and are important to memory and learning. The *hippocampus* is not where memories are stored (Leisman and Koch, 2000), but it aids in the formation of memories after receiving the input from other brain regions. Individuals with damage of the *hippocampus* have problems with acquiring new knowledge but can recall information learned before their injury (Kemperman and Gage, 1999). Current information suggests that stem cells probably make new neurons in another part of the human brain and lie dormant in additional locations. Previous studies in rats have shown that some neurogenesis occurs throughout life not only in the *hippocampus* but also in the olfactory system. Stem cells also exist in other brain areas like the *septum*, which is involved in emotion and learning, the *striatum* and in the spinal cord. The cells outside the *hippocampus* and olfactory system do not seem to create new neurons under normal conditions. Although based on the volume of studies, neurobiologists have become clearly convinced that enriching the environment of mature rodents affects brain wiring in ways that emphasize brain function; they have scoffed at the idea that production of new nerve cells in the adult brain could be a contributing factor. New research now appears to show that different environments do affect neurogenesis. Kempermann and colleagues (1997) have shown that adult mice in enriched environments grew 60 percent more new *granule* cells in the *dentate gyrus* than do genetically identical control organisms. They also do better on a learning task that involves finding their way out of a pool of water. It is not thought that the new neurons are solely responsible for the behavioral improvement; changes in connectivity and the chemical production are also involved.

Even more evidence on environmental effects, especially increased physical activity, on development of new nerve cells was published in February 1999. New evidence reported that mice that run more have substantially more brain cells than do sedentary siblings. Again, researchers at Salk Institute for Biological Studies reported in *Nature-Neuroscience* (van Praag et al., 1999a, 1999b, 2000; Gage, 2000) that new cells nearly doubled in mice that regularly used a running wheel. It has been shown in the past that running increases chemical growth factors, also known as *trophic* factors in the *hippocampus*, which suggests that this might help facilitate cell growth in that area of the brain. Gage has stated he thinks that it may be running or other vigorous exercise which stimulate brain cell production in people as well. It was noted that running mice had approximately 7,000 cells per square millimeter, while sedentary mice showed about 4,000 cells per square millimeter. All of the studies show that physical activity and movement create chemical and physical changes to the brain that appears to increase intelligence. One of the most significant areas is the *hippocampus*, an area involved in learning new information and which facilitates storage of it as a long-term memory.

In regard to child development, these facts were emphasized in a major 1994 report—*Starting Points: Meeting the Needs of Our Youngest Children* (*Carnegie Task Force on Meeting the Needs of Young Children*, 1994). Beginning in the 1960s, scientists began to demonstrate that the quality and variety of the environment have direct impact on brain development. Today researchers around the world are amassing evidence that the role of the environment is even more important than earlier studies had suggested. For example, histological and brain scan studies of organisms show changes in brain structure and function as a result of variation in early experience. These findings are consistent with research in child development that has shown the first 18 months of life to be an important period of development. Studies of children raised in poor environments show that they have cognitive deficits of substantial magnitude by 18 months of age and that full reversal of these deficits may not be possible (Rosenzweig and Bennett, 1996).

INVOLUNTARY LACK OF PHYSICAL ACTIVITY

There are various situations where children are not allowed or permitted to be physically active. They are prohibited from interacting with their environment. These children have been shown to suffer from cognitive and behavioral problems due to a developmental delay of their brains. One of the situations where this exists is when newborns are raised in an institution. For example, camera crews from America were recently permitted to film children in some state-run orphanages in Romania. Hundreds of babies and small children, in rows of white cribs that looked like cages, were videotaped. The number of children was overwhelming to the State and they could barely provide food and necessities to the children, much less provide them with affection or even human contact. These children were almost never allowed out of their cribs to crawl or move about. Many of these children would just lie in their cribs, self-stimulating by rocking back and forth, or shaking their heads, which was thought to be a desperate attempt to provide stimulation to their own brain. The children rock on their backs or move their necks, probably because these areas provide the most stimulation to the brain. These children were human examples of the rats that had been placed in impoverished or IC cages. Many rushed to adopt these deprived children; the children all had physical examinations and were apparently physically healthy. However, a follow-up story a few years later revealed something different. Many of these children exhibited learning disabilities or behavior problems. A few of these children underwent PET studies which revealed that various areas of their brain were underactive especially the frontal areas (Holden, 1996; Chugani et al., 2001; Gunnar et al., 2001). What these children lack most is physical activity; limits on space and concerns about injury are factors that cause these children to be less physically active. In fact, many of these Romanian children found their way to occupational therapists practicing sensory integration, and once placed in a program of specific activities they showed improvement in abilities to learn and in their behavior.

There are other situations, less drastic, but that nonetheless restrict a child's movement and development. Although never objectively evaluated, Doman and Delacato in the 1960s wrote a paper on what they termed *patterning*, or the developmental requirement of a child's progress through different stages of creeping and crawling. Both Doman and Delacato (Delacato, 1963) thought that at first children must move their arms and legs freely on their backs. An important step is that a child develops the ability from this position to lift the head up and flex the trunk slightly. Then the child must learn to roll over, from here the child learns a belly crawl. The child now must be able to extend its head and hold it up. From here, a child then learns to get up on all fours and crawl until it can eventually stand up and walk. It is thought that these stages are important to the development of various higher centers of the developing nervous system. Doman thinks that if any of these stages are missed or if there is an injury to the brain during this progression, the child may become "stuck" at that point in development. Standing and walking appear to facilitate the beginning of development of the neocortex; soon after a child learns to walk, it will start to develop speech. It is thought that if a child does not stand and walk, the brain will not develop maximally.

While a good deal of controversy has surrounded this approach to remediation and its theoretical basis, recent studies examining the importance of locomotion for improving children's spatial search capabilities (Yan et al., 1998) have been published and lend some support for Doman and Delacato's theoretical position. This meta-analysis quantitatively summarized the developmental influence and the effects of locomotor experience as well as the benefits of locomotor practice, locomotor assistance, and active searching patterns on children's search performance. Based on specific criteria, a search of a database, and reference lists identified 19 studies including 1,029 children (510 boys and 519 girls) from 4 to 44 months of age. The authors note that various studies have shown the positive effect of motor activities on facilitating children's cognitive development (Kail, 1988). It is thought that from both a theoretical and practical viewpoint, understanding the influence of motor activities on child development is crucial for parents, educators, and researchers. The performance of spatial search refers to a skill or action that allows a child to successfully find hidden objects (Brenner et al., 1967). It has been reported that a child's early locomotor activities facilitate development of spatial ability (Thelen and Smith, 1994). Other studies have shown the positive effect of locomotor activities on spatial search skills in infancy (e.g., Kermoian and Campos, 1988; Bae and Bertenthal, 1992). Developmental changes in locomotor activities (from crawling to creeping and walking) improve not only locomotor experience but spatial search skills or spatial information processing (Brenner and Bryant, 1977; Acredolo and Goodwyn, 1984).

It has been suggested that because vision and posture facilitate perception of the environment, they are more important than locomotion in the development of spatial abilities (Adelson and Fraiberg, 1974; Savage-Rumbaugh, 1993). According to Brewer and Bryant (1977), and Acredolo and Goodwyn (1984), developmental changes occur in the spatial information processing in infants when their relatively static bodies start to move. It is speculated that initially infants try to locate an object by using visual information from an egocentric perspective. As infants become capable of locomotor movements, they try to discover the object by manipulating the spatial relationship between the object and their own body movements. Eventually as a child's locomotor patterns become mature, they are able to search or remember the location of objects by using multiple relationships among the objects themselves and the environment. Numerous authors support the position that when children have enriched locomotor experiences, more visual cues are available for searching activities, which is thought to improve performance of spatial search (Rosenzweig and Bennett, 1996; de Vries, 1999). Studies support the effect of locomotor training and practice, which appear to affect spatial cognitive development. Since no significant age differences have been found for training effects, training or practice seems equally important for subjects from infancy through early childhood.

In summary, numerous authors support the hypothesis that children's locomotor experiences enhance their spatial search skills and spatial cognitive development. Based on the studies reviewed, children with locomotor assistance, training, and active search patterns, perform better than those without assistance or training and those using passive searching patterns. This suggests that not only does a child need to be free to move and crawl about, but also if the parent actively helps the child or teaches the child how to move about its environment, it may facilitate the child's cognitive development. On the other hand, restricted mobility, not going through crawling, creeping, walking stages or lack of parental involvement during locomotor development, would all be expected to result in cognitive delays.

Cajal (1911) noted that the micro-neurons of the *cerebellum* develop in direct response to infants' activities. Misner (1969) raised the question as to whether mobility, especially crawling and creeping in infancy, is a necessary forerunner of higher efficiency in the more complex motor and communicative skills later in life. In this paper, he defines neurological organization and the precepts on which it is based. Neurological organization is defined as "the degree to which the central nervous system and more specifically the brain, provides the organism with all the capabilities necessary to relate it successfully to its environment." The precepts on which it is based are:

1. The acquisition of neurological function during embryonic and postnatal life occurs as a continuum.
2. In this continuum, one innovative accomplishment lays the foundation for development of the next.
3. The development of mental and coordinative capability can be retarded and stopped; it can also be stimulated and accelerated.
4. Important gaps or arrests may occur and distorting biases may form.

Using the concept of neurological organization as a hypothesis, the author predicts that children restricted during the first one or two years of postnatal life, as Navaho children normally are, would reveal neurological disorganization in the form of reduced motor and communicative skills in later life. A comparative study was performed of two American Indian tribes with respect to mobility during the first two years of life, and athletic and scholastic abilities during the grade and high school periods. Misner's findings revealed that children with restricted mobility during early life showed substantially lower performance capabilities in later periods. It is important to note that the two tribes studied lived so closely together that their reservations physically touched. The tribes shared almost identical cultural and economic circumstances. Yet, one tribe tested with an average IQ over 25 points higher than that of the other. The only difference was the amount of physical mobility during infancy.

Similar findings, based on extensive study of cultures at all levels of complexity, have shown that tribes or cultures allowing and promoting infants to creep and crawl tend to be more complex societies possessing higher technology, and some form of written language. On the other hand, cultures that restrict their infants from crawling, have no writing of their own, and can be taught to read only with great difficulty. It was also noted, that individuals raised in these cultures have difficulty focusing their eyes at arms length, the distance that people read, write, make arts and crafts and tools (Maccoby, 2000). This underdevelopment of cognitive and *ocular–motor* abilities can be attributed to underdevelopment of the *cerebellum* which is involved in learning, motor activities, and *ocular–motor* control. This hand–eye feedback is thought to develop as a child starts to creep and crawl. As a child crawls along the floor, it has been noted that as each hand moves forward, the infant's eyes focus on it. This process occurs side by side thousands of times during development. It obviously involves the *cerebellum*, which not only would help to initiate this activity, but would receive feedback from it as well.

Development of the *cerebellum*, especially starting with the midline, *vermis*, or spinocerebellum, is thought to play a major role in the development of the midline postural and *ocular–motor* muscles, which have been associated with right brain development and is crucial to behavioral development. Crawling and creeping may be so important to developing intelligence as well as verbal and nonverbal communication skills, because of its effect on *cerebellar* development, and the subsequent effect that the *cerebellum* appears to have in the development of the neocortex, *association* areas, and especially the *frontal* cortex. Ultimately the effects that voluntary and involuntary sedentary behavior have on the brain are the same although the initiating cause may be slightly different. Both, however, point to an important role of the parent to promote and facilitate early and continuous mobility especially natural forms of movement, and to discourage sedentary behavior. Other times during development, however, a child may be subjected to an unfavorable environment, may be injured, or become ill, and quite often this is recognized as a factor that may produce developmental delays.

INJURY AND ILLNESS

Injury or illness can retard the growth and development of the body and, of course, the brain. Injuries and illnesses can occur at any period of development and have long-lasting effects if not treated. We have broken down these periods into three major categories: (1) prenatal, (2) the birthing process itself, and (3) postnatal.

Prenatal Injury or Illness

Two major factors that can affect a fetus prematurely: (1) genetic disorders that are not generally nonconducive to life and usually end with spontaneous abortion, and (2) the condition and health of the mother. The mother provides the complete environment in which the child initially develops. Although development at this stage is primarily genetically driven, recent research has shown that the immediate environment that the infant develops in, inside, and even outside of the mother plays a major role in the prenatal and postnatal development of the child. In fact behaviorally speaking, there is very little difference between a newborn baby and a 32-week-old fetus. A new wave of research suggests that the fetus can feel and dream (*Psychology Today*, October 1998; Christos, 1990). Recent research shows that at 32 weeks of gestation, a fetus behaves the same as a newborn. Some interesting facts about prenatal life are emerging:

1. By 9 weeks, a fetus can hiccup (Pillai and James, 1990a; Goldkrand and Farkouh, 1991) and react to loud noises (Reron, 1992). By the end of the second trimester, it can hear (Reron, 1992).
2. It has been shown that the fetus experiences rapid eye movement (REM) sleep associated with dreaming (Okai et al., 1992).
3. The fetus apparently appreciates the mother's intake of food, first picking up the food tastes of their culture in the womb (Mennella et al., 2001; Temple et al., 2002)
4. The fetus can distinguish between its mother's voice and that of a stranger, (Ockelford et al., 1988).
5. Even a premature baby is aware, feels, responds, and adapts to its environment (Ronca et al., 1996).

The basis of human behavior according to researchers begins to develop very early—just weeks after conception. Well before a mother typically knows she is pregnant, her embryo's brain has already begun to develop. By the fifth week of embryonic life, the brain has already begun the most fantastic feat of human development, which is the creation of the deeply creased and convoluted cerebral cortex. At 9 weeks, the embryo's growing brain allows it to bend its body, hiccup, and react to loud noises. At the tenth week, it moves its arms, "breathes" amniotic fluid in and out, and opens its jaws and stretches. Before the first trimester is over, it yawns, sucks, and swallows as well as feels and smells. By the end of the second trimester, it can hear, and toward the end of the pregnancy, it can see. The fetus however at 32 weeks sleeps 90–95 percent of the time. Some of this time has been shown to be spent in REM sleep. Close to birth, the fetus sleeps 85–90 percent of the time. Awake or asleep, the fetus moves over 50 times or more each hour, flexing or extending its body, moving its head, face, and limbs. Why people from certain cultures grow up with specific tastes for food may have to do with the fetal environment. At 13–15 weeks, it is thought that a fetus' taste buds already look like adults. It is known that amniotic fluid that surrounds it can smell strongly of curry, cumin, garlic, onion, and other essences from the mother's diet. Whether a fetus can taste these flavors is not known; however, it has been found that a 53-week-old premature infant will suck harder on a sweetened nipple than on a plain rubber one.

Recently it has been shown that salt preference may be imprinted while still in the womb, and linked to the level of morning sickness experienced by their mothers (Crystal and Bernstein, 1995, 1999). Studying 16-week-old infants, the researchers found that babies whose mothers suffered moderate to severe nausea and vomiting in early pregnancy showed a greater preference for salt water solutions than did babies whose mother experienced little or no morning sickness. In earlier work, Crystal and Bernstein (1995) showed a similar pattern of salt preference among young adults. This demonstrates that whatever happens in the womb to influence food preferences persists at least through young adulthood.

There is no question about the ability of a fetus to hear. A very young premature infant at 24 or 25 weeks responds to sounds around it, therefore, the auditory system must have been functioning in the womb. It has been shown that fetal heart rate slows when the mother is speaking, suggesting that the fetus not only hears and recognizes sounds, but also is calmed by it. It has also been shown that the fetus along with the ability to hear has the ability to learn and remember. In the 1980s DeCasper and colleagues (DeCasper and Fifer, 1980; DeCasper and Prescott, 1984) invented a feeding device that allows a baby to suck faster or slower depending on the type of sound presented. Using this device, the group was able to show that within a few hours after birth, an infant already prefers its mothers voice to that of a stranger, which is thought to suggest that it learned and remembered the voice, although not explicitly but implicitly. DeCasper and others have found that newborns can not only distinguish their mother from a stranger speaking, but also prefer their mother's voice, especially the way it sounds filtered through amniotic fluid rather than through air. In addition, they prefer to hear their mother speaking in her native language than to hear her or someone else speaking in a foreign tongue.

Vision is the last sense to develop; a premature infant can see light and shape (Eswaran et al., 2002). Researchers have reported a distinct fetal reaction to flashes of light shined on the mother's belly (Timor-Tritsch, 1986; Boos et al., 1987; Leroy et al., 1987; Lecanuet and Schaal, 1996). When and how behavioral traits originate in the womb is now the subject of several studies. In the first study of fetal temperament, researchers measured heart rate and movement of 31 fetuses six times before birth and compared them to readings taken twice after birth. The study has been extended to include 100 more fetuses. Their results show that fetuses that are very active in the womb tend to be infants that are more irritable. Those with irregular sleep/wake patterns in the womb sleep more poorly as young infants (Junge, 1980; Emory and Noonan, 1984; Pillai and James, 1990b). In addition, fetuses with high heart rates become unpredictable inactive babies. It is thought that behavioral development or any higher brain development may be influenced by the fetal environment provided by the mother, her eating patterns, her movements, her sleep/wake cycles, and hormonal balances may all influence the early development of the brain. Likewise, a negative environment provided by the mother can affect early development of the brain so that a child may start out at birth already delayed.

Vulnerable periods during the development of the nervous system are sensitive to environmental insults because they are dependent on the temporal and regional emergence of critical developmental processes (i.e., proliferation, migration, differentiation, *synaptogenesis, myelination*, and *apoptosis*). Evidence from numerous sources (Rice and Barone, 2000) demonstrates that neural development extends from the embryonic period through adolescence. In general, the sequence of events is comparable among species, although the time scales are considerably different. Developmental exposure of lower organisms or humans to numerous agents (e.g., X-ray irradiation, *methylazoxymethanol*, ethanol, lead, methyl mercury, or *chlorpyrifos*) demonstrates that interference with one or more of these developmental processes can lead to developmental neurotoxicity. Different behavioral domains (e.g., sensory, motor, and various cognitive functions) are subserved by different brain areas. Although there are important differences between the rodent and human brain, analogous structures can be identified. Moreover, the ontogeny of specific behaviors can be used to draw inferences regarding the maturation of specific brain structures or neural circuits in rodents and primates, including humans. Furthermore, neurobehavioral disorders in humans may also be the result of interference with normal ontogeny of developmental processes in the nervous system. Of critical concern is the possibility that developmental exposure to neurotoxicants may result in an acceleration of age-related decline in function.

Toxins, nutrition, alcohol, drugs, smoking, as well as the health of the mother, and the activity of her neocortex are likely to determine the level of her stress hormones. Stress hormones or stress reaction seems to play a significant role in intelligence and behavioral development. Stress, diet, and toxins may combine to have a harmful effect on intelligence. A study by Bernice Devlin of the University of Pittsburgh suggests that genes may have less impact on IQ than previously thought (Devlin et al., 1997).

Histories of 300 hyperactive children were compared to a control group, which included the medical histories of 190 normal children (Hartsough and Lambert, 1985). The results showed several differences including that mothers of children with *ADHD* seem to have generally poorer health during pregnancy, are of younger age than average, are more likely to be having their first child, have more frequent reports of *toxemia* or *eclampsia*, and generally have labor lasting longer than 13 hr. Hyperactive children are twice as likely to have evidence of fetal distress, head injuries, or other birth injuries. More hyperactive children have been shown to be born at a gestational age of 10 months or later. In addition, hyperactive children more frequently have medical problems or physical malformations at birth. Other researchers have identified relationships between *ADHD* and prenatal or perinatal distress. There has also been demonstrated a relationship between *ADHD* and maternal alcohol abuse (Coles et al., 1997; Weinberg, 1997), heavy smoking (Jacqz-Aigrain et al., 2002) and drug abuse (Bandstra et al., 2001). However, it is important to note that not all hyperactive children, not even the majority have known maternal histories of stress and that the environment of the womb may account for much more. For instance, many children who appear to have been born hyperactive, excitable, or intellectually slower, have been said to be most likely a product of "bad genes," especially because the mother or father may have a history of learning disability or behavioral problems like *ADHD*. However, if the mother or father are less physically active, more sedentary, or overweight, if they eat poorly and do not exercise, are *hypoxic* because they smoke, these factors will likely affect the mothers *cerebellum, thalamus*, and neocortical activity. Therefore, this may affect the mothers autonomic, immune, hormonal, and emotional systems. This may also affect the developing fetus and this affect will continue outside the womb, since the parents are providing their environment to the neonate and child. Therefore, genes may not be as important in these cases even if neurobehavioral disorders seemingly run in families, since all family members are functioning and developing in the same impoverished environment.

An increasing body of literature shows that there is an interesting association between maternal smoking during pregnancy and *ADHD* in their children. In a recent study (Milberger et al., 1996), a strong and significant association was found between cigarette smoking in mothers during pregnancy and *ADHD* in their children. This same study also showed an association between maternal smoking during pregnancy and IQ scores in the children. These results were reported to be consistent with studies that linked maternal cigarette smoking during pregnancy and long-term behavioral and cognitive problems in children and lower organisms (Butters and Goldstein, 1973; Denson et al., 1975; Dunn et al., 1977; Sexton, 1978; Nichols and Chen, 1981; Rantakallio, 1983; Naeye and Peters, 1984; Hagino and Lee, 1985; Krist-Jansson et al., 1989; Sexton et al., 1990; Weitzman et al., 1992; Richardson and Day, 1994). A number of studies have shown a greater susceptibility of the male fetus to intrauterine and other perinatal circumstances contributing to central nervous system disorders (Nesbitt, 1957; Montague, 1962; Hynd and Semmid-Clikeman, 1989a, 1989b; O'Callaghan et al., 1992). Milberger and colleagues (1998) examined whether the association between *ADHD* and maternal smoking during pregnancy would be seen in their high-risk brothers and sisters. Siblings of *ADHD* probands are considered to be at high risk for *ADHD*, comorbid psychiatric disorders, and impairments in intellectual functioning (Morrison and Stewart, 1971; Cantwell, 1972; Welner et al., 1977; Biederman et al., 1992; Faraone et al., 1993a, 1993b; Faraone and Biederman, 1994a, 1994b). Results of this study (Milberger et al., 1998) showed that 47 percent of the high-risk siblings with *ADHD* had a history of maternal smoking during pregnancy compared with 24 percent of the siblings without *ADHD*. This positive association remained significant after adjusting for socioeconomic status, parental IQ, and parental *ADHD* status. Lower IQ scores were found in those high-risk siblings whose mothers smoked during pregnancy compared with those whose mothers did not smoke. The results are consistent with a body of literature linking maternal smoking during pregnancy and impaired behavioral and cognitive functioning in their offspring (Butters and Goldstein, 1973; Denson et al., 1975; Sexton, 1978; Rantakillo, 1983; Naeye and Peters, 1984; Fogelman and Manor, 1988; Sexton et al., 1990; Makin et al., 1991).

The reason for the association between *ADHD* and cigarette smoking is not known; cigarette smoking, however, has been shown to increase the level of *carboxyhemoglobin* in both maternal and fetal blood, resulting in fetal *hypoxia* (Fielding, 1985). Another explanation is that nicotine receptors modulate *DA* activity. *Dopaminergic* dysregulation is thought to be involved in the underlying pathophysiology of *ADHD*. It has been stated that several lower organism and human studies support the *nicotine receptor hypothesis* linking *ADHD* and cigarette smoking (Hagino and Lee, 1985; Marks et al., 1992, 1993; Slotkin et al., 1993; Freedman et al., 1994). Maternal smoking has been linked to low birth weight (Landesman-Dwyer and Emanuel, 1979; Kline et al., 1989) that has been associated with *ADHD* (Breslau et al., 1996). Another study correlating parental smoking with behavioral problems in offspring (Wakschlag et al., 1997), shows that *conduct disorder*, a diagnosis designating serious antisocial behavior, is more than four times as common in boys whose mothers smoked more than 10 cigarettes per day during pregnancy. The 6-year study involved 177 pre-teenage boys. It has also been shown that nutrition, medication, and other environmental chemicals may affect the fetus as well.

Prenatal Stress

Asymmetry of the brain, as well as the development of the brain of the fetus has been shown to be affected by maternal stress responses. Cortical asymmetry in males has been demonstrated at birth (Diamond, 1983) and at 6 days of age (Diamond et al., 1980). Males *gonadectomized* at birth lose their usual right-greater-than-left asymmetry and are left-greater-than-right asymmetric in some cortical regions, whereas females *ovariectomized* at birth show an opposite shift in cortical asymmetry to the right (Diamond et al., 1979). Exposing female rats to stressful conditions during pregnancy lowers fetal serum *testosterone* in male progeny (Ward and Weisz, 1980) but not in females (Ward and Weisz, 1984). Prenatal maternal stress causes male offspring to shift from right-greater-than-left to left-greater-than-right in somatosensory cortical asymmetry (Anderson et al., 1986). The fact that both prenatal stress and day 1 *gonadectomy* alter cortical asymmetry indicates that the critical period for establishing asymmetry begins prior to birth and extends into early postnatal life (Lewis and Diamond, 1995). It has been noted that although the cortex remains plastic in adulthood to influences like environmental enrichment (Smith, 1934; Diamond et al., 1964; Bennett et al., 1974) it has been shown that these interventions affect cortical thickness bilaterally more often than asymmetrically (Diamond, 1985). It is interesting to note that males seem to be affected more than females to maternal stress that reduces the development of the right hemisphere. This may be an important factor in that more males appear to suffer from right brain deficits like *ADHD*, since the male brain normally is larger on the right than the left.

A study conducted at the University of California (Wadhwa et al., 1996, 2001, 2002) suggests that psychological stress in pregnant women may result in temperamental and behavioral difficulties in her offspring. This research involved 120 infants and toddlers, relating maternal stress during pregnancy to resultant behavior using interviews and psychological testing. Some scientists think that this and other studies may help relate some behaviors to early pre-birth trauma.

What we see is that anything that may affect the function of the brain and the regulation of the autonomic, immune, and hormonal function of the mother may have effects on a developing fetus. A history of sedentary activity before or after the pregnancy or anything that may cause an imbalance of activation of the neocortex increasing the mothers *sympathetic* nervous system and circulating stress hormone levels, may have significant effects on intelligence and behavior. *Oxidative stress* or nutritional deficiencies as well as environmental toxins like drugs and alcohol may have a direct effect on the fetus or an indirect effect by damaging the mother's central nervous system.

The Birthing Process

It has been widely recognized in recent scientific literature that the birth process itself very often results in injury to the baby especially to the cervical spine. These frequent injuries often go unrecognized by physicians who are unfamiliar with how to diagnose and/or treat these types of birth-related injuries. These injuries if untreated may result in significant developmental and functional neurological problems that may persist throughout the individual's life. These injuries often result in decreased motion of cervical vertebrae and muscles which provide a large amount of ongoing stimulus to *cerebellum, thalamus* and brain, more in fact than any other single area of the body. Robbing the brain of this stimulation and information, usually unilaterally, can cause significant imbalances of brain activity, which has been shown to be the foundation of all brain function. Although humans have gained much selective advantage from achieving a bipedal posture, especially its apparent impact on the development of the neocortex, we did give up one advantage with *quadrapedalism*. In achieving a bipedal position, we humans needed to reposition the pelvic bones resulting in a smaller birth canal. This posed several problems, since human babies have relatively large heads, and yet our brains still need to grow four times as large as any other primate. The main problem we wish to consider is that the bipedal posture results in the fact that traveling through the birth canal may be the most dangerous journey a human baby makes.

Modern medicine has not been able to make the journey much safer. In fact, a study (Kiely, 1991) indicates that 10 percent of all babies that die at birth in the United States die from neurological damage to the *brainstem* caused by the medically assisted delivery process. This translates into 27,000 preventable infant deaths that occur every year from an injury known as *decerebration*. *Decerebration* has been said to be a form of stillbirth that results from using the head as a lever to forcibly extract the baby from the womb by using manual traction or mechanical extractors. *Decerebration* in a gentle way of saying that a professional person pulled a little too hard on the infant's head during delivery and disconnected the spinal cord from the brain resulting in death. It is reasonable to assume that if a trained medical professional is accidentally causing such severe tissue damage, then an even larger percentage may be causing subclinical soft tissue damage of the cervical spine. However, since this is never actually examined after birth and there may not be any immediate overt symptoms, this trauma is usually overlooked.

Birth trauma resulting in altered mobility of vertebral segments or muscle tone can affect development of subcortical and neocortical structures (Yates, 1967; Khaibullina, 1980; DeToledo and Haddad, 1999). This can produce a number of symptoms like deficiencies of IQ or learning disabilities, behavioral problems, autonomic, sensory, or motor dysfunction (Upledger, 1978; Herman et al., 1985; Sarmiere and Bamburg, 2002). Many of these symptoms especially learning disabilities and behavioral problems may be associated with a dysfunction of afferentation of the cervical spine, which was a result of birth trauma. When we consider the high percentage of children who probably suffer this type of trauma, it may actually be one of the largest causes of these higher cognitive and behavioral problems.

Some of the symptoms that may be recognized are known as *torticollis*, unilateral *microsomy, C-scoliosis*, and motoric asymmetries which are often accompanied by unilateral maturation of hip joints and slowed motor development. This complex has been referred to as *kinematic imbalance due to suboccipital strain (KISS syndrome)* (Biedermann, 1991). It is noted by Biedermann that the pathogenic importance of asymmetric posture and motion in small children is oftentimes played down or not recognized. He also notes that parents frequently complain about the bent position of their baby, and they are told that it will go away, and he thinks that this is not true in many cases. Biedermann further refers to a study in which more than 600 children (age <24 months) were treated for asymmetric posture. Of the original group of children, 135 babies referred between 1981 and 1990 were examined. The case histories included: tilt posture of head, *torticollis, opisthotonos*, head held in retro-flexion, unable to lie on back, uniform sleeping posture, the child cries if the mother tries to change its position, asymmetric motor patterns, asymmetric posture of trunk and extremities sometimes combined with a tilted head position reminiscent of persisting asymmetric tonic neck reflex, sleeping disorders (the baby wakes up crying every hour), extreme sensitivity of the neck, cranial scoliosis, swelling of one side of the facial soft tissue, blockage of the *ilbosocral* points, asymmetries of the *gluteus* muscles, asymmetric development and range of motion of hips, and fever of unknown origin, loss of appetite, and other symptoms of central nervous system disorder.

It was noted that the case histories of these babies reveal birth trauma in a higher percentage than found in the general population, and Biedermann reports his findings shown in Table 10.1.

Researchers and clinicians have considered intra-cerebral damage as the primary cause of abnormal posture and asymmetric development (Bobath, 1976). It has been pointed out

TABLE 10.1 Symptoms of Birth Trauma as Reported by Biedermann

Condition	Percent of Population
Mobility of the cervical spine reduced more than 30 percent	74.6
Torticollis	38.6
C-scoliosis	35.1
Unilateral microsomy	41.2
Retarded development of hip joints	43.9
Opisthotonos	43.9
Mobility of vertebral spine reduced more than 30 percent	10.5
Feet deformities	10.5
Pathological reflexes	3.2

that these theories point to the central nervous system as the first link and the muscle as the end result of this pathological process. Biederman (1992) states, "The immense pathogenic potential of the *proprioceptive afferences* of the sub-occipital" region has until now been widely underestimated." Gutmann described what he referred to as the *atlas blockage syndrome* (Gutmann, 1970, 1971; Vele and Gutmann, 1971). He focused on the pathogenic significance of the *cervical-occipital* junction. Seifert (1974) examined a random sample of over 1,000 newborns and found 11 percent with blockages in this region. In this group of 119 babies, 78 percent showed asymmetries of the vertebral spine (*scobotic posture*). It is thought that most of these children did not present with obvious symptoms since most were excluded from the study. Biederman (1992) notes that since it has been shown that 6 percent of British primary school children have significant disorders of their visuomotor system (Brenner et al., 1967), he states that, "Head stabilization ... is a complex process involving interaction of reflexes elicited by *vestibular*, visual, and *proprioceptive* signals." He further indicates that most of the afferent *proprioceptive* signals originate from the cranial–cervical junction. Therefore, anything that may interfere or interrupt these afferent impulses will have much more extensive consequences in a nervous system in formation, which depends on appropriate stimuli to organize itself. "Most of the cerebral development lies still before the newborn," state Buchmann and Bülow (1989). Thus, development begins at the head (Flemming, 1979).

During the delivery, the head is rotated about 90° and pressed against the trunk by the contractions of the uterine muscles. It is thought that a majority of newborns suffer from micro-trauma of *brainstem* tissue in the *peri-ventricular* areas (Frymann, 1966; Chagnon and Blery, 1982). Frymann has reported on the often missed or forgotten traumas of early childhood and its impact on perceptual-motor development. It is thought that traumatization of the sub-occipital structures inhibits functioning of the *proprioceptive* feedback loops therefore; motor development cannot proceed normally (Biederman, 1991). Since the majority of *proprioceptive* input projects to the *cerebellum* and follows *cerebellar-thalamo* cortical loops to the cortex, this input has been shown to be important to the development and function of the cortex. The *cerebellum* also coordinates *ocular–motor* activity and balances *vestibular*, visual, and *proprioceptive* input. Asymmetric motor activity would result in asymmetric *cerebellar* and cerebral activation. This imbalance of muscle tone, which produces *torticollis*, exacerbates the *proprioceptive* and *ocular–motor* deficits and perpetuates the problem. Subclinical decreased vertebral mobility may also result in significant deficits of the *cerebellum* and cerebrum. Children may show only minor symptoms in the first few months of life (e.g., temporary fixation of the head in one position) and appear to "recover" spontaneously, but by age 5 or 6, they may suffer from headaches, postural problems, or diffuse disorders, including being unable to concentrate (Biedermann, 1991). Dislocation of the atlas to the right in relation to the *occiput* is found more frequently. This is thought to be in accordance with Jirout (1980) who described similar asymmetries found in the majority of adults, largely between C_1 and C_2.

The Neck and Movement in Relation to Brain Development

Leary and Hill (1996) have reported on the presence of movement disturbance symptoms in individuals with *autism*. Typically, these symptoms have been seen as peripheral to *autism* or as belonging to a co-occurring syndrome. Some authors have dismissed these symptoms as having no apparent impact on the presence of behaviors defined as the core characteristics of *autism*. The authors considered the relation between symptoms of movement disturbance and symptoms of *autism* and included speculative and exploratory analyses of shared symptoms. Their analyses point out the difficulties posed by current definitions of *autism*. Leary and Hill propose that symptoms of movement disturbance can affect a person's experience of life and how he or she may be perceived by others.

Teitelbaum and colleagues (1998) studied 17 *autistic* children, each of whom demonstrated disturbances of movement that could be detected clearly at the age of 4–6 months, and sometimes even at birth. They used the *Eshkol–Wachman Movement Analysis System* in combination with still-frame videodisc analysis to study videos obtained from parents of children who had been diagnosed as *autistic* by conventional methods, usually around 3 years old. The videos showed their behaviors when they were infants, long before they had been diagnosed as *autistic*. The movement disorders varied from child to child. Disturbances were revealed in the shape of the mouth and in some or all of the milestones of development including, lying, righting, sitting, crawling, and walking. Their findings support the view that movement disturbances play an intrinsic part in the phenomenon of *autism* and likely *autistic spectrum disorders* including *ADHD* and that they are present at birth.

Green and associates (2002) attempted the objective measurement and extent and severity of motor impairment in children with *Asperger's Syndrome* (AS) to determine whether the motor difficulties experienced by such children differed in any way from those classified as having a *Specific Developmental Disorder of Motor Function* (*SDD-MF*). The *Movement Assessment Battery for Children* (Henderson and Sugdend, 1992) provided a standardized test of motor impairment. A *Gesture Test* based on that by Sharon Cermak and colleagues (1980) was used to assess the child's ability to mime the use of familiar tools and to imitate meaningless sequences of movements. All the children with AS turned out to meet the criterion for a diagnosis of motor impairment. However, no evidence of group differences in the pattern of impairment was found. This study is consistent with others suggesting a high prevalence of clumsiness in AS. The authors also find support for the widespread prevalence of motor impairment in developmental disorders and the problems such comorbidity poses for attempts to posit discrete and functionally coherent impairments underlying distinct syndromes.

Rinehart and colleagues (2001) note that *autism* and AS have long been associated with movement abnormalities, although the neurobehavioral details of these abnormalities remain poorly defined. Clumsiness has traditionally been associated with *Asperger's*, but not *autism*. Others have suggested that both groups demonstrate a similar global motor delay. In their study, movement preparation or movement execution was examined by means of a motor reprogramming task. The results indicated that individuals with *autism* and AS have atypical movement preparation with an intact ability to execute movement. An atypical deficit in motor preparation is noted in *Asperger's*, whereas movement preparation is characterized by a "lack of anticipation" in *autism*. The differences in movement preparation profiles in these disorders suggest differential involvement of the *frontal-striatal* region, in particular, the *supplementary motor area* and *anterior cingulate*.

Early neuro-developmental pathogenesis in *autism* potentially affects emerging functional maps, but little imaging evidence is available to support this notion. In an attempt to remedy the situation, Muller and associates (2001) studied eight male *autistic* and eight matched normal subjects, using functional MRI during visually paced finger movement, compared to a control condition (visual stimulation in the absence of motor response). Results demonstrate activation in contralateral *perirolandic* cortex, *basal ganglia*, and *thalamus*, bilateral *supplementary motor area*, and ipsilateral *cerebellum* for both groups. However, activations are less pronounced in *autistics*. Direct group comparisons demonstrate greater activation in *perirolandic* and *supplementary motor areas* in the control group and greater activation (or reduced deactivation) in *posterior* and *prefrontal* cortices in the *autism* group. Intraindividual analyses further show that strongest activations are consistently located along the contralateral *central sulcus* in control subjects but occur in locations differing from individual to individual in the *autism* group. The authors conclude that there exists abnormal individual variability of functional maps and less distinct regional activation/deactivation patterns in *autism*. The observations relate to known motor impairments in *autism* and are compatible with the general hypothesis of disturbances of functional differentiation in the *autistic* cerebrum affecting motor function.

By assessing motor function in adult *autistics*, we may be able to glean a better understanding of the nature of motor deficit developmentally in children with neurobehavioral disorders. Kinesiological analysis of the gaits of 21 *autistic*, 15 normal, and 5 hyperactive-aggressive children revealed that the *autistic* patients had: (1) reduced stride lengths, (2) increased stance times, (3) increased hip flexion at "toe-off," and (4) decreased knee extension and ankle dorsiflexion at ground contact. In many respects, the gait differences between the *autistic* and normal subjects resemble differences between the gaits of *Parkinson's* patients and of normal adults. These results are compatible with the view that the *autistic* syndrome may be associated with specific dysfunction of the motor system affecting, among other structures, the *basal ganglia* (Vilensky et al., 1981). However, Hallett and colleagues (1993) attempted to assess gait in patients with *autism*. Five adults with *autism* and five healthy, age-matched control subjects were evaluated by them. Clinical and biomechanical assessment revealed mild clumsiness in four patients and upper limb posturing during gait in three patients. The velocity of gait, step length, cadence, step width, stance time, and vertical ground reaction forces were normal in all patients. The only significant abnormality was decreased range of motion of the ankle. Some patients exhibited slightly decreased knee flexion in early stance. Clinically, the gait appeared to be irregular in three patients, but the variability was not significantly increased. The findings in adult patients with *autism* indicate a nonspecific, neurological disturbance involving the motor system. The normal velocity of gait and the normal step length argue against a *Parkinson's*-type disturbance, whereas the clinical picture suggests a disturbance of the *cerebellum*.

Bram and Meir (1977) had noted an important link existing between the development of motor control and language development. Diagnostic and therapeutic work with *autistic* children reveals a high incidence of gross motor dysfunction along with language deficits. A study to evaluate the effectiveness of behavior modification techniques in eliciting speech from nursery-age *autistic* children yielded results to confirm the importance of the relationship. Results from a case study of one *autistic* child demonstrated that the child most frequently vocalized when motorically quiet, and engaged in motor activity when not speaking and especially prior to speech. Data from 30 half-hour behavior modification sessions and an intensive clinical interview indicated that limiting the child's gross motor activity is effective in increasing the frequency of vocalization. Kern and colleagues (1984) have noted that the experimental literature indicates that physical exercise can positively influence both appropriate and inappropriate behaviors in *autistics*, including the children's stereotypic behaviors, they studied whether the specific type of exercise (i.e., mild vs. vigorous) would differentially affect subsequent stereotyped behaviors. Their results demonstrate that (1) 15 min of mild exercise (ball playing) has little or no influence on the *autistic* children's subsequent stereotyped responding, and (2) 15 min of continuous and vigorous exercise (jogging) is always followed by reductions in stereotyped behaviors.

Children especially infants, should they make it through the birth canal intact, are still extremely susceptible to neck trauma. Babies and children have extremely large heads for their body size and they sit on very small little necks. The muscles of the neck are extremely immature and weak and cannot support the head. This makes it easy for a child to suffer a whiplash injury to the cervical spine. Injury can occur from an adult picking a child up improperly, pulling a child, hitting a child, or from falls where children might strike their heads. If a child hits its head hard enough to produce a cut or bruise to the head, face, or mouth, they most likely will also suffer a whiplash injury to the neck. This may result in subtle or severe unilateral muscle spasm. Unilateral muscle spasms of the neck may result in an asymmetric posture or head tilt. This may cause or also be a result of unilateral *ocular–motor* palsies. Asymmetric motor activity is a potential significant problem because it will most likely result in asymmetric activation of the cerebral cortex and especially the *prefrontal* cortex of the *frontal lobe*. Any motor dysfunction usually indicates a dysfunction of the motor portion of the *frontal lobe*. We have seen that non-motor dysfunction (cognitive and behavioral) of the *frontal lobe* appears to parallel the motor dysfunction.

In adults, whiplash injury has been associated with a constellation of symptoms. Bogduk and colleagues (1994) in an extensive clinical review of whiplash injuries, provide the following definition: "the term whiplash has been applied to the mechanism of injury, to the injury resulting from the mechanism, (whiplash injury) and to the syndrome of neck pain with or without other symptoms following such an injury (whiplash syndrome)." The following represents a list of symptoms most commonly reported in various studies, neck pain, headache, visual disturbance, dizziness, auditory and *vestibular* symptoms, weakness of muscles, often accompanied by consistent reflex and sensory signs as well as motor disturbances of both upper and lower extremity, concentration and memory disturbances, and psychological symptoms. Although some of these injuries may also be related to mild traumatic brain injury, the majority of symptoms are probably a result of abnormal afferentation of neck joints and muscles. Bogduk also states:

> in the acute phase, most patients exhibit features of muscular pain, but are destined to recover. Among them, however are patients with disc and *zygapophysical* joint injuries that cannot be seen on plain radiographs and are elusive even on CT or MRI. These injuries do not cause neurological signs and exhibit no known *pathognomonic* clinical features. Consequently they are not diagnosed or recognized.

Infection and Immunity

Although musculoskeletal injuries and their potential neurological consequences may be by far the greatest risk, childhood illness is also a potential problem. Acute or chronic infections or allergies can not only cause direct damage, but can result in neurological developmental delays. Illness again can result in an extended period of convalescence and sedentary behavior. Sedentary behavior can result in hemispheric deficits but infection or allergies can result in specific neurological deficits.

Perinatal exposure to infectious agents and toxins is linked to the pathogenesis of neuropsychiatric disorders, but the mechanisms by which environmental triggers interact with developing immune and neural elements to create neurodevelopmental disturbances are poorly understood. Hornig and associates (1999) describe a model for investigating disorders of central nervous system development based on neonatal rat infection with *Borna disease virus*, a *neurotropic* non-cytolytic *RNA* virus. Infection results in abnormal righting reflexes, hyperactivity, inhibition of open-field exploration, and stereotypic behaviors. Architecture is markedly disrupted in the *hippocampus* and *cerebellum*, with reduction in *granule* and *Purkinje* cell numbers. Neurons are lost predominantly by *apoptosis*, as supported by increased *mRNA* levels for *pro-apoptotic* products, decreased *mRNA* levels for the *anti-apoptotic bcl-x*, and in situ labeling of fragmented *DNA*. Although inflammatory infiltrates are observed transiently in *frontal* cortex, *glial* activation is prominent throughout the brain and persists for several weeks in concert with increased levels of pro-inflammatory *cytokine mRNAs* and progressive *hippocampal* and *cerebellar* damage. These functional and neuropathological abnormalities resemble human neurodevelopmental disorders suggesting interactions of environmental influences with the developing central nervous system.

Fatemi and associates (2002b) investigated the role of maternal exposure to *human influenza virus* (*H1N1*) in mice on Day 9 of pregnancy on *pyramidal* and *non-pyramidal* cell density, *pyramidal* nuclear area, and overall brain size in Day 0 neonates and 14-week-old progeny and compared them to sham-infected cohorts. *Pyramidal* cell density increased significantly by 170 percent in Day 0 infected mice vs. controls. *Pyramidal* cell nuclear size decreased significantly by 29 percent in exposed newborn mice vs. controls. Fourteen-week-old exposed mice continued to show significant increases in both *pyramidal* and *non-pyramidal* cell density values vs. controls, respectively. Similarly, *pyramidal* cell nuclear size exhibited 37–43 percent reductions when compared to control values; these were statistically significant vs. controls. Brain and ventricular area measurements in adult exposed mice also showed significant increases and decreases, respectively, vs. controls. Ventricular brain ratios exhibited 38–50 percent decreases in exposed mice vs. controls. While the rate of *pyramidal* cell proliferation per unit area decreased from birth to adulthood in both control and exposed groups, *non-pyramidal* cell growth rate increased only in the exposed adult mice. These data show for the first time that prenatal exposure of pregnant mice on Day 9 of pregnancy to a sublethal intranasal administration of influenza virus has both short-term and long-lasting deleterious effects on developing brain structure in the progeny as evident by altered *pyramidal* and *non-pyramidal* cell density values; atrophy of *pyramidal* cells despite normal cell proliferation rate and final enlargement of brain. Moreover, abnormal *cortico-genesis* is associated with development of abnormal behavior in the exposed adult mice.

Shi and colleagues (2003) note that maternal viral infection is known to increase the risk for schizophrenia and *autism* in the offspring. Using this observation in an organism model, they find that respiratory infection of pregnant mice with the human influenza virus yields offspring that display highly abnormal behavioral responses as adults. As in *schizophrenia* and *autism*, these offspring display deficits in *prepulse inhibition* (*PPI*) in the acoustic startle response. Compared with control mice, the infected mice also display striking responses to the acute administration of *antipsychotic* (*clozapine* and *chlorpromazine*) and *psychomimetic* (*ketamine*) drugs. Moreover, these mice are deficient in exploratory behavior in both open-field and novel-object tests and they are deficient in social interaction. At least some of these behavioral changes likely are attributable to the maternal immune response itself. That is, maternal injection of the synthetic double-stranded RNA *polyinosinic-polycytidylic* acid causes a *PPI* deficit in the offspring in the absence of virus. Therefore, maternal viral infection has a profound effect on the behavior of adult offspring, probably via an effect of the maternal immune response on the fetus.

Jyonouchi and colleagues (2001) at the *University of Minnesota* determined innate and adaptive immune responses in children with developmental regression and *autism spectrum disorders* (*ASD*), developmentally normal siblings, and controls with *lipopolysaccharide* (*LPS*), a stimulant for innate immunity and *peripheral blood mononuclear cells* (*PBMCs*). ASD patients produced greater than 2 SD above the control mean values of *TNF-alpha, IL-1beta*, and/or *IL-6* produced by control *PBMCs*. *ASD PBMCs* produced higher levels of pro-inflammatory/counter-regulatory *cytokines* without stimuli than did controls. With stimulants of *phytohemagglutinin* (*PHA*), tetanus, *IL-12p70*, and *IL-18*, *PBMCs* from 47.9 to 60 percent of the *ASD* patients produced greater than 2 SD above the control mean values of *TNF-alpha* depending on stimulants. The results indicate excessive innate immune responses in a number of *ASD* children.

Juul-Dam and associates (2001) examined various pre-, peri-, and neonatal factors in *autistic* participants and in *pervasive developmental disorder—not otherwise specified* (*PDD-NOS*) participants compared with the incidence of each factor to that of the normal population. Twenty-eight pre-, peri-, and neonatal factors were examined in these two groups using both medical records and parental interviews. Although most of the factors showed comparable incidences between the index and control groups, several factors showed statistically significant differences. The *autism* group was found to have a significantly higher incidence of uterine bleeding, a lower incidence of maternal vaginal infection, and less maternal use of contraceptives during conception when compared with the general population. Similarly, the *PDD-NOS* group showed a higher incidence of *hyperbilirubinemia* when compared with the general population. The results of this study support a consistent association of unfavorable events in pregnancy, delivery, and the neonatal phase

Wilkerson and associates (2002) attempted to examine the relationship between perinatal complications of pregnancy and neurobehavioral disorders in their children. In their study, the biological mothers of 183 *autistic* children and 209 normals completed the *Maternal Perinatal Scale (MPS)*, a maternal self-report that surveys complications of pregnancies and medical conditions of the mother. Performing a discriminant function analysis, the researchers examined perinatal complications as predictors between the *autistic* and normal subjects. Using the MPS, 65 percent of the *autistic* cases were correctly grouped. The results further indicated significant differences on three of the ten factors of the MPS, in particular, Gestational Age, Maternal Morphology, and Intrauterine Stress. When considered in an item-by-item fashion, five items were found to significantly predict group membership (prescriptions taken during pregnancy, length of labor, viral infection, abnormal presentation at delivery, and low birth weight). Finally, three maternal medical conditions were found to be highly significant and to contribute to the separation between groups including urinary infection, high temperatures, and depression.

Stella Chess in the 1970s (1977) performed a longitudinal study of 243 children with congenital *rubella*. In this sample, a high rate of *autism* was observed. Examination of the data suggests that the *rubella* virus was the primary etiological agent. She had hypothesized that the course of *autism* was that of a chronic infection in which recovery, chronicity, improvement, worsening, and delayed appearance of the *autistic* syndrome were all found. Other *rubella* consequences such as blindness, deafness, and cardiac and neuromuscular defects remained present. While the degree of mental retardation initially was related to the outcome of *autism*, shifts in mental retardation over time did not correlate significantly for the group with shift in the *autistic* symptoms. Matarazzo (2002) described two cases of children who at first developed normally, but before the age of 3 developed *autistic* symptoms following the reactivation of a chronic *oto-rhino-laryngologic* infection. The clinical and laboratory data of the cases support the etiological hypothesis of an autoimmune process. *ACTH*, prescribed in the first months of the disease, cured one case. The other patient, who was 2 years old when *autistic* symptoms appeared, and was treated only 6 years later, showed a partial but definitive improvement with the immunosuppressive treatment.

PARENTAL PHYSICAL AND ENVIRONMENTAL INFLUENCES

While it now clearly understood that poor parenting and lack of parental warmth has nothing whatsoever to do with the causation of ASD and is clearly not the cause of most behavior problems like *ADHD* or learning disabilities, the relation between parenting practice as a means of providing enrichment for the developing infant, impaired or not, should be addressed.

Fleming and colleagues (1999) note that the optimal coordination between the new mammalian mother and her young involves a sequence of behaviors on the part of each that ensures that the young will be adequately cared for and show healthy physical, emotional, and social development. This coordination is accomplished by each member of the relationship having the appropriate sensitivities and responses to cues that characterize the other. Among many mammalian species, new mothers are attracted to their infants' odors and some recognize them based on their odors; they also respond to their infants' vocalizations, thermal properties, and touch qualities. Together these cues ensure that the mother will nurse and protect the offspring and provide them with the appropriate physical and stimulus environment in which to develop. The young, in turn, orient to the mother and show a suckling pattern that reflects sensitivity to the mother's odor, touch, and temperature characteristics. Fleming and colleagues emphasize the importance of learning and plasticity in the formation and maintenance of the mother–young relationship and mediation of these experience effects by the brain and its neurochemistry.

Mark Smith and colleagues (1997), and Michael Meaney and associates (2001a, 2001b) have found in newborn lab rats that separation led to substantial increase in the number of brain cells that die normally during development. The untouched, unloved organisms had twice the rate of dead neurons than did those who were stroked by their mothers. Joseph (1999) examined the effects of early environmental influences on neural plasticity, the *limbic system*, and social and emotional development. He indicated that deprivation or abnormal rearing conditions induce severe disturbance in all aspects of social and emotional functioning, and affect the growth and survival of dendrites, axons, synapses, interneurons, neurons, and *glia*. The *amygdala, cingulate*, and *septal nuclei* develop at different rates, which correlate with the emergence of wariness, fear, selective attachments, and play behavior. These immature *limbic* nuclei are *experience-expectant*, and may be differentially injured depending on the age at which they suffer deprivation. The medial *amygdala* and later the *cingulate* and *septal nuclei* are the most vulnerable during the first 3 years of life. If denied sufficient stimulation, these nuclei may atrophy, develop seizure-like activity or maintain or form abnormal synaptic interconnections, resulting in social withdrawal, pathological shyness, explosive and inappropriate emotionality, and an inability to form normal emotional attachments. While neurobehavioral disorders are clearly not the result of parenting relationships with the infant, the disorders themselves will exacerbate the already compromised developing brain and nervous system.

Zhang and associates (2002) note that prolonged separation from the mother can interfere with normal growth and development and is a significant risk factor for adult psychopathology. In rodents, separation of a pup from its mother increases the behavioral and endocrine responses to stress for the lifetime of the organism. They investigated whether maternal deprivation could affect brain development of infant rats via changes in the rate of cell death as measured by *DNA* fragments using *terminal transferase (ApopTag)*. At postnatal day 12, the number of cells undergoing cell death approximately doubled in the cerebral cortex, *cerebellar* cortex and in several white matter tracts following 24 hr of maternal deprivation. Deprivation strongly increased the number of *ApopTag*-labeled cells. Stroking the infant rats only partially reversed the effects of maternal deprivation. Increased cell death in white matter tracts correlated with an induction of nerve growth factor previously associated with *oligodendrocyte* cell death. Cell birth was either unchanged or decreased in response to deprivation. Their results indicate that maternal deprivation can alter normal brain development by increasing cell death of neurons and *glia*, and provides a potential mechanism by which early environmental stressors may influence subsequent behavior. In examining the effects of maternal deprivation in rats (Grove et al., 2001) in the dorsomedial *hypothalamic nucleus (DMH), peri-fornical region (PFR)*, in addition to the *arcuate nucleus (ARH)*, significant effects are noted in the *ARH* in response to maternal deprivation suggesting to the authors that these neuronal populations respond to signals of energy balance. Also noted are effects in the *DMH* and *PFR*, which are differentially regulated by maternal deprivation and factors associated with maternal separation.

Mary Carlson, a Harvard researcher, has spent 3 years studying the impact of neglect on Romanian orphans living in state run wards. She has chronicled (Carlson and Earls, 1997) abnormal levels of stress hormones in these children, who profoundly lack affection. She has noted that it has been about 40 years since Harry Harlow conducted the classic experiments removing monkeys from their mothers and observing that the experience forever changed them. When she had seen news clips of Romanian children in orphanages, she was reminded of her Harlow's monkeys. She found that the ocular–motor or gaze and behavior of the abandoned infants looked much like the rhesus monkeys and chimps separated from their mothers that she had studied in laboratories throughout her career. She collected saliva from the Romanian orphans and studied the production of stress hormones to get a chemical picture of how the children are responding to their environment. She had also become aware of a yearlong project instituted by the University of North Carolina psychologist Joseph Sparling to study the effects of a social enrichment on Romanian children in orphanages. Half of the 30 orphans selected for the study were randomly assigned to live in a special environment with a caregiver for every four children and an intense learning program. Each child was held, cuddled, and spoken to by name. The other children lived life as usual, in large rooms of 20 children to one caretaker. Sparling allowed Carlson to collect saliva from the groups tested, by this time it had already been 6 months since the study was completed. The children were back in the normal population of crowded rooms. She would also study other orphans and Romanian children living at home with their parents. They collected saliva from children morning, noon, and night, on two separate days. They returned three times for further studies. *Cortisol* is a hormone that is implicated in many everyday functions including the body's stress response. In the morning, it is thought to mobilize glucose, a main source of fuel for the brain. A normal person's level of *cortisol* is highest in the morning and it should decrease as the day goes on. Carlson found that the Romanian orphans were nothing like normal children. Their morning levels were suppressed and noon and evening levels were high. Other studies have shown that children score highest on tests of mental and motor abilities when *cortisol* levels are highest in the morning. She reportedly found similar suppressed levels when she looked at Romanian toddlers living the week in day care centers. However, when she tested these same children at home on the weekend, their levels were more normal albeit slightly higher than average. Those children with the most elevated *cortisol* levels at noon, according to Carlson, were the same ones who had the lowest behavioral scores. High *cortisol* is thought to interfere with growth and regulation of the immune and reproductive system, and is regulated primarily by the right hemisphere. Carlson thinks that human touch or tactile sensation or the lack of it plays an important role in regulating the *cortisol* system in the brain. The children in the enrichment program seemed to have more normal levels of *cortisol*, when they reached 3 years old, a time when everyone in the orphanage began receiving educational instruction. Carlson indicated that the relationship between *cortisol* level and general behavior in both groups suggests that the positive effects of stimulation, as well as the negative effects of deprivation and stress, may have persistent neural and behavioral consequences for human infants.

McGill psychologist Michael Meaney (Lupien et al., 2000; Lupien and Lepage, 2001) thinks touch is vital for children to develop normally. His research suggests that *corticotrophin-releasing hormone* (*CRH*) is intimately tied to the amount of separation the infant experiences. He found that organisms exposed to long periods of maternal deprivation in early life have substantially higher levels of *CRH* in the *hypothalamus* and therefore have a greater response to stress. Meaney's group thinks that, the amount of maternal licking during the first 10 days of life is highly correlated with the production of *CRH* in the *hypothalamus* of the adult offspring. *CRH* has also been strongly implicated in depression. It is part of the *hypothalamus–pituitary–adrenal* axis, which is thought to be overactive in depressed individuals, and *cortisol* has been shown to be high. The *hypothalamus* is part of the *limbic system* and is inhibited by the neocortex especially the *frontal lobe*. Tactile sensation or touch is processed through the *cerebellum* and *thalamus*, and receives information via the *dorsal column*. These same pathways are traversed by *proprioceptive* input, which has also been called *subconscious touch*. *Proprioception* is stimulated by movement and gravity.

Deprivation of a human mother's touch can have severe consequences on both a child's developing nervous system and production of stress hormones. Researchers at the University of California at Irvine (de Quervain et al., 1998) have recently shown that elevated levels of a stress hormone *corticosterone*, which is similar to *cortisol*, hinders the ability of the rats to find their way back to a hidden target. They published their results in the journal *Nature*. The University of California researchers first taught rats to swim to a plastic platform hidden lust beneath the waters surface in a steel tank. They then gave the rats a small electric shock and tested how well they were able to swim back to the platform after 2 min, 30 min, and 4 hr. The rats were equally able after 2 min and 4 hr but at least 50 percent less successful after 30 min. The 30-min trial corresponded to a peak level of the stress hormone *cortico*sterone, which was secreted in response to the electric shock. The researchers indicated that they were finding physical evidence of physical changes in the brain caused by stress in infancy.

Human babies have by far the largest period of parental dependence and much of our gestation continues outside the womb. Therefore, if a child does not receive loving contact (tactile/sensory input) and consistent attention, the positive emotion or happiness he would normally feel when a parent or a friend smiles at him is diminished. Also positive emotions like happiness are generated by the left *prefrontal* cortex (Davidson, 1993), therefore, development of this area of the brain may be dependent on the frequent stimulation of parental touch and affection. It is thought that it is impossible to have a healthy attachment without a healthy reward system. It is thought that the *DA* system is involved in this reward system. They are several ways in which children respond to abuse and neglect. Some children disassociate, others become hypervigilant and some shift from one state to another. The children who disassociate withdraw into a fantasy world.

There is an organism model for dissociative states referred to as *immobilization*, in which rats prevented from moving became frozen with fear. Their brains respond by pumping out endogenous opiates, shutting down the release of stress hormones and lowering heart rate. It is interesting that this state occurs with decreased mobility of joints or muscles. Other children respond to abuse by becoming hypervigilant. This has the exact opposite effect on the body—the heart rate raises, the stress system activates. Both of the responses are opposite but may have the same underlying mechanisms. Similar to *hypokinetic* and *hyperkinetic* disorders, the latter is the result, whereas the former is the brain's initial response to lack of stimulation. It is important to note that abuse and neglect are forms of stimulus deprivation, especially positive stimulus like touch, auditory development, and positive emotional feedback. Research by Bruce Perry (Perry and Pollard, 1998), supports the clinical observations that severely abused children have physically smaller brains and lower IQs than those who were not exposed to trauma. Perry has noted by MRI that the *frontal lobes* of these children are especially small. He notes that these children do not have cortical atrophy, commenting on one child in particular he notes, "His MRI was read by a neuroradiologist as *cortical atrophy*, but his *frontal* cortex did not atrophy. It did not grow and shrink, it just never grew" (Diamond and Hopson, 1998). Perry relates to the case of a 12-year-old girl who had been starved during early childhood and thought to be anorexic. This young girl had been diagnosed with having *ADHD*, depression, and *schizoaffective* disorder. Perry theorized that the girl was not anorexic, but was suffering from what he called a *preadolescent failure to thrive*—a wasting syndrome usually seen in babies who do not get enough attention or affection. They are missing the physical, auditory, and visual stimulation needed to develop the brain. A young child's brain can be likened to a growing plant; it needs water and sunlight to grow properly. If it is placed in a dim dry area, it will grow much more slowly and it will be more frail and immature looking. A child needs stimuli like a plant needs water, without which the brain cells will not grow thicker and healthier, they will not sprout new branches or dendrites to connect with other neurons, they will not stimulate the growth of *glial* cells, which help support a thick healthy cells, and the end result is a physically smaller fragile brain, especially neocortex that cannot perform higher functions optimally. The neocortex can also not inhibit the *limbic system* structures like the *amygdala* and *hypothalamus* so the child may have little to no control over emotions and may be impulsive, angry, or violent. This is the alteration of *approach and avoidance* behavior, where some children have too much *avoidance* and lack of *approach*, typical of left-brain deficit, and are withdrawn. Others have decreased *avoidance* and increased *approach*, typical of right-brain deficits and may grow up aggressive. The most primitive parts of the human brain like the *limbic system* and *autonomic* control systems develop first in a child. The neocortex, which will take over the decision-making, will gain primacy later in development as it develops. Until it does, much of a child's decisions are directed by the *amygdala*, making decisions emotion rather than knowledge-based. Fear and anger are the regulating factors that govern much emotion and basic biological needs.

Genetics and Neurotransmitters

Among the most common theories of ASDs are that the primary cause is genetic, and that this genetic problem somehow produces a neurotransmitter deficiency, or "chemical imbalance." Many researchers now think that much of human behavior and many disorders of behavior are genetically driven. The most recent tool of study is quantitative genetics, a statistical approach that explores the causes of variation of traits among individuals. Studies comparing the performance of twins and adopted children on certain tests of cognitive skills, for example, can assess the relative contributions of nature and nurture. The term typically used is *heritability*, a standard measure of the genetic contribution to differences among individuals. *Heritability* tells us what proportion of individual differences in a population, which is known to variance, can be attributed to genes. *Heritability* is a way of explaining what makes people different, not what constitutes a given individual's intelligence. Early research that involved trying to estimate the *heritability* of specific cognitive abilities started with family studies.

The largest family study performed on specific cognitive abilities was done in Hawaii in the 1970s. The study involved more than 1,000 families and sibling pairs. The study examined the correlation between relatives on tests of verbal and spatial ability. Because children share half their genes with each parent and with their brothers and sisters, the highest correlation in test scores that could be expected on genetic basis alone would be 0.5. The Hawaiian study showed that the actual correlation for the verbal and spatial tests were an average 0.25 (DeFries et al., 1974, 1976a, 1976b, 2002). This demonstrates that family members are more alike than unrelated individuals. However, these correlations alone do not show whether cognitive abilities run in families because of genetics or environmental effects.

Researchers have been examining the genetic basis for *ADHD* and ASDs. Recently several investigators have reported that the DRD_4, located on chromosome 11 was linked to *ADHD*. The gene, which comes in several forms, seems to be involved in impulsivity, risk taking, and novelty-seeking behavior. Five separate laboratories have confirmed the findings that 30 percent of people with *ADD* have a certain form of the gene. The genetic formula repeats itself in DRD_4. Normally there are about four repeats. In risk takers including many people with *ADD*, researchers are finding a seven-repeat version. It is thought that the DRD_4 gene determines whether the cell fires in response to *DA*. People with a seven-repeat version have cells in specific regions of the brain that are less sensitive to *DA*, which means it may take a higher level of the neurotransmitter to have the same effect found in normal cell function. It also could mean that these people are more susceptible to anything that may reduce *DA* production (Jiang et al., 2000, 2001; Thapar et al., 2000; Kirley et al., 2002).

Reviewers in the 1960s and early 1970s were skeptical about any substantial role for genetic factors in the etiology of *autism*. A realization that the 2 percent rate of *autism* in siblings (as estimated at that time) was far above the general population base rate, and that this suggested a possible high genetic liability, led to the first small-scale twin study of *autism*. The replicated evidence from both twin and family studies undertaken in the 1970s and 1980s indicated both strong genetic influences and the likelihood that they applied to a phenotype that was much broader than the traditional diagnostic category of *autism*. Medical and chromosomal findings also indicated genetic heterogeneity. Advances in molecular genetics have now led to genome-wide scans of affected relative pair samples with a positive log of the odds to base 10 score for a location on chromosome 7. The major remaining research challenges and the likely clinical benefits that should derive from genetic research are now considered in relation to both current knowledge and that anticipated from research over the next decade.

Andres (2002) presents a review of organism models of PDD and *Autism*. The prevalence is about 7/10,000 taking a restrictive definition and more than 1/500 with a broader definition, including all the PDD. The importance of genetic factors has been highlighted by epidemiological studies showing that *autistic* disorder is one of the most genetic neuropsychiatric disorders. The relative risk of first relatives is about 100-fold higher than the risk in the normal population and the concordance in monozygotic twin is about 60 percent. Different strategies have been applied on the track of susceptibility genes.

The systematic search of linked loci led to contradictory results, in part due to the heterogeneity of the clinical definitions, to the differences in the DNA markers, and to the different methods of analysis used. An oversimplification of the inferred model is probably also a cause of disappointment. While more work is necessary to give a clearer picture, one region emerges more frequently: the long arm of chromosome 7. Several candidate genes have been studied and some have given indications of association: the *Reelin gene and the Wnt2* gene. *Cytogenetical* abnormalities are frequent at *15q11–13*, the region of the *Angelman* and *Prader–Willi* syndrome. Imprinting plays an important role in this region and no candidate gene has yet been identified in *autism*. Biochemical abnormalities have been found in the *serotonin* system. Association and linkage studies gave no consistent results with some *serotonin* receptors and in the transporter, although it seems interesting to go further in the biochemical characterization of the *serotonin* transporter activity, particularly in platelets, easily accessible. Two monogenic *diseases have been associated with autistic disorder: tuberous sclerosis* and *Fragile X*. A better knowledge of the pathophysiology of these disorders can help to understand *autism*. Different other candidate genes have been tested, positive results await replications in other samples, but are now slowly beginning to emerge. Organism models have been developed, generally by knocking out the different candidate genes. Behavior studies have mainly focused on anxiety and learning paradigms. Another group of models results from surgical or toxic lesions of candidate regions in the brain, in general during development. The tools to analyze these organisms are not yet standardized, and an important effort needs to be undertaken.

Genetic research in *autistic spectrum disorder* is still exploratory but the accumulating data is slowly beginning to reveal that genetic factors, yet undetermined, play a significant role in the etiology of the conditions associated with *autism*. There have been numerous full genome-wide searches that have been recently performed to localize susceptibility regions within the genome (Philippe et al.,

1999; Risch et al., 1999). These recent studies have been reviewed by Gutknecht (2001). What is now just beginning to emerge is that strong evidence exists for a genetic etiology of *ADHD*. Smalley and colleagues (2000) used affected sib-pair analysis in 203 families to localize the first major susceptibility locus for *ADHD* to a 12-cM region on chromosome *16p13* building upon an earlier genome-wide scan of this disorder. The region overlaps that highlighted in three genome scans for *autism* and physically maps to a 7-Mb region on *16p13*. These findings suggest that variations in a gene on *16p13* may contribute to common deficits found in both *ADHD* and *autism*. Fisher and associates (2002) of the *Wellcome Trust Centre for Human Genetics* at Oxford University have noted that molecular genetic studies of *ADHD* have previously focused on examining the roles of specific candidate genes, primarily those involved in *dopaminergic* pathways. Fisher's group is the first to have performed systematic genome-wide linkage scans for loci influencing *ADHD* in 126 affected sib pairs. A survey of regions containing 36 genes that have been proposed as candidates for *ADHD* indicated that 29 of these genes, including *DRD4* and *DAT1*, could be excluded. Only three of the candidates—*DRD5, 5HTT*, and *CALCYON*—coincided with sites of positive linkage identified by their screen. Two of the regions highlighted in their study, *2q24* and *16p13*, coincided with the top linkage peaks reported in other recent genome-scan studies of *autistic* sib pairs.

Numerous studies have shown that there is consistent evidence implicating genetic factors in the etiology of *autism*. In some cases, chromosomal abnormalities have been identified. One type of these abnormalities includes gaps and breaks non-randomly located in chromosomes, denominated *fragile sites* (FS). Arrietta and associates (2002) cytogenetically analyzed a group of *autistic* individuals and a normal population, and examined the *FS* found in both samples with the aim of (1) comparing their *FS* expression, (2) ascertaining whether any *FS* could be associated with their *autistic* sample, and (3) examining if there are differences between individual and pooled-data analyses. Their results show statistically significant differences in the spontaneous expression of breakages between patients and controls.

Auranen and associates (2002) recently attempted to identify genetic loci for ASD by performing a two-stage genome-wide scan in 38 Finnish families. The detailed clinical examination of all family members revealed infantile *autism*, but also *AS* and developmental *dysphasia*, in the same set of families. The most significant evidence for linkage was found on chromosome *3q25–27, 1q21–22*, and *7q*, which overlap with the previously reported candidate regions for infantile *autism* and *schizophrenia*, Association, *linkage disequilibrium* (LD), and *haplotype* analyses, provide evidence for shared ancestor alleles among affected individuals. Havlovicova and colleagues (2002) analyzed a sample of 20 patients with ASDs. The patients have been subjected to clinical genetic examination, cytogenetic analysis, and DNA analysis of the *FMR1* gene. In the sample studied, they observed various degrees of mental retardation (18/20), high frequency of complications during pregnancy (10/20), and delivery (10/20), as well as psychiatric disorders, behavioral abnormalities and suicides among the relatives, and increased head circumference and unusually formed ears in the probands. Three patients had different chromosomal aberrations or variants *(t(21;22), inv(9)*, and *inv(10))*. One patient harbored expansion of the *trinucleotide* repeat sequence in the *FMR1* gene on the full mutation level characteristic for the *fragile X* syndrome, and one patient was suspected of suffering from *Rett* syndrome. Their observations confirm and extend the results reported in the literature including microcephaly, ear malformations, indicative of aberrations in the *HOXA1* gene pathway, the occurrence of chromosomal inversions recurrent in *autism*, and peculiarities in the pedigrees of the patients.

Menold and associates (2001) note that *gamma-amino-butyric acid* (*GABA*) is the major inhibitory neurotransmitter in the brain, acting through the *GABAA* receptors. The *GABAA* receptors are comprised of several different homologous subunits, forming a group of receptors that are both structurally and functionally diverse. Three of the *GABAA* receptor subunit genes (*GABRB3*, *GABRA5*, and *GABRG3*) form a cluster on chromosome *15q11-q13*, in a region that has been genetically associated with *autism*. These investigators examined 16 *single nucleotide polymorphisms* (*SNPs*) located within *GABRB3*, *GABRA5*, and *GABRG3* for *LD* in 226 *autistic* families. They tested for *LD* using the *Pedigree Disequilibrium Test* (*PDT*). *PDT* results gave significant evidence that *autism* is associated with two *SNPs* located within the *GABRG3* gene suggesting that the *GABRG3* gene or a gene nearby contributes to genetic risk in *autism*. Among the regions of potential genetic involvement in ASD includes the *glutamate receptor 6* (*GluR6* or *GRIK2*) gene (Jamain et al., 2002). Glutamate is the principal excitatory neurotransmitter in the brain and is directly involved in cognitive functions such as memory and learning. Also, using the *transmission disequilibrium test* (*TDT*), which indicated a significant maternal transmission dysequilibrium and a significant association between *GluR6* and *autism*, mutation screening was performed in 33 affected individuals, revealing *several nucleotide polymorphisms* (*SNPs*), including one amino acid change (*M867I*) in a highly conserved domain of the *intracytoplasmic C-terminal* region of the protein. This change is found in 8 percent of the *autistic* subjects and in 4 percent of the control population. This change seems to be more maternally transmitted than expected to *autistic* males and when all is taken together, Jamain and colleagues' data suggests that *GluR6* is in LD with *autism*.

ASDs have been described as possessing cytogenetic abnormalities in the same region as the *Prader–Willi/Angelman* syndrome region (*15q11–13*), as well as in the 155CA-2 regions related to the GABA *type-A receptor beta3 subunit gene* (*GABRB3*). One study performed an association analysis with *155CA-2* using the *TDT* in a set of 80 *autistic* families (Buxbaum et al., 2002). The authors found an association between *autistic* disorder and *155CA-2* in the families they studied, supporting a role for genetic variants within the *GABA* receptor gene complex in *15q11-13* in *autism*.

Torres and colleagues (2002) in Utah evaluated possible contributions of *HLA-DRB1* alleles to ASD in 103 families of Caucasian descent. The *DR4* allele occurred significantly more often in probands than controls, whereas the *DR13, 14* alleles occurred less often in probands than controls. The *disequilibrium test* indicated that the *autistic* probands inherited the *DR4* allele more frequently than expected from the fathers and revealed that fewer *DR13* alleles than expected were inherited from the mother by *autistic* probands. The conclusion is that the *DR4* and *DR13* alleles are linked to *autism* with transmission by parent-of-origin.

Regarding brain structural anomalies that have been previously discussed, Fatemi and colleagues (2002b) attempted to examine the genetic basis of the observed *cerebellar* pathology including *Purkinje* cell atrophy. They hypothesized that cell migration and *apoptotic* mechanisms may account for observed *Purkinje* cell abnormalities. *Reelin* is an important secretory *glycoprotein* responsible for normal layering of the brain. *Bcl-2* is a regulatory protein responsible for control of programmed cell death in the brain. *Autistic* and normal control *cerebellar* cortices matched for age, sex, and *postmortem interval* (*PMI*) were prepared by these investigators for *SDS-gel* electrophoresis and Western blotting using specific *anti-Reelin* and *anti-Bcl-2* antibodies. Their results showed 43–44 percent reductions in *autistic cerebellum* compared with controls. Quantification of *Bcl-2* levels showed a 34–51 percent reduction in *autistic cerebellum*. Measurement of *beta-actin* in the same homogenates did not differ significantly between groups. Their results demonstrate for the first time that dysregulation of *Reelin* and *Bcl-2* may be responsible for some of the brain structural and behavioral abnormalities observed in *autism*.

We can conclude thus far that ASDs are likely caused, in part, by inheritance of multiple interacting susceptibility alleles. To further identify these inherited factors, Yu and associates (2002) performed an important study in which linkage analysis of 105 families with two or more affected sibs was performed. Segregation patterns of short tandem repeat poly-morphic markers from four chromosomes revealed null alleles at four marker sites in 12 families that were the result of deletions ranging in size from 5 to >260 kb. In one family, a deletion at marker *D7S630* was complex, with two segments deleted (37 and 18 kb) and two retained (2,836 and 38 bp). Three families had deletions at *D7S517*, with each family having a different deletion (96, 183, and >69 kb). Another three families had deletions at *D8S264*, again with each family having a different deletion, ranging in size from <5.9 kb to >260 kb. At a fourth marker, *D8S272*, a 192-kb deletion was found in five families. Unrelated subjects and additional families without *autism* were screened for deletions at these four sites. Families screened included 40 families from *Centre d'Etude du Polymorphisme Humaine* and 28 families affected with learning disabilities. Unrelated samples were 299 elderly control subjects, 121 younger control subjects, and 248 subjects with *Alzheimer*'s disease. The deletion allele at *D8S272* was found in all populations screened. For the other three sites, no additional deletions were identified in any of the groups without *autism*. Thus, these deletions appear to be specific to *autism* kindred's and are potential *autism*-susceptibility alleles. An alternative hypothesis is that *autism*-susceptibility alleles elsewhere cause the deletions detected here, possibly by inducing errors during meiosis.

Catecholamine Theory

For some time now, the overriding theory as to the basis of brain dysfunction in children and adults has been focused on the production or lack of adequate production of neurotransmitters. The main group of neurotransmitters that has been implicated in neurobehavioral disorders of childhood has been the *catecholamines* especially the *monoamines*. The original reason for the *catecholamine hypothesis* were findings that in patients being treated for *tuberculosis* who were administered *monoamine oxidase inhibitors*, specifically *iproniazid*, these agents acted clinically as mood elevators or antidepressants. Soon after, it was found that this class of compounds also produced marked increases in brain *amine* levels. Also *reserpine*, a powerful tranquilizer and an effective *antihypertensive*, depletes brain *amines* (serotonin as well as *catecholamines*) and produces a significant depressed state in about 20 percent of those treated and even suicide attempts in some people. In addition, several *reuptake inhibitors* like *Ritalin*, which are specific for *DA* and several *SSRI*s such as *Prozac* or *Paxil* have been shown to have clinically positive effects in some individuals with *ASD*s. The assumption that followed was that there must be a deficit in the amount of neurotransmitter being produced. Further, it had been thought that this could only result from genetic sources. More recently, however it has become apparent that the "chemical imbalance" theory may not be accurate. Theories have moved in a more functional direction. Neurotransmitters and their receptors are produced by the nerve cell in response to activation. The fact that there may be less neurotransmitter produced is more likely a result of decreased activation of a pathway and therefore less neuronal activity results in decreased need for neurotransmitters, and a decrease in the cellular machinery to produce neurotransmitters. It does not mean that the cells cannot produce the neurotransmitter; there may not be a deficit of the neurotransmitter but of environmental stimuli necessary to activate neuronal pathways. Artificially increasing the amount of neurotransmitters through certain drugs may "trick" the cells into believing that there is increased activation and may temporarily reduce the symptoms. However, since it is not actually activation, there are no long-term improvements in the deficits being treated by these chemicals.

DA acts as a key neurotransmitter in the brain. Numerous studies have shown its regulatory role for motor and *limbic* functions.

However, in the early stages of *Parkinson's disease (PD)*, alterations of executive functions also suggest a role for *DA* in regulating cognitive functions. Some other diseases, which can also involve *DA* dysfunction, such as *schizophrenia* or *ADHD* in children, as shown from the ameliorative action of *dopaminergic* antagonists and agonists, respectively, also show alteration of cognitive functions. Experimental studies show that selective lesions of the *dopaminergic* neurons in rats or primates can actually provide cognitive deficits, especially when the meso-*cortico-limbic* component of the *dopaminergic* systems is altered. Data from the experiments also have shown significant alteration in attentional processes, thus raising the question of direct involvement of *DA* in regulating attention. Since the *dopaminergic* influence is mainly exerted over the *frontal lobe* and *basal ganglia*, it has been suggested that cognitive deficits express alteration in these subcortical brain structures closely linked to cortical areas, more than simple deficit in *dopaminergic* transmission. This point is still a matter of debate but, undoubtedly, *DA* acts as a powerful regulator of different aspects of cognitive brain functions. In this respect, normalizing *DA* transmission will contribute to improve the cognitive deficits. Ontogenic and phylogenetic analysis of *dopaminergic* systems can provide evidence for a role of *DA* in the development of cognitive general capacities. *DA* can have a trophic action during maturation, which may influence the later cortical specification, particularly of *pre-frontal* cortical areas. Moreover, the characteristic extension of the *dopaminergic* cortical innervation in the rostro-caudal direction during the last stages of evolution in mammals can also be related to the appearance of progressively more developed cognitive capacities. Such an extension of cortical *DA* innervation could be related to increased processing of cortical information through the *basal ganglia*, either during the course of evolution or development. *DA* has thus to be considered as a key neuro-regulator which contributes to behavioral adaptation and to anticipatory processes necessary for preparing voluntary action consequent upon intention. All together, it can be suggested that a correlation exists between *DA* innervation and expression of cognitive capacities. Altering the *dopaminergic* transmission therefore, could contribute to cognitive impairment of the kind seen in ASDs.

However, according to Cooper and colleagues (1996), the extent to which products of a specific gene are expressed in brain is a matter of the demands placed on neurons or *glial* cells by the incoming information. The environment, which is both physical and social, is the primary source of these external stimuli. This information is then converted into synaptic signals, which are in turn transduced into intracellular second messengers, and ultimately into altered cytoplasmic and nuclear signals that through regulation of transcription, determine which genes are turned on or kept off. Therefore, it has been noted that the degree to which our genotype is reflected in our functional form or phenotype comes under environmental influence. The effects of the environment are likely to be cumulative and perhaps vary with the stages of development, and these influences on early brain function may not be apparent until much later in life. They note that a neuron sees the world through the information it receives from synaptic afferents. The synaptic afferents receive their information from receptors.

The cumulative results of the synaptic activity modify gene expression by way of the nuclear actions of metabolic end products, like intracellular second messengers, ions, and immediate early genes leading to transcriptional regulation, which are triggered by synaptic activity. Changes in gene expression, that result from the influence of external events or from changes in the internal environment of the brain, alter the neurons phenotype such as the rate of utilization of its own transmitter or receptors and subsequently the operation of the circuits that it is connected and participates with.

Therefore, the phenotypic expression of the genotype is dependent on synaptic activity, which in turn is dependent on receptor activity, which is dependent on environmental stimuli. Alteration of gene expression can be a result of decreases in environmental stimuli. Therefore, a purely biochemical approach to such a complex system does not make sense. These sentiments are also shared by Liotti and Tucker (1995). "The idea that a psychiatric disorder can be traced to a neurochemical abnormality is simplistic, and may lead to a functional understanding only as the neuromodulator systems are seen within the context of the parts they play in the brains multileveled motivational circuitry."

In an article discussing the current feelings in regard to depression (Marano, 1999, March/April), "it overturns the widespread belief that depression is 'just' a chemical imbalance. Yes, neurotransmitters like *serotonin* function abnormally in depression—but so do lots of other things." Davidson states in the same article, "the idea that there is a global derangement of the *serotonin* or *norepinephrine* system is not sustainable in the light of recent brain imaging data. What distinguishes depressed from non depressed individuals are patterns of regional brain function, differences in specific circuits." This is where the future is pointing—treatment must not be directed at chemical intervention that is not specific, but with ways of utilizing specific environmental and cognitive stimuli that are targeted toward specific circuits. This thinking is opening the way for a new acceptance of old approaches to learning disabilities and neurobehavioral disorders—nondrug therapies. Davidson points out "non-pharmacological treatments may exert quite specific biological effects in being able to affect certain select brain regions." For example, he states, "the deficit in activation of the *prefrontal* cortex that we and others have identified in depression may be something that can be changed with cognitive therapy, or possibly even meditation. What we realize is that the two primary theories of genetics and chemical imbalance needs to be redefined to a more functional cause and approach to treatment. For some reason people seem to reject the notion that most disorders are societal or lifestyle driven, especially so-called 'mental' or 'psychological disorders'."

To emphasize this point, Robert Sapolsky, a noted neuroendocrinologist at Stanford University, who studies the cellular mechanisms of stress-induced diseases, published an editorial in *USA Today* (February 23, 1999) addressing this very concept. He commented on the emphasis that there will be technological breakthroughs in medical research and treatment. He states,

I heartily endorse this given that I do that sort of research (gene therapy for neurological diseases). Nevertheless, I was disappointed by one omission in their antidote to doomsday. They said barely a word about preventative medicine. By this, I do not mean preventing illness with some scientific marvel of say altering the genome of a fetus to prevent it from even getting a disease. I am talking about the down-Homey version of preventive medicine—those platitudes about eating right, exercising, and avoiding vices. A majority of Westernized diseases are related to lifestyle, and a couple of ounces of preventative measures in these realms would have a staggering impact on our health (and its costs).

He continues

... I'm not sure if a top priority is more research on preventive medicine, since we already know a lot about that and no one listens anyway. My point is a top priority is research should be figuring out why no one listens. ...Why do we have trouble believing that big outcomes are the end product of lots of small steps?

Autonomic, Hormonal, and Immune System Involvement

We have discussed the normal asymmetric localization of autonomic control centers in the human brain. We have shown the relationship between decreased brain autonomic function resulting in increased heart rate, blood pressure, arrhythmias, and possibly sudden cardiac death. Schwartz and colleagues (1982) have found evidence of a *sympathetic* imbalance to the left in *sudden infant death syndrome (SIDS)*, more recently it was reported that more than one third of all cases of *SIDS* may result from a heart rhythm abnormality, the largest *SIDS* study ever conducted concludes. The latest research conducted in Italy singles out a heart defect that can trigger cardiac arrest. The study involved 33,043 infants and took 19 years to complete. Physicians examined babies with electrocardiograms a few days after birth and then followed them for a year. Babies with a heart rhythm defect called a prolonged QT interval were 41 times more likely than usual to die from *SIDS*. Half of the 24 babies in the study who died of *SIDS* had this condition compared with none of those who died of other causes. The study was conducted by Peter John Schwartz and colleagues (1998) from Italy's University of Pavia. They think that no less than 30–35 percent of infants who subsequently die of *SIDS* can be expected to have a prolonged QT interval in the first week of life. The QT interval is the time between when the left ventricle begins to beat and when it is ready to beat again. When the interval is too long, life-threatening rhythm disturbances can result.

In the United States, about 1 of every 1,000 babies dies of *SIDS*. Yanowitz and colleagues (1966) showed that right *stellate* stimulation increases HR, lowers T-wave amplitude, and does not change the QT interval, while left *stellate* stimulation does not change heart rate, increases T-wave amplitude, and increases the QT interval. Changes in T-wave amplitude and prolongation of the QT interval are clinically relevant as they reflect repolarization phenomenon associated with life-threatening arrhythmias and Sudden Cardiac Death (SDC) (Zipes, 1991; Schwartz et al., 1998). Therefore, a decrease in left-sided neocortical inhibition of the left-sided *sympathetic* system may result in arrhythmia and death. Other left-sided neo-cortical decreases are associated with depression and learning disabilities or low intelligence. There are interesting correlations between depression and heart disease as well as intelligence and heart disease. A study by Elias et al. (1995) reports a strong correlation between risk factors for heart disease and cognitive disabilities. In a 15-year study of almost 1,800 volunteers, it was discovered that as they got older their abilities to think seemed to decrease in proportion to the number of risk factors they possessed for heart disease.

These factors included smoking, diabetes, high blood pressure, and obesity. None of the subjects had cognitive impairments at the start of the study, but as they aged, a series of neurophysiological tests began to show differences that had occurred gradually over the years. Each additional risk factor showed a strong corresponding amount of cognitive decline.

Depression according to Davidson (Davidson et al., 1999; Marano, 1999, March/April) shows that the left side of the *prefrontal* cortex is crucial to establishing and maintaining positive feelings, while the right is associated with negative ones. Depressed people appear to have a power failure of the left *prefrontal* cortex. The failure has been shown in electrical studies of the brain as well as PET scans to indicate decreased blood flow and metabolism. The depressed just do not activate the machinery necessary to process positive emotions or respond to positive stimuli. Specifically the left *prefrontal* cortex is critical in producing what Davidson calls, "pre-goal attainment positive affect," this is what typically could be described as eagerness, the emotion that comes about as we approach a desired goal or stay tuned to rewards. The result is a lowered capacity for pleasure, lack of motivation, and loss of interest.

If the *prefrontal* cortex controls depression by decreasing its activation, the *amygdala* controls the severity of the depression by its negative output. Davidson and colleagues (2002) have shown that blood flow is increased in the *amygdala* the more an individual is depressed. The *amygdala* also determines how firmly a negative event is held in memory. Ordinarily as the left *prefrontal* cortex is activated, it simultaneously inhibits the *amygdala* and decreases the flow of negative emotion. In depressed individuals, the general failure of activation of the left *prefrontal* cortex allows the *amygdala* to fire unchecked and this overwhelms the individual with dread, fear, and other negative feelings.

Drevets (Drevets, 1988, 1994, 1998a, 1998b, 1999; Drevets et al., 1997) found that depressed individuals not only had decreased activity of *prefrontal* cortex, but that their *prefrontal* cortices are actually smaller than nondepressed individuals. Drevets specifically found that depressed subjects have a significantly smaller volume of a section of the left *prefrontal* cortex known as the ventral *anterior cingulate*. This area was found to be approximately 40 percent smaller in depressed patients. This area, also known as the *subquenual* cortex, is one of the few cortical areas that connect to the *hypothalamus*, which controls the body's stress response. Depression is now not just seen as a disorder from the neck up, but it affects several body systems. It has been shown to lead to heart disease in otherwise healthy individuals and magnifies the deadliness of existing cardiac problems. In addition, it accelerates changes in bone mass that leads to *osteoporosis*.

There are a number of autonomic and hormonal factors associated with depression including *tachycardia, arrhythmia*, hypertension, greater tendency for the blood to clot, higher blood levels of *cortisol*, increased secretion of growth hormone which is thought to shift *cholesterol* balance toward dangerous low-density lipoproteins, increased blood levels of *norepinephrine*, and increased fat accumulation. Michelson and colleagues (1996) was quoted in regard to osteoporosis "as compared with the normal woman, the mean bone density in the woman with past or current depression was 6.5 percent lower at the spine, 13.6 percent lower at the femoral neck, 13.6 percent lower at *Ward's triangle* and 10.8 percent lower at the *trochanter*." Depression has also been shown to be associated with a smaller size of the *hippocampus*.

We have seen that *neuroimmunological* activity is asymmetrically distributed. In one study, Bardos and colleagues (1981) found that natural killer (NK) cell activity of mouse spleen cells, thought to be a measure of spontaneous resistance against tumor cells was significantly impaired following left cortical lesions. This was not seen with right-sided lesions. Renoux and associates (1987) examined a broad range of immune-related functions in mice and found that unilateral neocortical ablation of the dorsal and lateral aspects of the *frontal, parietal,* and *occipital* regions affected the development of lymphoid tissue specifically reduced spleen and thymus weight in left-lesioned mice and significantly increased in right-lesioned mice. In addition, they found that the number of splenic *T* (*THY-1+*) cells was reduced in left-lesioned mice to about 50 percent of sham-operated mice, whereas it was not changed in right-lesioned mice. Additionally, left cortical lesions significantly decreased the ability of splenic *lymphocytes* to be stimulated by T-cell specific *mitogens PHA* and *CONA*; however, right cortical lesions significantly enhanced lymphproliferolic responses to these *mitogens*. These studies as well as others suggest that *T*-cell mediated immunity is directed by the asymmetrically balanced activity of both hemispheres thought to play opposite roles in neuro-immuno-modulation.

Left hemisphere activity appears to increase responsiveness of several *T*-cell dependent immune parameters; right hemisphere activity seems to be primarily immuno-suppressive. Therefore, an imbalance of activity to either side can disrupt normal immune response. Decreased left-side activity appears to decrease immune responses. Children with left-sided decreases may be prone to chronic infections, like *otitis media*, that do not respond to antibiotics and which are common in learning-disabled children. Children with right-sided decrease may have overactive immune responses, these children may be prone to autoimmune disorders and chronic allergies, for example, allergic asthma, a common finding in learning-disabled or *ADHD* children.

Sweda and colleagues (1989) found that patients with *Sydenham's chorea* frequently demonstrate *obsessive–compulsive disorder* (*OCD*) symptoms. In addition, it is known that *OCD* patients display choreiform movements (Hollander et al., 1990). *Sydenham's chorea* is a manifestation of rheumatic fever and is thought to be mediated by *streptococcal* antibodies that cross-react with the brain. Therefore, patients with *Sydenham's chorea* have antibodies that cross-react with neuronal cells of human *basal ganglia* and are thought to be absorbed by *membranes of certain types of streptococci* (Bronze and Dale, 1993). OCD may be a result of decreased activity of the right hemisphere; this would possibly result in heightened autoimmune response. Cerebral control of hormonal activity has been shown to be primarily under the influence of the *hypothalamus*, which in turn is under the control of the *limbic system*, the *amygdala*, and the *frontal* cortex (Feldman, 1989; Schmidt and Thews, 1989). Studies that involve organisms which are lesioned in the medial *preoptic* area, lower *brainstem*, or medial basal *hypothalamus*, were shown to have a two to three times higher mortality rate if the lesions were on the right as opposed to the left side (Gerendai, 1984). These results suggest that the right brain is more crucially involved in vital functions than the left. The asymmetry in cortical regulation of *cortisol* secretions in humans has been studied (Wittling and Pfluger, 1990). Results show that cerebral control of *cortisol* secretions in emotion-related situations is under the primary control of the right hemisphere, whereas the left hemisphere seems to play a relatively minor role (Reineit and Wittling, 1980).

It is known that chronically elevated *cortisol* levels are associated with the development of several diseases such as *atherosclerosis* (Troxler et al., 1977) and *neoplastic* disease (Sklar and Anisman, 1981). However, it also appears crucial that an adequate level of *cortisol* is present to ensure an undisturbed functioning of the organism resulting in increased resistance to stress (Manck et al., 1984), increased immunoreactivity (Cupps and Fauci, 1982; Dunn, 1989), and preventing disease (Manck et al., 1984). Therefore, *cortisol* secretion is increased in a number of stress-related situations of physical or psychological nature, ranging from emotional threats and challenges to work-stress, noxious stimuli, antigen challenge, and illness. It is therefore assumed that the physiological function of these stress-induced increases in *corticosteroid* levels is primarily to protect the organism against its own normal defense reactions that are activated by stress and, if not decreased or inhibited, may result in damage to the organism (Munck et al., 1984) by causing suppressive effects in just about every aspect of the immune and inflammatory response, as well as by its effect on metabolic, renal, endocrine, and neural functions (Meyer, 1985).

Cortisol is thought to prevent defense responses from overreacting and jeopardizing homeostasis. Therefore, a chronically elevated as well as a chronically low *cortisol* level may be associated with abnormal physiological function and a multitude of health problems (Wittling, 1995). Therefore, it appears that an increase in neocortical activity will increase *cortisol* levels. This may also explain why a decreased left frontal activation, as seen in depressed patients, may result in increased levels of *cortisol*. Decreased left hemisphere activation will likely decrease the inhibitory balance in the right hemisphere thereby allowing abnormally high levels of *cortisol*. This combined with the increased firing of the IML on the ipsilateral left side may also result in increased *corticosteroid* response.

A wide range of studies in man and other species suggest that early compromise of immunological tolerance (both maternal–fetal and self) may lead to severe and varied cognitive deficits including those commonly noted in *PDD*, particularly *autism*. A range of immunological injury hypotheses for the genesis of the *PDD* has developed over the past 30 years. Numerous studies have established a strong and clear relationship between *ASDs* and immune system compromise (Stubbs and Crawford, 1977; Weizman et al., 1982; Todd, 1986; Warren et al., 1986, 1987, 1990a, 1990b, 1996; Ferrari et al., 1988; Root-Bernstein and Westall, 1990; Yonk et al., 1990; Gupta et al., 1996) and are now currently gaining significant support.

Both genetic and environmental factors contribute to the pathogenesis of a wide variety of neurobehavioral disorders, including *autism*, mental retardation, and schizophrenia. Some heritable disorders approach 100 percent penetrance; nonetheless, even in these disorders, subtle aspects of clinical disease expression may be influenced by the environment. In other disorders with genetic influences, exogenous factors, and the time points during nervous system development at which they are introduced, modulate the expression of disease. Elucidation of the mechanisms guiding this intricate interplay between host response genes, environmental agents, and the neurobehavioral context within which these interactions occur, is necessary to understand the continuum of clinical outcomes. We will attempt to review the evidence that immune factors may contribute to the pathogenesis of neurobehavioral disorders, and identify processes by which neural circuitry may be compromised.

ASDs are strongly associated with autoimmune dysfunction in the patients' relatives. To evaluate the frequency of autoimmune disorders, as well as various prenatal and postnatal events in *autism*, Comi and colleagues (1999) surveyed the families of 61 *autistic* patients and 46 healthy controls using questionnaires. The mean number of autoimmune disorders was greater in families with *autism*; 46 percent had two or more members with autoimmune disorders. As the number of family members with autoimmune disorders increased from one to three, the risk of *autism* was greater, with an odds ratio that increased from 1.9 to 5.5, respectively. In mothers and first-degree relatives of *autistic* children, there were more autoimmune disorders (16 and 21 percent) as compared to controls (2 and 4 percent), with odds ratios of 8.8 and 6.0, respectively. The most common autoimmune disorders in both groups were Type 1 *diabetes*, adult *rheumatoid arthritis, hypothyroidism,* and systemic *lupus erythematosus.* Forty-six percent of the *autism* group reported having relatives with *rheumatoid* diseases, as compared to 26 percent of the controls. Prenatal maternal urinary tract, upper respiratory and vaginal infections, *asphyxia*, prematurity, and seizures were more common in the *autistic* group, although the differences were not significant. Thirty-nine percent of the controls, but only 11 percent of the *autistic* group, reported allergies. An increased number of autoimmune disorders suggest that in some families with *autism*, immune dysfunction could interact with various environmental factors to play a role in *autism* pathogenesis.

Perinatal exposure to infectious agents and toxins is linked to the pathogenesis of neuropsychiatric disorders, but the mechanisms by which environmental triggers interact with developing immune and neural elements to create *neurobehavioral* disturbances are poorly understood. Hornig and associates (1999) reporting in the *Proceedings of the National Academy of Sciences,* describe a model for investigating disorders of central nervous system development based on neonatal rat infection with *Borna* disease virus, a neurotropic noncytolytic *RNA* virus. Infection results in abnormal righting reflexes, hyperactivity, inhibition of open-field exploration, and stereotypic behaviors. Architecture is markedly disrupted in the *hippocampus* and *cerebellum*, with reduction in *granule* and *Purkinje* cell numbers. Neurons are lost predominantly by apoptosis, as supported by increased *mRNA* levels for pro-apoptotic products, decreased *mRNA* levels for the anti-apoptotic products, and in situ labeling of fragmented *DNA*. Although inflammatory infiltrates were observed transiently in *frontal* cortex, glial activation is prominent throughout the brain and persists for several weeks in concert with increased levels of pro-inflammatory *cytokine mRNA*s and progressive *hippocampal* and *cerebellar* damage. The resemblance of these functional and neuropathological abnormalities to human *neurobehavioral* disorders suggests the utility of this model for defining cellular, biochemical, histological, and functional outcomes of interactions of environmental influences with the developing central nervous system.

Singh and associates (1991, 1993) examined, based on a possible pathological relationship of autoimmunity to *autism*, antibodies reactive with myelin basic protein (*anti-MBP*) were investigated in the sera of *autistic* children. Approximately 58 percent sera of *autistic* children were found to be positive for *anti-MBP* and were found to be significantly different from the controls (9 percent positive), which included age-matched children with normal health, idiopathic mental retardation, *Down's* syndrome (DS), and normal adults. Two of the most consistently observed biological findings in *autism* are increased *serotonin* levels in the blood and immunological abnormalities (including autoreactivity with tissues of the central nervous system). Warren and Singh (1996) investigated the relationship between these two sets of observations and confirmed associations of the *major histocompatibility complex* (*MHC*) with *autism*. Since the *MHC* is known to regulate the immune system and is associated with autoimmune disorders, they studied serum *serotonin* levels in 20 *autistic* subjects with or without *MHC* types previously found to be associated with *autism*. They found a positive relationship between elevated *serotonin* levels and the *MHC* types previously associated with *autism*. Immune factors such as autoimmunity have been implicated in the genesis of *autism*, a *neurobehavioral* disorder. Singh (1996) further measured the plasma levels of *interferon-alpha* (*IFN-alpha*), *interferon-gamma* (*IFN-gamma*), *interleukin-12* (*IL-12*), *interleukin-6* (*IL-6*), *tumor necrosis factor-alpha* (*TNF-alpha*), and *soluble intercellular adhesion molecule-1* (*sICAM-1*) in *autistic* patients and age-matched normal controls, since autoimmune response involves immune activation. The levels of *IL-12* and *IFN-gamma* were significantly higher in patients as compared to controls. However, *IFN-alpha, IL-6, TNF-alpha*, and *sICAM-1* levels did not significantly differ between the two groups. Singh suggested that *IL-12* and *IFN-gamma* increases indicate antigenic stimulation of *Th-1* cells pathogenetically linked to autoimmunity in *autism*.

Daniels and colleagues (1995) found that 22 *autistic* subjects had an increased frequency of the extended or ancestral *MHC* haplotype. They confirmed this observation by studying 23 additional randomly chosen *autistic* subjects, most of their parents and 64 unrelated normal subjects. In agreement with earlier findings, the *MHC haplotype* was determined to be associated with *autism*. In combining the data from both studies, they found that a substantial fragment of the extended *haplotype* was represented in 40 percent of the *autistic* subjects and/or their mothers as compared to about 2 percent of unrelated subjects. They concluded that one or more genes of the *MHC* are involved in the development of some cases of *autism*.

The emerging concept of opioid peptides as a new class of chemical messengers of the neuroimmune axis and the presence of a number of immunological abnormalities in *autism* prompted Scifo and colleagues (1996) to correlate biological (hormonal and immunological) determinations and behavioral performances during treatment with the potent opiate antagonist, *naltrexone*. Twelve *autistic* patients ranging from 7 to 15 years, diagnosed according to *Diagnostic and Statistical Manual of the American Psychiatric Association.* (*DSM-III-R*), entered a double-blind crossover study with *naltrexone* at the doses of 0.5, 1.0, and 1.5 mg/kg every 48 hr. The behavioral evaluation was conducted using the specific *BSE* and *CARS* rating scales. *Naltrexone* treatment produced a significant reduction of the *autistic* symptomatology in 7 out of 12 children. The behavioral improvement was accompanied by alterations in the distribution of the major *lymphocyte* subsets, with a significant increase of the *T-helper-inducers* (*CD4+CD8−*) and a significant reduction of the *T-cytotoxic-suppressor* (*CD4−CD8+*) resulting in a normalization of the *CD4/CD8* ratio. Changes in NK cells and activity were inversely related to plasma *beta-endorphin* levels. The authors suggest that the mechanisms underlying *opioid*-immune interactions are altered in *autistic* children and that an immunological screening may have prognostic value for the pharmacological therapy with *opiate antagonists*.

Given the now extensive evidence that *autism* is often accompanied by abnormalities in the *inflammatory response system*, products of the system such as pro-inflammatory *cytokines* may induce some of the behavioral symptoms of *autism*, such as social withdrawal, resistance to novelty, and sleep disturbances. Croonenberghs and associates (2002) measured the production of *interleukin* (*IL*)-*6, IL-10,* the *IL-1* receptor antagonist (*IL-1RA*), *interferon* (*IFN*)-*gamma,* and *tumor necrosis factor* (*TNF*)-*alpha* by whole blood and by the serum concentrations of *IL-6*, the *IL-2* receptor (*IL-2R*), and *IL-1RA*. They found a significantly increased production of *IFN-gamma* and *IL-1RA* and a trend toward a significantly increased production of *IL-6* and *TNF-alpha* in *autistic* children. There were no significant differences in the serum concentrations of *IL-6, IL-2R,* and *IL-1RA* between *autistic* and normal children. Their results suggest that *autism* may be accompanied by an activation of the *monocytic* (increased *IL-1RA*) and *Th-1*-like (increased *IFN-gamma*) arm of the *inflammatory response system,* with increased production of proinflammatory cytokines playing a role in the pathophysiology of *autism*. Connolly and colleagues (1999) likewise found evidence supporting the notion that autoimmunity plays a role in the pathogenesis of language and social developmental abnormalities of *ASDs,* supporting Croonenberghs' findings.

Jyonouchi and associates (2001) determined innate and adaptive immune responses in children with developmental regression and *ASDs,* developmentally normal siblings, and controls. With *LPS,* a stimulant for innate immunity, *PBMCs* from 59/71 *autistic spectrum* patients produced greater than 2 SD above the control mean values of *TNF-alpha, IL-1beta,* and/or *IL-6* produced by control *PBMCs. Autistic spectrum PBMCs* produced higher levels of proinflammatory/counter-regulatory *cytokines* without stimuli than controls. Their results indicate excessive innate immune responses in *autistic spectrum* children. In examining the relation between immune and gastrointestinal function in *autistic* children, Jyonouchi and associates (2002) further examined children with *ASD*. They measured *IFN-gamma, IL-5,* and *TNF-alpha* production against representative *gliadin,* cow's milk protein, and soy by *PBMCs* from *autistic spectrum* and control children. They concluded that immune reactivity is associated with apparent gastrointestinal inflammation in *autistic spectrum* children that may be partly associated with aberrant innate immune response.

NUTRITION AND *AUTISTIC SPECTRUM* DISORDERS IN AN EVOLUTIONARY CONTEXT

Diet has been recognized as a major factor in the development of the large human brain, therefore, it is practical to assume that changes in the modern diet of humans may also play a role in dysfunction of the brain. It is no coincidence that the emergence of the first real human, a member of the genus *Homo* was apparently the first to adapt a broader approach to gathering food, which included meat. This was thought to be in contrast to earlier primates and *hominids* whose diet was thought to consist primarily of vegetation. This species was *Homo habilis* thought to have appeared in Africa about 2.3 million years ago. Two critical features of this species were of course its larger brain size about 600–750 ml and regular toolmaking ability. *Homo erectus* compared to *H. habilis* became much more sophisticated toolmakers. This was significant because the use of tools to obtain meat opened up a completely new area of ecological possibilities for humans. From this time on, *Hominids* would never be the same, once they learned how to harness the power of a concentrated and nutritious food source, meat. Among the many advantages of meat was that it reduced the demand on their digestive systems, which were necessarily large to process and digest vegetation, which was relatively low in nutrition. Meat on the other hand provided a much more powerful and efficient source of energy and nutrients. This may explain why the human gut is the only energy-using organ that is significantly small in relation to body size compared to other mammals, whereas the brain is significantly large. Our gut, including our stomach and intestines, is about half the size expected, based on other mammals. Small guts are thought to reflect a diet that would consist of high quality and easy to digest food. This change in gut relative to body size first occurs with *H. erectus*. This is thought to be a critical step, which allowed energy resources and blood flow to be channeled to the brain to help fuel the increased mass and reduce the demand for these same resources from the gut. It would most likely not be possible to have both a big brain and a big gut; the demand for energy from both would be too great. This also shows how the brain and gut are evolutionarily and, therefore, developmentally tied together. This may partly explain the presence of severe gut developmental dysfunction in certain children in the *autistic* spectrum.

Meat provided mothers with high quality food for developing children and continued to fuel their rapid growth. However it is thought that it was not just meat consumption that allowed for a smaller gut but also the consumption of fat and bone marrow along with the meat that truly provided a powerful and easy to digest source of energy and raw material. This is thought to have created a positive feedback loop that as we started to eat meat, we became more intelligent. This, in turn, allowed us to develop more sophisticated ways of obtaining meat as well as other rich, easily absorbable, and digestible foods. This, of course, would complement the increasingly more sophisticated motor activities that were previously acquired through bipedalism and which would continue as demands for further speed and strength would grow as *hominids* became more sophisticated hunters rather than just scavengers. Therefore, we can see that from the beginning, brain growth, digestion, diet, and motor activity were intimately tied together and all must be considered as potential factors in brain dysfunction. In fact, as human's diet changed, the onset of many modern illnesses were evidenced for the first time. In fact, human bodies and brains have actually been getting smaller for the last 10,000 years. *Cro-Magnon* males were thought to have averaged 6 ft tall whereas modern man averages 5 ft 8 in., and our brains are about 10 percent smaller as well. One possible explanation why this occurred is that tools that are more efficient and better hunting techniques reduced pressures which would have selected for larger size and strength. This was thought to have accentuated even further with the advent of farming, which began about 10,000 years ago. It is also possible that human's farming became more efficient as populations started to increase.

With the availability of a reliable food supply nearby, the first sedentary societies were created, with their villages, towns, and eventually their cities and civilization's manipulation of plants and organisms started, independently and within a relatively short time, in several areas of the world. Wheat was cultivated in the Middle East, rice in China, maize in South America, and sorghum, millet, and yams in West Africa. A major driving force in this change was the alteration that was then going on in the climate. The last Ice Age had ended, and sea levels were rising. Researchers such as Theya Molleson (Molleson and Oakley, 1966), of London's Natural History Museum, have traced the effects of agriculture's development on the bones of ancient people, such as those from Abu Hureyra, a *Neolithic* (New Stone Age) settlement in Syria. There, it has become clear, that the grind of everyday life quite literally marked the anatomies of the world's first farmers and, largely, their wives. Molleson (1994) has studied the bones of 162 individuals who lived at Abu Hureyra from about 11,500 to 7,500 years ago, and found injuries indicative of demanding physical activity but only among the farming people, not among their hunter-gatherer predecessors. There was vertebrae damage, severe *osteoarthritis* in toes, curved and buttressed femurs, and knees with bony extensions. At first, Molleson blamed sport or dance. "But crippled ballerinas seemed unlikely during the Neolithic period," she adds. Then the cause became apparent. "With the advent of agriculture, men cultivated the job of grain preparation." Women had to kneel roll stones to crush corn. "Kneeling for many hours strains the toes and knees, whereas grinding puts additional pressure on hips and the lower back," Molleson writes. The result: damaged discs and crushed vertebrae. It is clear therefore, that agriculture and the primary foods cultivated had a devastating effect on health.

A study of Native American skeletons has shown a significant impact on their health when they changed their diet to one consisting primarily of maize. Healthy hunter-gatherers were suddenly turned into sickly farmers. Tooth cavities jumped 7-fold; children's teeth defects reveal that their mothers were badly undernourished; anemia quadrupled in frequency; *tuberculosis*, yaws, *osteoarthritis*, and *syphilis* afflicted large numbers of the population; and mortality rates jumped. Almost a fifth of the population died in infancy. Far from being one of the blessings of the New World, corn was a public health disaster. What has become a common theme around the world is repeated throughout history. As farming populations increase, the health of individuals decreases dramatically. In other words, we get our calories cheaper but we give up nutritional value. The establishment of dense populations of people, who could exist on grain and rice stores, triggered many of the world's deadliest epidemics, diseases that thrived among cramped, underfed peoples. Tuberculosis, leprosy, cholera, and malaria all appeared in the wake of farming. Similarly, smallpox, the plague, and measles only manifested themselves with the arrival of cities. This trend continued through the Middle Ages and was associated with increasingly smaller stature. It is thought that medieval people were smaller because they ate so poorly and did not reach their genetically preprogrammed size; this is why it appears that we have grown since. However, we are smaller than our ancient predecessors are. *Homo sapiens*, as hunter-gatherers, clearly lived on a wide variety of foods: butter, milk, and cheese played no part in our eating habits for most of our history, for they only arrived on the scene with the advent of the Agricultural Revolution. These are "unnatural" products, high in fats and other cholesterol-raising constituents.

More importantly, when we examine the modern eating practices of western children, they exist primarily on a self-imposed diet consisting predominantly of wheat and dairy in every combination. Children exist on a cereal, macaroni and cheese, cream cheese and bagel, and pizza. Most children with behavioral disorders appear to avoid vegetables and fruits like the plague. Studies done by the *American Dietetic Association* (1999), the *U.S. Department of Health and Human Services*, and the *National Cancer Institute*, indicate that a large percentage of children in the United States are not obtaining the RDA for nutrients from their food. In 1997, Munoz and associates studied 3,307 children in the United States (age 2–19) to determine the number of children meeting national recommendations for food group intake I and found that only 1 percent met all the recommendations. Furthermore, 64 percent studied failed to meet the minimum RDA requirements for vegetable intake and, of the 36 percent that actually met these requirements, a quarter of all the vegetables they consumed were in the form of French fries. This poor diet combined with sedentary behavior is also a factor in obesity. We are probably programmed to store fat reserves against possible lean times. The leading cause of death to humans throughout history up until recent times has been famine. The result is overweight Westerners. We may be conditioned to minimize physical activity when it is not absolutely necessary—an adaptation to conserve food stores.

Besides stimulation, the brain needs fuel provided largely by oxygen and glucose. Glucose is broken down from our food, but there are also hosts of other nutrients that appear to be important to the brain and body's function. A brain that has ample stimulation but too little fuel will not be able to take advantage of that stimulation adequately. Without fuel, cells will fatigue and function inadequately, being unable to make new proteins to build new branches, or make and repair organelles like *mitochondria*, which produce energy for the cell. Without an adequate fuel supply, cells can undergo *oxidative stress* and be damaged or die because of *free radical* production.

Zeisel (1986) indicates that diet clearly influences neurotransmission. This can be important in grossly undernourished children. It can also be important in children in whom normal homeostatic mechanisms governing food intake are bypassed. Subtle differences in behavior can occur with physiological variations in food intake. Components of foods can also be used as drugs. Starvation can impair neuronal maturation and can have lasting effects upon behavior and intellectual performance. The extent of impact of starvation upon the brain depends upon whether under nutrition occurred during a critical phase in brain development. Short-term fasting has small, but significant, effects upon intellectual performance. Even when gross malnutrition is not present, subtle changes in diet may modulate brain function. *Tryptophan, tyrosine*, and *choline* in the diet are used as precursors for neuronal synthesis of *serotonin, dopamine, norepinephrine*, and *acetylcholine*, respectively. It is likely that the brain's sensitivity to certain components of the diet exists to permit monitoring of food intake by the central nervous system. Other components of the diet that may affect behavior include food additives, sugar, and caffeine. Food additives may exacerbate hyperactive symptoms in children with *ADD/ADHD*.

The early clinical and laboratory studies of John Dobbing (Katz et al., 1982; Dobbing, 1984, 1985a, 1985b; Dobbing and Sands, 1985) show that starvation during pregnancy and nursing can cause brain damage. If severe malnutrition happens in the first trimester, the result can be neural tube malformations such as *spina bifida, anencephaly*, or *microcephaly*. Malnutrition in the second trimester is likely to result in too few neurons, since nerve cells proliferate during this period. Malnutrition in the third trimester affects synaptic connections, brain growth, and the branching of dendrites (Morgan and Gibson, 1991; Diamond, 1998). Significantly underweight infants and children starved of protein and calories have small head size, lowered IQ scores, problems hearing, speaking, coordinating hand–eye movements, learning disabilities, and problems forming normal relationships. Studies in Latin America, Africa, and the United States also show that malnourished children had lower intelligence and school scores than children who had proper nutrition from the same neighborhood (JaredDiamond, 1990, 1998). While it has been thought that brain damage was primarily responsible for these findings, research has shown that the environment is equally important, and depending on its stimulation value, can help to either repair the malnourished brain or make it worse.

Cells that are adequately supplied with fuel but with no stimulation will die. If a cell is stimulated it will improve its own production of energy, but if there is no fuel it will also die. In a series of experiments with pregnant rats and their offspring, Diamond (1998) provided mother rats with high-protein pellets during the first period of pregnancy, and then half of them were switched to a low-protein pellet. Once the baby rats were born, they also were fed half high-protein and half low-protein diets. They also split the high-protein group in half, with one half living in a standard rat cage, and the other half in enriched cages. Their results showed that pups that were protein-deprived in the womb weighed much less and their cerebral cortex was much thinner than pups whose mothers were fed high protein. After weaning, it was found that the deprived pups could be rehabilitated with higher protein food; however it was noted that those rehabilitated in an enrichment cage developed a thicker cortex than the well-fed pups that were placed in a standard cage. The smartest pups with the thickest cortex were the ones with both good nutrition and a stimulating environment (Diamond, 1998). This is similar to the findings of Volkow and associates (2000) regarding *DA* levels. She found that the two factors that affected *DA* levels were lack of physical activity with high caloric intake or poor diet. This resulted in significant decreases in *DA* and *DA* receptor contents. Decrease in *DA* was highly correlated with decreases in cognitive and motor skills.

Most children today have low-protein, high-caloric diets combined with decreases in physical and mental stimulation. It appears, however, that if we increase stimulation and change nutrition, the children can recover quickly. In another study related by Diamond and Hopson (1998), research by Ernesto Pollitt, a pediatrics professor at the University of California at Davis is discussed. Pollitt and colleagues from Central America (Pollitt, 1994; Perez-Escamilla and Pollit, 1995; Watkins and Pollitt, 1996; Watkins et al., 1996) studied 2,000 children in four Guatemalan villages for almost 10 years. In two of the villages, they supplied the mothers and children with hot maize gruel full of protein and vitamins to supplement their normal meals. In two other villages, the mothers and children received a high-calorie vitamin-filled fruit drink but no extra protein. They found that over 8 years, the fetuses, infants, and children on protein supplements grew and developed various skills quicker, they were more energetic, and showed better social and emotional gains than children eating less protein. Both groups were exposed to an enrichment program, which included learning games, social skills, and preschool activities; both groups did better on this program. The well-nourished children however were able to play, experiment, and explore their environment more completely and this is thought to have promoted higher IQs and learning ability in this group.

In growing children, diets that are high in carbohydrates and low in protein are detrimental. This will also decrease essential fatty acids that have also been tied to brain function. Nutrition appears to play an important role in governing behavior in children with learning or attention problems. According to recent research, some children cannot produce essential fatty acids (EFAs) important for brain and nerve cell function. Jacqueline Stordy (Morgan and Stordy, 1995) theorized that low levels of certain nutritional fats impede a child's ability to focus attention, and affect learning ability. An estimated three million children suffer from *ADD*. Stordy thinks that *ADD* results from an inability to turn many of the EFAs in the diet to much needed long chain fatty acids that are the building blocks of nerve cells. Stordy suggests that the substance most lacking is *docosahexaenoic acid* (*DHA*), which is found naturally in the body and in foods like fish. The body normally converts EFAs, found in seeds and nuts to *DHA* and other long chain fatty acids. Stordy thinks children with *ADD* are unable to convert EFAs into *DHA*. In studying the effectiveness of DHA in children with learning disabilities, it has been demonstrated that *DHA* is found in rod cells in the eye necessary for night, peripheral vision, and movement detection. Oral supplements provided to dyslexics improved vision according to Stordy (2000) and others (Greatrex et al., 2000), which is thought to assist children to process information better.

Burgess and Stevens (Stevens et al., 1995; Burgess et al., 2000) of Perdue University studied 53 children diagnosed with *ADHD* and 43 controls. They noted that fatty acid levels were much lower in children with *ADHD* than in the controls. They also noted that subjects with lower *omega-3* fatty acid levels were more likely to exhibit behavioral problems. They also thought that these findings are supported by studies that show abnormal behavior in lower organisms deficient in *omega-3* fatty acids. Interestingly, despite their lower plasma levels of fatty acids, the *ADHD* group consumed more total and *polyunsaturated* fats than controls leading researchers to think there may be another factor altering metabolism. This may indicate that a nutritional deficit is not the primary factor but that decreased brain function and altered metabolism affect the way that a child can digest and absorb food from the gut.

In another similar study (Mitchell et al., 1987), 48 hyperactive children were compared with 49 age- and sex-matched controls. Significantly, more hyperactive children were noted to have auditory, visual, language, reading, and learning difficulties and the birth weight of hyperactive children was significantly lower than that of controls. In another study from Mitchell's group, serum *essential fatty acid* levels (*EFA*) were measured in 44 hyperactive children and 45 controls. The levels of *docosahexaenoic, dihomogammalinolesic*, and *arachidonite* acids were significantly lower in hyperactive children than in controls. Decreases in EFA levels have also been linked to depression and *schizophrenia*. Another study reports that *EFA* can reduce symptoms of *schizophrenia* by 25 percent (Peet et al., 1997), less than half an ounce per day is enough to produce the effect.

Goldfischer and colleagues (1973) showed that children with *Zellerweger cerebro-hepato-renal syndrome* lacked demonstrable *peroxisomes* in liver and kidney. It is now thought that *Zellerweger syndrome* (*ZS*)-type disturbances went from being conceived of as a multiple congenital anomaly syndrome to a new one of a new class of genetic disorders called *metabolically induced dysplasias*. *Peroxisominal* disorders are described as disorders that can be divided into two main categories. An X-linked *adrenoleukodystrophy* (*ALD*) and *adrenomyelo-neuropathy* (*AMN*) in which *peroxisome* biogenesis results from a defective organelle assembly and therefore multiple *peroxisome* enzyme abnormalities, mostly defects in *beta oxidation* of *very long chain fatty acids* (VLCFA) and from the synthesis of essential lipids such as *plasmalogens* (Lazarow and Moser, 1995).

DHA is an important constituent of retinal photoreceptor cell membranes and may play an important role in membranes of other nervous system cells (Neuringer et al., 1986). Therefore, reduced levels of *DHA* may have pathogenetic significance. Over the past 5 years, Martinez and associates (Martinez et al., 1993; Martinez, 1996) have reported on clinical neurophysiological as well as biochemical improvement in patients treated with *DHA*. Martinez and Vasquez (1998) present interesting results showing that supplementation with *DHA* results in improved central nervous system myelination as documented on T_2 weighted images obtained through MRI scans. In four of the children examined, the pattern of white matter disturbances on initial scans was consistent with *hypomyelination* or *dysmyelination*. Follow-up scans on three patients taken after 8 and 22 months show white matter patterns that are within normal limits for their age. A fourth child's scan shows contrast enhancement within the frontal white matter, a finding pathognomonic for inflammatory *demyelination* seen in classic X-linked *ALD*. Two years later the contrast enhancement was no longer present, and some decrease in the area of *demyelination* was noted. Also during the period of supplementation (which ranged from 1.5 to 3 years), clinical improvement was observed most notably increasing visual ability and to a lesser degree, resolution of axial *hypotonia*. Concomitantly, *DHA* levels became normal, liver dysfunction improved, *erythrocyte plasmalogen* levels rose, and *VLCFAs* were reduced.

The latest research suggests that *DHA* may decrease the effects of stress as well as depression, *dementia, ADHD, schizophrenia*, alcoholism, *bipolar disorder*, and postpartum depression. As 40–60 percent of the brain is fat, mostly in the form of *myelin*, which insulates brain cells, the better the insulation, the faster the cells can conduct impulses. The healthier the membrane, the better it will react to stimulation with less fatigue. This may ultimately increase the frequency of firing of brain cells promoting growth and plasticity. Not surprisingly then *DHA* is critical for early development of the brain, especially the 3 months prior to and following delivery, the time of most rapid brain growth. A deficiency of *DHA* may result in some degree of impairment. Premature infants have been shown to have lower levels of *DHA* compared with full-term babies, because they are denied the full 9 months of nutrients from their mother. Giving premature infants *DHA* supplements has resulted in higher scores at 1 year of age, suggesting a link between *DHA* and intelligence. It is also thought that bottle-feeding can also result in low levels of *DHA*. Infant formula does not contain *DHA*, even though this *omega-3* fatty acid is present in breast milk. It is thought possible that this hidden deficiency may be one of the reasons that in later years, bottle-fed infants have been shown to score lower on standardized tests of reading, visual interpretation, sentence completion, nonverbal skills, and math (Anderson et al., 1999).

In examining the relationship of nutrition with neurobehavioral disorders of childhood in an evolutionary context, we know that diet has been demonstrated to be significantly involved in *ASDs*. The impact of diet on the health and disease of society is well documented and the impact is profound. Diet also has been shown to affect the expression of genes. Biological anthropologist Barry Bogin (Bogin and MacVean, 1981a, 1981b; Bogin, 1998) reviews the effect of nutrition on societies. He thinks that nutrition's effect on society is an expression of plasticity. He defines plasticity as "the ability of many organisms, including humans, to alter themselves and their behavior or even their biology in response to changes in the environment." He further states:

> we tend to think that our bodies get locked into their final form by our genes, but in fact we alter our bodies as the conditions around us shift, particularly as we grow during childhood. Plasticity is as much a product of evolution's fine-tuning as any particular gene, and it makes just as much evolutionary good sense. Rather than being able to adapt to a single environment, we can change our bodies, thanks to plasticity, to cope with a wide range of environments. Combined with genes we inherit from our parents, plasticity accounts for what we are and what we can become.

Bogin notes that the concept of plasticity among anthropologists was first clearly defined in 1969 by Gabriel Lasker, biological anthropologist at Wayne State University. At that time, most scientists thought that genetics was the main factor in the make up of a person and their offspring. An example of this was the ability of adults in certain human societies to drink milk. It is known that as children, humans all produce the enzyme *lactase*, which is needed to break down the sugar *lactose* in a mother's milk. In many humans, the *lactase* gene slows down significantly as we reach adolescence, possibly due to another regulating gene. When this happens, we cannot digest milk. *Lactase* intolerance, which results in intestinal gas and diarrhea, effects between 70 and 90 percent of African and Native Americans, Asians, and people who are from the Mediterranean area. However, people of Central and Western European descent and the Fulani of West Africa are able to drink milk with little difficulty. It is thought that this is due to the fact that people descended from these societies have a long history of raising goats and cattle and there was therefore a benefit of being able to consume milk and milk products. Therefore, natural selection over time altered the regulation of their *lactase* gene, allowing it to function into adulthood. Bogin notes that these adaptations take centuries to establish while Lasker pointed out two other types of adaptations that do not require as much time to have an effect. In one instance, if humans face a cold winter with a paucity of heat available, their metabolic rates will increase over the time frame of only a few weeks so they can produce more heat, only to decrease again when the heat of summer returns. Lasker's other model indicates that adaptation can be considered the irreversible lifelong modification of people as they develop or their "plasticity." Humans, because they take a significantly long time to reach adulthood and live in so many varied environments are among the world's most adaptable species in physical form and behavior. Bogin uses as a prime example, the great range in height of humans, which is a subject he has studied for 25 years. He asks us to consider these statistics. In 1850, Americans were the tallest people in the world with men averaging 5'6" in height. However, 150 years later, American men average 5'8" in height, but have declined in the world standing and are now the third tallest people in the world. The tallest are presently the Dutch. However, in 1850 the Dutch averaged only 5'4" inches and were the shortest men in Europe, whereas today they average 5'10". Bogin questions what happened. "Did all the short Dutch sail over to the United States? Did the Dutch back in Europe get an infusion of 'tall genes'?" He thinks that neither of these two answers is true. He explains that the chief factor is that in both America and the Netherlands, the quality of life improved, but more so for the Dutch people and height increased as a result. The evidence for this can be found in studies of how height is determined. It is the product of plasticity in our childhood and in our mothers childhood as well. For example, if a young girl is undernourished and subsequently has poor health, the growth of her body and her reproductive system is also diminished. With a decrease in the amount of raw materials, she cannot create more cells to produce a larger body and the foods she does have are utilized in repairing already existing cells and tissues from the damage produced by the disease. Her smallest size as an adult is the product of a compromise her body has made during her growth and development.

We can also see that the same process would affect the growth and function of the developing brain and nervous system and the brain would be expected to be retarded in its development. Bogin further pointed out that,

> such a woman can pass on her short stature to her child, but genes have nothing to do with it for either of them. If she becomes pregnant, her small reproductive system probably would not be able to supply a normal level of nutrients and oxygen to her fetus. This harsh environment reprograms the fetus to grow more slowly than it would if the woman was healthier so she is more likely to give birth to a smaller baby.

Statistically it is noted that low birth weight infants (less than 5.5 pounds) will be most likely to continue their program of slow growth throughout childhood, especially if they grow up in a nutritionally deprived environment as well. Therefore, by the time they are teenagers, they are more likely to be significantly shorter than individuals who are of normal birth weight.

Studies of identical twins in certain cases have unequal access to the nutrients supplied by the mother's placenta. It has been shown that the twin with access to a smaller percentage of placenta is often born with low birth weight, whereas the other twin is normal. This difference can persist throughout their life span. Therefore, research shows that the average height of any group can be used as a measurement of the health of their society in general. Bogin further points out that after the turn of the century, the United States and the Netherlands began to protect the health of their people by purifying drinking water, installing sewer systems, regulating the safety of food, and most importantly providing better health care and diets to children. This, in twins, in turn produced taller children and adults. Differences in Dutch and American societies therefore determine the difference in height we see today. The Dutch provided public health benefits to all people including their poor, whereas in the United States, improved health was enjoyed by the wealthier citizens who could afford it. The poor on the other hand frequently lacked adequate housing, sanitation, and health care with differences between economic classes seen at birth. In 1994, 2 percent of Dutch babies were born at low birth weight, whereas the United States had 7 percent. For White Americans, the rate was 5.7 percent and for Black Americans, the rate was significant at 13.3 percent. Studies show that poor Americans are shorter than affluent, by about one inch. Bogin thinks that this explains our drop in height to third place worldwide.

Bogin relates another example of a common belief that the central African tribe, the *Tutsi* (or *Watusi*), is reportedly thought to be tall. However, this "myth" when examined reveals that today's *Tutsi* men average 5'7" and that they have maintained that height for over 100 years. Therefore, when much smaller European men first encountered the *Tutsi*, these men appeared extremely tall and therefore this 2–3 in. difference led to their reputation as giants. Although today, while the *Tutsi* could indeed be as tall as the Dutch, due to poverty conditions in Rwanda or Burundi where they live and with frequent warfare, their height has not increased. However, it has been shown that if the *Tutsi* and other Africans migrate to Western Europe or North America at younger ages, they attain greater adult height than Africans who remain in Africa.

In contrast, the shortest people are the *Pigmies* who average 4'9" tall for adult men with part of the reason being genetic and another factor being environmental. *Pigmies* who live in the forests of Central Africa appear to be undernourished while *Pigmies* who live on farms and ranches outside of the forest are better fed and are taller as well. Another group of people who were also called *Pigmies* are a group of extremely short people in New Guinea, who were thought to be genetically shorter. However, it was found that these individuals ate a diet deficient in iodine and other essential nutrients. When these people were provided with vitamin and mineral supplements, their supposedly genetically short size disappeared in their children, who grew to more normal height. It is also noted, "another way for the so called *Pigmies* to stop being *Pigmies* is to immigrate to the

United States." Bogin notes that his own research on the Mayans, refugees who arrived in the United States in the 1980s, shows that the average increase in height of the first generation of the immigrants was 2.2," which means that these individuals underwent one of the largest single generation increases in height ever recorded.

Height of course is only one example of rapid plasticity. Bogin also points out that it may play a key factor in determining people at risk for developing certain diseases. An example is *Parkinson's* disease. Ralph Garruto, a medical researcher and biological anthropologist at the National Institute of Health, has been studying the role of the environment and human plasticity in *Parkinson's* and *ALS*. Garruto and his team (Garruto et al., 1999; Plato et al., 2003) went to the islands of Guam and New Guinea where rates of both diseases are 50–100 times higher than in the United States. In fact, among the native people these two diseases kill 20 percent of the people over the age of 25. They found that both diseases are linked to a shortage of calcium in the diet. This deficiency eventually leads to the body absorbing too much of the aluminium present in the diet and causes severe damage to the body and brain. Garruto's team discovered that 70 percent of the people they examined in Guam had some brain damage, but only 20 percent progressed to manifesting *Parkinson's* disease or *ALS*. It appears that there exists a certain degree of genetic adaptability in the individuals studied. Bogin concluded:

> maybe *Lou Gehrig's* disease and *Parkinson's* disease, as well as many others, including some cancers, are not our genetic doom but a product of our development, just like variation in human height. Maybe the danger of these diseases will prove in time as illusory as the notion that *Tutsi* are giants or the *Mazza* are *Pigmies* or Americans still the tallest of tall.

These facts lay the foundation of how important the role of nutrition and environmental factors are to the growth and development of a child's body and especially their brain.

Of interest to our discussion is the relationship between *Parkinson's* and *basal ganglia* dysfunction and its relationship to the *cerebellum, thalamus,* and *frontal lobe. Parkinson's* is a *hypokinetic* disorder, as opposed to *hyperkinetic* disorders like *ADD, OCD, Tourette's,* and *Huntington's* disease. Therefore, if environmental nutritional deficiencies can produce *Parkinson's*, there is no reason to think that these factors could not also produce similar conditions, especially since *hyperkinetic* disorders may simply represent a different stage of neurological location. In addition, *Parkinson's* is thought to be produced by a deficiency of *DA* production by the *substantia nigra zona compacta. ADD* and other behavioral disorders have also shown a relationship between their symptoms and decreased *DA* production. We have noted how quickly changes in nutrition and environment can impact a society as a whole; only these factors can properly explain the rapid and dramatic increases that we have witnessed over the past few decades of the incidence of learning disabilities and *ADD*.

Wakefield (2002) in a recent review have indicated that there is growing awareness that primary gastrointestinal pathology may play an important role in the inception and clinical expression of *autistic spectrum* disorders. In addition to frequent gastrointestinal symptoms, children with *autism* often manifest complex biochemical and immunological abnormalities. The gut–brain axis is central to certain encephalopathies of extra-cranial origin, hepatic encephalopathy being the best characterized. Commonalities in the clinical characteristics of hepatic encephalopathy and a form of *autism* associated with developmental regression in an apparently previously normal child, accompanied by immune-mediated gastrointestinal pathology, have led to the proposal that there may be analogous mechanisms of toxic encephalopathy in patients with liver failure and some children with *autism*. Aberrations in *opioid* biochemistry are common to these two conditions, and there is evidence that *opioid* peptides may mediate certain aspects of the respective syndromes. The importance of identification of the pathways involved will allow therapeutic targets for this *autistic* phenotype which may include: modification of diet and enterocolonic microbial milieu in order to reduce toxin substrates, improve nutritional status, and modify mucosal immunity; anti-inflammatory/ immunomodulatory therapy; and specific treatment of dysmotility, focusing, for example, on the pharmacology of local *opioid* activity in the gut.

Gut motility and absorption is another autonomic regulatory factor that is under central autonomic regulation. We have seen that the heart has a natural pacemaker that is under regulation of the brain. Recently it has been shown that the stomach and intestine also have a pacemaker, but until recently, no one knew for sure where it was. However, a team of researchers at McMaster University (Huizinga et al., 1997, 2000; Thomsen et al., 1998) has pinpointed its location. They think that a group of star-shaped cells that underlie the gut muscles can generate gentle waves of excitation that control the rate at which the muscles contract and relax moving food through the intestine. Huizinga and colleagues thought that without this pacemaker, the gut muscles would go into a tonic-contraction, similar to what is seen in a spastic colon. The pacemaker cells are known as the intestinal cells of Cajal. On the other hand, *gastroparesis* results as a failure of the gut muscle to contract properly. It is not certain although a good possibility that this pacemaker, like that of the heart, is under control of the brain. We know that other aspects of neurological control over gut motility are controlled by a balance of the *parasympathetic* division of the *autonomic* nervous system headed by the *nucleus tractus solitarius* (*NTS*) and the *vagus* nerve. The *NTS* is under direct control of the neo-cortex, which balances out and inhibits the *sympathetic* division of the *autonomic* nervous system to maintain a proper balance of activity essential for normal digestion.

Digestive disorders and other metabolic problems have been associated with *autism* and *PDD*. In one study, researchers examined a series of children with chronic *enterocolitis* and *PDD* (Wakefield et al., 1998). Twelve children with ages ranging from 3 to 10, with a history of normal development followed by loss of acquired skills, including language, together with diarrhea and abdominal pain were referred to the pediatric gastroenterology unit. These children then underwent gastroenterological, neurological, and developmental assessment, and their developmental records were reviewed. The onset of the behavior symptoms was associated by the parents with measles, mumps, and rubella vaccination in 8 of the 12 children, with measles infection in one child, and *otitis media* in another. All 12 children had intestinal abnormalities, ranging from lymphoid nodular *hyperplasia* to aphthoid ulceration. Histology showed patchy chronic inflammation in the colon in 11 children and reactive *ileal lymphoid hyperplasia* in 7, but no *granulomas*. Behavioral disorders included autism (nine), *disintegrative psychosis* (one), and possible postural or *vaccinal encephalitis* (two). There were no obvious neurological abnormalities on MRI and EEG. Lab results included significantly raised urinary *methylmalonic* acid, low *hemoglobin*, and low serum *IgA*. In discussing the results, the authors note that the intestinal and behavioral pathologies may have occurred together by chance. However, the uniformity of the intestinal changes and the fact that previous studies have found intestinal dysfunction in children with *autistic spectrum* disorders, are thought to suggest that the connection is real and reflects a unique disease process. They also note that Asperger first recorded the link between *celiac disease* and behavior psychosis (Asperger, 1961).

Walker-Smith and Andrews (1972) showed low concentration of Alpha I *antitrypsin* in children with typical *autism*. In addition, D'Eufemia and colleagues (1996) identified abnormal intestinal permeability, seen in small intestine *enteropathy*, in 4,390 of a group of *autistic* children with no gastrointestinal symptoms, but not seen in controls. The authors conclude that based on these studies, together with their own results, including *anemia* and *IgA* deficiency, the results support the hypothesis that the consequences of an inflamed or dysfunctional intestine may play a part in behavior changes in some children.

Torrente and associates (2002) reported *lymphocytic colitis* in children with regressive *autism*. Where *epithelial* damage is prominent while comparing *duodenal* biopsies in 25 children with regressive *autism* to 11 with *celiac* disease, 5 with *cerebral palsy* and mental retardation, and 18 histologically normal controls. Immunohistochemistry was performed for *lymphocyte* and *epithelial* lineage and functional markers. Standard histopathology showed increased *enterocyte* and Paneth cell numbers in the *autistic* children. Immunohistochemistry demonstrated increased *lymphocyte* infiltration in both *epithelium* and *lamina propria* with upregulated crypt cell proliferation, compared to normal and cerebral palsy controls. *Intraepithelial lymphocytes* and *lamina propria* plasma cells were lower than in *celiac* disease, but *lamina propria T*-cell populations were higher and crypt proliferation similar. Most strikingly, *IgG* deposition was seen on the basolateral epithelial surface in 23/25 *autistic* children and was not seen in the other conditions. These findings demonstrate a novel form of enteropathy in *autistic* children, in which increases in mucosal *lymphocyte* density and *crypt* cell proliferation occur with *epithelial IgG* deposition suggesting to these authors the existence of an autoimmune lesion.

Furlano and colleagues (2001), in a clinical study, attempted to characterize an *ileal lymphoid nodular hyperplasia* (*LNH*) lesion to determine whether *LNH* is specific for *autism*. Colonoscopy was performed in 21 consecutively evaluated children with *autistic spectrum* disorders and bowel symptoms. Blinded comparison was made with 8 children with histologically normal *ileum* and *colon*, 10 developmentally normal children with *ileal LNH*, 15 with *Crohn's* disease, and 14 with *ulcerative colitis*. Immunohistochemistry was performed for cell lineage and functional markers, and histochemistry was performed for *glycosaminoglycans* and *basement membrane* thickness. Histology demonstrated *lymphocytic colitis* in the *autistic* children, less severe than classical inflammatory bowel disease. However, *basement membrane* thickness and mucosal *gamma delta* cell density were significantly increased above those of all other groups including patients with *inflammatory bowel disease*. Epithelial, but not *lamina propria, glycosaminoglycans* were disrupted. Furlano and associates concluded that immunohistochemistry confirms a distinct *lymphocytic colitis* in *autistic spectrum* disorders in which the epithelium appears particularly affected. This is consistent with increasing evidence for gut epithelial dysfunction in *autism*.

Horvath and colleagues (1999) also examining gastrointestinal abnormalities in children with *autism* evaluated the structure and function of their upper gastrointestinal tract by means of gastrointestinal endoscopy with biopsies, intestinal and pancreatic enzyme analyses, and bacterial and fungal cultures. Histological examination revealed grade II or I *reflux esophagitis* in 25/36 cases, chronic *gastritis* in 15/36, and chronic *duodenitis* in 24/36. The number of Paneth's cells in the duodenal crypts was significantly elevated in *autistic* children compared with non-*autistic* control subjects. Low intestinal carbohydrate digestive enzyme activity was reported in 21/36 children. Seventy-five percent of the *autistic* children had increased *pancreaticobiliary* fluid output after intravenous *secretin* administration. Nineteen of the twenty-one patients with diarrhea had significantly higher fluid output than those without diarrhea. The investigators concluded that gastrointestinal disorders, especially *reflux esophagitis* and *disaccharide malabsorption*, might contribute to the behavioral problems of *autistics*. The observed increase in *pancreaticobiliary* secretion after *secretin* infusion suggests an upregulation of *secretin* receptors in the *pancreas* and liver.

Sun and associates (1999a, 1999b) at the University of Florida, publishing in the journal *Autism*, indicate that *autism* may be linked to an individual's inability to properly break down a protein found in milk. The digestive problem might actually lead to the disorder's symptoms, according to these researchers. When the milk protein is not broken down, it produces *exorphins* or morphine-like compounds that are then taken up by areas of the brain known to be involved in *autism* where they cause cells to dysfunction. Their findings in lower organisms suggest an intestinal flaw according to Cade who is testing the theory in humans (Sun and Cade, 1999). Preliminary findings from that study show 95 percent of 81 *autistic* children studied had 100 times the normal levels of the milk protein in their blood and urine. When these children were put on a milk-free diet, at least 8 out of 10 no longer had symptoms of *autism*.

In an effective overview of causal variables associated with *autistic spectrum* disorder, McGuiness has overviewed the literature in summary form, which is provided in Table 10.2 (cf. www.biometricdiagnostics.com/content).

TABLE 10.2   Nutritional Perspectives on the *Autistic Spectrum* Child

I. **Physical health profile of the *autistic* child strongly tends toward:**

1. **Gastrointestinal abnormality**
   - **Malabsorption** (*J Autism/Childhood Schizo,* 1971; *1*(1), 48–62)
     - Frequently reports alcoholic stools, undigested fibers, positive Sudans
     - 85 percent of *autistics* meet criteria for malabsorption (B. Walsh, 500 pts)
   - **Maldigestion**—elevated urinary peptides
     - P. Shattuck (*Brain Dysfunct,* 1990; *3*: 338–45 and 1991; 4: 323–4)
     - K. L. Reichelt (*Develop Brain Dysfunct,* 1994; *7:* 71–85, and others)
     - Z. Sun and R. Cade (*Autism,* 1999a; *3:* 85–96 and 1999b; *3:* 67–83)
     - Low intestinal carbohydrate digestion 58 percent and increased pancreatic output with secretin in diarrhea subgroup K. Horvath (*J Pediatrics,* 1999; *135:* 559–63)
   - **Microbial Overgrowth**—fungal, bacterial, and viral
     - William Shaw, *Biological Basis of Autism and PDD,* 1989
     - E. Bolte on *Clostridium* (*Med Hypoth,* 1998; *51:* 133–44)
     - P. Shattock and A. Broughton IAG elevations
     - W. Walsh and W. McGinnis pyrrole elevations
     - Andrew Wakefield, *Lancet,* 1998; *351:* 637–44)
     - T. J. Borody, Center for Digestive Diseases, New S. Wales, Austral.
   - **Undischarged Paneth cells**
     - K. Horvath (*J Pediatrics,* 1999; *135:* 559–63)
   - **Abnormal intestinal permeability**
     - P. DEufemia (*Acta Pediatr,* 1995; *85:* 1076–9)
   - **Endoscopic abnormality** (K. Horvath, *J Pediatrics,* 1999; *135:* 559–63)
     - Chronic esophagitis 69 percent
     - Chronic duodenitis 67 percent
     - Chronic gastritis 42 percent
   - **GI symptoms reported by parents:** diarrhea, constipation, gas, belching, probing, visibly undigested food, and need for rubs

2. **Compromised immunity**
   - **Recurrent infections**
     - Euro Child/*Adolesc Psych,* 1993; *2*(2): 79–90
     - *J Autism Dev Disord,* 1987; *17*(4): 585–94
   - **Abnormal indices**
     - *T*-cell deficiency (*J Autism/Childhood Schizo,* 1977; *7:* 49–55)
     - Reduced *NK* cell activity (*J Ann Acad Child Psyc,* 1987; *26:* 333–35)
     - Low or absent IgA (*Autism Develop Dis,* 1986; *16:* 189–197)
     - Low C4B levels (*Clin Exp Immunol,* 1991; *83:* 438–40)
   - **Skewed ("elevated") viral titers** increasing grassroots reports, V. Singh, University of Michigan

3. **Detoxification weakness**
   - **Phase II depression** (S. Edelson and D. S. Cantor *Toxicol Industr Health,* 1998; *14*(4): 553–63)
     - Sulfation low in 15 of 17 (mean 5 vs. nl 10–18)
     - Glutathione conjugation low in 14 of 17 (mean 0.55 vs. 1.4–2.9)

Table 10.2 *Continued*

- ○ Glucuronidation low in 17 of 17 (mean 9.6 vs. 26.0–46.0)
- ○ Glycine conjugation low in 12 of 17 (15.4 vs. 30.0–53.0)
- **Sulfation deficit** (*Biol Psych,* 1999; *1*: 46(3): 420–4)
- **Peroxisomal malfunction** (P. Kane, *J Orthomolec Med,* 1997; *12–4*: 207–18 and 1999; *14–2*: 103–9)
- **Higher blood lead levels in** *autism* and documented *response to EDTA chelation* (*Am J Dis Child,* 1976; *130*: 47–8)
- **Apparent temporal association** *autism* onset and lead exposure (*Clin Pediatr,* 1988; *27*(1): 41–4)
4. **Abnormal nutritional profile in children with** *autism*
    - **Lower serum magnesium** than controls (Mary Coleman, *The Autistic Syndromes,* 1976; 197–205)
    - **Lower RBC magnesium** than controls (Findling et al., 1997)
    - **Low activated B6** (P5P) in 42 percent. Autistic group also **higher in serum copper** (*Nutr. Behav,* 1984; *2:* 9–17)
    - **Low EGOT (functional B6) in 82** percent and all 12 subjects low in 4 *amino acids* (tyrosine, carnosine, lysine, hydroxylysine). Dietary analysis revealed below-RDA intakes in *zinc* (12 of 12 subjects), *calcium* (8 of 12), *vitamin D* (9 of 12), *vitamin E* (6 of 12) and *vitamin A* (6 of 12) (Pfeiffer et al., 1995)
    - **B6 and magnesium therapeutic efficacy**—multiple positive studies (start with *Am J Psych,* 1978; *135*: 472–5)
    - **Low derivative** *Omega-6* **RBC membrane levels** 50 of 50 autistics assayed through Kennedy Krieger had GLA and DGLA below mean. Low *Omega-3* less common (may even be elevated) (*J Orthomolecular Med,* 1997; *12*(4))
    - **Low methionine level's** not uncommon (Observation by J. Pangborn)
    - **Below normal glutamine** (14 of 14), **high glutamate** (8 of 14) (*Invest Clin,* 1996 June; *37*(2): 112–28)
    - **Higher copper/zinc** ratios in autistic children (*J Appl Nutr,* 1997; *48*: 110–18)
    - **Reduced sulfate conjugation** and **lower plasma sulfate** in *autistics*. (*Dev Brain Dysfunct,* 1997; *10*: 40–3) (*J Environ Nutr Med,* Mar 2000)
    - **B12 deficiency** suggested by elevated urinary methylmalonic acid (*Lancet,* 1998; *351*: 637–41)
    - **Hypocalcinurics improve with calcium supplementation** Lower hair calcium in autistics reported (*Dev Brain Dysfunct,* 1994; *7*: 63–70)
    ARI parent survey for therapeutic responses by *autistic* children:
    - 50 percent improved with zinc (6 percent worsened)
    - 49 percent improved with vitamin C
    - 46 percent improved with magnesium and B6 (5 percent worsened)
    - 58 percent improved with calcium (Later survey 42 percent)

II. **Physical health profile in** *ADHD*
1. **Gastrointestinal abnormality**
    - *Colicky infants* and *older children diarrhea-prone* (V. Colquhoun HACSG, Sussex UK 1987)
    - *Severe stomach aches* (*Am J Clin Nutr,* 1995; *62*: 761–8)
    - *Elevated stool creosols* (*Lancet,* December 7, 1985)
    - *Ileal lymphoid nodular hyperplasia* (*Lancet,* July 18, 1998)
    - *Urinary peptide elevations*—P. Shattock and A. Broughton
    - *Urinary organic acids elevations*—W. Shaw
    - *IAG elevations*—A. Broughton
    - *Parasitosis* 67 percent—M. Lyon and J. Cline

2. **Compromised immunity**
    - *More infections and antibiotics* (*Am J Clin Nutr,* 1995; *62*: 761–8)
    - *Low complement C4B* (*J Am Acad Child Adolesc Psych,* 1995; *34*(8): 1009–14)

3. **Detoxification weakness**
    - *Low-level lead exposure induces hyperactivity in rats* (*Science,* 1973; *182*(116): 1022–4)
    - *Marked improvement in 7 of 13 chelated for "non-toxic" lead levels* (*Am J Psych,* 1976; *133*(10): 1155–8)
    - *Neonatal and maternal hair lead predict LD at age 6* (*Lancet,* 1987; *2:* 285)
    - *Hair lead levels correlate* with teacher-rated and physician-diagnosed *ADHD* (*Arch Environ Health,* 1996; *51*(3): 214–20)
    - *Striking chelation results* in 50 Vancouver children (*Turning Lead into Gold* (paperback), Nancy Hallaway and Ziggert Strauts, 1996)

Table 10.2 *Continued*

4. **Abnormal nutritional profile in *ADHD***
   - **Zinc deficiency**
     - Lower urinary, serum, nail, and hair zinc than controls plus quick drop in serum and salivary zinc with double-blind tartrazine. United Kingdom. (*J Nutr Med*, 1990; *1*: 51–7)
     - Plasma, erythrocytes, urine, and hair lower than controls. Poland (*Psychiatr Pol*, 1994; *28*(3): 345–53)
     - Zinc deficiency in *attention-deficit hyperactivity disorder*. Israel. (*Biol Psychiatry*, 1996; *40*(12): 1308–10)
     - Serum zinc—and free fatty acids—lower. Turkey. (*J Child Psychol Psychiatry*, 1996; *37*(2): 225–7)
     - *In vitro* study demonstrates decreased loss of fatty acids from mesenteric phospholipids with perfusion of physiological zinc. Canada. (*Can J Physiol Pharmacol*, 1990 *68*(7): 903–7)
   - **Fatty acid deficiency**
     - Lower serum DHA, DGLA, and AA in hyperactives than controls. (*Clin Pediatr*, 1987; *26*(8): 406–11)
     - Double-blind administration of evening primrose oil to a subgroup of prior study was associated with improved parent ratings for attention and excess motor activity compared to placebo. (*J Abn Child Psychol*, 1987; *15*(1): 75–9)
     - Evening primrose oil (GLA) 1 gram/day improved 53 of 79 hyperactive children selected as a subgroup on the basis of mood swings. The most striking improvement was noted in children with sleep disorders, crying spells and family history of alcohol or bipolar. (Muriel Blackburn, Crawley Hospital, Sussex, UK)
     - Lower plasma DHA, EPA, and AA, and lower RBC AA in *ADHD* than controls (*Am J Clin Nutr*, 1995; *62*: 761–8)
     - (Same group as above correlated greater tendency to behavioral problems with lower total plasma *Omega-3*, more colds and antibiotics with lower total *Omega-6*. *Physiol Behav* 59(4/5): 915–920, 1996)
     - Zinc and Evening primrose oil the mainstay for thousands of successes claimed by the HACSG, Sussex England (Personal Communication, Vicky Colquhoun, 1997)
   - **Magnesium deficiency**
     - Magnesium deficiency measured in 95 percent of 116 Polish children with *ADHD*: 78 percent low hair, 59 percent low RBCs, 34 percent low serum. (*Magnesium Res*, 1997; *10*(2):143–8)
     - Double-blind administration of 200 mg elemental magnesium per day to 25 of the above group produced measurable decrease in hyperactivity over 6 months compared to control. (*Magnesium Res*, 1997 *10*(2): 149–156)
   - **Iron deficiency**
     - Preliminary study showed improved behavior in nonanemic hyperactives given 5 mg/kg/day of iron for 30 days, with significant increase in serum ferritin. (*Neuropsychobiology*, 1997; *35*(4): 178–80)
     - Lower iron plasma, RBC, urine, and hair levels in 50 hyperactives (*Psychiatr Pol*, 1994; *28*(3): 343–53)
   - **Calcium deficiency**
     - Plasma, RBC, urine, and hair calcium in 50 hyperactive Polish children lower than controls. (*Psychiatry Pol*, 1994; *28*(3): 343–353.
   - **B6 in ADHD**
     - B6 to hyperactives with low *serotonin* levels resulted in normal *serotonin* levels and behavior. (*Pediatrics*, 1975; *55*: 437–41)
     - B6 to 6 hyperactives with low *serotonin* levels increased *serotonin* and reduced hyperactivity better than Ritalin in double-blind cross-over. Benefit carried over into the following placebo period, but not with Ritalin. (*Biol Psychiatry*, 1979; *14*(5): 741–51)
     - Significant subgroup of patients with *ADHD* (and *autism*) found to have pyrroluria by Bill Walsh (Pfeiffer Treatment Center, Napperville, IL) and Hugh Riordan (BioCenter, Wichita, KS). Good clinical track record for response to generous B6 and Zinc in thousands of pyrroluric patients. (Walsh also finds biotin very useful in "slender malabsorber group")
   - **B12 in *ADHD***
     - Elevated urinary methylmalonic acid and early reports of response to oral B12 from John Linnell, research director at The Childrens Medical Charity, UK. Some reports of response to B12 shots.

From W. R. McGinnis, www.biometricdiagnostics.com/content.

The first step toward treatment of the problems with children is recognizing the causes and educating parents, teachers, and clinicians on how to prevent them. Besides this, the next chapter will emphasize the theoretical rationale based on our understanding of brain-behavior relationships for the multitude of treatment options available.

11

Therapeutic Theory and Strategy

We have now reviewed how symptoms are produced and what are the underlying causes of decreases in stimulation to the brain in individuals with neurobehavioral disorders in the context of evolutionary aspects of bipedalism. At first glance, the answer to how intervention strategies should be developed seems simple—increase stimulation and provide the proper fuel. In practice, however, it is not quite that straightforward. The purpose of this chapter is to provide the theoretical rationale for therapeutic and intervention strategies currently available, but not to provide a "how to" manual, the subject of a subsequent volume. We need now to consider how we increase stimulation and what is the best fuel.

TREATMENT RATIONALE

The current thinking is that only two options for treatment exist: one being medication which makes up approximately 75 percent of recommendations and the other being psychological or behavioral counseling which makes up the other 25 percent. Most teachers, as well as clinicians are not aware of the options or the possible effectiveness and scientific rationale for their use. In fact, many other alternative treatment options have been available for decades that are effective and safe. Some of these options have merit and will be examined here in the light of recent brain, behavioral, pharmacological, biochemical, and genetic research.

Behavioral Intervention Strategies

Cognitive Behavioral Therapy

This form of therapy has been used on adults for many fears or *phobias, anxiety disorders, panic attacks*, and *post traumatic stress disorder*. Developed in the 1960s by psychiatrist Aaron Beck (1967, 1976), this form of therapy gradually allows those with these disorders to talk about, physically approach, and ultimately experience the very things that terrify them. It is thought that this particular form of therapy deliberately sets up a program of repeated programmed self-awareness exercises to rewire connections in the brain and form helpful new memories, just as repetitive practicing of the piano gradually creates a memory of motor skills. Common cognitive distortions without a neurobehavioral disorder are likely to have adaptive evolutionary value. Cognitive distortions are natural consequences of using fast track defensive algorithms that are sensitive to threat. In various contexts, especially those of threat, humans have evolved to think adaptively rather than logically. Hence, cognitive distortions are not strictly errors in brain functioning and it can be useful to inform patients that "negative thinking" may be dysfunctional but is a reflection of basic brain design and not personal irrationality. The neuropathology underlying attention deficit hyperactivity disorder (*ADHD*) and neurobehavioral disorders in general most consistently points to dysfunction in cortical-*striatal* pathways leading to inactivation, or insufficient engagement, of *frontal* and *prefrontal lobes*. By implication, there may be functional disconnection between the anterior and posterior higher cortical regions, instead of a fixed dysfunction in either one. Given this premise, reconnection of these systems via cognitive interventions constitutes a logical remedial approach in the treatment of *ADHD* and neurobehavioral disorders through integrative cognitive and neuropsychological interventions.

Related to neurobehavioral disorders, however, studies have shown that Cognitive Behavioral Therapy (CBT) is just as effective as drugs in many instances (Hinshaw et al., 1984; Barlow et al., 2000; Ninan et al., 2000; Hunter et al., 2002). We have discussed the involvement of the *thalamus* in the manifestation of developmental neurobiologic abnormalities. Rosenberg and colleagues (2000) had recently studied the pathophysiology of *obsessive–compulsive disorder* (*OCD*) treated by CBT. They had reported increased *thalamic* volume in treatment-naive pediatric *OCD* patients versus case-matched healthy comparison subjects that decreased to levels comparable to control subjects after effective *paroxetine* therapy. No study prior to theirs had measured neuroanatomic changes in the *thalamus* of *OCD* patients at the onset of illness before and after CBT. Volumetric magnetic resonance imaging studies were conducted in 11 psychotropic drug-naive 8–17-year-old children with *OCD* before and after 12 weeks of effective CBT monotherapy. They reported no significant change in *thalamic* volume in *OCD* patients before and after CBT suggesting that reduction in *thalamic* volume after *paroxetine* therapy may be specific to *paroxetine* treatment and not the result of a general treatment response or spontaneous improvement. However, recent research has demonstrated that CBT for *OCD* can systematically modify cerebral metabolic activity in a manner, which is significantly related to clinical outcome. There does exist a substantial body of research supporting an involvement of neural circuitry connecting the *orbitofrontal* cortex, *cingulate gyrus*, and *basal ganglia* in the expression of the symptoms of *OCD*. Data has been reported (Schwartz et al., 1998) which expands upon previous work demonstrating effects of CBT on functional interactions between *limbic cortex* and the *basal ganglia* reflecting the interactive nature of the relationships between cognitive choice, behavioral output, and brain activity.

CBT gradually stimulates the neocortex by increasingly having the patients physically interact with their environment. We have seen that children exposed to violence or neglect have similar symptoms as *post traumatic stress disorder*, where the *amygdala* and *limbic system* are overactive. In children, this results from a delayed development of the *prefrontal* cortex and the *amygdala* remains the primary site of emotionally based information processing. The rise of physical activity engages the muscle and joint receptors and the *cerebellum*, which is the initiator of all human learning, cognitive or social. The *cerebellum* also is the largest source of stimulation to the *thalamus* and the *prefrontal* cortex as well as the *basal ganglia*. As the person physically interacts with the environment, it induces the *cerebellum* to promote developments of the *prefrontal* cortex, which then inhibits the *amygdala* and *limbic system*, which reduces the fear and stress responses. The *prefrontal* cortex now allows the individual to have perception that is more appropriate and awareness of the reaction of others to their actions as they learn what is socially appropriate behavior. We also think that the recall of the painful memories at the same time helps to link these in time and space or synchronize these memories with other areas, if the now more efficiently functioning neocortex forms new associative cognitive patterns for the previously inhibited emotions. The individual would be theoretically forming new memories that are less disturbing than the previous associations.

Wykes and colleagues (2002) at the *Institute of Psychiatry* in London had recently performed an evaluation of the effects on the brain of CBT by means of functional magnetic resonance imaging (*f*MRI). The authors examined the effects on brain activity as a result of engaging in CBT. Three groups (patients receiving control therapy or CBT and a healthy control group) were investigated in a repeated measures design using the two-back test. Data obtained by *f*MRI and a broad assessment of executive functioning was completed at baseline and post-treatment. Brain activation changes were identified after accounting for possible task-correlated motion artifact. The *f*MRI analyses indicate that the control group shows decreased activation but the two patient groups show significant increases in activation over time. The patient group that receives successful CBT has significantly increased brain activation in regions associated with working memory, particularly the *frontal* cortical areas. The results are the first ever indication that brain activation changes in a seriously disabled group of patients with *schizophrenia* can be associated clearly with psychological rather than pharmacological therapy.

A study was performed in which the effects of *citalopram* and CBT on *regional cerebral blood flow* (*rCBF*) were explored in *social phobia* by means of *positron emission tomography* (PET) (Furmark et al., 2002). *rCBF* was assessed in 18 previously untreated patients with *social phobia* during an anxiogenic public speaking task. Patients were matched for sex, age, and phobia severity, based on social anxiety questionnaire data, and randomized to *citalopram* medication, CBT group therapy, or a waiting-list control group. Scans were repeated after 9 weeks of treatment or waiting time. The outcome was assessed by subjective and psychophysiological state anxiety measures and self-report questionnaires. The questions were re-administered after one year. The results indicate that symptoms improved significantly and roughly equally with *citalopram* and CBT, whereas the waiting-list group remains unchanged. Within both treated groups, and in responders regardless of treatment approach, improvement is accompanied by a decreased *rCBF*-response to public speaking bilaterally in the *amygdala, hippocampus*, and the *periamygdaloid, rhinal*, and *parahippocampal* cortices. Between-group comparisons confirm that *rCBF* in these regions decreases significantly more in treated groups than in control subjects, and in responders than non-responders, particularly in the right hemisphere. The degree of *amygdala-limbic* attenuation is associated with clinical improvement a year later. The authors conclude that common sites of action for *citalopram* and CBT of *social anxiety* are observed in the *amygdala, hippocampus*, and neighboring cortical areas that subserve bodily defense reactions to threat.

CBT can be employed as a systematic effort to assist brain-impaired individuals in developing ways to compensate for cognitive deficits. Brett and Laatscha (1998) had evaluated children with acquired nervous system brain injury receiving biweekly CBT sessions for 20 weeks in a school setting. Treatments were provided by trained schoolteachers under the supervision of psychologists specializing in cognitive rehabilitation. Students were evaluated pre- and post-treatment using neuropsychological tests. After treatment, the students demonstrated a significant increase in general memory ability. These gains were mostly due to increases in verbal learning ability according to the authors. The theoretical rationale for CBT intervention has been practically adapted to attempt to teach meta-cognitive executive thinking strategies to children with disorders of executive function. The intervention is based on the notion that some children with disorders of executive function have disorders of higher-level language, which predispose them to the executive impairments demonstrated. The teaching and reinforcing of meta-cognitive thinking strategies may well help advance verbal mediation of complex tasks and self-regulation of behavior in children with neurobehavioral disorders. Despite the growing literature on developmental executive disorders, little has been written about interventions that may enable the children to acquire some of the requisite adaptive skills.

Utilizing performance on intelligence testing spanning 20 years, Bellus and colleagues (1998) performed a study evaluating changes in cognitive functioning of a severely brain injured individual, who had been placed in a long-term psychiatric hospital and treated in an intensive behavioral rehabilitation program. Results found that the patient demonstrated a significant improvement in overall verbal and nonverbal cognitive functioning during treatment. These improvements were maintained for a 1-year period. The authors suggest that the use of "low tech," small group interventions, within intensive behavioral rehabilitation programs, may lead to the recovery of cognitive functioning for individuals who are significantly cognitively and socially impaired. Applying the same rationale to children with acquired brain injury and with difficulties in problem-solving and social adjustment, Suzman and associates (1997) provided case studies and a series of multiple baseline experiments examining the effects of a multi-component CBT on the remediation of problem-solving deficits in five children with acquired brain injury. Results indicate that the training program resulted in a substantial decrease in errors on a computerized problem-solving task used to monitor problem-solving performance during baseline and treatment. In addition, significant improvements are found on two of four standardized measures of problem-solving abilities.

Frolich and associates (2002) have indicated that in the past, cognitive behavioral treatment concepts failed to demonstrate their clinical effectiveness in the treatment of *ADHD* children. They combined *CBT* with a special focus on self-instructional and self-management skills with subsequent *parent management training* (*PMT*) in order to reduce academic problems and oppositional/aggressive behavior. Eighteen children with a diagnosis of *ADHD* combined type and *Oppositional Defiant Disorder* participated in the study. The effects of a 12-week treatment phase (6 weeks *CBT*; 6 weeks *PMT*) were compared with a preceding 4-week baseline. Core symptoms of *ADHD*, conduct, and homework problems were assessed by weekly administration of parent and teacher questionnaires. *CBT* was found effective in reducing the core symptoms of *ADHD* and conduct problems at home and in school. PMT resulted in a further amelioration of the cited symptoms. These investigators conclude that *CBT* is an important component in the treatment of *ADHD* if aspects of generalization are considered. *PMT* is a useful adjunct to *CBT* due to its effectiveness in situations where children still have problems of self-guidance.

Stevenson and colleagues (2002) attempted to systematically examine the efficacy of a CBT for management of adult *ADHD*. Their CBT program was designed to target problems commonly associated with adult *ADHD*, namely, attention problems, poor motivation, poor organizational skills, impulsivity, reduced anger control, and low self-esteem. In a randomized, controlled trial, a representative sample of adults with *ADHD* who were both medicated and non-medicated were assigned to either CBT or a waiting list control. CBT was delivered in an intensive format with eight 2-hour, weekly sessions with support people who acted as coaches, and participant workbooks with homework exercises. Participants who completed CBT reported reduced *ADHD* symptoms, improved organizational skills, and reduced levels of anger. Clinically significant improvements in *ADHD* symptoms and organizational skills were maintained one year after the intervention. The study's authors conclude that CBT provides a practical way of enhancing daily functioning for adults with *ADHD*.

Wilmshurst (2002) examined youth with severe *emotional* and *behavioral disorders* (*EBD*) by randomly assigning them for three months of intensive treatment to a 5-day residential program or a community-based alternative, family preservation program.

Programs differed not only in method of service delivery (residential unit vs. home-based), but also in treatment philosophy (solution focused brief therapy vs. CBT). Results confirm high rates of comorbidity in this population for externalizing and internalizing disorders. A significant Treatment × Program interaction was evident for internalizing disorders. At 1-year follow-up, significantly higher percentages of youth from the family preservation program revealed a reduction of clinical symptoms for *ADHD*, as well as, general anxiety and depression, whereas significant proportion of youth from the 5-day residential program demonstrated clinical deterioration and increased symptoms of anxiety and depression. These investigators conclude that greater emphasis be placed on research linking treatment to specific symptom clusters, especially highly comorbid clusters based on neuropsychological test performance.

Eye Movements as Reflective of Neurococognitive Processes Employed in Behavioral Therapeutic Intervention

Our eyes usually move in brief motions called *saccades*. Between the *saccades*, they focus on the objects that we see (Leisman, 1976a). Although our eyes move several times per second, we perceive the world only during the brief moments of fixation (Leisman, 1976a). In essence, we can think of our visual experience as a rapid sequence of still life photographs. However, the fixations occur too rapidly for us to notice the interval between the snapshots (Leisman, 1976a; Haber, 1978).

The saccadic eye movements, generated during a visual oddball task, of *autistic* children, normal children, children with *ADHD*, and dyslexic children were examined to determine whether *autistic* children differed from these other groups in saccadic frequency (Kemner et al., 1998). *Autistic* children made more *saccades* during the presentation of frequent stimuli (than normal and *ADDH* children), and between stimulus presentations. In addition, unlike the normal and dyslexic groups, their saccadic frequency did not depend on stimulus type. This abnormal pattern of *saccades* may negatively influence the ability to attend to stimuli, and thereby learning processes.

Goldberg and colleagues (2002) have recently employed ocular motor paradigms to examine whether or not *saccades* are impaired in individuals with *high functioning autism* (HFA). They recorded eye movements in patients with HFA and in normal adolescents on anti-saccade, memory-guided saccade (MGS), predictive saccade, and gap/overlap tasks. Compared with the normal subjects, patients with HFA had a significantly higher percentage of directional errors on the anti-saccade task, a significantly higher percentage of response suppression errors on a MGS task, and a significantly lower percentage of predictive eye movements on a predictive saccade task. They also showed longer latencies on a MGS task and for all conditions tested on a gap/null/overlap task (fixation target extinguished before, simultaneously, or after the new peripheral target appeared). When the latencies during the gap condition were subtracted from the latencies in the overlap condition, there was no difference between patients and normals. These authors conclude that abnormalities in ocular motor function in patients with HFA provide evidence for the involvement of a number of brain regions in HFA including the dorsolateral *prefrontal* cortex, the frontal eye fields, the *basal ganglia*, and *parietal lobes*.

It is known that *autistic* children demonstrate abnormal gaze behavior toward human faces as observed in daily-life situations (van der Geest et al., 2001, 2002a, 2002b). van der Geest and associates (2002a) investigated this process in two fixation time studies. They selected a group of high-functioning *autistic* children (including a group of sub-threshold Pervasive developmental disorder not otherwise specified [PDD-NOS] children) who were compared with a group of normal children, with respect to their fixation behavior for photographs of human faces. Using an infrared eye-tracking device, fixation times for the whole face and for the facial elements of faces were compared between the two groups. The first study dealt with faces having an emotional expression. The second study dealt with neutral faces presented either upright or upside-down. Results indicated that *autistic* children have the same fixation behavior as normal children for upright faces, with or without an emotional expression. Furthermore, results of the second study showed that normal children spent less time looking at upside-down faces, but that the fixation times of *autistic* children were not influenced by the orientation of the faces. These results plead against the notion that the abnormal gaze behavior in everyday life is due to the presence of facial stimuli per se. Furthermore, the absence of a face orientation effect in *autistic* children might be a reflection of a lack of holistic processing of human faces in *autism* and is a problem that may be addressed by therapeutic intervention using eye movements. van der Geest and associates (2002b) additionally noted that *autistic* children have a problem in processing social information and that several studies on eye movements have indeed found indications that children with *autism* show particularly abnormal gaze behavior in relation to social stimuli even though previous studies did not allow for precise gaze analysis. In their study (van der Geest et al., 2002b), the looking behavior of *autistic* children toward cartoon-like scenes that included a human figure was measured quantitatively using an infrared eye-tracking device. The fixation behavior of *autistic* children was found to be similar to that of their age- and IQ-matched normal peers. Their results do not support the notion that *autistic* children have a specific problem in processing socially loaded visual stimuli as reflected by eye tracking. In addition, there is no indication for an abnormality in gaze behavior in relation to neutral objects. It is suggested that the often-reported abnormal use of gaze in everyday life is not related to the nature of the visual stimuli but to other factors, like social interaction. Therefore, intervention strategies employing eye movements as a therapeutic vehicle are theoretically useful.

These same investigators (van der Geest et al., 2001) hypothesized that children with *autism* have deficits in attentional (dis-) engagement mechanisms. A saccadic gap-overlap task was used to study visual engagement and disengagement in 16 HFA children of about 10 years of age and 15 age- and IQ-matched normal control children. Subjects were asked to make saccadic eye movements from a fixation point to a suddenly appearing target as fast as possible. The saccadic reaction time was compared in two conditions: (1) the overlap condition, in which the fixation point was continuously visible, and (2) the gap condition, in which the fixation point was turned off 200 ms before the target appeared. Although no differences between the groups in either condition was observed, the gap effect (i.e., the difference in saccadic reaction time between the overlap condition and the gap condition) was smaller in the *autistic* group than in the control group. They concluded that *autistic* children show a lower level of attentional engagement.

Ruffman and colleagues (2001) addressed these issues by studying social understanding in *autism* employing eye gaze as a measure of core insights. Twenty-eight children with *autism* and 33 mentally handicapped children were given two tasks tapping social understanding and a control task tapping probability understanding. For each task, there was a measure of eye gaze (where children looked when anticipating the return of a story character or an object) and a verbal measure (a direct question). They found that eye gaze was better than verbal performance at differentiating children with *autism* from children with other mental handicaps. Children with *autism* did not look to the correct location in anticipation of the story character's return in the social tasks, but they did look to the correct location in the nonsocial probability task. These investigators also found that within the *autistic* group, children who looked to the correct location the least were rated as having the most severe *autistic* characteristics. Further, they found that whereas verbal performance correlated with general language ability in the *autistic* group, eye gaze did not. They argue that eye gaze probably taps unconscious but core insights into social behavior and as such is better than verbal measures at differentiating children with *autism* from mentally handicapped controls. Additionally, eye gaze taps either spontaneous processes of simulation or rudimentary pattern recognition, both of which are less based in language, and the social understanding of children with *autism* is probably based mostly on verbally mediated theories whereas control children also possess more spontaneous insights indexed by eye gaze.

Manifestations of core social deficits in *autism* are more pronounced in everyday settings than in explicit experimental tasks. To bring experimental measures in line with clinical observation, Klin and associates (2002) reported a novel method of quantifying atypical strategies of social monitoring in a setting that simulate the demands of daily experience. While viewing social scenes, eye-tracking technology measured visual fixations in 15 cognitively able males with *autism* and 15 age-, sex-, and verbal IQ-matched control subjects. The investigators coded fixations on four regions: mouth, eyes, body, and objects. Statistical analyses compared fixation time on regions of interest between groups and correlation of fixation time with outcome measures of social competence (i.e., standardized measures of daily social adjustment and degree of *autistic* social symptoms). They found significant between-group differences for all four regions with the best predictor of *autism* being reduced eye region fixation time. Fixation on mouths and objects was significantly correlated with social functioning: increased focus on mouths predicted improved social adjustment and less *autistic* social impairment, whereas more time on objects predicted the opposite relationship. When viewing naturalistic social situations, individuals with *autism* demonstrate abnormal patterns of social visual pursuit consistent with reduced salience of eyes and increased salience of mouths, bodies, and objects. Fixation times on mouths and objects but not on eyes are strong predictors of degree of social competence.

From an evolutionary standpoint gaze is an important component of social interaction. The function, evolution, and neurobiology of gaze processing are therefore of interest in the context of neurobehavioral disorders of childhood. The role of social gaze has changed considerably for primates compared to other organisms. This change may have been driven by morphological changes to the face and eyes of primates, limitations in the facial anatomy of other vertebrates, changes in the ecology of the environment in which primates live, and a necessity to communicate information about the environment, emotional, and mental states. The eyes represent different levels of signal value depending on the status, disposition, and emotional state of the sender and receiver of such signals. There are regions in the monkey and human brain, which contain neurons, that respond selectively to faces, bodies, and eye gaze. The ability to follow another individual's gaze direction is affected in individuals with *autism* and other neurobehavioral disorders as we have seen, as well as following particular localized brain lesions. We can hypothesize that gaze following is "hard-wired" in the brain, and may be localized within a circuit linking the superior *temporal sulcus, amygdala*, and *orbitofrontal* cortex. A more complete review is provided by Emery (2000). This being the case, intervention strategies employing eye gaze as a vehicle should clearly be employed in children with neurobehavioral disorders.

Supporting the notion of eye gaze involvement, Howard and colleagues (2000) at the *University of Liverpool* reported a convergence of behavioral and neuroanatomical evidence in support of an *amygdala* hypothesis of *autism*. They found that *high-functioning autistics* (*HFA*) show neuropsychological profiles characteristic of the effects of *amygdala* damage, in particular selective impairment in the recognition of facial expressions of fear, perception of eye-gaze direction, and recognition memory for faces. Using quantitative magnetic resonance imaging analysis techniques, they found that the same individuals also show abnormalities of medial *temporal lobe*, notably bilaterally enlarged *amygdala* volumes. These results combine to suggest that developmental malformation of the *amygdala* may underlie the social-cognitive impairments characteristic of *HFA*. While they attribute these findings to incomplete neuronal pruning in early development, these data further support a primary underlying involvement of eye gaze.

In attempting to explain further why gaze-shift impairment is part of the clinical picture in neurobehavioral disorders of childhood including *autism*, we should recall that recent imaging and clinical studies have challenged the concept that the functional role of the *cerebellum* is exclusively in the motor domain. Townsend and associates at the *University of California-San Diego* (1999) presented evidence of slowed covert orienting of visual-spatial attention in patients with developmental *cerebellar* abnormality (at least 90 percent of all postmortem cases of *autism* reported to date have *Purkinje* neuron loss), and in patients with *cerebellar* damage acquired from tumor or stroke. In spatial cuing tasks, normal control subjects across a wide age range were able to orient attention within 100 ms of an attention-directing cue. Patients with *cerebellar* damage showed little evidence of having oriented attention after 100 ms but did show the effects of attention orienting after 800–1,200 ms. These effects were demonstrated in a task in which results were independent of the motor response. In this task, smaller *cerebellar* vermal lobules VI–VII (from magnetic resonance imaging) were associated with greater attention-orienting deficits. Although eye movements may also be disrupted in patients with *cerebellar* damage, abnormal gaze shifting cannot explain the timing and nature of the attention-orienting deficits. These data are consistent with evidence from organism models that suggest damage to the *cerebellum* disrupts both the spatial encoding of a location for an attentional shift and the subsequent gaze shift. These data are also consistent with a model of *cerebellar* function in which the *cerebellum* supports a broad spectrum of brain systems involved in both non-motor and motor function.

Disturbances in the orbital *prefrontal* cortex and its ventral *striatal* target fields have been identified in neuroimaging studies of *OCD*. In organism models and studies of patients with lesions to this brain circuitry, a selective disturbance in the ability to suppress responses to irrelevant stimuli has been demonstrated. Such a deficit in response suppression might underlie the apparent inhibitory deficit suggested by the symptoms of *OCD*. Although *OCD* commonly emerges during childhood or adolescence, few studies have examined psychotropic-naive pediatric patients near the onset of illness to find the possible role of atypical developmental processes in this disorder. Oculomotor tests were administered to 18 psychotropic medication-naive, non-depressed patients with *OCD* aged 8.8–16.9 years and 18 case-matched healthy comparison subjects to assess the following three well-delineated aspects of *prefrontal* cortical function: the ability to suppress responses, the volitional execution of delayed responses, and the anticipation of predictable events. A significantly higher percentage of response suppression failures were observed in patients with *OCD*, particularly in younger patients compared with their case-matched controls. No significant differences between patients with *OCD* and controls were observed on other *prefrontal* cortical functions. Severity of *OCD* symptoms was related to response suppression deficits. A basic disturbance of behavioral inhibition in *OCD* was detected that may underlie the repetitive symptomatic behavior that characterizes the illness (Rosenberg et al., 1997).

Mostofsky and colleagues (2001a) assessed saccadic eye movements in boys with *Tourette's syndrome* with and without *ADHD*, comparing performance with that of an age-matched group of male controls. Three different *saccade* tasks (*prosaccades, antisaccades,* and MGS) were used to examine functions necessary for the planning and execution of eye movements, including motor response preparation, response inhibition, and working memory. The study included 14 boys with *Tourette's syndrome* without *ADHD*, 11 boys with *Tourette's syndrome* and *ADHD*, and 10 male controls. Mostofsky and associates found that the latency of *prosaccades* was prolonged in boys with *Tourette's syndrome* (both with and without *ADHD*) compared with controls. Variability in *prosaccade* latency was greater in the groups of boys with *Tourette's syndrome* and *ADHD* compared with both the *Tourette's syndrome*-only and control groups. Response inhibition errors on both the *antisaccade* task (directional errors) and MGS task (anticipatory errors) were increased in boys with *Tourette's syndrome* and *ADHD* compared with those with *Tourette's syndrome*-only. There were no significant differences among the three groups in accuracy of MGS. Mostofsky's oculomotor findings suggest that *Tourette's syndrome* is associated with delay in initiation of motor response as evidenced by excessive latency on *prosaccades*. Signs of impaired response inhibition and variability in motor response appear to be associated with the presence of *ADHD*.

Abnormalities of executive function are observed consistently in children with *ADHD*, and it is hypothesized that these arise because of disruption of a behavioral inhibition system. In examining contextual abnormalities of saccadic inhibition in children with *ADHD*, Cairney and associates (2001) compared executive and inhibitory functions between unmedicated and medicated children with *ADHD*, age-matched healthy children, and healthy adults. Executive functions were measured using a test of spatial working memory shown previously to be sensitive to *ADHD* and to stimulant medication. Inhibitory functions were measured using an ocular motor paradigm that required individuals to use task context to control the release of fixation. Context was set according to the probability that a target would appear at either of the two locations. In one block, targets appeared on 80 percent of trials. In the other block, targets appeared on 20 percent of trials. The ability to control the release of fixation was inferred from the *fixation-offset effect*, or the difference in *saccade* latency when the current fixation is offset 200 ms prior to the onset of the *saccade* target (gap condition), compared with when there is no offset (overlap condition). Although the healthy children made more errors on the spatial working memory task than the healthy adults, there was no difference between the two groups in their ability to control fixation using context. Both showed a larger *fixation-offset effect* when target probability was low. The unmedicated *ADHD* group made more errors on the spatial working memory test than the healthy children, although spatial working memory performance was normal in the medicated *ADHD* group. However, both the unmedicated and medicated *ADHD* groups were unable to modulate the *fixation-offset effect* according to context, and this was due to their inability to voluntarily inhibit *saccades* when there was a low target probability. These data suggest that the context-based modulation of fixation release is not controlled by the same systems that control executive function. Furthermore, deficits in executive function and inhibitory control appear independent in children with *ADHD* with others (Ross et al., 1994; Karatekin and Asarnow, 1998, 1999; Castellanos et al., 2000; Gould et al., 2001; Mostofsky et al., 2001b) having found similar effects in both males and females.

Bergmann (1998) speculates that the *hippocampus* and *amygdala* are involved in using eye movements as a rehabilitation tool. He notes that it is thought that these two structures are involved in much of the brain's learning and remembering. Therefore where the *amygdala* "retains the emotional flavor of memory, the *hippocampus* retains the dry facts." He also notes that inhibition of the *amygdala* is thought to arise from the left *prefrontal* cortex (Ledoux, 1986). It is noted that the *prefrontal* cortex is the area of the brain responsible for *working memory*. Therefore, strong emotions that are generated in the *amygdala* may create neural static, decreasing the ability of the *prefrontal* cortex to maintain *working memory* (Selemon et al., 1995). Sensory information from visual, auditory, and olfactory areas, are first sent to the *thalamus* and then monosynaptically to the *amygdala*. A second pathway from the *thalamus* projects to the neocortex; this arrangement allows the *amygdala* to react before the neocortex. The neocortex processes the information through several brain circuits before it finally perceives and responds (LeDoux, 1986). It has been suggested that eye gaze interventions resynchronize the activity of the two hemispheres, by way of the alternating stimulus which may mimic the activity of the pacemaker function within the cortex that may be suppressed. Bergmann concludes "stated more specifically, eye movement retraining gradually shifts the brain activity from amygdaline hyperactivity to activation of greater neocortical function."

Eye movement training can be powerful but difficult to perform for many children with learning or behavioral problems. Eye movement training that is more specific to the side of the *cerebellar* or cerebral deficit has been clinically found to be more effective. As children become more coordinated with their eye movements, there usually is improvement in learning and behavior. A direct neurological connection exists between the neck and extraocular muscles; weakness of neck muscles can be tested by examining for weakness and fatigueability of eye muscles. It has been noted that many children with neurobehavioral disorders have difficulty or cannot cross their eyes (convergence) in the midline (Eden et al., 1994). The extra-ocular muscles are analogous to midline postural muscles, with weakness in each reflecting a bilateral limitation. The child may be able to adduct one eye and not the other, this usually represents a unilateral weakness or neurological imbalance, and the weakness is often found on the same side as a neocortical decrease in activation. *Saccadic dysmetria* especially *hypometria* is more indicative of *cerebellar* deficit. *Saccadic dysmetria* with the child looking up to the right or down and left is usually associated with right *cerebellar* lesions, whereas up and left- and down and right *dysmetria* is associated with left *cerebellar* lesions. These findings need to be correlated with other signs and symptoms. There are currently some interesting studies examining eye movement intervention strategies in neurobehaviorally involved children using functional imaging techniques that should reveal or confirm more specifically its effectiveness and mechanism of function.

Biofeedback

Biofeedback training according to practitioners teaches one how to consciously control the autonomic nervous system using biofeedback devices in order to alleviate stress, migraine headaches, asthma, high BP, and a host of other health conditions. Neurofeedback or EEG biofeedback is the form of biofeedback typically used on children with epilepsy and neurobehavioral conditions like hyperactivity, *ADHD*, learning disability, etc. It is thought that neurofeedback seems to work by interacting in the area of frequency. Frequency in this context is the rate at which electrical activity proceeds through the nervous system. Human brain activity is measured by EEG in basic frequency ranges. These frequency ranges are reflective of states of consciousness and each frequency range is a component of a bounded continuum ranging from death through various states of sleep and ultimately to excitement and seizures. These are described in Table 11.1 and Fig. 11.1 below.

With decreasing states of consciousness, the EEG frequency slows and the amplitude increases. Additionally, with psychological milestones adapted from Piaget, one can notice a parallel growth of brain weight and the average frequency of EEG background activity from posterior regions of the scalp as reflected in Fig. 11.2.

Even though these measures of frequency by EEG are now considered relatively crude, they do provide a window into excitability within the brain. Researchers thought that problems arose when the operating speed of someone's brain is either too low (underaroused) or too high (over-aroused). This is similar to the concept of *hyperkinetic* state or *hypokinetic* state. There is speculation that arousal levels may be the major component in a whole host of disorders. The goal of neurofeedback therefore is said to be to stabilize the brain and nervous system so that it does not fluctuate easily between over-arousal and under-arousal. Viewing the brain from this perspective returns us to some original theories of arousal that were popular in the 1950s. It was then thought that two main states existed, stability and arousal (Leisman, 1980).

TABLE 11.1 EEG Frequency Ranges

Waveform	Frequency Range (in Hz)	Amplitude (in μV)	Occurrence
Gamma rhythm	30–50		Excitement
Beta rhythm	18–30	< 10	Alert/eyes open, arousal, anterior scalp
Alpha rhythm	8–13	0–40	Adults, older children, relaxed wakefulness/eye closed, parietal, occipital temporal regions
Mu rhythm	7–11	0–20	Asymmetric, asynchronous between two sides at times unilateral, central parietal, attenuates with contralateral extremity movement, thought of movement, or tactile stimulation; no reaction to eye opening and closing
Theta rhythm	4–7	40–60	Childhood, light sleep, temporal areas through adolescence
Delta rhythm	0.5–4	40–200	Sleep
Delta rhythm	0.5–3	40–200	Infancy, deep sleep, coma
Lambda & K complex & sleep spindles	Not defined solely in terms of rhythm		Deep sleep

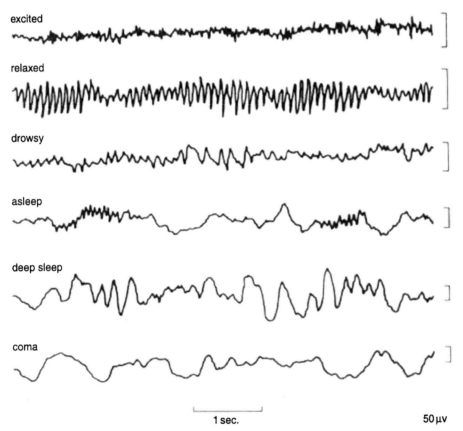

Fig. 11.1. Neural activity (EEGs) and the corresponding behavioral states accompanying arousal levels including those associated with an awake excited person, the alpha rhythm associated with relaxation or with eyes closed, the slowing in frequency associated with a drowsy condition, the slow high amplitude waves of sleep, the larger slow waves associated with deep sleep, and the further slowing of EEG waves associated with coma.

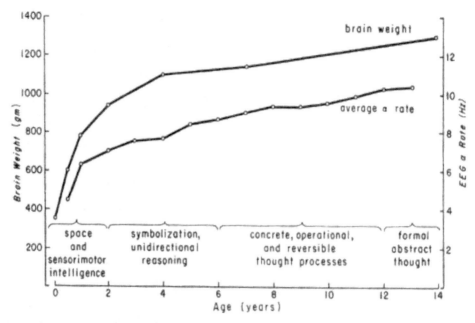

Fig. 11.2. Parallel growth of brain weight and average frequency of EEG background activity from posterior regions of the scalp. Psychological maturation milestones adapted from Piaget (1952). (From Stevens et al., 1968.)

According to this theory, optimal idling speed for the human brain is about 14 Hz. Therefore, if the brain's major activities are of speed lower than 8–13 Hz, an individual may feel tired and might seek stimulation from coffee or stimulating behavior. Individuals might suffer from depression, *ADD*, and mild *dissociative* disorder. Alternatively, overarousal might provoke an individual to feel unsettled and might then seek out alcohol to decrease arousal level or need medication to calm down, the situations being akin to depression vs. mania or left vs. right hemisphere dysfunction. Anxiety, hyper-vigilant stress, and obsessive behavior are thought to be all symptoms of over-arousal.

In the 1960s, neurofeedback was considered a revolutionary way to examine the mind and its capabilities. One of the early pioneers of biofeedback was M. B. Sterman, who was one of the first to experiment with a type of beta wave called *sensory-motor rhythm* (*SMR*) in the 12–15 Hz range. More specifically, a number of research studies confirmed the identification of a 12–14 Hz rhythm in the EEG of a number of species observed over the *Rolandic* (*sensory-motor*) cortex. This rhythm is associated with inhibition of motor activity (Chase and Harper, 1971; Howe and Sterman, 1972; Sterman, 1977). It was labeled *SMR* for its location in the sensory motor cortex. The rhythm has also been identified in humans. An increase in *SMR* in the EEG of cats by operant conditioning was subsequently demonstrated (Sterman and Wyrwicka, 1967; Wyrwicka and Sterman, 1968). Similar findings were found in primates. One effect of such training in cats and in humans was to increase the incidence and duration of *Rolandic sleep spindles*, which occur in the identical spectral band (12–15 Hz) as the waking *SMR* and in the same location. This is accompanied by more sustained periods of quiet sleep in both normal subjects and insomniacs (Sterman et al., 1970). It was also noted that paraplegics and quadriplegics exhibit larger than normal amounts of *sleep spindles* and reduced amounts of low frequency (4–7 Hz) EEG activity. In addition, patients with spinal cord injuries exhibit a relative dearth of epileptic behavior. Additionally, cats with cervical *dorsal column* transection exhibit a heightened threshold for drug-induced seizures. In one case it was noted in an epileptic patient who experienced upper *cervical* cord compression, following the injury his seizure activity disappeared (Shouse and Sterman, 1982). These findings suggest the fundamental relation between the incidence of *SMR* rhythm and of motor-induced seizures. Reduction of activity in the 4–7 Hz frequency band has also been demonstrated in monkeys during sleep, after administration of four anticonvulsant drugs. This suggests that excessive low frequency amplitude is indicative of insufficient cortical control and is a concomitant of susceptibility to seizure onset.

Following this postulation, it was found in 1969 that after training for enhanced *SMR* rhythm in cats, threshold for seizure onset was increased for chemically induced seizures (Fairchild, 1969; Sterman, 1976). Following this, EEG feedback training in poorly controlled epileptics showed reports of seizure reduction. Sterman achieved an average 66 percent reduction in seizure incidence in four epileptics using *SMR* enhancement training in combination with inhibition of excessive slow activity in 6–9 Hz region (Sterman, 1974). Wyler and associates (1976) showed that enhancement of EEG activity above 14 Hz and suppression of activity below 10 Hz was effective in seizure reduction. Synchronization of the EEG worsened seizure incidents whereas a desynchronization of EEG improved it (Wyler et al., 1976).

Lubar and Bahler (1976) took Sterman's work even further and in a different direction. They had noticed that hyperactivity decreased in patients treated for epilepsy and based on this created the protocol now used for treatment of *ADHD*. This is not considered entirely unexpected since hyperactivity may also be regarded in terms of insufficient motor inhibition and since EEG observables are similar in general to *interictal* epileptiform activity consisting of a relative increase of low frequency activity and a relative dearth of intermediate frequency activity (*SMR* and *beta* activity). The first study of the EEG biofeedback effectiveness with hyperactivity in the absence of seizure history was reported by Lubar and Shouse (1976). A number of behaviors associated with hyperactivity were monitored and significant changes were observed for 8 of 13 behavioral categories. The EEG training was shown to be more effective than the use of stimulant medication (*Ritalin* alone).

A study by Lubar and Lubar (1984) extends the technique to attentional defects and learning disabilities. The appropriateness of doing this is based primarily on observations that more than 60 percent of the cases of learning disabilities exhibit EEG abnormalities (Muehl et al., 1965). The experimental protocol was complimented with training in the 15–18 Hz region associated with EEG activation in general and with arousal and focus. Changes in EEG were documented with power spectral density measurements, which were compared with those of normal subjects. The EEG biofeedback was also accompanied by academic training. Acquisition of the desired EEG characteristics was observed in all six subjects under study. Significant improvement in academic performance was also documented for all of the subjects. It has been noted that *ADD* and *ADHD* have a strong neurological basis with increased *beta* activity occurring over central and frontal portions of the cortex and decreased beta activity occurring centrally, posteriorly, and sometimes even frontally. *Beta* activity has been associated with daydreaming and a lack of *beta* activity with poor ability to concentrate and to complete tasks.

Biofeedback treatment for *ADD* and learning disability typically involves approximately 50 sessions initially carried out 2–3 times a week and then phased out over a period. A session may consist of playing some kind of computer game in which a smiling *Pacman* gobbles up globs or a balloon that tries to float up to the sky while the patient's brain waves are continuously monitored. Each time the brain wave finds their way to the optimal state set by the clinician, the patient is rewarded with positive feedback. *Pacman* eats his enemy with pleasant tones. After anywhere from 5–50 sessions the brain seems to be able to find the optimal state on its own. Examining this information based on our previously described neurological model, we can see that consistent with these EEG studies, decreased activity or arousal of the brain, especially over *somatosensory* and *frontal* cortices, is associated with *epilepsy, ADD, ADHD*, and learning disabilities. In addition, increasing activity to the *somatosensory* cortex has been shown to significantly improve these conditions. EEG activity is known to be produced by the ascending activation system from the *thalamus*. Synchronization results from decreased *thalamic* activity and decreased desynchronization from increased *thalamic* activity. These studies also show that there is a relationship with activity from input from the *dorsal column*, its effect on EEG activity, and with the production of abnormal EEG activity. In addition, the *dorsal column* projections to the *cerebellum* and *thalamus* and subsequently to the *somatosensory* cortex and *frontal lobe* seem to be the basis of *gamma* oscillations and overall brain baseline activity. Therefore, this would seem to confirm that decreased *cerebellar* and *thalamic* activation of the brain as a whole and the *frontal lobe* specifically is associated with decreased stability and function of the brain pathways and their processing capability. EEG may be an effective remediation tool because cognitive activation of *frontal lobe* will activate ascending pathways from the *cerebellum* and *thalamus*, as well as descending frontal projections to the *brainstem reticular formation* and *basal ganglia*; and will increase muscle tone with subsequent feedback through spine *cerebellar* and *dorsal column* pathways to *cerebellar-thalamic*; and cortical projections. The benefit of EEG is that it is a relatively low-level output, which will not exceed the metabolic rate of neuronal pathways. However, neurofeedback therapists do not consider hemispheric differences and therapy does not appear to be specific to one hemisphere. In addition, if the problem arises from the musculoskeletal system and its lack of feedback as in a neck injury, this intervention strategy will not be effective in effecting change in the region of primary involvement.

Whereas stimulating motor activity may be more specialized to a specific hemisphere and found to be more easily used in children, especially ones demonstrating hyperactive behavior and attentional deficits.

Sensory-Motor Intervention Strategies

Therapeutic Use of Light

Levitan and associates (1999) at the *University of Toronto* report that evidence exists from clinical, epidemiological, and neuroimaging studies that *ADHD* and seasonal affective disorder (*SAD*) have several features in common. They assessed *SAD* symptoms in adults with *ADHD*. One hundred and fifteen individuals attending an adult *ADHD* clinic in Toronto, Canada were asked to complete the Seasonal Pattern Assessment Questionnaire. The rate of *SAD* in the overall clinic sample was estimated at 19.1 percent, a prevalence rate significantly greater than rates reported in large population surveys at similar latitudes.

Decreasing or increasing the amount of light, varying the color of light, or blocking out light from certain fields of vision, have been used for many different conditions. These conditions include forms of depression, anxiety, learning disabilities, sleep disorders, etc. One of the most recognized disorders that have been treated with the use of light is *SAD*. In a recent New York City survey, more than one third of responding adults reported at least mild winter malaise: 6 out of 100 reported severe depression (Caldwell, 1999). It is thought that *SAD*, as well as other similar disorders are due to the effect that light has on regulation of biological or *circadian* clocks.

The first experiments of biological clocks date back to 1911 and were conducted by Karl Von Frisch, an Austrian zoologist. Von Frisch discovered that minnows did not respond to ambient light perceived through their eyes. Von Frisch thought that something deep inside the brain itself could respond directly to light. He later traced this effect to the minnows *pineal* gland, which we now know to be the source of *melatonin*. In the brain, a recently discovered group of cells known as the *suprachiasmatic nucleus* of the hypo*thalamus* or *SCN* is thought to be the basis of biological clocks (Reppert and Schwartz, 1984; Honma et al., 2002). In mammals, it appears to be remarkably reliable. Even when removed from an experimental organism and placed in a dish, it continues to keep time on its own for approximately a day. The *SCN* is divided into two structures. One is in the right hemisphere and one is in the left, just below and behind the eyes. Each part of the SCN is made up of approximately 10,000 densely packed neurons. Recent research in mice suggests that mammals have a set of special photoreceptors in their eyes, which react to light signals and carry them directly to the *SCN*. These photoreceptors are thought to be different from the rods and cones used to perceive light stimulating the retina. It is thought that light helps to reset the biological clocks. Light in the morning is thought to set the clock ahead and in the evening, backwards. Whether the running of these biological clocks is an innate quality of cells or a product of some unseen force, like gravity, is not known. Nevertheless, disruption of light stimuli can disrupt these clocks.

Many other areas of the brain are affected by light. We can recall our earlier discussion of the intricacies of the development and function of the ventral and dorsal visual systems and their role in cognitive perception and *limbic system* function. We recall that the dorsal visual system is associated more with right hemisphere function, which is involved in global or low frequency stimulation. The ventral visual system is associated more with left hemisphere function, specific for a local visual function or higher frequency stimulation. These systems are powerfully connected to cognitive and emotional functions.

Modulating the frequency of the stimulus can change the firing rate of the *thalamus* and the brain as a whole. However, different frequencies of the same stimulus may have asymmetric effects on the brain. Modulating the frequency of stimulus also takes into account the metabolic rate of brain cells. Interestingly, it has been noted that red, which is a low frequency source, would be expected to slow down the neocortex and affect more of the right brain, increasing *sympathetic* responses. Blue light would be expected to speed up the firing rate of the neocortex, (thereby inhibiting *sympathetic* and exciting *parasympathetic* responses) which also may be more specific to the left brain and in fact, has been shown to increase *parasympathetic* functions (Vel'khover and Elfimov, 1995). It has also been noted that pale blue paint on the walls of schools appears to decrease hyperactivity in children, where as pale yellow on the walls of schools appears to improve concentration and learning abilities.

Altering the balance of light or vision from one hemisphere to the other has also been shown to have powerful psychological affects. Dr Frederick Schiffer, a psychiatrist, has found that using a pair of glasses that block vision to either the right or left hemispheres can help alleviate anxiety or depression (Schiffer, 1997, 1998). Schiffer thinks that these glasses that work to relieve anxiety and distress are a remarkable testimony to the link between the eyes and the two sides of the brain and a variety of psychological problems. Schiffer attempted a simple experiment on himself. When he covered one eye and part of the other, he detected a slight difference in his clarity of thought. Schiffer then made safety glasses that covered one eye completely and half of the other. He had 70 patients suffering from severe anxiety wear the glasses and measured the effect on a one to ten scale, by which anxiety is calculated. Sixty percent of those patients had a 1-point difference (improvement on the scale) and 23 percent had a 2-point difference. Schiffer also had the patients wear glasses that covered the other eye. In interviews, 40 of the patients said they felt more anxiety when they wore the glasses. A study of a control group of college students, who were not in therapy, also found measurable changes in their feelings of anxiety and changes in brain wave patterns with the glasses. Although no significant knowledge base at present exists in the application of light to the remediation of neurobehavioral disorders of childhood, the existing theory supports its further investigation.

Various forms of visual imagery have been found to help improve intelligence, motor performance, and behavior. Guided imagery is one form of visualization. Guided imagery has the subject create internal scenarios and mental pictures that evoke positive physical responses. Imagery is reported to improve the immune system (Gruzelier, 2002a), reduce stress (Gruzelier, 2002b), and slow HR (Oishi et al., 2000).

Leiner and associates (1991) have noted that several studies of ideation or mental imaging show significant blood flow changes to the *cerebellum* and *frontal lobe*. If individuals imagine a motor act like playing tennis or a cognitive act like mental calculation, both show significant increases in different areas of the *cerebellum* and neocortex. On the other hand, it has been shown that those who imagine shooting a basketball display as much improvement in that skill after a week as those who actually physically practice shooting a basketball. This would lead us to think that whether imagined or actually done, motor acts must activate similar areas of the *cerebellum* that create functional improvement in motor control.

In the last decade, there has been a dramatic increase in research effectively integrating cognitive psychology, functional neuroimaging, and behavioral neurology. This new work is typically conducting basic research into aspects of the human mind and brain. Parsons (2001), in one study employed object recognition, mental motor imagery, and mental rotation paradigms, to clarify the nature of a cognitive process; imagined spatial transformations used in shape recognition. Among other implications of the study was that recognizing a hand's handedness or imagining one's body movement depends on cerebrally lateralized sensory-motor structures and deciding upon handedness depends on exact match shape confirmation. In a second study, using cutaneous, tactile, and auditory pitch discrimination paradigms, it was suggested that the *cerebellum* has non-motor sensory support functions upon which optimally fine sensory discriminations depend. Mental imagery, the generation and manipulation of mental representations in the absence of sensory stimulation, is a core element of numerous cognitive processes. Numerous investigators have recently investigated the cortical mechanisms underlying imagery and spatial analysis in the visual domain using event-related ƒMRI during the mental clock task (Formisano et al., 2002) and ƒMRI (Knauff et al., 2002). The time-resolved analysis of cortical activation from auditory perception to motor response reveals a sequential activation of the left and right posterior parietal cortex, suggesting that these regions perform distinct functions in imagery tasks.

Knauff and colleagues found that in the absence of any correlated visual input, reasoning activates an *occipital-parietal-frontal* network, including parts of the *prefrontal* cortex (*Brodmann's area (BA) 6, 9*), the *cingulate gyrus (BA 32)*, the superior and inferior parietal cortex (*BA 7, 40*), the *precuneus (BA 7)*, and the *visual association cortex (BA 19)*. Because *reasoners* envisage and inspect spatially-organized mental models to solve deductive inference problems, we do have a basis for concluding that imagery has the capacity for effecting change in brain state. In one of the view imaging studies to date on clinical applications of imagery as a therapeutic tool, Marks and associates (2000) investigated subjective imagery in *OCD* before and after exposure therapy using ƒMRI. A small randomized study was performed, with controls for type and order of mental imagery and for treatment condition (exposure therapy guided by a computer or by a therapist, or relaxation guided by audio-tape). Before and after treatment, during ƒMRI scanning, patients imagined previously rehearsed scenarios that evoked an urge to ritualize, non-*OCD* anxiety, or a neutral state, and rated their discomfort during imagery. The method evoked greater discomfort during *OCD* imagery and anxiety (non-*OCD*) imagery than during neutral imagery. Discomfort was reduced by canceling imagery. Discomfort during *OCD* imagery (but not during anxiety non-*OCD* imagery) fell after exposure therapy but not after relaxation. The results showed differences between *OCD* and non-*OCD* images and their change after successful treatment, and confirmed clinical suggestions that canceling images reduced *OCD* discomfort.

Therapeutic Use of Sound

Dr Oliver Sacks, author of *An Anthropologist on Mars* (1995) and *Awakenings* (1973) has written that music therapy is a tool of "great power" for Alzheimer's and *Parkinson's* patients "because of its unique capacity to reorganize cerebral function where it has been damaged." "There's an overlap in brain mechanisms in the neurons used to process music, language, mathematics, and abstract reasoning," says Dr Mark Tramo, a neuroscientist at Harvard Medical School (cf. Tramo and Bharucha, 1991; Patel et al., 1998; Tramo, 2001; Tramo et al., 2001, 2002). "We think a hand full of neuronal codes is used by the brain, so exercising the brain through music strengthens other cognitive skills. It's a lot like saying if you exercise your body by running, you enhance your ability not only to run but also to play soccer or basketball."

Various techniques use sound and music as their primary mode of therapy. These techniques have reportedly been effective for children with learning disabilities and behavioral problems. There are those who think that sound and music can affect dysfunction in the nervous system through both its calming and energizing effects on the brain and CNS. As a clinical therapy it is reported to be used in hospitals, schools, and psychological treatment programs to reduce stress or lower BP, alleviate pain, overcome various learning disabilities, improve movement and balance, and promote endurance and strength.

Campbell, author of the book *The Mozart Effect* (1997), has researched the effect of music and its therapeutic affect (Rauscher et al., 1993). He thinks that since development of the first musical instruments between 43,000–82,000 years ago, humans knew that music created special effects. He suggests that music in the form of song and dance preceded speech in humans and was the first form of language. He states that research has shown that two-thirds of the inner ear cilia (hair cells) resonate only at the higher music frequencies of 13,000–20,000 Hz. To indicate that the application of the so-called *Mozart Effect* is without controversy would be an understatement especially considering there have been numerous reports of an inability to replicate the effect (McCutcheon, 2000).

Hughes and Fino (2000) had performed a study reported in *Clinical Electroencephalography*. The goal of this study was to determine distinctive aspects of Mozart music that may explain the *Mozart Effect*, specifically, the decrease in seizure activity. As many as 81 musical selections of Mozart, but also 67 of J. C. Bach, 67 of J. S. Bach, 39 of Chopin, and 148 from 55 other composers were computer analyzed to quantify the music in search of any distinctive aspect and later to determine the degree to which a dominant periodicity could be found. Long-term periodicity (especially 10–60 s, mean and median of 30 s), was found often in Mozart music but also that of the two Bachs, significantly more often than the other composers and was especially absent in the control music that had no effect on epileptic activity in previous studies. Short-term periodicities were not significantly different between Mozart and the Bachs vs. the other composers. The conclusion is that one distinctive aspect of Mozart music is long-term periodicity that may resonate well within the cerebral cortex and also may be related to coding within the brain. Thompson and Andrews (2000) in reporting on the *Mozart Effect* in which the claim was made that people perform better on tests of spatial abilities after listening to music composed by Mozart; examined whether the *Mozart Effect* is a consequence of between-condition differences in arousal and mood. Participants completed a test of spatial abilities after listening to music or sitting in silence. The music was a Mozart sonata (a pleasant and energetic piece) for some participants and an Albinoni adagio (a slow, sad piece) for others. These investigators also measured enjoyment, arousal, and mood. Performance on the spatial task was better following the music than the silence condition but only for participants who heard Mozart.

The two music selections also induced differential responding on the enjoyment, arousal, and mood measures. Moreover, when such differences were held constant by statistical means, the *Mozart Effect* disappeared. Thompson's findings provide compelling evidence that the *Mozart Effect* is an artifact of arousal and mood. We are, here, less concerned about Mozart as a composer and more about the effects of sound in effecting change in brain and cognitive function.

Thompson and Andrews (2000) in their paper provide an overview of the theoretical underpinnings of the *Tomatis Method*, along with a commentary on other forms of sound/music training and the need for research. A public debate was sparked over the *Mozart Effect*. This debate has turned out to be unfortunate because the real story is being missed. The real story starts with Alfred Tomatis. Dr Tomatis was the first to develop a technique using modified music to stimulate the rich interconnections between the ear and the nervous system to integrate aspects of human development and behavior. The originating theories behind the *Tomatis Method* are reviewed by Thomson and Andrews to describe the ear's clear connection to the brain and the nervous system. The neuropsychology of sound training describes how and what the *Tomatis Method* affects. The 50 years of clinical experience and anecdotal evidence amassed by Tomatis, show that sound stimulation can provide a valuable remediation and developmental training tool for individuals with neurobehavioral disorders.

In Norway, in the 1980s, educators used music therapy for children with severe physical and mental disabilities. They found that music reduced muscle contraction in patients with severe spastic conditions, increased range of motion in their spines, arms, hips, and legs. These effects would suggest effects not only on the neocortex, but the *brainstem reticular formation* as well as the *cerebellum*. Since music has powerful effects on the hair cells in the *vestibular apparatus*, it would also be expected to have effects on the *olivary complex* and the *cerebellum*. This, in fact, may be its primary effect.

Studies of *Alzheimer's* patients show that there is a pathological involvement of the visual (O'Mahoney et al., 1994), primary auditory paths (Hinton et al., 1986; Blanks et al., 1991; O'Mahoney et al., 1994), and olfactory (Esiri and Wilcock, 1984; Talmo et al., 1989). In a study of auditory system degeneration associated with *Alzheimer's Disease* (*AD*), Sinha and associates (1993), demonstrated marked degenerative changes in the ventral nucleus of the *medial geniculate body*, the central nucleus of the *inferior colliculus*, and the *primary auditory cortex* in 9 *AD* patients matched with controls on neuropathological examination. German and colleagues (1987), showed *neurofibrillary* degeneration in several *brainstem* nuclei, other than the *locus coeruleus* and dorsal *Raphe nucleus*, including a structure known as the *pedunculo-pontine nucleus* that forms part of the *reticular activating system*. O'Mahoney and colleagues (1994) studied patients with *AD* and examined the *primary auditory pathway* using *brainstem* auditory evoked responses and middle latency response. They concluded from their results, that dysfunction of the *primary auditory pathway* in patients with mild to moderate *AD* supports the hypothesis that impairment of auditory function and arousal are primary features of *AD*. In addition, Shigeta and associates (1993) have shown that the rate of improvement of the electro-oculogram rapid eye movement frequency (a sensitive direct indicator of level of arousal) (Amadeo and Shagass, 1973) is positively correlated with the rate of improvement in the *dementia scale* score in treated patients with senile *dementia*. This suggests that intellectual dysfunction in *dementia* is partly the result of reduced arousal. Therefore, music can increase arousal of the brain through primary auditory pathways. This may improve intelligence and *frontal lobe* function in neurobehavioral disorders of childhood. The *cerebellum* and *thalamus* may be responsible for the majority of this brain arousal. The auditory effects of listening to music as well as the motor effects of playing music can increase *cerebellar, thalamic*, and neocortical function.

A common intervention that has been employed with children with *ADD*, learning disabilities, and *central auditory processing* (*CAP*) deficits involves attempts to influence activity in the *cerebellar-vestibular* system (*CVS*) by auditory stimulation in frequency ranges generating *hyperacusis* and discomfort. Ann Blood and colleagues (Blood et al., 1999; Blood and Zatorre, 2001), at *McGill University* in Montreal, conducted the first scientific studies on music's emotional impact on the brain. They employed PET to examine cerebral blood flow (*CBF*) changes related to effective responses to music. Ten volunteers were scanned while listening to six versions of a novel musical passage varying systematically in degree of dissonance. Reciprocal CBF covariations were observed in several distinct *paralimbic* and neocortical regions as a function of dissonance and of perceived pleasantness/unpleasantness. The findings suggest that music may recruit neural mechanisms similar to those previously associated with pleasant/unpleasant emotional states, but different from those underlying other components of music perception, and other emotions such as fear. In a subsequent study, Blood's group again employed PET to study neural mechanisms underlying intensely pleasant emotional responses to music. CBF changes were measured in response to subject-selected music that elicited the highly pleasurable experience of "shivers-down-the-spine" or "chills." Subjective reports of chills were accompanied by changes in HR, electromyogram, and respiration. As intensity of these chills increased, CBF increases and decreases were observed in brain regions thought to be involved in reward/motivation, emotion, and arousal, including ventral *striatum, midbrain, amygdala, orbitofrontal* cortex, and ventral medial *prefrontal* cortex. These brain structures are known to be active in response to other euphoria-inducing stimuli, such as food, sex, and drugs of abuse. This finding links music with biologically relevant, survival-related stimuli via their common recruitment of brain circuitry involved in pleasure and reward.

Others (Penhune et al., 1998; Riecker et al., 2000; Bodner et al., 2001; Parsons, 2001; Satoh et al., 2001; Schlaug, 2001; Langheim et al., 2002) have observed significant increase in the function of the *cerebellum* in studies that examine the neuroanatomy of expert musicians as they listen to music. It has been known for sometime that children develop an early appreciation for music. In the first year of life, children have been shown to pay attention more acutely to stimuli with a harmonic structure, and it seems that they learn music in the same way as language, with one note exponentially acquiring new ones (Garat, 1993). In the past few years, researchers have begun to map areas of the brain involved in performing music or while silently reading scores. However, no previous studies have examined the emotions elicited during a musical piece.

Blood and her colleagues at *McGill*, decided to target the emotional response to music, by studying ten adults from ages 19–43, as they listened to musical notes that either clashed or had a harmonic tone. They designed the experiment using a single melody and adding on six versions from a very pleasant sounding piece to something that a two-year-old would bang out on the piano. They measured blood flow in the brain during these experiences to see if they would find a difference. According to Blood, the discordant notes triggered activity in the *parahippocampal gyrus*, an area near the *temporal lobe* that plays an important role in processing sensory memory. When the music was pleasing to the subjects, the investigators found significant activity in the lower region of the *frontal lobe*. Responses were primarily found on the right side of the brain. This activation is thought by Blood and colleagues as an indication of the emotional responses to music. These brain regions were also different from the region activated when musicians read a score, or were asked to pick out mistakes in musical pieces. In 1992, researchers at the *Montreal Neurological Institute* published the first study using PET to identify areas of the brain active during musical tasks. The late Justine Sergent and her colleagues studied musicians as they read scores or performed, and found many areas of the brain were involved in converting the written score into finger movement.

Lawrence Parsons and his colleagues (reviewed in Parsons, 2001), advancing Sergent's studies, found that specific tasks called upon during the musical experience rely on different areas of the brain. In one study, subjects were eight right-handed faculty conductors. The conductors were instructed to focus on errors in melody, harmony, or rhythm in a Bach Choral. The errors appeared one to every two beats and the musicians were instructed to take notice but not to perform any motor responses. Each task was shown to produce very different patterns of brain activity. Melody activated both the left and right hemispheres in the *temporal lobe*, while harmony and rhythm triggered activity more in the left hemisphere. Harmony did not activate the *temporal lobe* at all. Each of the tasks also activated right *fusiform gyrus*. This same area in the left hemisphere has been linked to visual processing of words. Researchers suspect that the right *fusiform gyrus* may have evolved to regulate information on musical notes and passages. It has been noted that stroke patients who have lost language function may be able to gain some verbal improvement, by singing words, which Parsons' thinks may be facilitated by this region of the brain. Listening for errors in harmony, melody, and rhythm also activates the *cerebellum* even though the conductors were not moving, indicating the *cerebellum*'s involvement in cognition.

Schlaug (2001), of *Beth Israel Deaconess Medical Center* in Boston, reported that skilled male musicians he studied have larger *cerebella* than average. He employed CT scans to compare 32 right-handed male musicians with 24 right-handed men with no musical training. Schlaug and his colleagues (1995) had previously reported that male musicians have larger *corpus callosa* and larger *primary motor cortices* in the *frontal lobe*. They have not found similar differences in the *cerebellar* volumes between male musicians and non-musicians. Other studies have shown *functional* brain changes in individuals who have mastered certain skills or suffered brain damage; this is the fist study that has identified *structural* changes in the brain that are associated with a learned skill. This indicates that the *cerebellum* is involved in both motor and cognitive function and in the processing and the production of music.

An interesting parallel exists between brain changes associated with the acquisition of a musical skill (Schlaug, 2001) and languages acquisition (Leiner et al., 1991). Leiner and associates have noted that the lateral areas of the *cerebellum* are activated during the cognitive aspects of speech production, whereas, the medial *cerebellar* cortex is activated during the motor function of speech. If music were indeed the first form of human language, then we would expect to find a parallel between music and language processing in the *cerebellum, thalamus*, and *frontal lobe*. It is also interesting to note that musicians have been found to demonstrate structural changes in the *cerebellum, frontal lobe* (*motor cortex*), and the *corpus callosum*. Since the *cerebellum* is thought to act as an association cortex external to the neocortex, especially between the dorsal and ventral language areas, it may assist in integrating function of the hemispheres. This being the case, we may expect to see an enlargement in the *corpus callosum* associated with hemispheric integration. Since both the *corpus callosum* and the *cerebellum* may be involved in the temporal synchronization of multiple areas, this may be the reason for the observed enlargement.

This may also explain the results of a recent study by Chan and her colleagues (1998) and others (Patel et al., 1998), that indicate that children who spend a few years learning to play a musical instrument, also develop better verbal skills compared with those who never studied music. These findings seem to be consistent with the brain scans that have shown the left *planum temporale* to be larger in musicians than in individuals without extensive training in instrumental music. Chan and associates studied 60 female college students from the University of Hong Kong, of whom 30 had at least six years of training with western musical instruments before the age of 12 and the other 30 had no formal training. The students were tested for verbal memory by attempts to remember lists of words. The researchers stated, "We found that adults with music training learn significantly more words than those without any music training." However, they found adults with and without music training were not significantly different in their ability to remember visual images such as words written on paper.

According to one neuroscientist, Evan Balaban, "There has been a long tradition of researchers trying to segregate speech and muscle in the brain. There is now evidence from several studies that they (music and speech), may have more to do with each other than was previously thought." If musically trained individuals had structural changes (the *planum*) as well as functional language changes that non-musicians do not have, the main differences in ability between them would be the motor acts associated with playing music and possibly the breathing associated with wind instruments (Patel and Balaban, 2001a, 2001b). Both of these differences in ability and the functional changes associated with them can be attributed to the effects of the *cerebellum*. This would also demonstrate that motor activities have a carry-over effect on cognitive function of the *frontal lobe*, such as in the case of verbal memory.

Musically trained brains respond to randomly heard musical tones in fundamentally different ways than those who are untrained. This effect is apparently more pronounced among those who take music lessons before the age of six. Recent studies suggest (Pantev et al., 1998, 2001; Pascual-Leone, 2001; Schlaug, 2001) that when a piano tone is played, either more neurons are activated or the neurons are responding in a more synchronized fashion. No change occurs in those without musical training. In one study, German researchers asked 20 musicians from a local conservatory to watch a cartoon, while either pure non-musical tones or piano tunes were presented. They measured electrical activity in the brain during the activity. A control group of non-musicians heard the same tones. The results of this study conducted by Christo Pantev and colleagues at the *University of Münster* found that musically trained brains showed about 25 percent more activity than non-musicians. What was also interesting is that the effect was observed only when piano tones were presented to musicians. Pure tones, which are not musical, had no apparent effect. In addition, those who began lessons before their sixth birthday seemed to have the strongest response. It was also noted that the brains of musicians respond differently to piano tones (Pantev et al., 1998).

Penhune et al., (1998), at *Montreal Neurological Institute* reported that there exists a difference in brain activities measured by PET scans in musicians with perfect pitch (the ability to hear a tone and name it) compared to those with relative pitch (the ability to recognize the difference between a major and a minor note). The recorded activity differences were noted in the left *frontal* cortex in those with perfect pitch. When those with relative pitch were asked to choose between a major and minor note, this region in the *frontal lobe* also became active. The authors concluded that those with perfect pitch were processing the information more efficiently. The left *frontal lobe* is the motor speech area, shown to be connected with activation of the right *cerebellum*. With this recent increase in research showing the effects of music on the brain, music as therapy has gained wider acceptance in more mainstream centers.

Therapeutic Use of Olfaction

The sense of smell can be used as a powerful stimulus to the brain. A number of recent investigations have suggested a significant role for olfactory stimulation in the alteration of cognition, mood, and social behavior. These orthodox investigations have a common, if uneasy, relationship with the holistic practice of so-called *aromatherapy*. In children and adults, various studies have shown improvement in learning abilities and emotional disturbances, as well as effects on BP and stress responses. One study using peppermint oil has shown improvement in cognitive function on children as compared to controls (Soussignan et al., 1995). In their study Soussignan and associates examined the facial responsiveness of ten mutic children with *pervasive developmental disorder (PDD)* and ten normal children matched for sex and chronological age who were all covertly videotaped while presented with a set of odors contrasted in hedonic valence. Hedonic ratings of the stimuli were obtained from both the group of normal subjects and a panel of adults. Two methods were used to measure facial responses in the same subjects. The first method consisted of an analysis of facial movements with the Facial Action Coding System. Results show that *PDD* and normal subjects displayed distinct action units in response to unpleasant odors. *PDD* subjects typically displayed muscular actions indexing negative experience, while normal subjects showed more smiles. With the second method, odor-elicited facial behavior was rated by a panel of observers, who were asked to judge whether the subjects were exposed to a pleasant, neutral, or unpleasant smell. The facial responses to unpleasant odors were classified more accurately in *PDD* than in normal subjects. These findings suggest a functional ability to sense the hedonics attached to odors, but a deficit of socialization of hedonic facial displays in developmentally disordered subjects.

Murphy et al., (2001) examined 105 young adults with *ADHD* who were compared with a control group of 64 normals on 14 measures of executive function and olfactory identification. The *ADHD* group performed significantly worse on 11 measures with no Group × Sex interaction, contrary to the findings of Ceccarelli and colleagues (2002) in rats. No differences were found in the *ADHD* group as a function of *ADHD* subtype or comorbid oppositional defiant disorder. After IQ was controlled, some group differences in verbal working memory, attention, and odor identification were no longer significant, whereas those in inhibition, interference control, nonverbal working memory, and other facets of attention remained so. The deficits in olfactory identification seen in neurobehavioral disorders in part can serve as a basis for therapeutic intervention in this modality as well as others.

Further support for the use of olfactory stimulation as part of an overall intervention strategy in neurobehavioral disorders of childhood comes from the study of Levy and colleagues (1999) who noted that memory for odors induces brain activation as measured by fMRI. Functional MRI brain scans were obtained in 21 normal male and female subjects and in two patients with *hyposmia* or diminished sense of smell in response to the imagination of odors of banana and peppermint and to the actual smells of the corresponding odors of *amyl acetate* and *menthone*, respectively. In normal subjects, brain activation in response to imagination of odors was significantly less than that in response to the actual smell of these odors, and activation following imagination of banana odor was significantly greater in men than in women, for the actual smell of the odor of *amyl acetate*. The ratio of brain activation by imagination of banana to activation by actual *amyl acetate* odor was about twice as high in women as in men. Before treatment, in patients with *hyposmia*, brain activation in response to odor imagination was greater than after presentation of the actual odor itself. After treatment, in patients with *hyposmia* in whom smell acuity returned to or toward normal, brain activation in response to odor imagination was not significantly different quantitatively from that before treatment; however, brain activation in response to the actual odor was significantly greater than that in response to imagination of the corresponding odor. Brain regions activated by both odor imagination and actual corresponding odor were similar and consistent with regions known to respond to odors. Their study indicates that (1) odors can be imagined and similar brain regions are activated by both imagined and corresponding actual odors; (2) imagination of odors elicits quantitatively less brain activation than do actual smells of corresponding odors in normal subjects; (3) absolute brain activation in men by odor imagination is greater than in women for some odors, but on a relative basis, the ratio for odor imagination to actual smell in women is twice that in men; (4) odor imagination, once the odor has been experienced, is present, recallable, and capable of inducing a relatively constant degree of brain activation even in the absence of the ability to recognize an actual corresponding odor.

Henkin and Levy (2001) looked further at the nature of the lateralization of brain activation to imagination and smell of odors also using *f*MRI finding a left hemispheric localization for pleasant and right hemispheric localization for unpleasant odors. Functional MRI brain scans were obtained in 24 normal subjects in response to imagination of banana and peppermint odors and in response to smell of corresponding odors of *amyl acetate* and *menthone*, respectively, and of *pyridine*. The results indicated that in normal subjects, activation generally occurs in the direction of left (L) to right (R) brain hemisphere in response to banana and peppermint odor imagination and to smell of corresponding odors of *amyl acetate* and *menthone*. There are no overall hemispheric differences for *pyridine* odor. Localization of all lateralized responses indicates that anterior *frontal* and *temporal* cortices are the brain regions most involved with imagination and smell of odors. Imagination and smell of odors perceived as pleasant generally activate the dominant or L to R brain hemisphere. Smell of odors perceived as unpleasant generally activates the contralateral or R to L brain hemisphere. According to these authors, predominant L to R hemispheric differences in brain activation in normal subjects occur in the order *amyl acetate* > *menthone* > *pyridine*, consistent with the hypothesis that pleasant odors are more appreciated in left hemisphere and unpleasant odors more in right hemisphere. Anterior *frontal* and *temporal* cortex regions previously found activated by imagination and smell of odors accounted for most hemispheric differences.

Henkin and Levy (2002), in a more recent study again employed *f*MRI to define brain activation in response to odors and imagination ("memory") of odors and tastes in patients who never recognized odors (congenital *hyposmia*). These authors studied nine patients with congenital *hyposmia* as they responded to odors of *amyl acetate, menthone*, and *pyridine*, to imagination ("memory") of banana and peppermint odors, and to salt and sweet tastes. Functional MRI brain scans were compared with those in normal subjects and patients with acquired *hyposmia*. The authors found that brain activation in response to odors was present in patients with congenital *hyposmia*, but activation was significantly lower than in normal subjects and in patients with acquired *hyposmia*. Regional activation localization was in anterior *frontal* and *temporal* cortex similar to that in normal subjects and patients with acquired *hyposmia*. Activation in response to presented odors was diverse, with a larger group exhibiting little or no activation with localization only in anterior *frontal* and *temporal* cortex and a smaller group exhibiting greater activation with localization extending to more complex olfactory integration sites. "Memory" of odors and tastes elicited activation in the same central nervous system regions in which activation in response to presented odors occurred, but responses were significantly lower than in normal subjects and patients with acquired *hyposmia* and did not lateralize. Odors induced CNS activation in patients with congenital *hyposmia*, which distinguishes olfaction from vision and audition since neither light nor acoustic stimuli induce CNS activation. Henkin and Levy concluded that odor activation localized to anterior *frontal* and *temporal* cortex consistent with the hypothesis that olfactory pathways are hard-wired into the *CNS* and that further pathways are undeveloped with primary olfactory system *CNS* connections but lack secondary connections. However, some patients exhibited greater odor activation with response localization extending to *cingulate* and *opercular* cortex, indicating some olfactory signals impinge on and maintain secondary connections consistent with similar functions in vision and audition. Activation localization of taste "memory" to anterior *frontal* and *temporal* cortex is consistent with *CNS* plasticity and cross-modal *CNS* reorganization as described for vision and audition. Thus, there are differences and similarities between olfaction, vision, and audition; the differences are dependent on the unique qualities of olfaction, perhaps due to its diffuse, primitive, and fundamental role in survival. These studies add further support to employing odor intervention strategies in neurobehaviorally-involved children in programs of differential hemispheric activation.

The effect of the olfactory system on the *limbic system* is profound especially when we consider the evolutionary development of the brain. The *limbic system* is intimately connected to the *rhinencephalon* or primitive "nose brain." Therefore, we would expect that odors or pheremonal activity would have direct effect on emotions, autonomic regulations, and through effects on the *parahippocampal* complex, on memory acquisition. Although it has been generally accepted that the sense of smell is the only sense that is not related to the *thalamus*, there has been recent evidence that a *hypothalamic-thalamo-cortical* circuit mediates pheremonal influences on eye and head movement (Risold and Swanson, 1995). Through this mechanism, pheremonal activity is thought to regulate attentional mechanisms. Risold and Swanson used a method for simultaneous iontophoretic (movement of ions as a result of an applied electric field) injections of anterograde tracer *phaseolus vulgaris leukoagglutinin* using retrograde trace of flora gold to characterize in rat a *hypothalamic-thalamo-cortical* pathway ending in a region thought to regulate attentional mechanisms by way of eye and head movement. The investigators thought that the relevant medial *hypothalamic* nuclei receive pheremonal information from the *amygdala* and project to specific parts of the *thalamic* nucleus *reuniens* and antero-medial nucleus, which then projects to a specific lateral part of the *retrosplenial* area (or medial *visual* cortex). They note that this area receives convergent input from the lateral posterior *thalamic nucleus* and projects to the *superior colliculus*. In addition, bi-directional connections with the *hippocampal formation* suggests that activity in this circuit is modified by previous experience. They further note that there are striking parallels with *basal ganglia* circuitry. In discussion of their results, they note that their evidence suggests that the rostral medial zone nuclei of the *hypothalamus* participate in a *thalamo-limbic* projection similar to the classic *mammillo-thalamic limbic* projections from the caudal medial zone and that the former receives olfactory information and modulates well established attentional mechanisms involving eye and head movement.

With regard to intra-cortical projections of the *retrosplenial* area, these are divided into three streams. One major stream extends rostrally to end in the anterior *cingulate* caudal *pre-limbic* and ventral lateral *orbital* areas. They note that their double injection results suggest that the first two areas project back to the *retrosplenial* area. This is of interest because in the rat, anterior and to a lesser extent caudal *pre-limbic* areas are thought to be associated with the frontal eye fields along with the adjacent secondary motor areas mainly because they project to several *brainstem* regions involved in ocular motor control including the *superior colliculus* (Leichnetz, 1989; Cooper and Phillipson, 1993). It is also noted that the anterior *cingulate* and the secondary *motor* areas receive inputs from the lateral posterior *thalamic nuclei* (Risold and Swanson, 1995) and the *medial dorsal nucleus No. 17* (Thompson and Robertson, 1987). The anterior *cingulate* area receives input from the *lateral dorsal nucleus No. 20* (van Groen and Wyss, 1992) and the rostral *nucleus reuniens*.

Risold and Swanson (1995) suggest that information arriving at the rostral medial *hypothalamus* from *pheremonal cortex* (in the cortical medial *amygdala*) projects to the mid-brain motor regions by descending pathways, as well as to parts of the cerebral cortex involved in regulating eye and head movements by ascending pathways to the rostral *nucleus reuniens* and ventral *intermedial nucleus*. They note that the *hippocampal formation* participates in conceptually similar circuitry involving the caudal medial *hypothalamus* (*mammillary body*), which is thought to give rise to the *mammillo-thalamic* and *mammillo-tegmental* tracks. Iso-cortical regions project to the *basal ganglia*, which in turn generate descending projections to midbrain motor regions and ascending projections to secondary motor cortical regions by way of the ventral anterior *thalamic nucleus*.

In summary Risold and Swanson (1995) state that their model predicts that the rostral *nucleus reuniens* and ventral anterior *medial nucleus* projecting to the retrosplenial area pathway, conducts pheremonal information to a polymodal cortical mid-brain pathway eliciting attentional motor responses involved in the procurement phase of appropriate motivated or goal-directed behavior (Swanson and Morgenson, 1981). We know that goal-directed behavior is a function of the *prefrontal* cortex and *approach and avoidance* behavior. Olfactory stimulation therefore would be expected to increase *frontal* cortex activation through its effect on *orbital*-frontal and *frontal eye fields*, as well as secondary *motor* cortex. Olfactory stimulation affects *limbic* structures like the *amygdala* and *hypothalamus*, which regulate emotional and autonomic responses and which are inhibited by *frontal* cortical activity. It can also influence learning and memory through its effect on the *hippocampal formation* and *intra-hippocampal* circuit. The *intra-hippocampal* circuit plays a critical role in short-term episodic or declarative memory (Risold and Swanson, 1995).

Olfactory stimulation also effects the anterior and posterior *cingulate* areas, which have been implicated in several aspects of spatial memory (Sutherland et al., 1988). By affecting attentional mechanisms of eye and head movements, it would also be expected that there may be influence on *cerebellar* activity either through effects on *ocular-motor* or *brainstem* motor nuclei. Therefore, the use of olfactory therapy or pheremonal activity has a neurophysiological basis for affecting both learning abilities and behavioral and emotional disorders.

Integrated Sensory-Motor Intervention Strategies

According to practitioners, Occupational Therapy (OT) is a health profession concerned with improving a person's occupational performance. In a pediatric setting, the Occupational Therapist deals with children whose occupations are usually players, preschoolers, or students. The Occupational Therapist evaluates a child's performance in relation to what is developmentally expected for that age group. If there is a discrepancy between developmental expectations and functional ability, the Occupational Therapist looks at a variety of perceptual and neuromuscular factors, which influence function.

A. Jean Ayres is credited with having first identified sensory integrative dysfunction, which is defined as an irregularity or disorder in brain function that makes it difficult to integrate sensory input effectively. It is thought that sensory integrated dysfunction may be present in motor, learning, social, emotional, speech, language, or attention disorders. Ayres thought that proprioceptive input is extremely important to the function of the sensory system and the brain as a whole. She identified gravity as an important input to the central nervous system because of its constancy of input. She thought that the primary source of this proprioceptive and gravitational input was from the vestibular apparatus of the inner ear and the *vestibular system*. She called this the *cerebellar vestibular system*. She thought that this system was a primary force in brain development. This was insightful considering the paucity of research to then support her theories of the development and function of the brain. Her observations and results were impressive enough that now Occupational Therapy with its developmental early intervention focus is universally adopted.

The *vestibular apparatus* and its receptors do not vary their sensitivity or influence the brain directly. The balance and sensitivity of the apparatus is set by the function of the *cerebellum* and the function of the *cerebellum* is a product of the service of four major pathways: (1) the *visual system*, (2) the *proprioceptive system* from muscles and joints, especially the cervical spine, (3) the *vestibular system*, and (4) the cerebral cortex. Since the cerebral cortex is just forming in a developing child and the *vestibular* and *visual systems* are relatively constant, the *proprioceptive system* is by far the most important to the *cerebellum* and its effects on the *thalamus* and the neocortex.

Ayres observed that children with learning and neurobehavioral problems exhibit what she termed *sensory defensiveness*. It was thought that this *sensory defensiveness* was the result of an over-activation of our protective senses. It was noticed that some children had decreased responses to sensory stimuli and some appeared to have increased sensitivity to sensory input. We now have a better way of understanding and explaining these observations and realize that both are the product of decreased sensory input to the *cerebellum, thalamus*, and neocortex. The *cerebellum* has two halves as does the *cerebrum*. These two halves must be balanced in their activation. If they are not, the hemisphere with decreased activity may initially be less sensitive to incoming sensory stimuli with an increased threshold of activation because the neurons are less active. However, over time, this decreased activation causes the cells to shift closer to threshold as a compensatory mechanism. This makes the child more sensitive to stimuli that effect the dysfunctional half of the *cerebellum*. Tactile, proprioceptive, extraocular, and vestibular input indeed results in over-firing of the *cerebellum*. The child will experience lower threshold to touch, movement of the head, neck, or body expressed as motion sickness, disordered eye movements, or visual perceptual disturbances. Cerebral activity associated with cognition or emotion also can make the *cerebellum* fire aberrantly and the cells, which have less endurance or fatigueability to this input, may cause these cells to produce free radicals and result in oxidative stress injury to these same cells (Forster et al., 1996; Joseph et al., 1998). In the *basal ganglia* this may produce *hyperkinetic* and/or *hypokinetic* behavior through the same process. In the cerebrum, we recognize this as epilepsy, epileptiform activity, or spontaneous firing of neurons (Joseph et al., 1998, 1999; Heim et al., 2000).

Ayres (1972a; 1972b) observed the symptoms of this process and describes several types of sensory defensiveness.

1. *Tactile defensiveness*. Children with tactile defensiveness avoid letting others touch them and would rather touch others. They frequently fuss or resist hair washing or cutting. They may act as if their life is being threatened when being bathed or having clothes changed. These children are often irritated by some types of clothes, clothing labels, or new clothes. They may dislike being close to others or avoid crowds. They can be agitated by people accidentally bumping into them. They often do not like to get their hands or feet dirty. They may seem unnecessarily rough to people. Some may bump or crash into things on purpose as a way of seeking sensation or seen under responsiveness to certain sensations or pain.

2. *Oral defensiveness*. Some children dislike or avoid certain textures or types of foods. They may be over or under sensitive to spicy or hot foods, avoid putting objects in their mouth and/or intensely dislike teethbrushing or face-washing. Sometimes have a variety of feeding problems since infancy.

3. *Gravitational insecurity*. This appears to be an irrational fear of change in position or movement. These children are often fearful of having their feet leave the ground or having their head tip backwards.

4. *Postural insecurity*. This is a fear and avoidance of certain movement activities due to poor postural mechanisms.

5. *Visual defensiveness*. This may involve an over sensitivity to light and visual distractibility. With this problem, children may avoid going outside in certain light and/or need to wear hats or sunglasses to block out light. They may startle more easily and/or avert their eyes or seem to avoid eye contact.

6. *Auditory defensiveness*. This reflects an over sensitivity to certain sounds and may involve irritable or fearful responses to noises like vacuum cleaners, motors, fire alarms, etc. Children sometimes make excessive amounts of noise to block out sounds (cf. Zentner and Kagan, 1996). Other symptoms can include unusual sensitivity to taste and/or smell (cf. preceding section).

When we understand how the *cerebellum* functions and how it affects the *thalamus* and cerebral cortex, we will then be able to explain more fully all of the symptoms of *autistic spectrum disorders* as a primary deficit or imbalance of *cerebellar*-cortical activation. We remember that the primary output of the *cerebellar* cortical cells or *Purkinje* cells is inhibitory to the *cerebellar* output nuclei. The *cerebellum* also controls motor coordination, balance, postural stability, and *extraocular* eye movements. It also activates the *thalamus*, which relays all sensory input to the cerebrum. Decreased activation of the *cerebellum* results in its dysfunction even though it may be more sensitive to input and may cause decreased stability of *thalamus* and cerebrum, even though the overall level of stimulation is decreased. This decreased threshold or increased signal-to-noise ratio may cause the cells to fire prematurely, they may reach oxidative stress earlier. This increased sensitivity is a product of decreased activation and is perceived as unpleasant by the child. This explains why a child with the same problem can present differently with one being over-reactive to certain stimuli and another being under-reactive. The underlying problem is the same, a lack of the central nervous system being properly activated. The same lack of stimulation can produce *hyperkinetic* behavior, while another may present with *hypokinetic* behavior.

Ayres devised a number of ways of treating these problems of *sensory defensiveness*. In Occupational Therapy, the approach to treatment primarily involves *vestibular, proprioceptive*, and *tactile* stimulation along with behavior modification techniques. Examples of some of these treatments for particular problems are:

1. *Tactile defensiveness*. OT treatment approaches include applying rapid and firm pressure touch to arms, hands, back, legs, and feet with a non-scratching brush with many bristles. The brushes recommended are specific plastic surgical scrub brushes. This is followed by gentle joint compression to the shoulders, elbows, wrists, hips, knees, ankles, and sometimes fingers and feet. This treatment is recommended because the results are effective for short periods. Occupational Therapists note that if these procedures are applied consistently over time, the *defensiveness* is permanently reduced or even eliminated. Deep pressure touch, compression, or traction to the joints, and heavy muscle action together is a special combination to reduce or eliminate *sensory defensiveness*. (Summation of sensation that is neurologically experienced in a short period over a large body space.)

2. *Oral defensiveness*. OT treatment of applying heavy pressure across the roof of the mouth and giving input to the *temporomandibular* joint. Oral motor activities are also used that involve biting or resistive sucking of a knot on the end of fruit roll-ups, beef jerky, etc. Occupational Therapists use small straws, sports bottles, plastic tubes, etc. for sucking as well as mouth toys that involve sucking and blowing.

3. *Gravitational and postural insecurity*. Treatment includes jumping on bouncing surfaces, trampolines, bed mattresses, or on the floor with jarring action, jumping and crashing into piles of pillows or beanbag chairs, bouncing while sitting on an inflated ball, play wrestling, swinging on suspended tire inner tubes, "frog" sling swings, wet hammocks, platform swings, and "bungy" cords. Climbing and crawling over and under large pillows, beanbag chairs, jungle gyms, rocks, trees, up stairs, on hands and knees through obstacle courses made of furniture, balance activities, walking on a balance beam, rocker and wobble boards, fine motor coordination, handwriting, and peg board drawing.

A significant number of outcome studies have indicated the effectiveness of this approach to treatment along with support of Ayres' (1972) concept of *sensory defensiveness* (Ayres, 1980; Larson, 1982; Hamill, 1984; Royeen, 1985; Case-Smith, 1991; Baranek and Berkson, 1994; Sakamoto, 1994; Baranek et al., 1997; Humphries et al., 1997). In general though, Occupational Therapy techniques do not utilize a specific approach based on asymmetric hemispheric function or deficits.

Theories of Physical-Mechanical Interventions

The Effects of Physical Exercise on Cognitive Performance

If there is one activity that seems to be the "magic bullet" against almost every disease or disorders, it is exercise, especially aerobic exercise. It seems almost every day a new study shows exercise to reduce the risk and severity of a new disease from cancer to the common cold to depression, exercise seems to be the one thing that prevents or cures them all. There have been many theories proposed as to why exercise has such dramatic health benefits. Some think it is because of its effect on the heart and cardiovascular system. Some think because of its effect on the endocrine system, while others think it affects the immune system. The fact is that it affects all of these systems but aerobic exercises most impact the efficient functioning of the *central nervous system*. When one modality affects all of the systems of the body, it must be because of a primary effect on the brain. As we have seen, autonomic, immune, endocrine, cognitive, emotional, and sensory systems are all asymmetrically distributed in the brain. Exercise therefore must have a generalized effect on all brain functions. As we know, the primary source of activation of the brain is through the motor system, therefore, high-frequency low-intensity activity of the motor system will have powerful effects on the global activation, arousal, and attention of the *cerebellum, thalamus, basal ganglia*, and cerebrum. Aerobic exercise affects all muscles of the body including the involuntary postural anti-gravity muscles, as well as the voluntary muscles of the extremities and trunk. It also increases the efficiency of the cardiovascular system to deliver blood and oxygen to the brain, and increases the capacity of the lungs to take in oxygen. We would expect, therefore, that exercise would be helpful in improving a child's ability to learn and control behavior and to focus attention. Lack of physical activity would be expected to cause the opposite affect.

Researchers at the *Salk Institute for Biological Studies* in La Jolla, California, demonstrated that mice that regularly exercise on running wheels had twice the number of new brain cells compared to sedentary mice. One of the study's authors, Fred Gage, has said that, "More people in my lab have started running since we found this result." The studies published in the *Proceedings of the National Academy of Sciences* and in *Nature, Neuroscience* (van Praag et al., 1999a, 1999b) are remarkable in several ways. Gauge's Lab demonstrated that humans along with mice and non-human primates do grow new brain cells after birth. In a previous study, the Salk researchers had found that those mice who had "enriched environment" with a tunnel, toys, and an exercise wheel grew more cells than those in regular lab cages (Kempermann and Gage, 1999; van Praag et al., 2000). What's more, in the area of new brain cell growth, the *hippocampus* is associated with learning and memory. Researchers thought that it might not just be running per se, but exercise in general that causes the growth of new brain cells. Does growing more brain cells mean the running mice are necessarily smarter? van Praag and Gauge have said it is reasonable to think so because previous studies on "enriched environment" mice showed that they perform better on learning tasks. Exercise has been shown to enhance cognitive function and to help stroke victims recover from brain injury (Levy et al., 2001; Shepherd, 2001).

The type of exercise is important and the combination of physical activity and mental focus or "purposeful" activity at the same time or close together, appear to yield the greatest changes. Nudo and associates (1996) documented plastic changes in the functional topography of *primary motor cortex (M1)* that are generated in motor skill learning in the normal, intact primate. The investigators employed intra-cortical micro-stimulation mapping techniques to derive detailed maps of the representation of movements in the distal forelimb zone of *M1* of squirrel monkeys, before and after behavioral training on two different tasks that differentially encouraged specific sets of forelimb movements. After training on a small-object retrieval task, which required skilled use of the digits, their evoked-movement digit representations expanded, whereas their evoked-movement wrist/forearm representational zones contracted. These changes were progressive and reversible. In a second motor skill exercise, a monkey pronated and supinated the forearm in a key (eyebolt)-turning task. In this case, the representation of the forearm expanded, whereas the digit representational zones contracted. Their results show that *M1* is alterable by use throughout the life of an organism. These studies also reveal that after digit training there was an areal expansion of dual-response representations, that is, cortical sectors over which stimulation produced movements about two or more joints. Movement combinations that were used more frequently after training were selectively magnified in their cortical representations. This close correspondence between changes in behavioral performance and electrophysiologically defined motor representations indicates that a neurophysiological correlate of a motor skill resides in *M1* for at least several days after acquisition. The finding that cocontracting muscles in the behavior come to be represented together in the cortex argues that, as in sensory cortices, temporal correlations drive emergent changes in distributed motor cortex representations.

Tantillo and associates (2002) had performed a study examining the effects of exercise on children with *ADHD* evaluated by studying the rate of spontaneous eye blinks, the *acoustic startle eye blink response* (*ASER*), and motor impersistence. The children evaluated, both male and female, were between 8 and 12 years old all meeting the DSM-III-R criteria for *ADHD*. All children in their study ceased *methylphenidate* medication 24 hours before and during each of three daily conditions. After a maximal treadmill walking test to determine cardio-respiratory fitness (VO[2peak]), each child was randomly assigned to counterbalanced conditions of treadmill walking at an intensity of 65–75 percent VO(2peak) or quiet rest. Responses were compared with a matched group of control participants. Boys with *ADHD* had increased spontaneous blink rate, decreased *ASER* latency, and decreased motor impersistence after maximal exercise. Girls with *ADHD* had increased *ASER* amplitude and decreased *ASER* latency after sub-maximal exercise. The authors' findings suggest an interaction between sex and exercise intensity that is not explained by physical fitness, activity history, or selected personality attributes. Their findings support the employment of vigorous exercise programs as adjuvant in the management of the behavioral features of *ADHD*.

Elliot and associates (1994) examined the effects of antecedent exercise conditions on maladaptive and stereotypic behaviors in six adults with both *autism* and moderate-to-profound mental retardation. The behaviors were observed in a controlled environment before and after exercise and non-exercise conditions. From the original group of participants, two were selected subsequently to participate in aerobic exercise immediately before performing a community-integrated vocational task. Only antecedent aerobic exercise significantly reduced maladaptive and stereotypic behaviors in the controlled setting. Neither of the less vigorous antecedent conditions did. When aerobic exercise preceded the vocational task, similar reductions were observed. There were individual differences in response to antecedent exercise. These authors note that the use of antecedent aerobic exercise to reduce maladaptive and stereotypic behaviors of adults with both *autism* and mental retardation is supported.

Similar results were reported by Rosenthal-Malek and Mitchell (1997) in an investigation of the reduction of self-stimulatory behaviors in adolescents with *autism* after vigorous exercise. Celiberti and colleagues (1997) in a detailed case study of an *autistic* boy also examined the differential and temporal effects of two levels of antecedent exercise (walking versus jogging) on his self-stimulatory behavior. The exercise conditions were applied immediately before periods of academic programming. Maladaptive self-stimulatory behaviors were separately tracked, enabling identification of behaviors that were more susceptible to change (e.g., physical self-stimulation and "out of seat" behavior) versus those that were more resistant (e.g., visual self-stimulation). Examination of temporal effects indicated a decrease in physical self-stimulation and "out of seat" behavior, but only for the jogging condition. In addition, sharp reductions in these behaviors were observed immediately following the jogging intervention and gradually increased but did not return to baseline levels over a 40 min period.

We now know that exercise has benefits for overall health as well as for cognitive function. Recent studies using organism models have been directed towards understanding the neurobiological bases of these benefits. It is now clear that voluntary exercise can increase levels of brain-derived *neurotrophic* factor (*BDNF*) and other growth factors stimulate neurogenesis, increased resistance to brain insult, improve learning, and mental performance. Recently, high-density *oligonucleotide* microarray analysis has demonstrated that, in addition to increasing levels of *BDNF*, exercise mobilizes gene expression profiles that would be predicted to benefit brain plasticity processes. Thus, exercise can provide a simple means to maintain brain function and promote brain plasticity (Cotman and Berchtold, 2002).

Lieberman and colleagues (2002) reporting in the *American Journal of Clinical Nutrition* note that the brain requires a continuous supply of glucose to function adequately. During aerobic exercise, peripheral glucose requirements increase and carbohydrate supplementation improves physical performance. The brain's utilization of glucose also increases during aerobic exercise. However, the effects of energy supplementation on cognitive function during sustained aerobic exercise are not well characterized. The investigators examined the effects of energy supplementation, as liquid carbohydrate, on cognitive function during sustained aerobic activity. Young, healthy men were randomly assigned to 1 of 3 treatment groups. The groups received a 6 percent (by vol) carbohydrate (35.1 kJ/kg), 12 percent (by vol) carbohydrate (70.2 kJ/kg), or placebo beverage in 6 isovolumic doses, and all groups consumed two meals (3200 kJ). Over the 10-hour study, the subjects performed physically demanding tasks, including a 19.3-km road march and two 4.8-km runs, interspersed with rest and other activities. Wrist-worn vigilance monitors, which emitted auditory stimuli (20/h) to which the subjects responded as rapidly as possible, and a standardized self-report mood questionnaire were used to assess cognitive function. These investigators found that vigilance consistently improved with supplemental carbohydrates in a dose-related manner; the 12 percent carbohydrate group performed the best and the placebo group, the worst. Mood-questionnaire results corroborated the results from the monitors; the subjects who received carbohydrates reported less confusion, and greater vigor than did those who received the placebo. Supplemental carbohydrate beverages enhance vigilance and mood during sustained physical activity and interspersed rest. In addition, ambulatory monitoring devices can continuously assess the effects of nutritional factors on cognition as individuals conduct their daily activities or participate in experiments. These approaches have not been employed in studying neurobiological involved children.

In an interesting study reported in the *Journal of Physiology, London* by Thornton and associates (2001) at the Laboratory of Physiology at *Oxford University*, PET was used to identify the neuroanatomical correlates underlying "central command" during imagination of exercise under hypnosis, in order to uncouple central command from peripheral feedback. Three cognitive conditions were used: imagination of freewheeling downhill on a bicycle (no change in HR, or ventilation, V(I)), imagination of exercise, cycling uphill (increased HR by 12 percent and V(I) by 30 percent of the actual exercise response), or volitionally driven hyperventilation to match that achieved in the second condition (no change in HR). The researchers found significant activations in the right dorso-lateral *prefrontal* cortex, *supplementary motor areas* (*SMA*), the right premotor area (*PMA*), supero-lateral *sensorimotor* areas, *thalamus*, and bilaterally in the *cerebellum*. In the second condition, significant activations were present in the *SMA* and in lateral *sensorimotor* cortical areas. The *SMA/PMA*, dorso-lateral *prefrontal* cortex, and the *cerebellum* are concerned with volitional motor control, including that of the respiratory muscles. The neuroanatomical areas activated suggest that a significant component of the respiratory response to "exercise," in the absence of both movement feedback and an increase in CO_2 production, can be generated by what appears to be a behavioral response.

Lardon and Polich (1996) examined the electrophysiologic effects of physical exercise by comparing groups of individuals who engage in regular intensive physical exercise (12+ hr/week) to control subjects (2+ hr/week). EEG activity was recorded under eyes open/closed conditions to assess baseline differences between these groups. Spectral power was less for the exercise compared to the control group in the *delta* band, but greater in all other bands. Mean band frequency was higher for the exercise compared to controls in the *delta*, *theta*, and *beta* bands. Some differences in scalp distribution for power and frequency between the exercise and control groups were found. The findings suggest that physical exercise substantially affects resting EEG and again supports the effects of exercise on brain function.

Traditionally the view has been that there is a separation between motor skills and cognitive ability. However, the same pathways and same global increased activation of the areas involved with motor skill also underlie the areas that form the foundation of cognitive ability. However, the brain is activity dependent, therefore even though the potential to learn is enhanced through motor training, if a specific cognitive skill is not trained, it will not adequately develop. In an interview with Steven Keele (Keele and Posner, 1968; Keele et al., 1985, 1988; Keele and Ivry, 1990), who is associated with the Psychology Department at the *University of Oregon* and whose research has helped bring the study of motor control into the mainstream of cognitive psychology, addresses this question. When he is asked why the topic of motor control does not get discussed much in textbooks of cognitive psychology, he states "… the basic neglect may derive from the implicit but faulty assumptions that psychologists have about motor control. They often think that motor control is an add on something beyond cognition itself …." He continues, "… some scientists view the learning or control of a series of actions as being relatively primitive compared to such notable human achievements as declarative memory, the ability to reason, or language." He goes on to relate that, "Almost 50 years ago the great neuropsychologist, Carl Lashley, pointed out a remarkable fact about human skill that should disabuse one of the notions that motor control and motor skill are less "human" than other remarkable capacities of our species. Humans speak, they type, they sign, they write and perform intricate motor skills. In the domain of music people play the fiddle, may dance to it, and they may sing or hum along. People build cabinets, knit, and blow fine glassware. These diverse motor activities are beyond the realm of other organisms and suggest that motor capabilities are related to other intellectual capabilities. Indeed some psychologists such as Jerome Bruner have suggested that "even human language capability is an outgrowth of capabilities involved to create new motor sequences."

"Extensive evidence suggests that knowledge is acquired as a result of extensive practice, thousands of hours of highly dedicated practice is key in separating the most successful people in various motor and non-motor skill domains from the rest of us. This perspective grew initially out of analysis of chess expertise, but also has been found to apply to muscle performance and basketball." Keele concluded, "…The surprising idea that stands out in the expertise literature is that the extraordinary motor capabilities of humans are best understood as an extension of their extraordinary cognitive abilities." When we examine "geniuses" through out history, we can see that artists and sculptors like da Vinci and Michelangelo, musicians like Bach and Mozart, were geniuses not only in their cognitive ability but also in their motor skill as well, to paint or play music.

The question is, does the constant practice of developing a motor skill like painting, or playing an instrument create the cognitive genius or vice versa? Motor skills develop first. We know motor skills develop first in a child but if they focus on the motor skill to the exclusion of all else, then they will not perform well in other areas of life. However, if motor skill is used as a tool to develop brain areas, and then academic and social pursuits are diligently taught, the child will learn those activities better and faster. The key is balance and in an otherwise normal child who is behind, and in a child who is developmentally delayed, the fastest and most effective way to increase the rate of development of their brain function may be through motor activity and motor development. If there is a delay in motor skill development, then there will be a subsequent delay in their cognitive and emotional development as well.

Occupational Therapists think that hyperactive children often have persistent tonic neck reflexes. This is a normal reflex present in young children and they think that if this reflex persists in older children, it is not only a sign of poor neurological organization, but makes it difficult and uncomfortable for the child to sit normally. Occupational Therapists note that many children who are hyperactive are also fidgety, sit in unusual postures, especially a slumped posture or hook their feet under the chair for support. They may tend to stand when eating or doing homework and they may experience fatigue of their neck and postural muscles, which becomes painful and affects the child's ability to concentrate. Occupational Therapists have designed a series of crawling exercises and claim that these intervention techniques are effective in alleviating the *ADD* symptoms and improve academic performance and behavior. These techniques emphasize the importance of the motor system's effect on the neurological development of a child's brain and a subsequent improvement in learning and behavior. While the theory does not take the *cerebellum*, differential hemispheric activation, and their effects on developing brain function into consideration, the theory does emphasize the role of postural muscles in the manifestation of the observed symptoms. Most children with learning disability and *ADD* do have poor postural tone, indicative of *cerebellar, thalamic*, and *frontal lobe* dysfunction. Therefore, any activity emphasizing proximal and postural coordination will increase feedback to these brain regions. These activities may be more significant to right brain development and therefore would be expected to improve symptoms of right brain deficit, including hyperactivity, *ADD*, and behavioral problems.

The use of spinal manipulation for purposes of promoting health and curing disease dates back almost as far as recorded history to the Greeks and Egyptians. The primary effect of spinal manipulation is on the brain and nervous system. Another premise of D. D. Palmer (B. J. Palmer, 1906) was that the main conduit or path to affect the brain is the musculoskeletal system, especially the spine, and particularly the upper cervical spine. In addition, we now see based on a preponderance of scientific information that this observation is indeed correct. The majority of all sensory input arises from somatosensory receptors of the musculoskeletal system and the largest percentage of that amount comes from the receptors of the spinal muscles and joints located in the upper cervical spine receptors (Leisman, 1976a). Through the ability of these spinal receptors and based on their upright orientation and transduction of gravitational forces into electrochemical impulses that constantly bombard the brain by way of the *dorsal column* and *spino-cerebellar* and nonspecific *thalamic* pathways, they provide the baseline activation or arousal on which, in part, other brain activity is based. Through specific pathways, the same somatosensory receptors can affect specific cortical areas that are involved with higher functions of perception, cognition, and emotional behavior, especially in the *frontal lobe*.

However, with manifestation of symptoms, especially musculoskeletal symptoms, which make up the vast majority of health complaints of humans, they are primarily symptoms of neurological dysfunction and are best treated by affecting the nervous system directly. This can be achieved by use of spinal manipulation, joint mobilization, exercise, and by stimulating the brain through a variety of environmental stimuli, such as sound, light, heat, cold, odors, tactile sensation, and cognitive activities, as indicated earlier in the chapter. Virtually all those with neurobehavioral disorders of childhood also demonstrate dysfunction of their motor-sensory system. Either this dysfunction of the motor-sensory system may in fact be a primary cause of the brain dysfunction or the brain dysfunction may be the primary cause of the motor-sensory dysfunction. Either way, the motor-sensory systems, which include the postural muscles and joints of the spine and neck are dysfunctional.

Therefore, no matter what the primary source, an intervention strategy for the motor-sensory dysfunction ought to result in an improvement of brain function and vice versa. This is especially true in the *frontal lobe* where we have seen that both motor and non-motor functions can be measured and a dysfunction of one is reflective of an equal dysfunction of the other. Therefore, an improvement of *frontal lobe* motor function associated with an improvement in a child's motor function capacities, such as muscle tone, coordination, mobility, strength, and endurance, should also be reflective of an improvement in non-motor functions of the *frontal lobe* such as cognitive, emotional, and behavioral. By directing and including diagnosis and treatment of musculoskeletal system function, we develop tools of measuring and affecting *central nervous system* status.

An example of the relationship between musculoskeletal complaints relating to higher brain function has been presented in a recent article in *Spine* (Luoto et al., 1999). In this paper, the authors examine the mechanisms explaining the association between lower back pain and deficits in information processing. Low back trouble, chronic pain in general and depression has been associated with impaired cognitive functions and slow reaction times. It is known that the preferred hand performs significantly better than the non-preferred hand in motor tasks. The authors hypothesize that chronic low back pain hampers the functioning of short-term memory in a way that leads the preferred hand to lose its advantage over the non-preferred hand, but that the advantage would be restored during rehabilitation. Reaction times for the preferred and non-preferred upper limbs were tested in 61 healthy control and 68 low back pain patients. A multi way *analysis of covariance* was used to examine the group, handedness, and rehabilitation effects on reaction time. A significant interaction among group, handedness, and rehabilitation was found. At the beginning, the reaction times for the preferred hand were faster among the control subjects, but not among the patients with low back pain. After the rehabilitation, the preferred hand was faster among both the control subjects and the patients with low back trouble. During the rehabilitation, back pain, psychological distress, and general disability decreased significantly among the patients with chronic low back trouble. The results support the hypothesis that chronic low back pain and disability impedes the functioning of short-term memory, resulting in decreased speed of information processing among patients with chronic low back symptoms.

Recent studies (Byl et al., 1996, 1997; Byl and Melnick, 1997; Byl and McKenzie, 2000; Blake et al., 2002) report on a theory that suggests that a dysfunction in the way the brain receives and processes information from the body, may trigger so-called writer's and musician's cramps. Researchers at *The University of California* and *San Francisco School of Medicine* say that the debilitating disorder also called *focal dystonia* of the hands stems from pushing the brain past its ability to learn quick repetitive movements. When the brain signals become "jumbled" these researchers think the muscles spasm and stiffen. Nancy Byl, Professor Physical Therapy, and Michael Merzenich, Professor of Otolaryngology, based their studies on previous research that explains the mechanism of how messages are wired to the brain. Studies explain how tactile receptors or nerves upon the skin speed signals to an area of the brain called *sensory maps*, which undergo rewiring or plasticity with each learning experience. Researchers read these maps and identify zones that correspond to individual fingers and parts of the fingertips.

According to ongoing studies conducted by Byl, Merzenich, and colleagues on monkeys first published in 1996, rapid repetitive movements result in degeneration of the brain's *sensory map* that leads to muscle spasm and impaired muscle tone. They suggest that during successive movements, the brain is forced to process too numerous sensations and muscle commands. This gives rise to faulty movements they say that causes fingers and hands to spasm. Standard therapy is used to treat *focal dystonia* including anti-*Parkinson's* drugs, muscle relaxants, and injections of *botulinum toxin* (*Botox*) to weaken problematic muscles. The authors feel that this treatment approach may be inappropriate. Instead, they suggest retraining therapy that consists of exercises to help patients fine-tune their tactile senses. This should, they think, help diffuse overloaded *sensory maps* so that they can discriminate sensations better.

Byl noted that after 12 weeks of retraining therapy, 14 of 16 patients with severe *focal dystonia* of the hand who were not helped by standard therapies reported improvement in function and were able to return to work. A brain scan taken on one of the patients showed that the *sensory maps* appeared more neatly arranged. Although many think that there are primary biomechanical factors that produce repetitive strain type injuries, Byl thinks that *focal dystonia* of the hand is more likely to occur if a person is exposed to biomechanical risk factors like a high level of repetitive movements and small fingers spread. She maintains that a significant factor in *focal dystonia* is a disorganization of the *sensory maps* adding that *Botox* and other treatments simply "quiet symptoms." She further states, "The nervous system is responsive to repetitive behaviors but we have always assumed that those modifications of the nervous system from repetitive movements would have a positive outcome, that it would make one smarter and be able to test more accurately.

But what we are saying is we have identified a dysfunctional reorganization of the sensory brain that seems to be associated with the disability disorder negative outcome." *Focal dystonia*, as described by Byl and associates, can be considered a primary dysfunction in the motor system, including the *basal ganglia, thalamus, cerebellum,* and *frontal lobe*. Focal dystonia or *hypokinetic* behaviors may be isolated to the *sensory-motor* cortices. Lower back and neck pain are also oftentimes considered repetitive strain injuries and the same mechanism may apply. Hypermobility of the spinal joints may also produce improper repetitive sensory input that may rewire *sensory maps* to produce fatigueability or *oxidative stress* to brain cells. This may result in a *focal dystonia* of the spinal muscles with effects on the sensory-motor cortices. These painful muscle spasms may either be a product of the central nervous system's irregular activation or may also result from abnormal repetitive feedback to the cortex.

Kelly and his colleagues (2000) had employed a mental rotation reaction-time paradigm to measure the effects of upper cervical manual spine adjustments on cortical processing. Thirty-six subjects with clinical evidence of upper cervical joint dysfunction participated in the study. Subjects in the experimental group received a high-velocity, low-amplitude upper cervical adjustment. A non-intervention group was used to control for improvement in the mental rotation task because of practice effects. Reaction time was measured for randomly varying angular orientations of an object appearing as either normal or mirror-reversed on a computer screen. The average decrease in mental rotation reaction time for the experimental group was 98 ms, a 14.9 percent improvement, whereas the average decrease in mental rotation reaction time for the control group was 58 ms, an 8 percent improvement. The difference in scores after the intervention time were significantly greater for the experimental group compared with the control group. The results demonstrate significant improvement in a complex reaction-time task after an upper cervical adjustment providing support for the notion that upper cervical spinal manipulation may affect cortical processing.

Carrick (1997) attempted to ascertain whether manipulation of the cervical spine is associated with changes in brain function. He employed physiological cortical maps reflecting brain activity before and after manual manipulation of the cervical spine in 500 subjects. Subjects were divided into six groups and each subject underwent manipulation of the second cervical segment. Blinded examiners obtained reproducible pre- and post-manipulative cortical maps. Brain activity was demonstrated by reproducible circumferential measurements of cortical hemispheric blind-spot maps before and after manipulation of the second cervical motion segment. Carrick found that manipulation of the cervical spine on the side of an enlarged cortical map is associated with significantly increased contralateral cortical activity. Manipulation of the cervical spine on the side opposite an enlarged cortical map is associated with decreased cortical activity with strong statistical significance. Manipulation of the cervical spine was specific for changes in only one cortical hemisphere. Carrick concluded that accurate reproducible maps of cortical responses could be used to measure the neurological consequences of spinal joint manipulation. As cervical manipulation activates specific neurological pathways, manipulation of the cervical spine may be associated with an increase or a decrease in brain function depending upon the side of the manipulation and the cortical hemishpericity of a patient. Carrick's essential point is that there exists an asymmetric distribution of the visual field loss and cerebral dysfunction. He also shows that specific manipulation of the cervical spine on the side opposite that of the functional cortical deficit specifically increases the function of that hemisphere of the brain. He also shows manipulation of the same side can increase the dysfunction. This study therefore supports the notion that spinal manipulation may be effective in facilitating the restoration of aspects of brain function, but that it can be specifically directed toward one hemisphere or the other.

In the context of the foregoing discussion, attempts to evaluate manual and manipulative therapies in the treatment of children with *ADHD* have been performed (Giesen et al., 1989). Investigators evaluated the effectiveness of the treatment for reducing activity levels of hyperactive children by these intervention strategies. Data collection included independent evaluations of behavior using unique wristwatch type devices to mechanically measure activity while the children completed tasks simulating schoolwork. Further evaluations included electrodermal tests to measure autonomic nervous system activity. Clinical evaluations to measure improvement in spinal biomechanics were also completed. Placebo care was given prior to intervention. Five of seven children showed improvement in mean behavioral scores from placebo care to treatment. Four of the seven showed improvement in arousal levels, and the improvement in the group as a whole was highly significant. Agreement between tests was also high in this study. For all seven children, three of the four principal tests used to detect improvement agreed either positively or negatively (parent ratings of activity, motion recorder scores, electrodermal measures, and X-rays of spinal distortions). While the behavioral improvement taken alone can only be considered suggestive, the strong interest agreement can be taken as support for the efficacy of this intervention strategy.

Metabolic-Physiological Intervention Strategies

Nutrition

Diet
Although we live in affluent countries, our children, as we have shown, are suffering from malnutrition and lack of environmental stimuli, adequate physical activity, and the average child's diet is deficient in basic nutrition. Children are getting fatter and eating more, but are still deficient in the proper nutrients that they need for their brains to develop and function optimally. Another factor is that if the brain is not functioning or

activated properly, the neurological regulation of the digestive system, the ability to secrete acids with which to break down food, the ability to absorb food and nutrients, as well as the circulation to the intestines and brain, are all decreased. Therefore, even with proper diet or even dietary supplementation, a child may be undernourished, as the brain requires both fuel and stimulation. Without fuel or with too much stimulation the brain will dysfunction. Without stimulation, the body cannot break down and absorb food and fuel; both the body and brain work together.

A low calorie diet is not good for the development of the child, pre- or post-natally. On the other hand, a high calorie diet, especially combined with physical inactivity, produces obesity, decrease in both *dopamine* and *dopamine* receptors (Popa et al., 1988), and both cognitive (Rodin et al., 1977) and motor dysfunction (Petrolini et al., 1995; Maffeis et al., 1997). Small decreases or restriction of caloric intake appears to work best on the brain. Caloric restriction experiments on primates and other organisms have extended their life span by up to 50 percent (Greenberg and Boozer, 2000; Ingram et al., 2001). Ray Walford has conducted a number of studies on caloric restriction. His studies indicate that slight caloric restriction prevented the decline of *dopamine* receptors in the brain cells of organisms. In another study, Walford showed that function of dendrites of brain cells was also improved by caloric restriction. The caloric restriction that Walford recommended is not extreme, it is reported that optimal caloric intake is from 1,500 to 2,000 calories a day, which is about 500 to 1,000 calories less than most men eat (Walford et al., 1995). Therefore, slight modifications in diet can have demonstrable changes in learning disabilities and behavioral problems and overall brain development and function.

Laura Stevens (Stevens et al., 1995; Burgess et al., 2000) has studied biochemical factors affecting children with *ADHD*. She notes that as children use about 45 raw materials (vitamins, minerals, amino acids, essential fatty acids, water, and carbohydrates) to make thousands of other chemicals that make up their bodies and brains, if they do not take in the right nutrients and in the appropriate amounts, their bodies and brains cannot develop and function properly. She notes that in computerized 3-day records for 100 children in her *ADHD* study, she determined that sugary drinks frequently replace milk at meals, fruits and vegetables are scarce, and sugar and fatty snacks replace healthy foods. Many children were found to be taking in much less than the recommended dietary allowances of several nutrients, calcium, magnesium, iron and zinc, and vitamins B-6 and C. She recommends some basic principles that she has found to be effective. The first principle is variety. Different fruits, vegetables, grains, meats, fish, etc. helps to guarantee that a child will get all of the nutrients they need for proper brain function.

The basic food pyramid can be used as a guide for what a child should eat every day. At the bottom of the food pyramid are breads, cereal, rice, and pasta. One should therefore ensure to choose whole grain products. A child requires the most servings of these foods daily. At the top of the pyramid are fats, oils, and sweets. Sweets and certain fats should be used sparingly and should be picked carefully. Efforts should be taken to reduce "bad fats." These consist of saturated and hydrogenated, and partially hydrogenated fats. On the other hand, "good fats," which are a source of essential fatty acids, should be included every day in a child's diet. Essential acids are extremely important for proper brain function. They are necessary to be consumed in a child's diet because the body does not produce these important nutrients. Rich sources of essential fatty acids can be found in pure soy, walnuts, canola oils, beans, tofu, and cold water fish. A child should have at least five servings of fruits and vegetables daily. Also, included should be two to three servings of lean meat and poultry, fish, eggs, dry beans, and nuts. A child should also have two to three servings of low fat milk, yogurt, and cheese daily. If a child is sensitive or intolerant of milk and dairy products, they will need a calcium supplement because it is difficult to get enough calcium in the diet without dairy. However, if a child eats too many dairy products, they may not eat enough of other foods they require. Milk is an excellent source of calcium, protein, and vitamin D, but is a poor source of iron and zinc. If a child consumes too many dairy products, they may be at risk for iron deficiency anemia, because milk contains little iron and the calcium may inhibit the absorption of iron from other foods.

Based on reports from caregivers, case studies, and observation of patients with schizophrenia and children with severe behavioral disorders, Dohan hypothesized (Dohan, 1966a, 1966b, 1969, 1970, 1979, 1980a, 1980b, 1988; Dohan et al., 1969, 1984) that gluten and dairy foods might worsen these behaviors. He noted that in many cases, a restricted diet could lead to significant improvement or recovery from these disorders. For several years, the biochemical explanation for this phenomenon remained unclear. However, several other studies seemed to bear out this observation, and in 1981, using more advanced laboratory technology, Karl Reichelt found and reported (Reichelt et al., 1981) abnormal peptides in the urine of schizophrenics and *autistics*. Peptides are pieces of proteins that are not completely broken down into individual amino acids. Reichelt has observed that these peptides, which are 4, 5, or 6 amino acids long, have sequences that match those of *opioid* peptides (*casomorphin* and *gliadomorphin*). The known dietary sources of these *opiate* peptides are *casein* (from milk) and *gliadin* or gluten (from cereal grains). He has since conducted several studies examining this finding (Seim and Reichelt, 1995; Reichelt et al., 1996; Pedersen et al., 1999; Knivsberg et al., 2002, 2001), as have several other researchers (Buckley, 1998; Garvey, 2002), including Paul Shattock (Whiteley and Shattock, 2002) in England. The best evidence for this correlation lies in the many case reports of improvement or recovery of children with *autism* on this diet.

O'Banion and associates (1978) examined the effect of particular foods on levels of hyperactivity, uncontrolled laughter, and disruptive behaviors in an 8-year-old *autistic* boy. The floor of the child's room was taped off into six equal-sized rectangles to measure general activity level. Frequency data were recorded on screaming, biting, scratching, and object throwing. A time-sample technique was used to record data on laughing. Data were gathered during four phases. During an initial 4-day period, the child was fed a normal American diet. A 6-day fasting period followed, during which time only spring water was allowed. The third phase lasted 18 days and involved the presentation of individual foods. During the final phase of the study, the child was given only foods that had not provoked a reaction in the third phase. Results showed that foods such as wheat, corn, tomatoes, sugar, mushrooms, and dairy products were instrumental in producing behavioral disorders with this child.

Lucarelli and colleagues (1995) also examined the hypothesis that food peptides might be able to determine toxic effects at the level of the central nervous system by interacting with neurotransmitters. They note, as have others, that a worsening of neurological symptoms can be demonstrated in *autistic* patients after the consumption of milk and wheat. These investigators examined the effects of a cow's milk-free diet (or other foods which gave a positive result after a skin test) in 36 *autistic* patients. They looked for immunological signs of food allergy in *autistic* patients on a free-choice diet. They noticed a significant marked improvement in the behavioral symptoms of patients after a period of 8 weeks on an elimination diet and found high levels of *IgA* antigen specific antibodies for *casein, lactalbumin*, and *beta-lactoglobulin*, and *IgG* and *IgM* for *casein*. The levels of these antibodies were significantly higher than those of a control group, which consisted of 20 healthy children. Their results supported the hypothesis that a relationship exists between food allergy and infantile *autism*. Reichelt's group (Knivsber et al., 2001) performed dietary intervention applied to 15 subjects with *autistic* syndromes. Pathological urine patterns and increased levels of peptides were found in their 24-hr urine samples. The peptides, some of which are probably derived from *gluten* and *casein*, were thought to have a negative pharmacological effect on attention, brain maturation, social interaction, and learning. Dietry intervention was reported to significantly reduce the *autistic* symptoms.

Knivsber and associates (2002) have noted that in *autistic spectrum disorders*, urinary peptide abnormalities, derived from *gluten, gliadin*, and *casein* exist. They reflect processes with opioid effect. These investigators evaluated the effects of gluten and casein-free diet for children with *autistic* syndromes and urinary peptide abnormalities. Randomly selected diet and control groups with 10 children in each group participated with the study lasting 1 year. The development for the group of children on diet was significantly better than for the controls. In a review, this same group of researchers (Knivsber et al., 2001) has noted that since the 1980s various studies on dietary intervention have been published. The scientific studies include both groups of participants as well as single cases, and beneficial results are reported in all, but one study. While some studies are based on urinary peptide abnormalities, others are not. The reported results are, however, identical; reduction of *autistic* behavior increases social and communicative skills, and *autistic* traits reappear after the diet has been broken.

Nutritional therapies are varied. There are dietary changes, nutritional supplements, herbal medicines, and homeopathy that are the most commonly used and accepted forms of therapy for children and adults. We will discuss all of these approaches in regard to treatment of neurobehavioral disorders of childhood.

Essential fatty acids
Farooqu and Horrocks (2001) reporting in the *Journal of Molecular Neuroscience*, state that *plasmalogens* are *glycerol-phospholipids* of neural membranes, rate-limiting enzymes, are in the *peroxisomes* and are induced by *docosahexaenoic acid* (*DHA*). The authors suggest that deficiencies of *DHA* and *plasmalogens* occur in *peroxisomal* disorders, like *Alzheimer's* disease, depression, and *ADHD*. The authors claim that this situation may be responsible for abnormal signal transduction associated with learning disability, cognitive deficit, and visual dysfunction. These abnormalities in the signal-transduction process can be partially corrected by supplementation with a diet enriched with *DHA* or essential fatty acids.

The symptoms of essential fatty acid deficiencies include:

- Excessive thirst
- Frequent urination
- Dry skin
- Dry hair
- Dandruff
- Soft and brittle nails
- Small hard little white bumps on the backs of the arms, elbows, or thighs.

There are two essential fatty acids relevant to our discussion: *linolenic acid,* an *omega-6* fatty acid and *alpha-linolenic* acid, an *omega-3* fatty acid. These are the precursor molecules for making long chain fatty acids such as *dihommo-gamma-linolenic* acid (*DGLA*), *arachadonic* acid, *eccosapentaenoic* acid (*EPA*), and *DHA*. Omega-6 and Omega-3 fatty acids are essential because they help make up the cell membrane around every cell including brain cells. The balance of these essential fatty acids helps to determine the fluidity of the membrane and the ability of molecules to enter and exit the cell. This balance also affects the ability of molecules to bind to receptors in the membrane. These long chain fatty acids are also critical because the body converts them to *prostaglandins* and other important molecules that help cells communicate with each other. *Arachadonic* acid and *DHA* are more concentrated in the brain and retina than in other cells. They therefore play a crucial role in brain and nervous system function. Studies have suggested that approximately 40 percent of boys with *ADHD* have significantly lower levels of *Omega-3* and *6* fatty acids than controls with normal behavior (Mitchell et al., 1987). However, boys with *ADHD* have few symptoms of fatty acid deficiencies and have blood levels comparable to boys in the control group.

It is thought that the reason for lower levels could include lower dietary intake or a metabolic block in the *omega-3* and *6* fatty acid pathways.

Ways to increase intake of essential fatty acids, especially *omega-3*, is to use cold process soy or canola oils for homemade salad dressing. These oils can be included into spaghetti sauce, soup, etc. or used to make a pasta salad, also baking with them is good. Frying with these oils should be avoided because the molecules do not sustain their structure with high heat and oxygen. Flaxseed oil is an even more concentrated source of *omega-3* fatty acids. Beans are also a good source of essential fatty acids, especially soy, navy, and kidney. Tofu and cold water fish, like salmon, tuna, mackerel, and sardines are a very good source of *omega-3* fatty acids.

Food allergies and behavior
Well-designed studies reported in the late 1980s and early 1990s have reported that food sensitivities are indeed a major factor for many children with *ADHD* (Egger, 1985; Kaplan et al., 1989; Carter, 1993; Boris and Mandel, 1994; Rowe and Rowe, 1994; Breaky, 1997). It is possible that a brain dysfunction, especially right brain deficit, will alter the body's immune response making it over responsive. It may also alter the environment of the intestinal tract, which may create a sensitivity or allergic response to certain foods and environmental allergies. There has been extensive research examining the brain-intestinal connection. It is fairly well documented that the *GALT* or *gut associated lymphoid tissue* is associated with an individual's absorption and immune responsiveness. It is also thought that various individuals who may be overly sensitive, have specific nutritional deficiencies, or who may suffer from poor brain regulation, may demonstrate a dysfunctional *GALT*. The gastrointestinal system, as a result, becomes less efficient at filtering allergens and larger molecules that may be associated with allergic responses. Infection may pass more easily across the stomach lining and enter the blood stream. Therefore, sensitivity may be secondary to the decreased activation or stimulation of motor or other sensory input (Henderson et al., 1995; Uhlig et al., 1997). It is also possible that the decreased *parasympathetic* activation associated with higher-level neocortical activity and development may be deficient and therefore, inhibition of the *sympathetic* control mechanisms are decreased (Mikkelsen et al., 1981; Girardi et al., 1995). This is associated with decreased blood flow to the gut, which may make the gut tissue break down, may eventually produce injury to the gut tissue, such as ulcers, but will also reduce the ability to absorb food and nutrients from the gut itself. It also will decrease the *parasympathetic* system's ability to appropriately secrete acid, to break down foods effectively, and may reduce some of the peristaltic activity in muscle contraction. These sensitivities may decrease with therapy or exercise, but nonetheless, these food allergies and sensitivities are important factors to consider.

Symptoms of food sensitivities include:

- Pale sallow complexion
- Puffiness, dark circles under eyes
- Recurrent stomachaches, constipation, or soiling
- Headaches including migraines
- Leg and muscle aches
- Bed wetting
- Chronic stuffy, itchy, running noses
- Recurrent ear infections
- Fatigue
- Mental sluggishness, "spaciness" or inability to concentrate.

The most common foods and additives that cause reactions are:

- Artificial colors and flavors
- Chocolate
- Sugar
- Milk and milk products
- Eggs
- Corn
- Wheat and rye
- Citrus (oranges, lemons, grapefruits, etc.)
- Legumes (peas, beans, peanuts, and soy).

A child may be sensitive to more than one of these as well as other foods. Any food the child craves is suspect. One way to test a child for food sensitivity is to give a child a serving of a suspected food. Then totally avoid those particular foods for 5–12 days and take note if the child's behavior or learning ability improves. At the end of the 5–12 days, reintroduce each food one at a time in its pure form such as a glass of milk, popcorn cooked in corn oil, or a poached egg. Artificial colors can be tested in a similar fashion by placing several drops of different color food dyes in water or tolerated juice and then note any reaction. One should not reintroduce more than one suspected food a day. If the child reacts to a specific food, allow time for the reaction to subside before reintroducing a different suspected food. Once identified, reaction-producing foods should be avoided for approximately 2–3 months and then reintroduced in small amounts if tolerated. It is also recommended that if a child has a severe food reaction, try giving them one or two tablets of *Alka Seltzer* cold, the aspirin-free type, in a glass of water. The child may show a dramatic improvement. This should not be used on a long-term basis due to the high concentration of sodium. If a child cannot tolerate milk or dairy products, it is important to supplement their diet with about 500–600 mg of uncolored, unsweetened calcium carbonate in liquid or chewable tablets for as long as they avoid dairy products.

Sugar and chocolate are the most common foods that are associated with behavior problems in children with *ADHD*. More sophisticated blood tests can be conducted for children who are suspected of nutritional deficiency, sensitivities, and fatty acid deficiencies. *Autistic* children in particular may have a deficiency of one essential fatty acid and too much of others, so specific tests are needed to prescribe specific supplementation.

Sugar and artificial sweetners
If one thinks that a child is sensitive to sugar, then one might attempt a no sugar diet for a couple of weeks. A no sugar diet avoids sugar, corn syrup, fructose, dextrose, honey, and maple syrup. The child's symptoms may increase for the first few days without sugar. Then reintroduce sugar and observe their behavior before and after. Even if a child were not sensitive to sugar one would want to keep their consumption of sugar low. Foods that are high in sugar are usually low in important essential nutrients, as well as usually contain artificial colors and flavors, chocolate, and saturated, hydrogenated, and partially hydrogenated fats. In addition, if a child eats a large amount of sugar containing foods, this reduces their intake of high-density nutritious foods. Some suggestions to help avoid sugar in the diet, substitute all—fruit juices, fruit drinks or punches, with 100 percent unsweetened orange, grape, grapefruit, or tomato juice. One should choose frozen and canned unsweetened fruits and pure fruit juices as well as fresh fruits for desserts and use small amounts of all fruit jams. A nutritious breakfast should include high protein food (eggs, homemade sausage, fresh fruit, whole grain cereal or bread, and low fat milk). Most western children start the day with sugary cereal. Not only are these products high in sugar content, they also contain artificial colors and preservatives, which can all cause adverse reactions.

Artificial sweeteners. Some children with *ADHD* do not tolerate the artificial sweetener *aspartame* (Equal, Nutrataste, or Nutrasweet). Other children will tolerate small amounts. A child may be able to tolerate several ounces of sugar-free 7-Up or Sprite on special occasions. These citrus-based drinks are all sweetened with *aspartame* and are not artificially colored. A parent can make their own soda by using club soda with concentrated unsweetened juices. Another note on *aspartame* is that *aspartate* is a precursor to the most powerful excitatory neurotransmitter in the brain, *glutamate*. It is thought by some that by ingesting *aspartame* the breakdown into *aspartate* may be associated with an overproduction of this type of neurotransmitter which could produce over-excitation or even excito-toxicity of the nervous system. *Saccharin* and *acesulfame* are other popular artificial sweeteners that are chemically different from *aspartame*. A child may tolerate small amounts of these, although some children have mild intestinal distress or diarrhea from too much. In general, artificial sweeteners should be avoided.

Yeast sensitivities
Candida albicans is a common and usually harmless yeast that normally lives in the intestinal and urinary tracts in humans. It usually does not produce difficulty in a healthy child with a properly functioning brain and immune system. However, a child who has chronic infection, which may be associated with decreased function of the left hemisphere and in turn associated with immune responsiveness (Rich and McKeever, 1990; Flannery and Liederman, 1995; Phillips et al., 1995; Clow et al., 2003) may expose the child to many rounds of antibiotics. The chronic use of antibiotics provides a less than optimum environment for both bacteria normally present and necessary for effective digestion and mutant forms of the bacteria. Normal bacteria not only aid in digestion, but also create a certain pH of the intestine that helps suppress the growth of yeast. When the bacteria are suppressed, the yeast can grow unchecked and *Candida albicans* is therefore thought to play a role in a number of health problems. These include recurrent infection, fatigue, irritability, hyperactivity, and other neurological symptoms, like short attention span, "spaciness," and depression. It is also possible that some of these symptoms will also reflect a decrease in brain activation, especially of the left hemisphere that may be the prime cause of or be associated with other diseases or disorders co-existing with the yeast infection.

Inferential data of the likelihood of yeast infection can be obtained by determining whether a child received four or more prescriptions for antibiotics in the preceding year; whether the child had been on a prolonged course of antibiotics for a 4-week period or longer; whether the presenting symptoms (not the infection) worsened after treatment with antibiotics; whether the child craves sugar, and if the child complains or appears to have persistent digestive problems, including gas, bloating, and constipation or diarrhea. Several simple treatments can be used to treat a child's yeast problem. Avoid foods that promote the growth of yeast, such as sugar, honey, corn syrup, maple syrup, etc. Other possible interventions include:

1. Anti-fungal medications that can be cautiously prescribed. *Nystatin* powder seems to be a safe anti-fungal prescription medication that kills yeast in the intestines. However, *Nystatin* in tablet form also contains food dyes, as many pills and vitamins do, which can cause a reaction in children that are sensitive. In addition, the liquid suspensions, like many liquid prescriptions and over-the-counter medications, contain sugar leaving only one choice for children that are sensitive to dyes and sugars and that is the powdered form of *Nystatin*. However, this has a slightly unpleasant taste, so it may be advisable to prepare the medication in clear gels and tablets and place the powder in the tablets if a child is able to swallow pills. A child's symptoms may also worsen in the first few days.
2. Foods that contain yeast should be avoided because it is thought that some children may be allergic or sensitive to yeast. These foods would include breads, dried fruit, and cheese.
3. One of the best ways to treat yeast problems besides continuing the use of antibiotics and increasing brain activation, especially left brain stimulation therapies, is to restore the natural bacterial environment by reintroducing the symbiotic bacteria. This can be ingested in the forms of *Lactobacillus acidophilus* and *planterum*.
4. Many infections that children experience are viral, few are bacterial. Recent studies have indicated that antibiotic abuse is rampant among physicians, with an estimated 50,000 prescriptions for antibiotics being unnecessarily written in 1 year. A large percentage of these are for *otitis media* infections that are often viral. Not only does this unnecessary medication promote yeast infection, it creates stronger antibiotic resistance to mutant bacteria.

Vitamin and mineral supplements

Although supplements are generally not viewed as being medication, they are reportedly effective and necessary in most children with *autistic spectrum disorders*. Their use should be viewed in much the same way as the use of medication. This means that although a child may benefit from these treatments, most of the time, like medication, these supplements and remedies are treating the symptoms and not the root causes of the problem. If a child's body and nervous system are functioning and developing properly and the child receives a sufficient and nutritious diet, the child should not require the use of supplements, herbs, or homeopathic remedies. The most important factor in brain development is natural environmental stimulation. When the child is deprived of such stimulation, the brain does not develop and function adequately and it will not adequately regulate the breakdown and absorption of food. Even if the diet is proper, the child may ultimately be deficient in various vitamins and minerals. In some cases, a child may have a genetic predisposition to a certain mineral or vitamin deficiency and therefore that individual child may require supplementation to compensate for the inherited decrease in production, absorption, or inability to produce the relevant substances on their own. In these cases, the deficiencies may exacerbate the symptoms and even create additional symptoms. In this case, nutritional supplementation may be helpful to compensate for the inability of the body to absorb these nutrient forms from the diet or to produce them themselves. When this is properly prescribed and monitored, it may have anywhere from a mild to dramatic effect in alleviating the symptoms that a child has and improve learning ability and behavior. This may lead some to conclude that nutrition deficiency therefore is the primary cause of the symptoms and supplementing the diet is the cure. However, if after a course of supplementation, which is then discontinued, symptoms return, then the root problem is probably related to something else and nutritional intervention is palliative. The treatment can be continued, but the search and therapy for the underlying cause must be continued.

Many who use nutrition as a therapy are quick to criticize the use of medication because it is not "natural" and is riskier than the use of vitamins, minerals, or herbs. This may be true, but the philosophy of giving a child a pill or a powder to cure a problem can be as damaging if it takes the focus off the root problem, which is likely decreased physical activity, combined with poor nutrition, and inadequate environmental stimulation.

The maintenance of health by ensuring a proper balance of nutrients, vitamins, minerals, and amino acids in the body as an organized modality of intervention is an emerging clinical science. Extensive research in recent years has shown the far reaching impact of various nutrients, including vitamin C, vitamin B-6, vitamin B-12, magnesium, calcium, and others, have specificity for brain and neuronal growth and for nervous system function. Nutrients and supplements that have been shown, under controlled circumstances, to enhance brain function and cognitive abilities and improve behavior include:

1. *Vitamin A*. A powerful antioxidant, which helps to protect the membranes of brain cells, which can be damaged by *oxidative stress* and *free radical* production. It also aids circulation. It should be taken with *beta carotene*, another source of *vitamin A*. For most children a daily dose would be approximately 10,000 units.

2. *Vitamins B*. A group of vitamins essential for neuronal growth and stability. The four B vitamins that are most essential for brain function are *B-12, B-6, B-1*, and *folic acid*. B-12 deficiency is thought to result in cognitive defects or learning difficulties, poor memory, a decrease in reasoning skills, as well as behavioral symptoms. A deficiency of *B-12* is 300 percent more common in individuals who do not take vitamin supplements. A daily dose of *B-12* is anywhere between 100 to 1,000 mg, depending on age and size of the child. *B-6* is thought to aid in converting blood sugar into glucose, which is the brain's only fuel. It also protects blood vessels. *B-6* improves memory and is thought to boost the

efficiency of the immune system. For children, an appropriate dose is approximately 50 to 80 mg daily. *B-1 (thiamine)* is important for metabolic processes in the brain and peripheral nervous system. It is also a powerful *antioxidant* and it improves the ability of *B-6* and vitamin *E* to destroy free radicals. *B-1* deficiency can result in behavior and memory impairment and is implicated in Korsokoff's psychosis. A daily dose of 50 to 80 mg is considered appropriate. Vitamin *B-3 (niacin)* is used to manufacture neurotransmitters, convert carbohydrates to glucose, and lower cholesterol level. It also appears to have a calming effect in children because it is thought to increase the effects of the inhibitory neurotransmitter *GABA*. Niacin can cause a mild flushing and burning sensation of the skin, however another form of *B-3, niacinamide*, does not. A daily dosage of 100 mg is considered appropriate. Vitamin *B-5 (pantothenic* acid) is thought to be crucial to the synthesis of neurotransmitter *acetylcholine*, implicated in memory and the efficient function of the *autonomic* nervous system. *B-5* also assists in forming a protective sheath around the spinal cord. A severe deficiency of *B-5* can produce paralysis and certainly motor symptoms. A daily dose of 100 mg is considered appropriate. *Folic acid* has been noted to be helpful in decreasing depression, even in low doses, and in improving cerebral circulation. Behavioral and emotional symptoms appear to be significantly higher in people with low *folic acid* levels. *Folic acid* is particularly helpful at breaking down the chemical *homocysteine*, which is toxic to brain cells and nervous tissue. A daily dose of 400 mg is recommended.

3. *Vitamin C*. It is so important to brain function that its levels in the brain are almost 15 times higher than they are outside the brain. Vitamin *C* is thought to be the most powerful of all the *antioxidants*. It also enhances the antioxidant effect of other nutrients, especially vitamin *E*. Vitamin *C* is important in the manufacture of several neurotransmitters including *acetylcholine, dopamine*, and *norepinephrine*, all thought to be deficient in children with *ADHD*. Vitamin *C* also improves the responses of the immune system and arterial function. This vitamin should be taken twice a day; the dose is 500 mg each. Vitamin *E* is thought to be a powerful antioxidant and therefore protects brain cells from damage from *oxidative stress* and free radicals. Therefore, it can not only help prevent brain dysfunction, but can promote and improve neuronal development and function. Studies have shown that vitamin *E*, when taken with *selenium* can improve mood and cognitive function in people. Vitamin *E* also improves immune function. A daily dose is 400 Iu and because vitamin *E* is fat soluble, it can accumulate to toxic levels.

4. *Calcium*. Supplements are essential when a child dislikes milk or dairy products or is sensitive to dairy products. Calcium may also have a calming effect on behavior. The dosage is 500 mg of calcium carbonate liquid or chewable tablets. Too much calcium can decrease the absorption of other minerals.

5. *Magnesium*. It is thought to aid in the metabolism of neurons. Magnesium is also a powerful *antioxidant* that increases the effect of vitamin *E*. If a child is hyperactive, irritable, has difficulty sleeping, muscle twitching, and wets the bed, *magnesium* may be helpful. *Magnesium* is an important cofactor in a multitude of chemical reactions in the body. Refined foods are often deficient in *magnesium* content. Good food sources are whole grains, nuts, seeds, seafood, fresh vegetables, and fruits. *Magnesium* supplements may improve a child's behavior and decrease allergies. Magnesium has a calming effect on many children with *ADHD*. Two recent studies suggested (Kozielec, 1997; Starobrat-Hermalin, 1997) that *magnesium* deficiency is common in children with *ADHD* and that supplementation with *magnesium* "200 mg a day" resulted in significant decreases in hyperactivity. *Magnesium chloride* and *magnesium citrate* are two well-absorbed forms. Dosage of 200 to 600 mg a day is appropriate.

6. *Zinc*. It is also an essential mineral that works as a cofactor in many of the body's metabolic pathways. *Zinc* is thought to play a particularly important role in brain metabolism. It is considered part of an *antioxidant*

"chain reaction" that destroys free radicals. It also increases the strength and stability of neurons protecting them from damage. If a child has a loss of appetite, slow growth, and white spots on their fingernails, they may have a *zinc* deficiency. Good food sources include eggs, liver, shellfish, wheat germ, beef, dark meat, turkey, nuts, and seeds. Taking 10 mg of *zinc* together with a multi-mineral supplement for 2 months may improve behavior. Too much *zinc* can decrease the absorption of other vital minerals.

7. *Selenium.* It is thought to be the most powerful *antioxidant* mineral. It is especially effective in preventing the oxidation of fats. This is particularly important in the brain, since approximately 60 percent of the brain is composed of fat. *Selenium* also aids the immune system and is thought to improve circulation. *Selenium* has been shown to produce an anti-anxiety effect in some individuals. *Selenium* content is reduced in refined foods. Good food sources are seafood, liver, and meat. Grains are a good source only if they are grown in *selenium* rich soils. Dosage should be 100 mg in a preparation of *sodium selenate*. This dosage should not be exceeded as *selenium* may be toxic in large amounts.

In an exceedingly ambitious attempt to enroll parents of *autistic* children in evaluating treatment intervention strategies, Table 11.2 below presents the results of parent's behavioral ratings of drug and diet interventions including both the benefits and adverse effects. These ratings have been collected from 1967 to the present and represent the responses of 21,500 parents who have completed questionnaires for the *Autism Research Institute*, albeit under non-controlled conditions. For the purposes of Table 11.2, the parents responses are given on a six-point scale which had been combined into three categories: "made worse" (ratings 1 and 2), "no effect" (ratings 3 and 4), and "made better" (ratings 5 and 6). The "Better-Worse" column gives the number of children who "Got Better" for each one who "Got Worse." There are three sections: Drugs, Biomedical/Non-Drug/Supplements, and Special Diets.

Psychopharmacology

Pharmacotherapy and Autism

We have indicated throughout the work that neurobehavioral disorders have a remarkable comorbidity with many disorders of childhood (Jensen et al., 1997, 1999). As many as two thirds of elementary school-aged children with *ADHD*, referred for clinical evaluation, have at least one other diagnosable psychiatric disorder (Cantwell, 1996). Concomitant disorders include: conduct disorder (Klein et al., 1997; Riggs et al., 1998), oppositional defiant disorder (Eiraldi et al., 1997), learning disorders, anxiety disorders (Diamond et al., 1999), and mood disorders, especially depression (Garrison et al., 1997). The genetic, neurochemical, and neurophysiological underpinnings have been explained throughout the text for both *ADHD* and other neurobehavioral disorders on the *autistic* spectrum. No wonder targeted drug intervention is not available. Table 11.2, while highly subjective, reports parents' qualitative responses of symptom reduction, over many years.

Clearly emerging, as subjectively effective for improving *autistic* symptoms are select anticonvulsants, and for *ADHD*, stimulant drugs that improve behavior and learning ability in 60 to 80 percent of correctly diagnosed children. The primary drugs employed in treatment are methylphenidate, amphetamines, and pemoline, with buspirone, and buporion (Welburtin) used in refractory cases. Of these, methylphenidate accounts for 90 percent of prescribed medication.

There are several reasons for the use of anti-epileptic drugs in *autistic spectrum disorders*, including the high incidence of epilepsy in these individuals, the anecdotal reports suggesting an improvement of communication and behavior in *autistic* subjects with epileptic discharges, and the increased awareness that some disruptive behaviors may be manifestations of an associated affective disorder. Di Martino and Tuchman (2001) report a study on the current use of anti-epileptic drugs in the treatment of *autism*, and on the association of affective disorders with

TABLE 11.2 Autism Research Institute Parent Ratings of Behavioral Effects of Biomedical Treatment Interventions (1967–2003)

Biomedical Treatment Interventions	Behavioral Effects				
	Got Worse (%)[a]	No Effect (%)	Got Better (%)	Better: Worse	No. of Cases[b]
Biomedical/Non-Drug/Supplements					
Vitamin A	2	59	39	22:1	334
Calcium[d]	2	62	35	14:1	988
Cod liver oil	3	51	46	14:1	411
Colostrum	6	58	37	6.7:1	163
Detox. (chelation)	3	28	70	27:1	116
Digestive enzymes	4	44	52	14:1	314
Di-methyl-glycine (DMG)	7	51	42	5.9:1	4547
Fatty acids	4	44	51	12:1	299
5 HTP	11	55	35	3.3:1	66
Folic acid	4	55	41	11:1	1100
Food allergy Treatment	4	37	59	14:1	290
Magnesium	6	65	29	5.2:1	288
Melatonin[e]	10	33	57	5.9:1	302
Pepcid	9	61	30	3.2:1	64
SAMe	25	46	29	1.1:1	28
St. Johns Wort	11	67	22	2.0:1	46
Tri-methyl-glycine (TMG)	14	42	44	3.1:1	182
Transfer factor	18	51	31	1.7:1	39
Vitamin B3	5	55	41	9.0:1	487
Vitamin B6 alone	7	64	29	4.1:1	590
Vitamin B6 & Magnesium	4	49	46	11.1:1	5079
Vitamin C	2	59	39	16:1	1306
Zinc	3	55	43	17:1	835
Special Diets					
Candida diet	3	45	52	18:1	605
Feingold diet	2	47	51	23:1	645
Gluten/Casein-free diet	4	33	64	18:1	724
Removed chocolate	1	50	49	36:1	1491
Removed eggs	2	61	37	21:1	882
Removed milk products/dairy	2	51	48	30:1	4950
Removed sugar	2	51	47	24:1	3392
Removed wheat	2	53	46	26:1	2701
Rotation diet	2	50	47	20:1	678
Drugs					
Aderall	39	28	34	0.9:1	285
Amphetamine	47	28	25	0.5:1	1174
Anafranil	31	37	31	1.0:1	351
Antibiotics	30	59	11	0.4:1	1617
Antifungals[c]: diflucan	7	42	51	7.2:1	185
Antifungals[c]: nystatin	5	48	47	10:1	727
Atarax	26	53	21	0.8:1	443
Benadryl	24	51	25	1.1:1	2512
Beta blocker	18	49	33	1.8:1	236
Buspar	26	45	30	1.2:1	281
Chloral hydrate	41	37	22	0.5:1	375
Clonidine	21	31	48	2.2:1	1090
Clozapine	44	39	16	0.4:1	79

TABLE 11.2 Continued

	Behavioral Effects				
Biomedical Treatment Interventions	Got Worse (%)[a]	No Effect (%)	Got Better (%)	Better: Worse	No. of Cases[b]
Cogentin	19	53	28	1.4:1	149
Cylert	45	35	21	0.5:1	580
Deanol	15	55	29	1.9:1	195
Depakene: behavior	25	43	32	1.3:1	871
Depakene: seizures	12	30	57	4.6:1	569
Desipramine	38	25	38	1.0:1	61
Dilantin: behavior	28	48	24	0.9:1	1049
Dilantin: seizures	14	36	51	3.8:1	377
Felbatol	26	45	29	1.1:1	38
Fenfluramine	21	51	28	1.4:1	453
Halcion	37	30	33	0.9:1	43
Haldol	37	27	35	0.9:1	1119
IVIG	13	45	42	3.2:1	31
Klonapin: behavior	28	33	38	1.4:1	156
Klonapin: seizures	38	50	12	0.3:1	26
Lithium	27	42	31	1.1:1	384
Luvox	28	36	37	1.3:1	120
Mellaril	28	38	33	1.2:1	2023
Mysoline: behavior	44	40	15	0.3:1	131
Mysoline: seizures	19	58	23	1.2:1	57
Naltrexone	22	46	32	1.5:1	200
Paxil	27	28	45	1.7:1	192
Phenergan	30	44	26	0.9:1	244
Phenobarbitol: behavior	47	37	16	0.3:1	1052
Phenobarbitol: seizures	17	43	40	2.4:1	458
Prolixin	34	34	33	1.0:1	83
Prozac	31	33	36	1.2:1	975
Risperidal	19	28	53	2.8:1	401
Ritalin	44	26	29	0.7:1	3540
Secretin: intravenous	8	43	49	6.2:1	217
Secretin: transdermal	12	47	41	3.6:1	78
Stelazine	28	44	27	1.0:1	415
Tegretol: behavior	24	45	31	1.3:1	1345
Tegretol: seizures	12	33	55	4.5:1	721
Thorazine	36	40	24	0.7:1	897
Tofranil	30	37	33	1.1:1	698
Valium	36	41	23	0.7:1	788
Zarontin: behavior	34	43	22	0.7:1	129
Zarontin: seizures	21	51	29	1.4:1	87
Zoloft	33	31	36	1.1:1	212

[a] "Worse" refers only to worse behavior. Drugs, but not nutrients, typically also cause physical problems if used long-term.
[b] No. of cases is cumulative over several decades, so does not reflect current usage levels (e.g., Haldol is now seldom used).
[c] Antifungal drugs are used only if *autism* is thought to be yeast-related.
[d] Calcium effects are not due to dairy-free diet; statistics are similar for milk drinkers and non-milk drinkers.
[e] Caution: While melatonin can benefit sleep and behavior, its long-term effects on puberty are unknown.

Note: *For seizure drugs:* The first line shows the drug's behavioral effects; the second line shows the drug's effects on seizures.

epilepsy and *autism*. The evidence supporting the hypothesis that there may be a subgroup of *autistic* children with epilepsy and affective disorders that preferentially respond to anti-epileptic drugs is, according to these authors preliminary. The authors note that evidence exists that *autism, epilepsy*, and affective disorders commonly co-occur, and that they may share a common neurochemical substrate, which is the common target of the psychotropic mechanism of action of different anti-epileptic drugs. This may then explain the data reported in Table 11.2. The authors summarize the literature on this topic and present an overview in Table 11.3 below.

ADHD and *autistics* seem to have different responses to stimulant medication. Stimulants such as *amphetamines* usually result in increases in activity, irritability, explosiveness, and stereotypic behavior when administered to *autistic* patients. An exception to this lack of effectiveness of stimulants is the moderate to high functioning *autistics* who have *ADHD*-like symptoms. This type of patient does benefit from stimulant medication.

TABLE 11.3 Anti-epileptics in Patients with *Autism* and Related Conditions

Patient	Age	Diagnosis	Seizure	Abnormal EEG	AED	Epilepsy Control	Behavior Improvement	Reference Citation
3	Child	Asperger ADHD (2) BPD (1)	–	–	VPA	/	+	[21]
2	Child	AUT	–	+	VPA	+	+	[20]
3	Child	AUT	–	+	VPA	?	+	[18]
1	Child	PDD	–	+	VPA	+	+	[15]
2	Child	AUT	–	+	VPA	+	+	[12]
2	Adult	AUT BPD (1) RCD (1)	–	–	VPA	±	+	[14]
1	Child	Asperger	+	+	CBZ	±	–	[19]
2	Child	AUT TS	+ (1)	+	CBZ	+	+	[16]
2	Adolescent	AUT BPD	1+	1+	CBZ	+(1)	+	[13]
13	Child/adolescent	AUT (12) Rett (1)	+	+	LMG	*	+ (8)	[17]

*After 4-month treatment (including the 13 *autistic* patients): 5 of 50 (11%) were seizure free; 16 of 50 (36%) had more than 30% reduction of seizures; 24 of 50 (53%) had no improvement.

Abbreviations:
Abnormal EEG = Presence of focal spikes or sharp waves
ADHD = Attention-deficit-hyperactivity disorder
AED = Anti-epileptic drugs
AUT = Infantile autism
BPD = Bipolar disorder
CBZ = Carbamazepine
LMG = Lamotrigine
PDD = Pervasive developmental disorder
RCD = Rapid cycling disorder
TS = Tuberous sclerosis
VPA = Valproic acid
+ = Present
– = Not present
± = Temporal seizure control
Source: (From Di Martino, A. and Tuchman, R. F. (2001). Anti-epileptic drugs: Affective use in *autism spectrum disorders. Pediatric Neurology*, 25, 199–207.)

In attempting to overview the literature on the clinical pharmacology of autism treatment, we can group the findings into the categories of controlled studies of *dopamine antagonists*, of *5-HT agonists, fenfluramine,* and controlled studies of *naltrexone* summarized in Tables 11.4–11.6.

The administration of conservative doses of *haloperidol* to *autistic* children results in significant decreases in hyperactivity, negativism, and stereotypes and in some cases has been shown to facilitate learning. On followup, *haloperidol* is therapeutically effective for up to $4\frac{1}{2}$ years and has helped many *autistic* children remain with their families as well as remain in educational programs without producing any adverse effects on IQ. Although *haloperidol* appears to be useful in the treatment of *autism*, the utility of the drug is limited by the risk of *tardive dyskinesia*. It would seem appropriate that future research with *neuroleptics* in the treatment of *autism* be targeted at drugs that do not cause *tardive dyskinesia* such as *clozapine* or more practically one of its congeners that does not cause bone marrow suppression or sedation.

Fenfluramine, as indicated in Table 11.4, may beneficially affect *autistic* patients by its ability to decrease CNS *serotonin* or *beta-endorphin* levels. However, the studies suggest that only a minority of patients were benefited by the therapy. It seems that too small a dose of the drug is being utilized. Thus future studies and treatments should be aimed at using doses greater than 1.5–2.0 mg/kg/d.

In the late 1970s, it was noted that neonatal rats and chicks exposed to high levels of opiates showed autistic-like withdrawal after they were born. Opiate treated animals exhibit

TABLE 11.4 Controlled Studies of *DA* Antagonists in Autistics

Study	Dose (mg/d)	Population	Results
Anderson et al., 1984	HLP 0.5–3.0 mg/d Placebo 4 week x-over i.e., PHP vs. HPH	Autism 29 males, 11 females 2–7 yr mild–profound MR	Improvement: Conners, CGI, Children's Rating Scale Performance: on HLP @ level 20 IQ pts higher
Anderson et al., 1989	HLP 0.16–0.18 mg/kg/d Placebo 3 week x-over i.e., HPP vs. PHP vs. PPH	Autism 35 males, 10 females 2–8 yr borderline–profound MR	Improvement: Conners (temper outbursts), CGI, Children's Rating Scale (trend) Cognition: HLP didn't facilitate or adversely effect learning decrease in hyperactivity, tantrums, withdrawal, & stereotypes increase in calm & relatedness

TABLE 11.5 Controlled Studies of *5-HT* Agonist, *Fenfluramine*, in *Autistics*

Study	Dose (mg/d)	Population	Results
August et al., 1984	placebo × 2 weeks FFA 1.5 mg/kg/d × 16 weeks placebo × 2 weeks	10 autistics 8 controls matched for age & sex 5–13 yr	decreased 5-HT by 62% Conners—improvement in hyperactivity, affect, & distractibility No effect on WISC IQ score
Yarbrough et al., 1987	FFA 2 mg/kg/d (max = 120 mg/d) placebo x-over × 15 weeks each	Autism 17 males 3 females 9–28 yr IQ = 12–51	Real Life Rating Scale: no effect ADRs (65%)—agitation, tenseness, insomnia & withdrawal ADRs
Ross et al., 1987	FFA 1.5 mg/kg/d (max = 60 mg/d) placebo x-over × 16 weeks each	Autism 8 males 1 females 3–12 yr IQ = 31–83	decreased bE not significant (low dose = ?) 3 responders decreased echolalia, motor disturbances perseveration & increased attention, social awareness ADRs: lethargy, weight loss
Beeghly et al., 1987	FFA 1.5 mg/kg/d placebo x-over × 4 mo each	Autism 7 males 2 females 7–14 yr IQ = 36–112	7 completers decreased 5-HT by 58% Conners & Real Life Rating Scale no change 2 responders (highest blood levels)
Leventhal et al., 1993	FFA (max = 60 mg/d) vs. PLB over 62 weeks	n = 15	hyperactivity decreased according to parents but not teachers an equivocal response at best

unusual motor flurries much like autistic children's hyperactivity. Additionally, they exhibited other unusual postures and perseverative behaviors and fail to evince normal separation anxiety when removed from their mothers. It is speculated that disturbances in brain opioid levels may block psychosocial development at its earliest stages leading to failures in language acquisition and other idiosyncrasies in learning. The theory as to why abnormal opioid levels cause self-injury, and why opiate antagonist drugs may decrease self-injurious behavior (SIB) is explained by the hypothetical "addiction theory." The theory proposes that the purpose of SIB is to promote pain-induced release of endogenous opiates. The use of the opiate antagonist, *naltrexone* in the treatment of *autism* is reasonable since it antagonizes endogenous *opiate* receptor activity. There are a large number of uncontrolled reports supporting the effectiveness of *naltrexone* in the treatment of *autism*. However, the five controlled trials that are available are far less encouraging. From these data we can conclude that despite encouraging anecdotal reports, there are now 8 double-blind placebo controlled trials that conclude that *naltrexone* is at best minimally effective in the treatment of *autism*. Naltrexone should not be utilized as a first line drug in the treatment of *autism*.

Pharmacotherapy and obsessive–compulsive disorder

In 1997 the results of the Expert Consensus Panel for Obsessive Compulsive Disorder was published as a supplement to the Journal of Clinical Psychiatry. As a part of this panel 79 physicians, who were considered "experts" in the field of *OCD* treatment, were asked to complete a survey. Their responses were then

TABLE 11.6 Controlled studies of *naltrexone* in autistics

Study	Dose (mg/d)	Population	Results
Sandman 1988	NTX 25, 50, and 100 mg po vs. placebo	4 autistics with severe to profound MR 23–36 yr 4 males	SIB decreased possible correlation between NTX dose and SIB decrease
Campbell et al., 1990	NTX 0.5–1.0 mg/kg/d or placebo × 21 days	18 autistics 3–8 yr profound to borderline MR	Improvement on global rating but only slight improvement in hyperactivity & fidgeting No ADR on learning
Scifo et al., 1991	NTX 0.5, 1.0, 1.5 mg/kg/d, and placebo × 5 weeks	12 autistics 7–16 yr 10 males	decreased autistic symptoms 7/12 (58%) no correlation [b-E]: clinical condition.
Zingarelli et al., 1992	NTX 0.6–1.1 mg/kg/d or placebo x-over × 21 days each	8 Autism 6 SIB 5 males 3 females 19–39 yr	No effect on SIB or mannerisms bE levels went up
Campbell et al., 1993	NTX 0.5–1.0 mg/kg/d vs. placebo parallel × 21 days	41 autistics 2.9–7.8 yr 10 males	NTX = placebo decreased hyperactivity, no effect on SIB or learning
Kolmen et al., 1995	NTX 1.0 mg/kg/d vs. placebo crossover × 14 days each	13 autistics 3.4–8.3 yr 12 males	parent global, impulsivity/ hyperactivity, and restlessness improved only teacher global rating improved 8/13 improved in home, school, and clinic setting, i.e., 2/3 settings
Bouvard et al., 1995	NTX 0.5 mg/kg/d vs. placebo crossover × 28 days each	10 autistics 5–14 yr 5 males	only 4/10 had a "strong response" responders at baseline had elevated vasopression and serotonin levels responders had robust decrease in beta-E levels There is a carryover effect of NTX that requires 1 month washout in crossover design studies
Willemsen-Swinkels et al., 1995	NTX 50 or 100 mg/d vs. placebo crossover × 28 days each 4 week washout between txs	32 mental retardates 7 autistics 16 autistic + SIB 9 SIB 18–46 yr 27 males	a heterogeneous population that was unable to exhibit any beneficial effect of NTX on SIB and autistic behavior

tallied and used to design clinical treatment guidelines. As part of these guidelines, an algorithm for the treatment of *OCD* in the acute phase was designed. This algorithm concluded that for more severe *OCD* the treatment of choice in the adult population is *CBT* plus a *serotonin reuptake inhibitor (SRI)* or an *SRI* alone. For adolescents and children the therapy of choice is *CBT* with the addition of an *SRI* if the patient desires. Despite the use of the *SRI*'s for the treatment of *OCD*, only 40–60 percent of patients become clinically improved with adequate medication trials. Oftentimes, even the patients that do respond to medication are not completely symptom-free. The following is a summary of the clinical studies that have been done investigating the use of medication for the treatment of *OCD*. Of the *SRI*'s that are mentioned all but one, *venlafaxine*, have been approved by the *US Food and Drug Administration* for the treatment of *OCD*.

TABLE 11.7 Comparison of Clinical Trials for *OCD*

Study	Diagnostic Criteria	Design	Drug (mean dose/day) × duration	Efficacy Measures	Outcome
Clomipramine					
Marks et al., 1980	Non-specified Chronic patients $n = 40$	CMI vs. PLB	CMI (145 mg) × 36 wks	Individualized rating scale, HAMD, Wakefield Inventory, General adjustment scale	CMI > PLB on Wakefield ($p < 0.005$), HAMD ($p < 0.002$), anxiety ratings ($p < 0.03$), compulsions ($p < 0.02$), and leisure, family and social adjustment ($p < 0.03$)
Mavissakalian et al., 1985	DSM-III $n = 16$	CMI vs. PLB	CMI (228.5 mg) × 8 wks	OCNS, HAMD	43% CMI responders, 0% PLB responders
Flament et al., 1985	DSM-III $n = 23$ (children / adolescents)	CMI vs. PLB	CMI (141 ± 40 mg) × 10 wks	LOI-CV, OCRS, CPRS, NIMH-OC, BPRS, NIMH self rating scale	CMI > PLB LOI-CV ($p < 0.05$), OCRS ($p = 0.007$), NIMH-OC ($p = 0.02$)
Marks et al., 1988	DSM-III and ICD-9, $n = 25$	CMI vs. PLB	CMI (127–157 mg) × 27 wks	Individualized rating scale, HAMD, BDI, Wakefield Inventory, General adjustment scale	CMI > PLB for target ritual ($p < 0.04$), global rituals ($p = 0.01$), behavioral avoidance ($p = 0.04$), depression ($p = 0.03$), and social leisure ($p = 0.006$)
Jenike et al., 1989	DSM-III $n = 27$	CMI vs. PLB	CMI up to 300 mg × 10 wks	YBOCS, NIMH-OC, HAMD, global assessment of OC symptoms	CMI > PLB all measures except HAMD ($p < 0.001$)
Mavissakalian et al., 1990	DSM-III $n = 25$	CMI vs. PLB	CMI (273.1 ± 43.9 mg) × 10 wks	OCNS, MOCI, LOCQ, ZAS, BDI, YBOCS, NIMH-OC	CMI > PLB all measures except MOCI subscales ($p < 0.05$)
CCSG 1991	DSM-III $n = 520$	CMI vs. PLB	CMI (218.8–34.5 mg) × 10 wks	YBOCS, NIMH	CMI > PLB YBOCS ($p < 0.001$), NIMH ($p < 0.001$)
DeVeaugh-Geiss et al., 1992	DSM-III $n = 60$ (children / adolescents)	CMI vs. PLB	CMI up to 200 mg or 3 mg/kg × 8 wks	YBOCS, NIMH-OC	CMI > PLB YBOCS ($p < 0.05$), NIMH-OC ($p < 0.05$) at week 3
Fluvoxamine					
Perse et al., 1987	DSM-III $n = 16$	FVM vs. PLB	FVM up to 150 mg × 8 wks crossover	SCL-90, OCC, MOCI, HAMA, HAMD, BDI	FVM > PLB SCL-90 ($p < 0.46$), HAMD ($p = 0.007$), HAMA ($p = 0.018$), GRS ($p < 0.045$)
Goodman et al., 1989	DSM-III $n = 42$	FVM vs. PLB	FVM (255 ± 60 mg) × 6–8 wks	YBOCS, HAMD, HAMA, CGI	FVM > PLB YBOCS ($p < 0.05$) week 2, HAMD ($p < 0.05$) week 3
Cottraux et al., 1990	DSM-III $n = 44$	FVM vs. PLB	FVM up to 300 mg × 24 wks	Four target ritual ratings, duration of rituals, ritual	FVM > PLB ritual duration ($p = 0.02$)wk 8, FVM > PLB HAMD and

TABLE 11.7 *Continued*

Study	Diagnostic Criteria	Design	Drug (mean dose/day) × duration	Efficacy Measures	Outcome
				improvement, compulsion checklist, HAMD, MADRS, BDI	MADRS ($p < 0.05$) wk 24, FVM = PLB all scales ($p =$ ns) week 48
Jenike et al., 1990	DSM-III $n = 38$	FVM vs. PLB	FVM up to 300 mg × 10 wks	YBOCS, NIMH-OC, CGI	FVM > PLB YBOCS ($p = 0.03$) wk 6, NIMH-OC ($p = 0.05$) wk 4
Goodman et al., 1996	DSM-III-R $n = 156$	FVM vs. PLB	FVM (215–245 mg) × 10 wks	YBOCS, NIMH-OC, CGI	FVM > PLB all measures ($p < 0.05$) wks 4 and 6
Chouinard et al., 1990	DSM-III without depression $n = 87$	STL vs. PLB	STL (180 mg) × 8 wks	YBOCS, NIMH, MOC, CGI	STL > PL (NIMH score, YBOCS Total score, CGI) ($p < 0.05$)
Jenike et al., 1990	DSM-III without depression $n = 19$	STL vs. PLB	STL (200 mg) × 10 wks	YBOCS, NIMH, MOC, CGI	STL = PLB all measures
Griest et al., 1992	DSM-III-R $n = 325$	SRT vs. PLB	STL fixed dose (50 mg, 100 mg, 200 mg) × 12 wks	YBOCS, NIMH, CGI	STL > PLB all measures ($p < 0.02$) wks 2 and 4
Paroxetine					
Wheadon et al., 1993	DSM-III-R $n = 348$	PXT vs. PLB	PXT fixed dose (20 mg, 40 mg, 60 mg) × 12 wks	YBOCS, NIMH, CGI	PXT (40 mg, 60 mg) > PLB YBOCS ($p < 0.017$) wk 4, NIMH ($p < 0.017$) wk 3 PXT (40 mg, 60 mg) > PXT 20 mg YBOCS ($p < 0.017$), NIMH ($p < 0.017$) wk 12
Fluoxetine					
Riddle et al., 1992	DSM-III-R $n = 14$ (children / adolescents)	FLX vs. PLB	FLX up to 20 mg/day × 8 or 12 wks	CY-BOCS, CGI-OCD, CGAS, LOI-CV, RCMAS, CDI	FLX > PLB CGI-OC ($p = 0.01$)
Tollefson et al., 1994	DSM-III-R $n = 355$	FLX vs. PLB	FLX 20 mg, 40 mg, 60 mg × 13 wks	YBOCS, HAMD, CGI-Severity and Improvment scales, CPRS, PGI	FLX (all doses) > PLB YBOCS ($p < 0.001$)
Venlafaxine					
Zajecka et al., 1990	Case study	VLF	VLF 375 mg/day × 8 wks	NIMH-OC	N/A
Non-SRI					
Foa et al., 1987	Non-specified OCD patients $n = 40$	IMI vs. PLB	IMI (233 mg) × 6 wks	BDI, OC symptoms: feared situation and compulsive behaviors	IMI = PLB all measures IMI > PLB self rated fear ($p < 0.04$)
Pigott et al., 1992	DSM-III-R $n = 21$	TRZ vs. PLB	TRZ (235 ± 10 mg) × 10 wks	YBOCS, NIMH-OC, HAMD	TRZ = PLB all measures

TABLE 11.7 Continued

Study	Diagnostic Criteria	Design	Drug (mean dose/day) × duration	Efficacy Measures	Outcome
Comparator Trials					
Thoren et al., 1980	RDC $n = 24$	CMI vs. NTP	CMI and NTP 150 mg × 5 wks	CPRS, LOI, HIS-WIS, ISS, and MADRS	CMI = NTP in OCD scores: CMI (42%), NTP (21%), PLB (7%) CMI > PLB CPRS ($p < 0.05$) wk 5
Ananth et al., 1981	Non-specified OCD neurosis $n = 20$	CMI vs. AMI	CMI and AMI up to 300 mg × 4 wks	Psychiatric Questionnaire for OCN and CGI	CMI = AMI OCN Questionnaire (p = ns)
Insel et al., 1983	DSM-III $n = 13$	CMI vs. CGL	CMI up to 300 mg and CGL up to 30 mg × 6 wks crossover	CPRS, OCRS, NIMH-OC, HAMD, LOI, Profile of mood states, compulsion checklist, BDI, and NIMH-side effect questionnaire	CMI > CGL CPRS ($p < 0.05$) and OCRS ($p < 0.05$)
Volavka et al., 1985	DSM-III $n = 16$	CMI vs. IMI	CMI and IMI up to 300 mg × 12 wks	SROCPI, CRONS, OCRS, HAMD, and GEE	CMI > IMI on GEE ($p < 0.05$) and HAMD ($p < 0.05$)
Zohar and Insel, 1987	DSM-III $n = 10$	CMI vs. DMI	CMI (235 ± 76 mg) and DMI (290 ± 32 mg) × 6 wks crossover	NIMH-OC, CPRS-OC, HAMD	CMI > DMI CPRS-OC ($p < 0.05$) an NIMH-OC ($p = 0.03$) at wks 4 and 6
Leonard et al., 1988	Non-specified OCD $n = 21$ (children / adolescents)	CMI vs. DMI	CMI and DMI up to 3 mg/kg × 5 wks crossover	Global OCD, NIMH-OC, and global depression scale	CMI > DMI Global OCD ($p = 0.002$), NIMH-OC ($p = 0.004$), global depression scale ($p = 0.03$) at wk 3
Leonard et al., 1989	DSM-III $n = 49$ (children / adolescents)	CMI vs. DMI	CMI (150 ± 53 mg) and DMI (153 ± 55 mg) × 5 wks crossover	LOI-CV, OCRS, CPRS, NIMH-OC, BPRS, and HAMD	CMI > DMI all scales except LOI-CV ($p < 0.006$)
Piggott et al., 1990	DSM-III-R $n = 11$	CMI vs. FLX	CMI (209 ± 13 mg) and FLX (75 ± 4 mg) × 10 wks crossover	YBOCS, NIMH, HAMD	CMI = FLX all measures CMI > FLX side-effects ($p < 0.05$)
Pato et al., 1991	DSM-III-R $n = 20$	CMI vs. BUS	CMI (225 ± 49 mg) and BUS (58 ± 7 mg) × 6 wks crossover	YBOCS, YBOCS-severity, YBOCS-global, NIMH-OC, HAMD	CMI = BUS, % improved: CMI (67%), BUS (56%)

TABLE 11.7 *Continued*

Study	Diagnostic Criteria	Design	Drug (mean dose/day) × duration	Efficacy Measures	Outcome
Vallejo et al., 1992	DSM-III n = 26	CMI vs. PLZ	CMI up to 225 mg and PLZ up to 75 mg × 12 wks	LOI, VOPI, OCIC, MOCI, GES, HAMD, HAMA, EPI-N, EPI-E	CMI = PLZ for all measures except LOCI, MOCI, EPI-N, and EPI-E
Hewlett et al., 1992	DSM-III-R n = 28	CMI vs. DIPH vs. CLO vs. CZP	CMI (239 mg), DIPH (237 MG), CLO (0.83 mg), CZP (6.85 mg) × 6 wks crossover	YBOCS, HAMD, HAMA	CMI > DIPH (YBOCS, $p < 0.05$), CZP > DIPH (YBOCS, $p < 0.05$)
Freeman et al., 1994	DSM-III-R without depression n = 64	FVM vs. CMI vs. PLB	FVM and CMI 250 mg × 10 wks	YBOCS, NIMH score, CGI-I	FVM = CMI for all efficacy measures % reduction in YBOCS: CMI (31%), FVM (33%)
Koran et al., 1996	DSM-III-R n = 79	CMI vs. FVM	CMI (201 mg) and FVM (255 mg)	YBOCS, NIMH-OC, CGI, HAMD	CMI = FVM on all measures % responders: CMI (54%), FVM (56%)
Goodman et al., 1990	DSM-III-R n = 40	FVM vs. DMI	FVM (223 ± 48 mg) and DMI (214 ± 55 mg) × 8 wks	YBOCS, HAMD, CGI	FVM > DMI YBOCS ($p < 0.02$) wk 7
Prasad, 1984	Non-specified OCD neurosis n = 6	ZMD vs. IMI	doses not reported × 4 wks	LOI, MADRS	ZMD > IMI LOI (p-value not given)

The expert consensus guidelines for the treatment of *OCD* indicates that the *SRI*'s are the most effective for the treatment of OCD and they recommend all five *SRI*'s (*fluvoxamine, fluoxetine, clomipramine, sertraline*, and *paroxetine*) as first line treatment. If the patient fails to respond to the first medication prescribed, the panel recommended gradually increasing the dose to the maximum recommended by the manufacturer within 4–8 weeks from the start of treatment. If the patient experiences a partial response to the medication, the dose should be increased to the maximum by 5–9 weeks from the start of treatment. Overall, the experts think that a trial of 8–13 weeks of adequate medication treatment is necessary before switching to an additional agent or adding an augmenting agent. Additionally the panel suggested that *CBT* be offered to every patient with *OCD*.

Pharmacotherapy of Tourette's Syndrome

The most common approach to treatment is the use of low dose neuroleptics (*haloperidol, pimozide*, and *risperidone*) because of their effectiveness in reducing tics. Besides the *neuroleptics, clonidine*, and *nicotine* have been studied in controlled trials to assess their efficacy. Additionally, there are positive case reports suggesting that *clonazepam, beta-blockers*, and *calcium-channel blockers* are of potential utility in the treatment of

Tourette's. However, we are only considering the controlled efficacy drug trials in the treatment of *Tourette's* syndrome (Shapiro et al., 1989).

The controlled drug trials in the treatment of *Tourette's* syndrome are presented in Table 11.8. *Haloperidol* and *pimozide* are effective in 70 percent of patients. However, because of a more conservative approach of reserving it for severe or refractory cases because of its risk of *extra-pyramidal* symptoms (EPS), *risperidone* might be a useful alternative *neuroleptic.* However, controlled trials are still not completed. *Pimozide* appears to have a lower risk of *extra-pyramidal* symptoms compared to *haloperidol.* However, it should be remembered that many patients respond to *haloperidol* doses (1–2 mg/d) that are low enough such that EPS are not a problem. The use of *nicotine* on an as and when needed basis could potentate the action of *haloperidol* during stressful periods allowing the use of lower doses, which may reduce the occurrence of EPS.

TABLE 11.8 Controlled Pharmacotherapy Studies in *Tourette's* syndrome

Study	Dose	Population	Results
Shapiro et al., 1989	HLP 0.08 mg/kg/d (18), PMZ 0.18 mg/kg/d (20) placebo (19) × 6 weeks 55 in x-over f-up	57 TS (DSM-III)	HLP sl > PMZ > PLB ADRs: HLP = PMZ > PLB akinesia, depression, cognitive dulling, decreased motivation, increased appetite & weight PMZ QT$_c$ increased WNL
McConville et al., 1989	10 HLP 1–10 mg/d & tx-refractory 9 Tx- naive 2 mg nicotine gum 10 HLP 5 Tx-naive placebo gum 4 Tx-naive	19 TS (DSM-III) 8–46 yr 10 tx refractory 24–300 months 9 never tx'd 2–6 months	HLP patients: decreased frequency & severity × 60 min Gum patient: decreased frequency & severity × 30 min Placebo: no effect
Sandor et al., 1990	HLP 5.5 ± 4 mg/d, PMZ 9 ± 6 mg/d, no drugs 1–15 yr f-up	33 TS 9–50 yr	Moderate-marked relief: HLP 78%, PMZ 70% TSGS: HLP = PMZ > PLB ADRs: HLP higher dc rate, more EPS, neither ECG delta
Gadow et al., 1992	MPD 0.2., 0.6, and 1.0 mg/kg/d vs placebo × 2 wk each double blind crossover × 8 weeks washout > or = 1 week	ADHD + TS or chronic motortics n = 11 males 6–12 yr	MPD decreased vocal tics at 0.6 and 1.0 mg/kg/dose but not 0.2 motor tics not improved but best response was to the 0.6 mg/kg/d dose
Kurlan et al., 1993	FLX 20–40 mg/d vs. placebo × 4 months double-blind, placebo controlled parallel design no washout	OCD (17) TS (4) MDD (2) trichotillomania (1) 10–18 yr	FLX = placebo for OCD and TS symptoms 6/11 subjects took either haloperidol or clonidine during the course of the study
Goetz et al., 1987	CLD 7.5–15 mcg/kg/d vs. placebo crossover design, 1 week washout	TS n = 23 males, 7 females 8–62 yr	19/30 required neuroleptics no effect on tics
Leckman et al., 1991	CLD 3.2–5.7 mcg/kg/d vs. placebo parallel design, 2 week washout	TS n = 40 7–48 yr	motor tics improved the most vs. placebo, 26% vs. 11%

The value of the *SRI*'s to treat the obsessions and compulsions associated with *Tourette's* remains to be resolved. Clinically they appear effective despite the finding in a small controlled trial that they are not. Patients with *Tourette's* appear to be responsive to *alpha₂ agonists*, although not nearly to the extent of the *neuroleptics*. *Clonidine* or *guanfacine* potentially may be the drug of choice in patients with concomitant *ADHD* and tic disorders. The paucity of *clonidine*-induced adverse drug reactions is impressive. The *calcium-channel blockers* represent a promising new class of drugs. Although more research needs to be conducted in this area, a trial of *verapamil* or *nifedipine* could be considered in a refractory patient or one unwilling to take other drugs because of their side effect profile.

Pharmacotherapy of Attention-Deficit Hyperactivity Disorder

Stimulants
While CNS stimulant medications currently are the drugs of choice, *tricyclic antidepressants* are also useful. Stimulants (*norepinephrine* or *dopamine* agonists) have been shown to help the symptoms of *ADHD*. Parents, teachers, and clinicians rate 75 percent of children with *ADHD* to be improved on stimulants, compared to 18 percent of placebo-treated children (Green, 1992). Approximately 20–25 percent of those who respond poorly to one medication will respond positively to another (Dulcan, 1990). Importantly, the psychopharmacology literature provides no agreement about how much improvement is required for a child to qualify as a "clinical responder." Stimulants tend to decrease physical activity, particularly during times when children are expected to be less active such as during school. They decrease vocalizations, noise, disruptive activity, and improve handwriting. Stimulants improve compliance with adults' commands, improve attention span and short-term memory, and reduce distractibility and impulsivity. The double-blind controlled studies reviewed here include those that investigated the efficacy of stimulants, antidepressants, and *clonidine* in the treatment of *ADHD*. *Dopamine* agonists such as *L-DOPA, piribidal*, and *amantadine* are not effective.

There are seven double-blind placebo controlled *ADHD* efficacy studies, seven involving *methylphenidate*, three *dextroamphetamine*, three *caffeine*, and one *pemoline* (Gittelman-Klein, 1987; Green, 1992; Greenhill, 1992). One hundred and fifty two patients were studied. *Methylphenidate, dextroamphetamine*, and *pemoline* appear equally effective in treating *ADHD*. The three psychostimulants were found to be more effective than *caffeine* and placebo. The following may be concluded from the studies summarized in Table 11.9:

1. The psychostimulants are the first line drugs in the treatment of *ADHD*.
2. The stimulants *MPH, DAS, Adderall*, and *pemoline* are equivalent in *ADHD* efficacy.
3. *ADHD* symptoms that should improve include hyperactivity, attention span, impulsivity and self-control, compliance, physical and verbal aggression, social interactions with peers, teachers, and parents, and academic productivity. *ADHD* symptoms that may or may not improve include reading skills, social skills, learning (less improvement in this area than behavior), academic achievement, antisocial behavior or arrest rates. It is estimated that the behavioral effects are twice as large as the academic effects (Swanson et al., 1991).
4. The opinion that stimulants do not have any effect on classroom learning and performance is debatable. The only adequately controlled study that investigated this issue found that methylphenidate 0.3 mg/kg per dose improved performance on arithmetical and language tasks (Douglas et al., 1986).
5. Aggressive behavior including stealing and vandalism is improved by standard doses of *MPH* 0.3–0.6 mg/kg.
6. Stimulants decrease friction between sibs and peers and improve maternal–child interactions.

TABLE 11.9 **Controlled Studies of Stimulants in the Treatment of Childhood and Adolescent** *ADHD*

Study	Dose (mg/d)	Population	Results
Garfinkel et al., 1975	MPH 20 mg/d 0.26 mg/kg/d Caffeine 150 mg/d Placebo, @ AM and noon X-over × 10 days each	MBD 8 males, 6–10 yr	MPH > Caffeine = Placebo 1. CTR aggression & hyperactivity = most 2. Kagan decreased errors 3. Reitan enhanced motor steadiness
Huestis et al., 1975	MPH Å 40 mg/d DAS Å 20 mg/d Caffeine ≥300 mg/d Placebo, @ AM and noon x-over × 3 weeks	MBD 12 males, 6 females mean = 8.5 yr	MPH = DAS > Caffeine = Placebo 1. DHRS hyperkinesis improved
Arnold et al., 1978	MPH Å 30 mg/d DAS Å 15 mg/d Caffeine Å 240 mg/d @ 2/3 AM and 1/3 noon x-over × 3 weeks	MBD 22 males, 8 females mean = 8.5 yr	MPH = DAS > Caffeine = Placebo 1. DHRS hyperkinetic behavior reduced improved school work 2. 26/29 improved best response: 12 DAS, 10 MPH, 1 Caffeine
Whalen et al., 1987	MPH 0.3 mg/kg MPH 0.6 mg/kg placebo, × 2 days each x-over	ADD or *ADHD* (DSM-III) 21 males, 3 females 6–11 yr	Social Behavior (SB) + SB not effected by MPH − SB (rule breaking etc.) linear improvement with MPH dose Nonsocial Behavior (solitary play etc.) no change with MPH Thus MPH does not decrease sociability
Klorman et al., 1990	3 week titration to a Å MPH 0.6 mg/kg/d Placebo x-over	ADD (DSM-III) 42 males 6 females 12–18 yr	1. CTRS improved hyperactivity, inattention and oppositionality 2. TOTS improvement on task 3. Nowles Mood Scale improved mood and affect
Pelham et al., 1990	MPH 10 mg bid MPH SR 20 mg bid pemoline 56.25 mg bid DAS spansules 10 bid placebo × 3–6 days each crossover 8 week trial MPH 0.3 mg/kg/does bid	*ADHD* (DSM-III-R) 22 males 8–13 yr	• pemoline and DAS spansules were the most effective treatments • pemoline was effective within 2–3 days • pemoline cause more initial insomnia problems than the other drugs
Musten et al., 1997	MPH 0.5 mg/kg/dose bid and Placebo x-over 7–10 days of each	*ADHD* (DSM-III-R) n = 41 31 completed 4–6 yr	Cognitive improvements in attention and impulsivity, behaviors (parent rating), attentional abilities, productive work improved on MPH doses (no difference between 0.3 and 0.5 dose) vs. placebo

TABLE 11.9 *Continued*

Study	Dose (mg/d)	Population	Results
Pelham et al., 1999	1) placebo at 7:30 AM, 11:30 AM, and 3:30 PM 2) 0.3 mg/kg of MPH at 7:30 AM, 11:30 AM, and 3:30 PM 3) 0.3 mg/kg of MPH at 7:30 AM and 11:30 AM with 0.15 mg/kg at 3:30 PM 4) 0.3 mg/kg of MPH at 7:30 AM only 5) 0.3 mg/kg of Adderall at 7:30 AM and at 3:30 PM 6) 0.3 mg/kg of Adderall at 7:30 AM with 0.15 mg/kg received at 3:30 PM 7) 0.3 mg/kg of Adderall at 7:30 AM only. To ensure blinding, placebo capsules were given at 11:30 AM in the Adderall conditions and or applicable doses in the other conditions. Mean = 11.1 mg for the 0.3 mg/kg conditions (range = 6.25–17.5) and 5.5 for the 0.15 mg/kg conditions (range = 3.75–8.75)	ADHD (DSM-IV) $n = 21$ 31 completed 6–12 yr	A single morning dose of Adderall had behavioral effects throughout an entire school day period that were equivalent to standard twice-daily MPH dosing Adderall may be used as a long-acting stimulant for children for whom midday dosing is a problem
Ahmann et al., 2001	Adderall 0.15 mg/kg/dose po bid × 7 d; Placebo × 7d Adderall 0.3 mg/kg/dose po bid × 7 d; placebo x-over placebo × 7 d	ADHD (DSM-IV) $n = 154$ 115 completed 5–18 yr	55% (78/143) concurrent (parent/teacher) response rate. Placebo response rate not noted. It was not determined if there was dose response relationship for adverse effects or efficacy

TABLE 11.10 Controlled Studies of Tricyclic Antidepressants in the Treatment of *ADHD*

Study	Dose (mg/d)	Population	Results
Krakowski et al., 1965	AMT 20–120 mg/d mean = 40 mg/d placebo, × 1–9 months	MBD 36 males, 14 females 3–18 yr	AMT good to excellent response (21/24) placebo good to excellent response (2/26)
Gross-Tsur, 1973	DAS 2.5–10 mg/d MPH 5–20 mg/d IMP 10–50 mg HS Placebo × 1 week each dose dependent on age of child	259 MBD 2–18 yr	Based on best tx response pts then put on that therapy for an additional 4 months MPH 25% DAS 21% IMP 21% IMP = DAS/MPH 10% IMP + DAS or IMP + MPH 12%
Rapoport et al., 1974	MPH ≤30 mg/d Å 20 mg/d IMP ≤ 150 mg/d Å 80 mg/d placebo × 6 weeks	76 hyperactive males, 6–12 yr	MPH > IMP > placebo Conners, Kagan MPH > IMP > placebo MD global assessment dose too low?
Donnelly et al., 1986	DSP ≥100 mg/d Å 3.4 mg/kg/d placebo × 2 weeks, parallel	ADHD 29 males, 6–12 yr	Classroom behavior improved, i.e., impulsivity and hyperactivity Cognition and attention did not improve
Biederman, 1989	DSP ≤5.6 mg/kg/d placebo × 6 weeks	MBD 58 males, 4 females <12 yr (n = 42) ≥12 yr (n = 42)	DSP > placebo CABRS & CGI very much or much improved DSP 68% vs. placebo 10%

Patients most likely to respond to *MPH* are characterized as having low *ADHD* severity, low anxiety, higher IQs, being younger, and being highly inattentive (Buitelaar et al., 1995). The response from a single stimulant dose predicts the home and school behavioral response at 4 weeks.

Antidepressants

Tricyclic antidepressants. The use of antidepressants in the treatment of *ADHD* has spanned nearly 30 years originating with Krakowski's *amitriptyline* study (1965). An *ADHD* subgroup may respond better to the *TCA*'s than to stimulants. Logically this group experiences more depression and anxiety symptoms than the stimulant responders (Pliszka et al., 2000). Eight controlled studies are available for efficacy evaluation. Gross (1973) and Rapoport and colleagues (1974) conducted studies using the *tricyclic antidepressants (TCA) imipramine.* Zametkin and associates (1985) studied the use of the *monoamine oxidase inhibitors (MAOI), clorgyline,* and *tranylcypromine* in *ADHD.* Donnelly et al. (1986) and Biederman and associates (1989) studied the effects of *desipramine* and Casat and colleagues (1987), Clay et al. (1987), Barrickman et al. (1995), and Conners et al. (1996) studied the use of *bupropion.*

Eight double-blind placebo-controlled antidepressant efficacy studies have studied 488 patients. *Tricyclics* are more effective than placebo in treating *ADHD.* Compared to

TABLE 11.11 Controlled Studies of Bupropion in the Treatment of ADHD

Study	Dose (mg/d)	Population	Results
Simeon et al., 1986	BPR 135 mg/d × 8 weeks 4 wk placebo lead-in, 15 completers	ADHD 17 males 7–13 yr	BPR > placebo CGI and Conner's Rating Scales for Parents, Teachers, and Physicians
Casat et al., 1987	BPR 3–6 mg/kg/d, $n = 20$ placebo, $n = 10$ × 28 days	ADHD 25 males 5 females 6–12 yr	BPR > placebo CTQ-hyperactivity CGI (severity and improvement)
Clay et al., 1988	BPR 3–6 mg/kg/d, $n = 18$ placebo, $n = 10$ × 28 days	ADHD 27 males 1 female 6–12 yr	BPR > placebo CGI (severity and improvement) no improvement on CPQ or CTQ
Barrickman et al., 1995	BPR = 3.3 mg/kg/d MPH = —mg/kg/d $n = 15$ × 42 days each; crossover	ADHD 12 males 3 females 7–17 yr	BPR = MPH according to CGI, Iowa-Conners, CPT, Kagan MFF, & Auditory-Verbal Test

the psychostimulants, *imipramine* and *clomipramine* were found to be less effective than *methylphenidate* while a third study found *imipramine* equal in effectiveness to *methylphenidate* and *dextroamphetamine*.

Monoamine oxidase inhibitors. (*MAOI*) One controlled efficacy study (Zametkin et al., 1985) exists that compares the effectiveness of *MAOI* to *dextroamphetamine* in the treatment of 22 *ADHD* patients. The study found that 10 mg/d *MAOI* doses were as effective as DAS. However, the authors noted that they preferred not to utilize *MAOI* to treat *ADHD* because of the possibility that the patient or parent might unwittingly utilize a psychostimulant if they run out of *MAOI* medication. The small size of the study and adverse drug reaction considerations suggest that *MAOI*'s be best regarded as third line drugs in the treatment of *ADHD*.

Bupropion. It appears that *bupropion* is effective in the treatment of *ADHD* according to the findings of four double-blind placebo controlled trials that studied a total of 93 patients administered doses ranging between 3–6 mg/kg/d. The drug was found to be as effective as *MPH*. Thus, it would appear that the drug would be a second line drug with the *TCA* in *ADHD* treatment.

Alpha-2 agonists
Alpha-2 agonists appear to be useful for *ADHD* patients confounded by tic disorders, extreme hyperactivity, oppositional or conduct disorder, hyperarousal, or a poor response to stimulants. (Hunt, 1990). Sedation and hypotension limit the usefulness of *clonidine* in the treatment of *ADHD*. However, *guanfacine* is an *alpha-2 agonist* that has a more desirable pharmacokinetic profile. The drug has an 18-hr half-life in adults and it is less sedating and causes less hypotension than *clonidine*. In an open trial, Hunt et al. (1995) administered *guanfacine* to 13 *ADHD* patients (4–20 years). The mean therapeutic dose was 3.2 mg/d (mean = 0.09 mg/kg/d). The 31-item Conner's Questionnaire demonstrated a 33 percent decrease in *ADHD* severity. On average, the hyperactivity factor decreased 42 percent, the inattention factor decreased 37 percent, and the immaturity factor also decreased significantly.

Fluoxetine

One open trial has concluded that *fluoxetine* 27 mg/d was partially effective in the treatment of *ADHD*. However, since *fluoxetine* has not been contrasted to either placebo or stimulants in the treatment of *ADHD*, the drug would have to be considered a fourth line drug.

In conclusion and based on these data, since *MPD* has been demonstrated to be effective in improving behavior, cognition, and sociability of children with *ADHD*, it is obvious that this is the pharmacological treatment of choice. The choice of a second line drug includes the *TCA*, *MAOI*, and *bupropion*. The six controlled *TCA* studies suggest these agents are effective but whether they improve the three spheres of behavior, cognition, and sociability as the stimulants do remains to be proven. The *MAOI* improve behavior but their effect on sociability and cognition remains undefined. Like the *TCA* the *bupropion* data suggest that the drug is slightly less effective than the stimulants.

References

Abell, F., Krama, M., Ashburner, J., Passingham, R., Friston, K., Frackowiak, R., Happe, F., Frith, C., and Frith U. (1999, June 3). The neuroanatomy of autism: A voxel-based whole brain analysis of structural scans. *Neuroreport, 10*, 1647–1651.

Aboitiz, F. (1999). Comparative development of the mammalian isocortex and the reptilian dorsal ventricular ridge. Evolutionary considerations. *Cerebral Cortex, 9*, 783–791.

Abraham, S., Collins, G., and Nordsiek, M. (1971). Relationship of childhood weight status to morbidity in adults. *Public Health Reports, 86*, 273–284.

Ackermann, H., Wildgruber, D., Daum, I., and Grodd, W. (1998). Does the cerebellum contribute to cognitive aspects of speech production? A functional magnetic resonance imaging (fMRI) study in humans. *Neuroscience Letters, 247*, 187–190.

Acredolo, A. and Goodwyn, S. W. (1984). The role of self-produced movement and visual tracking in infant spatial orientation. *Journal of Experimental Child Psychology, 38*, 312–327.

Aftanas, L. I., Varlamov, A. A., Pavlov, S. V., Makhnev, V. P., and Reva, N. V. (2002). Time-dependent cortical asymmetries induced by emotional arousal: EEG analysis of event-related synchronization and desynchronization in individually defined frequency bans. *International Journal of Psychophysiology, 44*, 67–82.

Aghajanian, G. K. and Wang, R. Y. (1977). Habenular and other midbrain raphe afferents demonstrated by a modified retrograde tracing technique. *Brain Research, 122*, 229–242.

Aharon, I., Etcoff, N., Ariely, D., Chabris, C. F., O'Connor, E., and Breiter, H. C. (2001). Beautiful faces have variable reward value: fMRI and behavioral evidence. *Neuron, 32*, 537–551.

Ahem, D. K., Gorkin, L., Anderson, J. L., Tiemey, C., Hallstlrom, A., Ewart, C., Capone, R. J., Schron, E., Komfield, D., Herd, J. A., Richardson, D. W., and Follick, M. J. (1990). Biobehavioral variables and mortality or cardiac arrest in the cardiac arrhythmia pilot shidy (CAPS). *American Journal of Cardiology, 66*, 59–62.

Ahmann, P. A., Theye, F. W., Berg, R., Linquist, A. J., Van Erem, A. J., and Campbell, L. R. (2001). Placebo-controlled evaluation of amphetamine mixture-dextroamphetamine salts and amphetamine salts (Adderall): Efficacy rate and side effects. *Pediatrics, 107*, E10.

Aizikov, G. S., Kreidich, Y. V., and Grigoryn, R. A. (1991). Sensory interaction and methods of non-medicinal prophylaxis of space motion sickness. *Physiologist, 34* (Suppl.). Proceeding of the 12th Annual Meeting of IUPS Commission on Gravitational Physiology, S220–S223.

Alba, F., Ramirez, M., Iribar, C., Cantalejo, E., and Oscar, C. (1985). Asymmetrical distribution of aminopeptidase activity in the cortex of rat brain. *Brain Research, 368*, 158–160.

Albert, M. D. (1973). A simple test of visual neglect. *Neurology, 23*, 658–664.

Albert, M. L., Feldman, R. G., and Wills, A. L. (1974). The subcortical dementia of progressive supranuclear palsy. *Journal of Neurology, Neurosurgery, and Psychiatry, 37*, 121–130.

Alberti, A., Pirrone, P., Elia, M., Waring, R. H., and Romano, C. (1999). Sulphation deficit in "low-functioning" autistic children: A pilot study. *Biological Psychiatry, 46*, 420–424.

Albin, R. L., Young, A. B., and Penney, J. B. (1995). The functional anatomy of disorders of the basal ganglia. *Trends in Neuroscience, 18*, 63–64.

Alexander, G. E., Delong, M. R., and Strick, P. L. (1986). Parallel organization of functionally segregated circuits linking basal ganglia and cortex. *Annual Review of Neuroscience, 9*, 357–381.

Alexander, G. E., Crutcher, M. D., and Delong, M. R. (1990). Basal ganglia–thalamocortical circuits; parallel substrates for motor, oculomotor, (prefrontal) and (limbic) functions. In H. B. N. Uylungs, C. G. Van Eben, J. P.C. DeBruin, M. A. Corner, M. G. P. Feenstra (Eds.), *The prefrontal cortex, its structure, function, and pathology* (pp. 266–271). Amsterdam, the Netherlands: Elsevier Science.

Allen, G. I., Gilbert, P. F. C., and Yin, T. C. T. (1978). Convergence of cerebral inputs onto dentate neurons in monkey. *Experimental Brain Research, 32*, 337–341.

Allison, A. C. (1953). The morphology of the olfactory system in vertebrates. *Annual Review of Cell and Developmental Biology, 28*, 195–244.

Almada, S. J., Zonderman, A. B., Shekelle, R. B., Dyer, A. R., Daviglus, M. L., and Costa, P. T., Jr. (1991). Neuroticism and cynicism and risk of death in middle-aged men: The Western electric study. *Psychosomatic Medicine, 53*, 165–175.

Altman, J. and Bayer, S. (1985). Embyronic development of the rat cerebellum: III. Regional differences in the time of origin, migration and settling of purkinje cells. *Journal of Comparative Neurology, 231*, 42–65.

Alyward, E. H., Minshew, N. J., Field, K., Sparks, B. F., and Singh, N. (2002). Effects of age on brain volume and head circumference in autism. *Neurology, 59*, 175–183.

Amadeo, M. and Shagass, C. (1973). Brief latency click-evoked potentials during waking and sleep in man. *Psychophysiology, 10*, 244–250.

American Academy of Child and Adolescent Psychiatry. (1997). Practice parameters for the assessment and treatment of children, adolescents, and adults with attention-deficit/hyperactivity disorder. *Journal of the American Academy of Child and Adolescent Psychiatry, 36* (Suppl.), 85S–121S.

American Dietetic Association (1999). Position of the American dietetic association: Nutrition standards for child-care programs. *Journal of the American Dietetic Association, 99*, 981–988.

American Psychiatric Association. (1994). *Diagnostic and statical manual of mental disorders* (4th ed.). Washington DC: American Psychiatric Press.

Ananth, J., Pecknold, J. C., Van Den Steen, N., and Engelsmann, F. (1981). Double-blind comparative study of clomipramine and amitriptyline in obsessive neurosis. *Progress in Neuro-Psychopharmcology, 5*, 257–262.

Andersen, P. (1960). Interhippocampal impulses. II. Apical dendritic activation on CA1 neurons. *Acta Physiologiae Plantarum/Polish Academy of Sciences, Committee of Plant Physiology Genetics and Breeding, 48*, 178–208.

Andersen, R. E. (1995). Is exercise or increased activity necessary for weight loss and weight management? *Medicine and Exercise in Nutrition and Health, 4*, 57–59.

Andersen, R. E., Crespo, C. J., Bartlett, S. J., Cheskin, L. J., and Pratt, M. (1998). Relationship of physical activity and television watching with body weight and level of fatness among children: Results from the third national health and nutrition examination survey. *The Journal of the American Medical Association, 279*, 938–942.

Anderson, J. W., Johnstone, B. M., and Remley, D. T. (1999). Breast-feeding and cognitive development: A meta-analysis. *The American Journal of Clinical Nutrition, 70*, 525–535

Anderson, L. T., Campbell, M., Adams, P., Small, A. M., Perry, R., and Shell, J. (1989). The effects of haloperidol on discrimination learning and behavioral symptoms in autistic children. *Journal of Autism and Developmental Disorders, 19*, 227–239.

Anderson, L. T., Campbell, M., Grega, D. M., Perry, R., Small, A. M., and Green, W. H. (1984). Haloperidol in the treatment of infantile autism: Effects on learning and behavioral symptoms. *American Journal of Psychiatry, 141*, 1195–1202.

Anderson, R. H., Fleming, D. E., Rhees, R. W., and Kinghom, E. (1986). Relationships between sexual activity, plasma testosterone and the volume of the sexually dimorphic nucleus of the preoptic area in prenatally stressed and non-stressed rats. *Brain Research, 370*, 1–10.

Anderson, S. D., Basbaum, A. I., and Fields, H. L. (1977). Response of medullary raphe neurons to peripheral stimulation and to systemic opiates. *Brain Research, 123*, 363–368.

Andreasen, N. C., O'Leary, D. S., Arndt, S., Cizadlo, T., Hurtig, R., Rezai, K., Watkins, G. L., Ponto, L. L., and Hichwa, R. D. (1995). Short-term and long-term verbal memory: A positron emission tomography study. *Proceedings of the National Academy of Sciences, USA, 92*, 5111–5115.

Andres, C. (2002). Molecular genetics and animal models in autistic disorder. *Brain Research Bulletin, 57*, 109–119.

Andrew J., Fowler, C., and Harrison, M. J. (1982). Hemi-dystonia due to focal basal ganglia lesion after head injury and improved by stereotaxic thalamotomy. *Journal of Neurology, Neurosurgery, and Psychiatry, 45*, 276.

Angst, J. and Merikangas, K. (1997). The depressive spectrum: Diagnostic classification and course. *Journal of Affective Disorders, 45*, 31–40.

Annett, M. and Manning, M. (1989). The disadvantages of dextrality for intelligence. *British Journal of Psychology, 80*, 213–226.

Anokhin, K. V. and Rose, S. P. (1991). Learning-induced increase of immediate early gene messenger RNA in the chick forebrain. *European Journal of Neuroscience, 3*, 162–167.

Aoyama, F., Iida, J., Inoue, M., Iwasaka, H., Sakiyama, S., Hata, K., and Kishimoto, T. (2000). Brain imaging in childhood and adolescence-onset schizophrenia associated with obsessive-compulsive symptoms. *Acta Psychiatrica Scandinavica, 102*, 32–37.

Aram, D. M., Ekelman, B. L., and Whitaker, H. A. (1986). Spoken syntax in children with acquired hemisphere lesions. *Brain and Language, 27*, 75–100.

Arbuthnott, G. W. and Crow, T. J. (1971). Relation of contraversive timing to unilateral release of dopamine from the nigrostriatal pathway in rats. *Experimental Neurology, 30*, 484–491.

Arcia, E. and Gualtieri, C. T. (1994). Neurobehavioural performance of adults with closed-head injury, adults with attention deficit, and controls. *Brain Injury, 8*, 395–404.

Armstrong, E. (1980). A qualitative comparison of hominoid thalamus; III. A motor substrate the ventrolateral complex. *American Journal of Physical Anthropology, 52*, 405–419.

Arnold, A. P. and Bottjer, S. W. (1985). Cerebral lateralization in birds. In S. D. Glick (Ed.), *Cerebral lateralization in nonhuman species* (pp. 11–39). Orlando, FL: Academic Press.

Arnold, H. M., Burk, J. A., Hodgson, E. M., Sarter, M., and Bruno, J. P. (2002). Differential cortical acetylcholine release in rats performing a sustained attention task versus behavioral control tasks that do not explicitly tax attention. *Neuroscience, 114*, 451–460.

Arnold, L. E., Christopher, J., Muestis, R., and Smeltzer, D. J. (1978). Methylphenidate vs. dextroamphetamine vs. caffeine in minimal brain dysfunction. *Archives of General Psychiatry, 35*, 463–473.

Arnulf, I., Bejjani, B. P., Garma, L., Bonnet, A. M., Houeto, J. L., Damier, P., Derenne, J. P., and Agid, Y. (2000). Improvement of sleep architecture in PD with subthalamic nucleus stimulation. *Neurology, 55*, 1732–1734.

Arora, R. C. and Meltzer, H. Y. (1989a). Serotonergic measures in the brains of suicide victims: 5-HT2 binding sites in the frontal cortex of suicide victims and control subjects. *American Journal of Psychiatry, 146*, 730–736.

Arora, R. C. and Meltzer, H. Y. (1989b). 3H-Imipramine binding in the frontal cortex of suicides. *Psychiatry Research, 30*, 125–135.

Arrigoni, G. and De Renzi, E. (1964). Constructional apraxia and hemispheric locus of lesion. *Cortex, 1*, 170–197.

Arrieta, I., Nunez, T., Martinez, B., Perez, A., Telez, M., Criado, B., Gainza, I., and Lostao, C. M. (2002). Chromosomal fragility in a behavioral disorder. *American Journal of Human Genetics, 71*, 777–790.

Asanuma, C., Thach, W. T., and Jones, E. G. (1983). Distribution of cerebellar terminations and their relation to other afferent terminations in the ventral lateral thalamic region of the monkey. *Brain Research, 286*, 237–265.

Asberg, M., Schalling, D., Traskman-Bendz, L., and Wagner, A. (1987). Psychobiology of suicide, impulsivity and related phenomena. In H. Y. Meltzer (Ed.), *Psychopharmacology: The third generation of progress* (pp. 655–668). New York, NY: Raven Press.

Asbjornsenm, A., Hugdahl, K., and Hynd, G. (1990). The effects of head and eye turns on the right ear advantage in dichotic listening. *Brain and Language, 39*, 447–458.

Ashcraft, M. H., Yamashita, T. S., and Aram, D. M. (1992). Mathematics performance in left and right brain lesioned children and adults. *Brain and Cognition, 19*, 208–252.

Asperger, H. (1961). Die Psycotathologie DES Coeliakakranken Kindes. *Annales paediatrici. International Review of Pediatrics, 197*, 146–161.

Associated Press. (1989). *Army puts couch potatoes on alert*. http://www.mndaily.com/daily/gopherarchives/1989/04/18/Army_puts_couch_potatoes_on_alert..txt.

Astrino, D. O. and Rodahl, K. (1970). *Text book of work physiology*. New York: McGraw-Hill.

Astrup, A., Hill, J. O., and Saris, W. H. (2001). Dietary fat: At the heart of the matter. *Science, 293*, 801–804.

Attwell, P. J., Cooke, S. F., and Yeo, C. H. (2002). Cerebellar function in consolidation of a motor memory. *Neuron, 34*, 1011–1020.

Audinat, E., Gahwiler, B. H., and Knopfel, T. (1992). Synaptic potentials in neurons of the deep nuclei in olivo-cerebellar slice cultures. *Neuroscience, 49*, 903–911.

August, G. J., Raz, N., Papanicolaou, A. C., Baird, T. D., Hirsh, S. L., and Hsu, L. L. (1984). Fenfluramine treatment in infantile autism. Neurochemical, electrophysiological and behavioral effects. *Journal of Nervous and Mental Disease, 172*, 604–612.

Aumann, T. D., Rawson, J. A., and Horne, M. K. (1998). The relationship between monkey dentate cerebellar nucleus activity and kinematic parameters of wrist movement. *Experimental brain research. Experimentelle Hirnforschung. Experimentation cerebrale, 119*, 179–190.

Auranen, M., Vanhala, R., Varilo, T., Ayers, K., Kempas, E., Ylisaukko-Oja, T., Sinsheimer, J. S., Peltonen, L., and Jarvela, I. (2002, June). A genomewide screen for autism-spectrum disorders: Evidence for a major susceptibility locus on chromosome 3q25-27. *Casopis lekaru ceskych, 141*, 381–387.

Ayers, A. J. (1972a). *Sensory integration and learning disabilities*. Los Angeles, CA: Western Psychological Services.

Aylward, E. H., Minshew, N. J., Goldstein, G., Honeycutt, N. A., Augustine, A. M., Yates, K. O., Barta, P. E., and Pearlson, G. D. (1999). MRI volumes of amygdala and hippocampus in non-mentally retarded autistic adolescents and adults. *Neurology, 53*, 2145–2150.

Ayres, A. J. (1972b). Types of sensory integrative dysfunction among disabled learners. *The American Journal of Occupational Therapy, 26*, 13.

Ayres, A. J. and Tickle, L. S. (1980). Hyper-responsivity to touch and vestibular stimuli as a predictor of positive response to sensory integration procedures by autistic children. *American Journal of Occupational Therapy, 34*, 375–381.

Bagshaw, M. H., Kimble, D. P., and Pribram, K. H. (1965). The GSR of monkeys during orienting and habituation and after ablation of the amygdala, hippocampus, and inferotemporal cortex. *Neuropsychologia, 3*, 111–119.

Bai, D. L. and Bertenthal, B. I. (1992). Locomotor status and the development of spatial search skills. *Child Development, 63*, 215–226.

Bailey, A., Phillips, W., and Rutter, M. (1996). Autism: Towards an integration of clinical, genetic, neuropsychological, and neurobiological perspectives. *Journal of Child Psychology and Psychiatry, and Allied Disciplines, 37*, 89–126.

Bailey, C. H. and Kandel, E. R. (1993). Structural changes accompanying memory storage. *Annual Review of Physiology, 993*, 397–426.

Bakker, D. J. (1969). Ear asymmetry with monaural stimulation: Task influences. *Cortex, 5*, 36–42.

Bakker, D. J., Van der Flugt, H., and Claushuis, M. (1978). The reliability of dichotic ear asymmetry in normal children. *Neuroposychologia, 16*, 753–757.

Balaban, C. D. (2002). Neural substrates linking balance control and anxiety. *Physiology and Behavior, 77*, 469–475.

Balaban, C. D. and Thayer, J. F. (2001). Neurological bases for balance-anxiety links. *Journal of Anxiety Disorders, 15*, 53–79.

Bandstra, E. S., Morrow, C. E., Anthony, J. C., Accornero, V. H., and Fried, P. A. (2001). Longitudinal investigation of task persistence and sustained attention in children with prenatal cocaine exposure. *Neurotoxicology and Teratology, 23*, 545–549.

Banich, M. T. (1995). Interhemispheric processing: Theoretical considerations and empirical approaches. In R. J. Davidson and K. Hugdahl (Eds.), *Brain asymmetry*. Cambridge, MA: MIT Press.

Banich, M. and Belger, A. (1991). Interversus intrahemishpheric concordance of judgments in a non-explicit memory task. *Brain and Cognition, 15*, 131–137.

Banker, D. and Girvin, J. (1971). The ultrastructural features of the mammalian muscle spindle. *Journal of Neuropathology and Experimental Neurology, 30*, 155–195.

Banks, M. S. and Dannemiller, J. L. (1987). Infant visual psychophysics. In P. Salapatek and L. Cohen (Eds.), *Handbooks of infant perception*, (Vol. 1, pp. 115–184). Orlando, FL: Academic Press.

Baranek, G. T. (1999). Autism during infancy: A retrospective video analysis of sensory-motor and social behaviors at 9–12 months of age. *Journal of Autism and Developmental Disorders, 29*, 213–224.

Baranek, G. T. and Berkson, G. (1994). Tactile defensiveness in children with developmental disabilities: Responsiveness and habituation. *Journal of Autism and Developmental Disorders, 24*, 457–471.

Baranek, G. T., Foster, L. G., and Berkson, G. (1997). Tactile defensiveness and stereotyped behaviors. *American Journal of Occupational Therapy, 51*, 91–95.

Barbas, H. and Pandya, D. N. (1989). Architecture and intrinsic connections of the prefrontal cortex in the rhesus monkey. *Journal of Comparative Neurology, 286*, 353–375.

Bardos, P., Degenne, D., Lebranchu, Y., Biziere, K., and Renoux, G. (1981). Neocortical lateralization of NK activity in mice. *Scandinavian Journal of Immunology, 13*, 609–611.

Barker, D. (1974). The morphology of muscle receptors. In C. C. Hung (Ed.), *Handbook of sensory physiology* (pp. 90–95). New York: Springer-Verlag.

Barkley, R. A. (1997). Behavioral inhibition, sustained attention and executive functions: Constructing a unifying theory of ADHD. *Psychological Bulletin, 121*, 65–69.

Barlow, C., Ribaut-Barassin, C., Zwingman, T. A., Pope, A. J., Brown, K. D., Owens, J. W., Larson, D., Harrington, E. A., Haeberle, A. M., Mariani, J., Eckhaus, M., Herrup, K., Bailly, Y., and Wynshaw-Boris, A. (2000). ATM is a cytoplasmic protein in mouse brain required to prevent lysosomal accumulation. *Proceedings of the National Academy of Sciences, 97*, 871–876.

Barneoud, P., Neveu, P. J., Vitiello, R., and LeMoal, M. (1987). Functional heterogeneity of the right and left neocortex in modulation of the immune system. *Physiology and Behavior, 41*, 525–530.

Barnes, C. A. (1998). Spatial learning and memory processes: The search for their neurobiological mechanisms in the rat. *Trends in Neuroscience, 11*, 163–169.

Baron-Cohen, S., Allen, J., and Gillberg, C. (1992). Can autism be detected at 18 months? The needle, the haystack, and the chat. *The British Journal of Psychiatry; The Journal of Mental Science, 161*, 839–843.

Baron-Cohen, S., Allen, J., and Gillberg, C. (1996). Can autism be detected at 18 months? The needle, the haystack, and the CHAT. *Journal of Autism and Developmental Disorders, 26*, 173–178.

Baroni, G., Pedrocchi, A., Ferrigno, G., Massion, J., and Pedotti, A. (2001). Motor coordination in weightless conditions revealed by long-term microgravity adaptation. *Acta Astronautica, 49*, 199–213.

Barrickman, L. L., Perry, P. J., and Allen, A. J. et al. (1995). Bupropion versus methylphenidate in the treatment of attention deficit hyperactivity disorder. *Journal of the American Academy of Child and Adolescent Psychiatry, 34*, 649–657.

Basbaum, A. I., Clanton, C. H., and Fields, H. L. (1976). Opiate and stimulus-produced analgesia: Functional anatomy of a medullospinal pathway. *Proceedings of the National Academy of Sciences, USA, 73*, 4685–4688.

Bastian, H. C. (1898). A treatise on aphasia and other speech defects. London: H. K. Lewis.

Bates, E., O'Connell, B., Vaid, J., Sledge, P., and Oakes, L. (1986). Language and hand preference in early development. *Developmental Neuropsychology, 2*, 1–15.

Bauman, M. L. and Kemper, T. L. (1985). Histoanatomic observations of the brain in early infantile autism. *Neurology, 35*, 866–874.

Bauman, M. L. and Kemper, T. L. (1986). Developmental cerebellar abnormalities: A consistent finding in an early infatile autism. *Neurology, 36*, 190.

Baxter, L. R., Phelps, M. E., Mazziotta, J. G., Schwartz, J. M., Gerner, R. H., Selin, C. E., and Sumida, R. M. (1985). Cerebral metabolic rates for glucose in mood disorders: Studies with positron emission tomography and fluorodeoxiglucose F18. *Archives of General Psychiatry, 42*, 441–447.

Bear, D. M. (1983). Hemispheric specialization and the neurology of emotion. *Archives of Neurology, 40*, 195–202.

Beck, A. T. (1967). Depression: Clinical, experimental, and theoretical aspects. New York, NY: Hoeber.

Beck, A. T. (1976). Cognitive therapy and the emotional disorders. New York, NY: International Universities Press.

Beckstead, R. M. (1978). Afferent connections of the entorhinal area in the rat as demonstrated by retrograde cell-labeling with horseradish peroxidase. *Brain Research, 152*, 249–264.

Beeghly, J. H., Kuperman, S., Perry, P. J., Wright, G. J., and Tsai, L. Y. (1987). Fenfluramine treatment of autism: Relationship of treatment response to blood levels of fenfluramine and norfenfluramine. *Journal of Autism and Developmental Disorders, 17*, 541–548.

Bell, C. C., Finger, T. E., and Russell, C. J. (1981a). Central connections of the posterior lateral line lobe in mormyrid fish. *Experimental Brain Research, 42*, 9–22.

Bell, J., Gruenthal, M., Finger, S., and Mangold, R. (1981b). Effects of one- and two-stage lesions of the posterior hypothalamus on temperature regulation in the rat. *Brain Research, 219*, 451–455.

Bellak, L. (1994). The schizophrenic syndrome and attention deficit disorder: Thesis, antithesis, and synthesis? *The American Psychologist, 49*, 25–29.

Bellugi, U., Bihrle, A., Jernigan, T., Trauner, D., and Doherty, S. (1990). Neuropsychological, neurological, and neuroanatomical profile of Williams syndrome. *American Journal of Medical Genetics* (Suppl.), 115–125.

Bellus, S. B., Kost, P. P., Vergo, J. G., and Dinezza, G. J. (1998). Improvements in cognitive functioning following intensive behavioural rehabilitation. *Brain Injury, 12*, 139–145.

Benabarre, A., Vieta, E., Martinez-Aran, A., Reinares, M., Colom, F., Lomena, F., Martin, F., and Valdes, M. (2002). The somatics of psyche: Structural neuromorphometry of bipolar disorder. *Psychotherapy and Psychosomatics, 71*, 180–189.

Benabid, A. L., Pollak, P., Louveau, A., Henry, S., and de Rougemont, J. (1987). Combined (thalamotomy and stimulation) stereotactic surgery of the VIM thalamic nucleus for bilateral Parkinson disease. *Applied Neurophysiology, 50*, 344–346.

Ben-Ari, Y. (2002). Excitatory actions of GABA during development: The nature of the nurture. *Nature, Revue of Neuroscience, 3*, 728–739.

Benarroch, E. E., Zollman, P. J., Schmelzer, J. D., Nelson, D. K., and Low, P. A. (1992). Guanethidine sympathectomy increases substance P concentration in the superior sympathetic ganglion of adult rats. *Brain Research, 584*, 305–308.

Benke, T. A., Luthi, A., Isaac, J. T., and Collingridge, G. L. (1998). Modulation of AMPA receptor unitary conductance by synaptic activity. *Nature, 393*, 793–797.

Bennett, E. L., Diamond, M. C., Krech, D., and Rosenzweig, M. R. (1964). Chemical and anatomical plasticity of the brain. *Science, 146*, 610–619.

Bennett, E. L., Rosenzweig, M. R., Diamond, M. C., Morimoto, H., and Hebert, M. (1974). Effects of successive environments on brain measures. *Physiology and Behavior, 12*, 621–631.

Benson, D. (1979). Aphasia, alexia, agraphia. New York, NY: Churchill Livingstone Inc.

Benson, D. F. and Geschwind, N. (1968). Cerebral dominance and its disturbances. *Pediatric Clinics of North America, 15*, 759–769.

Benson, D. F. and Geschwind, N. (1970). Developmental Gerstmann syndrome. *Neurology, 20*, 293–298.

Benson, D. F. and Geschwind, N. (1983). Aphasia and related disorders: Clinical approach. In M. M. Mesulam (Ed.), *Principles of behavioral neurology*. Philadelphia: S. A. Davis Co.

Benson, D. F., Sheremata, W. A., Bouchard, R., Segarra, J. M., Price, D., and Geschwind, N. (1973). Conduction aphasia. A clinicopathological study. *Archives of Neurology, 28*, 339–346.

Benson, F. F. and Blumer, D. (1975). Psychiatric aspects of neurologic disease. New York, NY: Grune and Stratton.

Benton, A. L., Varney, N. R., and Hamsher, K. deS. (1978). Lateral differences in tactile directional perception. *Neuropsychologia, 16*, 109–114.

Benton, A. R., Hamsher, K. D. S., Varney, N., and Spreen, O. (1983). Contributions to neuropsychological assessment: A clinical manual. New York, NY: Oxford University Press.

Berardelli, A., Rothwell, J. C., Hallett, M., Thompson, P. D., Manfredi, M., and Marsden, C. D. (1998). The pathophysiology of primary dystonia. *Brain, 121*, 1195–1212.

Bergmann, U. (1998). Speculations on the neurobiology of EMDR. *Traumatology, 4*, article 2, http://www.fsu.edu/~trauma/

Berman, A. J., Berman, D., and Prescott, J. W. (1978). The effect of cerebellar lesions on emotional behavior in the rhesus monkey. In I. S. Cooper, M. Riklon, and R. Snider (Eds.), *The cerebellum, epilepsy and behavior* (pp. 227–284). New York: Plenum Press.

Berquin, P. C., Giedd, J. N., Jacobsen, L. K., Hamburger, S. D., Krain, A. L., Rapoport, J. L., and Castellanos, F. X. (1998). Cerebellum in attention-deficit hyperactivity disorder: A morphometric MRI study. *Neurology, 50*, 1087–1093.

Berntson, G. G. and Micco, D. J. (1976). Organization of brainstem behavioral systems. *Brain Research Bulletin, 1*, 471–483.

Berntson, G. G. and Torello, M. W. (1982). The paleocerebellum and the integration of behavioral function. *Journal of Comparative and Physiological Psychology, 10*, 2–12.

Besedovsky, H. O., del Rey, A. E., and Sorkin, E. (1985). Immune-neuroendocrine interactions. *Journal of Immunology, 135*, 750–754.

Beversdorf, D. Q. and Heilman, K. M. (1998). Facilitatory paratonia and frontal lobe functioning. *Neurology, 51*, 968–971.

Biedermann, H. (1991). Kopfgelenk-induzierte symmetriestorungen bei kleinkindern. *Kinderarzt, 22*, 475–1482.

Biederman, I. (1987). Recognition by components: A theory of human image understanding. *Psychological Review, 94*, 115–147.

Biederman, J., Baldessarini, R. J., and Wright, V. et al. (1989). A double-blind placebo controlled study of desipramine in the treatment of ADD: I. Efficacy. *Journal of the American Academy of Child and Adolescent Psychiatry, 28*, 777–784.

Biederman, J., Faraone, S. V., Keenan, K., Benjamin, J., Krifcher, B., Moore, C., Sprich, S., Ugaglia, K., Jellinek, M. S., Steingard, R., Spencer, T., Norman, D., Kolodny, R., Kraus, I., Perrin, J., Keller, M. B., and Tsuang, M. T. (1992). Further evidence for family-genetic risk factors in attention deficit hyperactivity disorder (ADHD): Patterns of comorbidity in probands and relatives in psychiatrically and pediatrically referred samples. *Archives of General Psychiatry, 49*, 728–738.

Biederman, J., Newcorn, J., and Sprich, S. (1991). Comorbidity of Attention Deficit Hyperactivity Disorder with conduct, expressive, anxiety and other disorders. *American Journal of Psychiatry, 148*, 564–577.

Bingel, U., Quante, M., Knab, R., Bromm, B., Weiller, C., and Buchel, C. (2002). Subcortical structures involved in pain processing: Evidence from single-trial fMRI. *Pain, 99*, 313–321.

Birnbaumer, N. and Schmidt, R. F. (1991). *Biologische Psychologie*. Berlin: Springer-Verlag.

Biziere, K., Guillaumin, J. M., Degenne, D., Bardos, P., Renoux, M., and Renoux, G. (1985). Lateralized neocortical modulation of the T-cell lineage. In R. Guillemin, M. Cohn, and T. Melnechuk (Eds.), *Neural modulation of immunity* (pp. 81–91). New York, NY: Raven Press.

Bjorklund, D. F. (1997). The role of immaturity in human development. *Psychological Bulletin, 122*, 153–169.

Black, A. H. (1975). Hippocampal electrical activity and behavior. In R. L. Isaacson and K. H. Pribram (Eds.), *The Hippocampus* (Vol. 2, pp. 129–167). New York: Plenum Press.

Blackstad, T. W. (1977). Notes sur l'hodologie comparative de structures rhinencephaliques (limbiques). Quelques relations mutuelles et neocorticales. In J. de Ajuriaguerra and R. Tissot (Eds.), *Rhinencephale, Neurotransmetteurs et Psychoses* (pp. 17–60) (Symp. Bel-Air V, Geneve, Sept. 1976). Geneve: Georg & Cie S. A., Paris: Masson & Cie.

Blake, D. T., Byl, N. N., and Merzenich, M. M. (2002). Representation of the hand in the cerebral cortex. *Behavioral and Brain Research, 135*, 179–184.

Blanks, J. C., Torigoe, Y., Hinton, D. R., and Blanks, R. H. (1991). Retinal degeneration in the macula of patients with Alzheimer's disease. *Annals of the New York Academy of Sciences, 640*, 44–46.

Blatt, G. J., Anderson, R. A., and Stoner, G. R. (1990). Visual receptive field organization and cortico-cortical connections of the lateral intrapareital area (area LIP) in the macaque. *The Journal of Comparative Neurology, 299*, 421–445.

Blatt, G. J., Fitzgerald, C. M., Guptill, J. T., Booker, A. B., Kemper, T. L., and Bauman, M. L. (2001). Density and distribution of hippocampal neurotransmitter receptors in autism: An autoradiographic study. *Journal of Autism and Developmental Disorders, 31*, 537–543.

Bleuler, E. (1911). Dementia Praecox or the Group of Schizophrenias. Translated by J. Zinkin (1950). New York, NY: International Universities Press.

Blin, J., Baron, J. C., and Dubois, B. et al. (1990). Positron emission tomography study in progressive supranuclear palsy: Brain hypometabolic pattern and clinicometabolic correlations. *Archives of Neurology, 47*, 747–752.

Blin, J., Ruberg, M., and Baron, J. C. (1992). Positron emission tomography studies. In I. Litvan, Y. Agit (Eds.), *Progressive supranuclear palsy: Clinical and research approaches* (pp. 155–168). New York, NY: Oxford University Press.

Blinkov, S. M. and Glezer, I. I. (1968). *The human brain in figures and tables. A quantitative handbook*. New York: Plenum Press.

Bliss, T. V. and Collingridge, G. L. (1993). A synaptic model of memory: Long-term potentiation in the hippocampus. *Nature, 361*, 31–39.

Bliss, T. V. and Gardner-Medwin, A. R. (1973). Long-lasting potentiation of synaptic transmission in the dentate area of the unanaestetized rabbit following stimulation of the perforant path. *Journal of Physiology (London), 232*, 357–374.

Blokland, A., Lieben, C., and Deutz, N. E. (2002). Anxiogenic and depressive-like effects, but no cognitive deficits, after repeated moderate tryptophan depletion in the rat. *Journal of Psychopharmacology, 16*, 39–49.

Blonder, L. X., Bowers, D., and Heilman, K. M. (1991). The role of the right hemisphere in emotional communication. *Brain, 114*, 1115–1127.

Blood, A. J. and Zatorre, R. J. (2001). Intensely pleasurable responses to music correlate with activity in brain regions implicated in reward and emotion. *Proceedings of the National Academy of Sciences, USA, 98*, 11818–11823.

Blood, A. J., Zatorre, R. J., Bermudez, P., and Evans, A. C. (1999). Emotional responses to pleasant and unpleasant music correlate with activity in paralimbic brain regions. *Nature, Neuroscience, 2*, 382–387.

Blumer, D. and Benson, D. (1975). Personality changes with frontal and temporal lesions. In D. F. Benson and F. Blumer (Eds.), *Psychiatric aspects of neurologic disease*. New York: Grune and Stratton.

Bobath, B. (1976). *Abnorme Haltungsreflexe bei Gehirnschaden*. Stuttgart: Thieme.

Bock, O., Fowler, B., and Comfort, D. (2001). Human sensorimotor coordination during spaceflight: An analysis of pointing and tracking responses during the "Neurolab" Space Shuttle mission. *Aviation, Space, and Environmental Medicine, 72*, 877–883.

Bodner, M., Muftuler, L. T., Nalcioglu, O., and Shaw, G. L. (2001). FMRI study relevant to the Mozart effect: Brain areas involved in spatial-temporal reasoning. *Neurological Research, 23*, 683–690.

Bogduk, N. (1994). Post whiplash syndrome. *Australian Family Physician, 23*, 2303–2307.

Bogin, B. (1998). From caveman cuisine to fast food: The evolution of human nutrition. Growth hormone & IGF research. *Official Journal of the Growth Hormone Research Society and the International IGF Research Society, 8* (Suppl.), 79–86.

Bogin, B. and MacVean, R. B. (1981a). Nutritional and biological determinants of body fat patterning in urban Guatemalan children. *Human Biology; An International Record of Research, 53*, 259–268.

Bogin, B. and MacVean, R. B. (1981). Body composition and nutritional status of urban Guatemalan children of high and low socioeconomic class. *American Journal of Physical Anthropology, 55*, 543–551.

Boineoud, P., Le Moal, M., and Neveu, P. J. (1990). Asymmetric distribution of brain monoamines in left and right-handed mice. *Brain Research, 520*, 317–321.

Boliek, C. A. and Obrzut, J. E. (1995). Perceptual laterality. In R. J. Davidson and K. Hugdahl (Eds.), *Brain asymmetry*. Cambridge, MA: MIT Press.

Boliek, C. A., Obrzut, J. E., and Shaw, D. (1988). The effects of hemispatial and asymmetrically focused attention on dichotic listening with normal and learning-disabled children. *Neuropsychologia, 26*, 417–423.

Bolte, E. R. (1998). Autism, and Clostridium tetani. *Medical Hypotheses, 51*, 133–144.

Bonde, E. (2000). Comorbidity and subgroups in childhood autism. *European Child & Adolescent Psychiatry, 9*, 7–10.

Boos, R., Gnirs, J., Auer, L., and Schmidt, W. (1987). Controlled acoustic and photic stimulation of the fetus in the last pregnancy trimester. *Zeitschrift fur Geburtshilfe und Perinatologie 191*, 154–161.

Boris, M. and Mandel, F. S. (1994). Foods and additives are common causes of the attention deficit hyperactive disorder in children. *Annals of Allergy, 72*, 462–468.

Borod, J. C. (1992). Interhemispheric and intrahemispheric control of emotion: A focus on unilateral brain damage. *Journal of Consulting and Clinical Psychology, 60*, 339–348.

Borod, J. C. and Caron, H. S. (1980). Facedness and emotion related to lateral dominance, sex and expression type. *Neuropsychologia, 18*, 237–241.

Borod, J. C., Carper, M., Naeser, M., and Goodglass, H. (1985a). Left-handed and right-handed aphasics with left hemisphere lesions compared on nonverbal performance measures. *Cortex, 21*, 81–90.

Borod, J. C., Koff, E., Lorch, M. P., and Nicholas, M. (1985b). The expression and perception of facial emotion in brain-damaged patients. *Neuropsychologia, 24*, 169–180.

Boucsein, W. (1992). Electrodermal activity. New York, NY: Plenum Press.

Bouvard, M. P., Leboyer, M., Launay, J. M., Recasens, C., Plumet, M. H., Waller-Perotte, D., Tabuteau, F., Bondoux, D., Dugas, M., and Lensing, P. (1995). Low-dose naltrexone effects on plasma chemistries and clinical symptoms in autism: A double-blind, placebo-controlled study. *Psychiatry Research, 58*, 191–201.

Bowcker, H., Wills, A., and Ceballos-Baumann, A. et al. (1996). The effect of ethanol on alcohol responsive essential tremor: A positron emission tomography study. *Annals of Neurology, 39*, 652–658.

Bower, J. M. and Kassel, J. (1990). Variability and tactile projection patterns to cerebellar folia crus II-A of the norway rat. *Journal of Comparative Neurology, 302*, 768–778.

Bowers, D., Bauer, R. M., Coslett, H. B., and Heilman, K. M. (1985). Processing of faces by patients with unilateral hemisphere lesions. I. Dissociation between judgments of facial affect and facial identity. *Brain and Cognition, 4*, 258–272.

Bowsher, D. (1967). Compared study of the thalamic projections of 2 localized zones of the bulbar and mesencephalic reticular formations. *Comptes rendus hebdomadaires des seances de l'Academie des sciences, Serie D: Sciences naturelles, 265*, 340–342.

Bracke-Tolkmitt, R., Linden, A., Canavan, A. G. M., Rockstrok, B., Scholz, E., Wessel, K., and Diener, H. C. (1989). The cerebellum contributes to mental skills. *Behavioral Neuroscience, 103*, 442–446.

Bradley, P., Horn, G., and Bateson, P. (1981). Imprinting. An electron microscopic study of chick hyperstriatum ventrale. *Experimental Brain Research, 41*, 115–120.

Bradshaw, J. L. and Nettleton, N. C. (1988). Monaural asymmetries. In K. Hugdahl (Ed.), *Handbook of dichotic listening* (pp. 45–69). New York, NY: John Wiley & Sons.

Braitenberg, V. and Atwood, R. P. (1958). Morphological observation on the cerebellar cortex. *Journal of Comparative Neurology, 109*, 1–34.

Braitenberg, V. and Kemali, M. (1970). Exceptions to bilateral symmetry in the epithalamus of lower vertebrates. *The Journal of Comparative Neurology, 138*, 137–146.

Bram, S. and Meir, M. (1977). A relationship between motor control and language development in an autistic child. *Journal of Autism and Developmental Disorders, 14*, 57–67.

Brandel, J. P., Hirsch, E. C., Malessa, S., Duyckaerts, C., Cervera, P., and Agid, Y. (1991). Differential vulnerability of cholinergic projections to the mediodorsal nucleus of the thymus in senile dementia of alzheimer type and progressive supranuclear palsy. *Neuroscience, 41*, 25–31.

Breaky, J. (1997). The role of diet and behaviour in childhood. *Journal of Pediatric and Child Health, 33*, 190–194.

Breiter, H. C. and Rosen, B. R. (1999). Functional magnetic resonance imaging of brain reward circuitry in the human. *Annals of the New York Academy of Sciences, 877*, 523–547.

Breiter, H. C., Rauch, S. L., Kwong, K. K., Baker, J. R., Weisskoff, R. M., Kennedy, D. N., Kendrick, A. D., Davis, T. L., Jiang, A., Cohen, M. S., Stern, C. E., Belliveau, J. W., Baer, L., O'Sullivan, R. L., Savage, C. R., Jenike, M. A., and Rosen, B. R. (1966). Functional magnetic resonance imaging of symptom provocation in obsessive-compulsive disorder. *Archives of General Psychiatry, 53*, 595–606.

Brenneman, D. E., Hauser, J., Neale, E., Rubinraut, S., Fridkin, M., Davidson, A., and Gozes, I. (1988). Activity-dependent neurotrophic factor: Structure-activity relationships of femtomolar-acting peptides. *Journal of Pharmacology and Experimental Therapeutics, 285*, 619–627.

Brenner, M. W., Gillman, S., Zangwill, O. L., and Farrell, M. (1967). Visuo-motor disability in schoolchildren. *British Medical Journal, 4*, 259–264.

Breslau, N., Brown, G. G., DelDotto, J. E., Kumar, S., Ezhuthachan, S., Andreski, P., and Hufnagle, K. G. (1996). Psychiatric sequelae of low birth weight at 6 years of age. *Journal of Abnormal Child Psychology, 24*, 385–400.

Brett, A. W. and Laatscha, L. (1998). Cognitive rehabilitation therapy of brain-injured students in a public high school setting. *Pediatric Rehabilitation, 2*, 27–31.

Brigge, J. F. and Reale, R. A. (1985). Auditory cortex. In A. Peters and E. G. Jones (Eds.), *Cerebral cortex*. New York, NY: Plenum Press.

Broadbent, D. E. (1954). The role of auditory localization in attention and memory span. *Journal of Experimental Psychology, 47*, 191–196.

Broadbent, D. E. (1958). *Perception and Communication*. New York, NY: Plenum.

Broadbent, D. E. (1965a). Applications of information theory and decision theory to human perception and reaction. *Progress in Brain Research, 17*, 309–320.

Broadbent, D. E. (1965b, October 22). Information processing in the nervous system. *Science, 150*, 457–462.

Broca, P. (1861a). Remarques sur le siege de la faculte du langage articule, suivies d'une observation d'aphemie (perte de la parole). *Bulletins de la Societe Anatomique, 36*, 330–357.

Broca, P. (1861b). Nouvelle observation d'aphemie produite par une lesion de la moitie posterieure des deux-ieme et h-oisieme circonvolutions frontales. *Bulletms de la Sociele Analomique, 6* (Serie II), 398–407.

Brodal, A. (1947). The hippocampus and the sense of smell. A review. *Brain, 70*, 179–222.

Brodal, A. (1981). *Neurological anatomy in relation to clinical medicine* (3rd ed.). New York: Oxford University Press.

Brodal, A., Taber, E., and Walberg, F. (1960). The raphe nuclei of the brain stem in the cat. II. Efferent connections. *Journal of Comparative Neurology, 114*, 239–259.

Brodal, P. (1978). Principles of organization of the monkey corticopontine projection. *Brain Research, 148*, 214–218.

Bronze, M. S. and Dale, J. D. (1993). Epitopes of streptococcal M proteins that evoke antibodies that cross react with human brain. *Journal of Immunology* (Baltimore, MD: 1950), *151*, 2820–2828.

Bronzino, J. D., Stern, W. C., Leahy, J. P., and Morgane, P. J. (1976). Power spectral analysis of EEG activity obtained from cortical and subcortical sites during the vigilance states of the cat. *Brain Research Bulletin, 1*, 285–294.

Brooks, D. J. (1994). PET studies in progressive supranuclear palsy. *Journal of Neural Transmission. Supplementum, 42*, 119–134.

Brooks, D. J., Ibanez, V., and Sawle, G. V. et al. (1990). Differing patterns of striatal 18 F-Dopa uptake in parkinson's disease, multiple system atrophy and progressive supranuclear palsy. *Annals of Neurology, 28*, 547–555.

Brooks, D. J., Ibanez, V., and Sawle, G. V. et al. (1992). Striatal D2 receptor status in patient's with parkinson's disease, striatonigral degeneration, and progressive supranuclear palsy, measured with 11C-Raclopride and positron emission tomography. *Annals of Neurology, 31*, 184–192.

Brooks, H., Jellinger, K., Braak, E., and Bohl, J. (1992). Allocortical neurofibrillary changes in progressive supranuclear palsy. *Acta Neuropathologica* (Berlin), *84*, 478–483.

Brooks, V. B. (1984). Cerebellar functions in motor control. *Human Neurobiology, 2*, 251–260.

Brootsky, S. J. and Kagan, J. (1971). Stability of the orienting reflex in infants to auditory and visual stimuli as indexed by cardiac declaration. *Child Development, 42*, 2066–2070.

Brown, C. S. (1991). Treatment of attention deficit hyperactivity disorder: A critical review. *DICP, The Annals of Pharmacotherapy, 11*, 1207–1213.

Brown, H. D. and Kosslyn, S. M. (1995). Hemispheric differences in visual object processing: Structural versus allocation theories. In R. J. Davidson and K. Hugdahl (Eds.), *Brain asymmetry* (pp. 77–97). Cambridge, MA: MIT Press.

Brown, J. R. and Cramond, J. K. (1974). Displacement effects of television and the child's functional orientation to media. In J. G. Blumler, E. Katz (Eds.), *The uses of mass communications: Current perspectives on gratifications research* (pp. 93–112). Beverly Hills, CA: Sage Publications.

Brown, L. L., Schneider, J. S., and Lidsky, T. I. (1997). Sensory and cognitive functions of the basal ganglia. *Current Opinion in Neurobiology, 7*, 157–163.

Brown, T. E. (2000). Emerging understandings of attention-deficit/hyperactivity disorder and comorbidities. In T. E. Brown (Ed.), *Attention-deficit disorders and comorbidities in children, adolescents, and adults*. Washington, DC: American Psychiatric Press.

Brownell, K. D. and Wadden, T. A. (1992). Etiology and treatment of obesity. *Journal of Consulting and Clinical Psychology, 60*, 505–517.

Bruce, V. and Young, A. W. (1986). Understanding face recognition. *British Journal of Psychology, 77*, 305–327.

Bruggencate, G. T. (1975). Functions of extrapyramidal systems in motor control. Supraspinal descending pathways. *Acta Physiologica, Pharmacologica et Therapeutica Latinoamericana: Organo de la Asociacion Latinoamericana de Ciencias Fisiologicas y [de]* la *Asociacion Latinoamericana de Farmacologia*, 4, 587–610.

Bruner, J. S. (1968). *Processes of cognitive growth: Infancy* (Heinz Werner Memorial Lecture Series, Vol. 3). Worcester, MA: Clark University Press.

Brunet, M., Guy, F., Pilbeam, D., Mackaye, H. T., Likius, A., Ahounta, D., Neauvilain, A., Blondel, C., Bocherens, H., Boisserie, J. R., DeBonis, L., Coppens, Y., Dejax, J., Denys, C., Duringer, P., Eisenmann, V., Fanone, G., Fronty, P., Geralds, D., Lehmann, T., Lihoreau, F., Louchart, A., Majamat, A., Merceron, G., Mouchelin, G., Otero, O., Pelaez Campomanes, P., Ponce, De Leon, M., Rage, J. C., Sapanet, M., Schuster, M., Sudre, J., Tassy, P., Valentine, X., Vignaud, P., Viriot, L., Zazzo, A., and Zollikofer, C. (2002). A new hominid from the upper miocene of chad central Africa. *Nature*, 418, 801.

Brutkowski, A. (1964). Prefrontal cortex and drive inhibition. In J. Warren and K. Akert (Eds.), *The frontal granular cortex and behavior*. New York, NY: McGraw-Hill Book Co.

Bryden, M. P. (1988a). Dichotic studies of the lateralization of affect in normal subjects. In K. Hugdahl (Ed.), *Handbook of dichotic listening: Theory, methods, and research*. Chichester, England: John Wiley & Sons.

Bryden, M. P. (1988b). Does laterality make any difference? Thoughts on the relation between cerebral asymmetry and reading. In D. L. Molfese and S. J. Segalowitz (Eds.), *Brain lateralization in children: Developmental implications* (pp. 509–525). New York, NY: Guilford Press.

Bryden, M. P. and MacRae, L. (1988). Dichotic laterality effects obtained with emotional words. *Neuropsychiatry, Neuropsychology, and Behavioral Neurology*, 1, 171–176.

Bryden, M. P., McManus, I. C., and Steenhuis, R. E. (1991). Handedness is not related to self-reported disease incidence. *Cortex*, 27, 605–611.

Bryden, M. P., Munhall, K., and Allard, F. (1983). Attentional biases and the right-ear effect in dichotic listening. *Brain and Language*, 18, 236–248.

Buchmann, J., and Bülow, B. (1989). Asymmetrische Frühkindliche Kopfgelenksbeweglichkeit. Berlin: Springer.

Buchsbaum, M. S., Wu, J., DeLisi, L. E., Holcomb, H., Kessler, R., Johnson, J., King, A. C., Hazlett, E., Langston, K., and Post, R. M. (1986). Frontal cortex and basal ganglia metabolic rates assessed by positron emission tomography with F2-deoxiglucose in affective illness. *Journal of Affective Disorders*, 10, 137–152.

Buck, R. (1985). Prime theory: An integrated view of motivation and emotion. *Psychological Review*, 92, 389–413.

Buck, R. (1988). *Human motivation and emotion*. New York, NY: John Wiley & Sons.

Buckenhau, K. E. and Yeomans, J. S. (1993). An uncrossed telepontine pathway mediates ipsiversive circling. *Experimental Brain Research*, 54, 11–22.

Buckley, P. F. (1998). The clinical stigmata of aberrant neurodevelopment in schizophrenia. *Journal of Nervous and Mental Disease*, 186, 79–86.

Buhusi, C. V. and Schmajuk, N. A. (1996). Attention, configuration, and hippocampal function. *Hippocampus*, 6, 621–642.

Buitelaar, J. K., Van der Gaag, R. J., and Swaab-Barneveld, H., and Kuiper, M. (1995). Prediction of clinical response to methylphenidate in children with attention-deficit hyperactivity disorder. *Journal of the American Academy of Child and Adolescent Psychiatry*, 34, 1025–1032.

Bunzow, J. R., van Tol, H. H., Grandy, D. K., Albert, P., Salon, J., Christie, M., Machida, C. A., Neve, K. A., and Civelli, O. (1988). Cloning and expression of a rat D2 dopamine receptor cDNA. *Nature*, 336, 783–787.

Burgess, J. R., Stevens, L., Zhang, W., and Peck, L. (2000). Long-chain polyunsaturated fatty acids in children with attention-deficit hyperactivity disorder. *American Journal of Clinical Nutrition*, 71 (Suppl.), 327S–330S.

Burgess, N. (2002). The hippocampus, space, and viewpoints in episodic memory. *Quarterly Journal of Experimental Psychology, A*, 55, 1057–1080.

Bush, G., Frazier, J. A., Rauch, S. L., Seidman, L. J., Whalen, P. J., Jenike, M. A., Rosen, B. R., and Biederman, J. (1999, June 15). Anterior cingulated cortex dysfunction in attention-deficit/hyperactivity disorder revealed by fMRI and the Counting Stroop. *Biological Psychiatry*, 45, 1545–1552.

Bushnell, M. C., Goldberg, M. E., and Robinson, D. L. (1981). Behavioral enhancement of visual responses in monkey cerebral cortex: I. Modulation of posterior parietal cortex related to selected visual attention. *Journal of Neurophysiology*, 46, 755–772.

Butler, N. R. and Goldstein, H. (1973). Smoking in pregnancy and subsequent child development. *British Medical Journal*, 4, 573–575.

Butter, C. M., Kirsch, N. L., and Reeves, G. (1990). The effect of lateralized dynamic stimuli on unilateral spatial neglect following right hemisphere lesions. *Restorative Neurology and Neuroscience*, 2, 39–46.

Butters, N. and Rosvold, H. E. (1968). Effect of septal lesions on resistance to extinction and delayed alternation in monkeys. *Journal of Comparative and Physiological Psychology*, 66, 389–395.

Buxbaum, J. D., Silverman, J. M., Smith, C. J., Greenberg, D. A., Kilifarski, M., Reichert, J., Cook, E. H., Jr., Fang, Y., Song, C. Y., and Vitale, R. (2002). Association between a GABRB3 polymorphism and autism. *Molecular Psychiatry*, 7, 311–316.

Buzsaki, G. (2002). Theta oscillations in the hippocampus. *Neuron*, 33, 325–340.

Byl, N. N. and McKenzie, A. (2000). Treatment effectiveness for patients with a history of repetitive hand use and focal hand dystonia: A planned, prospective follow-up study. *Journal of Hand Therapy*, 13, 289–301.

Byl, N. N. and Melnick, M. (1997). The neural consequences of repetition: Clinical implications of a learning hypothesis. *Journal of Hand Therapy, 10,* 160–174.

Byl, N. N., Merzenich, M. M., Cheung, S., Bedenbaugh, P., Nagarajan, S. S., and Jenkins, W. M. (1997). A primate model for studying focal dystonia and repetitive strain injury: Effects on the primary somatosensory cortex. *Physical Therapy, 77,* 269–284.

Byl, N. N., Merzenich, M. M., and Jenkins, W. M. (1996). A primate genesis model of focal dystonia and repetitive strain injury: I. Learning-induced dedifferentiation of the representation of the hand in the primary somatosensory cortex in adult monkeys. *Neurology, 47,* 508–520.

Cabeza, R. and Nyberg, L. (2000). Neural bases of learning and memory: Functional neuroimaging evidence. *Current Opinion in Neurology, 13,* 415–421.

Cabot, J. B., Wild, J. M., and Cohen, D. H. (1979). Raphe inhibition of sympathetic pre-ganglionic neurons. *Science, 203,* 184–186.

Cahn, W., Pol, H. E., Bongers, M., Schnack, H. G., Mandl, R. C., Van Haren, N. E., Durston, S., Koning, H., Van Der Linden, J. A., and Kahn, R. S. (2002). Brain morphology in antipsychotic-naive schizophrenia: A study of multiple brain structures. *British Journal of Psychiatry* (Suppl.), *43,* S66–S72.

Caine, E. D., McBride, M. C., and Chiverton, P., Bamford, K. A., Rediess, S., and Shiao, J. (1988). Tourette's syndrome in Monroe County school children. *Neurology, 38,* 472–475.

Cairney, S., Maruff, P., Vance, A., Barnett, R., Luk, E., and Currie, J. (2001). Contextual abnormalities of saccadic inhibition in children with attention deficit hyperactivity disorder. *Experimental Brain Research, 141,* 507–518.

Cajal, R. (1911). Histologie du systéme nerveux de l'homme et des vertébrés (2 Vols.). Paris: Maloine.

Calabrese, J. R., Kling, M. A., and Gold, P. W. (1987). Alterations in immunocompetence during stress, bereavement, and depression: Focus on neuroendocrine regulation. *American Journal of Psychiatry, 144,* 1123–1187.

Caldwell, M. (1999, July). Mind over time. *Discover Magazine,* 50–59.

Campbell, K., Engel, H., Timperio, A., Cooper, C., and Crawford, D. (2000). Obesity management: Australian general practitioners' attitudes and practices. *Obesity Research, 8,* 459–466.

Clark, K. (1988). Major depressive disorder predicts cardiac events in patients with coronary artery disease. *Psychosomatic Medicine, 50,* 627–633.

Cammisa, K. M. (1994). Educational kinesiology with learning disabled children: An efficacy study. *Perceptual and Motor Skills, 78,* 105–106.

Campain, R. and Minckler, J. (1976). A note on the gross configurations of the human auditory cortex. *Brain and Language, 3,* 318–323.

Campbell, D. G. (1997). *The Mozart effect: Tapping the power of music to heal the body, strengthen the mind and unlock the creative spirit.* New York, NY: Avon Books.

Campbell, M., Anderson, L. T., Small, A. M., Adams, P., Gonzalez, N. M., and Ernst, M. (1993). Naltrexone in autistic children: Behavioral symptoms and attentional learning. *Journal of the American Academy of Child and Adolescent Psychiatry, 32,* 1283–1291.

Campbell, M., Anderson, L. T., Small, A. M., Locascio, J. J., Lynch, N. S., and Choroco, M. C. (1990). Naltrexone in autistic children: A double-blind and placebo controlled study. *Psychopharmacology Bulletin, 26,* 130–135.

Cancelliere, A. E. and Kertesz, A. (1990). Lesion localization in acquired deficits of emotional expression and comprehension. *Brain and Cognition, 13,* 133–147.

Cantwell, D. P. (1972). Psychiatric illness in the families of hyperactive children. *Archives of General Psychiatry, 27,* 414–417.

Cantwell, D. P. (1996). Attention deficit disorder: A review of the past 10 years. *Journal of the American Academy of Child and Adolescent Psychiatry, 35,* 978–987.

Cantwell, D. P. and Baker, L. (1991). Association between attention deficit hyperactivity disorder and learning disorders. *Journal of Learning Disabilities, 24,* 88–95.

Caparros-Lefebvre, D., Blond, S., and Vermersch, P. Pecheux, N., Guieu, J. D., and Petit, H. (1993). Chronic thalamic stimulation improves tremor in levodopa induced dyskinesias in parkinson's disease. *Journal of Neurology, Neurosurgery, and Psychiatry, 56,* 268–273.

Caplan, P. J. and Kinsbourne, M. (1976). Baby drops the rattle: Asymmetry of duration of grasp by infants. *Child Development, 47,* 532–534.

Carlen, M., Cassidy, R. M., Brismar, H., Smith, G. A., Enquist, L. W., and Frisen, J. (2002). Functional integration of adult-born neurons. *Current Biology, 12,* 606–608.

Carlson, J. N., Fitzgerald, L. W., Keller, R. W., Jr., and Glick, S. D. (1991). Side and region dependent changes in dopamine activation with various durations of restraint stress. *Brain Research, 550,* 313–318.

Carlson, M. and Earls, F. (1997). Psychological and neuroendocrinological sequelae of early social deprivation in institutionalized children in Romania. *Biological Psychiatry, 48,* 976–980.

Carlson, M., Hubel, D. H., and Wiesel, T. N. (1986). Effects of monocular exposure to oriented lines on monkey striate cortex. *Brain Research, 390,* 71–81.

Carlson, S., Pertovaara, A., and Tanila, H. (1987). Late effects of early binocular visual deprivation on the function of Brodmann's area 7 of monkeys (Macaca arctoides). *Brain Research, 430,* 101–111.

Carlsson, M. L. (1999). Hypothesis: Is infantile autism a hypoglutamatergic disorder? Relevance of glutamate – serotonin interactions for pharmacotherapy. *Trends in Neurosciences, 22,* 273–280.

Carmon, A. and Nachshon, I. (1973). Ear asymmetry in perception of emotional words. *Neuropsychiatry, Neuropsychology, and Behavioral Neurology, 1,* 171–176.

Carrick, F. R. (1997). Changes in brain function after manipulation of the cervical spine. *Journal of Manipulative and Physiological Therapeutic, 20,* 529–545.

Carter, R. L., Hohenegger, M. K., and Satz, P. (1982). Aphasia and speech organization in children. *Science, 218,* 797–799.

Casat, C. D., Pleasants, D. Z., and Van Wyck Fleet, J. (1987). A double-blind trial of bupropion in children with attention deficit disorder. *Psychopharmacology Bulletin, 87,* 120–122.

Case-Smith, J. (1991). The effects of tactile defensiveness and tactile discrimination on in-hand manipulation. *American Journal of Occupational Therapy, 45,* 811–818.

Case-Smith, J. and Miller, H. (1999). Occupational therapy with children with pervasive developmental disorders. *American Journal of Occupational Therapy, 53,* 506–513.

Casler, J. G. and Cook, J. R. (1999). Cognitive performance in space and analogous environments. *International Journal of Cognitive Ergonomics, 3,* 351–372.

Castellano, M. A., Diaz-Palarea, M. D., Barroso, J., and Rodriguez, M. (1989). Behavioral lateralization in rats and dopaminergic system: Individual and population laterality. *Behavioral Neuroscience, 103,* 46–53.

Castellanos, F. X. (1999). Stimulants and Tic disorders: From dogma to data. *Archives of General Psychiatry, 56,* 337–338.

Castellanos, F. X., Giedd, J. N., Berquin, P. C., Walter, J. M., Sharp, W., Tran, T., Vaituzis, A. C., Blumenthal, J. D., Nelson, J., Bastian, T. M., Zijdenbos, A., Evans, A. C., and Rapoport, J. L. (2001). Quantitative brain magnetic resonance imaging in girls with attention-deficit/hyperactivity disorder. *Archives of General Psychiatry, 58,* 289–295.

Castellanos, F. X., Giedd, J. N., Eckburg, P., Marsh, W. L., Vaituzis, A. C., Kaysen, D., Hamburger, S. D., and Rapoport, J. L. (1994). Quantitative morphology of the caudate nucleus in attention deficit hyperactivity disorder. *American Journal of Psychiatry, 151,* 1791–1796.

Castellanos, F. X., Giedd, J. N., Marsh, W. L., Hamburger, S. D., Vaituzis, A. C., Dickstein, D. P., Sarfatti, S. E., Vauss, Y. C., Snell, J. W., Lange, N., Kaysen, D., Krain, A. L., Ritchie, G. F., Rajapakse, J. C. and Rapoport, J. L. (1996). Quantitative brain magnetic resonance imaging in attention-deficit hyperactivity disorder. *Archives of General Psychiatry, 53,* 607–616.

Castellanos, F. X., Marvasti, F. F., Ducharme, J. L., Walter, J. M., Israel, M. E., Krain, A., Pavlovsky, C., and Hommer, D. W. (2000). Executive function oculomotor tasks in girls with ADHD. *Journal of the American Academy of Child and Adolescent Psychiatry, 39,* 644–650.

Carter, C. M., Urbanowicz, M., Hemsley, R., Mantilla, L., Strobel, S., Graham, P. J., and Taylor, E. (1993). Effects of a few food diet in attention deficit disorder. *Archives of Disease in Childhood, 69,* 564–568.

Ceccarelli, I., Masi, F., Fiorenzani, P., and Aloisi, A. M. (2002). Sex differences in the citrus lemon essential oil-induced increase of hippocampal acetylcholine release in rats exposed to a persistent painful stimulation. *Neuroscience Letters, 330,* 25–28.

Cechetto, D. F., Ciriello, J., and Calaresu, F. R. (1983). Afferent connections to cardiovascular sites in the amygdala: A horseradish peroxidase study in the cat. *Journal of the Autonomic Nervous System, 8,* 97–110.

Cechetto, D. F. and Saper, C. B. (1990). Role of the cerebral cortex in autonomic functioning. In A. D. Loewy and K. M. Spyer (Eds.), *Central regulation of autonomic functioning.* New York, NY: Oxford University Press.

Celiberti, D. A., Bobo, H. E., Kelly, K. S., Harris, S. L., and Handleman, J. S. (1997). The differential and temporal effects of antecedent exercise on the self-stimulatory behavior of a child with autism. *Research in Developmental Disabilities, 18,* 139–50.

Cerati, D. and Schwartz, P. J. (1991). Single cardiac vagal fiber activity, acute myocardial ischemia, and risk for sudden death. *Circulation, 69,* 1389–1401.

Cermak, S., Coster, W., and Drake, L. (1980). Representational and non-representational gestures in boys with learning disabilities. *American Journal of Occupational Therapy, 34,* 19–26.

Cesaroni, L. and Garber, M. (1997). Exploring the experience of autism through firsthand accounts. *Biological Psychiatry, 42,* 1148–1156.

Chagnon, S. and Blery, M. (1982). Sprains and luxations of the cervical spine in children. Report on 17 cases. *Journal de Radiologie, 63,* 465–470.

Chakrabarti, S. and Fombonne, E. (2001). Pervasive developmental disorders in preschool children. *The Journal of the American Medical Association, 285,* 3093–3099.

Chan, A. S., Ho, Y. C., and Cheung, M. C. (1998). Music training improves verbal memory. *Nature, 396,* 128.

Changeux, J. P. (1980). Genetic determinism and epigenesis of the neuronal network: Is there a biological compromise between Chomsky and Piaget? In M. Piattelli-Palmarini (Ed.), *Language and learning: The debate between Jean Piaget and Noam Chomsky.* Cambridge, MA: Harvard University Press.

Changeux, J. P. (1981). The acetylcholine receptor: An "allosteric" membrane protein. In *Harvey Lectures,* (pp. 85–254, 75), San Diego: Academic Press.

Changeux, J. P. (1983). Concluding remarks: On the singularity of nerve cells and its ontogenesis. *Progress in Brain Research, 58,* 465–478.

Changeux, J. P. (1985). *Neuronal man: The biology of mind.* New York: Oxford University Press.

Changeux, J. P. and Dennis, S. G. (1982). Signal transduction across cellular membranes. *Neurosciences Research Program Bulletin, 20,* 267–426.

Chase, M. H. and Harper, R. M. (1971). Somatomotor and visceromotor correlates of operantly conditioned 12–14 C-SEC sensorimotor cortical activity. *Electroencephalography and Clinical Neurophysiology, 31,* 85–92.

Chatterjee, A. (1998a). Feeling frontal dysfunction: Facilitory paratonia and the regulation of motor behavior. *Neurology, 51*, 937–939.

Chatterjee, A. (1998b). Motor minds and mental models in neglect. *Brain and Cognition, 37*, 339–349.

Chatterjee, A., Mennemeier, M., and Heilman, K. M. (1992a). A stimulus-response relationship in unilateral neglect: The power function. *Neuropsychologia, 30*, 1101–1108.

Chatterjee, A., Mennemeier, M., and Heilman, K. M. (1992b). Search patterns in neglect. *Neuropsychologia, 30*, 657–672.

Chatterjee, A., Yapundich, R., Menneneier, M., Mountz, J. M., Inampundi, C., Pan, J. W., and Mitchell, G. W. (1997). Thalamic thought disorder. *Cortex, 33*, 419–440.

Chen, R., Cohen, L. G., and Hallett, M. (2002). Nervous system reorganization following injury. *Neuroscience, 111*, 761–773.

Chess, S. (1977). Evolution of behavior disorder in a group of mentally retarded children. *Journal of the American Academy of Child Psychiatry, 16*, 4–18.

Cheverad, J. M., Falk, D., Hildebolt, C., Moore, A. J., Helmkamp, R. C., and Vannier, M. (1990). Heritability and association of cortical petalias in rhesus macaques (*Macaca-mulatta*). *Brain, Behavior and Evolution, 35*, 368–372.

Chi, J. G., Dooling, E. C., and Gilles, F. H. (1977). Gyral development of the human brain. *Annals of Neurology, 1*, 86–93.

Child Health Alert (2002, January).

Chiu, H. S. (1995). Psychiatric aspects of progressive supranuclear palsy. *Annals of General Hospital Psychiatry, 17*, 135–143.

Chouinard, G., Goodman, W., Greist, J., Jenike, M., Rasmussen, S., White, K., Hackett, E., Gaffney, M., and Bick, P. A. (1990). Results of a double-blind placebo controlled trial of a new serotonin uptake inhibitor, sertraline, in the treatment of obsessive-compulsive disorder. *Psychopharmacology Bulletin, 26*, 279–284.

Choy, S. P. (2002). *Access and persistence: Findings from 10 years of longitudinal research on students*. Washington, DC: American Council on Education, Center for Policy Analysis.

Christianson, S. A., Nilsson, L. G., and Silfvenius, H. (1986). Initial memory deficits and subsequent recovery in two cases of head trauma. *Scandinavian Journal of Psychology, 28*, 267–280.

Christos, G. A. (1990). Infant dreaming and fetal memory: A possible explanation of sudden infant death syndrome. *Archives of Disease in Childhood, 65* (10 Spec No), 1072–1075.

Chronister, R. B., Sikes, R. W., and White, L. E., Jr. (1976). The septohippocampal system: Significance of the subiculum. In J. F. DeFrance (Ed.), *The septal nuclei* (pp. 115–132). New York: Plenum Press.

Chudler, E. H. and Dong, W. K. (1995). The role of the basal ganglia in nociception and pain. *Pain, 60*, 3–38.

Chugani, H. T., Behen, M. E., Muzik, O., Juhasz, C., Nagy, F., and Chugani, D. C. (2001). Local brain functional activity following early deprivation: A study of postinstitutionalized Romanian orphans. *Neuroimage, 14*, 1290–1301.

Chugani, D. C., Sundram, B. S., Behen, M., Lee, M.-L., and Moore, G. J. (1999). Evidence of altered energy metabolism in autistic children. *Progress in Neuropsychopharmacology & Biological Psychiatry, 23*, 635–641.

Church, C. C. and Coplan, J. (1997). The high-functioning autistic experience: Birth to preteen years. *Clinical Chemistry, 43*, 543–545.

Cipolla-Neto, J., Horn, G., and McCabe, B. J. (1982). Hemispheric asymmetry and imprinting: The effect of sequential lesions to the hyperstriatum ventrale. *Experimental brain research. Experimentelle Hirnforschung. Experimentation cerebrale, 48*, 22–27.

Clark, C. R., Geffen, G. M., and Geffen, L. B. (1987). Catecholamines and attention I: Animal and clinical studies. *Neuroscience and Biobehavioral Reviews, 11*, 341–352.

Classen, J., Gerloff, C., Honda, M., and Hallet, M. (1998). Integrative visuomotor behavior is associated with interroginally coherent oscillations in the human brain. *Journal of Neurophysiology, 79*, 1567–1573.

Clay, T. H., Gualtieri, C. T., and Evens, R. W. et al. (1988). Clinical and neuropsychological effects of the novel antidepressant bupropion. *Psychopharmacology Bulletin, 88*, 143–148.

Clemente, C. D. and Chase, M. H. (1973). Neurological substrates of aggressive behaviour. *Annual Review of Physiology, 35*, 329–356.

Clomipramine Collaborative Study Group. (1991). Clomipramine in the treatment of patients with obsessive-compulsive disorder. *Archives of General Psychiatry, 48*, 730–738.

Clow, A., Lambert, S., Evans, P., Hucklebridge, F., and Higuchi, K. (2003). An investigation into asymmetrical cortical regulation of salivary S-IgA in conscious man using transcranial magnetic stimulation. *International Journal of Psychophysiology, 47*, 57–64.

Cohen, D. J. (2001). Into life: Autism, Tourette's syndrome and the community of clinical research. *Israel Journal of Psychiatry and Related Sciences, 38*, 226–234.

Cohen, M., Hynd, G., and Hugdahl, K. (1992). Dichotic listening performance in subtypes of developmental dyslexia and a left temporal lobe brain humor contrast group. *Brain and Language, 42*, 187–202.

Cohen, R. M., Semple, W. E., Gross, M., Holocomb, H. J., Dowling, S. M., and Nordahl, T. E. (1988). Functional localization of sustained attention. *Neuropsychiatry, Neuropsychology and Behavioral Neurology, 1*, 3–20.

Cohen, D. J., Donnellan, A. M., and Rhea, P. (Eds.) (1987). *Handbook of autism and pervasive developmental disorders* (pp. 3–19). New York, NY: Wiley.

Coleman, M. (1976). *The autistic syndromes*. New York, NY: American Elsevier Publishing Company, pp. 197–205.

Coles, C. D., Platzman, K. A., Raskind-Hood, C. L., Brown, R. T., Falek, A., and Smith, I. E. (1997). A comparison of children affected by prenatal alcohol exposure and attention deficit, hyperactivity disorder. *Alcohol and Clinical Experimental Research, 21*, 150–161.

Colling-Saltin, A. S. (1978). Enzyme histochemistry on skeletal muscle of the human soetus. *Journal of Neurological Science, 39*, 185–196.

Collins, J. J. and De Luca, C. J. (1993). Open-loop and closed-loop control of posture: A random-walk analysis of center-of-pressure trajectories. *Experimental Brain Research, 95*, 308–318.

Comi, A. M., Zimmerman, A. W., Frye, V. H., Law, P. A., and Peeden, J. N. (1999). Familial clustering of autoimmune disorders and evaluation of medical risk factors in autism. *Journal of Child Neurology, 14*, 388–394.

Comings, D. E. and Comings, B. G. (1984). Tourette's syndrome and attention deficit disorder with hyperactivity: Are they genetically related? *Journal of the American Academy of Child Psychiatry, 23*, 138–146.

Comings, D. E. and Comings, B. G. (1985). Tourette syndrome: Clinical and Psychological aspects of 250 cases. *American Journal of Human Genetics, 37*, 435–450.

Comings, D. E. and Comings, B. G. (1986). Evidence for an X-linked modifier gene affecting the expression of Tourette syndrome and its relevance to the increased frequency of speech, cognitive, and behavioral disorders in males. *Proceedings of the National Academy of Sciences, USA, 83*, 2551–2558.

Comings, D. E. and Comings, B. G. (1987). A controlled study of Tourette syndrome. I. Attention deficit disorder learning disorders, and school problems. *American Journal of Human Genetics, 41*, 701–741.

Comings, D. E. and Comings, B. G. (1988). A controlled study of Tourette syndrome-revisited: A reply to the letter of Pauls et al. *American Journal of Human Genetics, 43*, 209–217.

Comings, D. E. and Comings, B. G. (1998). Tourette's syndrome and attention deficit disorder. In D. J. Cohen, R. D. Bruun, J. F. Leckman (Eds.), *Tourette's syndrome and Tic disorders: Clinical understanding and treatment* (pp. 120–135). New York: Wiley.

Conce, M. and Delwaide, P. J. (1984). Clinical neurophysiological assessment of parkinson's disease. In R. G. Hassler and J. F. Christ (Eds.), *Advances in neurology* (pp. 365–372). New York: Raven Press.

Condon, W. S. and Ogston, W. D. (1966). Sound film analysis of normal and pathological behavior patterns. *The Journal of Nervous and Mental Disease, 143*, 338–347.

Condon, W. S. and Sander, L. W. (1974). Neonate movement is synchronized with adult speech: Interactional participation and language acquisition. *Science, 183*, 99–101.

Conners, C. K., Casat, C., and Gualtieri, C. T. et al. (1996). Bupropion hydrochloride in attention-deficit disorder with hyperactivity. *Journal of the American Academy of Child and Adolescent Psychiatry, 35*, 1314–1321.

Connolly, A. M., Chez, M. G., Pestronk, A., Arnold, S. T., Mehta, S., and Deuel, K. (1999). Serum autoantibodies to brain in Landau-Kleffner variant, autism, and other neurologic disorders. *The Journal of Pediatrics, 134*, 607–613.

Connor, D. F. (2002). Preschool attention deficit hyperactivity disorder: A review of prevalence, diagnosis, neurobiology, and stimulant treatment. *Journal of Developmental and Behavioral Pediatrics, 23* (Suppl.), S1–S9.

Conrad, B., Benecke, R., Carnehl, J., Hohne, J., and Meinck, H. M. (1983). Pathophysiological aspects of human locomotion. *Advances in Neurology, 39*, 717–726.

Conrad, L. C. A., Leonard, C. M., and Pfaff, D. W. (1974). Connections of the median and dorsal raphe nuclei in the rat: An autoradiographic and degeneration study. *The Journal of Comparative Neurology, 156*, 179–206.

Conrad, L. C. A. and Pfaff, D. W. (1976a). Efferents from medial basal forebrain and hypothalamus in the rat. I. An autoradiographic study of the medial preoptic area. *The Journal of Comparative Neurology, 169*, 185–220.

Conrad, L. C. A. and Pfaff, D. W. (1976b). Efferents from the medial basal forebrain and hypothalamus in the rat. II. An autoradiographic study of the anterior hypothalamus. *The Journal of Comparative Neurology, 169*, 221–261.

Contreras-Vidal, J. L., Grossberg, S., and Bullock, D. A. (1997). Neural model of cerebellar learning for arm movement control: Cortico-spino-cerebellar dynamics. *Learning & Memory* (Cold Spring Harbor, NY) *3*, 475–502.

Cook, N. (1984). Homotopic callosal inhibition. *Brain and Language, 23*, 116–125.

Coons, P. M., Bowman, E. L., and Milstein, V. (1988). Multiple personality disorder: A clinical investigation of 50 cases. *Journal of Nervous and Mental Disease, 176*, 519–527.

Cooper, D. M. (1994). Evidence for and mechanisms of exercise modulation of growth an overview. *Medical Science in Sports and Exercise, 26*, 733–740.

Cooper, J. D. and Phillipson, O. T. (1993). Central neuroanatomical organisation of the rat visuomotor system. *Progress in Neurobiology, 41*, 209–279.

Cooper, R. L., Winslow, J. L., Govind, C. K., and Atwood, H. L. (1996). Synaptic structural complexity as a factor enhancing probability of calcium-mediated transmitter release. *Journal of Neurophysiology, 75*, 2451–2466.

Cooper, S. (1960). Muscle spindles and other muscle receptors. In G. Bourne (Ed.), *The structure and function of muscle* (pp. 331–420). New York: Academic Press.

Coppoletta, J. M. and Wolbach, M. D. (1933). Body length and organ weights of infants and children. *American Journal of Pathology, 9*, 55–70.

Corballis, M. C. (1983). *Human laterality.* New York, NY: Academic Press.

Corballis, M. C. (1989). Laterality and human evolution. *Psychological Review, 96,* 492–505.

Corballis, M. C. (1991). *The lopsided ape: Evolution of the generative mind.* Oxford, England: Oxford University Press.

Corbier, P., Roffi, J., and Rhoda, J. (1983). Female sexual behavior in male rats: Effect of hour of castration at birth. *Physiology and Behavior, 30,* 613–616.

Corley, K. M. and McComas, A. J. (1984). Contrasting effects of suspension on hind limb muscles in the Hamster. *Experimental Neurology, 85,* 30–40.

Coryell, J. (1985). Infant rightward asymmetries predict right-handedness in childhood. *Neuropsychologia, 23,* 269–271.

Coryell, J. and Henderson, A. (1979). Role of the asymmetrical tonic reflex in hand visualization in normal infants. *The American Journal of Occupational Therapy, 33,* 225–260.

Coryell, J. and Michel, G. F. (1978). How supine postural preferences of infants can contribute toward the development of handedness. *Infant Behavior and Developmental, 1,* 245–257.

Costa, L. D., Vaughan, H. G., Horowitz, M., and Ritter, W. (1969). Patterns of behavior deficit associated with visual spatial neglect. *Cortex, 5,* 242–263.

Cotman, C. W. and Berchtold, N. C. (2002). Exercise: A behavioral intervention to enhance brain health and plasticity. *Trends in Neurosciences, 25,* 295–301.

Cotterill, R. M. (2001). Cooperation of the basal ganglia, cerebellum, sensory cerebrum and hippocampus: Possible implications for cognition, consciousness, intelligence and creativity. *Progress in Neurobiology, 64,* 1–33.

Cottraux, J., Mollard, E., Bouvard, M., Marks, I., Sluys, M., Nury, A. M., Rouge, R., and Cialdella, P. (1990). A controlled study of fluvoxamine and exposure in obsessive-compulsive disorder. *International Clinical Psychopharmacology, 5,* 17–30.

Corteen, R. S. and Wood, B. (1972). Autonomic responses to shock-associated words in an unattended channel. *Journal of Experimental Psychology, 94,* 308–313.

Cowan, W. M., Guillery, R. W., and Powell, T. P. S. (1964). The origin of the mamillary peduncle and other hypothalamic connexions from the midbrain. *Journal of Anatomy (London), 98,* 345–363.

Courchesne, E. (1991). Neuroanatomic imaging in autism. *Pediatrics, 87,* 781–790.

Courchesne, E., Hesselink, J. R., Jernigan, T. L., and Yeung-Courchesne, R. (1987). Abnormal neuroanatomy in a nonretarded person with autism. Unusual findings with magnetic resonance imaging. *Archives of Neurology, 3,* 335–341.

Courchesne, E., Townsend, J., Akshoomoff, N. A., Saitoh, O., Yeung-Courchesne, R., Lincoln, A. J., James, H. E., Haas, R. H., Schreibman, L. and Lau, L. (1994a). Impairment in shifting attention in autistic and cerebellar patients. *Behavioral Neuroscience, 108,* 848–865.

Courchesne, E., Townsend, J., and Saitoh, O. (1994b). The brain in infantile autism: Posterior fosa structures are abnormal. *Neurology, 44,* 214–223.

Courchesne, E., Saitoh, O., Yeung-Courchesne, R., Press, G. A., Lincoln, A. J., Haas, R. H. and Schreibman, L. (1994c). Abnormality of cerebellar vermian lobules VI and VII in patients with infantile autism: Identification of hypoplastic and hyperplastic subgroups with MR imaging. *American Journal of Roentgenology, 162,* 123–130.

Courchesne, E., Yeung-Couchesne, R., Press, G. A., Hesselink, J. R., Andjernigan, T. L. (1988). Hypoplasia of cerebellar lobules VI and VII in autism. *New England Journal of Medicine, 318,* 1349–1354.

Cragg, B. G. and Hamlyn, L. H. (1959). Histologic connections and electrical and autonomic responses evoked by stimulation of the dorsal fornix in the rabbit. *Experimental Neurology, 1,* 187–213.

Cramer, K. S. and Sur, M. (1995). Activity-dependent remodeling of connections in the mammalian visual system. *Current Opinion in Neurobiology, 5,* 106–111.

Crelin, E. S. (1987). *The human vocal tract.* New York, NY: Vantage.

Crewther, D. P. and Crewther, S. G. (2002). Refractive compensation to optical defocus depends on the temporal profile of luminance modulation of the environment. *Neuroreport, 13,* 1029–1032.

Crinella, F., Eghbalieh, B., and Swanson, J. M. et al. (1997). *Nigrostriatal dopaminergic structures and executive functions.* Paper presented at the annual convention of the American Psychological Association, Chicago, IL.

Cristino, L. and Bullock, T. H. (1984). Cerebellum mediates modality-specific modulation of sensory responses of midbrain and forebrain in rat. *Proceedings of the National Academy of Sciences, USA, 81,* 2917–2920.

Critchley, M. (1966). *The parietal lobes.* New York, NY: Hafner.

Croonenberghs, J., Bosmans, E., Deboutte, D., Kenis, G., and Maes, M. (2002). Activation of the inflammatory response system in autism. *Neuropsychobiology, 45,* 1–6.

Crosby, E. C. and Humphrey, T. (1941). Studies of the vertebrate telencephalon. II. The nuclear pattern of the anterior olfactory nucleus, tuberculum olfactorium and the amygdaloid complex in adult man. *The Journal of Comparative Neurology, 74,* 309–335.

Crow, T. J., Ball, J., Bloom, S. R., Brown, R., Bruton, C. J., Colter, N., Frith, C. D., Johnstone, E. C., Owens, D. G., Roberts, G. W. (1989). Schizophrenia as an anomaly of development of cerebral asymmetry. A postmortem study and a proposal concerning the genetic basis of the disease. *Archives of General Psychiatry, 46,* 1145–1150.

Crystal, S. R. and Bernstein, I. L. (1995). Morning sickness: Impact on offspring salt preference. *Appetite, 30,* 297–307.

Crystal, S. R. and Bernstein, I. L. (1999). Infant salt preference and mother's morning sickness. *Integrative*

Physiological and Behavioral Science: The Official Journal of the Pavlovian Society, 67, 181–187.

Crystal, S. R., Bowen, D. J., and Bernstein, I. L. (1999). Morning sickness and salt intake, food cravings, and food aversions. *Science, 208,* 1174–1176.

Cullinen, B. E. (1997). Transactional literature discussions: Engaging students in the appreciation and understanding of literature. *Reading Teacher, 51,* 86–96.

Cummings, J. L. (1993). Frontal-subcortical circuits and human behavior. *Archives of Neurology, 50,* 873–880.

Cummings, J. L. (1995). Anatomic and behavioral aspects of frontal-subcortical circuits. *Annals of the New York Academy of Sciences, 769,* 1–13.

Cummings, J. L. and Cunningham, K. (1992). Obsessive-compulsive disorder in huntington's disease. *Biological Psychiatry, 31,* 263–270.

Cupps, T. R. and Fauci, A. S. (1982). Corticosteroid-mediated immunoregulation in man. *Immunological Reviews, 65,* 133–155.

Damasio, A. R. (1985). Prosopagnosia. *Trends in Neuroscience, 8,* 132–135.

Damasio, A. R. (1989a). The brain binds entities and events by multiregional activation from convergence zones. *Neural Computation, 1,* 123–132.

Damasio, A. R. (1989b). Time-locked regional retroactivation: A systems level proposal for the neural substrate of recall and recognition. *Cognition, 33,* 25–62.

Damasio, A. R. (1994). *Descarte's error.* New York, NY: G. P. Putnam.

Damasio, A. R. and Tranel, D. (1988). Domain-specific amnesia for social knowledge. *Society for Neuroscience, 14,* 1289.

D'Amelio, F., Fox, R. A., Wu, L. C., Daunton, N. G., and Corcoran, M. L. (1998a). Effects of microgravity on muscle and cerebral cortex: A suggested interaction. *Advances in Space Research, 22,* 235–244.

D'Amelio, F., Wu, L. C., Fox, R. A., Daunton, N. G., Corcoran, M. L., and Polyakov, I. (1998b). Hypergravity exposure decreases gamma-aminobutyric acid immunoreactivity in axon terminals contacting pyramidal cells in the rat somatosensory cortex: A quantitative immunocytochemical image analysis. *Journal of Neuroscience Research, 53,* 135–142.

Daniels, W. W., Warren, R. P., Odell, J. D., Maciulis, A., Burger, R. A., Warren, W. L., and Torres, A. R. (1996). Increased frequency of the extended or ancestral haplotype B44-SC30-DR4 in autism. *Neuropsychobiology, 32,* 120–123.

D'Antona, R., Baron, J. C., Samson, Y., Serdaru, M., Viader, F., Agid, Y., and Cambier, J. (1985). Subcortical dementia. Frontal cortex hypometabolism detected by positron tomography in patients with progressive supranuclear palsy. *Brain, 108,* 785–799.

Darian-Smith, C., Tan, A., and Edwards, S. (1999). Comparing thalamocortical and corticothalamic microstructure and spatial reciprocity in the macaque ventral posterolateral nucleus (VPLc) and medial pulvinar. *The Journal of Comparative Neurology, 410,* 211–234.

Daruna, J. H. and Morgan, J. E. (1990). Psychosocial effects on immune function: Neuroendocrine pathways. *Psychosomatics, 31,* 4–12.

Daum, I., Ackerman, H., Schugens, M. M., Reimold, C., Dichgans, J., and Birbaumer, N. (1993). The cerebellum and cognitive functions in humans. *Behavioral Neuroscience, 107,* 411–419.

Davey, G. L. C. (1987). *Cognitive processes and pavlovian conditioning.* Chichester, England: John Wiley & Sons.

Davidson, R. J. (1984). Affect, cognition and hemispheric specialization. In C. E. Izard, J. Kagan, and R. Zajonc (Eds.), *Emotion, cognition and behavior.* New York, NY: Cambridge University Press.

Davidson, R. J. (1987). Cerebral asymmetry and the nature of emotion: Implications for the study of individual differences and psychopathology. In R. Takahashi, P. Flor-Henry, J. Gruzelier, and S. Niwa (Eds.), *Cerebral dynamics, laterality and psychopathology.* New York, NY: Elsevier.

Davidson, R. J. (1988). EEG measures of cerebral asymmetry: Conceptual and methodological issues. *International Journal of Neuroscience, 39,* 71–89.

Davidson, R. J. (1992a). Anterior cerebral asymmetry and the nature of emotion. *Brain and Cognition, 20,* 125–151.

Davidson, R. J. (1992b). Prolegomenon to the structure of emotion: Gleanings from neuropsychology. *Cognition and Emotion, 6,* 245–268.

Davidson, R. J. (1993). Cerebral asymmetry and emotion: Conceptual and methodological conundrums. *Cognition and Emotion, 7,* 115–138.

Davidson, R. J., Abercrombie, H., Nitschke, J. B., and Putnam, K. (1999). Regional brain function, emotion and disorders of emotion. *Current Opinion in Neurobiology, 9,* 228–234.

Davidson, R. J., Abercrombie, H., Nitschke, J. B., and Putnam, K. (2002). Regional brain function, emotion and disorders of emotion. *Annual Review of Psychology, 53,* 545–574.

Davidson, R. J., Chapman, J. P., and Chapman, L. J. (1987). *Task-dependent EEG asymmetry discriminates between depressed and non-depressed subjects.* Paper presented at the Society for Psychophysiologic Research, Amsterdam.

Davidson, R. J., Chapman, J. P., Chapman, L. P., and Henriques, J. B. (1990a). Asymmetrical brain electrical activity discriminates between psychometrically-matched verbal and spatial cognitive tasks. *Psychophysiology, 27,* 528–543.

Davidson, R. J., Ekman, P., Saron, C. D., Senulis, J. A., and Friesen, W. V. (1990b). Approach withdrawal and cerebral asymmetry: Emotional expression and brain physiology I. *Journal of Personality and Social Psychology, 58,* 330–341.

Davidson, R. J. and Fox, N. A. (1989). Frontal brain asymmetry predicts infants response to maternal separation. *Journal of Abnormal Psychology, 98,* 127–131.

Davidson, R. J. and Hugdahl, K. (1995). *Brain asymmetry,* Boston, MA: MIT Press.

Davidson, R. J., Pizzagalli, D., Nitschke, J. B., and Putnam, K. (1994). Depression: Perspectives from affective neuroscience. *The Journal of Clinical Psychiatry, 55* (71–81 Suppl. A); discussion 82, 98–100.

Davidson, R. J., and Sutton, S. K. (1995). Affective neuroscience: The emergence of a discipline. *Current Opinion in Neurobiology, 5*, 217–224.

Davidson, R. J. and Tomarken, A. J. (1989). Laterality and emotion: An electrophysiological approach. In F. Boller and J. Grafman (Eds.), *Handbook of neuropsychology* (Vol. 3, pp. 419–441). Amsterdam: Elsevier.

Dawson, G. (1988). Cerebral lateralization in autism: Clues to its role in language and affective development. In D. L. Molfese and S. J. Segalowitz (Eds.), *Brain lateralization in children: Developmental implications* (pp. 437–461). New York, NY: Guilford Press.

Dawson, G. (1996). Brief report: Neuropsychology of autism: A report on the state of the science. *Journal of Autism and Developmental Disorders, 26*, 179–184.

Dawson, J. L. (1972). Effects of sex hormones on cognitive style in rats and men. *Behavioral Genetics, 2*, 21–42.

Dawson, M. E. and Schell, A. M. (1982). Electrodermal responses to attended or nonattended significant stimuli during dichotic listening. *Journal of Experimental Psychology: Human Perception and Performance, 8*, 82–86.

Derryberry, D. and Tucker, D. M. (1991). The adaptive base of the neural hierarchy: Elementary motivational controls on network function. In R. Dienstbier (Ed.), *Nebraska symposium on motivation*. Lincoln, NE: University of Nebraska Press.

Deacon, T. W. (1992). Cortical connections of the inferior arcuate sulcus cortex in the macaque brain. *Brain Research, 573*, 8–26.

De Campli, W. M. (1987). Medical problems associated with long duration space light. In G. L. Burdett and Gasos G. A. Sossen (Eds.), *The human crust in space*. 24th Goddard Memorial Symposium. Santiago: Univelt.

De Casper, A. J. and Fifer, W. P. (1980). Of human bonding: Newborns prefer their mothers' voices. *Science, 208*, 1174–1176.

De Casper, A. J. and Prescott, P. A. (1984). Human newborns' perception of male voices: Preference, discrimination, and reinforcing value. *Developmental Psychobiology, 17*, 481–491.

Decety, J. and Ingvar, D. H. (1990). Brain structures participating in mental simulation of motor behavior: A neuropsychological interpretation. *Acta Psychologica (Amsterdam), 73*, 13–34.

Decety, J., Jeannerod, N., and Prablanc, C. (1989). The timing of mentally represented actions. *Brain Research, 34*, 35–42.

Decety, J., Philippon, B., and Ingvar, D. H. (1988). RCBF landscapes lung motor performance and motor ideation of a graphic gesture. *European Archive of Psychiatry and Neurological Science, 238*, 33–39.

Decety, J., Sjoholm, H., Ryding, E., Stenberg, G., and Ingvar, D. H. (1990). The cerebellum participates in mental activity: Tomographic measurements of regional cerebral blood flow. *Behavioral Brain Research, 535*, 313–317.

Deckwerth, T. L., Elliott, J. L., Knudson, C. M., Johnson, E. M., Jr., Snider, W. D., and Korsmeyer, S. J. (1996). BAX is required for neuronal death after trophic factor deprivation and during development. *Neuron, 17*, 401–411.

DeFelipe, J. (2002). Cortical interneurons: From Cajal to 2001. *Progress in Brain Research, 136*, 215–238.

DeFerrari, G. M., Vanoli, E., Stramba-Badiale, M., Hull, S. S., Jr., Foreman, R. D., and Schwartz, P. J. (1991). Vagal reflexes and survival during acute myocardial ischemia in concious dogs with healed myocardial infarction. *American Journal of Physiology, 261*, H63–H69.

DeFries, J. C., Vandenberg, S. G., McClearn, G. E., Kuse, A. R., Wilson, J. R., Ashton, G. C., and Johnson, R. C. (1976a). Near identity of cognitive structure in two ethnic groups. *Nature, 261*, 131–133.

DeFries, J. C., Ashton, G. C., Johnson, R. C., Kuse, A. R., McClearn, G. E., Mi, M. P., Rashad, M. N., Vandenberg, S. C., and Wilson, J. R. (1976b). Parent-offspring resemblance for specific cognitive abilities in two ethnic groups. *Annual Review of Genetics, 10*, 179–207.

DeFries, J. C., Vandenberg, S. G., and McClearn, G. E. (2002). Genetics of specific cognitive abilities. *Neuropsychopharmacology, 27*, 607–619.

DeFries, J. C., Vandenberg, S. G., McClearn, G. E., Kuse, A. R., Wilson, J. R., Ashton, G. C., Johnson, R. C. (1974, January). Near identity of cognitive structure in two ethnic groups. *Science, 183*, 338–339.

Dejerine, J. and Gauckler, E. (1911). *Les Manifestations Functionelles des Psychoneuroses; Leur Traitement par la Psychotherapie*. Paris, France: Masson.

Delacato, C. H. (1963). *The diagnosis and treatment of speech and reading problems*. Springfield, IL: Charles C. Thomas.

Delgado, J. M. (1964). Free behavior and brain stimulation. *International Review of Neurobiology, 6*, 349–449.

Delgado, J. M. (1967). Aggression and defense under cerebral radio control. *UCLA Forum in Medical Sciences, 7*, 171–193.

De Long, G. R. (1999). Autism: New data suggest a new hypothesis. *Neurology, 52*, 911–916.

De Long, M. R. (1990). Primate models of movement disorders of basal ganglia origin. *Trends in Neuroscience, 13*, 281–285.

de Jong, P. T., de Jong, J. M., Cohen, B., Jongkees, L. B. (1977). Ataxia and nystagmus induced by injection of local anesthetics in the neck. *Annals of Neurology, 1*, 240–246.

Delwaide, P. J., Petin, J. L., and Maertens, A. (1991). de Noorthout Short-latency autogenic in addition in the patients with parkinsonian rigidity. *Annals of Neurology, 30*, 83–89.

Demeter, S., Rosene, D., and Van Hoesen, G. (1990). Fields of origin and pathways of the interhemispheric commissures in the temporal lobe of macaques. *Journal of Comparative Neurology, 302*, 29–53.

Demeurisse, G. and Capon, A. (1991). Brain activation during a linguistic task in conduction aphasia. *Cortex, 27*, 285–294.

De Montis, G. M., Olianas, M. C. Serra, G., Tagliamonte, A., and Scheel-Kruger, J. (1979). Evidence that in nigral gabaergic-cholinergic balance controls posture. *European Journal of Pharmacology, 53*, 181–190.

Denckla, M. and Rudel, R. (1978). Abnomalies of motor development in hyperactive boys. *Annals of Neurology, 3*, 231–233.

Denckla, M., Rudel, R., and Chapman, C. et al. (1985). Motor proficiency in dyslexic children with and without attention disorders. *Archives of Neurology, 42*, 228–231.

Denenberg, V. H., Gall, J. S., Berrebi, A. S., and Yutzey, D. A. (1986). Callosal mediation of cortical inhibition in the lateralized rate brain. *Brain Research, 397*, 327–332.

Denenberg, V. H., Garbanati, J., Sherman, G., Yutzey, D. A., and Kaplan, R. (1978). Infantile stimulation induced brain lateralization in rats. *Science, 201*, 1150–1152.

Denenberg, V. H., Hofmann, M., Garbanati, J. A., Shennan, G. F., Rosen, G. D., and Yutzey, D. A. (1980). Handling in infancy, taste aversion, and brain laterality in rats. *Brain Research, 200*, 123–133.

De Nil, L. F., Kroll, R. M., and Houle, S. (2001). Functional neuroimaging of cerebellar activation during single word reading and verb generation in stuttering and nonstuttering adults. *Neuroscience Letters, 302*, 77–80.

Dennis, M. (1980). Capacity and strategy for syntactic comprehension after left or right hemi-decortication. *Brain and Language, 10*, 287–317.

Dennis, M. and Kohn, B. (1975). Comprehension of syntax in infantile hemiplegics after cerebral hemidecortication: Left-hemisphere superiority. *Brain and Language, 2*, 475–486.

Dennis, M., Lovett, M., and Wiegel-Crump, C. A. (1981). Written language acquisition after left or right hemidecortication in infancy. *Brain and Language, 12*, 54–91.

Dennis, M. and Whitaker, H. A. (1976). Language acquisition following hemidecortication: Linguistic superiority of the left over the right hemisphere. *Brain and Language, 3*, 404–433.

Denny-Brown, D. (1958). The nature of apraxia. *Journal of Nervous and Mental Diseases, 126*, 9–32.

Denson, R., Nanson, J. L., and McWatters, M. A. (1975). Hyperkinesis and maternal smoking. *Canadian Psychiatric Association Journal, 20*, 183–187.

de Quervain, D. J., Roozendaal, B., and McGaugh, J. L. (1998). Stress and glucocorticoids impair retrieval of long-term spatial memory. *Child and Adolescent Psychiatric Clinics of North America, 7*, 33–51, viii.

De Quiros, J. B., and Schrager, O. (1978). *Neurophysiological fundamentals in learning disabilities*. San Rafael, CA: Academic Therapy.

De Renzi, E. (1982). *Disorders of space exploration and cognition*. Chichester, England: John Wiley & Sons.

De Renzi, E., Faglioni, P., and Previdi. (1978). Increased susceptibility of aphasics to a distractor task in the recall of verbal commands. *Brain and Language, 61*, 14–21.

De Renzi, E., Faglioni, P., and Spinnler, H. (1968). The performance of patients with unilateral brain damage on face recognition tasks. *Cortex, 4*, 17–34.

De Renzi, E., Faglioni, P., and Villa, P. (1977). Topographical amnesia. *Journal of Neurology, Neurosurgery and Psychiatry, 40*, 498–505.

De Renzi, E. and Spinnler, H. (1966). Visual recognition in patients with unilateral cerebral disease. *Journal of Nervous and Mental Diseases, 142*, 515–525.

de Schonen, S. M. and Mathivet, E. (1989). First come first served: A scenario about the development of hemispheric specialization in face recognition during infancy. *European Bulletin of Cognitive Psychology (CPC), 9*, 3–44.

Desmond, J. E., Gabrieli, J. D., Wagner, A. D., Ginier, B. L., and Glover, G. H. (1997). Lobular patterns of cerebellar activation in verbal working-memory and finger-tapping tasks as revealed by functional MRI. *Journal of Neuroscience, 17*, 9675–9685.

Desai, N. S., Cudmore, R. H., Nelson, S. B., and Turrigiano, G. G. (2002). Critical periods for experience-dependent synaptic scaling in visual cortex. *Nature, Neuroscience, 5*, 783–789.

DeToledo, J. C. and Haddad, H. (1980). Progressive scoliosis in early, non-progressive CNS injuries: Role of axial muscles. *Brain Injury, 13*, 39–43.

D'Eufemia, P., Celli, M., and Sinocciaro, R. L. et al. (1996). Abnormal intestinal permeability in children with autism. *Acta Taediaprica, 85*, 1076–1079.

Deuschl, D. and Krack, P. (1998). Tremors: Differential diagnoses, neurophysiology, and phenomenology. In J. Jankovic, E. Tolosa (Eds.), *Parkinson's disease and movement disorders* (pp. 419–451). Baltimore: Williams and Wilkins.

Deutsch, G., Papanicolaou, A. C., Bourbon, W. T., and Eisenberg, H. M. (1987). Cerebral blood flow evidence of right frontal activation in attention demanding tasks. *International Journal of Neuroscience, 36*, 23–28.

DeValois, R. L. and DeValois, K. K. (1988). *Spatial vision*. New York, NY: Oxford University Press.

DeVeaugh-Geiss, J., Moroz, G., Biederman, J., Cantwel, I. D., Fontaine, R., Greist, J. H., Reichler, R., Katz, R., and Landau, P. (1992). Clomipramine hydrochloride in childhood and adolescent obsessive-compulsive disorder—a multicenter trial. *Journal of the American Academy of Child and Adolescent Psychiatry, 31*, 45–49.

Devlin, B., Daniels, M., and Roeder, K. (1997). The heritability of IQ. *Nature, 388*, 468–471.

de Vries, M. W. (1999). Babies, brains and culture: Optimizing neurodevelopment on the savanna. *Behavioral Brain Research, 88*, 43–48.

Dewan, M. J., Pandurangi, A. K., Lee, S. H., Ramachandran, T., Levy, B. F., Boucher, M., Yozawitz, A., and Major, L. (1983). Cerebellar morphology in chronic schizophrenic patients: A controlled computed tomography study. *Psychiatry Research, 10,* 97–103.

Diamond, I. R., Tannock, R., and Schachar, R. J. (1999). Response to methylphenidate in children with ADHD and comorbid anxiety. *Journal of the American Academy of Child and Adolescent Psychiatry, 38,* 402–409.

Diamond, J. M. (1990). *The third chimpanzee: The evolution and future of the human animal.* New York, NY: Harper.

Diamond, J. M. (1998). *Guns, germs, and steel: The fates of human societies.* New York, NY: W. W. Norton.

Diamond, M. C. (1967). Extensive cortical depth measurements and neuron size increases in the cortex of environmentally enriched rats. *Journal of Comparative Neurology, 131,* 357–364.

Diamond, M. C. (1983). New data supporting cortical asymmetry differences in males and females. *Behavioral and Brain Sciences, 3,* 233–234.

Diamond, M. C. (1985a). Sex differences in the structure of the rat forebrain. *Brain Research Review, 12,* 235–240.

Diamond, M. C. (1985b). Rat forebrain morphology: Right-left, male-female, young-old, enriched-impoverished. In S. Glick (Ed.), *Cerebral lateralization in nonhuman species* (pp. 73–88). Orlando, FL: Academic Press.

Diamond, M. C. (1987). Sex differences in the rat forebrain. *Brain Research, 434,* 235–240.

Diamond, M. C. (2001). Response of the brain to enrichment. *Anais da Academia Brasileira de Ciencias, 73,* 211–220.

Diamond, M. C., Dowling, G. A., and Johnson, R. E. (1980). Morphologic cerebral cortical asymmetry in male and female rats. *Experimental Neurology, 71,* 261–268.

Diamond, M. C. and Hopson, J. (1998). *Magic trees of the mind: How to nurture your child's intelligence, creativity, and healthy emotions from birth through adolescence.* New York: Dutton.

Diamond, M. C., Ingham, C. A., Johnson, R. E., Bennett, E. L., and Rosenzweig, M. R. (1976). Effects of environment on morphology of rat cerebral cortex and hippocampus. *Journal of Neurobiology, 7,* 75–85.

Diamond, M. C., Johnson, R. E., and Ehlert, J. (1979). A comparison of cortical thickness in male and female ratsnormal and gonadectomized, young and adult. *Behavioral and Neural Biology, 26,* 485–491.

Diamond, M. C., Krech, D., and Rosenzweig, M. R. (1964). The effects of an enriched environment on the rat cerebral cortex. *The Journal of Comparative Neurology, 123,* 111–119.

Dichgans, J. and Diener, H. C. (1984). Clinical evidence for functional compartmentalization of the cerebellum. In J. R. Bloedel, J. Dichgans, and W. Precht (Eds.), *Cerebellar functions* (pp. 126–147). Berlin, Germany: Springer-Verlag.

Dietrichs, E. and Haines, D. E. (1989). Interconnections between hypothalamus and cerebellum. *Anatomy and Embryology, 179,* 207–220.

Dietz, W. H. and Gortmaker, S. L. (1985). Do we fatten our children at the television set? *Pediatrics, 75,* 807–812.

Dietz, W. H. and Strasburger, V. C. (1991). Children, adolescents and television. *Current Problems in Pediatrics, 1,* 8–31.

Di Martino, A. and Tuchman, R. F. (2001). Antiepileptic drugs: Affective use in autism spectrum disorders. *Pediatric Neurology, 25,* 199–207.

Dimond, S. J. and Farrington, L. (1977). Emotional response to films shown to the right or left hemisphere of the brain measured by heart rate. *Acta Psychologica, 41,* 255–260.

Ding, Y., Li. J., Lai, Q., Adam, S., Rafols, J. A., and Diaz, F. G. (2002). Functional improvement after motor training is correlated with synaptic plasticity in rat thalamus. *Neurological Research, 24,* 829–836.

Di Pietro, L., Mossberg, H. O., and Stunkard, A. J. (1994). A 40-year history of overweight children in Stockholm: Life-time overweight, morbidity, and mortality. *International Journal of Obesity, 18,* 585–590.

Dobbing, J. (1984). Infant nutrition and later achievement. *Nutrition Reviews, 42,* 1–7.

Dobbing, J. (1985a). Maternal nutrition in pregnancy and later achievement of offspring: A personal interpretation. *Early Human Development, 12,* 1–8.

Dobbing, J. (1985b). Infant nutrition and later achievement. *The American Journal of Clinical Nutrition, 41* (Suppl.) 477–484.

Dobbing, J. and Sands, J. (1973). Quantitative growth and development of human brains. *Archives of Disease in Childhood, 48,* 757–767.

Dobbing, J. and Sands, J. (1985). Cell size and cell number in tissue growth and development. An old hypothesis reconsidered. *Archives Francaises de Pediatrie, 42,* 199–203.

Dodge, K. A. (1993). Social-cognitive mechanisms in the development of conduct disorder and depression. *Annual Review of Psychology, 44,* 559–584.

Dohan, F. C. (1966a). Cereals and schizophrenia data and hypothesis. *Acta Psychiatrica Scandinavica, 42,* 125–152.

Dohan, F. C. (1966b). Wartime changes in hospital admissions for schizophrenia. A comparison of admission for schizophrenia and other psychoses in six countries during world war II. *Acta Psychiatrica Scandinavica, 42,* 1–23.

Dohan, F. C. (1969). Is celiac disease a clue to the pathogenesis of schizophrenia? *Mental Hygiene, 53,* 525–529.

Dohan, F. C. (1970). Coeliac disease and schizophrenia. *Lancet, 25,* 897–898.

Dohan, F. C. (1979). Schizophrenia and neuroactive peptides from food. *Lancet, 12,* 1031.

Dohan, F. C. (1980a). Celiac disease and schizophrenia. *New England Journal of Medicine, 302,* 1262.

Dohan, F. C. (1980b). Hypothesis: Genes and neuroactive peptides from food as cause of schizophrenia. *Advances in Biochemistry and Psychopharmacology, 22,* 535–548.

Dohan, F. C. (1988). Genetic hypothesis of idiopathic schizophrenia: Its exorphin connection. *Schizophrenia Bulletin, 14,* 489–494.

Dohan, F. C., Grasberger, J. C., Lowell, F. M., Johnston, H. T., Jr., and Arbegast, A. W. (1969). Relapsed schizophrenics: More rapid improvement on a milk- and cereal-free diet. *British Journal of Psychiatry, 115,* 595–596.

Dohan, F. C., Harper, E. H., Clark, M. H., Rodrigue, R. B., and Zigas, V. (1984). Is schizophrenia rare if grain is rare? *Biological Psychiatry, 19,* 385–399.

Domesick, V. B. (1969). Projections from the cingulate cortex in the rat. *Brain Research, 12,* 296–320.

Donnelly, M., Zametkin, A. J., and Rapoport, J. L., Ismond, D. R., Weingartner, H., Lane, E., Oliver, J., Linnoila, M., and Potter, W. Z. (1986). Treatment of childhood hyperactivity with desipramine: Plasma drug concentration, cardiovascular effects, plasma and urine catecholamine levels and clinical response. *Clinical Pharmacology and Therapeutics, 39,* 72–81.

Douglas, V. I., Barr, R. G., and O'Neill, M. E. et al. (1986). Short term effects of methylphenidate on the cognitive, learning and academic performance of children with attention deficit disorder in the laboratory and the classroom. *Journal of Child Psychology and Psychiatry, 27,* 191–211.

Dow, R. S. and Moruzzi, G. (1958). *The physiology and pathology of the cerebellum.* Minneapolis, MN: University of Minnesota Press.

Downer, J. L. (1962). Interhemispheric integration in the visual system. In V. B. Mountcastle (Ed.), *Interhemispheric relations and cerebral dominance* (pp. 87–100). Baltimore: Johns 5 Press.

Dray, A., Davies, J., Oakley, N. R., Tongroach, P., and Vellucci, S. (1978). The dorsal and medial raphe projections to the substantia nigra in the rat: Electrophysiological, biochemical and behavioural observations. *Brain Research, 151,* 431–442.

Drevets, W. C. (1988). Prefrontal cortical-amygdalar metabolism in major depression. *Encephale, 14,* 339–344.

Drevets, W. C. (1994). Geriatric depression: Brain imaging correlates and pharmacologic considerations. *The Journal of Clinical Psychiatry, 55* (Suppl.), 71–81; discussion 82, 98–100.

Drevets, W. C. (1997). Functional neuroimaging studies of depression: The anatomy of melancholia. *Nature, 386,* 824–827.

Drevets, W. C. (1998a). Functional neuroimaging studies of depression: The anatomy of melancholia. *Annual Review of Medicine, 49,* 341–361.

Drevets, W. C. (1998b). Geriatric depression: Brain imaging correlates and pharmacologic considerations. *Annual Review of Medicine, 49,* 341–361.

Drevets, W. C. (1999). Prefrontal cortical-amygdalar metabolism in major depression. *Annals of the New York Academy of Sciences, 877,* 614–637.

Drevets, W. C., Price, J. L., Simpson, J. R., Jr., Todd, R. D., Reich, T., Vannier, M., and Raichle, M. E. (1997). Subgenual prefrontal cortex abnormalities in mood disorders. *Nature, 386,* 824–827.

Drevets, W. C. and Raichle, M. E. (1992). Neuroanatomical circuits in depression: Implications for treatment mechanisms. *Psychopharmacology Bulletin, 28,* 261–274.

Drevets, W. C., Videen, T. O., Price, J. L., Preskom, S. H., Carmichael, S. T., and Raichle, M. E. (1992). A functional anatomical study of unipolar depression. *The Journal of Neuroscience, 12,* 3628–3641.

Duane, D. D. (1989). Neurobiological correlates of learning disorders. *Journal of the American Academy of Child and Adolescent Psychiatry, 28,* 314–318.

Duane, D. D. (Ed.) (1991). *The Reading Brain: The Biological Basis of Dyslexia.* Timonium, MD: York Press.

Dubinsky, R. and Hallett, M. (1987). Glucose hypermetabolism of the inferior olive in patients with the essential tremor. *Annals of Neurology, 22,* 118.

Duffy, F. H., Denckla, M. B., Bartels, P. H., Sandini, G., and Kiessling, L. S. (1980). Dyslexia: Automated diagnosis by computerized declassification of brain electrical activity. *Annals of Neurology, 7,* 421–428.

Duffy, F. H., Denckla, M. B., McAnulty, G. B., Holmes, J. A. (1988). Neurophysiological studies in dyslexia. *Research Publication of the Association of Research in Nervous and Mental Disease, 66,* 149–170.

Dulcan, M. K. (1990). Using psychostimulants to treat behavioral disorders of children and adolescents. *Journal of Child and Adolescent Psychopharmacology, 1,* 7–20.

Dunn, A. J. (1989). Psychoimmunology for the psychoneuroendocrinologist: A review of animal studies of nervous system-immune system interactions. *Psychoneuroendocrinology, 14,* 251–274.

Dunn, H., McBumey, A., Ingram, S., and Hunter, C. (1977). Maternal cigarette smoking during pregnancy and the child's subsequent development: II. Neurological and intellectual maturation to the age of $6\frac{1}{2}$ years. *Canadian Journal of Public Health, 68,* 43–50.

Dyachkova, L. N. (1991). Ultrastructural characteristics of plastic changes in the brain cortex of rats exposed to space flight. *The Physiologist, 34* (Suppl.), 185–186. In *Proceedings of the 12th Annual Meeting of IUPS Commission of Gravitational Physiology.*

Dykens, E. M. (2000). Psychopathology in children with intellectual disability. *Journal of Child Psychology and Psychiatry, and Allied Disciplines, 41,* 407–417.

Eccles, J. C. (1987). Mechanisms of learning in complex neural systems. In V. B. Mountcastle, F. Plum, and S. R. Geiger (Eds.), *Handbook of physiology—The nervous system V* (pp. 137–167). New York, NY: Oxford University Press.

Eccles, J. C., Ito, M., and Szentagothai, J. (1967). *The cerebellum as a neuronal machine.* New York: Springer.

Eccles, J. C., Sasaki, K., and Strata, P. (1967). Interpretation of the potential fields generated in the cerebellar cortex by a mossy fibre volley. *Experimental Brain Research, 3,* 58–80.

Eddy, D. R., Schiflett, S. G., Schlegel, R. E., and Shehab, R. L. (1998). Cognitive performance aboard the life and microgravity spacelab. *Acta Astronautica, 43,* 193–210.

Edelson, S. B. and Cantor, D. S. (1998a). Autism: Xenobiotic influences. *Toxicology and Industrial Health, 14,* 799–811.

Edelson, S. B. and Cantor, D. S. (1998b). Autism: Xenobiotic influences. *Toxicology and Industrial Health, 14,* 553–563.

Edelson, S. M., Rimland, B., Berger, C. L., and Billings, D. (1998). Evaluation of a mechanical hand-support for facilitated communication. *Journal of Autism and Developmental Disorders, 15,* 23–36.

Eden, G. F., Stein, J. F., and Wood, H. M. Wood, F. B. (1994). Differences in eye movements and reading problems in dyslexic and normal children. *Vision Research, 34,* 1345–1358.

Egger, J., Carter, C. M., Graham, P. J., Gumley, D., and Soothill, J. F. (1985). Controlled trial of oligoantigenic treatment in the hyperkinetic syndrome. *Lancet, 1,* 540–545.

Ehlers, S. and Gillberg, C. (1993). The epidemiology of Asperger syndrome: A total population study. *Journal of Child Psychology and Psychiatry, 34,* 1327–1350.

Ehret, G. (1987). Left hemisphere advantage in the mouse brain for recognizing ultrasonic communication calls. *Nature, 325,* 249–251.

Eidelberg, D. and Galaburda, A. M. (1982). Symmetry and asymmetry in the human posterior thalamus. I. Cytoarchitectonic analysis in normal persons. *Archives of Neurology, 39,* 325–332.

Eiraldi, R. B., Power, T. J., and Nezu, C. M. (1997). Patterns of comorbidity associated with subtypes of attention-deficit/hyperactivity disorder among 6- to 12-year-old children. *Journal of the American Academy of Child and Adolescent Psychiatry, 36,* 503–514.

Elbert, T., Flor, H., Birbaumer, N., Knecht, S., Hampson, S., Larbig, W., and Taub, E. (1994). Extensive reorganization of the somatosensory cortex in adult humans after nervous system injury. *Neuroreport, 20,* 2593–2597.

Elbert, T., Pantev, C., Wienbruch, C., Rockstroh, B., and Taub, E. (1995). Increased cortical representation of the fingers of the left hand in string players. *Science, 13, 270,* 305–307.

Elias, P. K., D'Agostino, R. B., Elias, M. F., and Wolf, P. A. (1995). Blood pressure, hypertension, and age as risk factors for poor cognitive performance. *Current Opinion in Neurobiology, 9,* 228–234.

Elliott, F. A. (1982). Neurological findings in adult minimal brain dysfunction and the dyscontrol syndrome. *Journal of Nervous and Mental Disease, 170,* 680–687.

Elliott, R. O., Jr., Dobbin, A. R., Rose, G. D., and Soper, H. V. (1994). Vigorous, aerobic exercise versus general motor training activities: Effects on maladaptive and stereotypic behaviors of adults with both autism and mental retardation. *Journal of Autism and Developmental Disorders, 24,* 565–576.

Ely, P. W., Graves, R. E., and Potter, S. M. (1989). Dichotic listening indices of right hemisphere semantic processing. *Neuropsychologia, 27,* 1007–1015.

Emery, N. J. (2000). The eyes have it: The neuroethology, function and evolution of social gaze. *Neuroscience and Biobehavioral Reviews, 24,* 581–604.

Emory, E. K. and Noonan, J. R. (1984). Fetal cardiac responding: A correlate of birth weight and neonatal behavior. *European Journal of Obstetrics, Gynecology, and Reproductive Biology, 10,* 239–246.

Engel, A., Konig, P., Kreiter, A., Schillen, T., and Singer, N. (1992). Temporal coding in the visual cortex: New vistas on integration in the nervous system. *Trends in Neurosciences, 15,* 218–226.

Engel, A., Konig, P., Kreiter, A., and Singer, W. (1991). Interhemispheric synchronization of oscillatory neuronal responses in cat visual cortex. *Science, 252,* 1177–1179.

Epstein, L. H., Saelens, B. E., Myers, M. D., and Vito, D. (1997). Effects of decreasing sedentary behaviors on activity choice in obese children. *Health Psychology, 16,* 107–113.

Erenberg, G., Cruse, R., and Rothner, D. (1986). Tourette syndrome: An analysis of 200 pediatric and adolescent cases. *Cleveland Clinic Quarterly, 53,* 127–131.

Ernst, M., Zametkin, A. J., Matochik, J. A., Pascualvaca, D., and Cohen, R. M. (1997). Low medial prefrontal dopaminergic activity in autistic children. *Lancet, 350,* 638.

Eriksson, P. S., Perfilieva, E., Bjork-Eriksson, T., Alborn, A. M., Nordborg, C., Peterson, D. A., and Gage, F. H. (1998). Neurogenesis in the adult human hippocampus. *Nature, Medicine, 4,* 1313–1317.

Eron, L. D. (1994). Theories of aggression. From drives to cognitions. In L. R. Huesmann (Ed.), *Aggressive behaviour: Current perspectives* (pp. 3–11). New York, NY: Plenum.

Esiri, M. M. and Wilcock, G. K. (1984). The olfactory bulbs in Alzheimer's disease. *Journal of Neurology, Neurosurgery, and Psychiatry, 47,* 56–60.

Esweran, H., Wilson, J., Preissl, H., Robinson, S., Vrba, J., Murphy, P., Rose, D., and Lowery, C. (2002). Magnetoencephalographic recordings of visual evoked brain activity in the human fetus. *Lancet, 360,* 779–780.

Evarts, E. V. (1971). Central control of movement. Feedback and corollary discharge: A merging of the concepts. *Neurosciences Research Program Bulletin, 9,* 86–112.

Ewald, P. W. (1993). The evolution of virulence. *Scientific American, 268,* 86–93.

Fairchild, M. D., Jenden, D. J., and Mickey, M. R. (1969). Discrimination of behavioral state in the cat utilizing

long-term EEG frequency analysis. *Electroencephalography and Clinical Neurophysiology, 27,* 503–513.
Falk, D. (1978). Cerebral asymmetry in old world monkeys. *Acta Anatomica, 101,* 334–339.
Falk, D., Hildebolt, C., Cheverud, J., Vannier, M., Helmkamp, R. C., and Konigsberg, L. (1990). Cortical asymmetries in frontal lobes of rhesus monkeys (Macaca mulatta). *Brain Research, 512,* 40–45.
Falkai, P., and Bogerts, B. (1992). Neurodevelopmental abnormalities in schizophrenia. *Clinical Neuropharmacology, 15* (Suppl.), 498A–499A.
Fallon, J. H. and Loughlin, S. E. (1987). Monoamine innervation of cerebral cortex and a theory of the role of monoamines in cerebral cortex and basal ganglia. In A. Peters and E. G. Jones (Eds.), *Cerebral cortex* (Vol. 6, pp. 41–127). New York, NY: Plenum Press.
Farah, M. J. (1984). The neurological basis of mental imagery: A componential analysis. *Cognition, 18,* 245–272.
Faraone, S. and Biederman, J. (1994a). Genetics of attention-deficit hyperactivity disorder. *Child and Adolescent Psychiatric Clinic of North America, 3,* 285–302.
Faraone, S. and Biederman, J. (1994b). Is attention deficit hyperactivity disorder familial? *Harvard Review of Psychiatry, 1,* 271–287.
Faraone, S., Biederman, J., Krifcher Lehman, B., Keenan, K., Norman, D., Seidman, L., Kolodny, R., Kraus, I., Perrin, J., and Chen, W. (1993a). Evidence for the independent familial transmission of attention deficit hyperactivity disorder and learning disabilities: Results from a family genetic study. *American Journal of Psychiatry, 150,* 891–895.
Faraone, S., Biederman, J., Krifcher Lehman, B., Spencer, T., Norman, D., Seidman, L., Kraus, I., Perrin, J., Chen, W., and Tsuang, M. T. (1993b). Intellectual performance and school failure in children with attention deficit hyperactivity disorder and in their siblings. *Journal of Abnormal Psychology, 102,* 616–623.
Faraone, S. V., Biederman, J., and Monuteaux, M. C. (2001). Attention deficit hyperactivity disorder with bipolar disorder in girls: Further evidence for a familial subtype? *Journal of Affective Disorders, 64,* 19–26.
Farooqu, A. A. and Horrocks, L. A. (2001). Plasmalogens, phospholipase A2, and docosahexaenoic acid turnover in brain tissue. *Journal of Molecular Neuroscience, 16,* 263–272; discussion 279–284.
Fatemi, S. H., Earle, J., Kanodia, R., Kist, D., Emamian, E. S., Patterson, P. H., Shi, L., and Sidwell, R. (2002a). Prenatal viral infection leads to pyramidal cell atrophy and macrocephaly in adulthood: implications for genesis of autism and schizophrenia. *The Journal of Neuroscience: The Official Journal of the Society for Neuroscience, 23,* 297–302.
Fatemi, S. H., Halt, A. R., Realmuto, G., Earle, J., Kist, D. A., Thuras, P., and Merz, A. (2002b). Purkinje cell size is reduced in cerebellum of patient with autism. *Cellular and Molecular Neurobiology, 22,* 171–175.

Feger, J. (1997). Updating the functional model of the basal ganglia. *Trends in Neuroscience, 20,* 152–153.
Feldman, S. (1989). Afferent neural pathways and hypothalamic neurotransmitters regulating adrenocortical secretion. In H. Weiner, I. Florin, R. Murison, and D. Hellhammer (Eds.), *Frontiers of stress research* (pp. 201–207). Lewiston, NY: Hans Huber.
Ferchmin, P. A., Eterovic, V. A., and Caputto, R. (1970). Studies of brain weight and RNA content after short periods of exposure to environmental complexity. *Brain Research, 20,* 49–57.
Ferrari, P., Marescot, M. R., Moulias, R., Bursztejn, C., Deville Chabrolle, A., Thiollet, M., Lesourd, B., Braconnier, A., Dreux, C., and Zarifian, E. et al. (1988). Immune status in infantile autism. Correlation between the immune status, autistic symptoms and levels of serotonin. *Encephale, 14,* 339–344.
Fielding, J. E. (1985). Smoking: Health effects and control. *New England Journal of Medicine, 313,* 491–498.
Fiez, J. A. (1996). Cerebellar contributions to cognition. *Neuron, 16,* 13–15.
Filipek, P. A., Accardo, P. J., Baranek, G. T., Cook, E. H., Jr., Dawson, G., Gordon, B., Gravel, J. S., Johnson, C. P., Kallen, R. J., Levy, S. E., Minshew, N. J., Ozonoff, S., Prizant, B. M., Rapin, I., Rogers, S. J., Stone, W. L., Teplin, S., and Tuchman, R. F. Volkmar, F. R. (1999). The screening and diagnosis of autistic spectrum disorders. *Journal of Autism and Developmental Disorders, 29,* 439–484.
Filipek, P. A. Semrud-Clikeman, M., Steingard, R. J., Renshaw, P. F., Kennedy, D. N., and Biederman, J. (1997). Volumetric MRI analysis comparing subjects having attention-deficit hyperactivity disorder with normal controls. *Neurology, 48,* 589–601.
Findling, R. L., Maxwell, K., Scotese-Wojtila, L., Huang, J., Yamashita, T., and Wiznitzer, M. (1997). High-dose pyridoxine and magnesium administration in children with autistic disorder: An absence of salutary effects in a double-blind, placebo-controlled study. *Journal of Autism and Developmental Disorders, 27,* 467–78.
Fiorelli, M., Blin, J., Bakchine, S., LaPlane, D., and Baron, J. C. (1991). PET studies of cortical diaschisic in patients with motor hemi-neglect. *Journal of the Neurological Sciences, 104,* 135–142.
Fisher, S. E., Francks, C., McCracken, J. T., McGough, J. J., Marlow, A. J., MacPhie, I. L., Newbury, D. F., Crawford, L. R., Palmer, C. G., Woodward, J. A., Del'Homme, M., Cantwell, D. P., Nelson, S. F., Monaco, A. P., and Smalley, S. L. (2002). A genomewide scan for loci involved in attention-deficit/hyperactivity disorder. *American Journal of Human Genetics, 70,* 1183–1196.
Fitts, P. M. and Peterson, J. R. (1964). Information capacity of discrete motor responses. *Journal of Experimental Psychology, 67,* 103–112.
Fitz-Ritson, D. (1982). The anatomy and physiology of the muscle spindle and its role in posture and movement: A review. *The Journal of the Canadian Chiropractic Association, 26,* 144–150.

Fitzgerald, L. W., Keller, R. W., Glick, S. D., and Carlson, J. N. (1989). The effects of stressor controllability on regional changes in mesocorticolimbic dopamine activity. *Society for Neuroscience Abstracts, 15*, 1316.

Flament, M. F., Rapoport, J. L., Berg, C. J., Sceery, W., Kilts, C., Mellstrom, B., and Linnoila, M. (1985). Clomipramine treatment of childhood obsessive-compulsive disorder: A double-blind controlled study. *Archives of General Psychiatry, 42*, 977–983.

Flanders, M., Daghestani, L., and Berthoz, A. (1999). Reaching beyond reach. *Experimental Brain Research, 126*, 19–30.

Flannery, K. A. and Liederman, J. (1995). Is there really a syndrome involving the co-occurrence of neurodevelopmental disorder talent, non-right handedness and immune disorder among children? *Cortex, 31*, 503–515.

Fleming, A. S., O'Day, D. H., and Kraemer, G. W. (1999). Neurobiology of mother–infant interactions: Experience and central nervous system plasticity across development and generations. *Child Psychiatry and Human Development, 29*, 189–208.

Flemming, I. (1979). *Normale emwickiung des sauglings-mit ihre Abweichungen*. Stuttgart, Germany: Thieme.

Flor-Henry, P. (1976). Lateralized temporal-limbic dysfunction and psychopathology. *Annals of the New York Academy of Sciences, 280*, 777–795.

Flor-Henry, P. (1986). Observations, reflections and speculations on the cerebral determinants of mood and on the bilaterally asymmetrical distributions of the major neurotransmitter systems. *Acta Neurologica Scandinavica, 74* (Suppl.), 75–89.

Foa, E. B., Steketee, G., Kozak, M. J., and Dugger, D. (1987). Imipramine and placebo in the treatment of obsessive-compulsives: The effect on depression and on obsessional symptoms. *Psychopharmacology Bulletin, 23*, 8–11.

Fogelman, K. R. and Manor, O. (1988). Smoking in pregnancy and development into early adulthood. *British Journal of Medicine, 297*, 1233–1236.

Fombone, E. (2003). The prevalence of autism. *Journal of the American Medical Association, 289*, 87–89.

Fombone, E., Roge, B., Claverie, J., Courty, S., and Fremolle, J. (1999). Microcephaly and macrocephaly in autism. *Journal of Autism and Developmental Disorders, 29*, 113–119.

Formisano, E., Linden, D. E., Di Salle, F., Trojano, L., Esposito, F., Sack, A. T., Grossi, D., Zanella, F. E., and Goebel, R. (2002). Tracking the mind's image in the brain I: Time-resolved fMRI during visuospatial mental imagery. *Neuron, 35*, 185–194.

Forster, M. J., Dubey, A., Dawson, K. M., Stutts, W. A., Lal, H., and Sohal, R. S. (1996). Age-related losses of cognitive function and motor skills in mice are associated with oxidative protein damage in the brain. *Proceedings of the National Academy of Sciences, USA, 93*, 4765–4769.

Fox, P. T., Raichle. M. E., and Thach, W. T. (1985). Functional mapping of the cerebellum with positron emission tomography. *Proceedings of the National Academy of Sciences, USA, 82*, 7462–7466.

Frank, J. and Levinson, H. N. (1973) Dysmetric dyslexia and dyspraxia: Hypothesis and study. *Journal of American Academy of Child Psychiatry, 12*, 690–701.

Frank, Y. and Pavlakis, S. G. (2001). Brain imaging in neurobehavioral disorders. *Pediatric Neurology, 25*, 278–287.

Frasure-Smith, N. (1991). In-hospital symptoms of psychological stress as predictors of long term outcome after acute myocardial infarction in men. *American Journal of Cardiology, 67*, 121–126.

Frasure-Smith, N., Lesperance, F., and Talajic, M. (1993). Depression following myocardial infarction impact on 6-month survival. *Journal of the American Medical Association, 270*, 1819–1825.

Frecska, E., Arato, M., Tekes, K., and Powchik, P. (1990). Lateralization of 3H-IMI binding in human frontal cortex. *Biological Psychiatry, 27*, 71.

Freedman, R., Adler, L., Bickford, P., Byerley, W., Coon, H., Cullum, C., Griffith, J., Harris, J., Leonard, S., Miller, C., Myles-Worsley, M., Nagamoto, H., Rose, G., and Waldo, M. (1994). Schizophrenia and nicotinic receptors. *Harvard Review of Psychiatry, 2*, 179–192.

Freeman, C. P. L., Timble, M. R., Deakin, J. F. W., Strokes, T. M., and Ashford, J. J. (1994). Fluvoxamine versus clomipramine in the treatment of obsessive-compulsive disorder: A multicenter, randomized, double-blind, parallel group comparison. *Journal of Clinical Psychiatry, 55*, 301–305.

Freeman, W. and Watts, J. W. (1947). Retrograde degeneration of the thalamus following free frontal lobotomy. *The Journal of Comparative Neurology, 86*, 65–93.

Freeman, W. and Watts, J. W. (1948). The thalamic projection to the frontal lobe. *The Journal of Nervous and Mental Disease, 27*, 200–209.

Freides, D., Barbati, J, van Kampen-Horowitz, L. J., Sprehn, G., Iversen, C., Silver, J. R., and Woodward, R. (1980). Blind evaluation of body reflexes and motor skills in learning disability. *Journal of Autism and Developmental Disorders, 10*, 159–171.

Frolich, J., Dopfner, M., Berner, W., and Lehmkuhl, G. (2002). Treatment effects of combined cognitive behavioral therapy with parent training in hyperkinetic syndrome. *Praxis der Kinderpsychologie und Kinderpsychiatrie, 51*, 476–493.

Fried, I., MacDonald, K. A., and Wilson, C. L. (1997). Single neuron activity in human hippocampus and amygdala during recognition of faces and objects. *Neuron, 18*, 753–765.

Frumkin, L. R., Ripley, H. S., and Cox, G. B. (1978). Changes in cerebral hemispheric lateralization with hypnosis. *Biological Psychiatry, 13*, 741–750.

Frymann, V. (1966). Relation of disturbances of craniosacral mechanisms to symptomatology of the newborn. *The Journal of the American Osteopathic Association, 65*, 1059.

Furlano, R. I., Anthony, A., Day, R., Brown, A., McGarvey, L., Thomson, M. A., Davies, S. E.

Berelowitz, M., Forbes, A., Wakefield, A. J., Walker-Smith, J. A., and Murch, S. H. (2001). Colonic CD8 and gamma delta T-cell infiltration with epithelial damage in children with autism. *Journal of Pediatrics, 138*, 366–372.

Furmark, T., Tillfors, M., Marteinsdottir, I., Fischer, H., Pissiota, A., Langstrom, B., and Fredrikson, M. (2002). Common changes in cerebral blood flow in patients with social phobia treated with citalopram or cognitive-behavioral therapy. *Archives of General Psychiatry, 59*, 425–433.

Fuster, J. M. (1990). Prefrontal cortex and the bridging of temporal gaps in the perception–action cycle. *Annals of the New York Academy of Sciences, 608*, 318–329; discussion 330–336.

Fuster, J. M. (1992). Neurophysiology of neocortical memory in the primate. Presented at the Annual Meeting of the International Neuropsychological Society, San Diego, CA.

Gaarder, K. R. (1975). *Eye movements, vision and behavior*. Washington DC: Hemisphere Publishing.

Gaddes, W. H. (1985). *Learning disabilities and brain function: A neuropsychological approach* (2nd ed.). New York, NY: Springer-Verlag.

Gadow, K. D., Nolan, E. E., and Sverd, J. et al. (1992). Methylphenidate in hyperactive boy with comorbid tic disorder: II. Short-term behavioral effects in school settings. *Journal of the American Academy of Child and Adolescent Psychiatry, 31*, 462–471.

Gaffney, G. R., Kuperman, S., Tsai, L. Y., and Minchin, S. (1988). Morphological evidence for brainstem involvement in infantile autism. *Biological Psychiatry, 24*, 578–586.

Gaffney, G. R., Tsai, L. Y., Kuperman, S., and Minchin, S. (1987). Cerebellar structure in autism. *American Journal of Diseases of Children, 141*, 1330–1332.

Gage, F. H. (2002). Neurogenesis in the adult brain. *Journal of Neuroscience, 22*, 612–613.

Gainotti, G. (1970). Emotional behavior of patients with right and left brain damage in neuropsychological test conditions. *Archivio di Psicologia, Neurologia, e Psichiatria, 31*, 457–480.

Gainotti, G. (1972). Emotional behavior and hemispheric side of the lesion. *Cortex, 8*, 41–55.

Gainotti, G. (1983). Laterality of affect. The emotional behavior of right and left brain-damaged patients. In M. S. Myslobodsky (Ed.), *Hemisyndromes: Psychobiology, neurology, psychiatry* (pp. 175–192). New York, NY: Academic Press.

Galaburda, A. M. (1980). La region de Broca: Observations anatomiques faites un siecle apres la mort de son decouvreur. *Revue de Neurologie (Paris), 136*, 609–616.

Galaburda, A. M. (1984). The anatomy of language: Lessons from comparative anatomy. In D. Caplan, A. R. Lecours, and A. Smith (Eds.), *Biological perspectives on language* (pp. 290–302). Cambridge: MIT Press.

Galaburda, A. M. (1994). Developmental dyslexia and animal studies: At the interface between cognition and neurology. *Cognition, 50*, 133–149.

Galaburda, A. M. and Bellugi, U. (2000). Multi-level analysis of cortical neuroanatomy in Williams syndrome. *Journal of Cognitive Neuroscience, 12* (Suppl.), 74–88.

Galaburda, A., Sherman, G., and Geschwind, N. (1985). Cerebral lateralization: Historical note on animal studies. In S. D Glick (Ed.), *Cerebral lateralization in nonhuman species* (pp. 1–10). Orlando, FL: Academic Press.

Galaburda, A. M., Sherman, G. F., Rosen, G. D., Aboitiz, F., and Geschwind, N. (1985). Developmental dyslexia: Four consecutive patients with cortical anomalies. *Annals of Neurology, 18*, 222–233.

Galin, D. (1974). Implications for psychiatry of left and right cerebral specialization. *Archives of General Psychiatry, 31*, 572–583.

Gao, J. H., Parsons, L. M., Bower, J. M., Xiong, J., Li, J., and Fox, P. T. (1996). Cerebellum implicated in sensory acquisition and discrimination rather than motor control. *Science, 272*, 545–547.

Garat, M. C. (1993). Speech and music. *Soins. Psychiatrie, 16–18*, 152–153.

Garber, H. J., Ananth, J. V., Chiu, L. C., Griswold, V. J., and Oldendorf, W. H. (1989). Nuclear magnetic resonance study of obsessive-compulsive disorder. *American Journal of Psychiatry, 46*, 1001–1005.

Gardner, H., Ling, P. K., Flamm, L., and Silverman, J. (1975). Comprehension and appreciation of humorous material following brain damage. *Brain, 98*, 399–412.

Garrison, C. Z., Waller, J. L., Cuffe, S. P., McKeown, R. E., Addy, C. L., and Jackson, K. L. (1997). Incidence of major depressive disorder and dysthymia in young adolescents. *Journal of the American Academy of Child and Adolescent Psychiatry, 36*, 458–465.

Garruto, R. M., Little, M. A., James, G. D., and Brown, D. E. (1999). Natural experimental models: The global search for biomedical paradigms among traditional, modernizing, and modern populations. *Proceedings of the National Academy of Sciences, USA, 96*, 10536–10543.

Gartner, J., Weintraub, S., and Carlson, G. A. (1997). Childhood-onset psychosis: Evolution and comorbidity. *American Journal of Psychiatry, 154*, 256–261.

Garver, D. L. and Zemlan, F. P. (1986). Receptor studies in diagnosis and treatment of depression. In A. J. Rush and K. Z. Altshuler (Eds.), *Depression: Basic mechanisms, diagnosis, and treatment* (pp. 143–170). New York, NY: Guilford Press.

Garvey, J. (2002). Diet in autism and associated disorders. *Journal of Family and Health Care, 12*, 34–38.

Gazzaniga, M. S. (1998). Brain and conscious experience. *Advances in Neurology, 77*, 181–92; discussion 192–193.

Gdowski, G. T. Belton, T., and McCrea, R. A. (2001). The neurophysiological substrate for the cervico-ocular reflex in the squirrel monkey. *Experimental Brain Research, 140*, 253–264.

Gedye, A. (1992). Anatomy of self-injurious, stereotypic, and aggressive movements: Evidence for involuntary explanation. *Journal of Clinical Psychology, 48,* 766–778.

Geffen, G. M., Sjoberg, G., Mason, C., Smyth, D., and Butterworth, P. (1992). *Manual of the Queensland University Auditory Word Memory Test (QUAWMT).* Brisbane, Australia: Queensland University.

Geffner, D., Lucker, J. R., and Gordon, A. (1996). Reviewing research and controversies surrounding AIT. *Advance for Speech-Language Pathologists and Audiologists, 6,* 20–21.

Geldmacher, D., Doty, L., and Heilman, K. M. (1991). Attentional bias in normal elderly subjects on a letter cancellation task. *Neurology, 41,* 236.

Geller, D. A., Biederman, J., and Griffin, S. et al. (1996). Comorbidity of juvenile obsessive-compulsive disorder with disruptive behavior disorders. *Journal of the American Academy of Child and Adolescent Psychiatry, 35,* 1637–1646.

Geller, D. A., Beiderman, J., and Reed, E. et al. (1995). Similarities in response to fluoxetine in the treatment of children and adolescents with obsessive compulsive disorder. *Journal of American Academic Child and Adolescence Psychiatry, 34,* 36–44.

Gemmell, C. and O'Mara, S. M. (2002). Plasticity in the projection from the anterior thalamic nuclei to the anterior cingulated cortex in the rat in vivo: Paired pulse facilitation, long term potentiation and short term depression. *Neuroscience, 109,* 401–406.

George, M. S., Ketter, T. A., Parekh, B. A., Gill, D. S., Huggins, T., Marangell, L., Pazaglia, P. J., and Post, R. (1994). Spatial ability in affective illness: Differences in regional brain activation during a spatial matching task. *Neuropsychiatry, Neuropsychology, and Behavioral Neurology, 7,* 143–153.

Gepner, B. and Mestre, D. R. (2002). Brief report: Postural reactivity to fast visual motion differentiates autistic from children with Asperger syndrome. *Journal of Autism and Developmental Disorders, 32,* 231–238.

Geraci, R. (1998). Don't worry yourself thick. *Men's Health, 13,* 61–63.

Gerendai, I. (1984). Lateralization of neuroendocrine control. In N. Geschwind and A. M. Galaburda (Eds.), *Cerebral dominance. The biological foundations* (pp. 167–178). Cambridge, MA: Harvard University Press.

German, D. C., White, C. L. 3rd, and Sparkman, D. R. (1987). Alzheimer's disease: Neurofibrillary tangles in nuclei that project to the cerebral cortex. *Neuroscience, 21,* 305–312.

Geschwind, N. and Galaburda, A. M. (1987). *Cerebral lateralizalion: Biological mechanisms, associations, and pathology.* Cambridge, MA: MIT Press.

Geschwind, N. and Levitsky, W. (1968). Human brain: Left-right asymmetries in temporal speech regions. *Science, 161,* 181–187.

Gesell, A. (1938). The tonic neck reflex in the human infant. *Journal of Pediatrics, 13,* 455–464.

Gesell, A. and Ames, L. B. (1950). Tonic neck reflex and symmetrotonic behavior. *Journal of Pediatrics, 35,* 165–178.

Ghaziuddin, M., Weidmer-Mikhail, E., and Ghaziuddin, N. (1998). Comorbidity of Asperger syndrome: A preliminary report. *Journal of Intellectual Disability Research, 42,* 279–283.

Gibbs, M. E. and Ng, K. T. (1977). Counteractive effects of norepinephrine and amphetamine on quabain-induced amnesia. *Pharmacology, Biochemistry, and Behavior, 6,* 533–537.

Giedd, J. N. (1999). Brain development, IX: Human brain growth. *American Journal of Psychiatry, 156,* 4.

Giedd, J. N., Blumenthal, J., Jeffries, N. O., Castellanos, F. X., Liu, H., Zijdenbos, A., Paus, T., Evans, A. C., and Rapoport, J. L. (1999a). Brain development during childhood and adolescence: A longitudinal MRI study. *Progress in Neuro-psychopharmacology & Biological Psychiatry, 23,* 571–588.

Giedd, J. N., Blumenthal, J., Jeffries, N. O., Rajapakse, J. C., Vaituzis, A. C., Liu, H., Berry, Y. C., Tobin, M., Nelson, J., and Castellanos, F. X. (1999b). Development of the human corpus callosum during childhood and adolescence: A longitudinal MRI study. *The American Journal of Psychiatry, 156,* 4.

Giedd, J. N., Castellanos, F. X., Casey, B. J., Kozuch, P., King, A. C., Hamburger, S. D., and Rapoport, J. L. (2001). Quantitative morphology of the corpus callosum in attention deficit hyperactivity disorder. *Proceedings of the National Academy of Sciences, USA, 98,* 11650–11655.

Giedd, J. N., Snell, J. W., Lange, N., Rajapakse, J. C., Casey, B. J., Kozuch, P. L., Vaituzis, A. C., Vauss, Y. C., Hamburger, S. D., Kaysen, D., and Rapoport, J. L. (1994). Quantitative magnetic resonance imaging of human brain development: Ages 4–18. *The American Journal of Psychiatry, 151,* 665–669.

Giedd, J. N., Snell, J. W., Lange, N., Rajapakse, J., Casey, B. J., Kozuch, P. L., Vaituzis, A. C., Vauss, Y. C., Hamburger, S. D., Kaysen, D., and Rapoport, J. L. (1996). Quantitative magnetic resonance imaging of human brain development: Ages 4–18. *Cerebral Cortex, 6,* 551–560.

Giesen, J. M., Center, D. B., and Leach, R. A. (1989). An evaluation of chiropractic manipulation as a treatment of hyperactivity in children. *Journal of Manipulative and Physiological Therapeutics, 12,* 353–363.

Gilbert, A. N. and Wisocki, C. J. (1992). Hand preference and age in the United States. *Neuropsychologia, 30,* 601–608.

Gillberg, C. (1989). Asperger syndrome in 23 Swedish Children. *Developmental Medicine and Child Neurology, 31,* 520–531.

Gillberg, C. L. (1992). The Emanuel Miller memorial lecture 1991. Autism and autistic-like conditions: Subclasses among disorders of empathy. *Journal of Child Psychology and Psychiatry, and Allied Disciplines, 33,* 813–842.

Gillberg, C. and Coleman, M. (1996). Autism and medical disorders: A review of the literature. *Developmental Medicine and Child Neurology, 38,* 191–202.

Gillberg, C. and Gillberg, I. C. (1989). Aspergers syndrome some epidemiological aspects: A research note.

Journal of Child Psychology and Psychiatry, 30, 631–638.

Gillberg, C., Rasmussen, P., Carlstrom G., Svenson, B. and Waldenstrom, E. (1982). Perceptual, motor and attentional deficits in six-year-old children. Epidemiological aspects. *Journal of Child Psychology and Psychiatry, 23,* 131–144.

Gillberg, C. and Svennerholm, L. (1987). CSF monoamines in autistic syndromes and other pervasive developmental disorders of early childhood. *The British Journal of Psychiatry; The Journal of Mental Science, 151,* 89–94.

Gillberg, I. C., Winnergard, I., and Gillberg, C. (1993). Screening methods, epidemiology and evaluation of intervention In DAMP in preschool children. *European Child Adolescence Psychiatry, 2,* 121–135.

Girardi, N. L., Shaywitz, S. E., Shaywitz, B. A., Marchione, K., Fleischman, S. J., Jones, T. W., and Tamborlane, W. V. (1995). Blunted catecholamine responses after glucose ingestion in children with attention deficit disorder. *Pediatric Research, 38,* 539–542.

Gittelman-Klein, R. (1987). Pharmacotherapy of childhood hyperactivity: An update. In H. Y. Meltzer (Ed.), *Psychopharmacology: The third generation of progress* (pp. 1215–1224). New York, NY: Raven Press.

Glass, L. and MacKay, M. C. (1988). *From clocks to chaos: The rhythms of life.* Princeton, NJ: Princeton University Press.

Glick, S. D. and Cox, R. D. (1978). Nocturnal rotation in normal rats: Correlation with amphetamine-induced rotation and effects of nigrostriatal lesions. *Brain Research, 150,* 149–161.

Glick, S. D., Crane, A., Jemssi, T., Fleisher, L., and Creen, J. (1975). Functional and neurochemical correlates of potentiation of striatal asymmetry by callosal section. *Nature, 254,* 616–617.

Glick, S. D., Jerussi, T. P., Waters, D. H., and Green, J. P. (1974). Amphetamine-induced changes in striatal dopamine and acetylcholine levels and relationship to rotation (circling behavior) in rats. *Biochemical Pharmacology, 23,* 3223–3225.

Glick, S. D., Meibach, R. C., Cox, R. D., and Maayani, S. (1980). Phencyclidine-induced rotation and hippocampal modulation of nigrostriatal asymmetry. *Brain Research, 196,* 99–107.

Glick, S. D., Meibach, R. C., Cox, R. D., and Maayani, S. (1983). Multiple and interrelated functional asymmetries in rat brain. *Life Sciences, 32,* 2215–2221.

Glick, S. D., Ross, D. A., and Hough, L. B. (1982). Lateral asymmetry of neurotransmitters in human brain. *Brain Research, 234,* 53–63.

Glick, S. D. and Shapiro, R. M. (1985). Functional and neurochemical asymmetries. In N. Geschwind and A. M. Galaburda (Eds.), *Cerebral dominance: The biological foundations* (pp. 147–166). Cambridge, MA: Harvard University Press.

Globus, A., Rosenzweig, M. R., Bennett, E. L., and Diamond, M. C. (1973). Effects of differential experience on dendritic spine counts in rat cerebral cortex.

Journal of Comparative and Physiological Psychology, 82, 175–181.

Gloor, P. (1960). Amygdala. In J. Field, H. W. Magoun, and V. E. Hall (Eds.), *Handbook of physiology* (Vol. II), Neurophysiology, Section I (pp. 1395–1420). Washington, DC: American Physiological Society.

Gloor, P. (1990). Experiential phenomena of temporal lobe epilepsy. Facts and hypotheses. *Brain, 113,* 1673–1694.

Goddard, G. V. (1980). Component properties of the memory machine: Hebb revisited. In P. W. Jusczyk and R. M. Klein (Eds.), *The nature of thought: Essays in honor of D. O. Hebb* (pp. 231–247). Hillsdale, NJ: Erlbaum.

Goetz, C. G., Tanner, C. M., and Wilson, R. S. et al. (1987). Clonidine and Gilles de la Tourette's syndrome: Double-blind study using objective rating methods. *Annals of Neurology, 21,* 307–310.

Goldberg, G. (1985). Supplementary motor area strucure and function: Review and hypothesis. *Behavioral and Brain Science, 8,* 567–616.

Goldberg, E. and Costa, L. D. (1981). Hemispheric differences in the acquisition and use of descriptive systems. *Brain and Language, 14,* 144–173.

Goldberg, M. C., Landa, R., Lasker, A., Cooper, L., and Zee, D. S. (2000). Evidence of normal cerebellar control of the vestibulo-ocular reflex (VOR) in children with high-functioning autism. *Journal of Autism and Developmental Disorders, 30,* 519–524.

Goldberg, M. C., Lasker, A. G., Zee, D. S., Garth, E., Tien, A., and Landa, R. J. (2002). Deficits in the initiation of eye movements in the absence of a visual target in adolescents with high functioning autism. *Neuropsychologia, 40,* 2039–2049.

Goldenberg, M. E. and Robinson, D. C. (1977). Visual responses of neurons in monkey inferior parietal lobule. The physiological substrate of attention and neglect. *Neurology, 27,* 350.

Goldfischer, S., Moore, C. L., Johnson, A. B., Spiro, A. J., and Valsamis, M. P. (1973). Peroxisomal and mitochondrial defects in the cerebro-hepato-renal syndrome. *Science, 182,* 62–64.

Goldin-Meadow, S., Nusbaum, H., Kelly, S. D., and Wagner, S. (2001). Explaining math: Gesturing lightens the load. *Psychological Science: A Journal of the American Psychological Society/APS, 12,* 516–522.

Goldkrand, J. W. and Farkouh, L. (1991). Vibroacoustic stimulation and fetal hiccoughs. *Journal of Perinatology, 11,* 326–329.

Goldman-Rakic, P. S. (1987a). Circuitry of the primate prefrontal cortex and regulation of behavior by representational memory. In F. Plum (Ed.), *Handbook of physiology. The nervous system, Higher cortical functions* (Vol. 5, pp. 373–417). Bethesda, MD: American Physiological Society.

Goldman-Rakic, P. S. (1987b). Circuitry of the frontal association cortex and its relevance to dementia. *Archives of Gerontology and Geriatrics, 6,* 299–309.

Goldman-Rakic, P. S. (1987c). Topography of cognition: Parallel distributed networks in primate association cortex. *Annual Review in Neuroscience, 11*, 137–156.

Goldman-Rakic, P. (1988). Topography of cognition: Parallel distributed networks in primate association cortex. *Annual Review of Neuroscience, 11*, 137–156.

Goldstein, K. (1939). *The organism*. New York, NY: American Books.

Goldstein, K. (1952). The effect of brain damage on the personality. *Psychiatry, 15*, 245–260.

Goleman, D. (1995). *Emotional intelligence*. New York: Bantam Books.

Golse, B., Debray-Ritzen, P., Durosay, P., Puget, K., and Michelson, A. M. (1978). Alterations in two enzymes; superoxide dismutase and glutathion peroxidase in developmental infantile psychosis (infantile autism). *Revista de neurologia, 134*, 699–705.

Gomez-Tortosa, E., Arias-Navalon, J. A., and Barrio-Alba, A. et al. (1996). Relation between frontal lobe blood flow and cognitive performance in huntington's disease. *Neurologia, 11*, 251–256.

Goodale, M. A. and Milner, A. D. (1992). Separate visual pathways for perception and action. *Trends in the Neurosciences, 15*, 20–25.

Goodman, C. S. and Shatz, C. J. (1993). Developmental mechanisms that generate precise patterns of neuronal connectivity. *Cell, 72* (Suppl.), 77–98.

Goodman, W. K., Kozak, M. J., Liebowitz, M., and White, K. L. (1996). Treatment of obsessive-compulsive disorder with fluvoxamine: A multicenter, double-blind, placebo-controlled trial. *International Clinical Psychopharmacology, 11*, 21–29.

Goodman, W. K., Price, L. H., Delgado, P. L., Palumbo, J., Krystal, J. H., Nagy, L. M., Rasmussen, S. A., Heninger, G. R., and Charney, D. S. (1990). Specificity of serotonin reuptake inhibitors in the treatment of obsessive-compulsive disorder. *Archives of General Psychiatry, 47*, 577–585.

Goodman, W. K., Price, L., and Rasmussen, S. et al. (1989). The Yale-Brown obsessive-compulsive scale. II. Validity. *Archives of General Psychiatry, 46*, 1012–1016.

Goodwin, R. S. and Michel, G. F. (1981). Head orientation position during birth and in infant neonatal period, and hand preference at nineteen weeks. *Child Development, 52*, 819–826.

Gordon, C. R., Gonen, A., Nachum, Z., Doweck, I., Spitzer, O., and Shupak, A. (2001). The effects of dimenhydrinate, cinnarizine and transdermal scopolamine on performance. *Journal of Psychopharmacology, 15*, 167–172.

Gore, M. J. (2000). Recognizing violence as a health issue pays off with successful interventions. *Medicine & Health, 54*, P1–P4.

Gottlieb, S. (2002). 1.6 million elementary school children have ADHD, says report. *BMJ (Clinical Research ed.), 324*, 1296.

Gould, T. D., Bastain, T. M., Israel, M. E., Hommer, D. W., and Castellanos, F. X. (2001). Altered performance on an ocular fixation task in attention-deficit/hyperactivity disorder. *Biological Psychiatry, 50*, 633–635.

Gracovetsky, S. and Farfan, H. (1986). The optimum spine. *Spine, 11*, 543–573.

Graf, P. and Schacter, D. L. (1985). Implicit and explicit memory for new associations in normals and amnesic patients. *Journal of Experimental Psychology: Learning, Memory, and Cognition, 11*, 501–518.

Grafman, J., Litvan, I., Gomez, C., and Chase, T. (1990). Frontal lobe function in progressive supranuclear palsy. *Archives of Neurology, 47*, 553–558.

Grafman, J. (1989). Plans, actions, and mental sets: Managerial knowledge units in the frontal lobes. In *Neuropsychology* (pp. 93–137). Hillsdale, NJ: Erlbaum.

Grafman, J., Litvan, I., Massaquoi, S., Stewart, M., Sirigu, A., and Hallett, M. (1992). Cognitive planning deficit in patients with cerebellar atrophy. *Neurology, 42*, 1493–1496.

Grafton, S. T., Mazziotta, J. C., Presty, S., Friston, K. J., Frackowiak, R. S., and Phelps, M. E. (1992). Functional anatomy of human procedural learning determined with regional cerebral blood flow and PET. *Journal of Neuroscience, 12*, 2542–2548.

Graham-Brown, T. (1911). The intrinsic factors in the act of progression in the mammal. *Proceedings of the Royal Society of London, B84*, 308–319.

Graham-Brown, T. (1914). On the nature of the fundamental activity of the nervous centres; together with an analysis of the conditioning of rhythmic activity in progression, and a theory of the evolution of function in the nervous system. *The Journal of Physiology, 49*, 18–46.

Graham-Brown, T. (1915) On the activities of the central nervous system of the unborn foetus of the cat, with a discussion of the question whether progression (walking, etc.) is a 'learnt' complex. *The Journal of Physiology, 49*, 208–215.

Gralton, E. J., James, D. H., and Lindsey, M. P. (1998). Antipsychotic medication, psychiatric diagnosis and children with intellectual disability: A 12-year follow-up study. *Journal of Intellectual Disability Research, 42*, 49–57.

Graves, R., Landis, T., and Goodglass, H. (1981). Laterality and sex differences for visual recognition of emotional and non-emotional words. *Neuropsychologia, 19*, 95–102.

Gray, C., Konig, P., Engel, A., and Singer, W. (1989). Oscillatory responses in cat visual cortex exhibit intercolumnar synchronization which reflects global stimulus properties. *Nature, 338*, 334–337.

Gray, C. and Singer, W. (1989). Stimulus-specific neuronal oscillations in orientation columns of cat visual cortex. *Proceedings of the National Academy of Sciences, USA, 86*, 1698–1702.

Graybiel, A. (1974). Visuo-cerebellar and cerebellovisual connections involving the ventral lateral geniculate nucleus. *Experimental Brain Research, 20*, 303–306.

Graybiel, A. (1995). Building action repertories: Memory and learning functions of the basal ganglia. *Current Opinion in Neurobiology, 5,* 733–741.

Greatrex, J. C., Drasdo, N., and Dresser, K. (2000). Scotopic sensitivity in dyslexia and requirements for DHA supplementation. *Lancet, 355,* 1429–1430.

Green, D., Baird, G., Barnett, A. L., Henderson, L., Huber, J., and Henderson, S. E. (2002). The severity and nature of motor impairment in Asperger's syndrome: A comparison with specific developmental disorder of motor function. *Journal of Child Psychology and Psychiatry, and Allied Disciplines, 43,* 655–668.

Green, W. (1992). Pharmacotherapy: Stimulants in attention-deficit hyperactivity disorder. *Child and Adolescent Psychiatric Clinics of North America, 1,* 411–427.

Greenberg, J. A. and Boozer, C. N. (2000). Metabolic mass, metabolic rate, caloric restriction, and aging in male Fischer 344 rats. *Mechanisms of Ageing and Development, 113,* 37–48.

Greenhill, L. L. (1992). Nonstimulant drugs in the treatment of ADHD. In Attention-Deficit Hyperactivity Disorder. *Child and Adolescent Psychiatric Clinics of North America, 1,* 449–465.

Greenough, W. T., Black, J. E., and Wallace, C. S. (1987). Experience and brain development. *Child Development, 58,* 539–559.

Greenough, W. T. and Chang, F. L. (1988). Dendritic pattern formation involves both oriented regression and oriented growth in the barrels of mouse somatosensory cortex. *Brain Research, 47,* 148–152.

Greenough, W. T., Hwang, H. M., and Gorman, C. (1985). Evidence for active synapse formation or altered postsynaptic metabolism in visual cortex of rats reared in complex environments. *Proceedings of the National Academy of Sciences, USA, 82,* 4549–4552.

Greenough, W. T., Larson, J. R., and Withers, G. S. (1985). Effects of unilateral and bilateral training in a reaching test on dendritic branching of neurons in the rat motor sensory forelimb cortex. *Behavioral and Neural Biology, 44,* 301–314.

Greenough, W. T. and Volkmar, F. R. (1973). Pattern of dendritic branching in occipital cortex of rats reared in complex environments. *Experimental Neurology, 40,* 491–504.

Greenwood, C. E. and Winocur, G. (1990). Glucose treatment reduces memory deficits in young adult rats fed high-fat diets. *Behavioral and Neural Biology, 53,* 74–87.

Greenwood, C. E. and Winocur, G. (1996). Cognitive impairment in rats fed high-fat diets: A specific effect of saturated fatty-acid intake. *Behavioral Neuroscience, 110,* 451–459.

Greenwood, C. E. and Winocur, G. (1996). Learning and memory impairment in rats fed a high saturated fat diet. *Behavioral Neuroscience, 110,* 451–459.

Greenwood, C. E. and Winocur, G. (2001). Glucose treatment reduces memory deficits in young adult rats fed high-fat diets. *Neurobiology of Learning and Memory, 75,* 179–189.

Griest, J., Chouinard, G., DuBoff, E., Halaris, A., Won, KIM, S., Koran, L., Liebowitz, M., Lydiard, R. B., Rasmussen, S., White, K., and Sikes, C. (1992). Double-blind comparison of three doses of sertraline and placebo in the treatment of outpatients with obsessive compulsive disorder. *Clinical Neuropharmacology, 15* (Suppl.), 316B.

Grigoriev, A. I. and Kozlovskaya, I. (1987). Man in space: 25 years of manned space light in the Soviet Union: Biomedical aspects. In S. Watanabe et al. (Eds.), *Biological sciences in space.* Tokyo: MYU Space Research.

Gross, M. D. (1973). Imipramine in the treatment of minimal brain dysfunction in children. *Psychosomatics, 14,* 283–285.

Gross-Glenn, K., Duara, R., Barker, W. W., Loewenstein, D., Chang, J. Y., Yoshii, F., Apicella, A. M., Pascal, S., Boothe, T., Sevush, S. et al. (1991). Positron emission tomographic studies during serial word-reading by normal and dyslexic adults. *Journal of Clinical and Experimental Neuropsychology, 13,* 531–544.

Grossman, M. A. (1980). Central processor for hierarchically structured material: Evidence from broca's aphasia. *Neuropsychologia, 18,* 299–308.

Grouse, L. D., Schrier, B. K., Bennett, E. L., Rosenzweig, M. R., and Nelson, P. G. (1978). Sequence diversity studies of rat brain RNA: Effects of environmental complexity on rat brain RNA diversity. *Journal of Neurochemistry, 30,* 191–203.

Grove, K. L., Brogan, R. S., and Smith, M. S. (2001). Novel expression of neuropeptide Y (NPY) mRNA in hypothalamic regions during development: Region-specific effects of maternal deprivation on NPY and Agouti-related protein mRNA. *Endocrinology, 42,* 4771–4776.

Gruber, D., Waanders, R., Collins, R. L., Wolfer, D. P., and Lipp, H. P. (1991). Weak or missing paw lateralization in a mouse strain with congenital absence of the corpus callosum. *Behavioral Brain Research, 46,* 9–16.

Gruzelier, J. H. (2002a). The role of psychological intervention in modulating aspects of immune function in relation to health and well-being. *International Review of Neurobiology, 52,* 383–417.

Gruzelier, J. H. (2002b). A review of the impact of hypnosis, relaxation, guided imagery and individual differences on aspects of immunity and health. *Stress, 5,* 147–163.

Gruzelier, J. H. and Venables, P. H. (1974). Bimodality and lateral asymmetry of skin conductance orienting activity in schizophrenics: Replication and evidence of lateral asymmetry in patients with depression and disorders of personality. *Biological Psychiatry, 8,* 55–73.

Guiard, Y. (1980). Cerebral hemispheres and selective attention. *Acta Psychologica, 46,* 4–61.

Guiard, Y. and Millerat, F. (1984). Writing posture in lefthanders: Inverters are hand crossers. *Neuropsychological, 22,* 535–538.

Guiard, Y. and Requin, J. (1978). Between-hand vs. within-hand choice: A single channel reduced capacity in the split-brain monkey. In J. Requin (Ed.), *Attention and performance VII* (pp. 391–410). Hillsdale, NJ: Lawrence Erlbaum Associates.

Gulcher, J. R., Jonsson, P., and Kong, A. et al. (1997). Mapping a familial essential tremor gene, FET1, to chromosome 3q13. *Nature Genetics, 17*, 84–87.

Gunnar, M. R., Morison, S. J., Chisholm, K., and Schuder, M. (2001). Salivary cortisol levels in children adopted from romanian orphanages. *Developmental Psychopathology, 13*, 611–628.

Guo, S. S., Roche, A. F., Chumlea, W. C., and Gardner, J. C. (1994). Mass index values for overweight at age 35. *The American Journal of Clinical Nutrition, 59*, 810–819.

Gupta, S., Aggarwal, S., and Heads, C. (1996). Dysregulated immune system in children with autism: Beneficial effects of intravenous immune globulin on autistic characteristics. *Journal of Autism and Developmental Disorders, 26*, 439–452.

Gupta, S., Aggarwal, S., Rashanravan, B., and Lee, T. (1998). Th1- and Th2- like cytokines in CD4+ and CD8+ T cells in autism. *Journal of Neuroimmunology, 85*, 106–109.

Gur, R. E., Mozley, P. D., Resnick, S. M., Shtasel, D. L., Kohn, M. I., Zimmerman, R. A., Herman, G. T., Atlas, S. W., Grossman, R. I., Erwin, R. J., and Gur, R. C. (1991). Magnetic resonance imaging in schizophrenia: I. Volumetric analysis of brain and cerebrospinal fluid. *Archives of General Psychiatry, 48*, 407–412.

Gutknecht, L. (2002). Full-genome scans with autistic disorder: A review. *Behavior Genetics, 32*, 397–412.

Gutmann, G. (1970). Anatomical justification of burdening the lumbar spine and pelvis in man. *Fysiatrickya reumatologicky vestnik, 48*, 88–92.

Gutmann, G. (1971). Traumatic asymmetry of cervical articular surfaces. Their diagnostic certainty and evaluative importance. *Zeitschrift fur Physiotherapie, 23*, 383–386.

Gutzmann, H. and Kuhl, K. P. (1987). Emotion control and cerebellar atrophy in senile dementia. *Archives of Gerontology and Geriatrics, 6*, 61–71.

Guyton, A. C. (1986). *Textbook of medical physiology.* Philadelphia, PA: W. B. Saunders.

Haber, R. N. (1978). Visual perception. *Annual Review of Psychology, 29*, 31–59.

Hachinsky, V. C, Oppenheimer, S. M., Wilson, J. X., Guiraudon, C., and Cechetto, D. F. (1992). Asymmetry of sympathetic consequences of experimental stroke. *Archives of Neurology, 49*, 697–702.

Hadders-Algra, M. and Towen, B. (1992). Minor neurological dysfunction is more closely related to learning difficulties than to behavioral problems. *Journal of Learning Disabilities, 25*, 649–657.

Haggard, M. and Parkinson, A. M. (1971). Stimulus and task factors as determinants of ear advantages. *Quarterly Journal of Experimental Psychology, 23*, 168–177.

Hagino, N. and Lee, J. (1985). Effect of maternal nicotine on the development of sites for [3h] nicotine binding in the fetal brain. *International Journal of Developmental Neuroscience, 3*, 567–571.

Haile-Selassie, Y. (2001) Late miocene hominids from the middle Awash, Ethiopia. *Nature, 412*, 178–181.

Hainaut, K. and Duchateau, J. (1989). Peripheral changes during muscle fatigue and their adaptation to training and disuse. *Biomedica Biochimica Acta, 48*, S525–S529.

Haines, D. E. and Dietrichs, E. (1987). On the organization of interconnections between the cerebellum and hypothalamus. In J. S. King (Ed.), *New concepts in cerebellar neurobiology* (pp. 113–149). New York: Allen R. Liss.

Haines, D. E., Dietrichs, E., and Sowa, T. E. (1984). Hypothalamo-cerebellar and cerebello-hypothalamic pathways: A review and hypothesis concerning cerebellar circuits which may influence autonomic centers and affective behavior. *Brain, Behavior and Evolution, 24*, 198–220.

Haines, D. E., May, P. G., and Dietrichs, E. (1990). Neuronal connections between the cerebellar nuclei and hypothalamus in macaca fascicularis: Cerebello-visceral circuits. *The Journal of Comparative Neurology, 299*, 106–122.

Halboni, P., Kaminski, R., Gobbele, R., Zuchner, S., Waberski, T. D., Herrmann, C. S., Topper, R., and Buchner, H. (2000). Sleep stage dependant changes of the high-frequency part of the somatosensory evoked potentials at the thalamus and cortex. *Clinical Neurophysiology, 111*, 2277–2284.

Halkjaer-Kristensen, J., and Ingeman-Hansen, T. (1985). Wasting of the human quadriceps muscle after knee ligament injuries. I. Anthropometric consequences. *Scandinavian Journal of Rehabilitation Medicine* (Suppl.) *13*, 5–55.

Hall, N. R. and Goldstein, A. L. (1985). Neurotransmitters and host defense. In R. Guillemin, M. Cohn, and T. Melnechuk (Eds.), *Neural modulation of immunity* (pp. 143–154). New York, NY: Raven Press.

Hallaway, N. and Strauts, Z. (1995). *Turning lead into gold: How heavy metal poisoning can affect your child and how to prevent and treat it.* Vancouver, Canada: New Star Books.

Hallett, M. (1986). Differential diagnosis of tremor. In R. J. Vinken, G. W. Bruyn, H. L. Klawans (Eds.), *Handbook of clinical neurology* (Vol. 49, Revised series V, pp. 583–589). Amsterdam: Elsevier Science.

Hallett, M. (1998). Neurophysiology of dystonia. *Archives of Neurology, 55*, 601–603.

Hallett, M. and Dubinsky, R. (1993). Glucose metabolism in the brain of patients with essential tremor. *Journal of Neurology Science, 114*, 45–48.

Hallett, M., Lebiecowska, M. K., Thomas, S. L., Stanhope, S. J., Denckla, M. B., and Rumsey, J. (1993). Locomotion of austic adults. *Archives of Neurology, 50*, 1304–1308.

Halliwell, J. W. and Solan, H. A. (1972). The effects of a supplemental perceptual training program on reading achievement. *Exceptional Child, 38*, 613–621.

Halsband, U., Ito, N., Tanji, J., and Freund, H. J. (1993). The role of premotor cortex and the supplementary motor area in the temporal control of movement in man. *Brain, 116,* 243–266.

Hamill, J. S. (1984). Tactile defensiveness. A type of sensory integrative dysfunction. *Journal of Rehabilitation, 50,* 95–96.

Hamilton, C. R. and Gazzaniga, M. S. (1964). Lateralization of learning of color and brightness discriminations following brain bisection. *Nature, 201,* 220.

Hamilton, C. R. and Lund, J. S. (1970). Visual discrimination of movement: Midbrain or forebrain? *Science, 170,* 1428–1430.

Hamilton, C. R. and Vermeire, B. A. (1983). Discrimination of monkey faces by split-brain monkeys. *Behavioural Brain Research, 9,* 263–275.

Hamilton, C. R. and Vermeire, B. A. (1988). Complementary hemispheric specialization in monkeys. *Science, 242,* 1691–1694.

Hamilton, N. G., Frick, R. B., Takahashi, T., and Hopping, M. W. (1983). Psychiatric symptoms and cerebellar pathology. *The American Journal of Forensic Psychiatry, 140,* 1322–1326.

Hamlin, R. L. and Smith, C. R. (1968). Effects of vagal stimulation on S-A and A-V nodes. *American Journal of Physiology, 215,* 560–568.

Hardan, A. Y., Minshew, N. J. and Keshavan, M. S. (2000). Corpus callosum in autism. *Neurology, 55,* 1033–1036.

Hardan, A. Y., Minshew, N. J., Harenski, K., and Keshavan, M. S. (2001). Posterior fossa magnetic resonance imaging in autism. *Journal of the American Academy of Child and Adolescent Psychiatry, 40,* 666–672.

Hardan, A. Y., Minshew, N. J., Mallikarjuhn, M., and Keshavan, M. S. (2001). Brain volume in autism. *Journal of Child Neurology. 16,* 421–424.

Hari, R. and Salmelin, R. (1997). Human cortical oscillations: A neuromagnetic view through the skull. *Trends in the Neurosciences, 20,* 44–49.

Harper, J. W. and Heath, R. G. (1973). Anatomic connections of the fastigial nucleus to the rostral forebrain in the cat. *Experimental Neurology, 39,* 285–292.

Harrington, D. L. and Haaland, K. Y. (1991). Hemispheric specialization for motor sequencing: Abnormalities in levels of programming. *Neuropsychologia, 29,* 147–163.

Harris, I. M., Fulham, M. J., and Miller, L. A. (2001). The effects of mesial temporal and cerebellar hypometabolism on learning and memory. *Journal of the International Neuropsychological Society, 7,* 353–362.

Harris, L. J. (1980). Left-handedness: Early theories, facts, and fancies. In J. Herron (Ed.), *Neuropsychology of left-handedness* (pp. 3–78). New York, NY: Academic Press.

Harris, L. J. and Fitzgerald, H. E. (1983). Postural orientation in human infants: Changes from birth to three months. In G. Young, S. J. Segalowitz, C. M. Corter, and S. E. Trehub (Eds.), *Manual specialization and the developing brain* (pp. 285–305). New York: Academic Press.

Harris, N. S., Courchesne, E., Townsend, J., Carper, R. A., and Lord, C. (1999). Neuroanatomic contributions to slowed orienting of attention in children with autism. *Brain Research. Cognitive Brain Research, 8,* 61–71.

Harter, M. R. (1991). Event-related potential indicies: Learning disabilities and visual processing. In J. E. Obmit and C. W. Hynd (Eds.), *Neuropsychological foundations of learning disabilities: A handbook of issues, methods, and practice* (pp. 437–473). San Diego, CA: Academic Press.

Harting, J. K. (1997). *The global brainstem.* http://www.medsch.wisc.edu/anatomy/bs97/text/bs/contents.htm

Hartsough, C. S. and Lambert, N. M. (1985). Medical factors in hyperactive and normal children: Prenatal, developmental, and health history findings. *American Journal of Orthopsychiatry, 55,* 190–201.

Hashimoto, T., Tayama, M., Murakawa, K., Yoshimoto, T., Miyazaki, M., Harada, M., and Kuroda, Y. (1995). Development of the brainstem and cerebellum in autistic patients. *Journal of Autism and Developmental Disorders, 25,* 1–18.

Haslam, R. H., Dalby, J. T., Johns, R. D., and Rademaker, A. W. (1981). Cerebral asymmetry in developmental dyslexia. *Archives of Neurology, 38,* 679–682.

Hassler, R. (1964). Spezifische und unspezifische Systeme, des, Menschen, Zwischenhirns. In W. Bargmann and J. P. Schabe' (Eds.), *Lectures on the diencephalon* (pp. 1–32). Amsterdam: Elsevier.

Hatta, T. and Koike, M. (1991). Left-hand preference in frightened mother monkeys in taking up their babies. *Neuropsychologia, 29,* 207–209.

Havlovicova, M., Propper, L., Novotna, D., Musova, Z., Hrdlicka, M., and Sedlacek, Z. (2002). Genetic study of 20 patients with autism disorders. *Casopis Lekaru Ceskych, 141,* 381–387.

Hawn, P. R. and Harris, L. J. (1983). Hand differences in grasp duration and reaching in two and five-month-old human infants. In G. Young, S. J. Segalowitz, C. M. Corter, and S. E. Trehub (Eds.), *Manual specialization and the developing brain* (pp. 71–92). New York: Academic Press.

Haywood, J., Rose, S. P., and Bateson, P. P. (1970). Effects of an imprinting procedure on RNA polymerase activity in the chick brain. *Nature, 228,* 373–375.

Hazeltine, E. and Ivry, R. B. (2002). Neuroscience. Can we teach the cerebellum new tricks? *Science, 296,* 1979–1980.

Healey, J. M. (1990). *Endangered minds: Why children don't think and what we can do about it.* New York, NY: Simon and Schuster.

Heath, R. G., Franklin, D. E., and Shraberg, D. (1979). Gross pathology of the cerebellum in patients diagnosed and treated as functional psychiatric disorders. *Journal of Neurological and Mental Disorders, 167,* 585–592.

Heath, R. G. and Harper, J. W. (1974). Ascending projections of the cerebellar fastigial nucleus to the hippocampus, amygdala and other temporal lobe sites: Evoked potential and other histologic studies in monkeys and cats. *Experimental Neurology, 45*, 268–287.

Heath, R. G., Llewellyn, R. C., and Rouchell, A. M. (1980). The cerebellar pacemaker for intractable behavioral disorders and epilepsy: Follow-up report. *Biological Psychiatry, 15*, 243–256.

Hebb, D. O. (1949). *The organization of behavior*. New York: Wiley.

Hebb, D. O. (1980). *Essay on mind*. Hillsdale, NJ: Lawrence Erlbaum Associates.

Hecaen, H. and Angelergues, R. (1962). Agnosia for faces (prosopoagnosia). *Archives of Neurology, 7*, 92–100.

Hécaen, H. (1972). *La neuropsychologie humaine*. Paris: Masson.

Hecht, K., Hai, N. V., Hecht, T., Moritz, V., and Woossmann, H. (1976). Correlations between hippocampus function and stressed learning and their effect on cerebro-visceral regulation processes. *Acta Biologica et Medica Germanica, 35*, 35–45.

Heilman, K. M. (1979). Neglect and related disorders. In K. M. Heilman and E. Valenstein (Eds.), *Clinical neuropsychology* (pp. 268–307). New York, NY: Oxford University Press.

Heilman, K. M. (1995a). Attention asymmetries. In R. J. Davidson and K. Hugdahl (Eds.), *Brain asymmetry*, (pp. 217–234). Cambridge, MA: The MIT Press.

Heilman, K. M. (1995b). Apraxia. In K. M. Heilman and E. Valenstein (Eds.), *Clinical neuropsychology* (3rd ed.). New York: Oxford University Press.

Heilman, K. M., Chatterjee, A., and Doty, L. (1993). Hemispheric asymmetries of spatial attention. *Journal of Clinical and Experimental Neuropsychology, 15*, 14.

Heilman, K. M. and Gilmore, R. L. (1998). Cortical influences in emotion. *Journal of Clinical Neurophysiology, 15*, 409–423.

Heilman, K. M., Pandya, D. N., and Geschwind, N. (1970). Trimodal inattention following parietal lobe abolations. *Transactions of the American Neurologic Association, 95*, 259–261.

Heilman, K. M. and Rothi, L. (1993). In K. M. Heilman and E. Valenstein (Eds.), *Apraxia. Clinical Neuropsychology* (pp. 141–163). New York, NY: Oxford University Press.

Heilman, K. M., Schwartz, H. D., and Watson, R. T. (1978). Hypoarousal in patients with neglect syndrome and emotional indifference. *Neurology, 28*, 229–232.

Heilman, K. M. and Valenstein, E. (1972). Frontal lobe neglect in man. *Neurology, 22*, 660–664.

Heilman, K. M. and Valenstein, E. (1979). Mechanisms underlying hemispatial neglect. *Annals of Neurology, 5*, 166–170.

Heilman, K. M. and Valenstein, E. (1985). *Clinical neuropsychology*. New York, NY: Oxford University Press.

Heilman, K. M., Valenstein, E., and Watson, R. T. (1983). Localization of neglect. In A. Kertesz (Ed.), *Localization in neurology* (pp. 471–492). New York, NY: Academic Press.

Heilman, K. M. and Watson, R. T. (1977). Mechanisms underlying the unilateral neglect syndrome. *Advances in Neurology, 18*, 93–106.

Heilman, K. M. and Watson, R. T. (1991). Intentional motor disorders. In H. S. Levin, H. M. Eisenberg, and A. L. Benton (Eds.), *Frontal lobe function and dysfunction* (pp. 199–213). New York: Oxford University Press.

Heilman, K. M., Watson, R. T., Valenstein, E., and Damasio, A. R. (1983). Localization of lesions in neglect. In A. Kertesz (Ed.), *Localization in neuropsychology* (pp. 445–470). New York, NY: Academic Press.

Heim, C., Zhang, J., Lan, J., Sieklucka, M., Kurz, T., Riederer, P., Gerlach, M., and Sontag, K. H. (2000). Cerebral oligaemia episode triggers free radical formation and late cognitive deficiencies. *European Journal of Neuroscience, 12*, 715–725.

Heinke, D. and Humphreys, G. W. (2003). Attention, spatial representation, and visual neglect: Simulation emergent attention and spatial memory in the selective attention for identification model (SAIM). *Psychological Review, 110*, 29–87.

Heinsbroek, R. P., van Haaren, F., Zantvoord, F., and van de Poll, N. E. (1987). Sex differences in response rates during random ratio acquisition: Effects of gonadectomy. *Physiological Behavior, 39*, 269–272.

Heller, W., Lindsey, D. L., Metz, J., and Farnum, D. M. (1990). Individual differences in right-hemisphere activation are associated with arousal and autonomic response to lateralized stimuli. *Journal of Clinical and Experimental Neuropsychology, 13*, 95.

Hellige, J. B. (1993). *Hemispheric asymmetry: What's right and what's left*. Cambridge, MA: Harvard University Press.

Hellige, J. B. and Michimata, C. (1989). Categorization versus distance: Hemispheric differences for processing spatial information. *Memory and Cognition, 17*, 770–776.

Henderson, S. and Sugdend, D. (1992). Movement assessment battery for children (Movement ABC) (San Antonio, TX): The Psychological Corporation. *Journal of Autism and Developmental Disorders, 31*, 79–88.

Heustis, R. D., Arnold, L. E., and Smeltzer, D. J. (1975). Caffeine versus methylphenidate and d-amphetamine in minimal brain dysfunction: A double blind comparison. *American Journal of Psychiatry, 132*, 868–870.

Hier, D. B., LeMay, M., Rosenberger, P. B., and Perlo, V. (1978). Developmental dyslexia: Evidence for a sub-group with reversed cerebral asymmetry. *Archives of Neurology, 35*, 90–92.

Henderson, Z. and Blakemore, C. (1986). Organization of the visual pathways in the newborn kitten. *Neuroscience Research, 3*, 628–659.

Henderson, J. M., Einstein, R., Jackson, D. M., Byth, K., and Morris, J. G. (1995). 'Atypical' tremor. *European Neurology, 35*, 321–326.

Hendrickson, C. W., Kimble, R. J., and Kimble, D. P. (1969). Hippocampal lesions and the orienting response. *Journal of Comparative and Physiological Psychology, 67*, 220–227.

Henkin, R. I. and Levy, L. M. (2001). Lateralization of brain activation to imagination and smell of odors using functional magnetic resonance imaging (fMRI): Left hemispheric localization of pleasant and right hemispheric localization of unpleasant odors. *Journal of Computer Assisted Tomography, 25*, 493–514.

Henkin, R. I. and Levy, L. M. (2002). Functional MRI of congenital hyposmia: Brain activation to odors and imagination of odors and tastes. *Journal of Computer Assisted Tomography, 26*, 39–61.

Henriques, J. B. and Davidson, R. J. (1991). Left frontal hypoactivation in depression. *Journal of Abnormal Psychology, 100*, 535–545.

Herman, R., Mixon, J., Fisher, A., Maulucci, R., and Stuyck, J. (1985). Idiopathic scoliosis and the central nervous system: A motor control problem. The Harrington lecture, 1983. Scoliosis Research Society. *Spine, 10*, 1–14.

Hewlett, W. A., Vinogradov, S., and Agras, S. (1992). Clomipramine, clonazepam, and clonidine treatment of obsessive-compulsive disorder. *Journal of Clinical Psychopharmacology, 12*, 420–430.

Higgins, J. A. (1997). Gene for essential tremor maps to chromosome 2p22-p25. *Movement Disorders, 12*, 859–864.

Hjorth-Simonsen, A. (1973). Some intrinsic connections of the hippocampus in the rat: An experimental analysis. *Journal of Comparative Neurology, 147*, 145–162.

Hikosaka, O. (1998). Neural systems for control of voluntary action—a hypothesis. *Advances in Biophysics, 35*, 81–102.

Hill, J. O. and Peters, J. C. (1998). Environmental contributions to the obesity epidemic. *Science, 280*, 1371–1374.

Hillyard, S. A. (1999). Event-related potentials and human information processing. In G. Adelman and B. H. Smith (Eds.), *Encyclopedia of neuroscience* (2nd ed.) (pp. 679–682). Amsterdam, The Netherlands: Elsevier.

Hillyard, S. A., Hink, R. F., Schwent, V. L., and Picton, T. W. (1973). Electrical signs of selective attention in the human brain. *Science, 182*, 177–180.

Hinshaw, S. P., Henker, B., and Whalen, C. K. (1984). Cognitive-behavioral and pharmacologic interventions for hyperactive boys: Comparative and combined effects. *Journal of Consulting and Clinical Psychology, 52*, 739–749.

Hinshelwood, J. (1900). Word-blindness and visual memory. *Lancet, 2*, 1564–1570.

Hinton, D. R., Sadun, A. A., Blanks, J. C., and Miller, C. A. (1986). Optic-nerve degeneration in Alzheimer's disease. *New England Journal of Medicine, 315*, 485–487.

Hiscock, M. and Kinsbourne. M. (1995). Phylogeny and ontogeny of cerebral lateralization. In R. Davidson and K. Hugdahl (Eds.), *Brain asymmetry* (pp. 535–578). Cambridge, MA: MIT Press.

Hjorth-Simonsen, A. (1971). Hippocampal efferents to the ipsilateral entorhinal area: An experimental study in the rat. *The Journal of Comparative Neurology, 142*, 417–438.

Hjorth-Simonsen, A. (1972). Projection of the lateral part of the entorhinal area to the hippocampus and fascia dentata. *The Journal of Comparative Neurology, 146*, 219–232.

Hockey, R. (1979). Stress and the cognitive components of skilled performance. In V. Hamilton and D. M. Warburton (Eds.), *Human stress and cognition: An information processing perspective*. Chichester, England: John Wiley & Sons.

Holden, C. (1996). Small refugees suffer the effects of early neglect. *Science, 274*, 1076–1077.

Holender, D. (1986). Semantic activation without conscious identification in dichotic listening, parafoveal vision, and visual masking: A survey and appraisal. *Behavioral and Brain Sciences, 9*, 1–66.

Hollander, E., Liebowitz, M. R., and DeCaria, C. (1989). Conceptual and methodological issues in studies of obsessive-compulsive and Tourette's disorders. *Psychiatric Developments, 4*, 67–296.

Hollander, E., Schiffman, E., Cohen, D., Rivera-Stein, M. A., Rosen, W., Gorman, J. M., Fyer, A. J., Papp, L., and Liebowitz, M. R. (1990). Signs of central nervous system dysfunction in obsessive-compulsive disorder. *Archives of General Psychiatry, 47*, 27–32.

Holmes, G. (1904). On certain tremor in organic cerebral lesions. *Brain, 27*, 325–375.

Holmes, G. (1917). The symptoms of acute cerebellar injuries due to gunshot injuries. *Brain, 40*, 461–535.

Holstein, G. R., Kukielka, E., and Martinelli, G. P. (1999). Anatomical observations of the rat cerebellar nodulus after 24 hr of spaceflight. *Journal of Gravitational Physiology: A Journal of the International Society for Gravitational Physiology, 6*, P47–P50.

Honma, S., Kawamoto, T., Takagi, Y., Fujimoto, K., Sato, F., Noshiro, M., Kato, Y., and Honma, K. (2002). Dec1 and Dec2 are regulators of the mammalian molecular clock. *Nature, 419*, 841–844.

Hoon, A. H., Jr. and Reiss, A. L. (1992). The mesial-temporal lobe and autism: Case report and review. *Developmental Medicine and Child Neurology, 34*, 252–259.

Hopkins, D. A. and Holstege, G. (1978). Amygdaloid projections to the mesencephalon and medulla oblongata in the cat. *Experimental Brain Research, 32*, 529–547.

Hopkins, W. D., Washbum, D. A., and Rumbaugh, D. M. (1989). Note on hand use to manipulation of joysticks by rhesus monkeys (Macaca mulatta) and chimpanzees (Pan troglodytes). *Journal of Comparative Psychology, 103*, 91–94.

Hornig, M., Weissenbock, H., Horscroft, N., and Lipkin, W. I. (1999). An infection-based model of neurodevelopmental damage. *Proceedings of the National Academy of Sciences, USA, 96*, 12102–12107.

Horvath, K., Papadimitriou, J. C., Rabsztyn, A., Drachenberg, C., and Tildon, J. T. (1999). Gastrointestinal abnormalities in children with autistic disorder. *The Journal of Pediatrics, 135*, 559–563.

Howard, M. A., Cowell, P. E., Boucher, J., Broks, P., Mayes, A., Farrant, A., and Roberts, N. (2000). Convergent neuroanatomical and behavioural evidence of an amygdala hypothesis of autism. *Neuroreport, 11*, 2931–2935.

Howe, J. (1999). Show me what you are thinking. *Psychology Today, 31*.

Howe, R. C. and Sterman, M. B. (1972). Cortical-subcortical EEG correlates of suppressed motor behavior during sleep and waking in the cat. *Electroencephalography and Clinical Neurophysiology, 32*, 681–695.

Howlin, P. (2000). Autism and intellectual disability: Diagnostic and treatment issues. *Journal of the Royal Society of Medicine, 93*, 351–355.

Hubel, D. H. and Wiesel, T. N. (1979). Brain mechanisms of vision. *Scientific American, 241*, 150–162.

Huddahl, K., Asbornsen, A., and Wester, K. (1993). Memory performance in Parkinson's disease. *Neuropsychiatry, Neuropsychology and Behavioral Neurology, 6*, 170–176.

Hudziak, J. and Todd, R. D. (1993). Familial subtyping of attention deficit hyperactivity disorder. *Current Medical Research and Opinion, 6*, 489.

Huettner, M. I. S., Rosenthal, B. L., and Hynd, G. W. (1989). Regional cerebral blood flow (rCBF) in normal readers: Bilateral activation with narrative speech. *Archives of Clinical Neuropsychology, 4*, 71–78.

Hugdahl, K. (1988) (Ed.). Handbook of dichotic listenting: Theory, methods and research. Chichester, England: John Wiley & Sons.

Hugdahl, K. (1992). Brain lateralization: Dichotic listening studies. In B. Smith and G. Adelman (Eds.), *Encyclopedia of neuroscience: Neuroscience year 2*. Boston, MA: Birkhauser.

Hugdahl, K. (1995). Dichotic listening: Probing temporal lobe functional integrity. In R. Davidson and K. Hugdahl (Eds.), *Brain asymmetry*. Cambridge, MA: MIT Press.

Hugdahl, K. and Andersson, B. (1987). Dichotic listening in children. In M. Tramontana and R. Hooper (Eds.), *Advances in Child Neuropsychology, 9*, 631–649.

Hugdahl, K. and Andersson, L. (1986). The forced-attention paradigm in dichotic listening to CV-syllables: A comparison between adults and children. *Cortex, 22*, 417–432.

Hugdahl, K. and Andersson, L. (1987). Dichotic listening and reading acquisition in children: A one-year follow-up. *Journal of Clinical and Experimental Neuropsychology, 9*, 631–649.

Hugdahl, K. and Brobeck, C. G. (1986). Hemispheric asymmetry and human electrodermal conditioning: The dichotic extinction paradigm. *Psychophysiology, 23*, 491–499.

Hugdahl, K., Satz, P., Mitrushina, M., and Miller, E. N. (1993). Left-handedness and old age: Do left-handers die earlier? *Neuropsychologia, 31*, 325–333.

Hughes, J. R. and Fino, J. J. (2000). The Mozart effect: Distinctive aspects of the music—a clue to brain coding? *Clinical Electroencephalography, 31*, 94–103.

Huguenard, J. R. and Prince, D. A. (1994). Intrathalamic rhythmicity studied in vitro: Nominal T-current nodulation causes robust anti-oscillatory effects. *The Journal of Neuroscience, 14*, 5485–5502.

Huizinga, J. D. Robinson, T. L., and Thomsen, L. (2000). The search for the origin of rhythmicity in intestinal contraction; from tissue to single cells. *Neurogastroenterology and motility, 12*, 3–9.

Huizinga, J. D. Thuneberg, L., Vanderwinden, J. M., and Rumessen, J. J. (1997). Interstitial cells of Cajal as targets for pharmacological intervention in gastrointestinal motor disorders. *Trends in Pharmacological Science, 18*, 393–403.

Hulme, C., Biggerstaff, A., Morgan, G., and McKinlay, I. (1982). Visual, kinaesthetic and cross-modal judgments of length by normal and clumsy children. *Developmental Medicine and Child Neurology, 24*, 461–471.

Humphreys, P., Kaufmann, W. E., and Galaburda, A. M. (1990). Developmental dyslexia in women: Neuropathological findings in three cases. *Annals of Neurology, 28*, 727–738.

Humphrey, T. (1972). The development of the human amygdaloid complex in the neurobiology of the amygdala. In B. E. Eleftheriou (Ed.), *The Physiology of Aggression and Defeat*. New York, NY: Plenum Press.

Humphries, T. W., Snider, L., and McDougall, B. (1997). Therapists' consistency in following their treatment plans for sensory integrative and perceptual-motor therapy. *American Journal of Occupational Therapy, 51*, 104–112.

Hunt, C. (1974). Muscle Receptors. In C. Hunt (Ed.), *Handbook of sensory physiology* (pp. 192–218). New York: Springer-Verlag.

Hunt, C. and Ottoson, D. (1976). Initial burst of primary endings of isolated mammalian muscle spindles. *Journal of Neurophysiology, 39*, 324–330.

Hunt, R. D. and Arsten, A. F. T. (1995). An open trial of guanfacine in the treatment of attention-deficit hyperactivity disorder. *Journal of the American Academy of Child Psychiatry, 34*, 50–54.

Hunt, R. D., Capper, L., and O'Connell, P. (1990). Clonidine in child and adolescent psychiatry. *Journal of Child and Adolescent Psychopharmacology, 1*, 87–102.

Hunter, M. S., Ussher, J. M., Cariss, M., Browne, S., Jelley, R., and Katz, M. (2002). Medical (fluoxetine) and psychological (cognitive-behavioural therapy) treatment for premenstrual dysphoric disorder: A study of treatment processes. *Journal of Psychosomatic Research, 53*, 811–817.

Huttenlocher, J., Levine, S., and Vevea, J. (1998). Environmental input and cognitive growth: A study using time-period comparisons. *Child Development, 69*, 1012–1029.

Hynd, G. W., Hall, J., Novey, E. S., and Eliopulos, D., Black, K., Gonzalez, J. J., Edmonds, J. E., Riccio, C., and Cohen, M. (1995). Dyslexia and corpus callosum morphology. *Archives of Neurology, 52*, 32–38.

Hynd, G. W., Hynd, C. R., Sullivan, H. G., and Kingsbury, T. B. 4th. (1987). Regional cerebral blood flow (rCBF) in developmental dyslexia: Activation during reading in a surface and deep dyslexic. *Journal of Learning Disabilities, 20*, 294–300.

Hynd, G. W. and Semrud-Clikeman, M. (1989a). Dyslexia and brain morphology. *Psychological Bulletin, 106*, 447–482.

Hynd, G. W. and Semrud-Clikeman, M. (1989b). Dyslexia and neurodevelopmental pathology: Relationships to cognition, intelligence and reading skill acquisition. *Journal of Learning Disabilities, 22*, 204–216.

Hynd, G. W. and Willis, W. G. (1988a). *Pediatric neuropsychology*. Needham Heights, MA: Allyn & Bacon.

Hynd, G. W. and Willis, W. G. (1988b). *Pediatric neuropsychology*. Orlando, FL: Grune & Stratton.

Hynd, G. W., Semrud-Clikeman, M., Lorys, A. R., Novey, E. S., and Eliopulos, D. (1990). Brain morphology in developmental dyslexia and attention deficit disorder/hyperactivity. *Archives of Neurology, 47*, 919–926.

Ikoma, K., Samii, A., Mercuri, B., Wassermann, E. M., and Hallett, M. (1996). Abnormal cortical motor excitability in dystonia. *Journal of Neurology, 46*, 1371–1376.

Ingram, D. K., Chefer, S., Matochik, J., Moscrip, T. D., Weed, J., Roth, G. S., London, E. D., and Lane, M. A. (2001). Aging and caloric restriction in nonhuman primates: Behavioral and in vivo brain imaging studies. *Annals of the New York Academy of Sciences, 928*, 316–326.

Ingvar, D. H. (1977). The cerebral ideogram. *Encephale, 3*, 5–33.

Ingvar, D. H. (1991). On ideation and 'ideography'. In J. Eccles (Ed.), *The principles of design and operation of the brain* (pp. 433–453). Berlin: Springer.

Ingvar, D. H. (1993). Language functions related to prefrontal cortical activity: Neurolinguistic implications. *Annals of the New York Academy of Sciences, 682*, 240–247.

Ingvar, D. H. (1994). The will of the brain: Cerebral correlates of willful acts. *Journal of Theoretical Biology, 171*, 7–12.

Inhoff, A. W., Diener, H. C., Rafal. R. D., and Ivry., R. B. (1989). The role of cerebellar structures in the execution of serial movements. *Brain, 112*, 565–581.

Innocenti, G. M. (1981). Growth and reshaping of axons in the establishment of visual callosal connections. *Science, 212*, 824–827.

Innocenti, G. M. and Frost, D. O. (1980). The postnatal development of visual callosal connections in the absence of visual experience or of the eyes. *Experimental Brain Research, 39*, 365–375.

Insel, T. R., Murphy, D. L., Cohen, R. M., Alterman, I., Kilts, C., and Linnoila, M. (1983). Obsessive-compulsive disorder: A double-blind trial of clomipramine and clorgyline. *Archives of General Psychiatry, 40*, 605–612.

Isaacson, R. L. (1975a). The myth of recovery from early brain damage. In N. R. Ellis (Ed.), *Aberrant development in infancy: Human and animal studies* (pp. 1–26). Hillsdale, NJ: Laurence Erlbaum Assoc.

Isaacson, R. L. (1975b). Memory processes and the hippocampus. In D. Deutsch and J. A. Deutsch (Eds.), *Short-Term memory* (pp. 313–337). New York: Academic Press.

Ito, M. (1984). *The cerebellum and neural control* New York: Raven Press.

Ito, M. (1990). A new physiological concept on cerebellum. *Revista de neurologia* (Paris), *146*, 564–569.

Ito, M. (1991). Neuro control as a major aspect of higher-order brain function. In J. Eccles (Ed.), *The principles of design and operation of the brain*. (pp. 281–293) Berlin: Springer.

Iverson, J. M. and Goldin-Meadow, S. (1998). Why people gesture when they speak. *Nature, 396*, 228.

Ivry, R. (1991). *High versus low temporal frequency analysis of auditory information in the cerebral hemispheres: An account of hemispheric asymmetries*. Paper presented at the Annual Meeting of the American Psychonomic Society, San Francisco, CA.

Ivry, R. B. and Keele, S. W. (1989). Timing functions of the cerebellum. *Journal of Cognitive Neuroscience, 1*, 136–152.

Ivry, R. B., Keele, S. W., and Diener, H. C. (1988). Timing functions of the cerebellum. *Experimental Brain Research, 73*, 167–180.

Ivry, R. B. and Lebby, P. (1993). Hemispheric differences in auditory perception are similar to those found in visual perception. *Psychological Science, 4*, 41–45.

Ivy, G. O. and Killackey, H. P. (1982). Ontogenetic changes in the projections of neocortical neurons. *Journal of Neuroscience, 2*, 735–743.

Iwai, E. and Yukie, M. (1987). Amygdalofugal and amygdalopetal connections with modality-specific visual cortical areas in macques (Macaca fuscata, M. mulatta, and M. fascicularis). *The Journal of Comparative Neurology, 261*, 362–387.

Izard, C. E., Kagan, J., and Zajonc, R. B. (1984). *Emotions, cognition, and behavior*. New York, NY: Cambridge University Press.

Jackson, J. H. (1878). On the affections of speech from disease of the brain. *Brain, 1*, 304–330.

Jacobson, C. D., Csernus, V. J., Shryne, J. E., and Gorski, R. A. (1981). The influence of gonadectomy, and rogen exposure, or a gonadal graft in the neonatal rat on the volume of the sexually dimorphic nucleus of the preoptic area. *Journal of Neuroscience, 1*, 1142–1147.

Jacobson, S. and Trojanowski, J. Q. (1975). Amygdaloid projections to prefrontal granular cortex in rhesus monkey demonstrated with horseradish peroxidase. *Brain Research, 100*, 132–139.

Jacqz-Aigrain, E., Zhang, D., Maillard, G., Luton, D., Andre, J., and Oury, J. F. (2002). Maternal smoking during pregnancy and nicotine and cotinine concentrations in maternal and neonatal hair. *BJOG: An International Journal of Obstetrics and Gynaecology, 109,* 909–911.

Jamain, S., Betancur, C., Quach, H., Philippe, A., Fellous, M., Giros, B., Gillberg, C., Leboye, M., and Bourgeron, T. (2002). Paris Autism Research International Sibpair (PARIS) Study. Linkage and association of the glutamate receptor 6 gene with autism. *Molecular Psychiatry, 7,* 302–310.

Jancke, L., Peters, M., Schlaug, G., Posse, S., Steinmetz, H., and Muller-Gartner, H. (1998). Differential magnetic resonance signal change in human sensorimotor cortex to finger movements of different rate of the dominant and subdominant hand. *Brain Research. Cognitive Brain Research. 6,* 279–284.

Jancke, L., Steinmetz, H., and Volkmann, J. (1992). Dichotic listening: What does it measure? *Neuropsychologia, 30,* 941–950.

Jaselskis, C. A., Cook, E. H., Jr., Fletcher, K. E., and Leventhal, B. L. (1992). Clonidine treatment of hyperactive and impulsive children with autistic disorder. *Journal of Autism and Developmental Disorders, 26,* 159–163.

Jason, G. W., Cowey, A., and Weiskrantz, L. (1984). Hemispheric asymmetry for a visuospatial task in monkeys. *Neuropsychologia, 22,* 777–784.

Jasper, H. H. and Droogleever-Fortuyn, J. (1947). Experimental studies of the functional anatomy of petit mal epilepsy. *Journal of the Association for Research in Nervous and Mental Disease, 26,* 272–298.

Jeffery, D. B., McLellan, R. W., and Fox, D. (1982). The development of children's eating habits: The role of televisions commercials. *Health Education, 9,* 78–93.

Jenike, M. A., Baer, L., Summergrad, P., Minichiello, W. E., Holland, A., and Seymour, R. (1990a). Sertraline in obsessive-compulsive disorder: A double-blind comparison with placebo. *American Journal of Psychiatry, 147,* 923–928.

Jenike, M. A., Baer, L., Summergrad, P., Weilburg, J. B., Holland, A., and Seymour, R. (1989). Obsessive-compulsive disorder: A double-blind, placebo-controlled trial of clomipramine in 27 patients. *American Journal of Psychiatry, 146,* 1328–1330.

Jenike, M. A., Hyman, S., Baer, L., Holland, A., Minichiello, W. E., Buttollph, L., Summergrad, P., Seymour, R., and Ricciardi, J. (1990b). A controlled trial of fluvoxamine in obsessive-compulsive disorder: Implications for a serotonergic theory. *American Journal of Psychiatry, 147,* 1209–1214.

Jenkins, I. H., Bain, P. G., Colebatch, P. D., Thompson, P. D., Findley, L. J., Frackowiac, R. S. Marsden, C. D., and Broolis, D. J. (1993). A positron emission tomography study of essential tremor: Evidence for overactivity of cerebellar connections. *Annals of Neurology, 34,* 82–90.

Jennigan, T. L., Zatz, L. M., Moses, J. A., and Cardellino, J. P. (1952). Computed tomography in schizophrenics and normal volunteers cranial asymmetries. *Archives of General Psychiatry, 39,* 771–773.

Jensen, K. F. and Killackey, H. P. (1987). Terminal arbors of axons projecting to the somatosensory cortex of the adult rat. I. The normal morphology of specific thalamocortical afferents. *Journal of Neuroscience, 7,* 3529–3543.

Jensen, P. S., Kettle, L., Roper, M. T., Sloan, M. T., Dulcan, M. K., Hoven, C., Bird, H. R., Bauermeister, J. J., and Payne, J. D. (1999). Are stimulants overprescribed? Treatment of ADHD in four U.S. communities. *Journal of the American Academy of Child and Adolescent Psychiatry, 38,* 797–804.

Jensen, P. S., Martin, D., and Cantwell, D. P. (1997). Comorbidity in ADHD: Implications for research, practice, and DSM-V. *Journal of the American Academy of Child and Adolescent Psychiatry, 36,* 1065–1079.

Jerison, H. J. (1985). Animal intelligence as encephalization. *Philosophical Transactions of the Royal Society of London. Series B: Biological Sciences, 308,* 21–35.

Jerison, Harry J. (1973). *Evolution of the brain and intelligence* (pp. 8–9). Academic Press.

Jernigan, T. L. and Bellugi, U. (1990). Anomalous brain morphology on magnetic resonance images in Williams syndrome and Down syndrome. *Archives of Neurology, 47,* 529–533.

Jernigan, T. L., Bellugi, U., Sowell, E., Doherty, S., and Hesselink, J. R. (1993). Cerebral morphologic distinctions between Williams and Down syndromes. *Archives of Neurology, 50,* 186–191.

Jerussi, T. P. and Glick, S. D. (1974). Amphetamine-induced rotation in rats without lesions. *Neuropharmacology, 14,* 283–286.

Jerussi, T. P. and Glick, S. D. (1976). Drug-induced rotation in rats without lesions: Behavioral and neurochemical indices of a normal asymmetry in nigrostriatal function. *Psychopharmacology, 47,* 249–260.

Jiang, S., Xin, R., Wu, X., Lin, S., Qian, Y., Ren, D., Tang, G., and Wang, D. (2000). Association between attention deficit hyperactivity disorder and the DXS7 locus. *American Journal of Medical Genetics, 96,* 289–292.

Jiang, S., Xin, R., Lin, S., Qian, Y., Tang, G., Wang, D., and Wu, X. (2001). Linkage studies between attention-deficit hyperactivity disorder and the monoamine oxidase genes. *American Journal of Medical Genetics, 105,* 783–788.

Jirout, J. (1980). The effect of unilateral dominance of the cerebral hemispheres on the radiographic appearance of the cervical spine. *Radiologie, 20,* 466–469.

Johnels, B. and Speg, G. (1980). The corpus striatum and the regulation of posture and locomotion. *Neuroscience Letters, 19* (Suppl.), 339–350.

Johnstone, J. R. and Mark, R. F. (1973). Corollary discharge. *Vision Research, 13,* 1621.

Jones, B. E. (1993). The organization of central cholinergic systems and their functional importance in sleep-waking states. *Progress in Brain Research, 98,* 61–71.

Jones, D. P. (1995). *Cell Biology of trauma.* In J. J. Lemasters and C. N. Oliver (Eds.). Boca Raton, FL: CRC Press.

Jones, E. G. (1985). *The thalamus*. New York: Plenum Press.

Jones, E. G. (1990). The role of afferent activity in the maintenance of primate neocortical function. *Journal of Experimental Biology, 153,* 155–176.

Jones, E. G. and Burton, H. (1976a). A projection from the medial pulvinar to the amygdala in primates. *Brain Research, 104,* 142–147.

Jones, E. G. and Burton, H. (1976b). Areal differences in the laminar distribution of thalamic afferents in cortical fields of the insular, parietal and temporal regions of primates. *The Journal of Comparative Neurology, 168,* 197–248.

Jones, E. G. and Powell, T. P. S. (1969a). Connexions of the somatic sensory cortex of the rhesus monkey. I. Ipsilateral cortical connections. *Brain, 92,* 477–502.

Jones, E. G. and Powell, T. P. S. (1969b). Connexions of the somatic sensory cortex of the rhesus monkey. II. Contralateral cortical connections. *Brain, 92,* 717–730.

Jones, E. G. and Powell, T. P. S. (1970a). An anatomical study of converging sensory pathways within the cerebral cortex of the monkey. *Brain, 93,* 793–820.

Jones, E. G. and Powell, T. P. S. (1970b). An electron microscopic study of the laminar pattern and mode of termination of afferent fibre pathways in the somatic sensory cortex of the cat. *Philosophical Transactions of the Royal Society of London. Series B: Biological Sciences, 257,* 45–62.

Jones, E. G. and Powell, T. P. S. (1970c). Connexions of the somatic sensory cortex of the rhesus monkey. III. Thalamic connexions. *Brain, 93,* 37–56.

Jones, M. B. and Szatmari, P. (2002). A risk-factor model of epistatic interaction, focusing on autism. *American Journal of Medical Genetics, 114,* 558–565.

Jones, S. M., Subramanian, G., Avniel, W., Guo, Y., Burkard, R. F., and Jones, T. A. (2002). Stimulus and recording variables and their effects on mammalian vestibular evoked potentials. *Journal of Neuroscience Methods, 30,* 23–31.

Jonsson, J. E. and Hellige, J. B. (1986). Lateralized effects of blurring: A test of the visual spatial frequency model of cerebral hemisphere asymmetry. *Neuropsychologia, 24,* 351–362.

Joseph, J. A., Denisova, N., Fisher, D., Bickford, P., Prior, R., and Cao, G. (1998). Age-related neurodegeneration and oxidative stress: Putative nutritional intervention. *Neurologic Clinics, 16,* 747–755.

Joseph, R. (1999). Environmental influences on neural plasticity, the limbic system, emotional development and attachment: A review. *Child Psychiatry and Human Development, 29,* 189–208.

Joubert, M., Eisenring, J. J., Robb, J. P., and Andermann, F. (1969). Familial agenesis of the cerebellar vermis: A syndrome of episodic hyperpnea, abnormal eye movements, ataxia, and retardation. *Neurology, 19,* 813–825.

Jouvet, M. (1972). The role of monoamines and acetylcholine-containing neurons in the regulation of the sleep-waking cycle. *Ergebnisse der Physiologie, Biologischen Chemie und Experimentellen Pharmakologie, 64,* 166–307.

Jouvet, M., Bobillier, P., Pujol, J. F., and Renault, J. (1966). Effects of lesions of the raphe system on sleep and cerebral serotonin. *Comptes rendus des seances de la Societe de biologie et de ses filiales, 160,* 2343–2346.

Jouvet, M., Bobillier, P., Pujol, J. F., and Renault, J. (1967). Permanent insomnia and diminution of cerebral serotonin due to lesion of the raphe system in cats. *Journal de Physiologie (Paris), 59,* 248.

Junck, L., Gilman, S., Rothley, J. R., Betley, A. T., Koette, R. A., and Hichwa, R. D. (1988). A relationship between metabolism in frontal lobes and cerebellum in normal subject studies with peg. *Journal of Cerebral Blood Flow and Metabolism, 8,* 774–782.

Junge, H. D. (1980). Behavioral states and state-related heart rate and motor activity patterns in the newborn infant and the fetus ante partum. A comparative study. III. Analysis of sleep state-related motor activity patterns. *European Journal of Obstetrics, Gynecology and Reproductive Biology, 10,* 239–246.

Juul-Dam, N., Townsend, J., and Courchesne, E. (2001). Prenatal, perinatal, and neonatal factors in autism, pervasive developmental disorder not otherwise specified, and the general population. *Pediatrics, 107,* E63.

Jyonouchi, H., Sun, S., and Le, H. (2001). Proinflammatory and regulatory cytokine production associated with innate and adaptive immune responses in children with autism spectrum disorders and developmental regression. *Pediatrics, 107,* E63.

Jyonouchi, H., Sun, S., and Itokazu, N. (2002). Innate immunity associated with inflammatory responses and cytokine production against common dietary proteins in patients with autism spectrum disorder. *Neuropsychobiology, 46,* 76–84.

Kaada, B. (1967). Brain mechanisms related to aggressive behavior. In C. D. Clemente and D. B. Lindsley (Eds.), *Aggression and defense. Neural Mechanisms and Social Patterns* (pp. 95–216). Berkeley: University of California Press.

Kaada, B. R. (1951). Somato-motor, autonomic and electrocorticographic responses to electrical stimulation of rhinencephalic and other structures in primates, cat and dog. A study of responses from the limbic, subcallosal, orbito-insular, piriform and temporal cortex, hippocampus-fornix and amygdala. *Acta Physiologica Scandinavica, 23,* 1–285.

Kaada, B. R. (1972). Stimulation and regional ablation of the amygdaloid complex with reference to functional representations. In B. E. Elftheriou (Ed.), *The neurobiology of the amygdala* (pp. 205–281). New York: Plenum Press.

Kaada, B. R., Andersen, P., and Jansen, J., Jr. (1954). Stimulation of the amygdaloid nuclear complex in unanesthetized cats. *Neurology, 4,* 48–64.

Kaas, J. H. (1991). Plasticity of sensory and motor maps in adult mammals. *Annual Review of Neuroscience, 14,* 137–167.

Kadesjo, B. and Gillberg, C. (1998). Attention deficits and clumsiness in Swedish 7-year-olds. *Developmental Medicine and Child Neurology, 40*, 796–804.

Kadesjo, B. and Gillberg, C. (2000). Tourette's disorder: Epidemiology and comorbidity in primary school children. *Journal of the American Academy of Child and Adolescent Psychiatry, 39*, 548–555.

Kaga, K., Suzuki, J. I., Marsh, R. R., and Tanaka, Y. (1981). Influence of labyrinthine hypoactivity on gross motor development of infants. *Annals of the New York Academy of Sciences, 374*, 412–420.

Kagan, J. (1971). *Change and continuity in infancy.* New York, NY: John Wiley.

Kagan, J. and Zajonc, R. (Eds.) (1988). *Emotion, cognition and behavior.* New York, NY: Cambridge University Press.

Kail, R. (1988). Developmental functions for speeds of cognitive processes. *Journal of Experimental Child Psychology, 45*, 339–364.

Kakei, S., Futami, T., and Shinoda, Y. (1996). Projection pattern of single corticocortical fibers from the parietal cortex to the motor cortex. *Neuroreport, 7*, 2369–2372.

Kalivas, P. W., Churchill, L., and Romanides, A. (1999). Involvement of the pallidal-thalamocortical circuit in adaptive behavior. *Annals of the New York Academy of Sciences, 877*, 64–70.

Kaluger, G. and Heil, C. L. (1970). Basic symmetry and balance their relationship to perceptual motor development. *Progress in Physical Therapy, 1*, 132–137.

Kamil, M. L., Mosenthal, P. B., Pearson, P. D., and Barr, R. (Eds.) (2000). *Handbook of reading research* (Vol. 3). Mahwah, NJ: Lawrence Erlbaum Associates.

Kandel, E. R., Schwartz, J. H., and Jessell, T. M. (1995). *Essentials of neural science and behavior.* Norwalk, CT: Appleton & Lange.

Kane, P. C. and Kane, E. (1997). Peroxisomal disturbances in autistic spectrum disorder. *Journal of Orthomolecular Medicine, 12*, 207–218.

Kang, D. H., Davidson, R. J., Coe, C. L., Wheeler, R. W., Tomarken, A. J., and Ershler, W. B. (1991). Frontal brain asymmetry and immune function. *Behavioral Neuroscience, 105*, 860–869.

Kang, Y. and Harris, L. (1993). *Social-cultural influences on handedness: A cross-cultural study of Koreans and Americans.* Paper presented at the TENNET (Theoretical and Experimental Neuropsychology) Meeting, Montral.

Kanner, L. (1943). Autistic disturbances of affective contact. *The Nervous Child, 2*, 217–250.

Kaplan, B. J., McNicol, J., and Conte, R. A. (1989). Moghadam HK. Dietary replacement in preschool-aged hyperactive boys. *Pediatrics, 83*, 7–17.

Karatekin, C. and Asarnow, R. F. (1998). Components of visual search in childhood-onset schizophrenia and attention-deficit/hyperactivity disorder. *Journal of Abnormal Child Psychology, 26*, 367–380.

Karatekin, C. and Asarnow, R. F. (1999). Exploratory eye movements to pictures in childhood-onset schizophrenia and attention-deficit/hyperactivity disorder (ADHD). *Journal of Abnormal Child Psychology, 27*, 35–49.

Keren, M., Manor, I., and Tyano, S. (2001). Attention deficit disorder in the preschool years: Its characteristics and course from infancy to toddlerhood. *Harefuah, 140*, 1021–1025.

Karni, A. and Bertini, G. (1997). Learning perceptual skills: Behavioral probes into adult cortical plasticity. *Current Opinion in Neurobiology, 7*, 530–535.

Katkin, E. S., Cestaro, V. L., and Weitkunat, R. (1991). Individual differences in cortical evoked potentials as a function of heartbeat detection ability. *International Journal of Neuroscience, 61*, 269–276.

Kato, H., Izumiyama, M., Koizumi, H., Takahashi, A., and Itoyama, Y. (2002, August). Near-infrared spectroscopic topography as a tool to monitor motor reorganization after hemiparetic stroke: A comparison with functional MRI. *Stroke, 33*, 2032–2036.

Katz, H. B., Davies, C. A., and Dobbing, J. (1982). Effects of undernutrition at different ages early in life and later environmental complexity on parameters of the cerebrum and hippocampus in rats. *The Journal of Nutrition, 112*, 1362–1368.

Kaufmann, W. E. and Galaburda, A. M. (1989). Cerebrocortical microdysgenesis in neurologically normal subjects: A histopathologic study. *Neurology, 39*, 238–244.

Keele, S. W. and Ivry, R. (1990). Does the cerebellum provide a common computation for diverse tasks? A timing hypothesis. *Annals of the New York Academy of Sciences, 608*, 179–207, discussion 207–211.

Keele, S. W., Cohen, A., Ivry, R., Liotti, M., and Yee, P. (1988). Tests of a temporal theory of attentional binding. *Journal of Experimental Psychology, Human Perception and Performance, 14*, 444–452.

Keele, S. W., Pokorny, R. A., Corcos, D. M., and Ivry, R. (1985). Do perception and motor production share common timing mechanisms? A correctional analysis. *Acta Psychologica* (Amsterdam), *60*, 173–191.

Keele, S. W. and Posner, M. I. (1968). Processing of visual feedback in rapid movements. *Journal of Experimental Psychology, 77*, 155–158.

Keller, A., Castellanos, F. X., Vaituzis, A. C., Jeffries, N. O., Giedd, J. N., and Rapoport, J. L. (2003). Progressive loss of cerebellar volume in childhood-onset schizophrenia. *American Journal of Psychiatry, 160*: 128–133.

Keller, A. D., Roy, R. S., and Chase, W. P. (1937). Extirpation of the neocerebellar cortex without electing so-called cerebellar signs. *American Journal of Physiology, 118*, 720–733.

Kelly, D. D., Murphy, B. A., and Backhouse, D. P. (2000). Use of a mental rotation reaction-time paradigm to measure the effects of upper cervical adjustments on cortical processing: A pilot study. *Journal of Manipulative and Physiological Therapeutics, 23*, 246–251.

Kelly, J. P. (1985). Anatomical basis of sensory perception and motor coordination. In E. R. Kandel and

J. H. Schwartz (Eds.), *Principles of neural science* (2nd ed., pp. 222–243). New York, NY: Elsevier.

Kemner, C., Verbaten, M. N., Cuperus, J. M., Camfferman, G., and van Engeland, H. (1998). Abnormal saccadic eye movements in autistic children. *Journal of Autism and Developmental Disorders, 28*, 61–67.

Kemper, T. L., Wright, S. J., Jr., and Locke, S. (1972). Relationship between the Septum and the Cingulate gyrus in Macaca mulatta. *Journal of Comparative Neurology, 146*, 465–478.

Kempermann, G. and Gage, F. H. (1999). Experience-dependent regulation of adult hippocampal neurogenesis: Effects of long-term stimulation and stimulus withdrawal. *Hippocampus, 9*, 321–332.

Kempermann, G., Kuhn, H. G., and Gage, F. H. (1997). Genetic influence on neurogenesis in the dentate gyrus of adult mice. *Proceedings of the National Academy of Sciences, USA, 94*, 10409–10414.

Keren, M., Manor, I., and Tyano, S. (2001). Attention deficit disorder in the preschool years: Its characteristics and course from infancy to toddlerhood. *Harefuah, 140*, 1021–1025.

Kermoian, R. and Campos, J. J. (1988). Locomotor experience: A facilitator of spatial cognitive development. *Child Development, 59*, 908–917.

Kern, L., Koegel, R. L., and Dunlap, G. (1984). The influence of vigorous versus mild exercise on autistic stereotyped behaviors. *Journal of Autism and Developmental Disorder, 14*, 57–67.

Keshner, E. A. and Peterson, B. W. (1988). Motor control strategies underlying head stabilization and voluntary head movements in humans and cats. *Progress in Brain Research, 76*, 329–339.

Kershner, J. and Micallef, J. (1992). Consonant-vowel lateralization in dyslexic children: Deficit or compensatory development? *Brain and Language, 43*, 66–82.

Kertesz, A. (1979). Visual agnosia. The dual deficit of perception and recognition. *Cortex, 15*, 403–419.

Khaibullina, F. G. (1980). Cerebral hemodynamic disorders in children who have had a birth injury to the cervical portion of the spinal cord and their dynamics in the pathogenetic therapy process. *Pediatrica, 3*, 51–52.

Khanna, S. (1988). Obsessive-compulsive disorder: Is there a frontal lobe dysfunction? *Biological Psychiatry, 24*, 602–613.

Kiely, J. L. (1999). Mode of delivery and neonatal death in 17,587 infants presenting by the breech. *Brain Injury, 13*, 39–43.

Kiess, W., Galler, A., Reich, A., Muller, G., Kapellen, T., Deutscher, J., Raile, K., and Kratzsch, J. (2001, February). Clinical aspects of obesity in childhood and adolescence. *Obesity Reviews: An Official Journal of the International Association for the Study of Obesity, 2*, 29–36.

Kim, H., and Levine, S. C. (1992). Variations in characteristic perceptual asymmetry: Modality specific and modality general components. *Brain and Cognition, 19*, 21–47.

Kim, H., Levine, S. C., and Kertesz, S. (1990). Are variations among subjects in lateral asymmetry real individual differences or random error in measurement? Putting variability in its place. *Brain and Cognition, 14*, 220–242.

Kim, J. J. and Diamond, D. M. (2002). The stressed hippocampus, synaptic plasticity and lost memories. *Nature Reviews, Neuroscience, 3*, 453–462.

Kim, S. G., Ugurbil, K., and Strick, P. L. (1994). Activation of a cerebellar output nucleus during cognitive processing. *Science, 265*, 949–951.

Kim, S. J., Veenstra-VanderWeele, J., Hanna, G. L., Gonen, D., Leventhal, B. L., and Cook, E. H., Jr. (2000). Mutation screening of human 5-HT(2B) receptor gene in early-onset obsessive-compulsive disorder. *Molecular and Cellular Probes, 14*, 47–52.

Kimura, D. (1961a). Some effects of temporal-lobe damage on auditry perception. *Canadian Journal of Psychology, 15*, 156–165.

Kimura, D. (1961b). Cerebral dominance and the perception of verbal stimuli. *Canadian Journal of Psychology, 15*, 166–171.

Kimura, D. (1967). Functional asymmetry of the brain in dichotic listening. *Cortex, 3*, 163–178.

Kimura, D. and Archibald, Y. (1974). Motor functions of the left hemisphere. *Brain, 97*, 337–350.

Kinsbourne, M. (1970). The cerebral basis of lateral asymmetries in attention. *Acta Psychologica, 33*, 193–201.

Kinsbourne, M. (1972). Eye and head timing indicates cerebral lateralization. *Science, 176*, 539–541.

Kinsbourne, M. (1973). Lateral interactions in the brain. In M. Kinsbourne and L. W. Smyth (Eds.), *Hemispheric disconnection and cerebral function* (pp. 239–259). Springfield, IL: Charles C. Thomas.

Kinsbourne, M. (1974a). Mechanisms of hemispheric interaction in man. In M. Kinsbourne and W. L. Smith (Eds.), *Hemispheric disconnection and cerebral function*. Springfield, IL: Charles C. Thomas.

Kinsbourne, M. (1974b). Direction of gaze and distribution of cerebral thought processes. *Neuropsychologia, 12*, 279–281.

Kinsbourne, M. (1975). Lateral interactions in the brain. In M. Kinsbourne and W. L. Smith (Eds.), *Hemispheric disconnection and cerebral function* (pp. 239–259). Springfield, IL: Charles C. Thomas.

Kinsbourne, M. (1978). Evolution of language in relation to lateral action. In M. Kinsbourne (Ed.), *Asymmetrical function of the brain* (pp. 553–566). Cambridge: Cambridge University Press.

Kinsbourne, M. (1987a). The material basis of mind. In L. Vaina (Ed.), *Matters of Intelligence: Conceptual structures in cognitive neuroscience* (pp. 407–427). Boston: D. Reidel Publishing Co.

Kinsbourne, M. (1987b). Mechanisms of unilateral neglect. In M. Jeannerod (Ed.), Neurophysiological and neuropsychological aspects of spatial neglect (pp. 69–86). Amsterdam: Elsevier Science Publishers.

Kinsbourne, M. (1988). Integrated field theory of consciousness. In A. Marcel and E. Bisiach (Eds.),

Consciousness in contemporary science (pp. 230–256). Oxford, England: Clarendon Press.

Kinsbourne, M. and Duffy, C. J. (1990). The role of dorsal/ventral processing dissociation in the economy of the primate brain. *Behavioral and Brain Sciences, 13*, 553–554.

Kinsbourne, M. and Hicks, R. E. (1978). Functional cerebral space: A model for overflow, transfer and interference effects in human performance. In J. Requin (Ed.), *Attention and performance* (Vol. 7, pp. 345–362). Hillsdale, NJ: Lawrence Erlbaum Associates.

Kinsbourne, M. and Hiscock, M. (1977). Does cerebral dominance develop? In S. J. Segalowitz and F. A. Gruber (Eds.), *Language development and neurological theory* (pp. 171–191). New York: Academic Press.

Kinsbourne, M. and Hiscock, M. (1983). The normal and deviant development of functional lateralization of the brain. In P. Mussen, M. Haith, and J. Campos (Eds.), *Handbook of child psychology* (4th ed.). New York, NY: John Wiley & Sons.

Kirley, A., Hawi, Z., Daly, G., McCarron, M., Mullins, C., Millar, N., Waldman, I., Fitzgerald, M., and Gill, M. (2002). Dopaminergic system genes in ADHD: Toward a biological hypothesis. *Neuropsychopharmacology, 27*, 607–619.

Kitterle, F. L., Hellige, J. B., and Christman, S. (1992, November). Visual hemispheric asymmetries depend on which spatial frequencies are task relevant. *Brain and Cognition, 20*, 308–314.

Klein, R. G., Abikoff, H., Klass, E., Ganeles, D., Seese, L. M., and Pollack, S. (1997). Clinical efficacy of methylphenidate in conduct disorder with and without attention deficit hyperactivity disorder. *Archives of General Psychiatry, 54*, 1073–1080.

Kleist, K. (1931). Die störungen der ichleistungen und ihre lokalisation in orbital-, innen- und zwischenhirn. *Monastsschreift fur Psychiatrie und Neurologie, 79*, 338–350.

Klin, A., Volkmar, F. R., Sparrow, S. S., Cicchetti, D. V., and Rourke, B. P. (1995). Validity and neuropsychological characterization of Aspergers syndrome: Convergence with non-verbal learning disabilities syndrome. *Journal of Child Psychology and Psychiatry, 36*, 1127–1140.

Klin, A., Jones, W., Schultz, R., Volkmar, F., and Cohen, D. (2002). Visual fixation patterns during viewing of naturalistic social situations as predictors of social competence in individuals with autism. *Archives of General Psychiatry, 59*, 809–816.

Klin, A., Sparrow, S. S., de Bildt, A., Cicchetti, D. V., Cohen, D. J., and Volkmar, F. R. (1999). A normed study of face recognition in autism and related disorders. *Advances in Neurology, 52*, 911–916.

Kline, J., Stein, Z., and Susser, M. (1989). *Conception to birth: Epidemiology of prenatal development*. New York: Oxford University Press.

Klein, R. G., Abikoff, H., Klass, E., Ganeles, D., Seese, L. M., and Pollack, S. (1997). Clinical efficacy of methylphenidate in conduct disorder with and without attention deficit hyperactivity disorder. *Archives of General Psychiatry, 54*, 1073–1080.

Kleist, K. (1931). Die störungen der ichleistungen und ihre lokalisation in orbital-, innen- und zwischenhirn. *Monastsschreift fur Psychiatrie und Neurologie, 79*, 338–350.

Kolb, B. and Taylor, L. (1981). Affective behavior in patients with localized cortical excisions: Role of lesion site and side. *Science, 214*, 89–91.

Klorman, R., Brumaghim, J. T., and Fitzpatrick, P. A. et al. (1990). Clinical effects of a controlled trial of methylphenidate on adolescents with attention deficit disorder. *Journal of the American Academy of Child and Adolescent Psychiatry, 29*, 702–709.

Knauff, M., Mulack, T., Kassubek, J., Salih, H. R., and Greenlee, M. W. (2002). Spatial imagery in deductive reasoning: A functional MRI study. *Brain Research. Cognitive Brain Research, 13*, 203–212.

Knivsber, A. M., Reichelt, K. L., Hoien, T., and Nodland, M. (2002). A randomised, controlled study of dietary intervention in autistic syndromes. *Nutritional Neuroscience, 5*, 251–261.

Knivsber, A. M., Reichelt, K. L., and Nodland, M. (2001). Reports on dietary intervention in autistic disorders. *Nutritional Neuroscience, 4*, 25–37.

Kobayashi, R. M., Palkovits, M., Jacobowitz, D. M., and Kopin, I. J. (1975). Biochemical mapping of the noradrenergic projection from the locus coeruleus. A model for studies of brain neuronal pathways. *Neurology, 25*, 223–233.

Koch, P. and Leisman, G. (1990). A continuum model of activity waves in layered neuronal networks: A neuropsychology of brain stem seizures. *International Journal of Neuroscience, 54*, 41–62.

Koch, P. and Leisman G. (1996). Wave theory of large-scale organization of cortical activity. *International Journal of Neuroscience, 86*, 179–196.

Koch, P. and Leisman, G. (2001a). A layered neural continuum architecture in attentional and seizure disorders. *International Journal of Neuroscience, 107*, 199–232.

Koch, P. and Leisman, G. (2001b). Effect of local synaptic strengthening on global activity-wave growth in the hippocampus. *Journal of Neuroscience, 108*, 1–2, 127–146.

Kohen-Raz, R. (1986). *Learning disabilities and postural control*. London: Freund.

Kohen-Raz, R. (1991, October). Application of tetra-ataziametric posturography in clinical and developmental diagnosis. *Perceptual and Motor Skills, 73*, 635–656.

Kohen-Raz, R. and Hiriartborde, E. (1979, June). Some observations on tetra-ataxiametric patterns of static balance and their relation to mental and scholastic achievements. *Perceptual and Motor Skills, 48*, 871–890.

Kohen-Raz, R., Volkmar, F. R., and Cohen, D. J. (1992). Postural control in children with autism. *Journal of Autism and Development Disorders, 22*, 419–432.

Kohler, C., Shipley, T., Srebro, B., and Harkmark, W. (1978). Some retrohippocampal afferents to the entorhinal cortex. Cells of origin as studied by the

HRP method in the rat and mouse. *Neuroscience Letters, 10,* 115–120.

Kolb, B. and Taylor, L. (1990). Neocortical substrates of emotional behavior. In N. L. Stein, B. Leventhal, and T. Trabasso (Eds.), *Psychological and biological approaches to emotion.* Hillsdale, NJ: Lawrence Erlbaum Associates.

Kolb, B. and Whishaw, I. Q. (2001). *Introduction to brain and behavior.* New York: Freeman-Worth.

Koller, W. C., Vetere-Overfield, B., and Barter, R. (1989). Tremor in early parkinson's disease. *Clinical Neuropharmacology, 12,* 293–297.

Kolmen, B. K., Feldman, H. M., Handen, B. L., and Janosky, J. E. (1995). Naltrexone in young autistic children: A double-blind, placebo-controlled crossover study. *Journal of the American Academy of Child and Adolescent Psychiatry, 34,* 223–231.

Kooistra, C. A. and Heilman, K. M. (1989). Hemispatial visual inattention masquerading as hemianopsia. *Neurology, 39,* 1125–1127.

Koran, L. M., McLeroy, S. L., Davison, J. R. T., Rasmussen, S. A., Hollander, E., and Jenike, M. A. (1996). Fluvoxamine versus clomipramine for obsessive-compulsive disorder: A double blind comparison. *Journal of Clinical Psychopharmacology, 16,* 121–129.

Korijak, Y. A. and Kozlovskaya, I. D. (1991). Influences of Anti-Orthostatic Rest (ABR) on Functional Properties of Neuromuscular System in Man. And this is present in Physiologist, 34 (Suppl.) S107–S109.

Korkman, M. and Pesonen, A. E. (1994). Comparison of neuropsychological test profiles of children with attention deficit-hyperactivity and/or learning disorder. *Journal of Learning Disabilities, 27,* 383–392.

Kornhuber, H. (1974). Cerebral cortex, cerebellum, and basal ganglia: An introduction to their motor functions. In F. Schmitt and F. Worden (Eds.), *The neurosciences: Third study program* (pp. 267–280). Cambridge, MA: MIT Press.

Kosik, K. S. (1998). Neurolab: Learning how the nervous system adapts to microgravity. *Neuroscience News, 5,* 36–38.

Kosslyn, S. M. (1987). Seeing and imaging in the cerebral hemispheres: A computational approach. *Psychological Review, 94,* 148–175.

Kosslyn, S. M., Chabris, C., Marsolek, C. J., and Koenig, O. (1992). Categorical versus coordinate spatial representations: Computational analyses and computer simulations. *Journal of Experimental Psychology: Human Perception and Performance, 18,* 562–577.

Kosslyn, S. M., Flynn, R. A., Amsterdam, J. B., and Wang, G. (1990). Components of high-level vision: A cognitive neuroscience analysis and accounts of neurological syndromes. *Cognition, 34,* 203–277.

Kosslyn, S. M., Sokolov, M. A., and Chen, J. C. (1989). The lateralization of BRIAN: A computational theory and model of visual hemispheric specialization. In D. Klahr and K. Kotovsky (Eds.), *Complex information processing comes of age.* Hillsdale, NJ: Lawrence Erlbaum Associates.

Kotloski, R., Lynch, M., Lauersdorf, S., and Sutula, T. (2002). Repeated brief seizures induce progressive hippocampal neuron loss and memory deficits. *Progress in Brain Research, 135,* 95–110.

Kowalski, D. P., Aw, T. Y., Park, Y., and Jones, D. P. (1992). Free radicals. *Biologie medicale, 12,* 205–212.

Kozielec, T. and Starobrat-Hermelin, B. (1997). Assessment of magnesium levels in children with attention deficit hyperactivity disorder (ADHD). *Magnesium Research: Official Organ of the International Society for the Development of Research on Magnesium, 10,* 143–148.

Krakowski, A. J. (1965). Amitriptyline in the treatment of hyperkinetic children: A double-blind study. *International Journal of Psychosomatics: Official Publication of the International Psychosomatics Institute, 6,* 355–360.

Kraut, R., Patterson, M., Lundmark, V., Kiesler, S., Mukopadhyay, T., and Scherlis, W. (1998). Internet paradox. A social technology that reduces social involvement and psychological well-being? *American Psychologist, 53,* 1017–1031.

Krech, H. (1960). *Einführung in die deutsche sprechwissenschaft/sprecherziehung (lehrbrief).* Berlin, DDR: Deutscher verlag der Wissenschaften.

Krettek, J. E. and Price, J. L. (1974). A direct input from the amygdala to the thalamus and the cerebral cortex. *Brain Research, 67,* 169–174.

Krettek, J. E. and Price, J. L. (1977a). The cortical projection of the medio-dorsal nucleus and adjacent thalamic nuclei in the rat. *The Journal of Comparative Neurology, 171,* 157–192.

Krettek, J. E. and Price, J. L. (1977b). Projections from the amygdaloid complex to the cerebral cortex and thalamus in the rat and cat. *The Journal of Comparative Neurology, 172,* 687–722.

Krettek, J. E. and Price, J. L. (1977c). Projections from the amygdaloid complex and adjacent olfactory structures to the entorhinal cortex and to the subiculum in the rat and cat. *The Journal of Comparative Neurology, 172,* 723–752.

Kristjansson, E., Fried, P., and Watkinson, B. (1989). Maternal smoking during pregnancy affects children's vigilance performance. *Drug and Alcohol Dependence, 24,* 11–19.

Kronfol, Z., Hamsher, K., deS, Digre, K., and Waziri, R. (1978). Depression and hemispheric functions: Changes associated with unilateral ECT. *British Journal of Psychiatry, 132,* 560–567.

Kubitz, K. A. and Landers, D. M. (1993). The effects of aerobic training on cardiovascular responses to mental stress: An examination of underlying mechanisms. *Journal of Sport and Exercise Psychology, 15,* 326–337.

Kubos, K. L., Pearlson, G. D., and Robinson, R. G. (1982). Intracortical kainic acid induces an asymmetrical behavioral response in the rat. *Brain Research, 239,* 303–309.

Kubos, K. L. and Robinson, R. G. (1984). Cortical undercuts in the rat produce asymmetrical behavioral response without altering catecholamine concentrations. *Experimental Neurology, 83,* 646–653.

Kuhl, D. E., Phelps, M. E., Markham, C. H., Metter, E. J., Riege, W. H., and Winter, J. (1982). Cerebral metabolism and atrophy in huntington's disease determined by 18 FDG, and computed tomographic scan. *Annals of Neurology, 12,* 425–434.

Kurlan, R., Como, P. G., and Deeley, C. et al. (1993). A pilot controlled study of fluoxetine for obsessive-compulsive symptoms in children with Tourette's syndrome. *Clinical Neuropharmacology, 16,* 167–172.

Kurtz, R. G. (1975). Hippocampal and cortical activity during sexual behavior in the female rat. *Journal of Comparative and Physiological Psychology, 89,* 158–169.

Kussmaul, A. (1877). A disturbance of speech. *Cyclopedia of Practical Medicine, 14,* 581–875.

Kutty, I. N. and Prendes, J. L. (1981). Psychosis and cerebellar degeneration. *The Journal of Nervous and Mental Disease, 169,* 390–391.

Kyuhou, S., Matsuzaki, R., and Gemba, H. (1997). Cerebello-cerebral projections onto the ventral part of the frontal cortex of the macaque monkey. *Neuroscience Letters, 230,* 101–104.

Lacey, B. C. and Lacey, J. I. (1974). Studies of heart rate and other bodily processes in sensorimotor behavior. In P. A. Obrist, A. Black, J. Brener, and I. DiCara (Eds.), *Cardiovascular psychophysiology: Current issues in response mechanisms. Biofeedback and methodology.* Chicago: Aldine-Atherton.

Ladavas, E. (1987). Is the hemispatial deficit produced by right parietal damage associated with retinal or gravitational coordinates? *Brain, 110,* 167–180.

LaHoste, G. J., Neveu, P. J., Mormede, P. and Le Moal, M. (1989). Hemispheric asymmetry in the effects of cerebral cortical ablations on mitogen-induced lymphoproliferation and plasma prolactin levels in female rats. *Brain Research, 483,* 123–129.

Lalonde, R. and Botez, M. L. (1990). The cerebellum and learning processes in animals. *Brain Research. Brain Research Reviews, 15,* 325–332.

Lalonde, R., Manseau, N., and Betez, M. I. (1988). Spontaneous alternation and exploration in stagger mutant mice. *Behavioral Brain Research, 27,* 273–276.

Lamantia, A. and Rakic, P. (1990). Cytological and quantitative characteristics of four cerebral commissures in the rhesus monkey. *Journal of Comparative Neurology, 291,* 520–537.

Lamb, M. R., Robertson, L. C., and Knight, R. T. (1989). Attention and interference in the processing of global and local information: Effects of unilateral temporal-parietal junction lesions. *Neuropsychologia, 27,* 471–483.

Lamb, M. R., Robertson, L. C., and Knight, R. T. (1990). Component mechanisms underlying the processing of hierarchically organized patterns: Inferences from patients with unilateral cortical lesions. *Journal of Experimental Psychology. Learning, Memory, and Cognition, 16,* 471–483.

Lance, J. W. and Schwab, R. S. (1963). Action, tremor and cogwheel phenomena in Parkinson's disease. *Brain, 86,* 95–110.

Landesman-Dwyer, S. and Emmanuel, I. (1979). Smoking during pregnancy. *Teratology, 19,* 119.

Landgren, M., Peterson, R., and Kjellman, B., and Gillberg, C. (1996). ADHD, DAMP and other neurodevelopmental/neuro-Psychiatric disorders in six-year-old children. Epidemiology and comorbidity. *Developmental Medicine and Child Neurology, 38,* 891–906.

Lane, R. D. and Jennings, J. R. (1995). Hemispheric asymmetry, autonomic asymmetry, and the problem of sudden cardiac death. In R. J. Davidson, K. Hugdahl, (Eds.), *Brain asymmetry,* (pp. 271–304). Cambridge, MA: MIT Press.

Lane, R. and Schwartz, G. (1987). Induction of lateralized sympathetic input to the heart by the CNS during emotional arousal: A possible neurophysiologic trigger of sudden cardiac death. *Psychosomatic Medicine, 49,* 274–284.

Landis, T., Assal, G., and Perret, E. (1979). Opposite cerebral hemispheric superiorities for visual associative processing of emotional facial expressions and objects. *Nature, 278,* 739–740.

Landis, T., Cunnings, J. L., Christen, L., Bogen, L., and Imhof, H. G. (1986). Are unilateral right posterior cerebral lesions sufficient to cause prosopagnosia? Clinical and radiological findings in six additional patients. *Cortex, 22,* 243–252.

Landis, T., Graves, R., and Goodglass, H. (1981). Dissociated awareness of manual performance on two different visual associative tasks: A "split-brain" phenomenon in normal subjects. *Cortex, 17,* 435–440.

Landis, T., Graves, R., and Goodglass, H. (1982). Aphasic reading and writing: possible evidence for right hemisphere participation. *Cortex, 18,* 105–112

Langheim, F. J., Callicott, J. H., Mattay, V. S., Duyn, J. H., and Weinberger, D. R. (2002). Cortical systems associated with covert music rehearsal. *Neuroimage, 16,* 901–908.

Lardon, M. T. and Polich, J. (1996). EEG changes from long-term physical exercise. *Biological Psychology, 27,* 19–30.

Larsell, O. and Jansen, J. (1972). *The comparative anatomy and histology of the cerebellum* (Vol. 3). Minneapolis: University of Minnesota Press.

Larson, K. A. (1982). The sensory history of developmentally delayed children with and without tactile defensiveness. *American Journal of Occupational Therapy, 36,* 590–596.

Larsen, J. P., Odegaard, H., Grude, T. H., and Hoien, T. (1989). Magnetic resonance imaging—a method of studying the size and asymmetry of the planum temporale. *Acta Neurologica Scandinavica, 80,* 438–443.

Larsen J. P., Hoien, T., Lundberg, I., and Odegaard, H. (1990). MRI evaluation of the size and symmetry of the planum temporal in adolescents with developmental dyslexia. *Brain and Language, 39,* 289–301.

Lassen, N. A. and Ingvar, D. H. (1990). Brain regions involved in voluntary movements as revealed by radioisotopic mapping of CBF or CMR-glucose changes. *Revue Neurologie* (Paris), *146*, 620–625.

Lassonde, M. (1986). The facilitatory influence of the corpus callosum on intrahemispheric processing. In F. Lepore, M. Ptito, and H. H. Jasper (Eds.), *Two hemispheres – one brain. Functions of the corpus callosum* (pp. 385–401). New York, NY: Academic Press.

Laurent, A., Goaillard, J. M., Cases, O., Lebrand, C., Gaspar, P., and Ropert, N. (2002). Activity-dependent presynaptic effect of serotonin 1B receptors on the somatosensory thalamocortical transmission in neonatal mice. *Journal of Neuroscience, 22,* 886–900.

Lauter, J. (1982). Dichotic identification of complex sounds: Absolute and relative ear advantage. *Journal of Acoustical Society of America, 71,* 701–707.

Lavielle, S., Tassin, J. P., Thierry, A. M., Blanc, G., Herve, D., Barthelemy, C., and Glowinski, J. (1979). Blockade by benzodiazepines of the selective high increase in dopamine turnover induced by stress in mesocortical dopaminergic neurons of the rat. *Brain Research, 168,* 585–594.

Lazarow, P. B. and Moser, H. W. (1995). Disorders of peroxisome biogenesis. In C. R. Scriver, A. L. Beaudet, W. S. Sly, and D. Valle (Eds.). *The metabolic and molecular basis of inherited disease* (pp. 2287–2324). New York: McGraw-Hill.

Lazarus, R. S. and McCleary, R. (1951). Autonomic discrimination without awareness: A study of subception. *Psychological Review, 58,* 113–122.

Leary, M. R. and Hill, D. A. (1996). Moving on: Autism and movement disturbance. *Mental Retardation, 34,* 39–53.

Leask, S. J. and Crow, T. J. (2001). Word acquisition reflects lateralization of hand skill. *Trends in Cognitive Neuroscience, 5,* 513–516.

Leaton, R. N. and Supple, W. S., Jr. (1986). Cerebellar vermis essential for long-term habituation of the acoustic startle response. *Science, 232,* 513–515.

Leboyer, M., Philippe, A., Bouvard, M., Guilloud-Bataille, M., Bondoux, D., Tabuteau, F., Feingold, J., Mouren-Simeoni, M. C., and Louney, J. M. (1999). Whole blood serotonin and plasma beta-endorphin in autistic probands and their first-degree relatives. *Biological Psychiatry, 45,* 158–163.

Lecanuet, J. P. and Schaal, B. (1996). Fetal sensory competencies. *European Journal of Obstetrics, Gynecology, and Reproductive Biology, 68,* 1–23.

Leckman, J. F., Hardin, M. T., and Riddle, M. A. et al. (1991). Clonidine treatment of Gilles de la Tourette's Syndrome. *Archives of General Psychiatry, 48,* 324–328.

Ledoux, J. (1986). Sensory systems and emotion: A model of affective processing. *Integrative Psychiatry, 4,* 237–248.

LeDoux, J. E. (1987). Emotion. In F. Plum (Ed.), *Handbook of physiology. The nervous system* (*Higher cortical functions*, Vol. 5, pp. 419–459). Bethesda, MD: American Physiological Society.

Ledoux, J. E. (1991). Emotion and the limbic system concept. *Concepts in Neuroscience, 2,* 169–199.

Leichnetz, G. R. (1989). Inferior frontal eye field projections to the pursuit-related dorsolateral pontine nucleus and middle temporal area (MT) in the monkey. *Visual Neuroscience, 3,* 171–180.

Leichnetz, G. R. and Astruc, J. (1976). The efferent projections of the medial prefrontal cortex in the squirrel monkey (Saimiri sciureus). *Brain Research, 109,* 455–472.

Leiner, H. C., Leiner, A. L., and Dow, R. S. (1986). Does the cerebellum contribute to menta. *Neuroscience, 100,* 443–454.

Leiner, H. C., Leiner, A. L., and Dow, R. S. (1987). Cerebrocerebellar learning loops in apes and humans. *Italian Journal of Neurological Sciences, 8,* 425–436.

Leiner, H. C., Leiner, A. L., and Dow, R. S. (1989). Reappraising the cerebellum: What does the hindbrain contribute to the forebrain? *Behaviorscience, 103,* 998–100.

Leiner, H. C., Leiner, A. L., and Dow, R. S. (1991). The human cerebro-cerebellar system: Its computing, cognitive, and language skills. *Behavioral and Brain Research, 44,* 113–128.

Leisman, G. (1973). Conditioning variables in attentional handicaps. *Neuropsychologia, 11,* 199–205.

Leisman, G. (1974). The relationship between saccadic eye movements and the alpha rhythm in attentionally handicapped patients. *Neuropsychologia, 12,* 209–218.

Leisman, G. (1976a). The role of visual processes in attention and its disorders. In G. Leisman (Ed.), *Basic visual processes and learning disability* (pp. 7–123). Springfield, IL: Charles C. Thomas.

Leisman, G. (1976b). The Neurophysiology of visual processing. In G. Leisman (Ed.), *Basic visual processes and learning disability* (pp. 124–187). Springfield, IL: Charles C. Thomas.

Leisman, G. (1980). Stability and flexibility on natural systems. *International Journal of Neuroscience, 11,* 153–155.

Leisman, G. (1989a). Cybernetic model of psychophysiologic pathways: II. Consciousness of effort and kinesthesia. *Journal of Manipulative and Physiological Therapeutics, 12,* 174–191.

Leisman, G. (1989b). Cybernetic model of psychophysiologic pathways: III. Impairment of consciousness of effort and kinesthesia. *Journal of Manipulative and Physiological Therapeutics, 12,* 257–265.

Leisman, G. (1989c). A neuropsychology of limb segment information transmission capacity. *Journal of Clinical and Experimental Neuropsychology, 10,* 320.

Leisman, G. (2002). Hemispheric coherence function in developmental dyslexia. *Brain and Cognition, 48,* 425–431.

Leisman, G. and Ashkenazi, M. (1980). Aetiological factors in dyslexia: IV. Cerebral hemispheres are functionally equivalent in developmental dyslexia. *International Journal of Neuroscience, 11,* 15–164.

Leisman, G. and Koch, P. (2000). Continuum model of mnemonic and amnesic phenomena. *Journal of the International Neuropsychological Society, 6,* 589–603.

Leisman, G. and Koch, P. (2003). Synaptic strengthening and continuum activity wave-growth in temporal sequencing during cognitive tasks. *International Journal of Neuroscience, 113*, 179–202.

Leisman, G. and Melillo, R. (2004). Functional brain organization in developmental dyslexia. In F. Columbus (Ed.), *Progress in Dyslexia Research.* New York, NY: Nova.

Leisman, G. and Vitori, R. (1990). Limb segment information transmission capacity infers integrity of spinothalamic tracts and cortical visual-motor control. *International Journal of Neuroscience, 50*, 175–183.

Leisman, G. and Zenhausern, R. (1982). Integratory systems deficits in developmental dyslexia. In R. N. Malatesha and L. Hartlage (Eds.), *Neuropsychology and cognition* (pp. 281–309). Amsterdam: Elsevier.

LeMay, M. (1984). Radiological, developmental, and fossil asymmetries. In N. Geschwind and A. M. Galaburda (Eds.), *Cerebral dominance: The biological foundations* (pp. 26–42). Cambridge, MA: Harvard University Press.

LeMay, M. and Culebras, A. (1972). Human brain morphologic differences in the hemispheres demonstrable by carotid arteriography. *New England Journal of Medicine, 287*, 168–170.

LeMay, M. and Geschwind, N. (1975). Hemispheric differences in the brains of great apes. *Brain Behavior and Evolution, 11*, 48–52.

Lempert, H. and Kinsbourne, M. (1985). Possible origin of speech in selective orienting. *Psychological Bulletin, 97*, 62–73.

Lenneberg, E. H. (1967). *Biological Foundations of Language.* New York, NY: Wiley.

Lennon, J. B. M., Shealy, C. N., Roger, K., Matta, W., Cox, R., and Simpson, W. F. (1994). Postural and respiratory modulation of autonomic function, pain and health. *American Journal of Pain Management, 4*, 36–39.

Leonard, H., Swedo, S., Rapoport, J. L., Coffey, M., and Cheslow, D. (1988). Treatment of childhood obsessive compulsive disorder with clomipramine and desmethylimipramine: A double-blind crossover comparison. *Psychopharmacology Bulletin, 24*, 93–95.

Leonard, H. L., Swedo, S. E., Rapoport, J. L., Koby, E. V., Lenane, M. C., Ceslow, D. L., and Hamburger, S. D. (1989). Treatment of obsessive-compulsive disorder with clomipramine and desipramine in children and adolescents: A double-blind crossover comparison. *Archives of General Psychiatry, 46*, 1088–1092.

Lepore, F., Phaneuf, J., Samson, A., and Guillemot, J. (1982). Interhemispheric transfer of visual pattem discriminations: Evidence for a bilateral storage of the engram. *Behavioral Brain Research, 5*, 359–374.

Leroy, B., Jeny, R., and Allouch, A. (1987). Fetal response to stimuli. *Revue Medicale de Liege, 42*, 853–858.

Leung, P. W. L. and Connolly, K. J. (1998). Do hyperactive children have motor organization and/or execution deficits? *Developmental Medicine and Child Neurology, 40*, 600–607.

Leventhal, B. L., Cook, E. H., Jr., Morford, M., Ravitz, A. J., Heller, W., and Freedman, D. X. (1993). Clinical and neurochemical effects of fenfluramine in children with autism. *Journal of Neuropsychiatry and Clinical Neuroscience, 5*, 307–315.

Leviel, V., Chesselet, M., Glowinski, J., and Cheramy, A. (1981). Involvement of the thalamus in the asymmetric effects of unilateral sensory stimuli on the two nigrostriatal dopaminergic pathways of the cat. *Brain Research, 223*, 257–272.

Levin, M., Thorlin, T., Robinson, K. R., Nogi, T., and Mercola, M. (2002). Asymmetries in $H+/K+$ -ATPase and cell membrane potentials compromise a very early step in left-right patterning. *Cell, 111*, 77–89.

Levin, S., Holzman, P. S., Rothenberg, S. J., and Lipton, R. B. (1981). Saccadic eye movements in psychotic patients. *Psychiatry Research, 5*, 47–58.

Levinson, H. N. (1980). *A solution to the riddle dyslexia.* New York: Springer-Verlag.

Levinson, H. N. (1984). *Smart but feeling dumb.* New York: Warner.

Levitan, R. D., Jain, U. R., and Katzman, M. A. (1999). Seasonal affective symptoms in adults with residual attention-deficit hyperactivity disorder. *Comprehensive Psychiatry, 40*, 261–267.

Levy, C. E., Nichols, D. S., Schmalbrock, P. M., Keller, P., and Chakeres, D. W. (2001). Functional MRI evidence of cortical reorganization in upper-limb stroke hemiplegia treated with constraint-induced movement therapy. *American Journal of Physical Medicine and Rehabilitation, 80*, 4–12.

Levy, F. (1991). The dopamine theory of attention deficit hyperactivity disorder. *The Australian and New Zealand Journal of Psychiatry, 25*, 277–283.

Levy, F. and Hobbes, G. (1988). The action of stimulant medication in attention deficit disorder with hyperactivity: Dopaminergic, noradrenergic. *Journal of the American Academy of Child and Adolescent Psychiatry, 27*, 802–805.

Levy, J. (1972). Lateral specialization of the human brain: Behavioral manifestation and possible evolutionary basis. In J. A. Kriger (Ed.), *The biology of behavior.* Corvallis, OR: Oregon State University Press.

Levy, J. (1985). Interhemispheric collaboration: Single mindedness in the asymmetric brain. In C. T. Best (Ed.), *Hemispheric function and collaboration in the child* (pp. 11–31). New York, NY: Academic Press.

Levy, J. (1990). Regulation and generation of perception in the asymmetric brain. In C. Trevarthen (Ed.), *Brain circuits and functions of the mind* (pp. 231–248). New York, NY: Cambridge University Press.

Levy, J., Heller, W., Banich, M. T., and Burton, L. A. (1983). Are variations among right-handed individuals in perceptual asymmetries caused by characteristic arousal differences between hemispheres? *Journal of Experimental Psychology: Human Perception and Performance, 9*, 329–359.

Levy, J. and Reid, M. (1976). Variations in writing posture and cerebral organization. *Science, 94*, 337–339.

Levy, L. M., Henkin, R. I., Lin, C. S., Hutter, A., and Schellinger, D. (1999). Odor memory induces brain activation as measured by functional MRI. *Journal of Computer Assisted Tomography, 23*, 487–498.

Levy, M. N. and Martin, P. (1984). Parasympathetic control of the heart. In W. C. Randall (Ed.), *Nervous control of cardiovascular function*. New York, NY: Oxford University Press.

Levy, M. N., Ng, M. D., and Zieske, H. (1966). Functional distribution of the peripheral cardiac sympathetic pathways. *Circulation Research, 14*, 650–661.

Lewine, J. D., Andrews, R., Chez, M., Patil, A. A., Devinsky, O., Smith, M., Kanner, A., Davis, J. T., Funke, M., Jones, G., Chong, B., Provencal, S., Weisend, M., Lee, R. R., and Orrison, W. W., Jr. (1999). Magnetoencephalographic patterns of epileptiform activity in children with regressive autism spectrum disorders. *Pediatrics, 104*, 405–418.

Lewis, D. W. and Diamond, M. C. (1995). The influences of gonadal steroids on the asymmetry of the cerebral cortex. In E. R. J. Davidson and K. Hugdahl (Eds.), *Brain asymmetry* (pp. 31–50). Cambridge, MA: MIT Press.

Lewis, P. R. and Shute, C. C. D. (1967). The cholinergic limbic system: Projections to hippocampal formation, medial cortex, nuclei of the ascending cholinergic reticular system and the subfornical organ and supraoptic crest. *Brain, 90*, 521–540.

Ley, R. G. and Bryden, M. P. (1982). A dissociation of right and left hemispheric effects for recognizing emotional tone and verbal content. *Brain and Cognition, 1*, 3–9.

Ley, R. G. and Bryden, M. P. (1983). Right hemisphere involvement in imagery and affect. In E. Perecman (Ed.), *Cognitive processing in the right hemisphere* (pp. 111–123). New York, NY: Academic Press.

Lhermitte, F. (1986). Human autonomy and the frontal lobes. Part II: Patient behavior in complex and social situations: The "environmental dependency syndrome." *Annals of Neurology, 19*, 335–343.

Lieberman, J. A. and Koreen, A. R. (1993). Neurochemistry and neuroendocrinology of schizophrenia: A selective review. *Schizophrenia Bulletin, 19*, 371–429.

Lieberman, H. R., Falco, C. M., and Slade, S. S. (2002). Carbohydrate administration during a day of sustained aerobic activity improves vigilance, as assessed by a novel ambulatory monitoring device, and mood. *American Journal of Clinical Nutrition, 76*, 120–127.

Liederman, J. and Kinsbourne, M. (1980). The mechanism of neonatal rightward turning bias: A sensory or motor asymmetry? *Infant Behavior and Development, 3*, 223–238.

Liederman, J. (1983). Is there a stage of left-sided precocity during early manual specialization? In G. Young, S. J. Segalowitz, C. M. Corter, and S. E. Trehub (Eds.), *Manual specialization and the developing brain* (pp. 321–330). New York: Academic Press.

Liederman, J. (1988). The dynamics of interhemispheric collaboration and hemispheric control. *Brain and Cognition, 36*, 193–208.

Liepmann, H. (1900). The Syndrome of Apraxia (motor Asymboly) based on a case of unilateral apraxia. In J. W. Brown (Ed.), *Agnosia and Apraxia: Selected Papers of Liepmann, Lange, and Potzl* (pp. 155–183). Hillsdale, NJ: Lawrence Erlbaum, 1988.

Liepmann H. (1905). Die Linke Hemisphare und das Handeln. *Munchener medizinische Wochenschrift, 49*, 2322–2326.

Lindboe, C. F. and Platou, C. S. (1984). Effect of immobilization of short duration on the muscle fibre size. *Clinical Physiology, 4*, 183–188.

Liotti, M., Sava, D., Rizzolatti, G., and Caffarra, P. (1991). Differential hemispheric asymmetries in depression and anxiety: A reaction time study. *Biological Psychiatry, 29*, 887–899.

Liotti, M., and Tucker, D. M. (1992). Right hemisphere sensitivity to arousal and depression. *Brain and Cognition, 18*, 138–151.

Liotti, M., and Tucker, D. M. (1995). Emotion in asymmetric corticolimbic networks. In R. J. Davidson and K. Hugdahl (Eds.), *Brain asymmetry* (pp. 389–423). Cambridge, MA: MIT Press.

Lippmann, S., Manshadi, M., Baldwin, H., Drasin, G., Rice, J., and Alrajeh, S. (1982). Cerebellar vermis dimensions on computerized tomographic scans of schizophrenic and bipolar patients. *American Journal of Psychiatry, 139*, 667–668.

Lipton, P. and Whittingham, T. S. (1979). The effect of hypoxia on evoked potentials in the in vitro hippocampus. *Journal of Physiology, London, 287*, 427–438.

Litvan, I., Mohr, E., Williams, J., Gomez, C., and Chase, T. N. (1991). Differential memory and executive functions in demented patients with Parkinson's and Alzheimer's disease. *Journal of Neurology, Neurosurgery, and Psychiatry, 54*, 25–29.

Litvan, I., Paulsen, J. S., Mega, M. S., and Cummings, J. L. (1998). Neuropsychiatric assessment of patients with hyperkinetic and hypokinetic movement disorders. *Archives of Neurology, 55*, 1313–1319.

Livingstone, M. S. and Hubel, D. H. (1987). Psychophysical evidence for separate channels for the perception of form, color, movement and depth. *Journal of Neuroscience, 7*, 3416–3468.

Livingstone, M. and Hubel, D. (1988). Segregation of form, color, movement and depth: Anatomy, physiology and perception. *Science, 240*, 740–750.

Llinas, R. R. (1990). *Thalamic oscillations and signaling* (pp. 50–55). New York: Wiley.

Llinas, R. R. (1995, January 12). Neurobiology. Thorny issues in neurons. *Nature, 373*, 107–108.

Llinas, R. R. (2001). *I of the vortex: From neurons to self.* Cambridge, MA: MIT Press.

Llinas, R. R., Grace, A. A., and Yarom, Y. (1991). In vitro neurons in mammalian cortical layer 4 exhibit intrinsic oscillatory activity in the 10- to 50-Hz frequency range. *Proceedings of the National Academy of Sciences, USA, 88*, 897–901.

Llinas, R. R., Walton, K., Hillman, D. E., and Sotelo, C. (1975). Inferior olive: Its role in motor learning. *Science, 190*, 1230–1231.

Lloyd, D. and Edwards, S. W. (1987). Temperature compensated ultradian rhythms in lower eukaryotes: Timers for cell cycles and circadian events? *Progress in Clinical Biological Research, 227A*, 131–151.

Lockhart, R. S. (1989). The role of theory in understanding implicit memory. In S. Lewandowsky, J. C. Dunn, and K. Kirsner (Eds.), *Implict memory: Theoretical issues.* Hillsdale, NJ: Lawrence Erlbaum Associates.

Loeber, R., Burke, J. D., and Lahey, B. B. (2002). What are adolescent antecedents to antisocial personality disorder? *Criminal Behaviour and Mental Health: CBMH, 12*: 24–26.

Lombard, J. (1998). Autism: A mitochondrial disorder? *Medical Hypotheses, 50*, 93–102.

Lotspeich, L. and Ciaranello, R. D. (1993). The neurobiology and genetics of infantile autism. *International Revue of Neurobiology, 35*, 87–129.

Lotspeich, L. (2001). *Personal communication.* http://www.xent.com/pipermail/fork/2001-December/007434.html.

Lou, J. S. and Bloedel, J. R. (1988). A study of cerebellar cortical involvement in motor learning using a new avoidance conditioning paradigm involving limb movement. *Brain Research, 445*, 171–174.

Lown, B. (1979). Sudden cardiac death: The major challenge confronting contemporary cardiology. *American Journal of Cardiology, 43*, 313–328.

Lown, B., DeSilva, R. A., Reich, P., and Murawski, B. J. (1980). Psychophysiologic factors in sudden cardiac death. *American Journal of Psychiatry, 137*, 1325–1335.

Lown, B., Verrier, R., and Rabinowitz, S. (1977). Neural and psychologic mechanisms and the problem of sudden cardiac death. *American Journal of Cardiology, 39*, 890–902.

Lovejoy, C. O., Johanson, D. C., and Coppens, Y. (1982). Hominid lower limb bones recovered from the hadar formation: 1974–1977 collections. *American Journal of Physical Anthropology, 57*, 679–700.

Lovejoy, O. C. (1981). The origin of man. *Science, 221*, 341–350.

Lowndes, M. and Stewart, M. G. (1994). Dendritic spine density in the lobus parolfactorius of the domestic chick is increased 24 h after one-trial passive avoidance training. *Brain Research, 654*, 129–136.

Lubar, J. F. and Bahler, W. W. (1976). Behavioral management of epileptic seizures following EEG biofeedback training of the sensorimotor rhythm. *Biofeedback and Self Regulation, 1*, 77–104.

Lubar, J. F. and Shouse, M. N. (1976). EEG and behavioral changes in a hyperkinetic child concurrent with training of the sensorimotor rhythm (SMR): A preliminary report. *Biofeedback and Self Regulation, 1*, 293–306.

Lubar, J. O. and Lubar, J. F. (1984). Electroencephalographic biofeedback of SMR and beta for treatment of attention deficit disorders in a clinical setting. *Biofeedback and Self-regulation, 9*, 1–23.

Lucarelli, S., Frediani, T., Zingoni, A. M., Ferruzzi, F., Giardini, O., Quintieri, F., Barbato, M., D'Eufemia, P., and Cardi, E. (1995). Food allergy and infantile autism. *Panminerva Medica, 37*, 137–141.

Luh, K. E., Rueckert, L. M., and Levy, J. (1991). Perceptual asymmetries for free viewing of several types of chimeric stimuli. *Brain and Cognition, 16*, 83–103.

Lund, J. S., Holbach, S. M., and Chung, W. W. (1991). Postnatal development of thalamic recipient neurons in the monkey striate cortex: II. Influence of afferent driving on spine acquisition and dendritic growth of layer 4C spiny stellate neurons. *Journal of Comparative Neurology, 309*, 129–140.

Luoto, S., Taimela, S., Hurri, H., and Alaranta, H. (1999). Mechanisms explaining the association between low back trouble and deficits in information processing. A controlled study with follow-up. *Spine, 24*, 255–261.

Lupien, S. J. and Lepage, M. (2001). Stress, memory, and the hippocampus: Can't live with it, can't live without it. *Behavioral and Brain Research, 127*, 137–158.

Lupien, S. J., King, S., Meaney, M. J., and McEwen, B. S. (2000). Child's stress hormone levels correlate with mother's socioeconomic status and depressive state. *Biological Psychiatry, 48*, 976–980.

Luria, A. R. (1966). *Higher cortical functions in man.* New York: Basic Books.

Luria, A. R. (1973). *The working brain.* New York, NY: Basic Books.

Luteijn, E. F., Serra, M., Jackson, S., Steenhuis, M. P., Althaus, M., Volkmar, F., and Minderaa, R. (2000). How unspecified are disorders of children with a pervasive developmental disorder not otherwise specified? A study of social problems in children with PDD-NOS and ADHD. *European Child and Adolescent Psychiatry, 9*, 168–179.

Luxenberg, J., Swedo, S., Flamem, M., Friedland, R., Rapoport, J., and Rapoport, S. I. (1988). Neuroanatomical abnormalities in obsessive-compulsive disorder detected with quantitative X-ray computed tomography. *American Journal of Psychiatry, 145*, 1089–1093.

Lynch, G. and Boudry, M. (1984). The biochemistry of memory: A new and specific hypothesis. *Science, 224*, 1057–1063.

Lynch, J. C. (1980). The functional organization of posterior parietal association cortex. *Behavioral Brain Science, 3*, 485–534.

Maccoby, E. E. (2000). Parenting and its effects on children: On reading and misreading behavior genetics. *Annual Review of Psychology, 51*, 1–27.

Machne, X. and Segundo, J. P. (1956). Unitary responses to afferent volleys in amygdaloid complex. *Journal of Neurophysiology, 19*, 232–240.

MacLean, P. D. (1975). An ongoing analysis of hippocampal inputs and outputs: Microelectrode and neuroanatomical findings in squirrel monkeys. In R. L. Isaacson and K. H. Pribram (Eds.), *The Hippocampus* (Vol. 1, pp. 177–211). New York: Plenum Press.

MacLean, P. D. (1985). Brain evolution relating to family, play, and the separation call. *Archives of General Psychiatry, 42*, 405–417.

MacLean, P. D. (1990). *The Triune brain in evolution. Role in paleocerebral functions.* New York: Plenum Press.

MacNeilage, P. F., Studdert-Kennedy, M. G., and Lindblom, B. (1987). Primate handedness reconsidered. *Behavioral and Brain Sciences, 10*, 247–303.

Madeau, S. E. and Heilman, K. M. (1991). Gaze-dependent hemianopia without hemispatial neglect. *Neurology, 41*, 1244–1250.

Maes, M., Vandewioude, M., Schotte, C., Martin, M., Blockx, P., Scharpe, S., and Cosyns, P. (1989). Hypothalamic-pituitary-adrenal and thyroid axis dysfunctions and decrements in the availability of L-tryptophan as biological markers of suicidal ideation in major depressed females. *Acta Psychiatrica Scandinavica, 80*, 13–17.

Maffeis, C., Zaffanello, M., and Schutz, Y. (1997). Relationship between physical inactivity and adiposity in prepubertal boys. *Journal of Pediatrics, 131*, 288–292.

Magnus, R. (1924). Korperstellung. Berlin: Springer.

Maiti, A. and Snider, R. S. (1975). Cerebellar control of basal forebrain seizures: Amygdala and hippocampus. *Journal of Epilepsia, 16*, 521–533.

Maki, B. E., Holliday, P. J., and Topper, A. K. (1994). A prospective study of postural balance and risk of falling in an ambulatory and independent elderly population. *Journal of Gerontology: Medical Sciences, 49*, M72–M84.

Makin, J., Fried, P. A., and Watkinson, B. (1991). A comparison of active and passive smoking during pregnancy: Long-term effects. *Neurotoxicology and Teratology, 13*, 5–12.

Malkova, L. and Mishkin, M. (2003). One-trial memory for object-place associations after separate lesions of hippocampus and posterior parahippocampal region in the monkey. *Journal of Neuroscience, 23*, 1956–1965.

Manck, A., Guyre, P. M., and Holbrook, N. J. (1984). Physiological functions of glucocorticoids in stress and their relation to pharmacological actions. *Endocrine Reviews, 5*, 25–44.

Manjiviona, J. and Prior, M. (1995). Comparison of Asperger syndrome and high-functioning autistic children on a test of motor impairment. *Journal of Autism and Developmental Disorders, 25*, 23–39.

Manuck, S. B. and Krantz, D. S. (1986). Psychophysiologic reactivity in coronary heart disease and essential hypertension. In V. A. Matthews, S. M. Weiss, T. Detre, T. M. Dembroski, B. Falkner, S. B. Manuck, and R. B. Williams Jr. (Eds.), *Handbook of stress, reactivity and cardiovascular disease* (pp. 11–34). New York, NY: John Wiley & Sons.

Marano, H. E. (1999, March). Depression: Beyond serotonin. *Psychology Today, 32*, 72–76.

Marcinkiewicz, M., Morlos, R., and Chretien, M. (1989). CNS connections with median Raphe nucleus. Retrograde tracing with WGA-Apo-HRP-gold complex in the rat. *Journal of Comparative Neurology, 289*, 11–35.

Mariani, M. A. and Barkley, R. A. (1997). Neuropsychological and academic functioning in preschool boys with attention deficit hyperactive disorder. *Developmental and Neuropsychology, 13*, 111–129.

Marien, P., Saerens, J., Nanhoe, R., Moens, E., Nagels, G., Pickut, B. A., Dierckx, R. A., and De Deyn, P. P. (1996). Cerebellar induced aphasia: Case report of cerebellar induced prefrontal aphasic language phenomena supported by spect findings. *Journal of Neurological Science, 144*, 34–43.

Marien, P., Engelborghs, S., Fabbro, F., and De Deyn, P. P. (2001). The lateralized linguistic cerebellum: A review and a new hypothesis. *Brain and Language, 79*, 580–600.

Marks, I. M., Lelliott, P., Basoglu, M., Noshirvani, H., Monteiro, W., Cohen, D., and Kasvikis, Y. (1988). Clomipramine, self-exposure and therapist-aided exposure for obsessive-compulsive rituals. *British Journal of Psychiatry, 152*, 522–534.

Marks, I. M., O'Dwyer, A. M., Meehan, O., Greist, J., Baer, L., and McGuire, P. (2000). Subjective imagery in obsessive-compulsive disorder before and after exposure therapy. Pilot randomised controlled trial. *British Journal of Psychiatry, 176*, 387–391.

Marks, I. M., Stern, R. S., Mawson, D., Cobb, J., and McDonald, J. (1980). Clomipramine and exposure for obsessive-compulsive rituals: I. *British Journal of Psychiatry, 136*, 1–25.

Marks, M. J., Grady, S. R., and Collins, A. C. (1993). Down regulation of nicotinic receptor function after chronic nicotine infusion. *Journal of Pharmacology and Experimental Therapeutics, 266*, 1268–1276.

Marks, M. J., Pauly, J. R., Gross, S. D., Deneris, E. S., Hermans-Borgmeyer, I., Heinemann, F., and Collins, A. C. (1992). Nicotine binding and nicotine receptor subunit RNA after chronic nicotine treatment. *Journal of Neuroscience, 12*, 2765–2784.

Marr, D. (1982). *Vision.* New York, NY: W.H. Freeman.

Marsloek, C. J., Kosslyn, S. M., and Squire, L. R. (1992). Form specific visual priming in the Right cerebral hemisphere. *Journal of Experimental Psychology: Learning, Memory, and Cognition, 18*, 492–508.

Martin, A., Kaufman, J., and Charney, D. (2000). Pharmacotherapy of earl-onset depression. Update and new directions. *Child and Adolescent Psychiatric Clinics of North America, 9*, 135–157.

Martin, J. B. and Riskind, P. N. (1992). Neurologic manifestations of hypothalamic disease. *Progress in Brain Research, 93*, 31–40; discussion 40–42.

Martin, M. (1979). Hemispheric specialization for local and global processing. *Neuropsychologia, 17*, 33–40.

Martin, P. and Albers, M. (1995). Cerebellum and Schizophrenia. *Schizophrenia Bulletin, 21*, 241–250.

Martin, R. C., Kretzmer, T., Palmer, C., Sawrie S., Knowlton, R., Faught, E., Morawetz., R., and Kuzinecky, R. (2002). Risk to verbal memory following anterior temporal lobectomy in patients with severe left-sided hippocampal sclerosis. *Archives of Neurology, 59*, 1895–1901.

Martin, W. R. W. and Raichle, M. E. (1983). Cerebellar blood flow and metabolism in cerebral hemisphere infarction. *Annals of Neurology, 14*, 168–176.

Martinez, M. (1996). Docosahexaenoic acid therapy in docosahexaenoic acid-deficient patients with disorders of peroxisomal biogenesis. *Lipids, 31* (Suppl.), S145–S152.

Martinez, M. and Vazquez, E. (1998). MRI evidence that docosahexaenoic acid ethyl ester improves myelination in generalized peroxisomal disorders. *Neurology, 51*, 26–32.

Martinez, M., Pineda, M., Vidal, R., Conill, J., and Martin, B. (1993). Docosahexaenoic acid—a new therapeutic approach to peroxisomal-disorder patients: Experience with two cases. *Neurology, 43*, 1389–1397.

Marzi, C. and Berlucchi, G. (1997). Right visual field superiority for accuracy of recognition of famous faces in normals. *Neuropsychologia, 15*, 751–756.

Mataga, N., Fujishima, S., Condie, B. G., Hensch, T. K. (2001). Experience-dependent plasticity of mouse visual cortex in the absence of the neuronal activity-dependent marker egr1/zif268. *Journal of Neuroscience, 21*, 9724–9732.

Matarazzo, E. B. (2002). Treatment of late onset autism as a consequence of probable autoimmune processes related to chronic bacterial infection. *World Journal of Biological Psychiatry, 3*, 162–166.

Matsushita, M. and Xiong, G. (2001). Uncrossed and crossed projections from the upper cervical spinal cord to the cerebellar nuclei in the rat, studied by anterograde axonal tracing. *Journal of Comparative Neurology, 432*, 101–118.

Mauritz, K. H., Schmitt, C., and Dichgans, J. (1981). Delayed and enhanced long latency reflexes as the possible cause of postural tremor in late cerebellar atrophy. *Brain, 104*, 97–116.

Mavissakalian, M. R., Jones, B., Olson, S., and Perel, J. M. (1990). Clomipramine in obsessive-compulsive disorder: Clinical response and plasma levels. *Journal of Clinical Psychopharmacology, 10*, 261–268.

Mavissakalian, M. R., Turner, S. M., Michelson, L., and Jacob, R. (1985). Tricyclic antidepressants in obsessive-compulsive disorder: Antiobsessional or antidepressive agents? II. *American Journal of Psychiatry, 142*, 572–576.

Mayberg, H. S. (1992). Neuroimaging studies of depression in neurological disease. In S. E. Starkstein and R. G. Robinson (Eds.), *Depression in neurologic disease*. Baltimore, MD: Johns Hopkins University Press.

Mayberg, H. S., Moran, T. H., and Robinson, R. G. (1990). Remote lateralized changes in cortical spiperone binding following focal frontal cortex lesions in the rat. *Brain Research, 516*, 127–131.

Mayberg, H. S., Robinson, R. G., Wong, D. W., Parikh, R., Bolduc, P., Starkstein, S. E., Price, T., Dannals, R. F., Links, J. M., Wilson, A. A., Ravert, H. T., and Wagner, H. N. (1988). PET imaging of cortical S2 serotonin receptors after stroke: Lateralized changes and relationship to depression. *American Journal of Psychiatry, U5*, 937–943.

Mayevsky, A. and Chance, B. (1975). Metabolic responses of the awake cerebral cortex to anoxia hypoxia spreading depression and epileptiform activity. *Behavioural Brain Research, 98*, 149–165.

Mayevsky, A., Doron, A., Manor, T., Meilin, S., Zarchin, N., and Ouaknine, G. E. (1996). Cortical spreading depression recorded from the human brain using a multiparametric monitoring system. *Brain Research, 740*, 268–274.

McCarthy, R. A. and Warrington, E. K. (1990). Object Recognition. In R. A. McCarthy and E. K. Warrington (Eds.), *Cognitive neuropsychology: A clinical introduction* (pp. 22–55). New York, NY: Academic Press.

McClurkin, J. W. and Optican, L. M. (1996). Primate striate and prestriate cortical neurons during discrimination. I. Simultaneous temporal encoding of information about color and pattern. *Journal of Neurophysiology, 75*, 481–495.

McComas, A. J. (1994). Human neuromuscular at adaptations that a company changes in activity. *Medicine and Science in Sports and Exercise, 26*, 1498–1509.

McConville, B. J., Sanberg, P. R., and Fogelson, M. H. et al. (1992). The effects of nicotine plus haloperidol compared to nicotine only and placebo nicotine only in reducing tic severity and frequency in Tourette's disorder. *Biological Psychiatry, 31*, 832–840.

McCormick, D. A. and Thompson, R. F. (1984). Cerebellum: Essential involvement in the classically conditioned eyelid response. *Science, 223*, 296–299.

McCutcheon, L. E. (2000). Another failure to generalize the Mozart effect. *Psychological Reports, 87*, 325–330.

McDougle, C. J., Kresch, L. E., Goodman, W. K., Naylor, S. T., Volkmar, F. R., Cohen, D. J., and Price, L. H. (1995). A case-controlled study of repetitive thoughts and behavior in adults with autistic disorder and obsessive-compulsive disorder. *Journal of Clinical Psychopharmacology, 12*, 322–327.

McEwen, B. S. (1997a). Possible mechanisms for atrophy of the human hippocampus. *Molecular Psychiatry, 2*, 255–62.

McEwen, B. S. (1997b). Hormones as regulators of brain development: Life-long effects related to health and disease. *Acta Paediatrica Supplement, 422*, 41–44.

McEwen, B. S. (1999a). Clinical review 108: The molecular and neuroanatomical basis for estrogen effects in the central nervous system. *Journal of Clinical Endocrinology and Metabolism, 84*, 1790–1797.

McEwen, B. S. (1999b). Stress and hippocampal plasticity. *Annual Revue of Neurosciences, 22*, 105–22.

McEwen, B. S. (2000a). Effects of adverse experiences for brain structure and function. *Biological Psychiatry, 48*, 721–731.

McEwen, B. S. (2000b). Protective and damaging effects of stress mediators: Central role of the brain. *Progress in Brain Research, 122*, 25–34.

McEwen, B. S. (2001). Plasticity of the hippocampus: adaptation to chronic stress and allostatic load. *Annals of the New York Academy of Sciences, 933*, 265–277.

McEwen, B. S. (2002). Protective and damaging effects of stress mediators: The good and bad sides of the response to stress. *Metabolism, 51*, 2–4.

McShane, S., Glaser, L., Greer, E. R., Houtz, J., and Diamond, M. C. (1988). Cortical asymmetry: A preliminary study: neurons-glia, female–male. *Experimental Neurology, 99*, 353–361.

Meador, K., Loring, D. W., Lee, G. P., Brooks, B. S., Thompson, W. O., and Heilman, K. M. (1988). Right cerebral specialization for tactile attention as evidenced by intracarotid sodium amytal. *Neurology, 38*, 1763–1766.

Meaney, M. J. (2001a). Maternal care, gene expression, and the transmission of individual differences in stress reactivity across generations. *Annual Revue of Neuroscience, 24*, 1161–92.

Meaney, M. J. (2001b). Nature, nurture, and the disunity of knowledge. *Annals of the New York Academy of Science, 935*, 50–61.

Mega, M. S. and Cummings, J. L. (1994). Frontal-subcortical circuit and neuropsychiatric disorders. *The Journal of Neuropsychiatry and Clinical Neurosciences, 6*, 358–370.

Meguro, K., Blaizot, X., Kondoh, Y., Le Mestric, C., Baron, J. C., and Chaviox, C. (1999). Neocortical and hippocampal glycose hypometabolism following neurotoxic lesions of the entorhinal and perirhinal cortices in the non-human primate shown as PET. Implications for Alzheimer's disease. *Brain, 122*, 1518–1531.

Meibach, R. C. and Siegel, A. (1977). Efferent connections of the septal area in the rat: An analysis utilizing retrograde and anterograde transport methods. *Brain Research, 19*, 1–20.

Mellgren, S. I., Harkmark, W., and Srebro, B. (1977). Some enzyme histochemical characteristics of the human hippocampus. *Cell and Tissue Research, 181*, 459–471.

Meltzer, H. Y. and Lowy, M. T. (1987). The serotonin hypothesis of depression. In H. Y. Meltzer (Ed.), *Psychopharmacolgy: The third generation of progress* (pp. 513–526). New York, NY: Raven Press.

Menold, M. M., Shao, Y., Wolpert, C. M., Donnelly, S. L., Raiford, K. L., Martin, E. R., Ravan, S. A., Abramson, R. K., Wright, H. H., Delong, G. R., Cuccaro, M. L., Pericak-Vance, M. A., and Gilbert, J. R. (2001). Association analysis of chromosome 15 gabaa receptor subunit genes in autistic disorder. *Journal of Neurogenetics, 15*, 245–259.

Mennella, J. A., Jagnow, C. P., and Beauchamp, G. K. (2001). Prenatal and postnatal flavor learning by human infants. *Pediatrics, 107*, E88.

Men's Heath. (1998, December). Don't worry yourself sick: Persistent stress can shrink your brain and expand your gut.

Merzenich, M. M., Schreiner, C., Jenkins, W., and Wang, X. (1993). Neural mechanisms underlying temporal integration, segmentation, and input sequence representation: Some implications for the origin of learning disabilities. *Annals of the New York Academy of Sciences, 682*, 1–22.

Messahel, S., Pheasant, A. E., Pall, H., Ahmed-Choudhury, J., Sungum-Paliwal, R. S., and Vostanis, P. (1998, January 23). Urinary levels of neopterin and biopterin in autism. *Neuroscience Letters, 241*, 17–20.

Mesulam, M. M. (1981). A cortical network for directed attention and unilateral neglect. *Annals of Neurology, 10*, 309–325.

Mesulam, M. M. (1985). Attention, confusional states and neglect. In M. M. Mesulam (Ed.), *Principles of behavioral neurology* (pp. 125–168). Philadelphia, PA: F. A. Davis.

Mesulam, M. M. (1998). From sensation to cognition. *Brain, 121*, 1013–1052.

Mesulam, M., Van Hesen, C. W., Pandya, D. N., and Geschwind, N. (1977). Limbic and sensory connections of the inferior parietal lobule (area PG) in the rhesus monkey: A study with a new method for horseradish peroxidase histochemistry. *Brain Research, 136*, 393–414.

Metherate, R., Cox, C. L., and Ashe, J. H. (1992). Cellular bases of neocortical activation: Modulation of neural oscillations by the nucleus basalis and endogenous acetylcholine. *Journal of Neuroscience, 12*, 4701–4711.

Meulenbroek, R. G. and Van Galen, G. P. (1988). Foreperiod duration and the analysis of motor stages in a line-drawing task. *Acta Psychologica, 69*, 19–34.

Meyer, J. S. (1985). Biochemical effects of corticosteroids on neural tissues. *Physiological Reviews, 65*, 946–1020.

Micco, D. J., Jr., McEwen, B. S., and Shein, W. (1979). Modulation of behavioral inhibition in appetitive extinction following manipulation of adrenal steroids in rats: Implications for involvement of the hippocampus. *Journal of Comparative and Physiological Psychology, 93*, 323–329.

Michel, G. F. (1981). Right-handedness: A consequence of infant supine head-orientation preference? *Science, 212*, 685–687.

Michel, F., Poncet, M., and Signoret, J. L. (1989). Les lesions responsables de la prosopagnosie sont-elles toujours bilaterales? *Revue Neurologique, 145*, 764–770.

Michelson, D., Stratakis, C., Hill, L., Reynolds, J., Galliven, E., Chrousos, G., and Gold, P. (1996). Bone mineral density in women with depression. *New England Journal of Medicine, 335*, 1176–1181.

Middleton, F. A. and Strick, P. L. (1994). Anatomical evidence for cerebellar and basal ganglia involvement in higher cognitive function. *Science, 266*, 458–461.

Middleton, F. A. and Strick, P. L. (2001). Cerebellar projections to the prefrontal cortex of the primate. *Journal of Neuroscience, 21*, 700–712.

Mikkelsen, E., Lake, C. R., Brown, G. L., Ziegler, M. G., and Ebert, M. H. (1981). The hyperactive child syndrome: Peripheral sympathetic nervous system function and the effect of d-amphetamine. *Psychiatry Research, 4*, 157–169.

Milberger, S., Biederman, J., Faraone, S. V., Chen, L., and Jones, J. (1996). Is maternal smoking during pregnancy a risk factor for attention deficit hyperactivity disorder in children? *The American Journal of Psychiatry, 153*, 1138–1142.

Milberger, S., Biederman, J., Faraone, S. V., and Jones, J. (1998). Further evidence of an association between maternal smoking during pregnancy and attention deficit hyperactivity disorder: Findings from a high-risk sample of siblings. *Journal of Clinical Child Psychology, 27*, 352–358.

Milner, B. (1968). Visual recognition and recall after right temporal-lobe excision in man. *Neuropsychologia, 6*, 191–209.

Minshew, N. J. (1996). Brief report: Brain mechanisms in autism: Functional and structural abnormalities. *Journal of Autism and Developmental Disorders, 26*, 205–209.

Mirsky, A. F., Rosvold, H. E., and Pribram, K. H. (1957). Effects of cingulectomy on social behavior in monkeys. *Journal of Neurophysiology, 20*, 588–601.

Mishkin, M. (1978). Memory in monkeys severely unpaired by combined but not separate removal of amygdala and hippocampus. *Nature, 273*, 297–298.

Mishkin, M., Malamut, B., and Bachevalier, J. (1984). Memories and habits: Two neural systems. In G. Lynch, J. L. McGaugh, and N. M. Weinberger (Eds.), *Neurobiology of learning and memory* (pp. 65–77). New York, NY: Guilford Press.

Mitchell, E. A., Aman, M. G., Turbott, S. H., and Manku, M. (1987). Clinical characteristics and serum essential fatty acid levels in hyperactive children. *Clinical Pediatrics* (Philadelphia), *26*, 406–411.

Miwa, H., Hatori, K., Kondo, T., Imai, H., and Mizuno, Y. (1996). Thalamic tremor: Case reports and implications of the tremor-generating mechanism. *Neurology, 46*, 75–79.

Mizuno, N. and Nakamura, Y. (1970). Direct hypothalamic projections to the locus coeruleus. *Brain Research, 19*, 160–163.

Mizuno, Y., Ohta, S., and Panaka, N. et al. (1989). Deficiencies in complex I subunits of the respiratory chain in Parkinson's disease. *Biochemical and Biophysical Research Communications, 163*, 1450–1455.

Mohammed, A. H., Zhu S. W., Darmopil, S., Hjerling-Leffler, J., Ernfors, P., Winblad, B., Diamond, M. C., Eriksson, P. S., and Bogdanovic, N. (2002). Environmental enrichment and the brain. *Progress in Brain Research, 138*, 109–33.

Molfese, D. L. (1989). The use of auditory evoked responses recorded from newborn infants predict later language skills. *Birth Defects Original Article Series, 25*, 4–62.

Molleson, T. (1994). The eloquent bones of Abu Hureyra. *Scientific American, 271*, 70–75.

Molleson, T. I. and Oakley, K. P. (1966). Relative antiquity of the Ubeidiya hominid. *Nature, 209*, 1268.

Molnar, Z. and Blakemore, C. (1995). How do thalamic axons find their way to the cortex? *Trends in Neuroscience, 18*, 389–397.

Molnar, Z., and Blakemore, C. (1999). Development of signals influencing the growth and termination of thalamocortical axons in organotypic culture. *Experimental Neurology, 156*, 363–93.

Monjan, A. A. and Peters, M. H. (1970). Cerebellar lesions and task difficulty in pigeons. *Journal of Comparative and Physiological Psychology, 72*, 171–176.

Montague, M. (1962). *Prenatal influences*. Springfield, IL: Thomas.

Moran, T. H., Sanberg, P. R., Kubos, K. L, Goldrich, M., and Robinson, R. G. (1984). Asymmetrical effects of unilateral cortical suction lesions: Behavioral characterization. *Behavioral Neuroscience, 98*, 747–752.

Moreno, H., Borjas, L., Arrieta, A., Saez, L., Prassad, A., Estevez, J., and Bonilla, E. (1992) Clinical heterogeneity of the autistic syndrome: A study of 60 families. *Investigacion Clinica, 33*, 13–31

Morgan, C. M., Tanofsky-Kraff, M., Wilfley, D. E., and Yanovski, J. A. (2002). Childhood obesity. *Child and Adolescent Psychiatric Clinics of North America, 11*, 257–278.

Morgan, G., and Gibson, K. R. (1991). Nutritional and environmental interactions in brain development. In K. R. Gibson and A. C. Petersen (Eds), Brain Maturation and Cognitive Development (pp. 91-106). New York: Aldine de Bruyter.

Morgan, J. and Stordy, J. (1995). Infant feeding practices in the 1990s. *Health Visit, 68*, 56–58.

Morgan, W. P. (1896). A case of congenital word-blindness. *British Medical Journal, 2*, 1978.

Morilak, D. A., Fornal C. A., and Jacobs, B. L. (1987). Effects of physiological manipulations on locus coeruleus neuronal activity in freely moving cats. II. Cardiovascular challenge. *Brain Research, 422*, 24–31.

Morris, G. S., Sifft, J. M., and Khalsa, G. K. (1988). Effect of educational kinesiology on static balance of learning disabled students. *Perceptual and Motor Skills, 67*, 51–54.

Morris, M., Bradley, M., Bowers, D., Lang, P., and Heilman, K. (1991). *Valence-specific hypo arousal following right temporal lobectomy*. Paper presented at the Annual Meeting of the International Neuropsychological Society, San Antonio, TX.

Morris, R. D., Hopkins, W. D., and Bolser-Gilmore, L. (1993). Assessment of hand preference in two language-trained chimpanzees. *Journal of Clinical and Experimental Neuropsychology, 15*, 487–502.

Morrison, D. C., Hinshaw, S. P., and Carte, E. T. (1985). Signs of neurobehavioral dysfunction in a sample of learning disabled children: Stability and concurrent validity. *Perceptual and Motor Skills, 61*, 863–872.

Morrison, J. R. and Stewart, M. A. (1971). A family study of the hyperactive child syndrome. *Biological Psychiatry, 3*, 9–195.

Morrow, L., Urtunski, O. B., Kim, Y., and Boller, F. (1981). Arousal responses to emotional stimuli and laterality of lesion. *Neuropsychologia, 19*, 65–71.

Moruzzi, G. and Magoun, H. W. (1949). Brainstem reticular formation and activation of the EEG. *Electroencephilography and Clinical Neurophysiology, 1*, 455–473.

Moscovitch, M. (1977). The development of lateralization of language functions and its relation to cognitive and linguistic development: A review and some theoretical speculations. In S. J. Segalowitz and F. A. Gruber (Eds.), *Language development and neurological theory* (pp. 193–211). New York: Academic Press.

Mosko, S. S., Haubrich, D., and Jacobs, B. L. (1977). Serotonergic afferents to the dorsal raphe nucleus: Evidence from HRP and synaptosomal uptake studies. *Brain Research, 119*, 269–290.

Mostofsky, S. H., Lasker, A. G., Singer, H. S., Denckla, M. B., and Zee, D. S. (2001a). Oculomotor abnormalities in boys with tourette syndrome with and without ADHD. *Journal of the American Academy of Child and Adolescent Psychiatry, 40*, 1464–1472.

Mostofsky, S. H., Lasker, A. G., Cutting, L. E., Denckla, M. B., and Zee, D. S. (2001b). Oculomotor abnormalities in attention deficit hyperactivity disorder: A preliminary study. *Neurology, 57*, 423–430.

Motter, B. C. and Mountcastle, V. B. (1981). The functional properties of the light-sensitive neurons of the posterior parietal cortex studied in waking monkeys: Sparing and opponent vector organization. *Journal of Neuroscience, 1*, 3–26.

Mountcastle, V. B. (1978). Brain mechanisms for directed attention. *Journal of the Royal Society of Medicine, 71*, 14–28.

Muehl, S., Knott, J. R., and Benton, A. L. (1965). EEG abnormality and psychological test performance in reading disability. *Cortex, 1*, 434–439.

Muley, S. A., Strother, S. C., Ashe, J., Frutiger, S. A., Anderson, J. R., Sidtis, J. J., and Rottenberg, D. A. (2001). Effects of changes in experimental design on PET studies of isometric force. *Neuroimage, 13*, 185–195.

Muller, R. A. (1996). A quarter of a century of place cells. *Neuron, 17*, 813–822.

Muller, R. A., Pierce, K., Ambrose, J. B., Allen, G., and Courchesne, E. (2001). A typical patterns of cerebral motor activation in autism: A functional magnetic resonance study. *Biological Psychiatry, 49*, 665–676.

Munck, A., Guyre, P. M., and Holbrook, N. J. (1984). Physiological functions of glucocorticoids in stress and their relation to pharmacological actions. *Endocrine Reviews, 5*, 25–44.

Munoz, K. A., Krebs-Smith, S. M., Ballard-Barbash, R. and Cleveland, L. E. (1997). Food intakes of US children and adolescents compared with recommendations. *Pediatrics, 100*, 323–329.

Munte, T. F., Altenmuller, E., and Jancke, L. (2002). The musician's brain as a model of neuroplasticity. *Nature Reviews, Neuroscience, 3*, 473–478.

Murakami, J. W., Courchesne, E., Press, G. A., Young-Courchesne, R., and Hesselink, J. R. (1989). Reduced cerebellar hemisphere size and its relationship to vermal hypoplasia in autism. *Archives of Neurology, 46*, 689–694.

Muris, P., Steerneman, P., Merckelbach, H., Holdrinet, I., and Meesters, C. (1998). Comorbid anxiety symptoms in children with pervasive developmental disorders. *The American Journal of Psychiatry, 152*, 772–777.

Murphy, J. T., Kwan, H. C., MacKay, W. A., and Wong, Y. C. (1975). Physiological basis of cerebellar dysmetria. *Canadian Journal of Neurological Science, 2*, 279–284.

Murphy, K., Corfield, D. R., Guz, A., Fink, G. R., Wise, R. J., Harrison, J., and Adams, L. (1997). Cerebral areas associated with motor control of speech in humans. *Journal of Applied Physiology, 83*, 1438–1447.

Murphy, K. R., Barkley, R. A., and Bush, T. (2001). Executive functioning and olfactory identification in young adults with attention deficit-hyperactivity disorder. *Neuropsychology, 15*, 211–220.

Musten, L. M., Firestone, P., and Pisterman, S. et al. (1997). Effects of methylphenidate on preschool children with ADHD: Cognitive and behavioral function. *Journal of the American Academy of Child and Adolescent Psychiatry, 36*, 1407–1415.

Nadeau, S. E. and Heilman, K. M. (1991). Gaze-dependent hemianopia without hemispatial neglect. *Neurology, 41*, 1244–1250.

Naeye, R. L. and Peters, E. C. (1984). Mental development of children whose mothers smoked during pregnancy. *Obstetrics and Gynecology, 64*, 601–607.

Nagaratnam, N., Phan, T. A., Barnett, C., and Ibrahim, N. (2002). Angular gyrus syndrome mimicking depressive pseudodementia. *Journal of Psychiatry and Neuroscience, 27*, 364–368.

Narabayashi, H. (1986). Tremor: Its generating mechanism and treatment. In R. J. Vinken, G. W. Bruyn, and H. L. Klawans (Eds.), *Handbook of clinical neurology* (Vol. 49, Revised Series V.) (pp. 597–607). Amsterdam: Elsevier Science.

Nasrallah, H. A., Jacoby, C. G., and McCalley-Whitters, M. (1981). Cerebellar atrophy in schizophrenia and mania. *Lancet, 1*, 1102.

Nasrallah, H. A., McCalley-Whitters, M., and Jacoby, C. G. (1982). Cortical atrophy in schizophrenia and mania: A comparative CT study. *Journal of Clinical Psychiatry, 43*, 439–441.

Nass, R., Gross, A., and Devinsky, O. (1998). Autism and autistic epileptiform regression with occipital spikes. *Developmental Medicine and Child Neurology, 40*, 453–458.

Nass, R., Gross, A., and Devinsky, O. (2000). Autism and autistic epileptiform regression with occipital spikes. *Journal of Autism and Developmental Disorders, 30*, 519–524.

Nass, R. and Koch, D. (1987). Temperamental differences in toddlers with early unilateral right and left-brain damage. *Developmental Neuropsychology, 3*, 93–99.

Nataan, R. (1982). Processing negativity: An evoked potential reflection of selective attention. *Psychological Bulletin, 92*, 605–640.

Nathan, P. W., Smith, M. C., and Deacon, P. (1990). The corticospinal tracts in man. Course and location of fibres at different segmental levels. *Brain, 113*, 303–324.

National Center for Educational Statistics. (2001). National Assessment of Educational Progress. http://nces.ed.gov/pubsearch/getpubcats.

Nauta, W. J. H. (1971). The problem of the frontal lobe: A reinterpretation. *Journal of Psychiatric Research, 8*, 167–187.

Nauta, W. J. H. (1977). An intricately patterned prefrontocaudate projection in the rhesus monkey. *The Journal of Comparative Neurology, 171*, 369–386.

Nauta, W. J. H. and Haymaker, W. (1969). Hypothalamic nuclei and fiber connections. In W. Haymaker, E. Anderson, and W. J. H. Nauta (Eds.), *The Hypothalamus* (pp. 136–209). Springfield, IL: Charles C. Thomas.

Neau, J. P., Arroyo-Anllo, E., Bonnaud, V., Ingrand, P., and Gil, R. (2000). Neuropsychological disturbances in cerebellar infarcts. *Acta Neurologica Scandinavica, 102*, 363–370.

Nelson, D. W. (2002). *Kids count data book.* Baltimore, MD: Annie E. Casey Foundation.

Nesbitt, R. (1957). *Perinatal loss in modem obstetrics.* Philadelphia: Davis.

Netter, F. H. (1962). *CIBA collection of medical illustrations. The nervous system* (Vol. 1). New York, NY: CIBA Pharmaceutical.

Neuringer, M., Connor, W. E., Lin, D. S., Barstad, L., and Luck, S. (1986). Biochemical and functional effects of prenatal and postnatal omega 3 fatty acid deficiency on retina and brain in rhesus monkeys. *Proceedings of the National Academy of Sciences, USA, 83*, 4021–4025.

Neveu, P. J. (1992). Asymmetrical brain modulation of immune reactivity in mice: A model for studying interindividual differences and physiological population heterogeneity? *Life Sciences, 50*, 1–6.

New York Times. (1988, March 15). Problem youth learn how to succeed in business. Section 1(Pt 2), p. 46.

New York Times. (1999). Fear itself. What we now know about how it works how it can be treated & what it tells us about our unconscious. http://open-mind.org/SP/Articles/6c.htm

Newman, R. J. and Rezea, H. (1979). Functional relationship between hippocampus in the cerebellum and electrophysiological study of the cat. *Journal of Physiology, 287*, 405–426.

Newsda. (1997, November 17). Fitness file: An exercise-hypertension link.

Newsday. (1998, April 14). Study links biology and violence. p. A06.

Newsday. (1998, December 22). C03.

Newsday. (2001, January 4). A22.

Ng, K. T., Gibbs, M. E., Crowe, S. F., Sedman, G. L., Hua, F., Zhao, W., O'Dowd, B., Rickard, N., Gibbs, C. L., and Sykova, E. (1991). Molecular mechanisms of memory formation. *Molecular Neurobiology, 5*, 333–350.

Nichols, P. L. and Chen, T. C. (1981). *Minimal brain dysfunction: A prospective study.* Hillsdale, NJ: Lawrence Eribaum Associates.

Nichols, T. R. and Houk, J. C. (1976). Improvement in linearity and regulation of stiffness that results upon actions of stretch reflex. *Journal of Neurophysiology, 30*, 466–441.

Nicogossian, A. E., Huntoon, C. L., and Pool, S. L., (Ed.) (1989) *Space physiology and medicine.* Philadelphia: Lea and Febiger.

Nicolelis, N. A. L., Linn, R. C. S., Woodward, B. J., and Chapin, J. K. (1993). Induction of immediate spatiotemporal changes in thalamic networks by peripheral block of ascending cutaneous information. *Nature* (London), *361*, 533–536.

Niemann, J., Winker, T., Gerling, J., Landwehrmeyer, B., and Jung, R. (1991). Changes of slow cortical negative DC-potentials during the acquisition of a complex finger motor task. *Experimental Brain Research, 85*, 417–422.

Nieto, F. J., Szklo, M., and Comstock, G. W. (1992). Childhood weight and growth rate as predictors of adult mortality. *American Journal of Epidemiology, 136*, 201–213.

Nieuwenhuys, R., Ten Donkelear, H. J., and Nicholson, C. (1998). *The Central Nervous System of Veterbrates* (Vol. 3). Berlin: Springer Verlag.

Ninan, P. T., Rothbaum, B. O., Marsteller, F. A., Knight, B. T., and Eccard, M. B. (2000). A placebo-controlled trial of cognitive-behavioral therapy and clomipramine in trichotillomania. *Journal of Clinical Psychiatry, 61*, 47–50.

Njiokiktjien, C., de Rijke, W., Dieker-van Ophem, A., and Voorhoeve, O. (1976). Stabilography as a diagnostic tool in child neurology. *Agressologie, 17* (Spec D), 41–48.

Noback, C. R. and Demarest, R. J. (1981). *The human nervous system: Basic principles of neurobiology* (3rd ed.). Philadephia, PA: Lea & Febiger.

Noden, D. M. (1992). Vetebrate craniofacial development: Novel approaches and new dilemmas. *Current Opinion in Genetic Development, 2*, 576–581.

Norman, R. M. (1945). Thalamic degeneration following bilateral premotor frontal lobe atrophy of the stru: Mpell type. *Journal of Neurology, Neurosurgery, and Psychiatry, 8*, 52–56.

Noterdaeme, M., Amorosa, H., Mildenberger, K., Sitter, S., and Minow, F. (2001). Evaluation of attention problems in children with autism and children with a specific language disorder. *European Child & Adolescent Psychiatry, 10*, 58–66.

Nottebohm, F. (1977). Asymmetries in neural control of vocalization in the canary. In S. Hamad, R. W. Doty, L. Goldstein, J. Jaynes, and G. Krauthamer (Eds.), *Lateralization in the nervous system* (pp. 23–44). New York: Academic Press.

Nudo, R. J., Milliken, G. W., Jenkins, W. M., and Merzenich, M. M. (1996). Use-dependent alterations of movement representations in primary motor cortex of adult squirrel monkeys. *Journal of Neuroscience, 16*, 785–807.

O'Banion, D., Armstrong, B., Cummings, R. A., and Stange, J. (1978). Disruptive behavior: A dietary approach. *Journal of Autism and Childhood Schizophrenia, 8*, 325–337.

O'Boyle, M. W., Alexander, J. E., and Benbow, C. P. (1991). Enhanced right hemisphere activation in the mathematically precocious: A preliminary EEG investigation. *Brain and Cognition, 17*, 138–153.

Obrzut, J. E. (1988). Defident lateralization in learning-disabled children: Developmental lag or abnormal cerebral organization? In D. L. Molfese and S. J. Segalowitz (Eds.), *Brain lateralization in children: Developmental implications* (pp. 567–589). New York, NY: Guilford Press.

Obrzut, J. E. (1991). Hemispheric activation and arousal asymmetry in learning disabled children. In J. E. Obrzut and G. W. Hynd (Eds.), *Neuropsychological foundations of learning disabilities: A handbook of issues, methods, and practice* (pp. 179–198). San Diego, CA: Academic Press.

Obrzut, J. E., Hynd, G. W., and Obrzut, A. (1983). Neuropsychological assessment of learning disabilities: A discriminant analysis. *Journal of Experimental Child Psychology, 35*, 46–55.

Obrzut, J. E., Mondor, T. A., and Uecker, A. (1993). The influence of attention on the dichotic REA in normal and learning-disabled children. *Neuropsychologia, 3*, 1411–1416.

O'Callaghan, E., Gibson, T., Colohan, H. A., Buckley, P., Walshe, D. G., Larkin, C., and Waddington, J. L. (1992). Risk of schizophrenia in adults born after obstetric complications and their association with early onset or illness: A controlled study. *British Medical Journal, 305*, 1256–1259.

Ockleford, E. M., Vince, M. A., Layton, C., and Reader, M. R. (1995). Responses of neonates to parents' and others' voices. *Early Human Development, 18*, 27–36.

Oganov, V. S. and Potapov, A. N. (1976). On the mechanisms of changes in skeletal muscles in the weightless environment. In G. Marechal and O. Takacs (Eds.), *Life science and space research*, (Vol. 14, pp. 137–143). Berlin: Academic-Verlag.

Ohman, A. (1983). The orienting response during pavlovian conditioning. In D. Siddle (Ed.), *Orienting and habituation perspectives in human research.* Chichester, England: John Wiley & Sons.

Oishi, K., Kasai, T., and Maeshima, T. (2000). Autonomic response specificity during motor imagery. *Journal of Physiological Anthropology and Applied Human Science, 19*, 255–261.

Ojeda, N., Ortuno, F., Arbizu, J., Lopez, P., Marti-Climent, J. M., Penuelas, I., and Cervera-Enguix, S. (2002). Functional neuroanatomy of sustained attention in schizophrenia: Contribution of parietal cortices. *Human Brain Mapping, 17*, 116–130.

Ojemann, G. A. (1977). Asymmetric function of the thalamus in man. In S. J. Dimond and D. A. Blizzard (Eds.), *Evolution and lateralization of the brain, Annals of the New York Academy of Sciences, 299*, 380–396.

Ojemann, G. A. and Creutzfeldt, O. D. (1987). Language in humans and animals: Contribution of brain stimulation and recording. In F. Plum (Ed.), *Handbook of physiology, Section I: The nervous system, Higher function of the brain* (Vol. 5, pp. 675–699). Bethesda, MD: American Physiological Society.

Okai, T., Kozuma, S., Shinozuka, N., Kuwabara, Y., and Mizuno, M. A. (1992). Study on the development of sleep–wakefulness cycle in the human fetus. *Early Human Development, 29*, 391–396.

Oke, A., Keller, R., Mefford, J., and Adams, R. N. (1978). Lateralization of norepinephrine in human thalamus. *Science, 2001*, 1411–1413.

Oke, A., Lewis, R., and Adams, R. N. (1980). Hemispheric asymmetry of norepinephrine distribution in rat thalamus. *Brain Research, 188*, 269–272.

O'Keefe, J. and Nadel, L. (1978). *The hippocampus as a cognitive map.* Oxford: Clarendon Press.

Oki, J., Miyamoto, A., Takahashi, S., and Takei, H. (1999). Cognitive deterioration associated with focal cortical dysplasia. *Pediatric Neurology, 20*, 73–77.

Oldfield, R. C. (1969). Handedness in musicians. *British Journal of Psychology, 60*, 91–99.

Oldham, J. M., Hollander, E., and Skodol, A. E. (Eds.) (1996). *Impulsivity and compulsivity.* Washington, DC: American Psychiatric press.

Olds, J. (1967). The limbic system and behavioral reinforcement. *Progress in Brain Research, 27*, 144–164.

Olds, J. (1969). The central nervous system and the reinforcement of behavior. *The American Psychologist, 24*, 114–132.

O'Mahony, D., Rowan, M., Feely, J., Walsh, J. B., and Coakley, D. (1994). Primary auditory pathway and reticular activating system dysfunction in Alzheimer's disease. *Neurology, 44*, 2089–2094.

O'Neill, M. and Jones, R. S. (1997). Sensory-perceptual abnormalities in autism: A case for more research? *Journal of Autism and Developmental Disorders, 27*, 283–293.

O'Neill, M. and Jones, R. S. (1998). Sensory-perceptual abnormalities in autism: A case for more research? *Journal of Autism and Developmental Disorders, 28*, 153–157.

O'Reilly, R. C. and Rudy, J. W. (2001). Conjunctive representations in learning and memory; Principles of cortical and hippocampal function. *Psychological Review, 108*, 311–345.

Ornitz, E. M., Atwell, C. W., Kaplan, A. R., and Westlake, J. R. (1991). Brain-stem dysfunction in autism. Results of vestibular stimulation. *Journal of Autism and Developmental Disorders, 21*, 303–313.

Ornstein, R. (1991). *Evolution of conciousness; The origins of the way we think.* Englewood Cliffs, NJ: Prentice Hall.

Orton, S. T. (1937). *Reading, writing and speech problems in children.* New York, NY: W. W. Norton & Co.

Otsuka, H., Harada, M., Mori, K., Hisaoka, S., and Nishitani, H. (1999). Brain metabolites in the hippocampus-amygdala region and cerebellum in autism; an 1H-MR spectroscopy study. *Neuroradiology, 41*, 517–519.

Ottenbacher, K. (1978). Identifying vestibular procession dysfunction in learning-disabled children. *The American Journal of Occupational Therapy, 32*, 217–221.

Ottersen, O. P. and Ben-Ari, Y. (1978a). Demonstration of a heavy projection of midline thalamic neurons upon the lateral nucleus of the amygdala of the rat. *Neuroscience Letters, 9*, 147–152.

Ottersen, O. P. and Ben-Ari, Y. (1978b). Pontine and mesencephalic afferents to the central nucleus of the amygdala of the rat. *Neuroscience Letters, 8*, 329–334.

Ottersen, O. P. and Ben-Ari, Y. (1979). Afferent connections to the amygdaloid complex of the rat and cat. Projections from the thalamus. *The Journal of Comparative Neurology, 187*, 401–424.

Pacquet, L. and Merikle, P. M. (1988). Global precendence in attended and nonattended objects. *Journal of Experimental Psychology. Human Perception and Performance, 14*, 89–100.

Page, T. and Moseley, C. (2002). Metabolic treatment of hyperuricosuric autism. *Progress in Neuropsychopharmacology and Biological Psychiatry, 26*, 397–400.

Page, T., Yu, A., Fontanesi, J., and Nyhan, W. L. (1997). Developmental disorder associated with increased cellular nucleotidase activity. *Proceedings of the National Academy of Sciences, USA, 94*, 11601–11606.

Palmer, B. J. (1906). *The science of chiropractic*. Davenport, Iowa: Palmer School of Chiropractic.

Palmer, S. E. (1977). Hierarchical structure in perceptual representations. *Cognitive Psychology, 9*, 441–474.

Palmer, S. S. and Hutton, J. T. (1995). Postural finger tremor exhibited by parkinson's patients and age matched subject. *Movement Disorders: Official Journal of the Movement Disorder Society, 10*, 658–663.

Pandya, D. M. and Kuypers, H. G. J. M. (1969). Corticocortical connections in the rhesus monkey. *Brain Research, 13*, 13–36.

Pandya, D. N., Seltzer, B., and Barbas, H. (1988). Input-output organization of the primate cerebral cortex. In G. Mitchell, J. Erwin, and D. R. Swindler (Eds.), *Comparative Primate Biology*. Neursociences (Vol. 4, pp. 39–80). New York, NY: A.R. Liss.

Pandya, D. N., Van Hoesen, G. W., and Domesick, V. B. (1973). A cingulo-amygdaloid projection in the rhesus monkey. *Brain Research, 61*, 369–373.

Pandya, D. N. and Yeteiran, E. H. (1985). Proposed neural circuitry for spatial memory in the primate brain. *Neuropsychologia, 22*, 109–122.

Pandya, D. N. and Yeterian, E. H. (1990). Architecture and connection of cerebral cortex: Implications for brain evolution and function. In A. B. Scheibel and A. S. Wechsler (Eds.), *Neurobiology of higher cognitive function* (pp. 53–84). New York: The Guilford Press.

Pantev, C., Engelien, A., Candia, V., and Elbert, T. (2001). Representational cortex in musicians. Plastic alterations in response to musical practice. *Annals of the New York Academy of Sciences, 930*, 300–314.

Pantev, C., Oostenveld, R., Engelien, A., Ross, B., Roberts, L. E., and Hoke, M. (1998). Increased auditory cortical representation in musicians. *Nature, 392*, 811–814.

Papathanassiou, D., Etard, O., Mellet, E., Zago, L., Mazoyer, B., and Tzourio-Mazoyer, N. A. (2000). A common language network for comprehension and production: A contribution to the definition of language epicenters with PET. *Neuroimage, 11*, 347–357.

Papez, J. W. (1937). A proposed mechanism of emotion. *Archives of Neurological Psychiatry, 38*, 725–743.

Pappas, C. C. and Brown, E. (1987). Learning to read by reading: Learning how to extend the functional potential of language. *Research in the Teaching of English, 21*, 160–174.

Pardo, J. V., Pardo, P. J., and Raichle, M. E. (1991). Human brain activation during dysphoria. *Society for Neuroscience Abstracts, 17*, 664.

Parkinson, J. K. and Mishkin, M. A. (1982). A selective mnemonic role for the hippocampus in monkeys: Memory for the location of objects. *Society for Neuroscience Abstracts, 8*, 23.

Parsons, L. M. (2001). Exploring the functional neuroanatomy of music performance, perception, and comprehension. *Annals of the New York Academy of Sciences, 930*, 211–231.

Pascual-Leone, A. (2001). The brain that plays music and is changed by it. *Annals of the New York Academy of Sciences, 930*, 315–329.

Pascual-Leone, A., Wassermann, E. M., Sadato, N., and Hallett, M. (1995a). The role of reading activity on the modulation of motor cortical outputs to the reading hand in braille readers. *Annals of Neurology, 38*, 910–915.

Pascual-Leone, A., Nguyet, D., Cohen, L. G., Brasil-Neto, J. P., Cammarota, A., and Hallett, M. (1995b). Modulation of muscle responses evoked by transcranial magnetic stimulation during the acquisition of new fine motor skills. 2: *Journal of Neurophysiology, 74*, 1037–1045.

Pascual, R. and Figueroa, H. (1996). Effects of preweaning sensorimotor stimulation on behavioral and neuronal development in motor and visual cortex of the rat. *Biology of the Neonate, 69*, 399–404.

Passingham, R. E. (1975). Changes in the size and organization of the brain in man and his ancestors. *Brain, Behavior and Evolution, 11*, 73–90.

Pasquier, D. A. and Reinoso-Suarez, F. (1976). Direct projections from hypothalamus to hippocampus in the rat demonstrated by retrograde transport of horseradish peroxidase. *Brain Research, 108*, 165–169.

Patel, A. D. and Balaban, E. (2001a). Cortical dynamics and the perception of tone sequence structure. *Annals of the New York Academy of Sciences, 930*, 422–424.

Patel, A. D. and Balaban, E. (2001b). Human pitch perception is reflected in the timing of stimulus-related cortical activity. *Nature, Neuroscience, 4*, 839–844.

Patel, A. D., Peretz, I., Tramo, M., and Labreque, R. (1998). Processing prosodic and musical patterns: A neuropsychological investigation. *Brain and Language, 61*, 123–144.

Paul, S. M., Heath, R. G., and Ellison, J. P. (1973). Histochemical demonstration of a direct pathway from the fastigial nucleus to the septal region. *Experimental Neurology, 40*, 798–805.

Pazo, J. H., Murer, G. M., and Segal, E. (1993). D1 and D2 receptors and circking behavior in rats with unilateral lesion of the entopeduncular nucleus. *Brain Research Bulletin, 30*, 635–639.

Pearlson, G. D., Kubos, K. L., and Robinson, R. G. (1984). Effect of anterior-posterior lesion location on the asymmetrical behavioral and biochemical response to cortical suction ablations in the rat. *Brain Research, 293*, 241–250.

Pearlson, G. D. and Robinson, R. G. (1981). Suction lesions of the frontal cerebral cortex in the rat induce asymmetrical behavioral and catecholaminergic responses. *Brain Research, 218*, 233–242.

Pearson, P. D. (1986). Twenty years of research in reading comprehension. In T. E. Raphael (Ed.), *Contexts for school-based literacy* (pp. 43–62). New York: Random House.

Pedersen, O. S., Liu, Y., and Reichelt, K. L. (1999). Serotonin uptake stimulating peptide found in plasma of normal individuals and in some autistic urines. *Journal of Peptide Research, 53*, 641–646.

Peet, M., Glen, I., and Horribin, D. F. (1997). *Phospholipid spectrum disorder in psychiatry*. Carnforth, UK: Marius Press.

Peled, O., Carraso, R., Globman, H., and Yehuda, S. (1997). Attention deficit disorder and hyperactivity—changes in hypothalamic function in hyperactive children: A new model. *Medical Hypotheses, 48*, 262–275.

Pelham, W. E., Gnagy, E. M., and Andrea, C. et al. (1999a). A comparison of morning-only and morning/late afternoon Adderall to morning-only, twice-daily, and three times-daily methylphenidate in children with ADHD. *Advances in Pediatrics, 104*, 1300–1311.

Pelham, W. E., Aronoff, H. R., and Midlam, J. K., et al. (1999b). A comparison of Ritalin and Adderall: Efficacy and time-course in children with ADHD. *Advances in Pediatrics, 103*, 805–806.

Pelham, W. E., Greenslade, K. E., and Vodde-Hamilton, M. et al. (1990). Relative efficacy of long-acting stimulants on children with ADHD: A comparison of standard methylphenidate, sustained-release methylphenidate, sustained-release dextroamphetamine, and pemoline. *Advances in Pediatrics, 86*, 226–237.

Penhune, V. B., Zattore, R. J., and Evans, A. C. (1998). Cerebellar contributions to motor timing: A PET study of auditory and visual rhythm reproduction. *Journal of Cognitive Neuroscience, 10*, 752–765.

Penin, X., Berge, C., and Baylac, M. (2002). Ontogenetic study of the skull in modern humans and the common chimpanzees: Neotenic hypothesis reconsidered with a tridimensional Procrustes analysis. *American Journal of Physical Anthropology, 118*, 50–62.

Penfield, W. G. (1975). *The mystery of the mind*. Princeton, NJ: Princeton University Press.

Pennington, B. F., Bennetto, L., and McAleer, O. et al. (1996). Executive functions and working memory. In G. R. Lyon and N. A. Krasnegor, (Eds.), *Attention, memory, and executive function* (pp. 327–348). Baltimore, MD: Paul H. Brookes.

Pennington, B. F. and Ozonoff, S. (1996). Executive functions and developmental psychopathology. *Journal of Child Psychology and Psychiatry, 37*, 51–87.

Perez-Escamilla, R. and Pollitt, E. (1995). Growth improvements in children above 3 years of age: The Cali Study. *The Journal of Nutrition, 125*, 885–893.

Perris, C., Gottfries, C. G., and von Knorring, L. (1979). Visual averaged evoked responses in psychiatric patients. Relationship to levels of 5-hydroxyindoleacetic acid, homovanilic acid and tryptophan in cerebrospinal fluid. *Journal of Psychiatric Research, 15*, 175–181.

Perry, B. D. and Pollard, R. (1998). Homeostasis, stress, trauma, and adaptation. A neurodevelopmental view of childhood trauma. *Child and Adolescent Clinics of North America, 7*, 33–51.

Perry, E., Walker, M., Grace, J., and Perry, R. (1999). Acetylcholine in mind: A neurotransmitter correlate of consciousness? *Trends in the Neurosciences, 22*, 273–280.

Perse, T. L., Greist, J. H., Jefferson, J. W., Rosenfeld, R., and Dar, R. (1987). Fluvoxamine treatment of obsessive-compulsive disorder. *American Journal of Psychiatry, 144*, 1543–1548.

Peschstedt, P. (1986). Suppression of ipsalateral auditory pathways increases with increasing task load in commissurotomy subjects. *Bulletin of Clinical Neurosciences, 51*, 73–76.

Peters, J. E., Romine, J. S., and Dykman, R. A. (1975). A special neurological examination of children with learning disabilities. *Developmental Medicine and Child Neurology, 17*, 63–78.

Peters, M. (1985). Constraints in the coordination of bimanual movements and their expression in skilled and unskilled subjects. *Quarterly Journal of Experimental Psychology, 37A*, 171–196.

Peters, M. (1988). The primate mouth as agent of manipulation and its relation to human handedness. *Behavioral and Brain Sciences, 11*, 729.

Peters, M. (1991). Laterality and motor control. In J. Marsh (Ed.), *Biological asymmetry and handedness* (CIBA Symposium 162) (pp. 300–311). London: John Wiley & Sons.

Peters, M. (1992). Cerebral asymmetry for speech and the asymmetry in path lengths for the right and left recurrent nerves. *Brain and Language, 43*, 349–352.

Peters, M. (1995). Handedness and its relation to other indices of cerebral lateralization. In R. J. Davidson and K. Hugdahl (Eds.), *Brain asymmetry*. Cambridge, MA: MIT Press.

Peters, M. and Filter, P. M. (1973). Performance of a motor task after cerebellar cortical lesions in rats. *Physiology and Behavior, 11*, 13–16.

Peterson, J. M. and Lansky, L. M. (1977). Left-handedness among architects: Partial replication and some new data. *Perceptual and Motor Skills, 45*, 1216–1218.

Peterson, S. F., Fox, P. T., Posner, M. I., Mintum, M. A., and Raichle, M. E. (1989). Positron emission tomographic studies in the processing of single words. *Journal of Cognitive Neuroscience, 1*, 153–170.

Petrie, B. N. and Peters, M. (1980). Handedness: Left/right differences in intensity of grasp response and duration of rattle holding in infants. *Infant Behavior and Development, 3*, 215–221.

Petrolini, N., Iughetti, L., and Bernasconi, S. (1995). Difficulty in visual motor coordination as a possible cause of sedentary behaviour in obese children. *International Journal of Obesity and Related Metabolic Disorders, 19*, 928.

Pettigrew, J. D. and Miller, S. M. (1998). A "sticky" interhemispheric switch in bipolar disorder? *Proceedings of the Royal Society of London. Series B. Biological Sciences, 265*, 2414–2418.

Pfeiffer, S. I., Norton, J., Nelson, L., and Shott, S. (1995). Efficacy of vitamin B6 and magnesium in the treatment of autism: A methodology review and summary of outcomes. *Journal of Autism and Developmental Disorders, 25*, 481–493.

Philippe, A., Martinez, M., Guilloud-Bataille, M., Gillberg, C., Rastam, M., Sponheim, E., Coleman, M., Zappella, M., Aschauer, H., Van Maldergem, L., Penet, C., Feingold, J., Brice, A., Leboyer, M., and van Malldergerme, L. (1999). Genome-wide scan for autism susceptibility genes. Paris Autism Research International Sibpair Study. *Human Molecular Genetics, 8*, 805–812.

Phillips, M. J., Weller, R. O., Kida, S., and Iannotti, F. (1995). Focal brain damage enhances experimental allergic encephalomyelitis in brain and spinal cord. *Neuropathology and Applied Neurobiology, 21*, 189–200.

Piaget, J. (1952). *The origins of intelligence in the child*. New York, NY: International Universities Press.

Pierce, E. T., Foote, W. E., and Hobson, J. A. (1976). The efferent connection of the nucleus raphe dorsalis. *Brain Research, 107*, 137–144.

Pigott, T. A., L'Heureux, F., Rubenstein, C. S., Bernstein, S. E., Hill, J. L., and Murphy, D. L. (1992). A double-blind, placebo controlled study of trazodone in patients with obsessive-compulsive disorder. *Journal of Clinical Psychopharmacology, 12*, 156–162.

Pigott, T. A., Pato, M. T., Berstein, S. E., Grover, G. N., Hill, J. L., Tolliver, T. J., and Murphy, D. L. (1990). Controlled comparisons of clomipramine and fluoxetine in the treatment of obsessive-compulsive disorder. *Archives of General Psychiatry, 47*, 926–932.

Pillai, M. and James, D. (1990a). Hiccups and breathing in human fetuses. *Archives of Diseases of Childhood, 65*, 1072–1075.

Pillai, M. and James, D. (1990b). Are the behavioural states of the newborn comparable to those of the fetus? *Early Human Development, 22*, 39–49.

Pinault, D. and Deschev, M. (1992). The origin of rhythmic subthreshold depolarizations in thalamic relay cells of rats under urethane anesthesia. *Psychological Research, 595*, 295–300.

Pitman, R. K., Green, R. C., and Jenike, M. A. et al. (1987). Clinical comparison of Tourette's disorder and obsessive-compulsive disorder. *American Journal of Psychiatry, 144*, 1166–1171.

Piven, J., Berthier, M. L., Starkstein, S. E., Nehme, E., Pearlson, G., and Folstein, S. (1990). Magnetic resonance imaging evidence for a defect of cerebral cortical development in autism. *The American Journal of Psychiatry, 147*, 734–739.

Piven, J. and Palmer, P. (1999). Psychiatric disorder and the broad autism phenotype: Evidence from a family study of multiple-incidence autism families. *Journal of Autism and Developmental Disorders, 29*, 499–508.

Pizzorusso, T., Fagiolini, M., Gianfranceschi, L., Porciatti, V., and Maffei, L. (2000). Role of neurotrophins in the development and plasticity of the visual system: Experiments on dark rearing. *International Journal of Psychophysiology, 35*, 189–196.

Plato, C. C., Garruto, R. M., Galasko, D., Craig, U. K., Plato, M., Gamst, A., Torres, J. M., and Wiederholt, W. (2003). Amyotrophic lateral sclerosis and parkinsonism-dementia complex of Guam: Changing incidence rates during the past 60 years. *American Journal of Epidemiology, 157*, 149–157.

Plaut, D. C. and Farah, M. J. (1990). Visual object representation: Interpreting neurophysiological data within a computational framework. *Journal of Cognitive Neuroscience, 2*, 320–343.

Plioplys, A. V., Hemmens, S. E., and Regan, C. M. (1990). Expression of a neural cell adhesion molecule serum fragment is depressed in autism. *The Journal of Neuropsychiatry and Clinical Neurosciences*, Fall, 2, 413–417.

Plioplys, A. V., Greaves, A., and Yoshida, W. (1989). Anti-CNS antibodies in childhood neurologic diseases. *Neuropediatrics, 20*, 93–102.

Pliszka, S. R., Browne, R. G., and Olvera, R. L., and Wynne, K. (2000). A double-blind, placebo-controlled study ad adderall and methylphenidate in the treatment of attention-deficit hyperactivity disorder. *Journal of American Academic Child and Adolescent Psychiatry, 39*, 619–626.

Poeck, K. (1969). Pathophysiology of emotional disorders associated with brain damage. In P. J. Vinken and G. W. Bruyn (Eds.), *Handbook of clinical neurology* (Vol. 3, pp. 343–376). New York, NY: Elsevier.

Pollitt, E. (1994). Poverty and child development: Relevance of research in developing countries to the

United States. *Child Development, 65*(2 Spec No), 283–295.

Polyakov, I. B., Drobyshev, I. V., and Krasnov, I. B. (1991). Morphological changes in the spinal chord and intervertebral ganglia of rats exposed to differential gravity levels. *Physiologist, 34* (Suppl.), S187–S188.

Popa, M., Stefanescu, A. M., Dumitriu, L., Simionescu, L., and Giurcaneanu, M. (1988). The assessment of the dopaminergic tonus by urinary determinations of homovanillic acid (HVA) and dihydroxyphenylacetic acid (DOPA) in normal, obese and GH-deficient short children. *Endocrinologie, 26*, 211–220.

Poppel, E. (1988). *Mindworks. Time and conscious experience.* Boston, MA: Harcourt Brace Jovanovich.

Poppel, E., Schill, K., and von Steinbuchel, N. (1990). Sensory integration within temporally neutral system states: A hypothesis. *Naturwissenschaften, 77*, 89–91.

Porac, C. and Coren, S. (1981). *Lateral preferences and human behavior.* New York, NY: Springer-Verlag.

Porter, R. J., Jr. and Berlin, C. I. (1975). On interpreting developmental changes in the dichotic right-ear advantage. *Brain and Language, 2*, 186–200.

Posner, M. I. (1986). A framework for relating cognitive to neural systems. *Electroencephalography Clinical Neurophysiology Supplement, 38*, 155–166.

Posner, M. I. and Cohen, Y. (1984). Components of visual orienting. In H. Bouma and D. G. Bouwhuis (Eds.), *Attention and performance X: Control of language processes.* Hillsdale, NJ: Lawrence Erlbaum Associates.

Posner, M. I., Inhoff, A. W., Friedrich, F. J., and Cohen, A. (1987). Isolating attentional systems: A cognitive-anatomical analysis. *Psychobiology, 15*, 107–121.

Posner, M. I., Petersen, S. E., Fox, P. T., and Raichle, M. E. (1988). Localization of cognitive operations in the human brain, *Science, 240*, 1627–1631.

Posner, M. I. and Raichle, M. E. (1994). *Images of mind.* New York: Scientific American Library.

Posner, M. I., Walker, L., Friedrich, F. J., and Rafal, R. D. (1984). Effects of parietal lobe injury on covert orienting of visual attention. *Journal of Neuroscience, 4*, 163–167.

Powell, E. W. (1973). Limbic projections to the thalamus. *Experimental Brain Research, 17*, 394–401.

Prasad, A. (1984). A double blind study of imipramine versus zimelidine in treatment of obsessive compulsive neurosis. *Pharmacopsychiatry, 17*, 61–62.

Prather, P., Jarmulowicz, L., Brownell, H., and Gardner, H. (1992). *Selective attention and the right hemisphere: A failure in integration, not detection.* Presented at the International Neuropsychology Society Meeting, San Diego, CA.

Prentice, A. M., Black, A. E., Coward, W. A., and Cole, T. J. (1996a). Energy expenditure in overweight and obese adults in affluent societies: An analysis of 319 doubly-labeled water measurements. *BMJ (Clinical Research ed.), 311*, 437–439.

Prentice, A. M., Goldberg, G. R., Murgatroyd, P. R., and Cole, T. J. (1996b). Physical activity and obesity: Problems in correcting expenditure for body size. *European Journal of Clinical Nutrition, 50*, 93–97.

Prentice, A. M. and Jebb, S. A. (1995). Obesity in Britain: Gluttony or sloth? *British Medical Journal, 11*, 437–439.

Prentice, A. M. and Paul, A. A. (1996). Fat and energy needs of children in developing countries. *International Journal of Obesity and Related Metabolic disorders: Journal of the International Association for the Study of Obesity, 20*, 688–691.

Pribram, K. (1986). The role of corticocortical connections. In F. Lepore, M. Ptito, and H. H. Jasper (Eds.), *Two hemispheres one brain: Functions of the corpus callosum* (pp. 523–540). New York, NY: Alan R. Liss.

Pribram, K. H., Chow, K. L., and Semmes, J. (1953). Limit and organization of the cortical projection from the medial thalamic nucleus in monkey. *The Journal of Comparative Neurology, 98*, 433–448.

Price, C., Wise, R., Ramsay, S., Friston, K., Howard, D., Patterson, K., and Frackowiak, R. (1992). Regional response differences within the human auditory cortex when listening to words. *Neuroscience Letters, 146*, 179–182.

Price, J. L. and Amaral, D. G. (1981). An autoradiographic study of the projections of the central nucleus of the monkey amygdala. *Journal of Neuroscience, 1*, 1242–1259.

Princeton Survey Research Associates. (1994). Prevention magazine's children's health index. *Prevention, 46*, 66–80.

Prizant, B. M. (1996). Brief report: Communication, language, social, and emotional development. *Journal of Autism and Developmental Disorders, 29*, 213–224.

Puidon Martin, J. (1967). Role of the vestibular system in the control of posture and movement in man. In A. V. S. de Rueck and J. Knight (Eds.), *Myotatic, kinesthtic and vestibular mechanisms* (pp. 92–96). London: Churchill.

Rademacher, J., Caviness, V. S., Jr., Steinmetz, H., and Galaburda, A. M. (1993). Topographical variation of the human primary cortices: Implications for neuroimaging, brain mapping, and neurobiology. *Cerebral Cortex, 3*, 313–329.

Rae, C., Karmiloff-Smith, A., Lee, M. A., Dixon, R. M., Grant, J., Blamire, A. M., Thompson, C. H., Styles, P., and Radda, G. K. (1998). Brain biochemistry in Williams syndrome: Evidence for a role of the cerebellum in cognition? *Neurology, 51*, 33–40.

Raine, A., Buchsbaum, M., and LaCasse, L. (1997). Brain abnormalities in murderers indicated by positron emission tomography. *Biological Psychiatry, 42*, 495–508.

Raine, A., Meloy, J. R., Bihrle, S., Stoddard, J., LaCasse, L., and Buchsbaum, M. S. (1998a). Reduced prefrontal and increased subcortical brain functioning assessed using positron emission tomography in predatory and affective murderers. *Behavioral Sciences & The Law, 16,* 319–332.

Raine, A., Phil, D., Stoddard, J., Bihrle, S., and Buchsbaum, M. (1998b). Prefrontal glucose deficits in murderers lacking psychosocial deprivation. *Neuropsychiatry, Neuropsychology, and Behavioral Neurology, 11,* 1–7.

Rajarethinam, R., DeQuardo, J. R., Miedler, J., Arndt, S., Kirbat, R., Brunberg, J. A., and Tandon, R. (2001). Hippocampus and amygdala in schizophrenia: Assessment of the relationship of neuroanatomy to psychopathology. *Psychiatry Research, 108,* 79–87.

Rakic, P. and Goldman-Rakic, P. S. (1982). The development and modifiability of the cerebral cortex. Overview. *Neurosciences Research Program Bulletin, 20,* 433–438.

Rakic, P., Bourgeois, J. P., and Goldman-Rakic, P. S. (1994). Synaptic development of the cerebral cortex: Implications for learning, memory, and mental illness. *Progress in Brain Research, 102,* 227–243.

Ramirez Rozzi, F. (1998). Enamel structure and development and its application in hominid evolution and taxonomy. *Journal of Human Evolution, 35,* 327–330.

Ramnani, N., Toni, I., Passingham, R. E., and Haggard, P. (2001). The cerebellum and parietal cortex play a specific role in coordination: A PET study. *Neuroimage, 14,* 899–911.

Ramsay, D. S. (1985). Fluctuations in unimanual hand preference in infants following the onset of duplicated syllable babbling. *Developmental Psychology, 21,* 318–324.

Randall, W. C. and Ardell, J. L. (1990). Nervous control of the heart: Anatomy and pathophysiology. In D. P. Zipes and J. Jalife (Eds.), *Cardiac electrophysiology.* Philadelphia, PA: W. B. Saunders.

Rantakallio, P. (1983). A follow-up study up to the age of 14 of children whose mothers smoked during pregnancy. *Acta Paediatrica Scandinavica, 72,* 747–753.

Rapoport, J. L., Quinn, P. O., and Bradbard, G. et al. (1974). Imipramine and methylphenidate treatment in hyperactive boys. *Archives of General Psychiatry, 30,* 789–793.

Rauch, S. L., van der Kolk, B. A., Fisler, R. E., Alpert, N. M., Orr, S. P., Savage, C. R., Fischman, A. J., Jenike, M. A., and Pitman, R. K. (1996). A symptom provocation study of posttraumatic stress disorder using positron emission tomography and script-driven imagery. *Archives of General Psychiatry, 53,* 380–387.

Rauch, S. L., Whalen, P. J., Shin, L. M., McInerney, S. C., Macklin, M. L., Lasko, N. B., Orr, S. P., and Pitman, R. K. (2000). Exaggerated amygdala response to masked facial stimuli in posttraumatic stress disorder: A functional MRI study. *Biological Psychiatry, 47,* 769–776.

Rauscher, F. H., Shaw, G. L., and Ky, K. N. (1993). Music and spatial task performance. *Nature, 365,* 611.

Rebai, M., Mecacd, L., Bagot, J., and Bonnet, C. (1986). Hemispheric asymmetries in the visual evoked potentials to temporal frequency: Preliminary evidence. *Perception, 15,* 589–594.

Reeves, A. G. and Hagamen, W. D. (1971). Behavioral and EEG asymmetry following unilateral lesions of the forebrain and midbrain of cats. *Electroencephalography and Clinical Neurophysiology, 39,* 83–86.

Reichelt, K. L. (1994). Biochemistry and psychophysiology of autistic syndromes. *Tidsskrift for den Norske laegeforening, 114,* 1432–1434.

Reichelt, K. L., Hole, K., Hamberger, A., Saelid, G., Edminson, P. D., Braestrup, C. B., Lingjaerde, O., Ledaal, P., and Orbeck, H. (1981). Biologically active peptide-containing fractions in schizophrenia and childhood autism. *Advances in Biochemistry and Psychopharmacology, 28,* 627–643.

Reichelt, K. L., Seim, A. R., and Reichelt, W. H. (1996). Could schizophrenia be reasonably explained by Dohan's hypothesis on genetic interaction with a dietary peptide overload? *Progress in Neuropsychopharmacology and Biological Psychiatry, 20,* 1083–1114.

Reiman, E. M., Fusselman, M. J., Fox, P. T., and Raichle, M. E. (1989). Neuroanatomical correlates of anticipatory anxiety. *Science, 243,* 1071–1074.

Reiman, E. M., Raichle, M. E., Butler, F. K., Hercovitch, P., and Robins, E. (1984). A focal brain abnormality in panic disorder, a severe form of anxiety. *Nature, 310,* 683–685.

Reinert, G. and Wittling, W. (1980). Klinische Psychologie: Konzepte und Tendenzen. In W. Wittling (Ed.), *Handbuch der klinischen Psychologie* (Vol. 1, pp. 14–80). Hamburg: Hoffmann & Campe.

Reis, D. J., Doba, N., and Nathan, M. A. (1973). Predatory attack, grooming, and consummatory behaviors evoked by electrical stimulation of cat cerebellar nuclei. *Science, 182,* 845–847.

Reiss, A. L., Aylward, E., Freund, L. S., Joshi, P. K., and Bryan, R. N. (1991). Neuroanatomy of fragile X syndrome: The posterior fossa. *Annals of Neurology, 29,* 26–32.

Reiss, A. L. and Freund, L. (1990). Fragile x syndrome DSM-III-R and autism. *Psychiatry, 29,* 885–891.

Reitan, R. M. and Boll, T. J. (1973). Neuropsychological correlates of minimal brain dysfunction. *Annals of the New York Academy of Sciences, 205,* 65–88.

Renoux, G. and Biziere, K. (1986). Brain neocortex lateralized control of immune recognition. *Integrative Psychiatry, 4,* 32–40.

Renoux, G., Biziere, K., Renoux, M., Bardos, P., and Degenne, D. (1987). Consequences of bilateral brain neocortical ablation on imuthiol-induced immunostimulation in mice. *Annals of the New York Academy of Sciences, 496,* 346–353.

Renoux, G., Biziere, K., Renoux, M., Guillaumin, J. M., and Degenne, D. (1983). A balanced brain asymmetry modulates T cell-mediated events. *Journal of Neuroimmunology, 5,* 227–238.

Repp, B. (1977). Measuring laterality effects in dichotic listening. *Journal of Acoustical Society of America, 42,* 720–737.

Reppert, S. M. and Schwartz, W. J. (1984). The suprachiasmatic nuclei of the fetal rat: Characterization of a functional circadian clock using 14C-labeled deoxyglucose. *Journal of Neuroscience, 4*, 1677–1682.

Reron, E. (1992). Kliniki Otolaryngologicznej AM, Krakowie [History of electrophysiological investigations of organs of hearing]. *Early Human Development, 29*, 391–396.

Reyes, M. G. and Gordon, A. (1981). Cerebellar vermis in schizophrenia. *Lancet, 2*, 700–701.

Ricardo, J. A. and Koh, E. T. (1978). Anatomical evidence of direct projections from the nucleus of the solitary tract to the hypothalamus, amygdala, and other forebrain structures in the rat. *Brain Research, 153*, 1–26.

Rice, D. and Barone, S., Jr. (2000). Critical periods of vulnerability for the developing nervous system: Evidence from humans and animal models. *Environmental Health Perspectives, 108*, 511–533.

Rich, D. A. and McKeever, W. F. (1990). An investigation of immune system disorder as a "marker" for anomalous dominance. *Brain and Cognition, 12*, 55–72.

Richardson, A. J. and Ross, M. A. (2000). Fatty acid metabolism in neurodevelopmental disorder: A new perspective on associations between attention-deficit/hyperactivity disorder, dyslexia, dyspraxia and the autistic spectrum. *Prostaglandins, Leukotrienes, and Essential Fatty Acids, 63*, 1–9.

Richardson, G. and Day, N. (1994). Detrimental effects of prenatal cocaine exposure: Illusion or reality? *Journal of the American Academy of Child and Adolescent Psychiatry, 33*, 28–34.

Richdale, A. L. (1999). Sleep problems in autism: Prevalence, cause, and intervention. *Developmental Medicine and Child Neurology, 41*, 60–66.

Riddle, M. A., Scahill, L., King, R. A., Hardin, M. T., Anderson, G. M., Ort, S. I., Smith, J. C., Leckman, J. F., and Cohen, D. J. (1992). Double-blind, crossover trail of fluoxetine and placebo in children and adolescents with obsessive-compulsive disorder. *Journal of the American Academy of Child and Adolescent Psychiatry, 31*, 1062–1069.

Riddoch, M. J. and Humphreys, G. (1983). The effect of cuing on unilateral neglect. *Neuropsychologia, 21*, 589–599.

Riecker, A., Ackermann, H., Wildgruber, D., Dogil, G., and Grodd, W. (2000). Opposite hemispheric lateralization effects during speaking and singing at motor cortex, insula and cerebellum. *Neuroreport, 11*, 1997–2000.

Riggs, P. D., Leon, S. L., Mikulich, S. K., and Pottle, L. C. (1998). An open trial of bupropion for ADHD in adolescents with substance use disorders and conduct disorder. *Journal of the American Academy of Child and Adolescent Psychiatry, 37*, 1271–1278.

Rinehart, N. J., Bradshaw, J. L., Brereton, A. V., and Tonge, B. J. (2001). Movement preparation in high-functioning autism and Asperger disorder: A serial choice reaction time task involving motor reprogramming. *Biological Psychiatry, 49*, 665–676.

Ringo, J. and O'Neill, S. (1993). Indirect inputs to ventral temporal cortex of monkey: The influence on unit activity of alerting auditory input, interhemispheric subcortical visual input, reward, and the behavioral response. *Journal of Neurophysiology, 70*, 2215–2225.

Rinn, W. E. (1984). The neuropsychology of facial expression: A review of the neurological and psychological mechanisms for producing facial expressions. *Psychological Bulletin, 95*, 52–77.

Risch, N., Spiker, D., Lotspeich, L., Nouri, N., Hinds, D., Hallmayer, J., Kalaydjieva, L., McCague, P., Dimiceli, S., Pitts, T., Nguyen, L., Yang, J., Harper, C., Thorpe, D., Vermeer, S., Young, H., Hebert, J., Lin, A., Ferguson, J., Chiotti, C., Wiese-Slater, S., Rogers, T., Salmon, B., Nicholas, P., and Myers, R. M. et al. (1999). A genomic screen of autism: Evidence for a multilocus etiology. *American Journal of Human Genetics, 65*, 493–507.

Risold, P. Y. and Swanson, L. W. (1995). Evidence for a hypothalamothalamocortical circuit mediating pheromonal influences on eye and head movements. *Proceedings of the National Academy of Sciences, USA, 92*, 3898–3902.

Risse, G. L., Rubens, A. B., and Jordan, L. S. (1984). Disturbances of long-term memory in aphasic patients: Comparison of anterior and posterior lesions, *Brain, 107*, 605–617.

Ritvo, E. R., Freeman, B. J., Scheibel, A. B., Duoung, T., Robinson, H., Guthrie, D., and Ritvo, A. (1986). Lower purkinje cell counts in the cerebella of four autistic subjects: Initial findings of the UCLA-NSAC autopsy research report. *The American Journal of Psychiatry, 143*, 862–866.

Rizzolatti, G., Umilta, C., and Berlucchi, G. (1971). Opposite superiorities of the right and left cerebral hemispheres in discriminative reaction time to physiognomical and alphabetical material. *Brain, 94*, 431–442.

Roberts, G. W. (1991). Schizophrenia: A neuropathological perspective. *British Journal of Psychiatry, 158*, 8–17.

Robinson, D. A. (1975). Tectal oculomotor connections. *Neuroscience Research Program Bulletin, 13*, 238–244.

Robinson, D. and Munne, W. H. et al. (1995). Reduced caudate nucleus volume in obsessive-compulsive disorder. *Archives of General Psychiatry, 52*, 393–398.

Robinson, D. L., Goldberg, M. E., and Stanton, G. B. (1978). Parietal association cortex in the primate: Sensory mechanisms and behavioral modulations. *Journal of Neurophysiology, 50*, 1415–1432.

Robinson, D. L. and Wurtz, R. H. (1976). Use of an extraretinal signal by monkey superior colliculus neurons to distinguish real from self-induced stimulus movement. *Journal of Neurophysiology, 39*, 852–870.

Robinson, F. R., Phillips, J. O., and Fuchs, A. F. (1994). Coordination of gaze shifts inprimates: Brainstem inputs to neck and extraocular motoneuron pools. *Journal of Comparative Neurology, 346*, 43–62.

Robinson, R. G. (1979). Differential behavior and biochemical efifects of right and left hemispheric cerebral infarction in the rat. *Science, 205*, 707–710.

Robinson, R. G. (1985). Lateralized behavioral and neurochemical consequences of unilateral brain injury in rats. In S. Glick (Ed.), *Cerebral lateralization in nonhuman species* (pp. 135–156) New York, NY: Academic Press.

Robinson, R. G. (1995). Mapping brain activity associated with emotion. *American Journal of Psychiatry, 152*, 327–329.

Robinson, R. G. and Bloom, F. E. (1977). Pharmacological treatment following experimental cerebral infarction: Implications for understanding psychological symptoms of human stroke. *Biological Psychiatry, 12*, 669–680.

Robinson, R. G., Boston, J. D., Starkstein, S. E., and Price, T. R. (1988). Comparison of mania and depression after brain damage: Causal factors. *American Journal of Psychiatry, 145*, 142–148.

Robinson, R. G. and Downhill, J. E. (1995). Lateralization of psychopathology in response to focal brain injury. In R. J. Davidson and K. Hugdahl (Eds.), *Brain asymmetry*. Cambridge, MA: MIT Press.

Robinson, R. G., Kubos, K. L., Starr, L. B., Rao, K., and Price, T. R. (1984). Mood disorders in stroke patients: Importance of location of lesion. *Brain, 107*, 81–93.

Robinson, R. G., Shoemaker, W. J., Schlumpf, M., Valk, T., and Bloom, F. E. (1975). Effect of experimental cerebral infarction in rat brain on catecholamines and behaviour. *Nature, 255*, 332–334.

Robinson, R. G. and Stitt, T. G. (1981). Intracortical 6-hydroxydopamine induces an asymmetrical behavioral response in the rat. *Brain Research, 213*, 387–395.

Robinson, R. G. and Szetela, B. (1981). Mood change following left hemispheric brain injury. *Annals of Neurology, 9*, 447–453.

Rodin, J., Slochower, J., and Fleming, B. (1977). Effects of degree of obesity, age of onset, and weight loss on responsiveness to sensory and external stimuli. *Journal of Comparative and Physiological Psychology, 91*, 586–597.

Rodriguez, E., George, N., Lachaux, J. P., Martinerie, J., Renault, B., and Varela, F. J. (1999). Perception's Shadow: Long-distance synchronization of human brainactivity. *Nature, 397*, 430–433.

Roeyers, H., Keymeulen, H., and Buysse, A. (1998). Differentiating attention-deficit hyperactivity disorder from pervasive developmental disorder not otherwise specified. *Journal of Learning Disabilities, 31*, 565–571.

Rogers, L. J. (1980). Ipsilateral and contralateral correlations between EEG and EP principal components. *Electroencephalography and Clinical Neurophysiology, 50*, 441–448.

Rogers, L. J. (1982). Light experience and asymmetry of brain distinction in chickens. *Nature, 297*, 223–225.

Rogers, M. C., Battit, G., McPeek, B., and Todd, D. (1978). Lateralization of sympathetic control of the human sinus node: ECG changes of stellate ganglion block. *Anesthesiology, 48*, 139–141.

Roland, P. E., Eriksson. L., Widen, L., and Stone-Elander, S. (1989). Changes in regional cerebral oxidative metabolism induced by lactile learning and recognition in man. *European Journal of Neuroscience, 1*, 3–18.

Rolls, E. T., Treves, A., Robertson, R. G., Georges-Francois, P., and Panzeri, S. (1998). Information about spatial view in an ensemble of primate hippocampal cells. *Journal of Neurophysiology, 79*, 1797–1813.

Rols, G., Tallon-Baudry, C., Girard, P., Bertrand, O., and Bullier, J. (2001). Cortical mapping of gamma oscillations in areas V1 and V4 of the macaque monkey. *Visual Neuroscience, 18*, 527–540.

Ronca, A. E., Abel, R. A., and Alberts, J. R. (1996). Perinatal stimulation and adaptation of the neonate. *Pediatrica Supplement, 416*, 8–15.

Root-Bernstein, R. S. and Westall, F. C. (1990). Serotonin binding sites. II. Muramyl dipeptide binds to serotonin binding sites on myelin basic protein, LHRH, and MSH-ACTH 4-10. *Brain Research Bulletin, 25*, 827–841.

Roschmann, R., Wittling, W., and Pfluger, M. (1992). Visumotorische Langzeitlateralisierung als Methode zur Untersuchung funktionaler Hemispharenasymmetrien (abstract). In L. Montada (Ed.), *Bericht uber den 38. Kongress der deutschen Gesellschaft fur Psychologie in Trier* (Vol. 1, p. 831). Gottingen: Hogrefe.

Rose, S. P. R. (1991). How chicks make memories: The cellular cascade from e-fos to dendritic remodelling. *Trends in Neuroscience, 14*, 390–396.

Rosen, G. D., Sherman, G. F., and Galaburda, A. M. (1989). Interhemispheric connections differ between symmetrical and asymmetrical brain regions. *Neuroscience, 33*, 525–533.

Rosenberg, D. R., Benazon, N. R., Gilbert, A., Sullivan, A., and Moore, G. J. (2000). Thalamic volume in pediatric obsessive-compulsive disorder patients before and after cognitive behavioral therapy. *Biological Psychiatry, 48*, 294–300.

Rosenberg, D. R., Dick, E. L., O'Hearn, K. M., and Sweeney, J. A. (1997). Response-inhibition deficits in obsessive-compulsive disorder: An indicator of dysfunction in frontostriatal circuits. *Journal of Psychiatry and Neuroscience, 22*, 29–38.

Rosenberg, D. R., Keshavan, M. S., O'Hearn, K. M., Dick, E. L., Bagwell, W. W., Seymour, A. B., Montrose, D. M., Pierri, J. N., and Birmaher, B. (1997). Frontostriatal measurement in treatment naïve children with obsessive-compulsive disorder. *Archives of General Psychiatry, 54*, 824–830.

Rosenberger, P. B. and Hier, D. B. (1980). Cerebral asymmetry and verbal intellectial deficits. *Annals of Neurology, 8*, 300–304.

Rosenhall, U., Nordin, V., Sandstrom, M., Ahlsen, G., and Gillberg, C. (1999). Autism and hearing loss. *Journal of Autism and Developmental Disorders, 29*, 349–357.

Rosenthal-Malek, A. and Mitchell, S. (1997). Brief report: The effects of exercise on the self-stimulatory behaviors and positive responding of adolescents with autism. *Journal of Autism and Developmental Disorders, 27*, 193–202.

Rosenzweig, M. R. (1957). Etudes sur l'association des mots. *l'Année Psychologique, 57*, 1, 23–32.

Rosenzweig, M. R. and Bennett, E. L. (1996). Psychobiology of plasticity: Effects of training and experience on brain and behavior. *Behavioral and Brain Research, 78*, 57–65.

Rosenzweig, M. R., Bennett, E. L., Martinez, J. L., Jr., Colombo, P. J., Lee, D. W., and Serrano, P. A. (1992). Studying stages of memory formation with chicks. In L. R. Squire and N. Butters (Eds.), *Neuropsychology of memory*. New York: Guilford.

Rosenzweig, M. R., Krech, D., Bennet, E. L., and Diamond, M. C. (1968). Modifying brain chemistry and anatomy by enrichment or impoverishment of experience. In G. Newton and S. Levine (Eds.), *Early experience and behavior* (pp. 258–297). Springfield, IL: Charles. C. Thomas.

Rosenstock. J., Field, T. D., and Greene, E. (1977). The role of mammillary bodies in spatial memory. *Experimental Neurology, 55*, 340–352.

Rosner, J. and Rosner, J. (1987). Comparison of visual characteristics in children with and without learning difficulties. *American Journal of Optometry and Physiological Optics, 64*, 531–533.

Ross, D. L., Klykylo, W. M., and Hitzemann, R. (1987). Reduction of elevated CSF beta-endorphin by fenfluramine in infantile autism. *Pediatric Neurology, 3*, 83–86.

Ross, E. D. (1981). The aprosodias: Functional-anatomical organization of the affective components of language in the right hemisphere. *Archives of Neurology, 38*, 561–569.

Ross, E. D., Kirkpatrick, J. B., and Lastimosa, A. C. (1979). Position and vibration sensations: Functions of the dorsal spinocerebellar tracts? *Annals of Neurology, 5*, 171–176.

Ross, R. G., Hommer, D., Breiger, D., Varley, C., and Radant, A. (1994). Eye movement task related to frontal lobe functioning in children with attention deficit disorder. *Journal of the American Academy of Child and Adolescent Psychiatry, 33*, 869–874.

Rossi, A., Stratta, P., Mancini, F., de Cataldo, S., and Casacchia, M. (1993). Cerebellar vermal size in schizophrenia: A male effect. *Biological Psychiatry, 33*, 354–357.

Rossor, M., Garrett, N., and Iversen, L. (1980). No evidence for lateral asymmetry of neurotransmitters in post-mortem human brain. *Journal of Neurochemistry, 35*, 743–745.

Rothwell, J. C., Obeso, J. A., Day, B. L., and Marsden, C. D. (1983). Pathophysiology of dystonias. *Advances in Neurology, 39*, 851–863.

Rouiller, E. M., Tanne, J., Moret, V., Kermadi, I., Boussaoud, D., and Welker, E. (1998). Dual morphology and topography of the corticothalamic terminals originating from the primary, supplementary motor, and dorsal premotor cortical areas in macaque monkeys. *Journal of Comparative Neurology, 396*, 169–185.

Roux, S., Bruneau, N., Garreau, B., Guerin, P., Adrien, J. L., Dansart, P., Gomot, M., and Barthelemy, C. (1997). Bioclinical profiles of autism and other developmental disorders using a multivariate statistical approach. *Biological Psychiatry, 42*, 1148–1156.

Rowe, K. S. and Rowe, K. J. (1994). Synthetic food coloring and behavior: A dose response effect in a double-blind, placebo-controlled, repeated-measures study. *Journal of Pediatrics, 125*, 691–698.

Royce, G. J. (1978a). Autoradiographic evidence for a discontinuous projection to the caudate nucleus from the centromedian nucleus in the cat. *Brain Research, 146*, 145–150.

Royce, G. J. (1978b). Cells of origin of subcortical afferents to the caudate nucleus: A horseradish peroxidase study in the cat. *Brain Research, 153*, 465–475.

Royeen, C. B. (1985). Domain specifications of the construct tactile defensiveness. *American Journal of Occupational Therapy, 39*, 596–599.

Rubens, A. B. (1985). Caloric stimulation and unilateral visual neglect. *Neurology, 35*, 1019–1024.

Ruff, H. A. and Lawson, K. R. (1990). Development of sustained, focused attention in young children during free play. *Developmental Psychology, 26*, 85–93.

Ruffman, T., Garnham, W., and Rideout, P. (2001). Social understanding in autism: Eye gaze as a measure of core insights. *Journal of Child Psychology and Psychiatry, 42*, 1083–1094.

Rumsey, J. M. (1985). Conceptual problem-solving in highly verbal, non-retarded autistic men. *Journal of Clinical Psychology, 48*, 766–778.

Rushton, J. L., Clark, S. J., and Freed, G. L. (2000). Pediatrician and family physician prescription of selective serotonin reuptake inhibitors. *Pediatrics, 105*, E82.

Russell, G. (1970). Discussion. In W. S. Fields and W. D. Willis (Eds.), *The cerebellum in health and disease* (p. 409). St. Louis MO: Warren H. Green.

Rutherford, O. M. and Jones, D. A. (1988). Contractile properties and fatigability of the human abductor pollicis and first dorsal interosseus: A comparison of the effects of two chronic stimulation patterns. *Journal of Neurological Science, 85*, 319–331.

Rutherford, M. D., Baron-Cohen, S., and Wheelwright, S. (2002). Reading the mind in the voice: A study with normal adults and adults with Asperger syndrome and high functioning autism. *Journal of Autism and Developmental Disorders, 32*, 189–194.

Ryding, E., Decety, J., Sjoholm, H., Stenberg, G., and Ingvar, D. H. (1993). Motor imagery activates the cerebellum regionally. A SPECT rCBF study with 99mTc-HMPAO. *Brain Research. Cognitive Brain Research, 1*, 94–99.

Sacheim, H. A., Greenberg, M. S., Weiman, A. L., Gur, R. C., Hungerbuhler, J. P., and Geschwind, M. (1982). Hemispheric asymmetry in the expression of positive and negative emotions: Neurologic evidence. *Archives of Neurology, 39*, 210–218.

Sacks, O. (1973). *Awakenings*. London, UK: Duckworth.

Sacks, O. (1995). *An anthropologist on Mars: Seven paradoxical tales*. New York, NY: Knopf.

Safer, M. A. and Leventhal, H. (1977). Ear differences in evaluating emotional tone of voice and verbal content.

Journal of Experimental Psychology: Human Perception and Performance, 3, 75–82.

Saint-Cyr, J. A., Taylor, A. E., and Nicholson, K. (1995). Behavior and the basal ganglia. *Advances in Neurology, 65*, 1–28.

Saitoh, O. and Courchesne, E. (1998). Magnetic resonance imaging study of the brain in autism. *Psychiatry and Clinical Neurosciences, 52* (Suppl.), S219–S222.

Sakai, K., Touret, M., Salvert, D., Leger, L., and Jouvet, M. (1977). Afferent projections to the cat locus coeruleus as visualized by the horseradish peroxidase technique. *Brain Research, 119*, 21–41.

Sakai, S. T., Inase, M., and Tanji, J. (2002). The relationship between MI and SMA afferents and cerebellar and pallidal efferents in the macaque monkey. *Somatosensory and Motor Research, 19*, 139–148.

Sakamoto, T. (1994). Psycho-educational approach for hyperactive children. *No To Hattatsu, 26*, 175–181.

Sakanaka, M., Shibasaki, T., and Lederis, K. (1986). Distribution and efferent projections of corticotropin-releasing factor-like immunoreactivity in the rat amygdaloid complex. *Brain Research, 382*, 213–238.

Salloway, S. and Cummings, J. (1994). Subcortical disease and neuropsychiatric illness. *Journal of Neuropsychiatry and Clinical Neuroscience, 6*, 93–99.

Salmon, D. P. and Butters, N. (1995). Neurobiology of skill and habit learning. *Current Opinion in Neurobiology, 5*, 184–190.

Sanders, A. F. (1983). Towards a Model of Stress and Human Performance. *Acta Psychologica, 53*, 61–97.

Sanders, C. E., Field, T. M., Diego, M., and Kaplan, M. (2000). The relationship of Internet use to depression and social isolation among adolescents. *Adolescence, 35*, 237–242.

Sandson, J. and Albert, M. L. (1984). Varieties of perseveration. *Neuropsychologia, 22*, 715–732.

Sandler, H. (1980). Effects of bedrest and weightlessness on the heart. In GH1 Bourne (Ed.), *Hearts and hearts–like organs* (Vol. 2, pp. 345–424). New York: Academic Press.

Sandman, C. A. (1988). B-endorphin disregulation in autistic and self-injurious behavior: A neurodevelopmental hypothesis. *Synapse, 2*, 193–199.

Sandor, P., Musisi, S., and Moldofsky, H. et al. (1990). Tourette's syndrome: A follow-up study. *Journal of Clinical Psychopharmacology, 10*, 197–199.

Sandson, J. and Albert, M. L. (1984). Varieties of perseveration. *Neuropsychologia, 22*, 715–732.

Sandyk, R., Kay, S. R., and Merriam, A. E. (1991). Atrophy of the cerebellar vermis: Relevance to the symptoms of schizophrenia. *International Journal of Neuroscience, 57*, 205–212.

Sanford, R. A. and Andy, O. J. (1969). Brain tumors. *Progress in Neurology and Psychiatry, 24*, 320–331.

Santini, M. and Ibata, Y. (1971). The fine structure of thin myelinated axons within muscle spindles. *Behavioural Brain Research, 33*, 289–302.

Saper, C. B., Loewy, A. D., Swanson, L. W., and Cowan, W. M. (1976). Direct hypothalamoautonomic connections. *Brain Research, 117*, 305–312.

Saper, C. D. (1987). Diffuse cortical projection system: Anatomical organization and role in cortical function. In F. Plum (Ed.), *Handbook of physiology, Section 1, The nervous system, higher functions of the brain* (Vol. V, Bethesda, MD, pp. 169–210): American Physiological Society.

Sarmiere, P. D. and Bamburg, J. R. (2002). Head, neck, and spines: A role for LIMK-1 in the hippocampus. *Neuron, 35*, 3–5.

Satoh, M., Takeda, K., Nagata, K., Hatazawa, J., and Kuzuhara, S. (2001). Activated brain regions in musicians during an ensemble: A PET study. *Brain research. Cognitive Brain Reseach, 12*, 101–108.

Satz, P. (1990). Developmental dyslexia: An etiological reformulation. In G. Pavlidis (Ed.), *Perspectives on dyslexia. Neurology, neuropsychology and genetics* (Vol. I, pp. 3–26). New York, NY: John Wiley & Sons.

Sava, D., Liotti, M., and Rizzolatti, G. (1988). Right hemisphere superiority for programming oculomotion: Evidence from simple reaction time experiments. *Neuropsychologia, 26*, 201–211.

Savage-Rumbaugh, E. S., Murphy, J., Sevcik, R. A., Brake, K. E., Williams, S. L., and Rumbaugh, D. M. (1993). Language comprehension in ape and child. *Monographs of the Society for Research in Child Development, 58*, 1–222.

Saxton, D. (1978). The behavior of infants whose mothers smoke in pregnancy. *Early Human Development, 2*, 363–369.

Schacter, D. L. (1987). Implicit memory: History and current status. *Journal of Experimental Psychology: Learning, Memory, and Cognition, 13*, 501–518.

Schaffer, C. E., Davidson, R. J., and Saron, C. (1983). Frontal and parietal electroencephalogram asymmetry in depressed and nondepressed subjects. *Biological Psychiatry, 18*, 753–762.

Scheibel, M. E. and Scheibel, A. B. (1966). The organization of the nucleus reticularis thalami: A golgi study. *Brain Research, 1*, 43–62.

Schiffer, F. (1997). Affect changes observed with right versus left lateral visual field stimulation in psychotherapy patients: Possible physiological, psychological, and therapeutic implications. *Comprehensive Psychiatry, 38*, 289–295.

Schiffer, F. (1998). Of two minds: The revolutionary science of dual-brain psychology. New York, NY: Free Press.

Schiffer, F., Zaidel, E., Bogen, J., and Chasan-Taber, S. (1998). Different psychological status in the two hemispheres of two split-brain patients. *Neuropsychiatry, neuropsychology, and behavioral neurology, 11*, 151–156.

Schildkraut, J. (1965). The catecholamine hypothesis on affective disorders: A review of supporting evidence. *American Journal of Psychiatry, 122*, 509–522.

Schlaug, G. (2001). The brain of musicians. A model for functional and structural adaptation. *Annals of the New York Academy of Sciences, 930*, 281–299.

Schlaug, G., Jancke, L., Huang, Y., Staiger, J. F., and Steinmetz, H. (1995). Increased corpus callosum size in musicians. *Neuropsychologia, 33*, 1047–1055.

Schmahmann, J. D. (1991). An emerging concept: The cerebellar contribution to higher function. *Archives of Neurology, 48,* 1178–1187.

Schmahmann, J. D. (1997). Rediscovery of an early concept. *International Review in Neurobiology, 41,* 3–27.

Schmahmann, J. D. and Pandya, D. N. (1990). Anatomical investigation and projections from thalamus to posterior parietal cortex in rhesus monkey: WGA-HRP and fluorescent tracer study. *The Journal of Comparative Neurology, 295,* 299–326.

Schmahmann, J. D. and Sherman, J. C. (1998). The cerebellar cognitive affective syndrome. *Brain, 121,* 561–579.

Schmid-Burgk, W., Becker, W., Diekmann, V., Jurgens, R., and Kornhuber, H. H. (1982). Disturbed smooth pursuit and saccadic eye movements in schizophrenia. *Archiv fur Psychiatrie und Nervenkrankheiten, 232,* 381–389.

Schmidt, R. F. and Thews, G. (Eds.) (1989). *Humanphysiology.* Berlin: Springer.

Schmidt, S. L., Manhaes, A. C., and DeMoraes, V. Z. (1991). The effects of total and partial callosal agenesis on the development of paw preference performance in the BALB/cCF mouse. *Brain Research, 545,* 175–182.

Schneider, L. H., Murphy, R. B., and Coons, E. E. (1982). Lateralization of striatal dopamine (D2) receptors in normal rats. *Neuroscience Letters, 33,* 281–284.

Schuler, A. L. (1995). Thinking in autism: Differences in learning and development, In K. A. Quill (Ed.) *Teaching Children with Autism* (pp. 11–32). Philadelphia, PA: Thomson Learning.

Schwartz, A. S., Marchok, P. L., Kreinick, C. J., and Flynn, R. E. (1979). The asymmetric lateralization of tactile extinction in patients with unilateral cerebral dysfunction. *Brain, 102,* 669–684.

Schwartz, J. M. (1998). Neuroanatomical aspects of cognitive-behavioural therapy response in obsessive-compulsive disorder. An evolving perspective on brain and behaviour. *British Journal of Psychiatry, 35* (Suppl.), 38–44.

Schwartz, P. (1984). Sympathetic imbalance and cardiac arrhythmias. In W. Randall (Ed.), *Nervous control of cardiovascular function.* New York, NY: Oxford University Press.

Schwartz, P. J., Montemerio, M., Facchini, J., Salice, P., Rosti, D., Poggio, G., and Giorgetti, R. (1982). The QT interval throughout the first six months of life: A prospective study. *Circulation, 66,* 496–501.

Schwartz, P. J. and Priori, S. G. (1990). Sympathetic nervous system and cardiac arrhythmias. In D. P. Zipes and J. Jalife (Eds.), *Cardiac electrophysiology.* Philadelphia, PA: W.B. Saunders.

Schwartz, P. J., Stramba-Badiale, M., Segantini, A., Austoni, P., Bosi, G., Giorgetti, R., Grancini, F., Marni, E. D., Perticone, F., Rosti, D., and Salice, P. (1998). Prolongation of the QT interval and the sudden infant death syndrome. *New England Journal of Medicine, 338,* 1709–1714.

Schweigart, G. and Eysel, U. T. (2002). Activity-dependent receptive field changes in the surround of adult cat visual cortex lesions. *European Journal of Neuroscience, 15,* 1585–1596.

Scifo, R., Batticane, N., and Quattropani, M. C. (1991). A double-blind trial with naltrexone in autism. *Brain Dysfunction, 4,* 310–317.

Scifo, R., Cioni, M., Nicolosi, A., Batticane, N., Tirolo, C., Testa, N., Quattropani, M. C., Morale, M. C., Gallo, F., and Marchetti, B. (1996). Opioid-immune interactions in autism: Behavioural and immunological assessment during a double-blind treatment with naltrexone. *Annali dell'Istituto Superiore di Sanita, 32,* 352–359.

Scott, R. B., Stoodley, C. J., Anslow, P., Paul, C., Stein, J. F., Sugden, E. M., and Mitchell, C. D. (2001). Lateralized cognitive deficits in children following cerebellar lesions. *Developmental Medicine and Child Neurology, 43,* 685–691.

Sears, L. L., Vest, C., Mohamed, S., Bailey, J., Ranson, B. J., and Piven, J. (1999). An MRI study of the basal ganglia in autism. *Progress in Neuro-Psychopharmacology & Biological Psychiatry, 23,* 613–624.

Segal, M. (1977). Afferents to the entorhinal cortex of the rat studied by the method of retrograde transport of horseradish peroxidase. *Experimental Neurology, 57,* 750–765.

Segal, M. (1979). A potent inhibitory monosynaptic hypothalamo-hippocampal connection. *Brain Research, 162,* 137–141.

Segal, M. and Landis, S. (1974). Afferents to the hippocampus of the rat studied with the method of retrograde transport of horseradish peroxidase. *Brain Research, 78,* 1–15.

Segalowitz, S. J. and Chapman, J. S. (1980). Cerebral asymmetry for speech in neonates: A behavioral measure. *Brain and Language, 9,* 281–288.

Segalowitz, S. J., Wagner, W. J., and Menna, R. (1992). Lateral versus frontal ERP predictors of reading skill. *Brain and Cognition, 20,* 85–103.

Segalowitz, S. L. and Bryden, M. P. (1983). Individual differences in hemispheric representation of language. In S. J. Segalowitz (Ed.), *Language functions and brain organization* (pp. 341–372). New York, NY: Academic Press.

Seidman, L. J., Benedict, K., and Biederman, J. et al. (1995a). Performance of ADHD children on the Rey-Osterrieth Complex figure: A pilot neuropsychological study. *Journal of Child Psychology Psychiatry, 36,* 1459–1473.

Seidman, L. J., Biederman, J., and Faraone, S. et al. (1995b). Effects of family history and comorbidity on the neuropsychological performance of ADHD children: Preliminary findings. *Journal of American Academic Children and Adolescence Psychiatry, 34,* 1015–1024.

Seidman, L. J., Biederman, J., and Faone, S. V. et al. (1997a). A pilot study of neuropsychological function in girls with ADHD. *Journal of American Academic Child and Adolescence Psychiatry, 36,* 366–373.

Seidman, L. J., Biederman, J., and Faraone, S. V. et al. (1997b). Toward defining a neuropsychology of attention deficit hyperactivity disorder: Performance of children and adolescents from a large clinically referred sample. *Journal of Consult Clinical Physiology*, 65, 150–160.

Seidman, L. J., Biederman, J., and Weber, W. et al. (1998). Neurological function in adults with attention deficit hyperactivity disorder. *Biological Psychiatry*, 44, 260–268.

Seifert, I. (1974). Functional aspects of C-scoliosis in infants. *Beitrage zur Orthopadie und Traumatologie*, 21, 265–271.

Seim, A. R. and Reichelt, K. L. (1995). An enzyme/brain-barrier theory of psychiatric pathogenesis: Unifying observations on phenylketonuria, autism, schizophrenia, and postpartum psychosis. *Medical Hypotheses*, 45, 498–502.

Seitz, R. J. and Roland, P. E. (1992). Learning of sequential finger movements in man: A combined kinematic and positron emission tomography (PET) study. *European Journal of Neuroscience*, 4, 154–165.

Selemon, L. D., Rajkowska, G., and Goldman-Rakic, P. S. (1995). Abnormally high neuronal density in the schizophrenic cortex. A morphometric analysis of prefrontal area 9 and occipital area 17. *Archives of General Psychiatry*, 52, 805–818, discussion 819–820.

Semmes, J. (1968). Hemispheric specialisation: A possible cue to mechanism. *Neuropsychologia*, 6, 11–25.

Semrud-Clikeman, M. and Hynd, G. W. (1990). Right hemispheric dysfunction in nonverbal learning disabilities: Social, academic, and adaptive functioning in adults and children. *Psychological Bulletin*, 107, 196–209.

Serdula, M. K., Ivery, D., Coates, R. J., Freedman, D. S., Williamson, D. F., and Byers, T. (1993). Do obese children become obese adults? *Preventive Medicine*, 22, 167–177.

Sergent, J. (1987a). Information processing and laterality effects for object and face perception. In G. W. Humphreys and M. S. Riddoch (Eds.), *Visual object processing: A cognitive neuropsychological approach* (pp. 145–173). Hillsdale, NJ: Lawrence Erlbaum Associates.

Sergent, J. (1987b). Failures to confirm the spatial-frequency hypothesis: Fatal blow or healthy complication? *Canadian Journal of Psychology*, 41, 412–428.

Sergent, J. (1989a). Structural processing of faces. In A. W. Young and H. D. Ellis (Eds.), *Handbook of face processing*. Amsterdam: Elsevier.

Sergent, J. (1995). *Brain asymmetry*. Cambridge: MIT Press.

Sergent, J. and Hellige, J. B. (1986). Role of input factors in visual-field asymmetries. *Brain and Cognition*, 5, 174–199.

Sergent, J., Ohta, S., and MacDonald, B. (1992a). Functional neuroanatomy of face and object processing: A PET study. *Brain*, 115, 15–29.

Sergent, J. and Poncet, M. (1990). From covert to overt recognition of faces in prosopagnosic patient. *Brain*, 113, 989–1004.

Sergent, J. and Signoret, J. L. (1992a). Functional and anatomical decomposition of face and processing: Evidence from prosopagnosia and PET study of normal subjects. *Philosophical Transactions of the Royal Society of London. Series B: Biological Sciences*, 335, 55–62.

Sergent, J. and Signoret, J. L. (1992b). Varieties of functional deficits in prosopagnosia. *Cerebral Cortex*, 2, 375–388.

Sergeant, J. (2000). The cognitive–energetic model: An empirical approach to attention-deficit hyperactivity disorder. *Neuroscience and Biobehavioral Reviews*, 24, 7–12.

Sergeant, J. A. and Van der Meere, J. (1990). Additive factor method applied to psychopathology with special reference to childhood hyperactivity. *Acta Psychologica*, 74, 377–395.

Servan-Schreiber, D., Bruno, R. M., Carter, C. S., and Cohen, J. D. (1998). Dopamine and the mechanisms of cognition: Part I, A neural network model predicting dopamine effects on selective attention. *Biological Psychiatry*, 43, 713–722.

Sexton, M, Fox, N., and Hebel, J. (1990). Prenatal exposure to tobacco. II. Effects on cognitive functioning at age three. *International Journal of Epidemiology*, 19, 72–77.

Shallice, T. (1982). Specific impairments of planning. Philosophical transactions of the Royal Society of London. *Series A: Mathematical and Physical Sciences*, 298, 199–209.

Shan, X., Aw, T. Y., Smith, E. R., Ingelman-Sundberg, M., Mannervik, B., Iyanagi, T., and Jones, D. P. (1992). Effect of chronic hypoxia on detoxication enzymes in rat liver. *Biochemical Pharmacolology*, 43, 2421–2426.

Shapiro, A. K., Shapiro, E. S., and Young, J. G. et al. (1988). *Gilles de la Tourette Syndrome*. New York: Raven.

Shapiro, E., Shapiro, A. K., and Fulpo, G. et al. (1989). Controlled study of haloperidol, pimozide, and placebo for the treatment of Gilles de la Tourette's Syndrome. *Archives of General Psychiatry*, 46, 722–730.

Shapiro, R. M., Glick, S. D., and Hough, L. B. (1986). Striatal dopamine uptake asymmetries and rotational behavior in unlesioned rats: Revising the model? *Psychopharmacology*, 89, 25–30.

Shapovalova, K. B. (1991). The role of central neurochemical mechanisms in regulation of posture adjustment and voluntary movement components in the dogs. *Physiologist*, 34 (Suppl.), S110–S113.

Shattock, P., Kennedy, A., Rowell, F., and Berney, T. (1990). Role of neuropeptides in autism and their relationships with classical neurotransmitters. *Brain Dysfunction*, 3, 328–345.

Shaywitz, S. E. and Shaywitz, B. A. (1984). Diagnosis and management of attention deficit disorder: A pediatric perspective. *Pediatric Clinics of North America*, 31, 429–457.

Shaywitz, S. E., Shaywitz, B. A., Pugh, K. R., Fulbright, R. K., Constable, R. T., Menel, W. E., Shankweiler, D. P., Liberman, A. M., Skudlarski, P., Fletcher, J. M., Katz, L., Marchione, K. E., Lacadie, C., Gatenby, C., and Gore, J. C. (1998). Function disruption in the organization of the brain for reading in dyslexia. *Proceedings of the National Academy of Sciences, USA, 95*, 2636–2641.

Shaw, W. (1997). *Biological treatments for Autism and PDD.* Overland Park, KS: Sunflower Press.

Shepherd, R. B. (2001). Exercise and training to optimize functional motor performance in stroke: Driving neural reorganization? *Neural Plasticity, 8*, 121–129.

Sherman, G. F., Garbanti, J. A., Rosen, G. D., Yutzey, D. A. and Denenberg, V. H. (1980). Brain and behavioral asymmetries for spatial preferences in rates. *Brain Research, 192*, 61–67.

Shi, L., Fatemi, S. H., Sidwell, R. W., and Patterson, P. H. (2003). Maternal influenza infection causes marked behavioral and pharmacological changes in the offspring. *Journal of Neuroscience, 23*, 297–302.

Shigeta, M., Nishikawa, Y., Shimizu, M., Usui, M., Hyoki, K., and Kawamuro, Y. (1993). Horizontal component of electro-oculogram as a parameter of arousal in dementia: Relationship between intellectual improvement and increasing arousal under pharmacotherapy. *Journal of Clinical Pharmacology, 33*, 741–746.

Shimizu, N., Ohnishi, S., Tohyama, M., and Maeda, T. (1974). Demonstration by degeneration silver methods of the ascending projection from the locus coeruleus. *Experimental Brain Research, 21*, 181–192.

Shin, L. M., Dougherty, D. D., Orr, S. P., Pitman, R. K., Lasko, M., Macklin, M. L., Alpert, N. M., Fischman, A. J., and Rauch, S. L. (2000). Activation of anterior paralimbic structures during guilt-related script-driven imagery. *Biological Psychiatry, 48*, 43–50.

Shute, C. C. D. and Lewis, P. R. (1967). The ascending cholinergic reticular system: Neocortical, olfactory and subcortical projections. *Brain, 90*, 497–520.

Siqueland, E. R. and Lipsitt, L. P. (1966). Conditioned head turning in human newborns. *Journal of Experimental Child Psychology, 4*, 356–357.

Shouse, M. N. and Sterman, M. B. (1982). Acute sleep deprivation reduces amygdala-kindled seizure thresholds in cats. *Experimental Neurology, 78*, 716–727.

Sidtis, J. J. (1980). On the nature of the cortical function underlying right hemisphere auditory perception. *Neuropsychologia, 18*, 321–330.

Sikes, R. W., Chronister, R. B., and White, L. E., Jr. (1977). Origin of the direct hippocampus anterior thalamic bundle in the rat: A combined horseradish peroxidase golgi analysis. *Experimental Neurology, 57*, 379–395.

Silberman, E. K. and Weingartner, H. (1986). Hemispheric lateralization of functions related to emotion. *Brain and Cognition, 5*, 322–353.

Simeon, J. G., Ferguson, H. B., and Van Wyck Fleet, J. (1986). Bupropion effects in attention deficit and conduct disorders. *Canadian Journal of Psychiatry, 31*, 581–585.

Singer, H. S., Reiss, A. L., Brown, J. E., Aylward, E. H., Shih, B., Chee, E., Harris, E. L., Reader, M. J., Chase, G. A., Bryan, R. N. et al. (1993). Volumetric MRI changes in basal ganglia of children with Tourette's syndrome. *Neurology, 43*, 950–956.

Singer, W. (1983). Neuronal mechanisms of experience-dependent self-organization of the mammalian visual cortex. *Acta Morphologica Hungarica, 31*, 235–259.

Singh, V. K. (1996). Plasma increase of interleukin-12 and interferon-gamma. Pathological significance in autism. *Journal of Neuroimmunology, 66*, 143–145.

Singh, V. K., Warren, R. P., Odell, J. D., and Cole, P. (1991). Changes of soluble interleukin-2, interleukin-2 receptor, T8 antigen, and interleukin-1 in the serum of autistic children. *Clinical Immunology Immunopathology, 61*, 448–455.

Singh, V. K., Warren, R. P., Odell, J. D., Warren, W. L., and Cole, P. (1993). Antibodies to myelin basic protein in children with autistic behavior. *Brain, Behavior, and Immunity, 7*, 97–103.

Sinha, U. K., Hollen, K. M., Rodriguez, R., and Miller, C. A. (1993). Auditory system degeneration in Alzheimer's disease. *Neurology, 43*, 779–785.

Skinner, J. E. and Reed, J. C. (1981). Blockade of frontocortical-brain stem pathway prevents ventricular fibrillation of ischemic heart. *American Journal of Physiology, 249*, H156–H163.

Sklar, L. S. and Anisman, H. (1981). Stress and cancer. *Psychological Bulletin, 89*, 369–406.

Slotkin, T. A., Lappi, S. E., and Seidler, F. J. (1993). Impact of fetal nicotine exposure on development of rat brain regions: Critical sensitive periods or effects of withdrawal? *Brain Research Bulletin, 31*, 319–328.

Small, S. L., Hlustik, P., Noll, D. C., Genovese, C., and Solodkin, A. (2002). Cerebellar hemispheric activation ipsilateral to the paretic hand correlates with functional recovery after stroke. *Brain, 125*, 1544–1557.

Smalley, S. L., McGough, J. J., Del'Homme, M., NewDelman, J., Gordon, E., Kim, T., Liu, A., and McCracken, J. T. (2000). Familial clustering of symptoms and disruptive behaviors in multiplex families with attention-deficit/hyperactivity disorder. *Journal of the American Academy of Child and Adolescent Psychiatry, 39*, 1135–1143.

Smith, M. A., Kim, S. Y., van Oers, H. J., and Levine, S. (1997). Maternal deprivation and stress induce immediate early genes in the infant rat brain. *Endocrinology, 138*, 4622–4628.

Smith, O. C. (1934). Action potentials from single motor units in voluntary contraction. *American Journal of Physiology, 108*, 629–638.

Snider, R. S. (1967). Functional alterations of cerebral sensory areas by the cerebellum. In C. A. Fox and R. S. Snider (Eds.), *The Cerebellum* (pp. 322–333). Amsterdam: Elsevier.

Snider, R. S. (1975). Cerebellar-ceruleus pathway. *Brain Research, 88*, 59–63.

Snider, R. S. and Snider, S. R. (1979). Commentary: Cerebellar lesions and psychiatric disorders. *Journal of Nervous and Mental Disease, 167*, 760–761.

Snider, R. S. and Maiti, A. (1976). Cerebellar contributions to the papez circuit. *Journal of Neuroscience Research, 2*, 133–146.

Snider, R. S., Maiti, A., and Snider, S. R. (1976). Cerebellar pathways to ventral midbrain and nigra. *Experimental Neurology, 53*, 714–728.

Snyder, J. S., Kee, N., and Wojowicz, J. M. (2001). Effects of adult neurigensis on synaptic-plasticity in the rat denate gyrus. *Journal of Neurophysiology, 85,* 2423–2431.

Sobotka, S. S., Davidson, R. J., and Senulis, J. A. (1992). Anterior brain electrical asymmetries in response to reward and punishment. *Electroencephalography and Clinical Neurophysiology, 83,* 236–247.

Soll, D. R. (1983). A new method for examining the complexity and relationships of "timers" in developing systems. *Developmental Biology, 95,* 73–91.

Solomon, P. R., Stowe, G. T., and Pendlbeury, W. W. (1989). Disrupted eyelid conditioning in a patient with damage to cerebellar afferents. *Behavioral Neuroscience, 103,* 898–902.

Sommer, M., Lang, N., Tergau, F., and Paulus, W. (2002). Neuronal tissue polarization induced by repetitive transcranial magnetic stimulation? *Neuroreport, 13,* 809–811.

Sommer, M. A. and Wurtz, R. H. (2002). A pathway in primate brain for internal monitoring of movements. *Science, 296,* 1480–1482.

Song, L. Y., Singer, M. I., and Anglin, T. M. (1998). Violence exposure and emotional trauma as contributions to adolescents' violent behaviors. *Archives of Pediatrics & Adolescent Medicine, 152,* 531–536.

Soussignan, R., Schaal, B., Schmit, G., and Nadel, J. (1995). Facial responsiveness to odours in normal and pervasively developmentally disordered children. *Chemical Senses, 20,* 47–459.

Spencer, T., Biederman, J., Harding, M., O'Donnell, D., Wilens, T., Faraone, S., Coffey, B., and Geller, D. (1998). Disentangling the overlap between Tourette's disorder and ADHD. *Journal of Child Psychology and Psychiatry, 39,* 1037–1044.

Spencer, T., Biederman, J., and Wilkens, T. (1996). Pharmacotherapy of attention deficit hyperactivity disorder across the cycle. *Journal American Academic Child Adolescent Psychiatry, 35,* 409–432.

Sperry, R. (1962). Some general aspects of interhemispheric integration. In V. B. Mountcastle (Ed.), *Interhemispheric relations and cerebral dominance* (pp. 43–49). Baltimore, MD: Johns Hopkins Press.

Spiegel, D., Cutcomb, S., Ren, C., and Pribram, K. (1985). Hypnotic hallucination alters evoked potentials. *Journal of Abnormal Psychology, 94,* 249–255.

Spitzer, R. L. Endicott, J., and Robins, E. (1978). Research diagnostic criteria: Rationale and reliability. *Archives of General Psychiatry, 35,* 773–782.

Spoont, M. R. (1992). Modulatory role of serotonin in neural information processing: Implications for human psychopathology. *Psychological Bulletin, 112,* 330–350.

Spoor, F., Wood., B., and Zonneveld, F. (1994). Implications of early hominid labyrinthine morphology for evolution of human bipedal locomotion. *Nature, 369,* 645–648.

Spoor, F. and Zonneveld, F. (1998). Comparative review of the human bony labyrinth. *American Journal of Physical Anthropology, 27* (Suppl.), 211–251.

Spreen, O. and Strauss, E. (1991). *A compendium of neuropsychological tests.* New York, NY: Oxford University Press.

Spurzheim, G. (1818). *Observations sur la folie ou Sur les dérangements des fonctions morales et intellectuelles de l'homme.* Paris, Treuttel et Würtz. http://pages.britishlibrary.net/phrenology2/spurzheim_ observations_ folie.rtf

Squire, L. (1991). Memory: Organization and locus of change. New York, NY: Oxford.

Starkstein, S. E., Fedoroff, P., Berthier, M. L., and Robinson, R. G. (1991). Manic-depressive and pure manic states after brain lesions. *Biological Psychiatry, 29,* 149–158.

Starkstein, S. E., Mayberg, H. S., Preziosi, T. J., Andrezejewski, P, Leiguarta, R., and Robinson, R. G. (1992). Reliability validity and clinical correlates of apathy in Parkinson's disease. *The Journal of Neuropsychiatry and Clinical Neurosciences, 4,* 134–139.

Starkstein, S. E., Petracca, G., Chemerinski, E., Teson, A., Sabe, L., Merello, M., and Leiguarda, R. (1998). Depression in classic versus akinetic-rigid Parkinson's disease. *Movement Disorders, 13,* 29–33.

Starkstein, S. E., Preziosi, T. J., Bolduc, P. L., and Robinson, R. G. (1990). Depression in Parkinson's disease. *Journal of Nervous and Mental Disase, 178,* 27–31.

Starkstein, S. E., Preziosi, T. J., Forrester, A. W., and Robinson, R. G. (1990). Specificity of affective and autonomic symptoms of depression in Parkinson's disease. *Journal of Neurology, Neurosurgery, and Psychiatry, 53,* 869–873.

Starkstein, S. E. and Robinson, R. T. (1988). Comparisons of patients with and without post-stroke major depression matched for size and location of lesion. *Archives of General Psychiatry, 45,* 247–252.

Starkstein, S. E. and Robinson, R. G. (1989). Affective disorders and cerebral vascular disease. *British Journal of Psychiatry, 154,* 170–182.

Starkstein, S. E., Robinson, R. T., and Price, T. R. (1987). Comparison of cortical and subcortical lesions in the production of post-stroke mood disorders. *Brain, 110,* 1045–1059.

Starobrat-Hermelin, B. and Kozielec, T. (1997). The effects of magnesium physiological supplementation on hyperactivity in children with attention deficit hyperactivity disorder (ADHD). Positive response to magnesium oral loading test. *Magnesium Research: Official Organ of the International Society for the Development of Research on Magnesium, 10,* 149–156.

State, M. W., Pauls, D. L., and Lechman, J. F. (2001). Tourette's syndrome and related disorders. *Child and Adolescent Psychiatry Clinics of North America, 10,* 317–331.

Steinbeck, K. S. (2001). The importance of physical activity in the prevention of overweight and obesity in childhood: A review and an opinion. *Obesity Reviews: An Official Journal of the International Association for the Study of Obesity, 2,* 117–130.

Stein, D. J. (1996). The neurobiology of obsessive-compulsive disorder, 2, 300–305.

Stephan, H. (1975). In W. Bargman (Ed.), *Aliocortex. Handbuch der mikroskopischen Anatomie des Menschen* (Vol. 4) Nervensyslem, Part 9. Berlin: Springer-Verlag.

Steriade, M. (1998). Corticothalamic networks, oscillations, and plasticity. *Advances in Neurology*, 77, 105–134.

Steriade, M. (2000). Corticothalamic resonance, states of vigilance and mentation. *Neuroscience*, 101, 243–276.

Steriade, M. (2001). Impact of network activities on neuronal properties in corticothalamic systems. *Journal of Neurophysiology*, 86, 1–39.

Steriade, M., Jones, E. G., and Llinas, R. R. (1990). *Thalamic oscillations and signaling* (pp. 50–55). New York: Wiley.

Steriade, M., Sakai, K., and Jouvet, M. (1984). Bulbothalamic neurons related to thalamocortical activation processes during paradoxical sleep. *Experimental Brain Research*, 54, 463–475.

Steriade, M., Timofeev, I., Grenier, F., and Durmuller, N. (1998). Role of thalamic and cortical neurons in augmenting responses and self-sustained activity: Dual intracellular recordings in vivo. *Journal of Neuroscience*, 18, 6425–6443.

Sterman, M. B. (1976). Effects of brain surgery and EEG operant conditioning on seizure latency following monomethylhydrazine intoxication in the cat. *Experimental Neurology*, 50, 757–765.

Sterman, M. B. (1977). Sensorimotor EEG operant conditioning: Experimental and clinical effects. *The Pavlovian Journal of Biological Science*, 12, 63–92.

Sterman, M. B., Howe, R. C., and Macdonald, L. R. (1970, February 20). Facilitation of spindle-burst sleep by conditioning of electroencephalographic activity while awake. *Science*, 167, 1146–1148.

Sterman, M. B. and Wyrwicka, W. (1967). EEG correlates of sleep: Evidence for separate forebrain substrates. *Brain Research*, 6, 143–163.

Stern, K. (1942). Thalamo-frontal projection in Man, *Journal of anatomy*, 76, 302–307.

Stevens, J. R. (1982). Neuropathology of schizophrenia. *Archives of General Psychiatry*, 39, 1131–1139.

Stevens, J. R., Sachdev, K., and Milstein, V. (1968). Behavior disorders of childhood and the electroencephalogram. *Archives of Neurology*, 18, 160–177.

Stevens, L. J., Zentall, S. S., Deck, J. L., Abate, M. L., Watkins, B. A., Lipp, S. R., and Burgess, J. R. (1995). Essential fatty acid metabolism in boys with attention-deficit hyperactivity disorder. *American Journal of Clinical Nutrition*, 62, 761–768.

Stevenson, C. S., Whitmont, S., Bornholt, L., Livesey, D., and Stevenson, R. J. (2002). A cognitive remediation programme for adults with attention deficit hyperactivity disorder. *The Australian and New Zealand Journal of Psychiatry*, 36, 610–616.

Steward, O. (1976). Topographic organization of the projections from the entorhinal area to the hippocampal formation of the rat. *The Journal of Comparative Neurology*, 167, 285–314.

Stokes, A., Bawden, H. N., Camfield, P. R., Backman, J. E., and Dooley, J. M. (1991). Peer problems in tourette's disorder. *Pediatrics*, 87, 936–942.

Stordy, B. J. (2000). Dark adaptation, motor skills, docosahexaenoic acid, and dyslexia. *The American Journal of Clinical Nutrition*, 71 (Suppl.), 323S–326S.

Story, M. T., Neumark-Stzainer, D. R., Sherwood, N. E., Holt, K., Sofka, D., Trowbridge, F. L., and Barlow, S. E. (2002). Management of child and adolescent obesity: Attitudes, barriers, skills, and training needs among health care professionals. *Pediatrics*, 110, 210–214.

Strasburger, V. C. (1992). Children, adolescents, and television. *Pediatrics in Review; American Academy of Pediatrics*, 13, 144–151.

Strassburger, T. L., Lee, H. C., Daly, E. M., Szczepanik, J., Krasuski, J. S., Mentis, M. J., Salerno, J. A., DeCarli, C., Schapiro, M. B., and Alexander, G. E. (1997). Interactive effects of age and hypertension on volumes of brain structures. *Stroke*, 28, 1410–1417.

Strauss, S. and Moscovitch, M. (1981). Perception of facial expressions. *Brain and Language*, 13, 308–332.

Stubbs, E. G., and Crawford, M. L. (1977). Depressed lymphocyte responsiveness in autistic children. *Journal of Autism and Childhood Schizophrenia*, 7, 49–55.

Studdert-Kennedy, M. and Shankweiler, D. P. (1970). Hemispheric specialization for speech perception. *Journal of the Acoustical Society of America*, 48, 579–594.

Studdert-Kennedy, M. and Shankweiler, D. (1981). Hemispheric specialization for language processes. *Science*, 211, 960–961.

Subrahmanyam, K., Kraut, R. E., Greenfield, P. M., and Gross, E. F. (2000). The impact of home computer use on children's activities and development. *Future Child*, 10, 123–144.

Sukov, W. and Barth, D. S. (2001). Cellular mechanisms of thalamically evoked gamma oscillations in auditory cortex. *Journal of Neurophysiology*, 85, 1235–1245.

Sullivan, K. and McKeever, W. (1985). *Loss of fluent speech in callosotomy: Patients who are discordant for speech and motor control dominances.* Presented at the 13th Annual Meeting of the International Neuropsychological Society, San Diego, CA.

Sun Zhongjie, J., Cade, R., Fregly, M. J., and Privette, R. M. (1999a). Beta-casomorphin induces Fos-like immunoreactivity in discrete brain regions relevant to schizophrenia and autism. *Autism*, 3, 67–84.

Sun Zhongjie, J. and Cade, R. (1999b). A peptide found in schizophrenia and autism causes behavioral changes in rats. *Autism*, 3, 85–96.

Supple, W. F. and Leaton, R. N. (1990). Cerebellar vermis: Essential for classically conditioned bradycardia in the rat. *Brain Research*, 509, 17–23.

Sussman, N. M., Gur, R. C., Gur, R. E., and O'Connor, M. J. (1983). Mutism as a consequence of callosotomy. *Journal of Neurosurgery*, 59, 514–519.

Sutherland, R. J., Whishaw, I. Q., and Kolb, B. (1988). Contributions of cingulate cortex to two forms of spatial learning and memory. *Journal of Neuroscience, 8*, 1863–1872.

Suzman, K. B., Morris, R. D., Morris, M. K., and Milan, M. A. (1997). Cognitive-behavioral remediation of problem solving deficits in children with acquired brain injury. *Journal of Behavior Therapy and Experimental Psychiatry, 28*, 203–212.

Swanson, L. W. (1978). The anatomical organization of septo-hippocampal projections. In *Functions of the septo-hippocampal system*. Ciba Foundation Symposium 58 (new series, pp. 25–43). Amsterdam: Elsevier Excerpta Medica.

Swanson, L. W. and Cowan, W. M. (1977). An autoradiographic study of the organization of the efferent connections of the hippocampal formation in the rat. *The Journal of Comparative Neurology, 72*, 49–84.

Swanson, L. W., and Mogenson, G. J. (1981). Neural mechanisms for the functional coupling of autonomic, endocrine and somatomotor responses in adaptive behavior. *Brain Research, 228*, 1–34.

Swedo, S. E., Rapopon, J. L., Cheslov, D. L. et al. (1989). High prevalence of obsessive-compulsive symptoms in patients with Sydenham's chorea. *American Journal of Psychiatry, 146*, 246–249.

Swett, J. and Eldren, E. (1960). Concussion in structure of stretch receptors in medial gastronemius and soleus muscles of the cat. *The Anatomical Record, 137*, 461–473.

Szatmari, P., Tuff, L., Finlayson, A. J., and Bartolucci, G. (1990). Asperger's syndrome and autism: Neurocognitive aspects. *Journal of the American Academy of Child and Adolescent Psychiatry, 29*, 130–136.

Szeligo, F. and Leblond, C. P. (1977). Response of the three main types of glial cells of cortex and corpus callosum in rats handled during suckling or exposed to enriched, control and impoverished environments following weaning. *Journal of Comparative Neurology, 172*, 247–263.

Tabary, J. C., Tabary, C., Tardieu, G., and Goldspink, G. (1972). Physiological and structural changes in the cat soleus muscle due to immobilization at different lengths of plaster cast. *The Journal of Physiology* (London), *224*, 231–244.

Taggart, P., Camithers, M., and Somerville, W. (1973). Electrocardiogram, plasma catecholamines and lipids, and their modifications by oxprenolol when speaking before an audience. *Lancet, 2*, 341–346.

Tanabe, H., Sawada, T., Inoue, N., Ogawa, M., Kuriyama, Y., and Shiraishi, J. (1987). Conduction aphasia and arcuate fascicus. *Acta Neurologica Scandinavica, 76*, 422–427.

Tanaka, T., Lange, H., and Naquet, R. (1975). Sleep, subcortical stimulation and kindling in the cat. *The Canadian Journal of Neurological Sciences, 2*, 447–455.

Tandon, R. and Greden, J. F. (1989). Cholinergic hyperactivity and negative schizophrenic symptoms. A model of cholinergic/dopaminergic interactions in schizophrenia. *Archives of General Psychiatry, 46*, 745–753.

Tanel, D. and Damasio, A. R. (1987). *Recognition of gender, age, and meaning of facial expression can be dissociated from recognition by prosopoagnosics*. Presented at the 39th Annual Meeting of the American Academy of Neurology, New York, NY.

Tannock, R. and Schachar, R. (1996). Executive dysfunction as an underlying mechanism of behavior and language problems in attention deficit hyperactivity disorder. In J. Beitchman, N. Cohen, M. Konstantareas and R. Tannock (Eds.), *Language, learning and behavior disorders* (pp. 128–155). Cambridge, UK: Cambridge University Press.

Tantam, D. (1991). Aspergers syndrome in adulthood. In U. Frith (Ed.), *Autism and asperger syndrome* (pp. 147–183). Cambridge, UK: Cambridge University Press.

Tantillo, M., Kesick, C. M., Hynd, G. W., and Dishman, R. K. (2002). The effects of exercise on children with attention-deficit hyperactivity disorder. *Medicine and Science in Sports and Exercise, 34*, 203–212.

Tartter, V. C. (1984). Laterality differences in speaker and consonant identification in dichotic listening. *Brain and Language, 23*, 74–85.

Tedeschi, J. and Felson, R. B. (1994). *Violence, aggression, and coercive actions*. Washington, DC: American Psychological Association.

Teitelbaum, P., Teitelbaum, O., Nye, J., Fryman, J., and Maurer, R. G. (1998). Movement analysis in infancy may be useful for early diagnosis of autism. *Proceedings of the National Academy of Sciences, USA, 95*, 13982–13987.

Temple, E. C., Hutchinson, I., Laing, D. G., and Jinks, A. L. (2002). Taste development: Differential growth rates of tongue regions in humans. *Brain Research. Developmental Brain Research, 135*, 65–70.

Thach, W. T., Jr., Goodkin, H. P., and Keating, J. G. (1992). The Cerebellum and the adaptive coordination of movement. *Annual Review of Neuroscience, 15*, 403–442.

Thapar, A., Harrington, R., Ross, K., and McGuffin, P. (2000). Does the definition of ADHD affect heritability? *Journal of the American Academy of Child and Adolescent Psychiatry, 39*, 1528–1536.

Thatch, W. T., Jr. (1980). The cerebellum. In D. D. Mount Casgle (Ed.), *Medical physiology* (14th ed., pp. 837–858). St. Louis: Mosby.

Thatcher, R. W., Krause, P. J., and Hrybyk, M. (1986). Cortico-cortical associations and EEG coherence: A two-compartmental model. *Electroencephalography and Clinical Neurophysiology, 64*, 123–143.

Thelen, E. and Smith, L. B. (1994). *A dynamic systems approach to the development of cognition and action*. Cambridge, MA: MIT Press.

Thierry, A. M., Tassin, J. P., Blanc, G., and Glowinski, J. (1976). Selective activation of the neocortical DA system by stress. *Nature, 263*, 242–244.

Thomas, E. W. (1969). *Brain-injured children*. Springfield, IL: Charles C. Thomas.

Thompson, B. M. and Andrews, S. R. (2000). An historical commentary on the physiological effects of music:

Tomatis, Mozart and neuropsychology. *Integrative Physiological and Behavioral Sciences, 35*, 174–188.

Thompson, P. M., Vidal, C., Giedd, J. N., Gochman, P., Blumenthal, J., Nicolson, R., Toga, A. W., and Rapoport, J. L. (2001). Mapping adolescent brain change reveals dynamic wave of accelerated gray matter loss in very early-onset schizophrenia. *Proceedings of the National Academy of Sciences, USA, 98*, 11650–11655.

Thompson, S. M. and Robertson, R. T. (1987). Organization of subcortical pathways for sensory projections to the limbic cortex. I. Subcortical projections to the medial limbic cortex in the rat. *Journal of Comparative Neurology, 265*, 175–188.

Thomsen, L., Robinson, T. L., Lee, J. C., Farraway, L. A., Hughes, M. J., Andrews, D. W., and Huizinga, J. D. (1998). Interstitial cells of Cajal generate a rhythmic pacemaker current. *Nature, Medicine, 4*, 845–851.

Thomsen, P. H. (2000). Obsessions: The impact and treatment of obsessive-compulsive disorder in children and adolescents. *Journal of Psychopharmacology, 14* (Suppl.), S31–S37.

Thoren, P., Asberg, M., Cronholm, B., Jornestedt, L., and Traskman, L. (1980). Clomipramine treatment of obsessive-compulsive disorder: A controlled clinical trial. *Archives of General Psychiatry, 37*, 1281–1285.

Thornton, J. M., Guz, A., Murphy, K., Griffith, A. R., Pedersen, D. L., Kardos, A., Leff, A., Adams, L., Casadei, B., and Paterson, D. J. (2001). Identification of higher brain centres that may encode the cardiorespiratory response to exercise in humans. *Journal of Physiology* (London), *533*, 8.

Thorpe, S. J., Rolls, E. T., and Maddison, S. (1983). The orbitofrontal cortex: Neuronal activity in the behaving monkey. *Experimental Brain Research, 490*, 93–115.

Tieman, D. G., McCall, M. A., and Hirsch, H. V. (1983). Physiological effects of unequal alternating monocular exposure. *Journal of Neurophysiology, 49*, 804–818.

Timor-Tritsch, I. E. (1986). The effect of external stimuli on fetal behaviour. *European Journal of Obstetrics, Gynecology, and Reproductive Biology, 21*, 321–329.

Todd R. D. (1986). Pervasive developmental disorders and immunological tolerance. *Psychiatric Development, 4*, 147–165.

Todd, J. T. and Perotti, V. J. (1999). The visual perception of surface orientation from optical motion. *Perception and Psychophysics, 61*, 1577–1589.

Tollefson, G. D., Rampey, A. H., Potvin, J. H., Jenike, M. A., Rush, A. J., Dominquez, R. A., Koran, L. M., Shear, M. K., Goodman, W., and Genduso, L. A. (1994). A multicenter investigation of fixed-dose fluoxetine in the treatment of obsessive-compulsive disorder. *Archives of General Psychiatry, 51*, 559–567.

Tomarken, A. J., Davidson, R. J., and Henriques, J. B. (1990). Resting frontal brain asymmetry predicts affective responses to films. *Journal of Personality and Social Psychology, 59*, 791–801.

Tomarken, A. J., Davidson, R. J., Wheeler, R. W., and Doss, R. (1992). Individual differences in anterior brain asymmetry and fundamental dimensions of emotion. *Journal of Personality and Social Psychology, 62*, 676–687.

Tomasch, J. (1969). The numerical capacity of the human cortico-ponto-cerebellar system. *Brain Research, 13*, 476–484.

Tonkonogy, J. and Goodglass, H. (1981). Language function, foot of the third frontal gyrus and rolandic operculum. *Archives of Neurology, 38*, 486–490.

Torack, R. M. and Morris, J. C. (1988). The association of ventral tegmental area histopathology with adult dementia. *Archives of Neurology, 45*, 497–501.

Toro, J., Cevera, M., and Osejo, E. et al. (1992). Obsessive-compulsive disorder in childhood and adolescence: A clinical study. *Journal of Child Psychology Psychiatry, 33*, 1025–1037.

Torrente, F., Ashwood, P., Day, R., Machado, N., Furlano, R. I., Anthony, A., Davies, S. E., Wakefield, A. J., Thomson, M. A., Walker-Smith, J. A., and Murch, S. H. (2002). Small intestinal enteropathy with epithelial IgG and complement deposition in children with regressive autism. *Molecular Psychiatry, 7*, 334, 375–382.

Torres, A. R., Maciulis, A., Stubbs, E. G., Cutler, A., and Odell, D. (2001). The transmission disequilibrium test suggests that HLA-DR4 and DR13 are linked to autism spectrum disorder. *Journal of Autism and Developmental Disorders, 31*, 529–535.

Torrey, E. F., Taylor, E. H., Bracha, H. S., Bowler, A. E., McNeil, T. F., Rawlings, R. R., Quinn, P. O., Bigelow, L. B., Rickler, K., and Sjostrom, K., et al. (1994). Prenatal origin of schizophrenia in a subgroup of discordant monozygotic twins. *Schizophrenia Bulletin, 20*, 423–432.

Townsend, J., Courchesne, E., Covington, J., Westerfield, M., Harris, N. S., Lyden, P., Lowry, T. P., and Press, G. A. (1999). Spatial attention deficits in patients with acquired or developmental cerebellar abnormality. *Journal of Neuroscience, 19*, 5632–5643.

Tracy, A. L., Jarrard, L. E., and Davidson, T. L. (2001). The hippocampus and motivation revisited: Appetite and activity. *Behavioral and Brain Research, 127*, 13–23.

Tramo, M. J. (2001). Biology and music. Music of the hemispheres. *Science, 291*, 54–56.

Tramo, M. J. and Bharucha, J. J. (1991). Musical priming by the right hemisphere post-callosotomy. *Neuropsychologia, 29*, 313–325.

Tramo, M. J., Cariani, P. A., Delgutte, B., and Braida, L. D. (2001). Neurobiological foundations for the theory of harmony in western tonal music. *Annals of the New York Academy of Sciences, 930*, 92–116.

Tramo, M. J., Shah, G. D., and Braida, L. D. (2002). Functional role of auditory cortex in frequency processing and pitch perception. *Journal of Neurophysiology, 87*, 122–139.

Tranel, D. and Damasio, A. R. (1985). Knowledge without awareness: An autonomic index of facial recognition by prosopoagnosics. *Science, 228*, 1453–1454.

Treisman, A. M., and Gelade, G. (1980). A feature integration theory of attention. *Cognitive Psychology, 12*, 97–136.

Trevarthen, C. (1984). Functional relations of disconnected hemispheres with the brain stem and with each other: Monkey and man. In M. Kinsbourne and W. Smith (Eds.), *Hemispheric disconnection and cerebral function*. Springfield, IL: Charles C. Thomas.

Trevarthen, C. (Ed.). (1990). *Brain circuits and functions of the mind. Essays in honor of Roger W. Sperry*. Cambridge, England: Cambridge University Press.

Trevarthen, C. B. (1968). Two mechanisms of vision in primates. *Psychologische Forschung, 31*, 299–337.

Tribble, D. L., Jones, D. P., and Edmondson, D. E. (1988). *Molecular Pharmacology, 34*, 413–420.

Troxler, R. G., Sprague, E. A., Albanese, R. A., and Thompson, A. J. (1977). The association of elevated plasma cortisol and early atherosclerosis as demonstrated by coronary angiography. *Atherosclerosis, 26*, 151–162.

Troiano, R. P., Flegal, K. M. (2001). Overweight children and adolescents: Description, epidemiology and demographics. *Pediatrics, 101*, 497–504.

Troiano, R. P., Flegal, K. M., Kuczmarski, R. J., Campbell, S. M., and Johnson, C. L. (1995). Overweight prevalence and trends for children and adolescents: The National Health and Nutrition Examination Surveys, 1963 to 1991. *Archives of Pediatric Adolescent Medicine, 149*, 1085–1091.

Tsai, L. Y. (1996). Brief report: Comorbid psychiatric disorders of autistic disorder. *Journal of Autism and Developmental Disorders, 29*, 439–484.

Tsokos, G. C. and Balow, J. E. (1986). Regulation of human cellular immune responses by glucocorticosteroids. In N. P. Plotnikoff, R. E. Faith, A. J. Murgo, and R. A. Good (Eds.), *Enkephalins and endorphins. Stress and the immune system* (pp. 159–171). New York, NY: Plenum Press.

Tucker, D. M. (1981). Lateral brain function, emotion and conceptualization. *Psychological Bulletin, 89*, 19–46.

Tucker, D. M. (1986). Neural control of emotional communication. In P. Blank, R. Buck, and R. Rosenthal (Eds.), *Nonverbal communication in the clinical context*. Cambridge, England: Cambridge University Press.

Tucker, D. M. (1991). Development of emotion and cortical networks. In M. Gunnar and C. Nelson (Eds.), *Minnesota symposium on child development: Developmental neuroscience*. New York, NY: Oxford University Press.

Tucker, D. M. and Derryberry, D. (1992). Motivated attention: Anxiety and the frontal executive functions. *Neuropsychiatry, Neuropsychology, and Behavioral Neurology, 5*, 233–252.

Tucker, D. M. and Frederick, S. L. (1989). Emotion and brain lateralization. In H. Wagner and A. Manstead (Eds.), *Handbook of social psychophysiology*. New York, NY: Wiley.

Tucker, D. M. and Liotti, M. (1989). Neuropsychological mechanisms of anxiety and depression. In F. Boller and J. Grafman (Eds.), *Handbook of neuropsychology*. Amsterdam: Elsevier.

Tucker, D. M., Stenslie, C. E., Roth, R. S., and Shearer, S. (1981). Right frontal lobe activation and right hemisphere performance decrement during a depressed mood. *Archives of General Psychiatry, 38*, 169–174.

Tucker, D. M. and Williamson, P. A. (1984). Asymmetric neural control systems in human self-regulation. *Psychological Review, 91*, 185–215.

Turkewitz, G., Gordon, E. W., and Birch, H. G. (1965). Head turning in the human neonate: Spontaneous patterns. *Journal of Genetic Psychology, 107*, 143–148.

Turkewitz, G., Moreau, T., Gordon, E. W., Birch, H. G., and Crystal, D. (1967). Relationships between prior head positions and lateral differences in responsiveness to somesthetic stimulation in the human neonate. *Journal of Experimental Child Psychology, 5*, 548–561.

Turner, A. M. and Greenough, W. T. (1985). Differential rearing effects on rat visual cortex synapses. I. Synaptic and neuronal density and synapses per neuron. *Brain Research, 329*, 195–203.

Uhlig, T., Merkenschlager, A., Brandmaier, R., and Egger, J. (1997). Topographic mapping of brain electrical activity in children with food-induced attention deficit hyperkinetic disorder. *European Journal of Pediatrics, 156*, 557–561.

Umilta, C., Bagnara, S., and Simion, F. (1978). Laterality effects for simple and complex geometrical figures, and nonsense patterns. *Neuropsychologia, 16*, 43–49.

Umilta, C., Rizzolatti, G., Marzi, C. A., Zamboni, G., Franzini, C., Camarda, R., and Berlucchi, G. (1974). Hemispheric differences in the discrimination of line orientation. *Neuropsychologia, 12*, 165–174.

Ungerleider, L. G. and Mishkin, M. (1982). Two cortical visual systems. In D. J. Ingle, M. H. Goodale, and R. J. W. Mansfield (Eds.), *The analysis of visual behavior*. Cambridge, MA: MIT Press.

Ungerstedt, U. (1971). Striatal dopamine release after amphetamine or nerve degeneration revealed by rotational behaviour. *Acta Physiologica Scandinavica. Supplementum, 367*, 49–68.

Ungerstedt, U. and Arbuthnott, G. W. (1970). Quantitative recording of rotational behavior in rats after 6-hydroxy-dopamine lesions of the nigrostriatal dopamine system. *Brain Research, 24*, 485–493.

Upledger, J. E. (1978). The relationship of craniosacral examination findings in grade school children with developmental problems. *Journal of the American Osteopathic Association, 17*, 760–776.

Ursin, H. and Kaada, B. R. (1960). Functional localization within the amygdaloid complex in the cat. *Electroencephalography and Clinical Neurophysiology, 12*, 1–20.

USA Today. (1999, February 23). Future good health will come to those who listen. p. A-15.

USA Today. (1999, May 3). Antidepressant use turns children into test subjects. p. D-09.

U.S. Department of Health and Human Services. (1996). *Physical activity and health: A report of the surgeon general*. Atlanta, GA: US Department of Health and Human Services, Centers for Disease Control and Prevention, National Center for Chronic Disease Prevention and Health Promotion.

U.S. Department of Health and Human Service, Administration for Children and Families. (1997). *The first progress report on head start program performance measures*. http://www.acf.hhs.gov/programs/core/ongoing_research/faces/faces_pubs_reports.html

U.S. Justice Department's Office of Juvenile Justice and Delinquency Prevention. (2000). http://ojjdp.ncjrs.org/pubs/alpha.html#J

U.S. News and World Report. (2000, June 19).

Vallar, C., Sandroni, P., Rusconi, M. L., and Barbieri, S. (1991b). Hemianopia, hemianesthesia, and spatial neglect: A study with evoked potentials. *Neurology, 41*, 1918–1922.

Vallar, G. (1990). Hemispheric control of articulatory speech output in aphasia. In G. Hammond (Ed.), *Cerebral control of speech and limb movement. Advances in psychology* (pp. 388–416). Amsterdam: North-Holland.

Vallar, G., Bottini, G., Sterzi, R., Passerini, D., and Rusconi, M. L. (1991a). Hemianesthesia, sensory neglect and defective access to conscious experience. *Neurology, 41*, 650–652.

Vallejo, J., Olivares, J., Marcos, T., Bulbena, A., and Menchon, J. M. (1992). Clomipramine versus phenelzine in obsessive-compulsive disorder: A controlled clinical trial. *British Journal of Psychiatry, 161*, 665–670.

van der Geest, J. N., Kemner, C., Camfferman, G., Verbaten, M. N., and van Engeland, H. (2001). Eye movements, visual attention, and autism: A saccadic reaction time study using the gap and overlap paradigm. *Biological Psychiatry, 50*, 614–619.

van der Geest, J. N., Kemner, C., Verbaten, M. N., and van Engeland, H. (2002a). Gaze behavior of children with pervasive developmental disorder toward human faces: A fixation time study. *Journal of Child Psychology and Psychiatry, 43*, 669–678.

van der Geest, J. N., Kemner, C., Camfferman, G., Verbaten, M. N., and van Engeland, H. (2002b). Looking at images with human figures: Comparison between autistic and normal children. *Journal of Autism and Developmental Disorders, 32*, 69–75.

van der Meere, J., van Baal, M., and Sergeant, J. (1989). The additive factor method: A differential diagnostic tool in hyperactivity and learning disability. *Journal of Abnormal and Child Psychology, 17*, 409–422.

van der Meere, J., Vreeling, H. J., and Sergeant, J. (1992). A motor presetting study in hyperactive, learning disabled and control children. *Journal of Child Psychology and Psychiatry, 33*, 1347–1354.

Van Buren, J. M. and Borke, R. C. (1972). *Variations and connections of the human thalamus* (Vol. I). New York: Springer.

van Groen, T. and Wyss, J. M. (1992). Connections of the retrosplenial dysgranular cortex in the rat. *Journal of Comparative Neurology, 315*, 200–216.

Van Hoesen, G. W. and Pandya, D. N. (1975). Some connections of the entorhinal (area 28) and peripheral (area 35) cortices in the monkey. *Trends in Neuroscience, 5*, 345–350.

van Praag, H., Christie, B. R., Sejnowski, T. J., and Gage, F. H. (1999a). Running enhances neurogenesis, learning, and long-term potentiation in mice. *Proceedings of the National Academy of Sciences, USA, 96*, 13427–13431.

van Praag, H., Kempermann, G., and Gage, F. H. (1999b). Running increases cell proliferation and neurogenesis in the adult mouse dentate gyrus. *Nature, Neuroscience, 2*, 266–270.

Van Praag, H., Kempermann, G., and Gage, F. H. (2000). Neural consequences of environmental enrichment. *Nature Reviews. Neuroscience, 1*, 191–198.

van Vreeswijk, C. (2000, May 29). Analysis of the asynchronous state in networks of strongly coupled oscillators. *Physical Review Letters, 84*, 5110–5113.

Vargha-Khadem, F., O'Gorman, A., and Watters, G. (1985). Aphasia and handedness in relation to hemispheric side, age at injury and severity of cerebral lesions during childhood. *Brain, 108*, 677–696.

Vastag, B. (2001). Pay attention: Ritalin acts much like cocaine. *Journal of the American Medical Association, 286*, 905–906.

Vauclair, J., Fagot, J., and Hopkins, W. D. (1993). Rotation of mental images in baboons when the visual input is directed to the left cerebral hemisphere. *Psychological Science, 4*, 99–103.

Veening, I. G. (1978a). Cortical afferents of the amygdaloid complex in the rat. An HRP study. *Neuroscience Letters, 8*, 191–195.

Veening, I. G. (1978b). Subcortical afferents of the amygdaloid complex in the rat. *Neuroscience Letters, 8*, 197–202.

Vele, F. and Gutmann, G. (1971). [Influencing of postural reflexes via the joints]. *Zeitschrift fur physiotherapie, 23*, 383–386.

Vel'khover, E. S. and Elfimov, M. A. (1995). The dependence of the parasympathetic reaction of the mesencephalic section of the brain stem to age-, sex- and pigment-reagent-related factors. *Zhurnal nevropatologii i psikhiatrii imeni S.S. Korsakova, 95*, 36–39.

Venables, P. H. (1989). The Emanuel Miller memorial lecture 1987. Childhood markers for adult disorders. *Journal of Child Psychology and Psychiatry, and Allied Disciplines, 30*, 347–364.

Verrier, R., Thompson, P., and Lown, B. (1974). Ventricular vulnerability during sympathetic stimulation: Role of heart rate and blood pressure. *Cardiovascular Research, 8*, 602–610.

Verrier, R. L. (1990). Behavioral stess, myocardial ischemia, and arrhythmias. In D. P. Zipes and J. Jalife (Eds.), *Cardiac Electrophysiology*. Philadelphia, PA: W. B. Saunders.

Verrier, R. L. and Dickerson, L. W. (1991). Autonomic nervous system and coronary blood flow changes related to emotional activation and sleep. *Circulation, 83*, 1181–1189.

Victor, M., Adams, R. D., and Mancall, E. L. (1959). A restricted form of cerebellar cortical degeneration occurring in alcoholic patients. *Archives of Neurology, 1*, 579–688.

Vignaud, P., Duringer, P., Mackaye, H. T., Likius, A., Blondel, C., Boisserie, J. R., De Bonis, L., Eisenmann, V., Etienne, M. E., Geraads, D., Guy, F., Lehmann, T., Lihoreau, F., Lopez-Martinez, N., Mourer-Chauvire, C., Otero, O., Rage, J. C., Schuster, M., Viriot, L., Zazzo, A., and Brunet, M. (2002). Geology and palaeontology of the upper miocene toros-menalla hominid locality, Chad. *Nature, 418*, 152–155.

Vilensky, J. A., Damasio, A. R., and Maurer, R. G. (1981). Gait disturbances in patients with autistic behavior: A preliminary study. *Archives of Neurology, 38*, 646–649.

Viviani, J., Turkewitz, G., and Karp. E. (1978). A relationship between laterality of functioning at 2 days and at 7 years of age. *Bulletin of the Psychonomic Society, 12*, 189–192.

Volavka, J., Neziroglu, F., and Yaryura-Tobias, J. A. (1985). Clomipramine and imipramine in obsessive-compulsive disorder. *Psychiatric Research, 14*, 83–91.

Volgushev, M., Pernberg, J., and Eysel, U. T. (2000). Comparison of the selectivity of postsynaptic potentials and spike responses in cat visual cortex. *European Journal of Neuroscience, 12*, 257–263.

Volkow, N. D., Levy, A., Brodie, J. D., Wolf, A. P., Cancro, R., van Gelder, P., and Henn, F. (1992). Low cerebellar metabolism in medicated patients with chronic schizophrenia. *American Journal of Psychiatry, 149*, 686–688.

Volkow, N. D., Logan, J., Fowler, J. S., Wang, G. J., Gur, R. C., Wong, C., Felder, C., Gatley, S. J., Ding, Y. S., Hitzemann, R., and Pappas, N. (2000). Association between age-related decline in brain dopamine activity and impairment in frontal and cingulate metabolism. *American Journal of Psychiatry, 157*, 75–80.

von Plessen, K., Lundervold, A., Duta, N., Heiervang, E., Klauschen, F., Smievoll, A. I., Ersland, L., and Hugdahl, K. (2002). Less developed corpus callosum in dyslexic subjects—a structural MRI study. *Neuropsychologia, 40*, 1035–1044.

Voronin, L. G., Semenov, B. F., Nikol'skaia, K. A., and Ozherelkov, S. V. (1980). Nature of proprioceptive information processing during learning in brain inoculated mice. *Doklady Akademii nauk/[Rossiiskaia akademii nauk], 251*, 500–504.

Wada, J. A., Clarke, R., and Hamm, A. (1975). Cerebral hemispheric asymmetry in humans. *Archives of Neurology, 32*, 239–246.

Wada, J. A. and Davis, A. E. (1977). Fundamental nature of human infant's brain asymmetry. *The Canadian Journal of Neurological Sciences. Le journal canadien des sciences neurologiques, 4*, 203–207.

Wadhwa, P. D., Culhane, J. F., Rauh, V., and Barve, S. S. (1991). Stress and preterm birth: Neuroendocrine, immune/inflammatory, and vascular mechanisms. *British Journal of Obstetrics and Gynaecology, 98*, 898–904.

Wadhwa, P. D., Culhane, J. F., Rauh, V., Barve, S. S., Hogan, V., Sandman, C. A., Hobel, C. J., Chicz-DeMet, A., Dunkel-Schetter, C., Garite, T. J., and Glynn, L. (2001). Stress, infection and preterm birth: A biobehavioural perspective. *Maternal and Child Health Journal, 5*, 119–125.

Wadhwa, P. D., Dunkel-Schetter, C., Chicz-DeMet, A., Porto, M., and Sandman, C. A. (2001 July). Prenatal psychosocial factors and the neuroendocrine axis in human pregnancy. *Paediatric and Perinatal Epidemiology, 15* (Suppl.), 17–29.

Wagner, H. N., Bums, D. H., Dannals, R. F., Wong, D. F., Langstrom, B., Duelfer, T., Frost, J. J., Ravert, H. T., Links, J. M., Rosenbloom, S. B., Lucas, S. E., Kramer, A. V., and Kuhlar, M. (1983). Imaging dopamine receptors in the human brain by positron tomography. *Science, 221*, 1264–1266.

Wakefield, A. J., Murch, S. H., Anthony, A., Linnell, J., Casson, D. M., Malik, M., Berelowitz, M., Dhillon, A. P., Thomson, M. A., Harvey, P., Valentine, A., Davies, S. E., and Walker-Smith, J. A. (1998). Ileal-lymphoid-nodular hyperplasia, non-specific colitis, and pervasive developmental disorder in children. *Lancet, 351*, 637–641.

Wakefield, A. J. (2002). The gut–brain axis in childhood developmental disorders. *Journal of pediatric gastroenterology and nutrition, 34* (Suppl.), S14–S17.

Wakschlag, L. S., Lahey, B. B., Loeber, R., Green, S. M., Gordon, R. A., and Leventhal, B. L. (1996). Maternal smoking during pregnancy and the risk of conduct disorder in boys. *Psychosomatic Medicine, 58*, 432–446.

Walford, R. L., Weber, L., and Panov, S. (1995). Caloric restriction and aging as viewed from Biosphere 2. *Receptor, 5*, 29–33.

Walker, B. B. and Sandman, C. A. (1979). Human visual evoked responses are related to heart rate. *Journal of Comparative Physiology and Psychology, 93*, 717–729.

Walker, B. B. and Sandman, C. A. (1982). Visual evoked potentials change as heart rate and carotid pressure change. *Psychophysiology, 19*, 520.

Walker-Smith, J. and Andrews, J. (1972). Alpha-1 antitrypsin, autism and coeliac disease. *Lancet, 2*, 883–884.

Wall, P. M. and Messier, C. (2001). The hippocampal formation—orbitomedial prefrontal cortex circuit in the attentional control of active memory. *Behavioral and Brain Research, 127*, 99–117.

Ward, A. A. (1948). The anterior cingular gyrus and personality. *Research Publications of the Association for Research in Nervous and Mental Diseases, 27*, 438–445.

Ward, I. L. and Weisz, J. (1980). Maternal stress alters plasma testosterone in fetal males. *Science, 207*, 328–329.

Ward, I. L. and Weisz, J. (1984). Differential effects of maternal stress on circulating levels of corticosterone, progesterone and testosterone in male and female rat fetuses and their mothers. *Endocrinology, 114*, 1635–1644.

Ward, J. P. (1991). Prosimians as animal models in the study of neural lateralization. In F. L. Kitterle (Ed.), *Cerebral laterality: Theory and research* (pp. 1–17). Hillsdale, NJ: Lawrence Erlbaum Associates.

Ward, O. B., Monaghan, E. P., and Ward, I. L. (1986). Naltrexone blocks the effects of prenatal stress on sexual behavior differentiation in male rats. *Pharmacology, Biochemistry and Behavior, 251*, 573–576.

Warren, R. P., Foster, A., and Margaretten, N. C. (1987). Reduced natural killer cell activity in autism. *Journal of the American Academy of Child and Adolescent Psychiatry, 26*, 333–335.

Warren, R. P., Margaretten, N. C., Pace, N. C., and Foster, A. (1986). Immune abnormalities in patients with autism. *Journal of Autism and Developmental Disorders, 16*, 189–197.

Warren, R. P., Singh, V. K., (1996). Elevated serotonin levels in autism: Association with the major histocompatibility complex. *Neuropsychobiology, 34*, 72–75.

Warren, R. P., Singh, V. K., Averett, R. E., Odell, J. D., Maciulis, A., Burger, R. A., Daniels, W. W., and Warren, W. L. Immunogenetic studies in autism and related disorders. *Molecular and Chemical Neuropathology, 28*, 77–81.

Warren, R. P., Cole, P., Odell, J. D., Pingree, C. B., Warren, W. L., White, E., Yonk, J., and Singh, V. K. (1990). Detection of maternal antibodies in infantile autism. *Journal of the American Academy of Child and Adolescent Psychiatry, 29*, 873–877.

Warrington, E. K. (1982). Neuropsychological studies of object recognition. *Philosophical Transactions of the Royal Society of London, 298*, 15–27.

Warrington, E. K. and James, M. (1967). An experimental investigation of facial recognition in patients with unilateral cerebral lesions. *Cortex, 3*, 317–326.

Watkins, W. E., Cruz, J. R., and Pollitt, E. (1996). The effects of deworming on indicators of school performance in Guatemala. *Transactions of the Royal Society of Tropical Medicine and Hygiene, 90*, 156–161.

Watkins, W. E. and Pollitt. E. (1996). Effect of removing ascaris on the growth of Guatemalan schoolchildren. *Pediatrics, 97*, 871–876.

Watson, R. T. and Heilman, K. M. (1982). Affect in subcortical aphasia. *Neurology, 32*, 102–103.

Watson, R. T., Heilman, K. M., Cauthen, J. C., and King, F. A. (1973). Neglect after cingulectomy. *Neurology, 23*, 1003–1007.

Watson, R. T., Heilman, K. M., Miller, B. D., and King, F. A. (1974). Neglect after mesencephalic reticular formation lesions. *Neurology, 24*, 294–298.

Watson, P. J. (1978). Non-motor functions of the cerebellum. *Psychological Bulletin, 85*, 944–967.

Weber, A. M. and Bradshaw, J. L. (1981). Levy and Reid's neuropsychological model in relation to writing/hand postural: An evaluation. *Psychological Bulletin, 90*, 74–88.

Wechsler, A. F. (1973). The effect of organic brain disease on recall of emotionally charged versus neutral narrative texts. *Neurology, 23*, 130–135.

Weinberg, N. Z. (1997). Cognitive and behavioral deficits associated with parental alcohol use. *Journal of the American Academy of child and Adolescent Psychiatry, 36*, 1177–1186.

Weinberger, D. R., Berman, K. F., Ladarola, N., Driesen, N., and Zec, R. F. (1988). Free frontal cortical blood flow and cognitive function in huntington's disease. *Journal of Neurology, Neurosurgery, and Psychiatry, 51*, 94–104.

Weinberger, D. R., Luchins, D. J., Morihisa, J., and Wyatt, R. J. (1982). Asymmetrical volumes of the right and left frontal and occipital regions of the human brain. *Annals of Neurology, 11*, 97–99.

Weinberger, D. R., Torrey, E. F., and Wyatt, R. J. (1979). Cerebellar atrophy in chronic schizophrenia. *Lancet, 1*, 718–719.

Weinberger, N. M. (1995). Dynamic regulation of receptive fields and maps in the adult sensory cortex. *Annual Review of Neuroscience, 18*, 129–158.

Weintraub, S. and Mesulam, M. M. (1987). Right cerebral dominance in spatial attention: Further evidence based on ipsilateral neglect. *Archives of Neurology, 44*, 621–625.

Weissenberger, A. A., Dell, M. L., Liow, K., Theodore, W., Frattali, C. M., Hernandez, D., and Zametkin, A. J. (2001, June). Aggression and psychiatric comorbidity in children with hypothalamic hemartomas and their unaffected siblings. *Journal of American Academic Child Adolescent Psychiatry, 40*, 696–703.

Weisz, J., Szilagyi, N., Lang, E., and Adam, G. (1992). The influence of monocular viewing on heart period variability. *International Journal of Psychophysiology, 12*, 11–18.

Weizman, A., Weizman, R., Szekely, G. A., Wijsenbeek, H., and Livni, E. (1982). Abnormal immune response to brain tissue antigen in the syndrome of autism. *American Journal of Psychiatry, 139*, 1462–1465.

Welch, K. and Stuteville, P. (1958). Experimental production of neglect in monkeys. *Brain, 81*, 341–347.

Welner, Z., Welner, A., Stewart, M., Palkes, H., and Wish, E. (1977). A controlled study of siblings of hyperactive children. *Journal of Nervous and Mental Disease, 165*, 110–117.

Welsh, J. P. (1998). Systemic harmaline blocks associative and motor learning by the actions of the inferior olive. *European Journal of Neuroscience, 10*, 3307–3320.

Welsh, J. P., Lang, E. J., Suglhara, I., and Llinas, R. (1995). Dynamic organization of motor control within the olivocerebellar system. *Nature, 374*, 453–457.

Werntz, D. A., Bickford, R. G., Bloom, F. E., and Shannahof-Khalsa, D. S. (1983). Alternating cerebral hemispheric activity and lateralization of autonomic nervous function. *Human Neurobiology, 2*, 39–43.

Wessel, K., Moschner, C., Wandinger, K. P., Kompf, D., and Heide, W. (1998). Oculomotor testing in the differential diagnosis of degenerative ataxic disorders. *Archives of Neurology, 55*, 949–956.

West, R. W. and Greenough, W. T. (1972). Effect of environmental complexity on cortical synapses of rats: Preliminary results. *Behavioral Biology, 7*, 279–284.

WGBH, Boston (2002). http://www.pbs.org/wgbh/pages/frontline/shows/teenbrain/interviews/todd.html

Wexler, B. E. and Halwes, T. (1983). Increasing the power of dichotic methods: The fused rhymed words test. *Neuropsychologia, 21*, 59–66.

Whalen, C. K., Swanson, J. M., and Henker, B. et al. (1987). Natural social behaviors in hyperactive children: dose effects of methylphenidate. *Journal of Consulting and Clinical Psychology, 55*, 187–193.

Whalen, P. J., Bush, G., McNally, R. J., Wilhelm, S., McInerney, S. C., Jenike, M. A., and Rauch, S. L. (1998a). The emotional counting Stroop paradigm: A functional magnetic resonance imaging probe of the anterior cingulate affective division. *Biological Psychiatry, 44*, 1219–1228.

Whalen, P. J., Rauch, S. L., Etcoff, N. L., McInerney, S. C., Lee, M. B., and Jenike, M. A. (1998b). Masked presentations of emotional facial expressions modulate amygdala activity without explicit knowledge. *Journal of Neuroscience, 18*, 411–418.

Whalen, R. (1993). Musculoskeletal adaption to mechanical forces on Earth and in space. *Physiologist, 36*, S127–S130.

Wheadon, D. E., Bushnell, W., and Steiner, M. (1993). *A fixed dose comparison of 20, 40, or 60 mg paroxetine to placebo in the treatment of obsessive-compulsive disorder.* Presented at the American College of Neuropharmacology Meeting, Puerto Rico.

Wheeler, R. W., Davidson, R. J., and Tomarken, A. J. (1993). Frontal brain asymmetry and emotional reactivity: A biological substrate of affective style. *Psychophysiology, 30*, 82–89.

Whitaker, R. C., Wright, J. A., Pepe, M. S., Seidel, K. D., and Dietz, W. H. (1997). Predicting obesity in young adulthood from childhood and parental obesity. *New England Journal of Medicine, 337*, 869–873.

White, B. (1973). *The first 3 years of life.* Englewood Cliffs, NJ: Prentice-Hall.

White, M. J. and Davies, C. T. M. (1984). The effects of the mobilization after lower leg fractures on the contractile properties of human triceps surae. *Clinical Science, 66*, 277–282.

Whiteley, P, and Shattock, P. (2002). Biochemical aspects in autism spectrum disorders: Updating the opioid-excess theory and presenting new opportunities for biomedical intervention. *Expert opinion on therapeutic targets, 6*, 175–183.

Whitlock, D. G. and Nauta, W. J. H. (1956). Subcortical projections from the temporal neocortex in macaca mulatta. *The Journal of Comparative Neurology, 106*, 183–212.

Wichman, T. and DeLong, M. (1996). Functional and pathophysiological models of the basal ganglia. *Current Opinion in Neurobiology, 6*, 751–758.

Wickelgren, I. (1997). Getting the brain's attention. *Science, 278*, 35–37.

Widom, C. S. (1989a). Child abuse, neglect, and adult behavior: Research design and findings on criminality, violence, and child abuse. *American Journal of Orthopsychiatry 59*, 355–367.

Widom, C. S. (1989b). Does violence beget violence? A critical examination of the literature. *Psychological Bulletin, 106*, 3–28.

Wiggs, L. and Stores, G. (1998). Behavioural treatment for sleep problems in children with severe learning disabilities and challenging daytime behaviour: Effect on sleep patterns of mother and child. *Journal of Sleep Research, 7*, 119–126.

Wilkerson, D. S., Volpe, A. G., Dean, R. S., and Titus, J. B. (2002). Perinatal complications as predictors of infantile autism. *The World Journal of Biological Psychiatry: The Official Journal of the World Federation of Societies of Biological Psychiatry, 3*, 162–166.

Wilkins, A. J., Shallice, T., and McCarthy, R. (1987). Frontal lesions and sustained attention. *Neuropsychologia, 25*, 359–366.

Williams, L. M., Phillips, M. L., Brammer, M. J., Skerrett, D., Lagopoulos, J., Rennie, C., Bahramali, H., Olivieri, G., David, A. S., Peduto, A., and Gordon, E. (2001). Arousal dissociates amygdala and hippocampal fear responses: Evidence from simultaneous fMRI and skin conductance recording. *Neuroimage, 14*, 1070–1079.

Williams, M. F. (2002). Primate encephalization and intelligence. *Medical Hypotheses, 58*, 284–290.

Williams, R. B. J., Haney, T. L., Lee, K. L., Kong, Y., Blumenthal, J. A., and Whalen, R. E. (1980). Type A behavior, hostility, and coronary atherosclerosis. *Psychosomatic Medicine, 42*, 539–549.

Willams, T. M. and Handford, A. G. (1986). Television and other leisure activities. In T. M. Williams (Ed.), *The impact of television: A national experiment in three commutities* (pp. 143–213). Orlando, FL: Academic Press.

Willemsen-Swinkels, S. H., Buitelaar, J. K., Nijhof, G. J., and van England, H. (1995). Failure of naltrexone to reduce self-injurious and autistic behavior in mentally retarded adults. *Archives of General Psychiatry, 52*, 766–773.

Williams, V. and Grossman, R. G. (1970). Ultrastructure of cortical synapses after the failure of presynaptic activity in ischemia. *Anatomical Research, 166*, 131–142.

Wilmshurst, L. A. (2002). Treatment programs for youth with emotional and behavioral disorders: An outcome study of two alternate approaches. *Mental Health Services Research, 4*, 85–96.

Winfree, A. T. (1983). Sudden cardiac death: A problem in topology. *Scientific American, 248*, 144–161.

Winfree, A. T. (1987). *When home breaks down.* Princeton, NJ: Princeton University Press.

Wing, L. (1981). Asperger syndrome: A clinical account. *Psychological Medicine, 11*, 115–129.

Wing, L. (1988). Autism: Possible clues to the underlying pathology Clinical facts. In L. Wing (Ed.), *Aspects of autism: Biological research* (pp. 1–10). London: Gaskell/National Autistic Society.

Wing, L. and Attwood, A. (1987). Syndromes of autism and atypical development. In D. J. Cohen, A. M. Donnellan, and P. Rhea (Eds.), *Handbook of autism and persuasive developmental disorders* (pp. 3–19). New York, NY: Wiley.

Wing, L. and Potter, D. (2002). The epidemiology of autistic spectrum disorders: Is the prevalence rising? *Mental Retardation and Developmental Disabilities Research Review, 8*, 151–161.

Winocur, G. and Greenwood, C. E. (1999). The effects of high fat diets and environmental influences on cognitive performance in rats. *Nature Neuroscience, 2*, 861–863.

Winstein, C. J. and Pohl, P. S. (1995). Effects of unilateral brain damage on the control of goal-directed hand movements. *Experimental Brain Research, 105*, 163–174.

Witelson, S. F. (1977). Developmental dyslexia: Two right hemispheres and none left. *Science, 195*, 309–311.

Witelson, S. F. (1985). The brain connection: The corpus callosum is larger in left-handers. *Science, 229*, 665–668.

Witelson, S. F. and Pallie, W. (1973). Left hemisphere specialization for language in the newborn: Neuroanatomical evidence of asymmetry. *Brain, 96*, 641–646.

Wittling, W. (1990). Psychophysiological correlates of human brain asymmetry: Blood pressure changes during lateralized presentation of an emotionally laden film. *Neuropsychologia, 28*, 457–470.

Wittling, W. (1995). Brain asymmetry in the control of autonomic-physiologic activity. In R. J. Davidson and K. Hugdahl (Eds.), *Brain asymmetry* (pp. 305–357), Cambridge, MA: MIT Press.

Wittling, W. and Pfluger, M. (1990). Neuroendocrine hemisphere asymmetries: Salivary cortisol secretion during lateralized viewing of emotion-related and neutral films. *Brain and Cognition, 14*, 243–265.

Wittling, W. and Roschmann, R. (1993). Emotion-related hemisphere asymmetry: Subjective emotional responses to laterally presented films. *Cortex, 29*, 431–448.

Wittling, W. and Schweiger, E. (1992). *Brain asymmetry in the regulation of autonomic arousal in emotion-related situations.* Presented at the International Neuropsychological Symposium, Schluchsee, Germany.

Wolffet, H., Michel, G. F., and Ovrut, M. et al. (1990). Race and timing of motor coordination in development dyslexia. *Development Psychology, 26*, 349–359.

Woods, B. T. and Teuber, H. L. (1978). Changing patterns of childhood aphasia. *Annals of Neurology, 32*, 239–246.

Woody-Ramsey, J. and Miller, P. H. (1988). The facilitation of selective attention in preschoolers. *Child Development, 59*, 1497–1503.

Wyke, B. (1979a). Cervical-articular contributions to posture and gait: Their relation to senile disequilibrium, *Age Aging, 8*, 251–258.

Wyke, B. (1979b). Conference on the aging brain. Cervical-articular contributions to posture and gait: Their relation to senile disequilibrium: *Age Aging, 8*, 251–258.

Wyke, B. D. (1975). *The neurological basis of movement: A developmental review.* In K. S. Holt (Ed.), *Movement and child development* (p. 1933). London: Heinemann.

Wyke, M. A. (1965). Comparative analysis of proprioception in the left and right arms. *The Quarterly Journal of Experimental Psychology, 17*, 149–157.

Wykes, T., Brammer, M., Mellers, J., Bray, P., Reeder, C., Williams, C., and Corner, J. (2002). Effects on the brain of a psychological treatment: Cognitive remediation therapy: Functional magnetic resonance imaging in schizophrenia. *British Journal of Psychiatry, 181*, 144–152.

Wyler, A. R., Lockard, J. S., Ward, A. A., Jr., and Finch, C. A. (1976). Conditioned EEG desynchronization and seizure occurrence in patients. *Electroencephalography and Clinical Neurophysiology, 41*, 501–512.

Wyrwicka, W. and Sterman, M. B. (1968). Instrumental conditioning of sensori-motor cortex EEG spindles in the waking cat. *Physiology and Behavior, 3*, 703–707.

Yakovlev, P. I. and Rakic, P. (1966). Patterns of decussation of bulbar pyramids and distribution of pyramidal tracts on two sides of the spinal cord. *Transactions of the American Neurological Association, 91*, 366–367.

Yamamoto, B. K. and Freed, C. R. (1984). Asymmetric dopamine and serotonin metabolism in nigrostriatal and limbic structures of the trained circling rat. *Brain Research, 297*, 115–119.

Yan, J. H., Thomas, J. R., and Downing, J. H. (1998). Locomotion improves children's spatial search: A meta-analytic review. *Perceptual and Motor Skills, 87*, 67–82.

Yanowitz, F., Preston, J. B., and Abildskov, J. A. (1966). Functional distribution of right and left stellate innervation to the ventricles: Production of neurogenic electrocardiographic changes by unilateral alteration of sympathetic tone. *Circulation Research, 28*, 416–428.

Yarbrough, E., Santat, U., Perel, I., Webster, C., and Lombardi, R. (1987). Effects of fenfluramine on autistic individuals residing in a state developmental center. *Journal of Autism and Developmental Disorders, 17*, 303–314.

Yates, P. O. (1967). The pathological basis for cerebral ischaemia. *Modern Trends in Neurology, 4*, 180–192.

Yeap, L. L. (1989). Hemisphericity and shident achievement. *International Journal of Neuroscience, 48*, 225–232.

Yeargin-Allsopp, M. and Boyle, C. (2002). Overview: The epidemiology of neurodevelopmental disorders. *Mental Retardation and Developmental Disabilities Research Review, 8*, 113–136.

Yeo, C. H. (1991). Cerebellum and classical conditioning of motor responses. *Annals of New York Academy of Sciences, 627*, 292–304.

Youmans, J. R. and Smith, A. H. (1991). Gravitational fields and aging. *Physiologist, 34* (Suppl.), S19–S22.

Young, A. B. and Penney, J. B. (1984). Neurochemical anatomy of movement disorders. *Neurological Clinics, 2*, 417–433.

Young, B. A., Penney, J. B., and Starosta-Rubinstein, S. et al. (1986). PET scan investigations of huntington's disease: Cerebral metabolic correlates of neurological features and functional decline. *Annals of Neurology, 20*, 296–303.

Young, G. and Gagnon, M. (1990). Neonatal laterality, birth stress, familial sinistrality, and left-brain inhibition. *Developmental Neuropsychology, 6*, 127–150.

Young, G., Segalowitz, S. J., Misek, P., Alp, I. E., and Boulet, R. (1983). Is early reaching left-handed? Review of manual specialization research. In G. Young, S. J. Segalowitz, C. M. Corter, and S. E. Trehub (Eds.), *Manual specialization and the developing brain* (pp. 13–32). London: Academic Press.

Yozawitz, A., Bruder, G., Sutton, S., Sharpe, L., Gurland, B., Fleiss, J., and Costa, L. (1979). Dichotic perception: Evidence for right hemisphere dysfunction in affective psychosis. *British Journal of Psychiatry, 135*, 224–237.

Yu, C. E., Dawson, G., Munson, J., D'Souza, I., Osterling, J., Estes, A., Leutenegger, A. L., Flodman, P., Smith, M., Raskind, W. H., Spence, M. A., McMahon, W., Wijsman, E. M., and Schellenberg, G. D. (2002). Presence of large deletions in kindreds with autism. *American Journal of Human Genetics, 71*, 100–115.

Zagon, I. S. and McLaughlin, P. J. (1983a). Increased brain size and cellular content in infant rats treated with an opiate antagonist. *Science, 221*, 1179–1180.

Zagon, I. S. and McLaughlin, P. J. (1983b). Naltrexone modulates growth in infant rats. *Life Sciences, 33*, 2449–2454.

Zajecka, J. M., Fawcett, J., and Guy, C. (1990). Coexisting major depression and obsessive compulsive disorder treated with venlafaxine. *Journal of Clinical Psychopharmacology, 10*, 152–153.

Zametkin, A., Rapoport, J. L., and Murphy, D. L., et al. (1985). Treatment of hyperactive children wit monoamine oxidase inhibitors: I. Clinical efficacy. *Archives of General Psychiatry, 42*, 962–966.

Zametkin, A. J. and Ernst, M. (1999). Problems in the management of attention-deficit-hyperactivity disorder. *The New England Journal of Medicine, 340*, 40–46.

Zanchetti, A. and Zoccolini, A. (1954). Autonomic hypothalamic outburst elicited by cerebellar stimulation. *Journal of Neurophysiology, 7*, 475–483.

Zappela, M. (2002). Early-onset Tourette syndrome with reversible autistic behaviour: A maturational disorder. *European Child and Adolescent Psychiatry, 11*, 18–23.

Zappia, J. V. and Rogers, L. J. (1983). Light experience during development affects asymmetry of forebrain function in chickens. *Developmental Brain Research, 11*, 93–106.

Zeisel, S. H. (1986). Dietary influences on neurotransmission. *Advances in Pediatrics, 33*, 23–47.

Zentner, M. R. and Kagan, J. (1996). Perception of music by infants. *Nature, 383*, 29.

Zhang, L. X., Levine, S., Dent, G., Zhan, Y., Xing, G., Okimoto, D., Kathleen Gordon, M., Post, R. M., and Smith, M. A. (2002). Maternal deprivation increases cell death in the infant rat brain. *Brain Research. Developmental Brain Research, 133*, 1–11.

Zilles, K. (1990). Cortex. In G. Paxinos (Ed.), *The human nervous system* (pp. 757–802). San Diego: Academic Press.

Zimmerberg, B., Glick, S. D., and Jerussi, T. P. (1974). Neurochemical correlate of a spatial preference in rats. *Science, 185*, 623–625.

Zingarelli, G., Ellman, G., Hom, A., Wymore, M., Heidorn, S., and Chicz-DeMet, A. (1992). Clinical effects of naltrexone on autistic behavior. *American Journal of Mental Retardation, 97*, 57–63.

Zipes, D. P. (1991). The long QT interval syndrome: A Rosetta Stone for sympathetic related ventricular tachyarrhythmias. *Circulation, 84*, 1414–1419.

Zoccolotti, P., Caltagirone, C., Benedetti, N., and Gainotti, G. (1986). Perturbation des reponses vegetatives aux stimuli emotionnels au cours des lesions hemispheriques unilaterales. *L'Emephale, 12*, 263–268.

Zoccolotti, P., Scabini, D., and Violani, C. (1981). Electrodermal responses in patients with unilateral brain damage. *Journal of Clinical Neuropsychology, 4*, 143.

Zohar, J. and Insel, T. R. (1987). Obsessive-compulsive disorder: Psychobiological approaches to diagnosis, treatment, and pathophysicology. *Biological Psychiatry, 22*, 667–687.

Zola-Morgan, S. and Squire, L. R. (1993). Neuroanatomy of memory. *Annual Review of Neuroscience, 16*, 547–563.

Zola-Morgan, S., Squire, L. R., Clower, R. P., and Rempel, N. L. (1993). Damage to the perirhinal cortex exacerbates memory impairment following lesions to the hippocampal formation. *Journal of Neuroscience, 13*, 251–265.

Zola-Morgan, S., Squire, L. R., and Mishkin, M. (1992). The neuroanatomy of amnesia: Amygdala-hippocampus vs. temporal stem. *Science, 218*, I337–I339.

Index

Acetylcholine, 103, 239, 260, 294, 347
Acoustic striata, 138
ADD *See* attention deficit disorder
Adductor pollicis, 250, 251
Adenylate cyclase, 149
ADHD *See* attention deficit hyperactive disorder
Adrenoleukodystrophy, 296
Adrenomyelo–neuropathy, 296
Agnosia, 164, 200
Akinesia, 220
Akinetic, 163, 200
Allocation theory, 133
Alzheimer's disease, 283, 321, 322
Ambient vision, 198
Aminergic, 50
Amines, 241, 283
Aminopeptidase, 118
Amygdala, 45, 61, 62, 71, 75, 88, 90, 94–101, 103, 111, 145, 147–149, 151, 153, 155, 159, 161–163, 169, 174, 199–201, 212, 232–236, 239, 248, 276, 279, 286, 288, 306, 307, 311, 313, 323, 328, 329
Amygdalofugal, 96
Anencephaly, 294
Angelman syndrome, 280, 282
Angular gyrus, 125, 218, 220
Anoxia, 197, 255
Anterior cingulate, 72, 77, 96, 98, 99, 158, 159, 207, 225, 230, 233, 271, 287, 328
Anterior commisure, 96, 98, 101
Anterior horn, 98
Anterolateral system, 109
Antihypertensive, 283
Antiorthostatic, 253
Antioxidant, 346–348
Antipsychotic, 274
Antisocial personality disorder, 237
Antitrypsin, 300
Anxiety disorders, 234, 235, 241, 246, 305, 348
Anxiety response, 97
Aphasic, 156
ApopTag, 276
Arachadonic acid, 342

Archicortex, 93, 94
Arcuate fasciculus, 173, 175
Ardipithecus ramidus kadabba, 17
Aromatherapy, 325
Artificial scotoma, 107
Ascending reticular activating system, 83, 182
Aspartame, 344
Asperger's syndrome, 1, 5, 9, 11–13, 237, 238, 271
Ataxia, 57, 58, 197
Atherosclerosis, 15, 150, 288
Atlas blockage syndrome, 270
Atria, 145
Attention deficit disorder, 9, 10, 177, 187, 210, 214, 299
Attention deficit hyperactive disorder, 1, 4, 5, 9–14, 45, 74, 80, 177, 208, 209, 211, 212, 217, 231–234, 237, 238, 240, 241, 244, 246, 266–268, 271, 275, 278, 279, 281, 284, 287, 294–296, 303, 304, 306, 308, 309, 312–314, 317, 318, 326, 333, 339, 340, 342–344, 347, 348, 351, 359–365
Attention window, 134, 135
Auditory association areas, 110
Australopithecus, 17, 18, 35, 36
Autism, 3–5, 9–13, 45, 75, 80, 177, 209–212, 214, 216, 223, 232, 237–239, 241, 270–272, 274, 275, 279–284, 288–292, 296, 299–303, 309–312, 331, 333, 341, 342, 346, 348, 350–354
Autistic spectrum disorders, 9, 11, 12, 45, 271, 279, 283, 284, 288, 291, 296, 299, 300, 301, 331, 342, 346, 348
Autonomic nervous system, 59, 103, 111, 146, 166, 168, 300, 314, 339, 347
AV node, 145, 151

Basal forebrain, 89
Basal ganglia, 2, 3, 14, 28, 29, 39, 42, 44, 45, 57, 65, 66, 69, 72–81, 83, 87, 90, 103, 109, 111–113, 122, 124, 128, 129, 132, 149, 151, 157, 159, 161, 169, 175, 178, 181, 198, 204–208, 210–212, 214, 215, 221, 224, 225, 230, 231, 234, 237–239, 241, 243, 245, 271, 272, 284, 287, 299, 306, 309, 318, 328–330, 332, 338
Basement membrane, 301
Beta-blockers, 358
Beta-endorphin, 290, 352

Beta oxidation of very long chain fatty acids, 296
Binding problem, 85
Biogenic amines *See* amines
Bipolar disorder, 296
Borderline personality disorder, 237
Borna disease virus, 273, 289
Brachiation, 17
Bradychardia, 145
Brain injured child syndrome, 231
Brainstem, 31, 40, 42, 44, 45, 49–52, 55, 57, 58, 61, 67, 76, 80–83, 88, 90, 96, 99, 100, 102, 103, 108, 145, 146, 149, 151, 158, 172–174, 182, 187, 191, 193, 205, 223, 230, 239, 268, 270, 288, 318, 322, 328, 329
Broca's aphasia, 125
Broca's area, 221
Bupropion, 363–365

Calcium, 43, 303, 304, 347, 350
Calcium-channel blockers, 358, 360
Callosal, 146, 171, 172
Callosum, 61, 71, 89, 94, 100, 101, 108, 119, 130, 138, 169, 171, 172, 203, 220, 249, 324
Candida albicans, 345
Carboxy-hemoglobin, 267
Casein, 341, 342
Catastrophic-depressive reaction, 152, 226
Catecholamine, 119, 195, 283
Catecholaminergic, 98
Caudate, 62, 69, 71, 72, 74, 75, 82, 90, 96, 97, 117, 157–159, 172, 208, 211, 230, 231, 233, 239
CBT *See* cognitive behavioral therapy
Celiac disease, 300, 301
Central auditory processing disorder, 202
Centralia lateralis, 167
Central nervous system, 43, 67, 78, 84, 86, 103, 108, 111, 132, 142, 147, 172, 181, 182, 195, 202, 209, 216, 244, 249, 250, 253, 257, 263, 267–270, 273, 289, 290, 294, 296, 327, 329, 331, 332, 337, 338, 341
Central sulcus, 108, 109, 190, 271
Cerebellar glomerulus *See* cerebellar
Cerebellum, 1–3, 13, 14, 16, 20, 28–31, 35, 38, 39, 43, 44, 46–52, 54–63, 65–67, 69, 72, 75–83, 87, 88, 90, 91, 98, 100, 103, 107, 109, 110–114, 124, 128–130, 132, 146, 148, 164, 175, 178, 180, 181, 183, 184, 187–191, 193–196, 202, 204–206, 209–214, 221–225, 229, 231, 233, 234, 239, 243, 245, 249, 257, 263, 266, 268, 270–273, 276, 278, 282, 289, 299, 306, 312, 314, 318, 320, 322–325, 329–332, 335, 336, 338
Cerebral palsy, 301
Cerebrocerebellum, 55, 56
Characteristic perceptual asymmetry, 227
Chimeric faces, 227
Chlorpromazine, 274
Choline *See* cholinergic
Cholinergic, 75, 83, 99, 101, 163, 213
Chordata, 117
Chronotrophic, 145
Cingulate gyrus, 67, 83, 88, 94, 97–101, 103, 145, 147, 156, 158, 162, 163, 194, 199, 231, 306, 320
Cingulotomy, 163, 200

Cingulum, 61
Citalopram, 307
Classical conditioning, 167, 168
Climbing fibers, 43, 49, 50
Clonazepam, 358
Clonic, 98
Clonidine, 14, 358–360, 364
Cochlea, 110
Cognitive-behavioral therapy, 358
Commissure, 71, 82, 94, 98, 100, 169
Conditioned stimulus, 167
Conduct disorder, 241, 244, 267, 348, 364
Conscious memory, 225, 235
Constructional apraxia, 163, 200
Corpus callosa *See* corpus callosum
Corpus callosum, 94, 130, 171, 172, 324
Corpus colliculus *See* corpus callosum
Corpus striatum, 119
Cortical atrophy, 224, 278
Cortical-striatal-thalamic, 233
Cortico-cortico, 29
Corticofugal, 167
Cortico-limbic, 155, 159, 164, 284
Corticospinal, 55, 128, 206
Corticosteroid, 150, 288
Corticotrophin-releasing hormone, 277
Cortisol, 149, 150, 248, 252, 254, 277, 278, 287, 288
Critical period, 106, 118, 136, 181, 258, 268
Crohn's disease, 301
Cross-temporal contingencies, 110, 111, 113, 214, 231
Cuneocerebellum, 54
Cytogenetical, 280
Cytokine, 273, 289

Darwinian medicine, 15
Decerebration, 268
Declarative memory, 222, 329, 335
Dementia, 80, 212, 224, 296, 322
Demyelination, 296
Dentate, 47, 54–56, 61, 63, 65, 72, 82, 94, 100, 128, 207, 260
Depolarization, 24, 87, 144, 145, 257
Descending medial longitudinal fasciculus, 51
DHA *See* docosahexaenoic acid
Diabetes, 246, 248, 249, 252, 286, 289
Diagonal band, 61
Dichotic listening, 121, 132, 138, 168, 173, 202, 203, 219, 227
Diencephalon, 69, 81, 89, 225
Dihaptic, 203
Dihommo-gamma-linolenic acid, 342
Dihomogammalino-lesic acid, 295
Direct transfer, 170, 174
Disaccharide malabsorption, 301
Disintegrative psychosis, 300
Docosahexaenoic acid, 295, 296, 304, 342
Dopamine, 2, 3, 13, 14, 62, 69, 72–80, 88, 98, 103, 119, 147, 149, 159, 172, 178, 208, 209, 213, 228, 234, 235, 238, 239, 246, 267, 278, 280, 281, 283, 284, 294, 299, 340, 347, 352, 360

Dorsal column, 77, 83, 87, 90, 107, 109, 113, 183, 190, 194, 243, 278, 316, 318, 336
Dorsi-flexors, 250
Down's syndrome, 223, 289
Dressing dyspraxia *See* dyspraxia
Duodenitis, 301
Dysarthria, 57, 221
Dyscalculia, 187, 216
Dysdiadchokinesia, 57
Dysgraphia, 187, 216
Dyslexia, 4, 9, 38, 180, 187, 213, 216–220, 246
Dysmetria, 57, 58, 211, 314
Dysmyelination, 296
Dysnergia, 57
Dysnomia, 187
Dyspraxia, 187, 200, 207–209, 216
Dysthymic, 155
Dystonia, 76–78, 80, 232, 337, 338

Eccosapentaenoic acid, 342
Echopraxia, 113
Eclampsia, 266
Electronystagmography, 213
Emboliform, 47, 55
Emotional brain, 95
Encephalization, 36, 42
Enhancement effect, 199
Enterocolitis, 300
Enterocyte, 301
Entorhinal, 75, 95, 100, 101, 103, 147, 159, 201, 234
Epinephrine, 144, 239
Epithelium, 301
Eshkol-Wachman Movement Analysis System, 271
Essential fatty acids, 295, 340, 342–344
Essential tremor, 78
Eukaryotes, 15
Exaptation, 21
Exorphins, 301
Extinction, 17, 164–166, 173, 192, 193

Facilitory paratonia *See* paratonia
Faraensis, 20
Fasciculus, 53, 61, 71, 77, 82
Fastigial nucleus, 47, 51, 54, 55, 58, 61, 62, 67, 98, 184, 205, 211–213, 222
Fastigobulbar, 54
Fear response, 97
Fixed action patterns, 44–46
Flocculonodular lobe, 51, 57
Focal dystonia, 338
Folic acid, 346
Forebrain, 58, 75, 89, 108, 123, 148, 213, 223, 259
Fornix, 82, 94, 98, 100, 101
Fragile X syndrome, 223, 281
Free radical, 255, 256, 293, 346
Frontal eye fields, 72, 185, 207, 309, 328, 329
Frontal lobe, 2, 3, 10, 39, 40, 59, 61, 63, 66, 67, 69, 72, 74–76, 79, 82, 88, 90, 97, 108–115, 120, 125, 144, 146, 147, 152, 158, 160, 163–167, 174, 175, 187, 194, 204–207, 209, 210–212, 214, 215, 217, 221, 222,
Frontal lobe *contd.*
224–226, 230–233, 239, 243, 245, 248, 273, 278, 284, 299, 318, 320, 322–325, 336–338
Funiculus, 90
Fusiform gyrus, 140, 141, 324

GABA, 47, 50, 73, 75, 84, 213, 281, 282, 347
GABAminergic, 213
Gamma fibers, 250
Gastritis, 301
Gastroparesis, 300
Gastrulation, 118
Generativity, 129, 131
Gerstmann's syndrome, 125, 218
Gestalt, 162, 198
Gigantocellularis, 77, 80
Gigenhalten, 113
Glial, 22, 27, 30, 38, 188, 189, 239, 273, 279, 284, 289
Globos nucleus, 47
Glutamate, 62, 117, 213, 239, 260, 282, 303, 344
Glutamatergic *See* glutamate
Gluten, 341, 342
Glycerol-phospholipids, 342
Glycoprotein, 282
Golgi cells, 48
Gonadectomized *See* gonadectomy
Gonadectomy, 118, 267
Gonadotrophic hormone, 98
Grandmother cells, 85
Granule, 47, 48, 50, 56, 100, 211, 212, 222, 239, 260, 273, 289
Granulomas, 300
Gray matter, 47, 71, 81, 93, 97, 98, 100, 101, 103
Gut associated lymphoid tissue, 343
Gyrus, 61, 94, 96, 98–100, 108, 109, 120, 121, 124, 125, 135, 138, 141, 163, 190, 191, 194, 217, 218, 220, 225, 260, 323, 324

Haloperidol, 352, 358, 359
Hemianesthesia, 165
Hemianopia, 165
Hemi-motor neglect, 206
Hemi-neglect, 202
Hemispatial neglect, 165, 172, 191, 192, 194
Hemispherectomy *See* hemisphericity
Hemisphericity, 3, 13, 38, 146, 155, 219, 339
Hemoglobin, 300
Hepatocytes, 255
Heritability, 279
Herpes simplex, 65, 72
Heschl's gyrus, 217
Hippocampus, 25, 26, 61, 62, 75, 88, 90, 93, 94, 98–103, 147, 148, 155, 159, 161–163, 169, 199–201, 212, 222, 225, 234, 235, 239, 260, 273, 287, 289, 307, 313, 328, 329, 332
Hominids, 17, 18, 20, 29, 38, 82, 291, 292
Homo erectus, 20, 21, 36, 37, 291
Homo habilis, 291
Homo rodolfensis, 36
Homo sapiens, 18, 20, 35, 129, 293

Human influenza virus, 274
Huntington's Disease, 73, 74, 75, 80, 299
5-hydroxyendoliacetic acid, 228
5-hydroxytryptamine acid, 148
Hyperactive child syndrome, 231
Hyperbilirubinemia, 275
Hyperkinesia, 72, 73, 205, 208, 253
Hyperkinetic, 45, 73, 74, 75, 77, 79, 80, 112, 178, 207, 208, 215, 216, 231, 232, 238, 278, 299, 314, 330, 331, 361
Hyperplasia, 300, 301
Hyperpolarization, 257
Hypoglossal, 117
Hypokinesia, 72, 73, 75, 79, 205, 207, 208, 215, 226, 278, 299, 314, 330, 331, 338
Hypometria, 314
Hypometric, saccades, 211
Hypomyelination, 296
Hypophysis, 103
Hyposmia, 326, 327
Hypothalamo-autonomic, 152
Hypothalamus, 45, 61, 62, 75, 83, 88, 90, 95–101, 103, 104, 118, 144–152, 158, 162, 169, 178, 230, 234, 248, 252, 276, 277, 279, 287, 288, 319, 328, 329
Hypothyroidism, 289
Hypotonia, 57, 190, 206, 207, 213, 240, 296
Hypoxia, 255, 256, 266, 267

Ideography, 60, 224
Ideomotor praxis, 164
Ilbosocral, 269
Ileal lymphoid, 300, 301
Ileum, 301
Implicit learning, 167, 168, 225
Implicit memory, 168, 173, 225
Inferior olivary nucleus, 43, 50, 51, 55, 78
Inferior olive, 30, 42, 43, 44, 46, 78, 79
Inferior-temporal cortices, 133
Inotrophic, 145
Intention tremor, 56, 57, 132
Interferon-alpha, 290
Interferon-gamma, 290
Interleukin-6, 290
Interleukin-12, 290
Intralaminar nucleus, 30, 63, 72
Intra-parietal sulcus, 161, 198
Ischemia, 142, 143

Joubert syndrome, 212

Ketamine, 274
Kinematic imbalance due to suboccipital strain, 269
Kinesthesia, 188
KISS syndrome *See* kinematic imbalance due to suboccipital strain
Klüver-Bucy syndrome, 162, 178, 235

Lactase, 297
Lamina propria, 301
Lateral geniculate nucleus, 51, 83, 109, 196

Learning disability, 9, 11, 57, 177, 206, 215, 216, 225, 237, 266, 314, 317, 336, 342
Le grand lobe limbique, 93
Lemniscus, 49, 53, 61, 82, 138
Lentiform nucleus, 71, 77
Light-sensitive neurons, 198
Limbic striatum, 75
Limbic system, 59, 61, 75, 84, 93–95, 99, 103–105, 111, 112, 147, 149, 158, 161–163, 166, 178, 194, 199, 201, 202, 212, 222, 225, 232, 234, 237, 248, 252, 276, 278, 279, 288, 306, 319, 328
Lingual gyrus *See* gyrus
Linking operations, 162, 198
Lipopolysaccharide, 274, 291
Locus coeruleus, 50, 61, 67, 75, 83, 88, 89, 90, 95, 96, 98, 99, 101, 103, 145, 322
Lordosis, 19, 256, 257
Lou Gehrig's disease, 299
Lumbricals, 250
Lupus erythematosus, 289
Luys, 69, 115
Lymph nodes, 150
Lymphocyte, 287, 301
Lymphocytic colitis, 301

Magnesium, 304, 340, 346, 347
Mammillary body, 98–101, 103, 328
Mania, 74, 75, 79, 103, 112, 115, 144, 152, 153, 158–160, 163, 208, 212, 226, 229, 230, 231, 233, 234, 237, 241, 316
Mazza, 299
Medulla, 50, 52, 97, 144, 151
Medullary lamina, 81
Medullary reticular formation, 90, 144
Melatonin, 69, 318, 350
Meniculate body, 138
Mesencephalic, 158, 159, 162, 166, 167, 213, 230, 234
Mesencephalon, 101, 147–149, 208, 212, 213
Mesolimbic, 62, 74, 75, 79, 160, 213
Metabolically induced dysplasias, 296
Methylazoxymethanol, 266
Methylmalonic acid, 300, 303, 304
Methylphenidate, 14, 333, 348, 360, 364
Microcephaly, 239, 281, 294
Microsomy, 269
Minimal brain dysfunction, 11, 209, 220, 231
Mitochondria, 27, 250, 254–256, 293
Monoamine oxidase inhibitors, 283, 363
Mooney faces, 85
Mossy fibers, 49, 50, 100
Motorneurons, 77, 251
Motor strip, 108, 110
Mozart Effect, 321, 322
MRNA, 273, 289
Mutism, 172, 173, 220
Myelination, 34, 171, 265, 296
Myogenic, 41

N-acetylaspartate, 66
Naltrexone, 118, 290, 352–354

Natural killer, 151, 154, 287, 290
Neo-cerebellum, 66, 128, 212, 193, 213, 221–223
Neolithic, 292
Neoplastic disease, 150, 288
Neostriatum, 29, 35, 69, 72, 73, 75, 79, 90, 113, 149
Neurobehavioral, 1–5, 9, 14, 40, 45, 179, 181, 245, 266,
    271, 272, 275, 283, 285, 288–290, 296, 305, 306, 308,
    311, 312, 314, 319, 322, 326, 330, 337, 342, 348
Neurogenic motricity, 42
Neuropile, 188
Nicotine, 267, 358, 359
Nicotine receptor hypothesis, 267
Nigrastriatal, 75
Nigrostriatal, 13, 73, 74, 119
Nigrostriatal bundle, 119
Non-verbal learning disabilities, 233
Noradrenergic, 14, 61, 67, 75, 76, 98, 101, 143, 147, 148,
    151, 213, 238
Noradrenergic tricyclic, 238
Norepinephrine, 50, 88, 98, 103, 119, 120, 144, 147, 148,
    159, 178, 228, 229, 239, 241, 285, 287, 294,
    347, 360
Normoxia, 255
Normoxic cells, 255
Notochord, 15
Nucleotide polymorphisms, 282
Nucleus accumbens, 61, 75, 98, 159, 233, 235
Nucleus ambiguous, 103
Nucleus intermedius, 88
Nucleus interpositus, 54, 55
Nucleus lovus coeruleus, 103
Nucleus medialis dorsalis, 82
Nucleus reticularis, 100, 167
Nucleus reticularis tegmenti pontis, 100
Nucleus tractus solitarius, 62, 144, 300
Nucleus tuberis, 103
Nystagmus, 58, 185, 187, 189, 190, 239

Oblique muscles, 51, 124, 164, 200, 217
Obsessive-compulsive disorder, 1, 4, 9, 10–12, 73–75, 112,
    177, 208, 232, 234, 237, 238, 241, 287, 299, 306, 312,
    320, 353, 355–359
Occipital lobe, 71, 82, 93, 108, 110, 125, 137, 141, 151,
    158, 164, 179, 196, 200, 201, 219, 220, 231, 252, 258,
    270, 287, 315, 320
Occipital petalia, 125
Occipital-temporal cortex, 133, 134, 140, 141, 164
OCD See obsessive-compulsive disorder
Olfactory bulb, 96, 97, 100
Olfactory cortex, 96, 155, 162, 163, 188
Oligodendrocyte, 276
Olivopontocerebellar, 58
Omega-3, 295, 296, 342, 343
Operculum, 125, 217
Opiate antagonists, 290
Opioid, 290, 299, 341, 342, 353
Oppositional paratonia, 113
Optokinetic nystagmus, 58
Orbito-frontal cortex, 72, 73, 103, 156, 158, 162, 166, 200,
    207, 208, 230, 231, 233, 237, 241

Organelles, 24, 27, 254, 293
Osmosicity, 254
Osteoarthritis, 292
Otitis media, 287, 300, 345
Oto-rhino-laryngologic, 275
Oxidative stress, 27, 254–257, 293, 330, 331, 338, 346, 347
Oxytocin, 103

Paedeogenesis, 30
paleocerebellum, 67
Paleo-cerebellum-limbic, 212
Paleocortex, 93, 94, 104, 164
Paleostriatum, 69, 79
Pallidum, 73
Pancreas, 301
Pancreatico-biliary, 301
Panic attacks, 62, 235, 305
Parabrachial nucleus, 96, 97
Parahippocampal gyrus, 94, 100, 141
Para-limbic, 155, 158, 159, 161, 230, 231, 234, 235
Paranthropus, 17
Parasubiculum, 95, 100
Parasympathetic, 103, 143–146, 150, 152, 256, 300, 319, 343
Paratonia, 113, 207
Paratonic rigidity, 113
Para-ventricular nucleus, 103
Parent cells, 105
Parent management training, 308
Parietal cortex, 61, 83, 109, 159, 161–163, 174, 190, 197,
    199, 201, 225, 320
Parietal lobe, 26, 83, 99, 109, 112, 125, 133, 157, 161, 163,
    165, 166, 172, 191–194, 197, 198, 200, 218
Parieto-occipital sulcus, 82, 108
Parkinson's disease, 13, 44, 73–80, 139, 157, 158, 160, 163,
    230, 272, 299, 321, 338
Paroxetine, 306, 358
Pars compacta, 69
Pars dorsalis, 77
Pars reticulata, 69, 72
Parvocellular, 55
Paxil, 283, 350
Pervasive developmental disorder (PPD), 10, 210, 274, 300,
    302, 309, 326
Pedigree Disequilibrium Test, 282
Peduncle-pontine, 83
Perceptual-motor disorder, 187
Periaqueductal gray matter, 98, 99
Periarcuate, 166, 193
Perirolandic cortex, 271
Peri-sylvian cortex, 217
Peri-ventricular, 270
Permian times, 17
Peroxisome, 296
Perseveration, 75, 90, 110, 112, 113, 119, 159, 162, 178,
    210, 214, 231–233, 237, 240, 353
Personality disorders, 168, 237
Pervasive developmental disorder, 1, 9, 10, 177, 210, 232,
    274, 280, 288, 300, 326
Phobias, 234, 305
Phytohemagglutinin, 274

Pigmies, 298, 299
Pimozide, 358, 359
Piriform, 93, 94, 100, 103
Pituitary, 103, 144, 150, 277
Planum temporale, 120, 124, 179, 217, 218, 220, 324
Plasmalogens, 342
Plasmologans, 296
Polyinosinic-polycytidylic acid, 274
Pontine nuclei, 51, 55, 101
Pontine reticulotegmental nucleus, 55
Pontis, 55
Positron emission tomography, 307, 323
Post-commissural fornix, 101
Post-ganglionic sympathetic, 250
Post traumatic stress disorder, 234, 235, 305, 306
Prader-Willi Syndrome, 280, 282
Praxis, 165
Prearcuate, 166, 193
Precuneus, 66, 67, 320
Prefrontal cortex, 13, 35, 63, 65, 67, 72, 74, 76, 77, 80, 81, 82, 88, 90, 91, 96, 97, 108, 110–113, 125, 128–130, 144, 145, 158, 160, 164, 165, 168, 178, 179, 181, 187, 194, 196, 201, 208, 210, 211, 214, 217, 224, 225, 229–235, 237, 238, 243, 246, 248, 273, 278, 285–287, 306, 309, 312, 313, 320, 323, 329, 334
Prefrontal lobes, 210, 306
Premotor, 40, 45, 46, 51, 55, 57, 72, 73, 77, 84, 97, 108, 110, 125, 128, 129, 132, 149, 164, 165, 191, 207, 208, 214, 217, 225, 334
Preoptic, 90, 96, 101, 149, 288
Presubiculum, 95, 99–101, 163
Pretectal, 99, 103
Primary auditory pathway, 322
Progressive lateralization hypothesis, 181, 182
Prokaryotes, 15
Proprioceptive, 29, 67, 184, 187, 188–190, 193, 195, 209, 250, 251, 257, 270, 278, 329, 330, 331
Prosencephalon, 105
Prosopagnosia, 140, 141, 163, 200
Prozac, 244, 283, 350
Pseudo hemi-anesthesia, 192
Psychomimetic, 274
Pulvinar, 81, 83, 96, 138, 148, 164, 200
Purkinje cells, 43, 47–50, 56–58, 62, 78, 211–213, 222, 239, 273, 282, 289, 312, 331
Putamen, 69, 71, 72, 74, 75, 90, 159, 225
Pyramidal, 69, 86, 100, 108, 128, 274, 359

Quadrapedalism, 268

Ramidus, 17, 18
Raphe nuclei, 50, 61, 67, 83, 88–90, 95, 96, 98, 101, 103, 145, 148, 228, 322
Receptive field, 107, 108, 136
Red nucleus, 50, 55, 57
Reelin gene, 280, 282
Reserpine, 283
Reticular activating system, 167, 322
Reticular formation, 57, 80, 83, 88, 97, 102, 103, 166, 167, 172, 185, 187, 318, 322

Reticular nucleus, 82, 84, 86
Reticulospinal tract, 51, 206
Retrosplenial, 163, 200, 328, 329
Reuptake inhibitors, 228, 244, 283
Rhinencephalon, 94, 105, 111, 328
Risperidone, 358
Ritalin, 8, 74, 236, 244, 283, 304, 317, 350
Rolandic lesions, 156
Rubella, 275, 300
Rubrospinal, 55, 206

Saccades, 58, 211, 309, 312, 313
Saccadic neurons, 198
Sahelanthropus tchadensis, 18
Schizo-obsessive spectrum disorders, 237
Schizophrenia, 3, 10, 209–211, 213, 235, 274, 281, 284, 288, 295, 296, 307, 341
Scoliosis, 206, 269
Scopolamine, 189
Secretin, 301
Selective attention, 75, 138, 147, 172, 195, 204
Selenium, 348
Sensory defensiveness, 330, 331
Sensory integration, 183, 209, 214, 261
Sensory-motor system, 204
Septal, 61, 75, 89, 94, 98, 99, 101, 158, 162, 213, 230, 276
Septofimbrial nucleus, 98
Septum, 75, 90, 95, 98–103, 147, 260
Serotonergic, 61, 90, 98, 118, 148, 151, 213, 228, 238
Serotonin, 50, 88, 89, 90, 103, 119, 147–149, 159, 178, 228, 229, 239, 241, 244, 280, 283, 285, 290, 294, 304, 352, 354
Shadantropus tchadereniss, 18
Sleep spindles, 315, 316
Smooth pursuit, 58, 211
Social indifference, 99
Social phobia, 307
Solitary tract, 95–97, 103
Soluble intercellular adhesion molecule-1, 290
Somatosensory, 57, 76, 78, 83, 87, 88, 90, 101, 107, 109, 112, 117, 118, 163, 165, 183, 187, 188, 190, 191, 194, 200, 204, 205, 215, 243, 267, 318, 336
Space dyslexia, 214
Spatial neglect, 112, 161, 192, 196
Spatial summation, 28–30
Spina bifida, 294
Spinocerebellum, 51, 54–56, 263
Stellate cells, 47, 48, 50, 56, 57, 145, 222, 286
Stellectomy, 143
Strabismus, 106, 206
Striatal, 57, 72, 73, 78, 79, 117, 119, 149, 158, 160, 172, 208, 233, 271, 306, 312
Stria terminalis, 96, 98
Subiculum, 95, 98, 100, 101
Subliminal implicit conditioning, 168
Substantia grisea centralis, 103
Substantia nigra, 69, 72–75, 77, 79, 88, 96–98, 119, 172, 299
Substantia nigra zona compacta, 299
Sudden infant death syndrome, 147, 285

Superior colliculus, 51, 72, 99, 103, 109, 328
Supplementary motor areas, 110, 129, 271, 334
Supplementary motor cortex, 108, 225
Supraoptic nucleus, 103
Surround inhibition, 50
Sydenham's chorea, 238, 287
Sylvian fissure, 110
Sympathetic, 90, 143–147, 150, 152, 168, 195, 252, 255–257, 268, 285, 286, 300, 319, 343
Syphilis, 293

Tactile, 60, 63, 115, 123, 181, 183, 184, 189–196, 209, 212, 216, 258, 259, 277, 278, 315, 320, 330, 331, 337, 338
Tardive dyskinesia, 352
t-butylhydroperoxide, 255
Tectum, 138
Tegmental area, 50, 53, 61, 62, 75, 88, 89, 98–101, 103, 159, 212, 234, 328
Tegmental nuclei of Gudden, 103
Telencephalon, 69, 94, 98, 105
Temporal gyrus, 135, 141
Temporal lobe, 93, 97, 100, 102, 110, 134, 136–138, 141, 145, 153, 160–162, 198, 200, 202, 204, 213, 217, 225, 235, 311, 323, 324
Temporal-parietal lesions, 125, 219
Temporal plana, 216
Temporal sulcus, 166, 193, 311
Terminal transferase, 276
Testosterone, 118, 267
Tetanus, 274
Thalamo-amygdala, 162
Thalamus, 2, 3, 25, 28–31, 35, 46, 51, 55, 57, 60, 61, 63, 65–67, 69, 71–84, 86–88, 90, 91, 96, 97, 99–103, 106, 107, 109–111, 113–115, 122, 124, 138, 146–148, 151, 158, 159, 162, 165, 167, 171, 178, 180–184, 187, 190, 194–196, 200, 202, 204, 205, 207, 212, 215, 221, 225, 227, 230, 233–235, 241, 243, 245, 251, 266, 268, 271, 278, 299, 306, 313, 318, 319, 322, 324, 328, 330–332, 335, 336, 338
Therapsids, 17
Third party convergence, 174
Thymus, 150, 287
Titubation, 57
Tomatis Method, 322
Tonic, 75, 98, 121, 152, 160, 171, 188, 206, 227, 250, 251, 269, 300, 336
Torticollis, 269, 270
Tourette's Syndrome, 9–11, 44, 73–75, 79, 177, 208, 209, 232, 234, 237, 238, 241, 299, 312, 358–360
Toxemia, 266
Transmission Disequilibrium Test, 282
Triassic times, 17
Trigeminal, 49, 52, 55, 61, 83
Trinucleotide, 281
Tryptophan, 294
Tuberculosis, 283, 292

Tuberinfundibular, 74
Tuberous sclerosis, 280
Tumor necrosis, 290
Tutsi, 298, 299
Tyrosine, 294, 303

Uncinate fasciculus, 54

Vaccinal encephalitis, 300
Vagus nerve, 62, 95, 97, 103, 144, 145, 300
Vasopressin, 98, 103
Ventral system, 133, 162, 199
Ventralis anterior, 82
Ventralis lateralis, 82
Ventralis posterior, 82
Ventral visual pathway, 161, 162, 164, 198, 200
Ventricular fibrillation, 142, 143, 145
Vermis, 51, 54, 55, 57, 58, 62, 66–78, 210–213, 222, 223, 231, 263
Vertigo, 187, 189, 190
Vestibular, 13, 20, 29, 42, 49, 51, 55, 58, 62, 67, 77, 128, 165, 184, 185, 187–190, 205, 209, 213, 214, 229, 239, 254, 270, 273, 322, 323, 329, 330, 331
Vestibular apparatus, 20, 51, 62, 165, 184, 185, 187, 229, 322, 329
Vestibular nuclei, 49, 51, 55, 62, 77, 128, 184, 185, 205
Vestibular nystagmus, 58
Vestibulocerebellum, 51, 165
Vestibulo-cochlear, 138
Vestibulospinal, 51, 55
Visceral brain, 95
Visual association cortex, 110, 320
Visual cortex, 83, 106, 107, 109, 110, 148, 188, 196, 197, 201, 218, 220, 235, 328
Visual neglect, 168, 196
Visual-spatial, 161, 200, 217, 233, 312
Visuo-spatial, 115, 124, 156, 197, 216
Vitamin A, 346
Vitamin B, 346
Vitamin C, 303, 347, 349

Ward's triangle, 287
Watusi, 298
Wernicke's aphasia, 124, 217
Wernicke's area, 173, 175, 201
William's syndrome, 66, 223
Wnt2 gene, 280
Wolfe's Law, 19
Working memory, 13, 66, 67, 130, 158, 209, 225, 233, 246, 307, 312, 313, 326

Xenopus, 118

Zellerweger cerebro-hepato-renal syndrome, 295
Zinc, 303, 304, 347, 349
Zygapophysical, 273

TN USA
009
00003B/7/P